Digital Control System Analysis and Design

Digital Control System Analysis and Design

CHARLES L. PHILLIPS
Professor

Department of Electrical Engineering
Auburn University

H. TROY NAGLE, JR.
Professor

Department of Electrical and Computer Engineering
North Carolina State University

PRENTICE-HALL, INC., Englewood Cliffs, N.J. 07632

Library of Congress Cataloging in Publication Data

Phillips, Charles L.
Digital control system analysis and design.

Includes bibliographical references and index.
1. Digital control systems. 2. Electric filters, Digital. 3. INTEL 8086 (Computer) I. Nagle, H. Troy, 1942– . II. Title.
TJ223.M53P47 1984 629.8′95 83-9531
ISBN 0-13-212043-7

Editorial/production supervision: Barbara Bernstein
Manufacturing buyer: Tony Caruso
Cover design: Edsal Enterprizes

Printed in the United States of America

10 9 8 7

ISBN 0-13-212043-7

Prentice-Hall International, Inc., *London*
Prentice-Hall of Australia Pty. Limited, *Sydney*
Editora Prentice-Hall do Brasil, Ltda., *Rio de Janeiro*
Prentice-Hall Canada Inc., *Toronto*
Prentice-Hall of India Private Limited, *New Delhi*
Prentice-Hall of Japan, Inc., *Tokyo*
Prentice-Hall of Southeast Asia Pte. Ltd., *Singapore*
Whitehall Books Limited, *Wellington, New Zealand*

To

Laverne, Susie, Chuck, and Carole

Julie

Contents

5 CLOSED-LOOP SYSTEMS 135

6 SYSTEM TIME-RESPONSE CHARACTERISTICS 161

7 STABILITY ANALYSIS TECHNIQUES 190

Preface

This book is intended to be used primarily as a text for a first course in discrete control systems and/or a first course in digital filters, at either the senior or first-year graduate level. Furthermore, the text is suitable for self-study by the practicing engineer.

This book is based on material taught at Auburn University, and in intensive short courses taught in both the United States and Europe. The practicing engineers who attended these short courses have influenced both the content and the direction of this book, resulting in more emphasis placed on the practical aspects of designing and implementing digital control systems. Also, the introduction of the microprocessor has greatly influenced the material of the book, with Chapter 13 devoted exclusively to microcomputer implementations.

Chapter 1 in this book presents a brief introduction and an outline of the text. Chapters 2 through 10 cover the analysis and design of discrete-time linear control systems. Some previous knowledge of continuous-time control systems is helpful in understanding this material. The mathematics involved in the analysis and design of discrete-time control systems is the z-transform and vector-matrix difference equations, with these topics presented in Chapter 2.

Chapter 3 is devoted to the very important topic of sampling signals; and the mathematical model of the sampler and data hold, which is basic to the remainder of the text, is developed here. The implications and the limitations of this model are stressed. In addition, analog-to-digital and digital-to-analog converters are discussed.

The next four chapters, 4, 5, 6, and 7, are devoted to the application of the mathematics of Chapter 2 to the analysis of discrete-time systems, with emphasis on digital control systems. Classical design techniques are covered in Chapter 8, with the frequency-response Bode technique emphasized. Modern design techniques are presented in Chapters 9 and 10. Throughout these chapters, practical computer-aided analysis and design are stressed. The required computer programs are given in Appendix I.

Chapters 11, 12, 13, and 14 are devoted to digital filters. In Chapter 11 the transformation of analog filters into discrete-time representations is presented. The properties of numerical integration techniques and their relation to sampled-data transformations are investigated. Chapter 12 demonstrates various structures for digital filters. Cascade and parallel arrangements are detailed.

Implementation of digital filters on microcomputers is the subject of Chapter 13. Assembly language programs for the INTEL 8086 and other 16-bit machines are included. The INTEL 2920 signal processor is also described. Several other signal processors and microcomputers are discussed.

Chapter 14 covers many of the theoretical aspects of digital filtering. Quantization effects on signal amplitude and filter coefficients are discussed. Quantization noise is examined and characterized. Limit cycles are investigated. The theoretical aspects are then employed in practical guidelines for implementing digital filters. Presented in Chapter 15 are case studies of three operational digital control systems.

At Auburn University, three courses based on the controls portion of this text, Chapters 2 through 10, have been taught. Chapters 2 through 8 are covered in their entirety in a one-quarter four credit-hour graduate course. Thus the material is also suitable for a three-semester-hour course. These chapters have also been covered in twenty lecture hours of an undergraduate course, but with much of the material omitted. The topics not covered in this abbreviated presentation are state variables, the modified z-transform, nonsynchronous sampling, steady-state accuracy, and closed-loop frequency response. A third course, which is a one-quarter three-credit course, requires one of the above courses as a prerequisite, and introduces the state variables of Chapter 2. Then the state-variable models of Chapter 4, and the modern design of Chapters 9 and 10, are covered in detail.

Also at Auburn University, a first course in digital filters has been taught using material from Chapters 2, 11, 12, 13, and 14. The course was offered to senior and beginning graduate students for three quarter hours credit and was organized around 28 lectures.

To further assist the user of this book, an instructor's manual containing problem solutions and possible course syllabuses has been developed. The authors feel that the problems at the end of the chapters are an indispensable part of the text, and should be fully utilized by all who study this text.

Finally, we gratefully acknowledge the many colleagues, graduate and undergraduate students, and staff members of the Electrical Engineering Department at Auburn University who have contributed to the development of this book. In particular, we wish to thank Professor Richard C. Jaeger for contributing the digital-to-analog and analog-to-digital sections of Chapter 3. We are especially indebted to

Professor J. David Irwin, our Department Head, for his aid and encouragement, and to the many typists in the Electrical Engineering Department, who, over a period of many years, have patiently typed and retyped this manuscript.

Auburn University

CHARLES L. PHILLIPS

H. TROY NAGLE, JR.

Digital Control System Analysis and Design

1

Introduction

1.1 OVERVIEW

This book is concerned with the analysis and design of closed-loop physical systems that contain digital computers. The computers are placed within the system to modify the dynamics of the closed-loop system such that a *more satisfactory* system response is obtained.

A closed-loop system is one in which certain system forcing functions (inputs) are determined, at least in part, by the response (outputs) of the system (i.e., the input is a function of the output). A simple closed-loop system is illustrated in Figure 1-1. The physical system (process) to be controlled is called the *plant*. The *sensor* (or sensors) measures the response of the plant, which is then compared to the desired response. This difference signal initiates actions that result in the actual response approaching the desired response, which drives the difference signal toward zero. Generally, an unacceptable closed-loop response occurs if the plant input is simply the difference in the desired response and the actual response. Instead, this difference signal must be processed (filtered) by another physical system, which is called a *compensator*, a *controller*, or simply a *filter*. One problem of the control system designer is the specification of the compensator.

An example of a closed-loop system is the case of a pilot landing an aircraft. For this example, in Figure 1-1 the plant is the aircraft and the plant inputs are the pilot's manipulations of the various control surfaces and of the aircraft velocity. The pilot is the sensor, with his or her visual perceptions of position, velocity, instrument indications, and so on, and with his or her sense of balance, motion, and so on. The desired response is the pilot's concept of the desired flight path. The compensation is

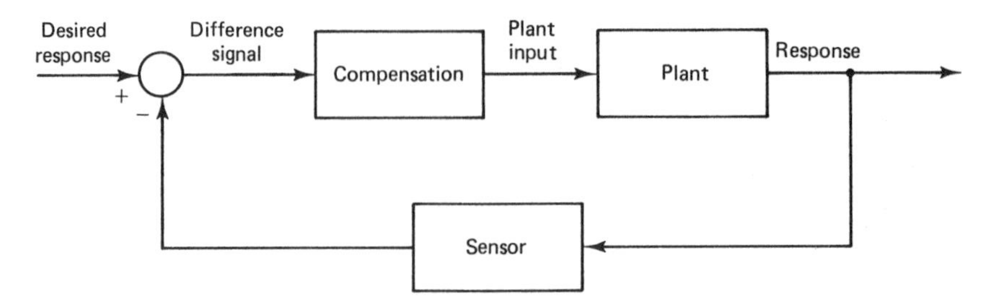

Figure 1-1 Closed-loop system.

the pilot's manner of correcting perceived errors in flight path. Hence, for this example, the compensation, the sensor, and the generation of the desired response are all functions performed by the pilot. It is obvious from this example that the compensation must be a function of plant (aircraft) dynamics. A pilot trained only in a fighter aircraft is not qualified to land a large passenger aircraft, even if he or she can manipulate the controls.

We will consider systems of the type shown in Figure 1-1, in which the sensor is an appropriate measuring instrument and the compensation function is performed by a digital computer. The plant has dynamics; we will program the computer such that it has *dynamics* of the same nature as those of the plant. Furthermore, although generally we cannot choose the dynamics of the plant, we can choose those of the computer such that, in some sense, the dynamics of the closed-loop system are *satisfactory*. For example, if we are designing an automatic aircraft landing system, the landing must be safe, the ride must be acceptable to the pilot and to any passengers, and the aircraft cannot be unduly stressed.

Both classical and modern control techniques of analysis and design are developed in this book. All control-system techniques developed are applicable to *linear time-invariant discrete-time* system models. A linear system is one for which the principle of superposition applies [1]. Suppose that the input of a system $x_1(t)$ produces a response (output) $y_1(t)$, and the input $x_2(t)$ produces the response $y_2(t)$. Then, if the system is linear, the principle of superposition applies and the input $x_1(t) + x_2(t)$ will produce the response $y_1(t) + y_2(t)$. All physical systems are inherently nonlinear; however, in many systems, if the system variables do not vary over too wide a range, the system responds in a linear manner.

When the parameters of a system are constant with respect to time, the system is called a time-invariant system. An example of a time-varying system is the booster stage of a space vehicle. As the fuel is consumed, the mass of the vehicle decreases with time.

A discrete-time system has signals that can change values only at discrete instants at time. We will refer to systems in which all signals can change continuously with time as *continuous-time*, or *analog*, systems. Even though the analysis and design techniques presented are applicable to linear systems only, certain nonlinear effects will be discussed.

The compensator, or controller, in this text is a digital filter. The filter implements a transfer function. The design of transfer functions for digital controllers is the subject of Chapters 2 through 11. Once the transfer function is known, algorithms for its realization must be programmed on a digital computer, or the algorithms must be implemented in special-purpose hardware. These subjects are detailed in Chapters 12, 13, and 14.

Presented next in this chapter is an example of a digital control system. Then the equations describing three typical plants that appear in closed-loop systems are developed. Finally, some basic results of Laplace transform theory are given.

1.2 DIGITAL CONTROL SYSTEM

The basic structure of a digital control system will be introduced through the example of an automatic aircraft landing system. The system to be described is the landing system that is currently operational on U.S. Navy aircraft carriers [2]. Only the simpler aspects of the system will be described.

Two-crew-member airliner flight deck. The digital electronics include an automatic flight control system (i.e., "automatic pilot"). (Courtesy of Boeing Airplane Company.)

The automatic aircraft landing system is depicted in Figure 1-2. The system consists of three basic parts: the aircraft, the radar unit, and the controlling unit. During the operation of this control system, the radar unit measures the approximate

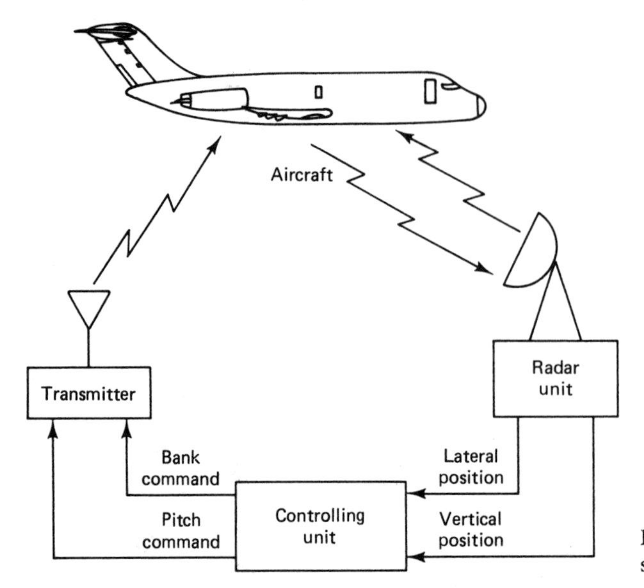

Figure 1-2 Automatic aircraft landing system.

vertical and lateral positions of the aircraft, which are then transmitted to the controlling unit. From these measurements, the controlling unit calculates appropriate pitch and bank commands. These commands are then transmitted to the aircraft autopilots, which in turn cause the aircraft to respond accordingly.

In Figure 1-2 the controlling unit is a digital computer. The lateral control system, which controls the lateral position of the aircraft, and the vertical control system, which controls the altitude of the aircraft, are independent (decoupled). Thus the bank command input affects only the lateral position of the aircraft, and the pitch command input affects only the altitude of the aircraft. To simplify the treatment further, only the lateral control system will be discussed.

A block diagram of the lateral control system is given in Figure 1-3. The aircraft lateral position, $y(t)$, is the lateral distance of the aircraft from the extended centerline of the landing area on the deck of the aircraft carrier. The control system attempts to force $y(t)$ to zero. The radar unit measures $y(t)$ every 0.05 s. Thus $y(kT)$ is the sampled value of $y(t)$, with $T = 0.05$ s and $k = 0, 1, 2, 3, \ldots$. The digital controller processes these sampled values and generates the discrete bank commands $\phi(kT)$. The data hold, which is on board the aircraft, clamps the bank command $\phi(t)$ constant at the last value received until the next value is received. Then the bank command is held constant at the new value until the following value is received. Thus the bank command is updated every $T = 0.05$ s, which is called the sample period. The aircraft responds to the bank command, which changes the lateral position $y(t)$.

Two additional inputs are shown in Figure 1-3. These are unwanted inputs, called *disturbances*, and we would prefer that they not exist. The first, $w(t)$, is the wind input, which certainly affects the position of the aircraft. The second disturbance input, labeled radar noise, is present since the radar cannot measure the exact position of the aircraft. The noise is the difference between the exact aircraft position and the

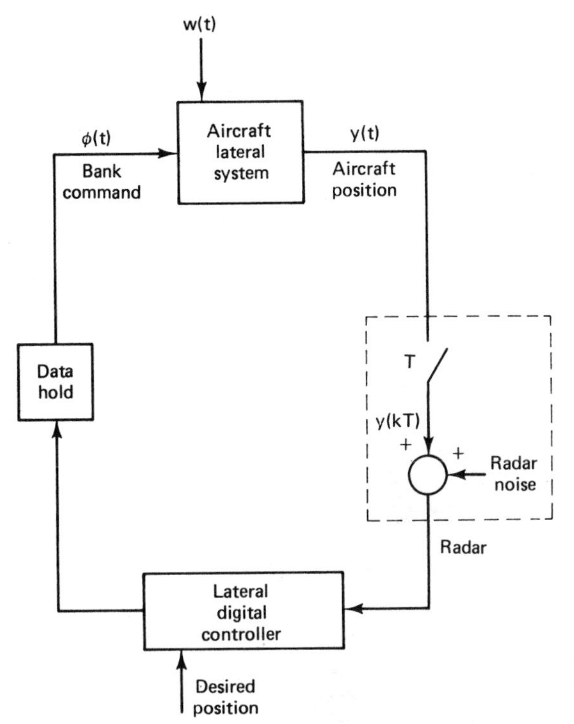

Figure 1-3 Aircraft lateral control system.

measured position. Sensor noise is always present in a control system, since no sensor is perfect.

The design problem for this system is to maintain $y(t)$ small in the presence of the wind and radar-noise disturbances. In addition, the plane must respond in a manner that is both acceptable to the pilot and does not unduly stress the structure of the aircraft.

In order to effect the design, it is necessary to know the mathematical relationships between the lateral position $y(t)$, the bank command input $\phi(t)$, and the wind input $w(t)$. These mathematical relationhips are referred to as the mathematical model, or simply the model, of the aircraft. For example, for the McDonnell-Douglas Corporation F4 aircraft, the model of lateral system is a ninth-order ordinary nonlinear differential equation [3]. For the case that the bank command $\phi(t)$ remains small in amplitude, the nonlinearities are not excited and the system model is a ninth-order ordinary linear differential equation.

The task of the control system designer is to specify the processing to be accomplished in the digital controller. This processing will be a function of the ninth-order aircraft model, the expected wind input, the radar noise, the sample period T, and the desired response characteristics. Various methods of digital controller design will be developed in Chapters 8, 9, 10, and 11.

The development of the ninth-order model of the aircraft is beyond the scope of this book. In addition, this model is too complex to be used in an example in this book. Hence, to illustrate the development of models of physical systems, the mathematical models of three simple, but common, control-system plants will be developed later in this chapter. Two of the systems relate to the control of position, and the third relates to temperature control.

1.3 THE CONTROL PROBLEM

We may state the control problem as follows. A physical system or process is to be accurately controlled through closed-loop, or feedback, operation. An output variable, called the response, is adjusted as required by an error signal. The error signal is a measure of the difference between the system response, as determined by a sensor, and the desired response.

Microcomputer-based measurement system and digital controller. (Courtesy of John Fluke Manufacturing Company.)

Generally, a controller, or filter, is required to process the error signal in order that certain control criteria, or specifications, will be satisfied. The criteria may involve, but not be limited to:

1. Disturbance rejection
2. Steady-state errors
3. Transient response
4. Sensitivity to parameter changes in the plant

Solving the control problem will generally involve:

1. Choosing sensors and any actuators required to drive the plant
2. Developing the plant, sensor, and actuator models (equations)

3. Designing the controller based on the developed models and the control criteria
4. Evaluating the design analytically, by simulation, and by system tests

1.4 SATELLITE MODEL

As the first example of the development of the mathematical model of a physical system, we will consider the attitude control system of a satellite. Assume that the satellite is spherical and has the thrustor configuration shown in Figure 1-4. Suppose that θ is the yaw angle of the satellite. In addition to the thrustors shown, thrustors will also control the pitch angle and the roll angle, giving complete three-axis control of the satellite. We will consider only the yaw-axis control systems, whose purpose is to control the angle $\theta(t)$.

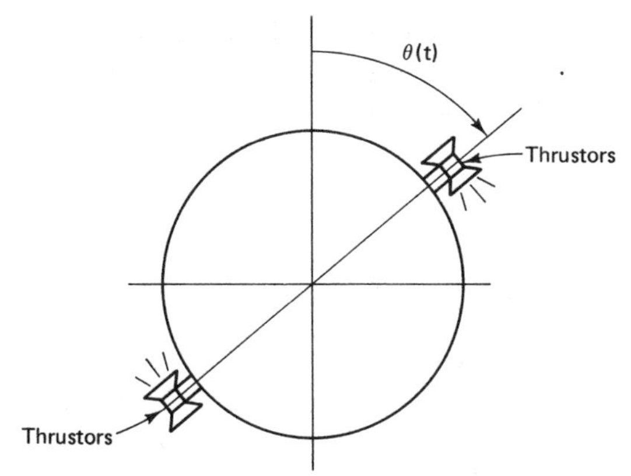

Figure 1-4 Satellite.

For the satellite, the thrustors, when active, apply a torque $\tau(t)$. The torque of the two active thrustors shown in Figure 1-4 tends to reduce $\theta(t)$. The other two thrustors shown tend to increase $\theta(t)$.

Since there is essentially no friction in the environment of a satellite, and assuming the satellite to be rigid, we can write

$$J\frac{d^2\theta}{dt^2} = \tau \tag{1-1}$$

where J is the satellite's moment of inertia in the yaw axis. Taking the Laplace transform of (1-1) yields

$$Js^2\Theta(s) = T(s) = \mathcal{L}[\tau(t)] \tag{1-2}$$

which can be expressed as

$$\frac{\Theta(s)}{T(s)} = G(s) = \frac{1}{Js^2} \tag{1-3}$$

The ratio of the Laplace transforms of the output variable $[\theta(t)]$ to input variable $[\tau(t)]$ is called the plant transfer function, and is denoted here as $G(s)$. A brief review of the Laplace transform is given in Section 1.7.

The model of the satellite may be specified by either the second-order differential equation of (1-1) or the second-order transfer function of (1-3). A third model is the state-variable model, which we will now develop. Suppose that we define the variables x_1 and x_2 as

$$x_1 = \theta \tag{1-4}$$

$$x_2 = \dot{x}_1 = \dot{\theta} \tag{1-5}$$

where $\dot{x}_1$ denotes the derivative of x_1 with respect to time. Then, from (1-1) and (1-5),

$$\dot{x}_2 = \ddot{\theta} = \frac{1}{J}\tau \tag{1-6}$$

We can now write (1-5) and (1-6) in vector-matrix form (see Appendix IV):

$$\begin{bmatrix} \dot{x}_1 \\ \dot{x}_2 \end{bmatrix} = \begin{bmatrix} 0 & 1 \\ 0 & 0 \end{bmatrix} \begin{bmatrix} x_1 \\ x_2 \end{bmatrix} + \begin{bmatrix} 0 \\ \frac{1}{J} \end{bmatrix} \tau \tag{1-7}$$

In this equation, x_1 and x_2 are called the state variables. Hence we may specify the model of the satellite in the form of (1-1), or (1-3), or (1-7). State-variable models of analog systems are considered in greater detail in Chapter 4.

1.5 SERVOMOTOR SYSTEM MODEL

In this section the model of a servo system (a positioning system) is derived. An example of this type system is an antenna tracking system. In this system, an electric motor is utilized to rotate a radar antenna which automatically tracks an aircraft. The error signal, which is proportional to the difference between the pointing direction of the antenna and the line of sight to the aircraft, is amplified and drives the motor in the appropriate direction so as to reduce this error.

A dc motor system is shown in Figure 1-5. The motor is armature controlled with a constant field. The armature resistance and inductance are R_a and L_a, respectively. We assume that the inductance L_a can be ignored, which is the case for many servomotors. The motor back emf e_m is given by [4]

$$e_m = K_b\omega = K_b\frac{d\theta}{dt} \tag{1-8}$$

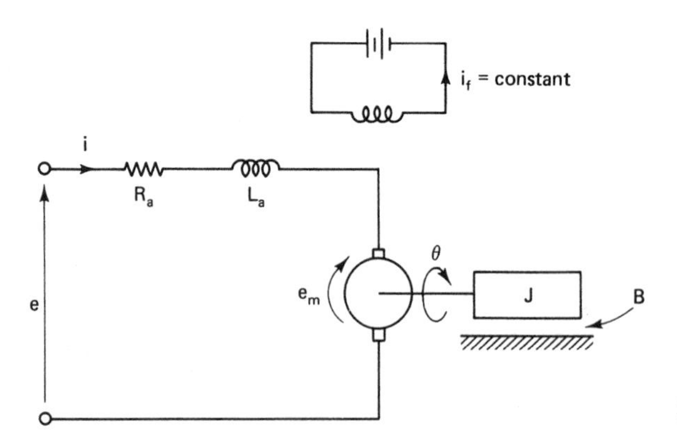

Figure 1-5 Servomotor system.

where θ is the motor shaft position, ω is the shaft angular velocity, and K_b is a motor-dependent constant. The total moment of inertia connected to the motor shaft is J, and B is the total viscous friction. Letting τ be the torque developed by the motor, we write

$$\tau = J\frac{d^2\theta}{dt^2} + B\frac{d\theta}{dt} \tag{1-9}$$

The developed torque for this motor is given by

$$\tau = K_T i \tag{1-10}$$

where i is the armature current and K_T is a constant. The final equation required is the voltage equation for the armature circuit.

$$e = iR_a + e_m \tag{1-11}$$

These four equations may be solved for $\theta(t)$, the output, as a function of $e(t)$, the input. First, from (1-11) and (1-8),

$$i = \frac{e - e_m}{R_a} = \frac{e}{R_a} - \frac{K_b}{R_a}\frac{d\theta}{dt} \tag{1-12}$$

Then, from (1-9), (1-10), and (1-12),

$$\tau = K_T i = \frac{K_T}{R_a}e - \frac{K_T K_b}{R_a}\frac{d\theta}{dt} = J\frac{d^2\theta}{dt^2} + B\frac{d\theta}{dt} \tag{1-13}$$

This equation may be written as

$$J\frac{d^2\theta}{dt^2} + \frac{BR_a + K_T K_b}{R_a}\frac{d\theta}{dt} = \frac{K_T}{R_a}e \tag{1-14}$$

which is the desired model. If the armature inductance were not negligible, this model would be third order.

If we take the Laplace transform of (1-14) and solve for the transfer function, the result is

$$\frac{\Theta(s)}{E(s)} = G(s) = \frac{K_T/R_a}{Js^2 + \dfrac{BR_a + K_T K_b}{R_a}s} = \frac{K_T/JR_a}{s\left(s + \dfrac{BR_a + K_T K_b}{JR_a}\right)} \tag{1-15}$$

Many of the examples of this text are based on this transfer function.

The state-variable model of this system is derived as in the preceding section. Let

$$\begin{aligned} x_1 &= \theta \\ x_2 &= \dot{\theta} \end{aligned} \tag{1-16}$$

Then, from (1-14),

$$\dot{x}_2 = \ddot{\theta} = -\frac{BR_a + K_T K_b}{JR_a}x_2 + \frac{K_T}{JR_a}e \tag{1-17}$$

Hence the state equations can be written as

$$\begin{bmatrix}\dot{x}_1 \\ \dot{x}_2\end{bmatrix} = \begin{bmatrix}0 & 1 \\ 0 & -\dfrac{BR_a + K_T K_b}{JR_a}\end{bmatrix}\begin{bmatrix}x_1 \\ x_2\end{bmatrix} + \begin{bmatrix}0 \\ \dfrac{K_T}{JR_a}\end{bmatrix}e \tag{1-18}$$

A closed-loop servo system, which in this case is an antenna pointing system, is shown in Figure 1-6a. The electric motor rotates the antenna and the sensor, which is a potentiometer. Hence the voltage out of the potentiometer, v_i, is directly proportional to the antenna position. The desired position is indicated by the input potentiometer. If the relative positions of the two potentiometers are different, the error signal is nonzero and a voltage is applied to the motor to cause rotation of the motor shaft. If the system operates satisfactorily, the output potentiometer then moves to the same relative position as that of the input potentiometer.

The system block diagram is given in Figure 1-6b. Since the error signal is normally a low-power signal, a power amplifier is required to drive the motor. However, this amplifier introduces a nonlinearity, since an amplifier has a maximum output voltage and can be saturated at this value. Suppose that the amplifier has a gain of 5 and a maximum output of 24 V. Then the amplifier input–output characteristic is as shown in Figure 1-6c. The amplifier saturates at an input of 4.8 V; hence, for an error signal larger than 4.8 V in magnitude, the system is nonlinear.

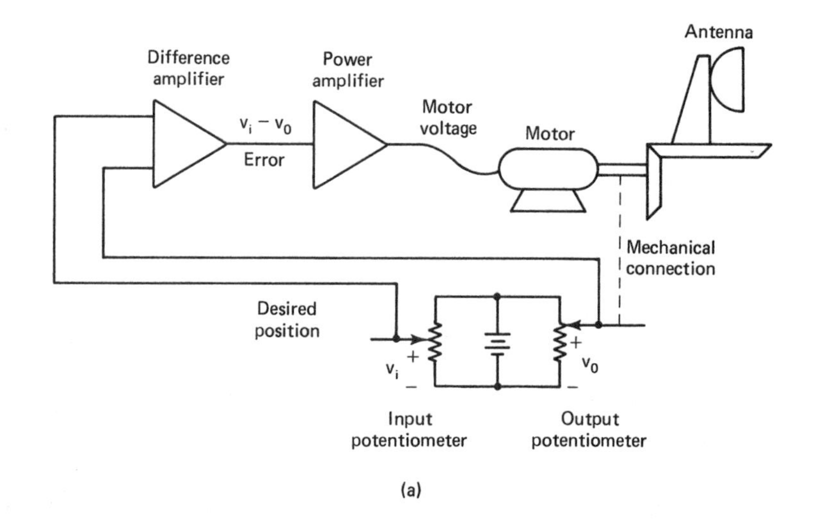

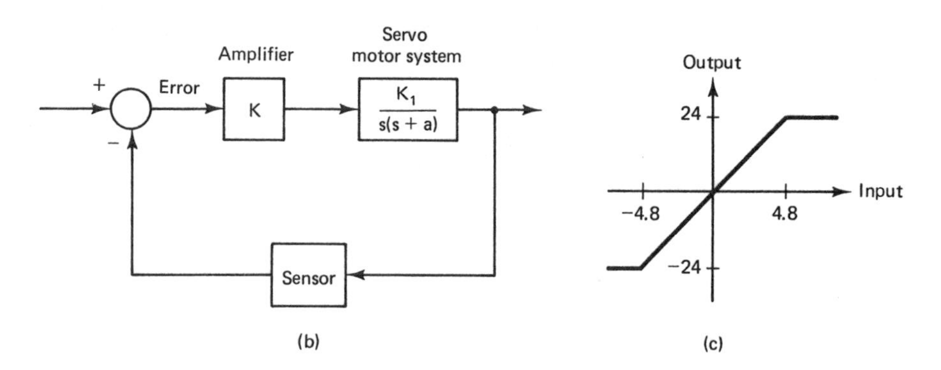

Figure 1-6 Servo control system.

In many control systems, we go to great lengths to ensure that the system operation is confined to linear regions. In other systems, we purposely design for nonlinear operation. For example, in this servo system, we must apply maximum voltage to the motor to achieve maximum speed of response. Thus for large error signals we would have the amplifier saturated in order to achieve a fast response.

The analysis and design of nonlinear systems is beyond the scope of this book; we will always assume that the system under consideration is operating in a linear mode.

1.6 TEMPERATURE CONTROL SYSTEM

As a third example of modeling, a thermal system will be considered. It is desired to control the temperature of a liquid in a tank. Liquid is flowing out at some rate, being replaced by liquid at temperature T_i as shown in Figure 1-7. A mixer agitates the liquid such that the liquid temperature can be assumed uniform at a value T throughout the tank. The liquid is heated by an electric heater.

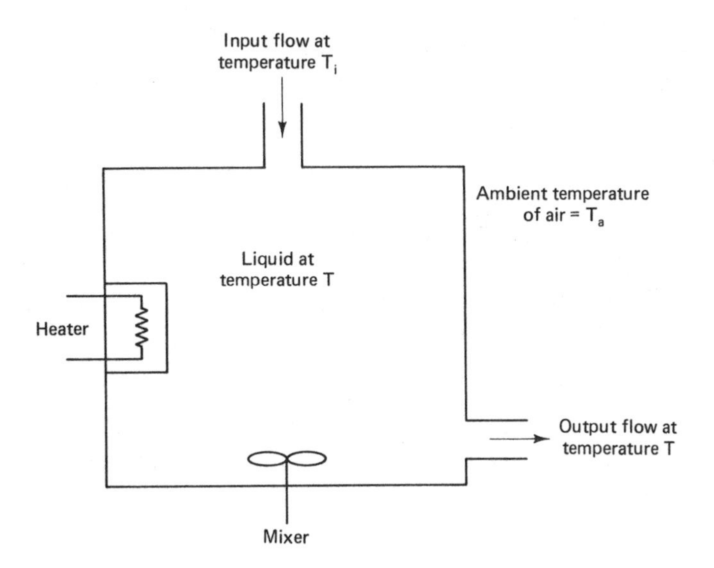

Figure 1-7 Thermal system.

We first make the following definitions:

Q_e = heat flow supplied by electric heater
Q_l = heat flow into liquid
Q_o = heat flow via liquid leaving tank
Q_i = heat flow via liquid entering tank
Q_s = heat flow through tank surface

By the conservation of energy, heat added to the tank of liquid must equal that lost from the tank. Thus

$$Q_e + Q_i = Q_l + Q_o + Q_s \tag{1-19}$$

Now [5]

$$Q_l = C\frac{dT}{dt} \tag{1-20}$$

where C is the thermal capacity of the liquid in the tank. Letting V equal the flow into and out of the tank (assumed equal) and H equal the specific heat of the liquid, we can write

$$Q_i = VHT_i \tag{1-21}$$

and

$$Q_o = VHT \tag{1-22}$$

The final equation needed is

$$Q_s = \frac{T - T_a}{R} \tag{1-23}$$

where R is the thermal resistance to heat flow through the tank surface and T_a is the ambient air temperature outside the tank.

Substituting (1-20) through (1-23) into (1-19) yields

$$Q_e + VHT_i = C\frac{dT}{dt} + VHT + \frac{T - T_a}{R} \tag{1-24}$$

This is a first-order linear differential equation with Q_e, T_i, and T_a as forcing functions. In terms of a control system, Q_e is the control input and T_i and T_a are disturbance inputs. If the liquid flow V is a function of time, (1-24) is a first-order linear time-varying differential equation.

To simplify the discussion, we will assume that V is constant. Taking the Laplace transform of (1-24) and solving for $T(s)$ yields

$$T(s) = \frac{Q_e(s)}{Cs + VH + (1/R)} + \frac{VHT_i(s)}{Cs + VH + (1/R)} + \frac{(1/R)T_a(s)}{Cs + VH + (1/R)} \tag{1-25}$$

Different configurations may be used to express (1-25) as a block diagram; one is given in Figure 1-8.

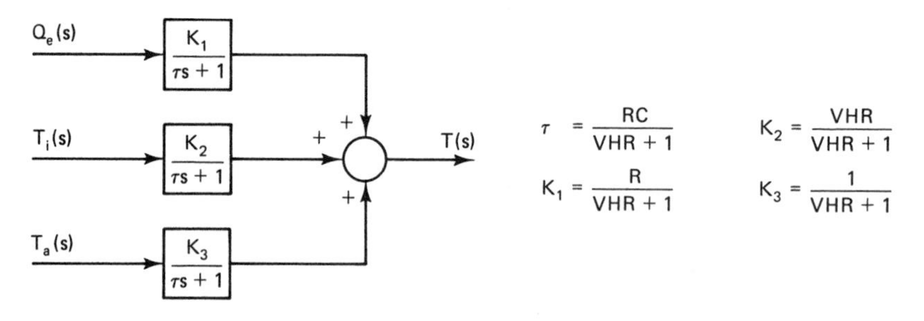

Figure 1-8 Block diagram of a thermal system.

If we ignore the disturbance inputs, the transfer-function model of the system is simple and first order. However, at some step in the control system design the disturbances must be considered. Quite often a major specification in a control system design is the minimization of system response to disturbance inputs.

1.7 THE LAPLACE TRANSFORM

A brief review of the Laplace transform is presented in this section. The Laplace transform is useful to the controls engineer in modeling a linear time-invariant analog system as a transfer function. The Laplace transform may also be used to find the time response of this type of system; however, we generally use simulations (machine solutions of the system equations) for this purpose.

The Laplace transform of a function $f(t)$ is defined as [6]

$$F(s) = \mathcal{L}[f(t)] = \int_0^\infty f(t)\epsilon^{-st}\,dt \tag{1-26}$$

and the inverse Laplace transform of $F(s)$ is then given by

$$f(t) = \frac{1}{2\pi j}\int_{\sigma-j\infty}^{\sigma+j\infty} F(s)\epsilon^{st}\,ds, \qquad j = \sqrt{-1} \tag{1-27}$$

In (1-27), σ is determined by the singularities of $F(s)$ [6]. We use (1-26) to construct a table of Laplace transform. Then, when possible, we use the table to find inverse transforms, rather than using (1-27).

As an example, we will derive the Laplace transform of the exponential function ϵ^{-at}. From (1-26),

$$\begin{aligned}\mathcal{L}[\epsilon^{-at}] &= \int_0^\infty \epsilon^{-at}\epsilon^{-st}\,dt = \int_0^\infty \epsilon^{-(s+a)t}\,dt \\ &= \frac{-\epsilon^{-(s+a)t}}{s+a}\bigg|_0^\infty \\ &= \frac{1}{s+a}, \qquad \mathrm{Re}\,(s+a) > 0\end{aligned} \tag{1-28}$$

where Re denotes the real part of the expression. A short table of commonly required transforms is given in Table 1-1.

In the development of analysis and design techniques for digital control systems, we will require several theorems from Laplace transform theory. As an example of these theorems, we will develop the final-value theorem. We wish to calculate, from $F(s)$, the final value of $f(t)$ [i.e., $\lim_{t\to\infty} f(t)$]. However, to derive this theorem, we must find the Laplace transform of the derivative of $f(t)$.

$$\mathcal{L}\left[\frac{df}{dt}\right] = \int_0^\infty \epsilon^{-st}\frac{df}{dt}\,dt \tag{1-29}$$

TABLE 1-1 LAPLACE TRANSFORMS

Name	Time function, $f(t)$	Laplace transform, $F(s)$
Unit impulse	$\delta(t)$	1
Unit step	$u(t)$	$\frac{1}{s}$
Unit ramp	t	$\frac{1}{s^2}$
nth-order ramp	t^n	$\frac{n!}{s^{n+1}}$
Exponential	ϵ^{-at}	$\frac{1}{s+a}$
Sine	$\sin \omega t$	$\frac{\omega}{s^2+\omega^2}$
Cosine	$\cos \omega t$	$\frac{s}{s^2+\omega^2}$
Damped sine	$\epsilon^{-at} \sin \omega t$	$\frac{\omega}{(s+a)^2+\omega^2}$
Damped cosine	$\epsilon^{-at} \cos \omega t$	$\frac{s+a}{(s+a)^2+\omega^2}$

This expression can be integrated by parts, with

$$u = \epsilon^{-st}, \qquad dv = \frac{df}{dt}\,dt$$

Thus

$$\begin{aligned} \mathcal{L}\left[\frac{df}{dt}\right] &= f(t)\epsilon^{-st}\Big|_0^\infty + s\int_0^\infty \epsilon^{-st} f(t)\,dt \\ &= 0 - f(0) + sF(s) \\ &= sF(s) - f(0) \end{aligned} \tag{1-30}$$

Now the final-value theorem can be derived. From (1-29),

$$\lim_{s\to 0} \int_0^\infty \epsilon^{-st}\frac{df}{dt}\,dt = \int_0^\infty \frac{df}{dt}\,dt = \lim_{t\to\infty} f(t) - f(0) \tag{1-31}$$

Hence, from (1-29), (1-30), and (1-31),

$$\lim_{t\to\infty} f(t) - f(0) = \lim_{s\to 0}\,[sF(s) - f(0)]$$

or

$$\lim_{t\to\infty} f(t) = \lim_{s\to 0} sF(s) \tag{1-32}$$

provided that the limit on the left side of the equation exists. A short table of theorems is given in Table 1-2.

An example illustrating the use of the Laplace transform will now be given.

TABLE 1-2 LAPLACE TRANSFORM THEOREMS

Name	Theorem
Derivative	$\mathcal{L}\left[\frac{df}{dt}\right] = sF(s) - f(0^+)$
nth-order derivative	$\mathcal{L}\left[\frac{d^n f}{dt^n}\right] = s^n F(s) - s^{n-1}f(0^+) - \cdots - f^{(n-1)}(0^+)$
Integral	$\mathcal{L}\left[\int_0^t f(t)\,dt\right] = \frac{F(s)}{s}$
Shifting	$\mathcal{L}[f(t-t_0)u(t-t_0)] = \epsilon^{-t_0 s}F(s)$
Initial value	$\lim_{t\to 0} f(t) = \lim_{s\to\infty} sF(s)$
Final value	$\lim_{t\to\infty} f(t) = \lim_{s\to 0} sF(s)$
Frequency shift	$\mathcal{L}[\epsilon^{-at}f(t)] = F(s+a)$

Example 1.1

Suppose that the servo system of Figure 1-5 has a transfer function of

$$\frac{\Theta(s)}{E(s)} = G(s) = \frac{1}{s(s+1)}$$

A unit step of 10 V is applied to the motor, and it is desired to find the motor velocity $d\theta/dt$. Now

$$\Theta(s) = G(s)E(s) = \frac{10}{s^2(s+1)}$$

Since this transform does not appear in Table 1-1, a partial-fraction expansion [6] must be used:

$$\Theta(s) = \frac{10}{s^2(s+1)} = \frac{10}{s^2} - \frac{10}{s} + \frac{10}{s+1}$$

Hence, from Table 1-1,

$$\theta(t) = 10t - 10 + 10\epsilon^{-t}$$

and

$$\frac{d\theta}{dt} = 10(1 - \epsilon^{-t})$$

Thus the motor speed exponentially approaches a constant value.

1.8 SUMMARY

This chapter has introduced the idea of a closed-loop control system, in particular the idea of a closed-loop digital control system. Next, the describing mathematical equations of three different physical systems were derived, to acquaint the reader with the mathematics required to analyze continuous-time (analog) systems. These systems were described by ordinary linear differential equations which are time invariant. The Laplace transform, which is useful in solving these equations, was then described briefly.

REFERENCES

1. M. Athans, M. L. Dertouzos, R. N. Spann, and S. J. Mason, *Systems, Networks, and Computations: Multivariable Methods*. New York: McGraw-Hill Book Company, 1974.
2. R. F. Wigginton, "Evaluation of OPS-II Operational Program for the Automatic Carrier Landing System," Naval Electronic Systems Test and Evaluation Facility, Saint Inigoes, Md., 1971.
3. C. L. Phillips, E. R. Graf, and H. T. Nagle, Jr., "MATCALS Error and Stability Analysis," Report AU-EE-75-2080-1, Auburn University, Auburn, Ala., 1975.
4. A. E. Fitzgerald and C. Kingsley, Jr., *Electric Machinery*. New York: McGraw-Hill Book Company, 1961.
5. J. D. Trimmer, *Response of Physical Systems*. New York: John Wiley & Sons, Inc., 1950.
6. D. K. Cheng, *Analysis of Linear Systems*. Reading, Mass: Addison-Wesley Publishing Company, Inc., 1959.

PROBLEMS

1-1. The satellite of Section 1.4 is connected in the closed-loop control system shown in Figure P1-1. The torque is directly proportional to the error signal. Derive the transfer function $\Theta(s)/\Theta_c(s)$, where $\theta_c(t)$ is the commanded attitude angle.

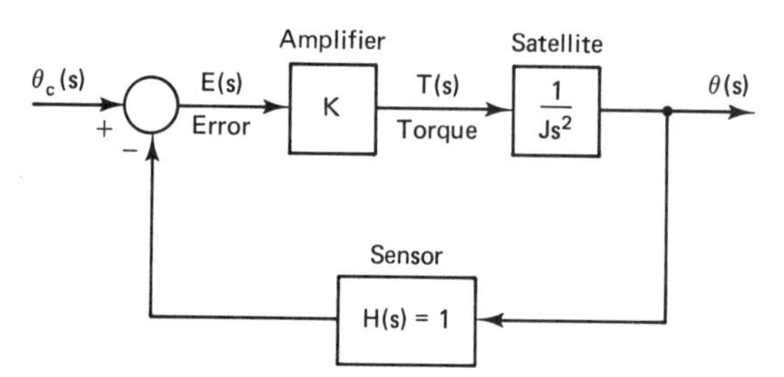

Figure P1-1 System for Problem 1-1.

1-2. In the system of Problem 1-1, $J = 0.5$ and $K = 10$ in appropriate units. The attitude of the satellite is initially at 0°. At $t = 0$, the attitude is commanded to 20°; that is, a 20° step is applied to the system at $t = 0$. Find the response $\theta(t)$.

1-3. Find the state-variable model of the system of Problem 1-1.

1-4. The antenna positioning system described in Section 1.5 is shown in Figure P1-4a. Let $\theta(t)$ be the antenna pointing angle, and suppose that the gears and the potentiometer power supply in the system are such that

$$v_0(t) = 0.05\theta(t)$$

where the units of $v_0(t)$ are volts and $\theta(t)$ is in degrees. Suppose that the gain of the power amplifier is 5 V/V, and with the motor voltage denoted as $e(t)$, the transfer function of the motor/antenna system is

$$\frac{\theta(s)}{E(s)} = \frac{10}{s(s+1)}$$

(a) The system block diagram is given in Figure P1-4b. Add the required gains and transfer functions to the block diagram.

(b) A step input of 30° is applied at the system input at $t = 0$. Find the response $\theta(t)$.

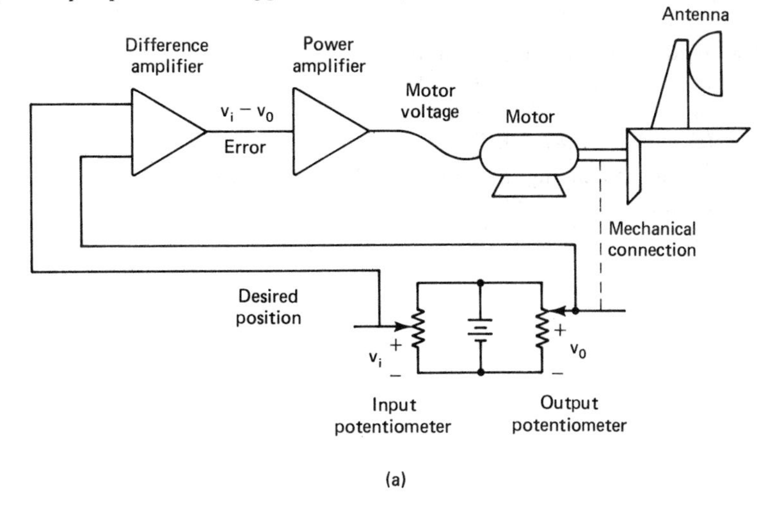

(a)

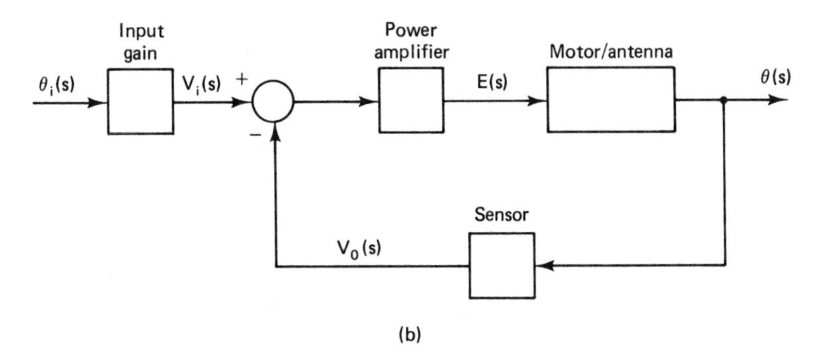

(b)

Figure P1-4 System for Problem 1-4.

1-5. Find the state-variable model for the system of Problem 1-4.

1-6. Show that the transfer function for the system of Figure P1-6 is

$$\frac{C(s)}{R(s)} = \frac{G(s)}{1 + G(s)H(s)}$$

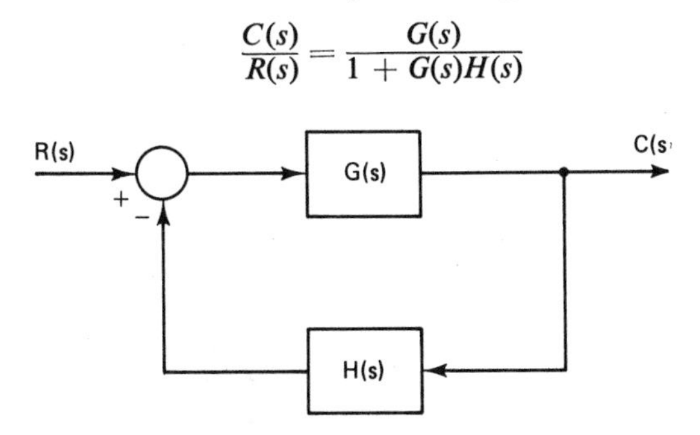

Figure P1-6 System for Problem 1-6.

1-7. Find the Laplace transform of:

(a) $f(t) = 3\epsilon^{-0.7t} \cos (5t)$

(b) $f(t) = \sin (3t + 30°)$

(c) $f(t) = t\epsilon^{-4t}$

1-8. Find the Laplace transform of:

(a) $f(t) = \cos (t - 1)u(t - 1)$

(b) $f(t) = \epsilon^{-(t-0.5)}u(t - 0.5)$

(c) $f(t) = \epsilon^{-t}u(t - 0.5)$

1-9. Using the defining integral of the Laplace transform, (1-26), derive the Laplace transform of:

(a) $f(t) = \epsilon^{-2t}$

(b) $f(t) = t$

2

Discrete-Time Systems and the z-Transform

2.1 INTRODUCTION

In this chapter two important concepts are introduced: a discrete-time system and the z-transform. In contrast to the continuous-time system whose operation is described or modeled by a set of differential equations, the discrete-time system is one whose operation is described by a set of difference equations. The transform method employed in the analysis of linear time-invariant continuous-time systems is the Laplace transform; in a similar manner, the transform used in the analysis of linear time-invariant discrete-time systems is the z-transform. The modeling of discrete-time systems by both transfer functions and state-variable equations is presented.

2.2 DISCRETE-TIME SYSTEMS

To illustrate the idea of a discrete-time system, consider the digital control system shown in Figure 2-1a. The digital computer performs the compensation function within the system. The interface at the input of the computer is an analog-to-digital (A/D) converter, and is required to convert the error signal, which is a continuous-time signal, into a form that can be readily processed by the computer. At the computer output a digital-to-analog (D/A) converter is required to convert the digital signals of the computer into a form necessary to drive the plant.

We will now consider the following example. Suppose that the A/D converter, the digital computer, and the D/A converter are to replace an analog, or continuous-

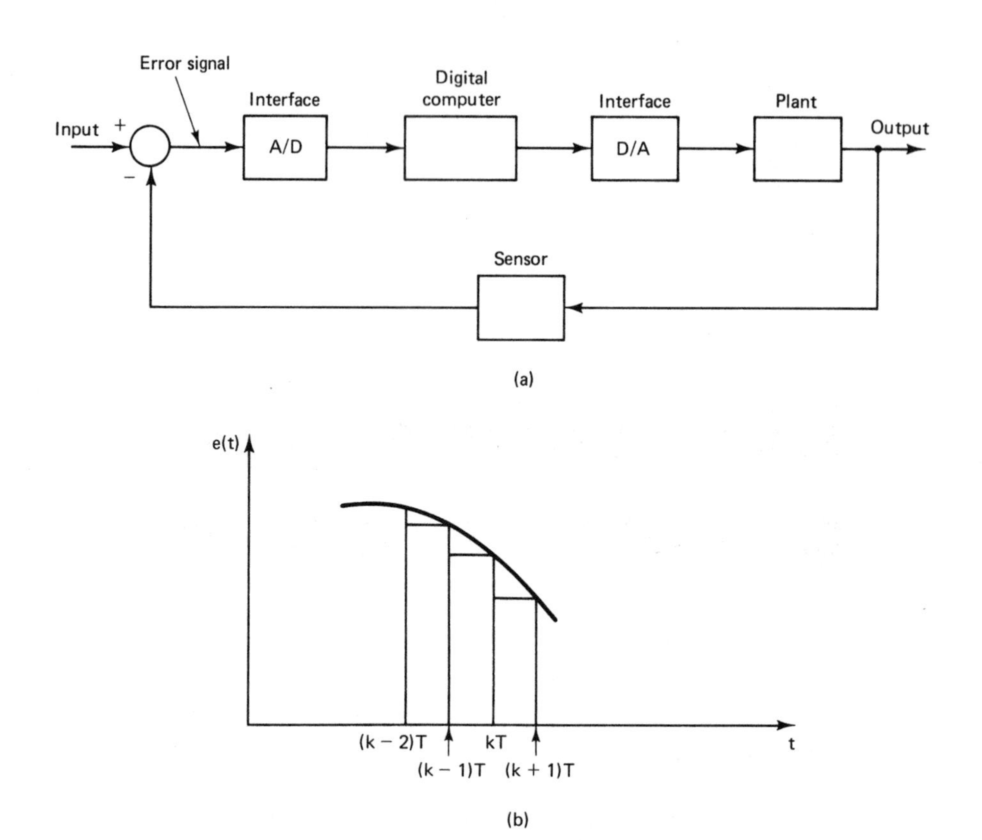

Figure 2-1 Digital control system.

time, proportional-integral (PI) compensator, such that the digital control system response has essentially the same characteristics as the analog system. (The PI controller is discussed in Chapter 8.) The analog controller output is given by

$$m(t) = K_P e(t) + K_I \int_0^t e(t)\, dt \tag{2-1}$$

where $e(t)$ is the controller input signal, $m(t)$ is the controller output signal, and K_P and K_I are constant gains determined by the design process.

Since the digital computer can be programmed to multiply, add, and integrate numerically, the controller equation can be realized using the digital computer. For this example, the rectangular rule of numerical integration, illustrated in Figure 2-1, will be employed. Of course, other algorithms of numerical integration may also be used. From Figure 2-1b, letting $x(t)$ be the numerical integral of $e(t)$, we can write

$$x(kT) = x[(k-1)T] + Te(kT) \tag{2-2}$$

where T is the numerical algorithm step size, in seconds. Then (2-1) becomes, for the digital compensator,

$$m(kT) = K_P e(kT) + K_I x(kT)$$

Equation (2-2) is a first-order linear *difference equation*. The general form of a first-order linear time-invariant difference equation is (with the T omitted for convenience)

$$x(k) = \alpha_1 e(k) + \alpha_0 e(k-1) - \beta_0 x(k-1) \tag{2-3}$$

This equation is first order since the signals from only the last sampling instant enter the equation. The general form of an nth-order linear difference equation is

$$\begin{aligned} x(k) = \alpha_n e(k) + \alpha_{n-1} e(k-1) + \cdots + \alpha_0 e(k-n) \\ - \beta_{n-1} x(k-1) - \cdots - \beta_0 x(k-n) \end{aligned} \tag{2-4}$$

It will be shown in Chapter 5 that, if the plant in Figure 2-1 is also linear and time invariant, the entire system may be modeled by a difference equation of the form of (2-4), which is generally of higher order than that of the controller. Compare (2-4) to a linear differential equation describing an nth-order continuous-time system.

$$\begin{aligned} y(t) = a_0 e(t) + a_1 \frac{de(t)}{dt} + \cdots + a_n \frac{d^n e(t)}{dt^n} \\ - b_1 \frac{dy(t)}{dt} - \cdots - b_n \frac{d^n y(t)}{dt^n} \end{aligned} \tag{2-5}$$

The describing equation of a linear, time-invariant analog (i.e., continuous-time) filter is also of the form of (2-5). The device that realizes this filter, usually an R-L-C network, perhaps with operational amplifiers, can be considered to be an analog computer programmed to solve (2-5). In a like manner, (2-4) is the describing equation of a linear, time-invariant discrete filter, which is usually called a digital filter. The device that realizes this filter is a digital computer programmed to solve (2-4), or a special-purpose computer built specifically to solve (2-4). Thus the digital computer in Figure 2-1 would be programmed to solve a difference equation of the form of (2-4), and the problem of the control system designer would be to determine (1) T, the sampling period; (2) n, the order of the difference equation; and (3) α_i and β_i, the filter coefficients, such that the system has certain desired characteristics. There are additional problems in the realization of the digital filter: for example, the computer wordlength required to keep system errors caused by round-off in the computer at an acceptable level. As an example, a digital filter (controller) was designed to compensate the attitude control system of the booster stage of the Saturn V rocket [1]. In this design, T was 0.05 s and the controller was of fifth order. The minimum wordlength required in the computer was found to be 32 bits, in order that system errors caused by round-off in the filter remain at an acceptable level.

In many applications other than control systems, digital filters have been designed to replace analog filters and the problems encountered are the same as those listed above.

2.3 TRANSFORM METHODS

In linear time-invariant continuous-time systems, the Laplace transform can be utilized in system analysis and design. For example, an alternative, but equally valid description of the operation of a system described by (2-5) is obtained by taking the

The NASA space shuttle utilizes a digital flight control system. (Courtesy of NASA.)

Laplace transform of this equation and solving for the transfer function:

$$\frac{Y(s)}{E(s)} = \frac{a_n s^n + \cdots + a_1 s + a_0}{b_n s^n + \cdots + b_1 s + 1} \tag{2-6}$$

In a similar manner, a transform will now be defined that can be utilized in the analysis of discrete-time systems modeled by difference equations of the form given in (2-4).

A transform is defined for number sequences as follows. The function $E(z)$ is defined as a power series in z^{-k} with coefficients equal to the values of the number sequence $\{e(k)\}$. This transform, called the z-transform, is then expressed as

$$E(z) = \mathfrak{z}[\{e(k)\}] = e(0) + e(1)z^{-1} + e(2)z^{-2} + \cdots \tag{2-7}$$

where $\mathfrak{z}(\cdot)$ indicates the z-transform operation. Equation (2-7) can be written in compact notation as

$$E(z) = \mathfrak{z}[\{e(k)\}] = \sum_{k=0}^{\infty} e(k)z^{-k} \tag{2-8}$$

The z-transform is defined for any number sequence $\{e(k)\}$, and may be used in the analysis of any type of system described by difference equations. For example, the z-transform is used in discrete probability problems, and for this case the numbers in the sequence $\{e(k)\}$ are discrete probabilities [2].

Equation (2-7) is the definition of the single-sided z-transform. The double-sided z-transform, sometimes called the generating function [3], is defined as

$$G[\{f(k)\}] = \sum_{k=-\infty}^{\infty} f(k)z^{-k} \tag{2-9}$$

Throughout this text, only the single-sided z-transform as defined in (2-7) will be used, and this transform will be referred to as the ordinary z-transform. If the sequence

$e(k)$ is generated from a time function $e(t)$ by sampling every T seconds, $e(k)$ is understood to be $e(kT)$ (i.e., the T is dropped for convenience).

Three examples will now be given to illustrate the z-transform.

Example 2.1

Given $E(z)$ below, find $\{e(k)\}$.

$$E(z) = 1 + 3z^{-1} - 2z^{-2} + z^{-4} + \cdots$$

We know, then, from (2-7), that the values of the number sequence $\{e(k)\}$ are

$$\begin{aligned} e(0) &= 1 \qquad & e(3) &= 0 \\ e(1) &= 3 \qquad & e(4) &= 1 \\ e(2) &= -2 \qquad & &\cdots \end{aligned}$$

Consider now the identity

$$\frac{1}{1-x} = 1 + x + x^2 + x^3 + \cdots, \qquad |x| < 1 \tag{2-10}$$

This power series is useful, in some cases, in expressing $E(z)$ in closed form, as will be illustrated in the next two examples.

Example 2.2

Given that $e(k) = 1$ for all k, find $E(z)$. By definition $E(z)$ is

$$E(z) = 1 + z^{-1} + z^{-2} + \cdots$$

Now, employing (2-10),

$$E(z) = \frac{1}{1 - z^{-1}} = \frac{z}{z-1}, \qquad |z^{-1}| < 1 \tag{2-11}$$

Note that $\{e(k)\}$ may be generated by sampling a unit step function. However, there are many other time functions that have a value of unity every T seconds, and thus all have the same z-transform.

Example 2.3

Given that $e(k) = \epsilon^{-akT}$, find $E(z)$. $E(z)$ can be written in power series form as

$$\begin{aligned} E(z) &= 1 + \epsilon^{-aT}z^{-1} + \epsilon^{-2aT}z^{-2} + \cdots \\ &= 1 + (\epsilon^{-aT}z^{-1}) + (\epsilon^{-aT}z^{-1})^2 + \cdots \end{aligned}$$

$E(z)$ can be put in closed form by applying (2-10), so that

$$E(z) = \frac{1}{1 - \epsilon^{-aT}z^{-1}} = \frac{z}{z - \epsilon^{-aT}}, \qquad |\epsilon^{-aT}z^{-1}| < 1$$

Note that, in this example, $\{e(k)\}$ may be generated by sampling the function $e(t) = \epsilon^{-at}$.

2.4 THEOREMS OF THE z-TRANSFORM

Several theorems of the z-transform will now be developed. These theorems will prove to be useful in the analysis of discrete systems.

Addition and Subtraction

Theorem. The z-transform of a sum of number sequences is equal to the sum of the z-transforms of the number sequences; that is,

$$\mathfrak{z}[\{e_1(k)\} \pm \{e_2(k)\}] = E_1(z) \pm E_2(z) \tag{2-12}$$

Proof. From the definition of the z-transform,

$$\mathfrak{z}[\{e_1(k)\} \pm \{e_2(k)\}] = \sum_{k=0}^{\infty} [e_1(k) \pm e_2(k)]z^{-k}$$

$$= \sum_{k=0}^{\infty} e_1(k)z^{-k} \pm \sum_{k=0}^{\infty} e_2(k)z^{-k} = E_1(z) \pm E_2(z)$$

Multiplication by a Constant

Theorem. The z-transform of a number sequence multiplied by a constant is equal to the constant multiplied by the z-transform of the number sequence:

$$\mathfrak{z}[a\{e(k)\}] = a\mathfrak{z}[\{e(k)\}] = aE(z) \tag{2-13}$$

Proof. From the definition of the z-transform,

$$\mathfrak{z}[a\{e(k)\}] = \sum_{k=0}^{\infty} ae(k)z^{-k} = a \sum_{k=0}^{\infty} e(k)z^{-k} = aE(z)$$

Example 2.4

In a manner similar to that shown above, the linearity property of the z-transform can be proved as follows. Let

$$\{e(k)\} = a\{e_1(k)\} + b\{e_2(k)\}$$

Then

$$\begin{aligned}\mathfrak{z}[\{e(k)\}] &= \mathfrak{z}[a\{e_1(k)\} + b\{e_2(k)\}] \\ &= \sum_{k=0}^{\infty} [ae_1(k) + be_2(k)]z^{-k} \\ &= a \sum_{k=0}^{\infty} e_1(k)z^{-k} + b \sum_{k=0}^{\infty} e_2(k)z^{-k} \\ &= a\mathfrak{z}[e_1(k)] + b\mathfrak{z}[e_2(k)] \\ &= aE_1(z) + bE_2(z)\end{aligned}$$

Real Translation

Theorem. Let n be a positive integer, and let $e(k)$ be zero for k less than zero. Further, let $E(z)$ be the z-transform of $\{e(k)\}$. Then

$$\mathfrak{z}[\{e(k - n)\}] = z^{-n}E(z) \tag{2-14}$$

and

$$\mathfrak{z}[\{e(k + n)\}] = z^{n}\left[E(z) - \sum_{k=0}^{n-1} e(k)z^{-k}\right] \tag{2-15}$$

In the use of the ordinary z-transform, it is understood that

$$e(k) = e(k)u(k)$$

where $u(k)$ is the unit step defined by

$$u(k) = \begin{cases} 0, & k < 0 \\ 1, & k \geq 0 \end{cases}$$

For (2-14), it is necessary that

$$e(k - n) = e(k - n)u(k - n)$$

and in (2-15),

$$e(k + n) = e(k + n)u(k)$$

With these definitions, the proof for (2-14) and (2-15) will be given.

Proof. From the definition of the z-transform,

$$\mathfrak{z}[\{e(k - n)\}] = e(-n) + e(-n + 1)z^{-1} + \cdots + e(0)z^{-n} + e(1)z^{-(n+1)} + \cdots$$
$$= z^{-n}[e(0) + e(1)z^{-1} + e(2)z^{-2} + \cdots] = z^{-n}E(z)$$

since $e(k) = 0$ for $k < 0$. Also,

$$\mathfrak{z}[\{e(k + n)\}] = e(n) + e(n + 1)z^{-1} + e(n + 2)z^{-2} + \cdots$$

By adding and subtracting terms, we obtain

$$\mathfrak{z}[\{e(k + n)\}] = z^n[e(0) + e(1)z^{-1} + \cdots + e(n)z^{-n} + e(n + 1)z^{-(n+1)} + \cdots$$
$$- e(0) - e(1)z^{-1} - \cdots - e(n - 1)z^{-(n-1)}]$$

or

$$\mathfrak{z}[\{e(k + n)\}] = z^n\left[E(z) - \sum_{k=0}^{n-1} e(k)z^{-k}\right]$$

Example 2.5

It was shown in Example 2.3 that

$$\mathfrak{z}[\{\epsilon^{-akT}\}] = \frac{z}{z - \epsilon^{-aT}}$$

Thus

$$\mathfrak{z}[\{\epsilon^{-a(k-3)T}\}] = \mathfrak{z}[\{\epsilon^{-a(k-3)T}u[(k - 3)T]\}] = z^{-3}\left[\frac{z}{z - \epsilon^{-aT}}\right] = \frac{1}{z^2(z - \epsilon^{-aT})}$$

where $u(kT)$ is the unit step. Also,

$$\mathfrak{z}[\{\epsilon^{-a(k+2)T}\}] = z^2\left[\frac{z}{z - \epsilon^{-aT}} - 1 - \epsilon^{-aT}z^{-1}\right]$$

Complex Translation

Theorem. Given that the z-transform of $\{e(k)\}$ is $E(z)$. Then

$$\mathfrak{z}[\{\epsilon^{ak}e(k)\}] = E(z\epsilon^{-a}) \tag{2-16}$$

Proof. From the definition of the z-transform,

$$\mathfrak{z}[\{\epsilon^{ak}e(k)\}] = e(0) + \epsilon^{a}e(1)z^{-1} + \epsilon^{2a}e(2)z^{-2} + \cdots$$
$$= e(0) + e(1)(z\epsilon^{-a})^{-1} + e(2)(z\epsilon^{-a})^{-2} + \cdots$$

or

$$\mathfrak{z}[\{\epsilon^{ak}e(k)\}] = E(z)\,|_{z=z\epsilon^{-a}} = E(z\epsilon^{-a})$$

Example 2.6

Given that the z-transform of $\{e(k)\} = \{k\}$ is $E(z) = z/(z-1)^2$, then the z-transform of $\{k\epsilon^{ak}\}$ is

$$E(z)\big|_{z=z\epsilon^{-a}} = \frac{z}{(z-1)^2}\bigg|_{z=z\epsilon^{-a}}$$

$$= \frac{z\epsilon^{-a}}{(z\epsilon^{-a}-1)^2} = \frac{\epsilon^a z}{(z-\epsilon^a)^2}$$

Initial Value

Theorem. Given that the z-transform of $\{e(k)\}$ is $E(z)$. Then

$$e(0) = \lim_{z\to\infty} E(z) \tag{2-17}$$

Proof. Since

$$E(z) = e(0) + e(1)z^{-1} + e(2)z^{-2} + \cdots$$

then (2-17) is seen by inspection.

Final Value

Theorem. Given that the z-transform of $\{e(k)\}$ is $E(z)$. Then

$$\lim_{n\to\infty} e(n) = \lim_{z\to 1}(z-1)E(z)$$

provided that the left-side limit exists.

Proof. Consider the transform

$$\mathfrak{z}[\{e(k+1)\} - \{e(k)\}] = \lim_{n\to\infty}\left[\sum_{k=0}^{n} e(k+1)z^{-k} - \sum_{k=0}^{n} e(k)z^{-k}\right]$$

$$= \lim_{n\to\infty}[-e(0) + e(1)(1-z^{-1}) + e(2)(z^{-1}-z^{-2}) + \cdots$$

$$+ e(n)(z^{-n+1}-z^{-n}) + e(n+1)z^{-n}]$$

Thus

$$\lim_{z\to 1}[\mathfrak{z}[\{e(k+1)\} - \{e(k)\}]] = \lim_{n\to\infty}[e(n+1) - e(0)]$$

Also, from the real translation theorem,

$$\mathfrak{z}[\{e(k+1)\} - \{e(k)\}] = z[E(z) - e(0)] - E(z)$$

$$= (z-1)E(z) - ze(0)$$

Equating the two expressions above, we obtain

$$\lim_{n\to\infty} e(n) = \lim_{z\to 1}(z-1)E(z)$$

provided that the left-side limit exists. It is shown in Chapter 7 that this limit exists provided that all poles of $E(z)$ are inside the unit circle, except for possibly a simple pole at $z = 1$.

Example 2.7

To illustrate the initial-value theorem and the final-value theorem, consider the z-transform of $e(k) = 1, k = 0, 1, 2, \ldots$. We have shown, in Example 2.2, that

$$E(z) = \mathfrak{z}[\{1\}] = \frac{z}{z-1}$$

Applying the initial-value theorem, we see that

$$e(0) = \lim_{z\to\infty} \frac{z}{z-1} = \lim_{z\to\infty} \frac{1}{1-1/z} = 1$$

Since the final value of $e(k)$ exists, we may apply the final-value theorem.

$$\lim_{k\to\infty} e(k) = \lim_{z\to 1} (z-1)E(z) = \lim_{z\to 1} z = 1$$

2.5 SOLUTION OF DIFFERENCE EQUATIONS

There are three basic techniques for solving linear time-invariant difference equations. The first method, commonly referred to as the classical approach, consists of finding the complementary and the particular parts of the solution [4], in a manner similar to that used in the classical solution of linear differential equations. This technique will not be discussed here; however, a technique similar to it is presented later in this chapter in the discussion of state variables. The second technique, which is a sequential procedure, is the method used in the digital-computer solution of difference equations and is illustrated by the following example. The third technique will be considered later.

Example 2.8

It is desired to find $m(k)$ for the equation

$$m(k) = e(k) - e(k-1) - m(k-1), \qquad k \geq 0$$

where

$$e(k) = \begin{cases} 1, & k \text{ even} \\ 0, & k \text{ odd} \end{cases}$$

and both $e(-1)$ and $m(-1)$ are zero. Then $m(k)$ can be determined by solving the difference equation first for $k = 0$, then for $k = 1$, $k = 2$, and so on. Thus

$$\begin{aligned} m(0) &= e(0) = 1 \\ m(1) &= e(1) - e(0) - m(0) = 0 - 1 - 1 = -2 \\ m(2) &= e(2) - e(1) - m(1) = 1 - 0 + 2 = 3 \\ m(3) &= e(3) - e(2) - m(2) = 0 - 1 - 3 = -4 \\ m(4) &= e(4) - e(3) - m(3) = 1 - 0 + 4 = 5 \end{aligned}$$

Note the sequential nature of the solution process. Using this approach, we can find $m(k)$ for any value of k. This technique is not practical, however, for large values of k, except when implemented on a digital computer. For the example above, a program segment, in FORTRAN, which solves the equation is

```
      M0 = 0.
      E0 = 0.
      E1 = 1.
      DO 1 I = 1, J
      M1 = E1 - E0 - M0
      M0 = M1
      E0 = E1
    1 E1 = 1. - E1
```

Note that the coding is such that MO represents $m(k-1)$, EO represents $e(k-1)$, M1 represents $m(k)$, and E1 represents $e(k)$. The first three FORTRAN statements initialize the sequential process, which begins with the fourth statement and extends to the last statement. A loop exists between the fourth and eighth statements and the first time through the loop M1 $= m(0)$, the second time through M1 $= m(1)$, and so on.

As a second example, consider the numerical integration of the first-order differential equation

$$\frac{dx(t)}{dt} = \dot{x}(t) = ax(t) + be(t)$$

For T small, $\dot{x}(t)$ may be approximated by the forward difference,

$$\dot{x}(t) \approx \frac{x(t+T) - x(t)}{T}$$

or, to an approximation,

$$\frac{x(t+T) - x(t)}{T} = ax(t) + be(t)$$

Solving this equation for $x(t+T)$, we obtain

$$x(t+T) = (1 + aT)x(t) + bTe(t)$$

Evaluating this equation for $t = kT$, we obtain the difference equation

$$x[(k+1)T] = (1 + aT)x(kT) + bTe(kT)$$

This technique of numerical integrations, which is known as Euler's method, leads directly to a difference equation. In fact, all numerical integration techniques may be expressed as difference equations [5], and can be programmed on a digital computer as illustrated in Example 2.8.

The third technique for solving linear time-invariant difference equations, which employs the use of the z-transform, will now be presented. Consider the following nth-order difference equation, where it is assumed that $\{e(k)\}$ is known.

$$\begin{aligned} m(k) + \beta_{n-1}m(k-1) + \cdots + \beta_0 m(k-n) \\ = \alpha_n e(k) + \alpha_{n-1}e(k-1) + \cdots + \alpha_0 e(k-n) \end{aligned} \tag{2-18}$$

The z-transform of (2-18), which results from the use of the real translation theorem (2-14), is

$$\begin{aligned} M(z) + \beta_{n-1}z^{-1}M(z) + \cdots + \beta_0 z^{-n}M(z) \\ = \alpha_n E(z) + \alpha_{n-1}z^{-1}E(z) + \cdots + \alpha_0 z^{-n}E(z) \end{aligned} \tag{2-19}$$

Note that the z-transform has changed the difference equation in (2-18) to the algebraic equation in (2-19). Solving the expression above for $M(z)$ yields

$$M(z) = \frac{\alpha_n + \alpha_{n-1}z^{-1} + \cdots + \alpha_0 z^{-n}}{1 + \beta_{n-1}z^{-1} + \cdots + \beta_0 z^{-n}} E(z) \tag{2-20}$$

$m(k)$ can be found by taking the inverse z-transform of (2-20). General techniques for determining the inverse z-transform are discussed in the next section.

Example 2.9

Consider the difference equation of Example 2.8.

$$m(k) = e(k) - e(k-1) - m(k-1)$$

The z-transform of this equation, obtained via the real translation theorem, is

$$M(z) = E(z) - z^{-1}E(z) - z^{-1}M(z)$$

or

$$M(z) = \frac{z-1}{z+1}E(z)$$

Since

$$e(k) = \begin{cases} 1, & k \text{ even} \\ 0, & k \text{ odd} \end{cases}$$

we see that

$$E(z) = 1 + z^{-2} + z^{-4} + \cdots = \frac{1}{1 - z^{-2}} = \frac{z^2}{z^2 - 1}$$

Thus

$$M(z) = \frac{z-1}{z+1}\frac{z^2}{z^2-1} = \frac{z^2}{z^2 + 2z + 1}$$

We can expand $M(z)$ into a power series by dividing the numerator of $M(z)$ by its denominator so as to obtain

$$\begin{array}{r|l} & 1 - 2z^{-1} + 3z^{-2} - 4z^{-3} + \cdots \\ z^2 + 2z + 1 & z^2 \\ & \underline{z^2 + 2z + 1} \\ & -2z - 1 \\ & \underline{-2z - 4 - 2z^{-1}} \\ & 3 + 2z^{-1} \\ & \underline{3 + 6z^{-1} + 3z^{-2}} \\ & -4z^{-1} - 3z^{-2} \\ & \cdots \end{array}$$

Therefore,

$$M(z) = 1 - 2z^{-1} + 3z^{-2} - 4z^{-3} + \cdots$$

and the values of $m(k)$ are seen to be the same as those found using the sequential technique in Example 2.8.

Thus far we have considered only difference equations for which the initial conditions are zero. The solution presented in (2-20) represents only the forced part of the response. In order to include initial conditions in the solution of (2-18), k is replaced with $k + n$. Then

$$\begin{aligned} m(k+n) + \beta_{n-1}m(k+n-1) + \cdots + \beta_0 m(k) \\ = \alpha_n e(k+n) + \alpha_{n-1}e(k+n-1) + \cdots + \alpha_0 e(k) \end{aligned} \tag{2-21}$$

From the real translation theorem (2-15),

$$\mathfrak{z}[\{m(k+i)\}] = z^i[M(z) - m(0) - m(1)z^{-1} - \cdots - m(i-1)z^{-(i-1)}] \tag{2-22}$$

Thus the z-transform of (2-21) can be found using (2-22), and all initial conditions are included in the solution. Note that if all initial conditions are zero, the z-transform of (2-21) yields the same results as given in (2-20). Given in Appendix I is a computer program, in FORTRAN, which solves (2-21).

2.6 THE INVERSE z-TRANSFORM

In order for the z-transform technique to be a feasible approach in the solution of difference equations, methods for determining the inverse z-transform are required. Four such methods will be given here.

Power Series Method

The power series method for finding the inverse z-transform of a function $E(z)$ which is expressed as the ratio of two polynomials in z involves dividing the denominator of $E(z)$ into the numerator such that a power series of the form

$$E(z) = e_0 + e_1 z^{-1} + e_2 z^{-2} + \cdots \tag{2-23}$$

is obtained. From the definition of the z-transform, it can be seen that the values of $\{e(k)\}$ are simply the coefficients in the power series. This technique was illustrated in Example 2.9. Another example will now be given.

Example 2.10

It is desired to find the values of $\{e(k)\}$ for $E(z)$ given by the expression

$$E(z) = \frac{z}{z^2 - 3z + 2}$$

Using long division, we obtain

$$\begin{array}{r|l}
 & z^{-1} + 3z^{-2} + 7z^{-3} + 15z^{-4} + \cdots \\
\hline
z^2 - 3z + 2 & z \\
 & \underline{z - 3 + 2z^{-1}} \\
 & \quad 3 - 2z^{-1} \\
 & \quad \underline{3 - 9z^{-1} + 6z^{-2}} \\
 & \qquad 7z^{-1} - 6z^{-2} \\
 & \qquad \underline{7z^{-1} - 21z^{-2} + 14z^{-3}} \\
 & \qquad\qquad 15z^{-2} - 14z^{-3} + \cdots \\
 & \qquad\qquad\quad \cdots\cdots\cdots
\end{array}$$

and therefore

$$\begin{aligned}
e(0) &= 0 & e(4) &= 15 \\
e(1) &= 1 & &\cdots, \\
e(2) &= 3 & e(k) &= 2^k - 1 \\
e(3) &= 7 & &\cdots.
\end{aligned}$$

In this particular case, the general expression for $e(k)$ as a function of k [i.e., $e(k) = 2^k - 1$] can be recognized. In general, this cannot be done using the power series method.

Partial-Fraction Expansion Method

In a manner similar to that employed with the Laplace transform, a function $E(z)$ can be expanded in partial fractions and then tables of known z-transform pairs can be used to determine the inverse z-transform. A table of commonly used z-transforms is given in Table 2-1, and a table of z-transforms based on sampled time functions is

TABLE 2-1 z-TRANSFORM TABLE

Number sequence, $\{e(k)\}$	z-transform, $E(z)$
$\{1\}$	$\dfrac{z}{z-1}$
$\{k\}$	$\dfrac{z}{(z-1)^2}$
$\{k^2\}$	$\dfrac{z(z+1)}{(z-1)^3}$
$\{a^k\}$	$\dfrac{z}{z-a}$
$\{ka^k\}$	$\dfrac{az}{(z-a)^2}$
$\{\sin ak\}$	$\dfrac{z \sin a}{z^2 - 2z\cos a + 1}$
$\{\cos ak\}$	$\dfrac{z(z - \cos a)}{z^2 - 2z\cos a + 1}$
$\{a^k \sin bk\}$	$\dfrac{az \sin b}{z^2 - 2az\cos b + a^2}$
$\{a^k \cos bk\}$	$\dfrac{z^2 - az\cos b}{z^2 - 2az\cos b + a^2}$

given in Table 2-2. Before proceeding with an example of the partial-fraction expansion method, consider the function

$$E(z) = \frac{z}{z-a} = 1 + az^{-1} + a^2z^{-2} + a^3z^{-3} + \cdots \tag{2-24}$$

Examination of the power series indicates that

$$\mathfrak{z}^{-1}\left[\frac{z}{z-a}\right] = \{a^k\} \tag{2-25}$$

where $\mathfrak{z}^{-1}[\cdot]$ indicates the inverse z-transform. This particular function is perhaps the most common z-transform encountered. Notice, from the z-transform table, that a factor of z is generally required in the numerator of the partial fractions. Hence the partial-fraction expansion should be performed on $E(z)/z$, which will result in the terms of $E(z)$ being of the same form as those in the tables.

TABLE 2-2 TABLE OF COMMONLY USED z-TRANSFORMS

Laplace transform, $E(s)$	Time function, $e(t)$	z-transform, $E(z)$
$\frac{1}{s}$	$u(t)$	$\frac{z}{z-1}$
$\frac{1}{s^2}$	t	$\frac{Tz}{(z-1)^2}$
$\frac{1}{s+a}$	ϵ^{-at}	$\frac{z}{z-\epsilon^{-aT}}$
$\frac{a}{s(s+a)}$	$1-\epsilon^{-at}$	$\frac{z(1-\epsilon^{-aT})}{(z-1)(z-\epsilon^{-aT})}$
$\frac{1}{(s+a)^2}$	$t\epsilon^{-at}$	$\frac{Tz\epsilon^{-aT}}{(z-\epsilon^{-aT})^2}$
$\frac{a}{s^2(s+a)}$	$t-\frac{1-\epsilon^{-at}}{a}$	$\frac{Tz}{(z-1)^2}-\frac{(1-\epsilon^{-aT})z}{a(z-1)(z-\epsilon^{-aT})}$
$\frac{a}{s^2+a^2}$	$\sin(at)$	$\frac{z\sin(aT)}{z^2-2z\cos(aT)+1}$
$\frac{s}{s^2+a^2}$	$\cos(at)$	$\frac{z(z-\cos(aT))}{z^2-2z\cos aT+1}$
$\frac{1}{(s+a)^2+b^2}$	$\frac{1}{b}\epsilon^{-at}\sin bt$	$\frac{1}{b}\left[\frac{z\epsilon^{-aT}\sin bT}{z^2-2z\epsilon^{-aT}\cos(bT)+\epsilon^{-2aT}}\right]$
$\frac{s+a}{(s+a)^2+b^2}$	$\epsilon^{-at}\cos bt$	$\frac{z^2-z\epsilon^{-aT}\cos bT}{z^2-2z\epsilon^{-aT}\cos bT+\epsilon^{-2aT}}$

Example 2.11

Consider the function $E(z)$ given in Example 2.10:

$$E(z)=\frac{z}{(z-1)(z-2)}$$

Hence

$$\frac{E(z)}{z}=\frac{1}{(z-1)(z-2)}=\frac{-1}{z-1}+\frac{1}{z-2}$$

Then

$$\mathfrak{z}^{-1}[E(z)]=\mathfrak{z}^{-1}\left[\frac{-z}{z-1}\right]+\mathfrak{z}^{-1}\left[\frac{z}{z-2}\right]$$

From (2-25), the value of $e(k)$ is

$$e(k)=-1+2^k$$

which is the same value as that found in Example 2.10.

not to be covered

Inversion-Formula Method

Perhaps the most general technique for obtaining the inverse of a z-transform is the inversion integral. This integral, derived via complex variable theory, is

$$e(k)=\frac{1}{2\pi j}\oint_{\Gamma} E(z)z^{k-1}\,dz, \qquad j=\sqrt{-1} \tag{2-26}$$

The reader interested in the derivation and ramifications of this expression is referred to Refs. 6 and 7. Basically, the expression is a line integral in the z-plane along the closed curve Γ. For our purposes Γ is any closed curve in the z-plane which encloses all of the finite poles of $E(z)z^{k-1}$.

Using the theorem of residues [8], we can evaluate the integral in (2-26) via the expression

$$e(k) = \sum_{\substack{\text{at poles} \\ \text{of } [E(z)z^{k-1}]}} [\text{residues of } E(z)z^{k-1}] \tag{2-27}$$

If the function $E(z)z^{k-1}$ has a simple pole at $z = a$, the residue is evaluated as

$$(\text{residue})_{z=a} = (z - a)E(z)z^{k-1}\big|_{z=a} \tag{2-28}$$

For a pole of order m at $z = a$, the residue is calculated using the expression

$$(\text{residue})_{z=a} = \frac{1}{(m-1)!}\frac{d^{m-1}}{dz^{m-1}}[(z-a)^m E(z)z^{k-1}]\bigg|_{z=a} \tag{2-29}$$

Example 2.12

Consider the function $E(z)$ from Examples 2.10 and 2.11:

$$E(z) = \frac{z}{(z-1)(z-2)}$$

Substituting this expression into (2-27) and (2-28) yields

$$e(k) = \frac{z^k}{z-2}\bigg|_{z=1} + \frac{z^k}{z-1}\bigg|_{z=2} = -1 + 2^k$$

and the result is seen to be the same as that obtained in the previous examples.

The following example illustrates the inversion-formula technique for a multiple-order pole.

Example 2.13

The function $E(z)$ below has a single pole of order 2 at $z = 1$.

$$E(z) = \frac{z}{(z-1)^2}$$

The inverse transform obtained using (2-29) is

$$\begin{aligned} e(k) &= \frac{1}{(2-1)!}\frac{d^{2-1}}{dz^{2-1}}\left[(z-1)^2\left(\frac{z}{(z-1)^2}\right)z^{k-1}\right]\bigg|_{z=1} \\ &= \frac{d}{dz}(z^k)\bigg|_{z=1} \\ &= kz^{k-1}\big|_{z=1} \\ &= k \end{aligned}$$

Discrete Convolution

The discrete convolution technique used for determining the inverse z-transform is analogous to the convolution integral employed in the use of Laplace transforms. Suppose that the function $E(z)$ can be expressed as the product of two functions, each of which, in general, will be simpler than $E(z)$; that is,

$$E(z) = E_1(z)E_2(z) \tag{2-30}$$

Further, let $E_1(z)$ and $E_2(z)$ be expressed as power series. Then

$$\begin{aligned} E(z) = [&e_1(0) + e_1(1)z^{-1} + e_1(2)z^{-2} + \cdots][e_2(0) + e_2(1)z^{-1} \\ &+ e_2(2)z^{-2} + \cdots] \end{aligned} \tag{2-31}$$

Direct multiplication of the two power series yields

$$E(z) = e_1(0)e_2(0) + [e_1(0)e_2(1) + e_1(1)e_2(0)]z^{-1} + [e_1(0)e_2(2) + e_1(1)e_2(1) + e_1(2)e_2(0)]z^{-2} + \cdots \tag{2-32}$$

Thus the general relationship for $e(k)$ is seen to be

$$\begin{aligned} e(k) &= e_1(0)e_2(k) + e_1(1)e_2(k-1) + \cdots + e_1(k)e_2(0) \\ &= \sum_{n=0}^{k} e_1(n)e_2(k-n) = \sum_{n=0}^{k} e_1(k-n)e_2(n) \end{aligned} \tag{2-33}$$

Equation (2-33) is the discrete convolution summation, and may be useful in determining the inverse z-transform of a function $E(z) = E_1(z)E_2(z)$ if $E_1(z)$ and $E_2(z)$ are initially expressed as power series.

Example 2.14

Once again consider the function $E(z)$ from Example 2.10:

$$E(z) = \frac{z}{(z-1)(z-2)} = E_1(z)E_2(z)$$

We now define

$$E_1(z) = \frac{z}{z-1} = 1 + z^{-1} + z^{-2} + \cdots$$

and

$$E_2(z) = \frac{1}{z-2} = z^{-1} + 2z^{-2} + (2)^2z^{-3} + \cdots$$

Then $e(k)$ can be formed directly from (2-33). For example, $e(3)$ is given by the expression

$$\begin{aligned} e(3) &= \sum_{n=0}^{3} e_1(n)e_2(3-n) \\ &= e_1(0)e_2(3) + e_1(1)e_2(2) + e_1(2)e_2(1) + e_1(3)e_2(0) \\ &= (1)(2^2) + (1)(2) + (1)(1) + (1)(0) = 7 \end{aligned}$$

The other values of $e(k)$ can be obtained in a similar manner and a quick check of them will show that they agree with the values obtained via the other inversion techniques.

2.7 SIMULATION DIAGRAMS AND FLOW GRAPHS

It has been shown that a linear time-variant discrete-time system may be represented by either a difference equation or a transfer function. A third representation commonly used is a simulation diagram. Simulation diagrams for discrete-time systems are presented in this section.

First the basic elements used to construct simulation diagrams for a system described by a linear difference equation are developed. Let the block shown in Figure 2-2a represent a shift register. Every T seconds, a number is shifted into the register, and at that instant, the number that was stored in the register is shifted out. Therefore, if we let $e(k)$ represent the number shifted into the register at $t = kT$, the number shifted out is $e(k-1)$. We let the symbolic representation of this memory device be as shown in Figure 2-2b. This symbol can represent any device that performs the foregoing operation.

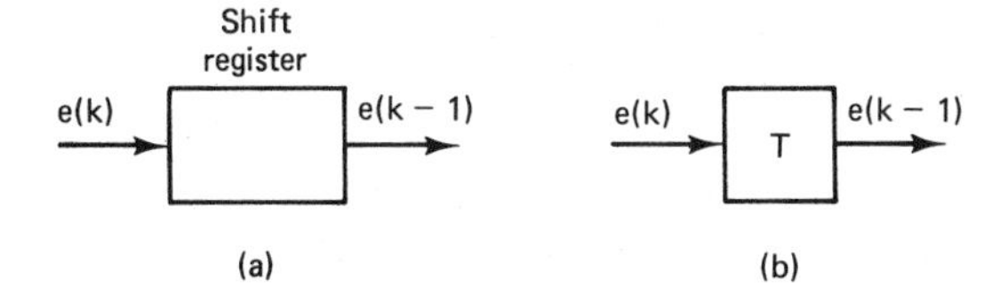

Figure 2-2 Ideal time-delay element.

An interconnection of these devices, together with devices that perform multiplication by a constant and summation, can be used to represent a linear time-invariant difference equation. For example, consider the difference equation used in Example 2.8:

$$m(k) = e(k) - e(k-1) - m(k-1) \tag{2-34}$$

A simulation diagram of this equation is shown in Figure 2-3.

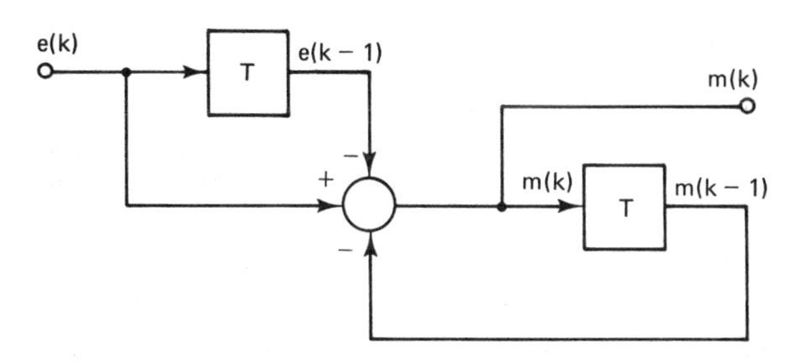

Figure 2-3 Simulation diagram for the difference equation used in Example 2.8.

Electronic devices may be constructed to perform all the operations shown in the figure. Now suppose that such a construction exists. Then in order to solve Example 2.8 using this constructed machine, the numbers in both memory locations (shift registers) are set to zero and the input $e(kT)$ is made equal to 1 at time instants $kT = 0$, $2T$, $4T$, . . . and $e(kT)$ is made equal to 0 at time instants $kT = T, 3T, 5T, \ldots$. The solution $m(k)$ then appears at the output terminal at $t = kT$. This machine would be a special-purpose computer, capable of solving only the difference equation (2-34). Recall that the computer program given in Example 2.8 also solves this difference equation, but in this case a general-purpose computer is used. The general-purpose computer software arranges the arithmetic registers, memory, and so on, to perform the operations depicted in Figure 2-3.

To include nonzero initial conditions in our special-purpose computer described above, replace k with $k + 1$ in both (2-34) and Figure 2-3. Then the value of $e(0)$ is placed in the register, in Figure 2-3, whose output is now $e(k)$, and the value of $m(0)$ is placed in the register (or memory location) whose output is now $m(k)$. The input values $e(1)$, $e(2)$, . . . are applied at the input terminal in Figure 2-3, and the output values $m(1)$, $m(2)$, . . . appear at the output terminal.

Recall that in the analog simulation of continuous systems, the basic element is the integrator. In the simulation of discrete systems, the basic element is the time delay (or memory) of T seconds.

A somewhat different but equivalent graphical representation of a difference equation is the signal flow graph. A block diagram, such as that illustrated in Figure 2-3, is simply a graphical representation of an equation, or a set of equations. The signal flow graph may also be used to graphically represent equations. The basic elements of a flow graph are the branches and the nodes. By definition, the signal out of a branch is equal to the branch gain (transfer function) times the signal into the branch. This is illustrated in Figure 2-4, which shows both the block-diagram representation and the flow-graph representation of a branch. Also, by definition, the signal at a node in a flow graph is equal to sum of the signals from all branches into that node. This concept is also shown in Figure 2-4. Thus a flow graph contains exactly the same information as a block diagram.

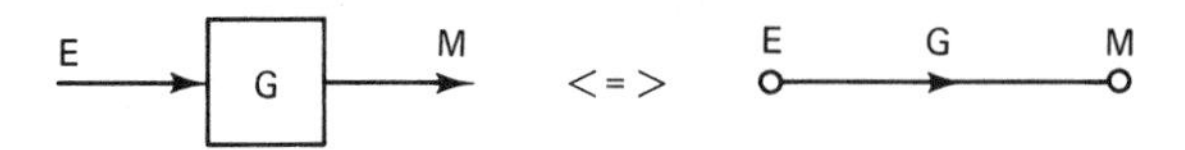

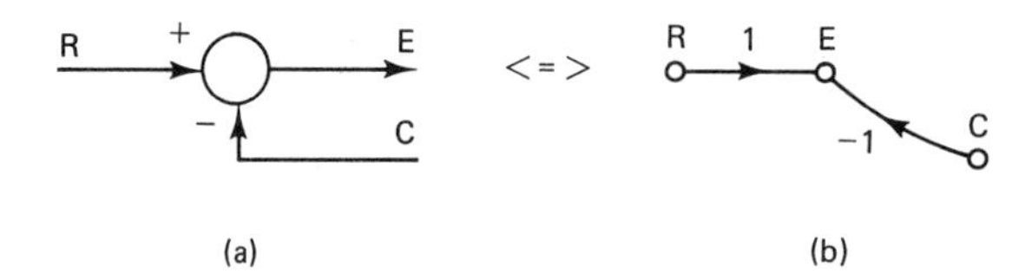

Figure 2-4 Equivalent block-diagram (a) and flow-graph (b) symbols.

Once again, consider the time-delay device shown in Figure 2-2b. The z-transform of the input $e(k)$ is $E(z)$, and the z-transform of the output $e(k - 1)$ is $z^{-1}E(z)$. Thus the transfer function of the time delay is z^{-1}. (Recall that the transfer function of an integrator is s^{-1}.) Consider again the system shown in Figure 2-3. A flow-graph representation of this system is as shown in Figure 2-5. The transfer function of this system may be obtained either from Figure 2-3 by block-diagram reduction or from Figure 2-5 by using Mason's gain formula. Those readers unfamiliar with Mason's gain-formula technique are referred to Appendix II. Basically, the gain formula is based on the geometry and signal flow directions of a flow graph. Since this architecture is exactly the same as that used in a block diagram, Mason's gain formula may also be applied directly to a block diagram.

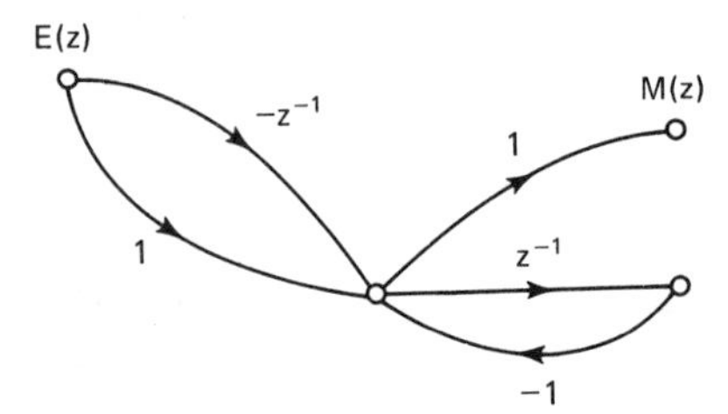

Figure 2-5 Signal flow graph for the system shown in Figure 2-3.

The application of Mason's gain formula to Figure 2-5 yields the transfer function

$$\frac{M(z)}{E(z)} = \frac{1 - z^{-1}}{1 + z^{-1}} = \frac{z - 1}{z + 1}$$

which is the same as that obtained in Example 2.9.

Consider now a general nth-order difference equation:

$$m(k) + \beta_{n-1}m(k-1) + \cdots + \beta_0 m(k-n) = \alpha_n e(k) + \alpha_{n-1}e(k-1) + \cdots + \alpha_0 e(k-n) \tag{2-35}$$

Taking the z-transform of this equation yields

$$M(z) + \beta_{n-1}z^{-1}M(z) + \cdots + \beta_0 z^{-n}M(z) = \alpha_n E(z) + \alpha_{n-1}z^{-1}E(z) + \cdots + \alpha_0 z^{-n}E(z) \tag{2-36}$$

This difference equation may then be represented by the transfer function

$$\frac{M(z)}{E(z)} = \frac{\alpha_n + \alpha_{n-1}z^{-1} + \cdots + \alpha_0 z^{-n}}{1 + \beta_{n-1}z^{-1} + \cdots + \beta_0 z^{-n}} \tag{2-37}$$

The system of (2-35) may be represented by the simulation diagram shown in Figure 2-6. The flow graph for this diagram is shown in Figure 2-7. Application of Mason's gain formula to this flow graph yields (2-37).

The simulation diagram of Figure 2-6 is only one of many that can be constructed to represent the transfer function of (2-37). The representation of Figure 2-6 is nonminimal in the sense that the diagram contains $2n$ delays, but an nth-order system can be represented with only n delays. This brings us to the topic of state variables for discrete systems, which is closely related to simulation diagrams. The remainder of this chapter is devoted to the topic of state variables.

2.8 STATE VARIABLES

In preceding sections we defined a discrete-time system as one that can be described by a difference equation. If the discrete-time system is linear and time invariant, we can also represent the system by a transfer function. For a linear time-invariant discrete-time system with input $E(z)$ and output $M(z)$, we can write

$$M(z) = G(z)E(z) \tag{2-38}$$

as illustrated in Section 2.7. Thus the discrete-time system may be represented by the block diagram of Figure 2-8a.

The more modern approach to the analysis and synthesis of discrete-time systems employs what is commonly called the state-variable method. In this approach the system is modeled as shown in Figure 2-8b. To be completely general, we must allow for the possibility of more than one input, and of more than one output. Thus, in Figure 2-8b, the variables u_i, $i = 1, \ldots, r$, are the external inputs which drive the system; the variables y_i, $i = 1, \ldots, m$, represent the system outputs or the system response; and the variables x_i, $i = 1, \ldots, n$ are the internal or state variables of the system. The state variables completely describe the dynamics of the system. In other words, they represent the minimum amount of information which is necessary to determine both the future states and the system outputs for given input functions; that is, given the system states, the system dynamics and the input functions, we can determine all subsequent states and outputs.

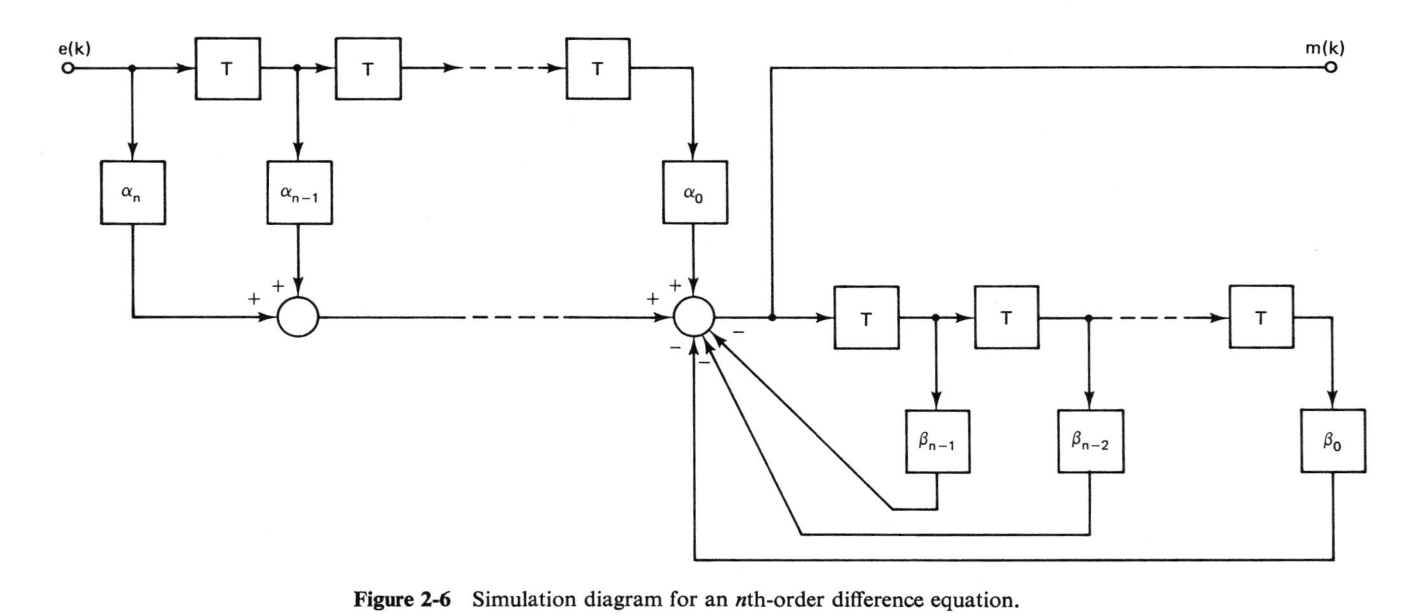

Figure 2-6 Simulation diagram for an *n*th-order difference equation.

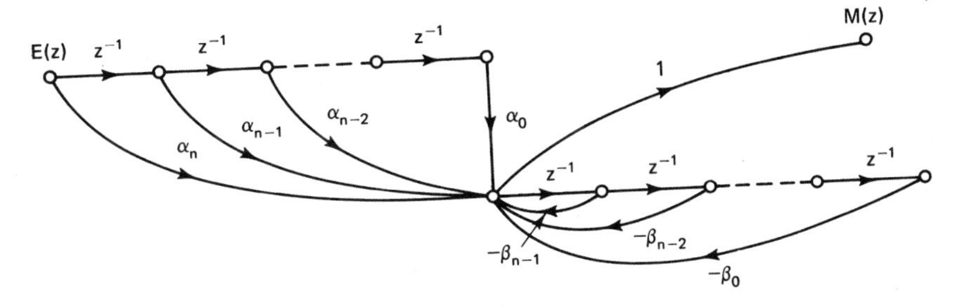

Figure 2-7 Flow graph for an nth-order difference equation.

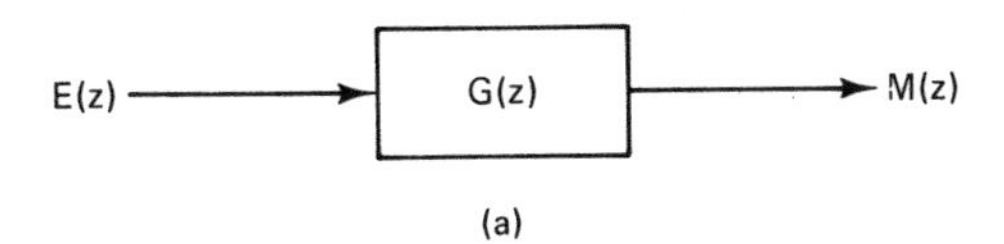

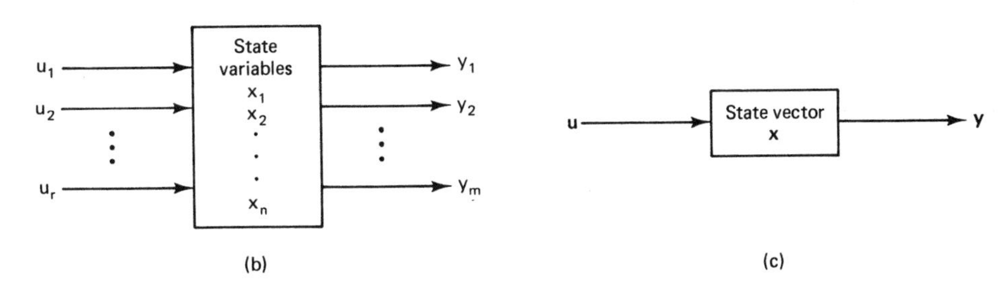

Figure 2-8 Representations of system dynamics: (a) z-transfer function representation; (b) state-variable representation; (c) state-vector representation.

For convenience we represent the system shown in Figure 2-8b by that shown in Figure 2-8c, where $\mathbf{u}$ is an input vector,

$$\mathbf{u} = \begin{bmatrix} u_1 \\ u_2 \\ \cdot \\ \cdot \\ \cdot \\ u_r \end{bmatrix}$$

$\mathbf{y}$ is an output vector,

$$\mathbf{y} = \begin{bmatrix} y_1 \\ y_2 \\ \cdot \\ \cdot \\ \cdot \\ y_m \end{bmatrix}$$

and $\mathbf{x}$ is a state vector containing the state variables,

$$\mathbf{x} = \begin{bmatrix} x_1 \\ x_2 \\ \cdot \\ \cdot \\ \cdot \\ x_n \end{bmatrix}$$

The set of values that the input vector $\mathbf{u}$ may assume as a function of discrete time k, written $\mathbf{u}(k)$, is called the input space of the system. The output space and state space are defined in a similar manner.

In general, the equation that describes the state of the system at any time $k + 1$ is given by the single-valued functional relationship

$$\mathbf{x}(k + 1) = \mathbf{f}[\mathbf{x}(k), \mathbf{u}(k)] \tag{2-39}$$

This equation simply states that the state $\mathbf{x}$ at time $k + 1$ is a function of the state and the input at the previous discrete-time increment k. The output response of the system is defined in a similar manner as

$$\mathbf{y}(k) = \mathbf{g}[\mathbf{x}(k), \mathbf{u}(k)] \tag{2-40}$$

If the system is linear, then equations (2-39) and (2-40) reduce to

$$\mathbf{x}(k + 1) = \mathbf{A}(k)\mathbf{x}(k) + \mathbf{B}(k)\mathbf{u}(k) \tag{2-41}$$

$$\mathbf{y}(k) = \mathbf{C}(k)\mathbf{x}(k) + \mathbf{D}(k)\mathbf{u}(k) \tag{2-42}$$

where $\mathbf{x}$ is an n-vector, $\mathbf{u}$ is an r-vector, $\mathbf{y}$ is an m-vector (as shown in Figure 2-8), and $\mathbf{A}(k)$, $\mathbf{B}(k)$, $\mathbf{C}(k)$, and $\mathbf{D}(k)$ are time-varying matrices of dimensions $n \times n$, $n \times r$, $m \times n$, and $m \times r$, respectively. If the system is time invariant, then the matrices in (2-41) and (2-42) are constant, and hence the equations reduce to

$$\mathbf{x}(k + 1) = \mathbf{A}\mathbf{x}(k) + \mathbf{B}\mathbf{u}(k) \tag{2-43}$$

$$\mathbf{y}(k) = \mathbf{C}\mathbf{x}(k) + \mathbf{D}\mathbf{u}(k) \tag{2-44}$$

State-variable modeling will now be illustrated by an example.

Example 2.15

It is desired to find a state-variable model of the system described by the difference equation

$$y(k + 2) = u(k) + 1.7y(k + 1) - 0.72y(k)$$

Let

$$x_1(k) = y(k)$$

$$x_2(k) = x_1(k + 1) = y(k + 1)$$

Then

$$x_2(k + 1) = y(k + 2) = u(k) + 1.7x_2(k) - 0.72x_1(k)$$

or, from these equations, we write

$$x_1(k + 1) = x_2(k)$$

$$x_2(k + 1) = -0.72x_1(k) + 1.7x_2(k) + u(k)$$

$$y(k) = x_1(k)$$

We may express these equations in vector-matrix form as

$$\mathbf{x}(k+1) = \begin{bmatrix} 0 & 1 \\ -0.72 & 1.7 \end{bmatrix} \mathbf{x}(k) + \begin{bmatrix} 0 \\ 1 \end{bmatrix} u(k)$$

$$y(k) = [1 \quad 0]\mathbf{x}(k)$$

Equations (2-43) and (2-44) are the state equations for a linear time-invariant system and usually represent the starting point in the analysis or synthesis of a discrete system. However, let us first examine the connection between this approach and the z-transform method. To do this we will show how to derive a set of discrete state-variable equations from the z-transform transfer function.

Given the transfer function

$$G(z) = \frac{a_{n-1}z^{n-1} + a_{n-2}z^{n-2} + \cdots + a_1 z + a_0}{z^n + b_{n-1}z^{n-1} + \cdots + b_1 z + b_0} \tag{2-45}$$

we can then write this expression in the form

$$\frac{Y(z)}{U(z)} = G(z) = \frac{a_{n-1}z^{n-1} + \cdots + a_1 z + a_0}{z^n + b_{n-1}z^{n-1} + \cdots + b_1 z + b_0} \frac{E(z)}{E(z)} \tag{2-46}$$

We now let

$$Y(z) = (a_{n-1}z^{n-1} + \cdots + a_1 z + a_0)E(z) \tag{2-47}$$

$$U(z) = (z^n + b_{n-1}z^{n-1} + \cdots + b_1 z + b_0)E(z) \tag{2-48}$$

At this point we recall from our earlier material the correspondence

$$\begin{aligned} E(z) &\longrightarrow e(k) \\ zE(z) &\longrightarrow e(k+1) \\ z^2E(z) &\longrightarrow e(k+2) \\ &\vdots \end{aligned}$$

Under this correspondence we define the state variables

$$\begin{aligned} x_1(k) &= e(k) \\ x_2(k) &= x_1(k+1) = e(k+1) \\ x_3(k) &= x_2(k+1) = e(k+2) \\ &\vdots \\ x_n(k) &= x_{n-1}(k+1) = e(k+n-1) \end{aligned} \tag{2-49}$$

From equations (2-48) and (2-49) we obtain the state equations

$$\begin{aligned} x_1(k+1) &= x_2(k) \\ x_2(k+1) &= x_3(k) \\ x_3(k+1) &= x_4(k) \\ &\vdots \\ x_n(k+1) &= -b_0 x_1(k) - b_1 x_2(k) - b_2 x_3(k) \cdots - b_{n-1}x_n(k) + u(k) \end{aligned}$$

which written in matrix form is

$$\begin{bmatrix} x_1(k+1) \\ x_2(k+1) \\ \cdot \\ \cdot \\ \cdot \\ \cdot \\ x_n(k+1) \end{bmatrix} = \begin{bmatrix} 0 & 1 & 0 & 0 & 0 & \cdots & 0 \\ 0 & 0 & 1 & 0 & 0 & \cdots & 0 \\ 0 & 0 & 0 & 1 & 0 & \cdots & 0 \\ & & & \cdot & & & \\ & & & \cdot & & & \\ & & & \cdot & & & \\ -b_0 & -b_1 & -b_2 & -b_3 & -b_4 & \cdots & -b_{n-1} \end{bmatrix} \begin{bmatrix} x_1(k) \\ x_2(k) \\ \cdot \\ \cdot \\ \cdot \\ \cdot \\ x_n(k) \end{bmatrix} + \begin{bmatrix} 0 \\ 0 \\ \cdot \\ \cdot \\ \cdot \\ 0 \\ 1 \end{bmatrix} u(k) \tag{2-50}$$

or simply

$$\mathbf{x}(k+1) = \mathbf{A}\mathbf{x}(k) + \mathbf{b}u(k) \tag{2-51}$$

Note that for single-input systems, the matrices **B** and **D** of (2-43) and (2-44) reduce to vectors **b** and **d**. The output equation obtained from (2-47) is

$$y(k) = [a_0 \quad a_1 \quad \cdots \quad a_{n-1}] \begin{bmatrix} x_1(k) \\ x_2(k) \\ \cdot \\ \cdot \\ \cdot \\ x_n(k) \end{bmatrix} \tag{2-52}$$

or simply

$$y(k) = \mathbf{c}\mathbf{x}(k) \tag{2-53}$$

Note that for single-output systems, the matrix **C** of (2-44) reduces to a row vector **c**, and for single-input, single-output systems the matrix **D** reduces to a single parameter d. As indicated in (2-53), the notation used for a vector (either row or column) is a boldface lowercase letter. Hence equations (2-51) and (2-53) are a set of state equations for the discrete system described by equation (2-45).

Another convenient and useful representation of the discrete system is the signal flow graph or the equivalent simulation diagram. These two forms can be derived from equation (2-45) by first dividing the numerator and denominator by z^n, which yields

$$G(z) = \frac{a_{n-1}z^{-1} + a_{n-2}z^{-2} + \cdots + a_1 z^{1-n} + a_0 z^{-n}}{1 + b_{n-1}z^{-1} + \cdots + b_1 z^{1-n} + b_0 z^{-n}} \tag{2-54}$$

Once again we can express this transfer function in the form

$$G(z) = \frac{Y(z)}{U(z)} = \frac{a_{n-1}z^{-1} + a_{n-2}z^{-2} + \cdots + a_1 z^{1-n} + a_0 z^{-n}}{1 + b_{n-1}z^{-1} + \cdots + b_1 z^{1-n} + b_0 z^{-n}} \frac{E(z)}{E(z)} \tag{2-55}$$

From this expression we obtain the two equations

$$Y(z) = (a_{n-1}z^{-1} + a_{n-2}z^{-2} + \cdots + a_1 z^{1-n} + a_0 z^{-n})E(z) \tag{2-56}$$

$$U(z) = (1 + b_{n-1}z^{-1} + \cdots + b_1 z^{1-n} + b_0 z^{-n})E(z) \tag{2-57}$$

Equation (2-57) can be rewritten in the form

$$E(z) = U(z) - b_{n-1}z^{-1}E(z) - \cdots - b_1 z^{1-n}E(z) - b_0 z^{-n}E(z) \tag{2-58}$$

A signal-flow-graph representation of the system described by equation (2-45) can be easily derived from equations (2-56) and (2-58) and is shown in Figure 2-9a.

Recall that the transfer function for a pure delay of T seconds is z^{-1}, where T is the discrete-time increment (i.e., T is the interval between sampling instants k, $k+1$, $k+2$, etc.). Using this relationship, the signal flow graph of Figure 2-9a can be immediately converted to the equivalent simulation diagram of Figure 2-9b. We see from (2-50) and (2-52) that the states are the register outputs, as noted in Figure 2-9b.

At this point the reader is encouraged to note very carefully the relationships

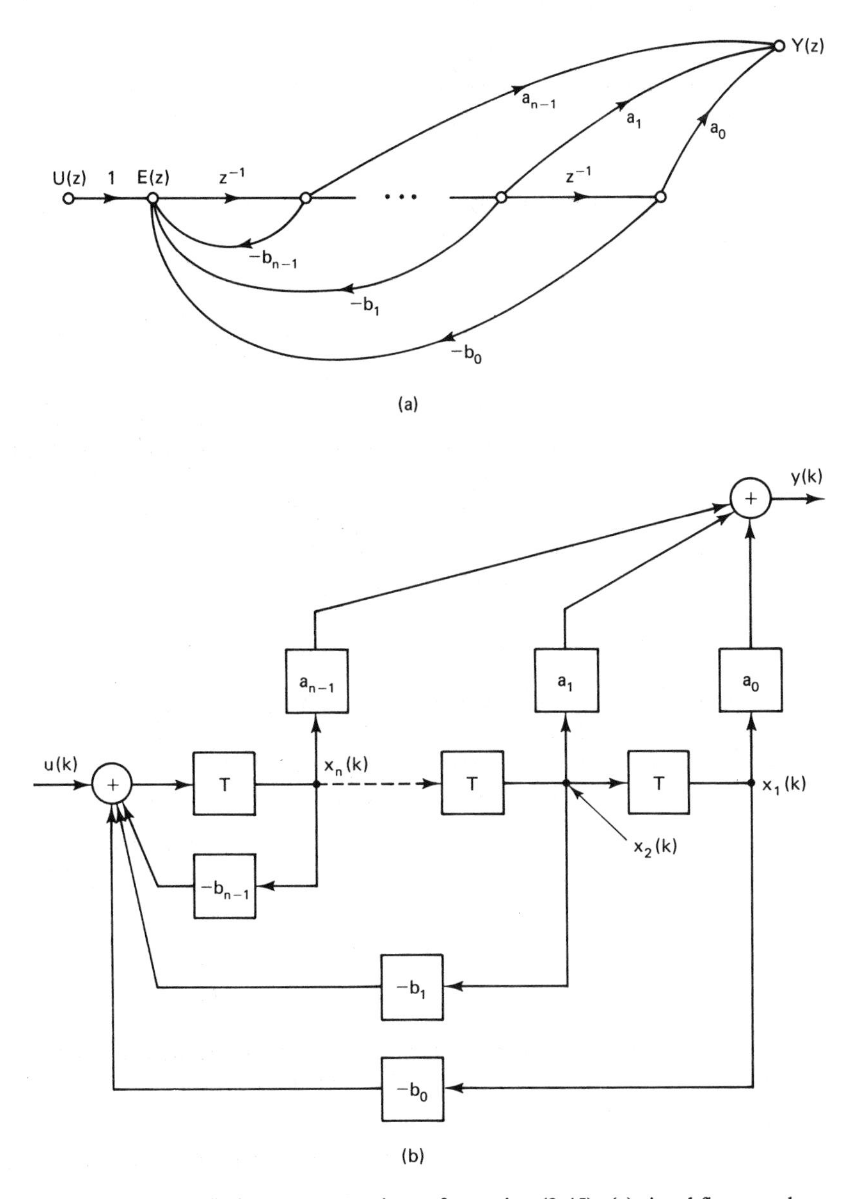

Figure 2-9 Equivalent representations of equation (2-45): (a) signal flow graph representation of equation (2-45); (b) simulation diagram equivalent to the flow graph in (a).

that exist between equations (2-50) and (2-52) and the diagrams in Figure 2-9. One who is very familiar with the correspondence between the equations and diagrams can derive a flow graph or simulation diagram by inspection. From the flow graph, one may use Mason's gain formula to reconstruct the original transfer function. The structure of Figure 2-9, together with equations (2-50) and (2-52), is called either the control canonical form or the phase-variable canonical form, and is useful in the design procedures presented in Chapter 9.

Example 2.16

Given the following function, we want to derive the signal flow graph and corresponding state equations:

$$G(z) = \frac{Y(z)}{U(z)} = \frac{z^2 + 2z + 1}{z^3 + 2z^2 + z + \frac{1}{2}}$$

Comparing this expression with (2-45) and Figure 2-9a indicates that the signal flow graph can be derived by inspection, as shown in Figure 2-10. Note that the states are

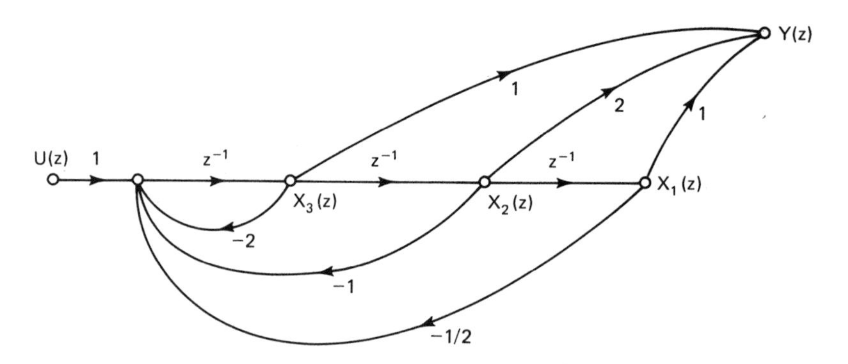

Figure 2-10 Flow graph for Example 2.16.

then chosen to be the register outputs. The state equations can now be derived from Figure 2-10 or from a comparison of the given transfer function with equations (2-45), (2-50), and (2-52). The resulting state equations are

$$\begin{bmatrix} x_1(k+1) \\ x_2(k+1) \\ x_3(k+1) \end{bmatrix} = \begin{bmatrix} 0 & 1 & 0 \\ 0 & 0 & 1 \\ -\frac{1}{2} & -1 & -2 \end{bmatrix} \begin{bmatrix} x_1(k) \\ x_2(k) \\ x_3(k) \end{bmatrix} + \begin{bmatrix} 0 \\ 0 \\ 1 \end{bmatrix} u(k)$$

$$y(k) = [1 \quad 2 \quad 1] \begin{bmatrix} x_1(k) \\ x_2(k) \\ x_3(k) \end{bmatrix}$$

The following example will demonstrate an important salient feature encountered in going from the transfer function to the state equations.

Example 2.17

Given the following transfer function, we wish to derive the state equations.

$$G(z) = \frac{Y(z)}{U(z)} = \frac{a_2 z^2 + a_1 z + a_0}{z^2 + b_1 z + b_0}$$

Therefore,

$$\frac{Y(z)}{U(z)} = \frac{a_2 + a_1 z^{-1} + a_0 z^{-2}}{1 + b_1 z^{-1} + b_0 z^{-2}} \frac{E(z)}{E(z)}$$

Following the previous development, the signal flow graph for this transfer function is shown in Figure 2-11. The states are shown on the flow graph, and the equations are seen to be

$$\begin{bmatrix} x_1(k+1) \\ x_2(k+1) \end{bmatrix} = \begin{bmatrix} 0 & 1 \\ -b_0 & -b_1 \end{bmatrix} \begin{bmatrix} x_1(k) \\ x_2(k) \end{bmatrix} + \begin{bmatrix} 0 \\ 1 \end{bmatrix} u(k)$$

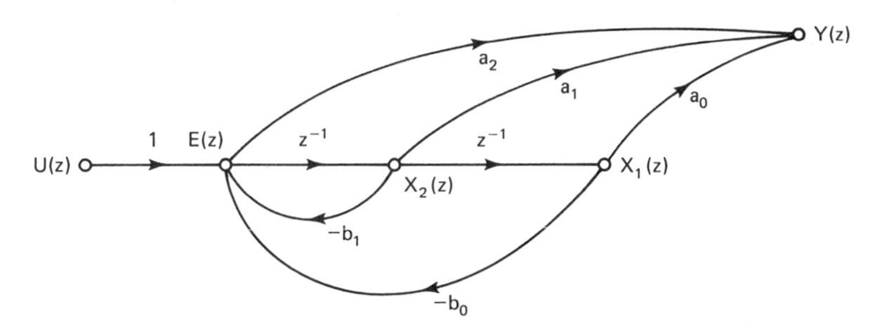

Figure 2-11 Flow graph for Example 2.17.

At this point it is important to note that the flow graph in Figure 2-11 is slightly different from the form presented earlier; that is, there is a direct path from $E(z)$ to $Y(z)$. From the flow graph the equation for the output is

$$Y(z) = a_0 X_1(z) + a_1 X_2(z) + a_2 E(z)$$

But

$$E(z) = U(z) - b_1 X_2(z) - b_0 X_1(z)$$

From these equations we obtain

$$Y(z) = a_2 U(z) + (a_0 - a_2 b_0) X_1(z) + (a_1 - a_2 b_1) X_2(z)$$

Hence the output state equation is

$$y(k) = [(a_0 - a_2 b_0) \quad (a_1 - a_2 b_1)] \begin{bmatrix} x_1(k) \\ x_2(k) \end{bmatrix} + a_2 u(k)$$

Whenever the numerator and denominator of the transfer function are of the same order, the output state equation must be handled as demonstrated above. Also, the output equation may be written by inspection from Figure 2-11.

Example 2.18

We wish to derive the state equations for the multivariable discrete system shown in Figure 2-12. The state equation can be obtained directly from the simulation diagram by inspection.

$$\begin{bmatrix} x_1(k+1) \\ x_2(k+1) \end{bmatrix} = \begin{bmatrix} 0.5 & 0 \\ 1 & -1 \end{bmatrix} \begin{bmatrix} x_1(k) \\ x_2(k) \end{bmatrix} + \begin{bmatrix} 1 & 1 \\ 0 & 1 \end{bmatrix} \begin{bmatrix} u_1(k) \\ u_2(k) \end{bmatrix}$$

$$\begin{bmatrix} y_1(k) \\ y_2(k) \end{bmatrix} = \begin{bmatrix} 1 & 2 \\ 0 & 1 \end{bmatrix} \begin{bmatrix} x_1(k) \\ x_2(k) \end{bmatrix} + \begin{bmatrix} 0 & 0 \\ 0 & 1 \end{bmatrix} \begin{bmatrix} u_1(k) \\ u_2(k) \end{bmatrix}$$

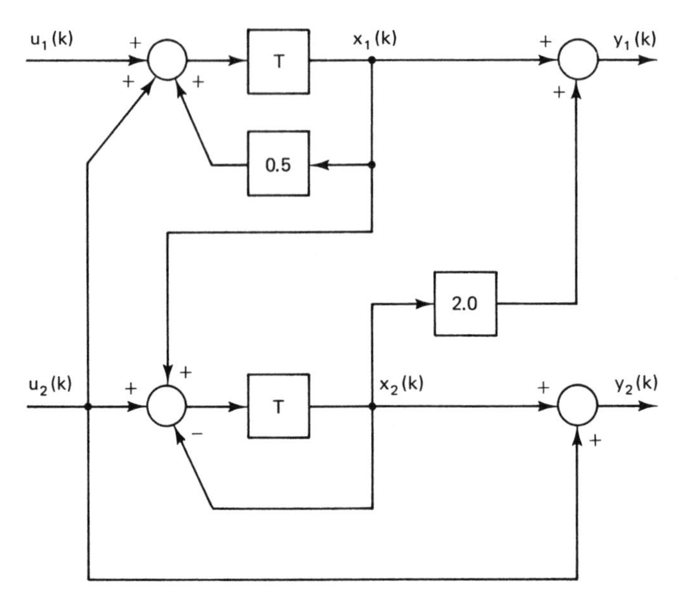

Figure 2-12 Simulation diagram for a multivariable system.

2.9 OTHER STATE-VARIABLE FORMULATIONS

We have just seen, in the preceding section, a technique for finding a state-variable formulation for a single-input, single-output discrete-time system, given either the system difference equation or the system transfer function. For a given system, there is no unique state-variable formulation. For a given transfer function, any simulation diagram with the system transfer function yields a valid state-variable model for the system. However, for certain analysis or design procedures, certain formulations present advantages with respect to calculations, as will be shown in later chapters. The following example illustrates the derivation of different formulations.

Example 2.19

Consider the difference equation of Example 2.15:

$$y(k+2) = u(k) + 1.7y(k+1) - 0.72y(k)$$

One state-variable model was derived in that example. Two additional models are derived here. Taking the z-transform of the difference equation yields

$$\frac{Y(z)}{U(z)} = \frac{1}{z^2 - 1.7z + 0.72}$$

This transfer function can be expressed as

$$\frac{Y(z)}{U(z)} = \left[\frac{1}{z-0.9}\right]\left[\frac{1}{z-0.8}\right]$$

A simulation diagram for this transfer function is given in Figure 2-13a. From this figure we write the state equations

$$\mathbf{x}(k+1) = \begin{bmatrix} 0.8 & 1 \\ 0 & 0.9 \end{bmatrix} \mathbf{x}(k) + \begin{bmatrix} 0 \\ 1 \end{bmatrix} u(k)$$

$$y(k) = [1 \quad 0]\mathbf{x}(k)$$

Also, the system transfer function can be expressed, through partial-fraction expansion, as

$$\frac{Y(z)}{U(z)} = \frac{10}{z-0.9} + \frac{-10}{z-0.8}$$

A simulation diagram for this transfer function is given in Figure 2-13b. From this figure we write the state equations

$$\mathbf{x}(k+1) = \begin{bmatrix} 0.9 & 0 \\ 0 & 0.8 \end{bmatrix} \mathbf{x}(k) + \begin{bmatrix} 1 \\ 1 \end{bmatrix} u(k)$$

$$y(k) = [10 \quad -10]\mathbf{x}(k)$$

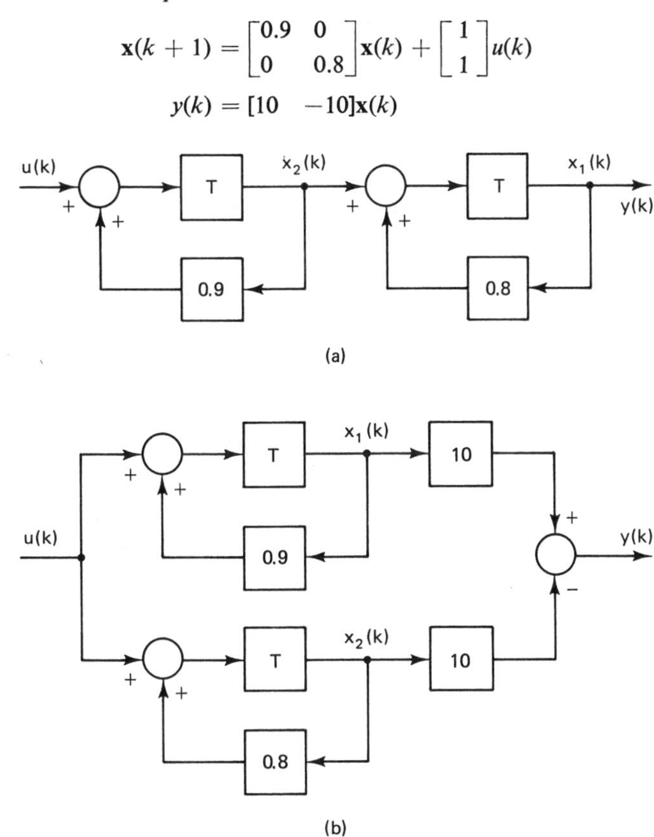

Figure 2-13 Simulation diagrams for Example 2.19.

In the example above, we first represented a higher-order transfer function as a product of simpler transfer functions. For the general case, we would express the transfer function $G(z)$ as

$$G(z) = G_{1_c}(z)G_{2_c}(z) \cdots G_{n_c}(z) \tag{2-59}$$

Each of the simpler transfer functions $G_{i_c}(z)$ is realized by the technique of the preceding section, and the realizations are then connected in cascade to realize $G(z)$. If $G(z)$ contains either complex poles or complex zeros, we may choose some of the $G_{i_c}(z)$ in (2-59) to be second order, to avoid computational difficulties with complex elements in the state-variable matrices.

Next, in Example 2.19, we represented a higher-order transfer function as the sum of simpler transfer functions through partial-fraction expansion. For the general case, we would express the transfer function $G(z)$ as

$$G(z) = G_{1_p}(z) + G_{2_p}(z) + \cdots + G_{n_p}(z) \tag{2-60}$$

Each of the simpler transfer functions $G_{i_p}(z)$ is realized by the technique of the preceding section, and the realizations are then connected in parallel to realize $G(z)$.

We have just seen three methods for arriving at different state models for a discrete-time system, with these models derived from the system transfer function. In fact, we can derive any number of different state models, given one state model of the system, through *similarity transformations*. This procedure will now be developed. Consider the state equation in the form

$$\begin{aligned} \mathbf{x}(k+1) &= \mathbf{Ax}(k) + \mathbf{Bu}(k) \\ \mathbf{y}(k) &= \mathbf{Cx}(\mathbf{k}) + \mathbf{Du}(k) \end{aligned} \tag{2-61}$$

Now, we can apply the linear transformation

$$\mathbf{x}(k) = \mathbf{Px}'(k) \tag{2-62}$$

where $\mathbf{P}$ is a constant $(n \times n)$ matrix, and $\mathbf{x}'(k)$ is the new state vector. Note that it is necessary that $\mathbf{P}^{-1}$, the inverse of $\mathbf{P}$, exists so that $\mathbf{x}'(k)$ can be determined from $\mathbf{x}(k)$. Substituting (2-62) into (2-61), we obtain the equation

$$\begin{aligned} \mathbf{x}'(k+1) &= \mathbf{P}^{-1}\mathbf{APx}'(k) + \mathbf{P}^{-1}\mathbf{Bu}(k) \\ \mathbf{y}(k) &= \mathbf{CPx}'(k) + \mathbf{Du}(k) \end{aligned} \tag{2-63}$$

These equations can be expressed as

$$\begin{aligned} \mathbf{x}'(k+1) &= \mathbf{A}_P\mathbf{x}'(k) + \mathbf{B}_P\mathbf{u}(k) \\ \mathbf{y}(k) &= \mathbf{C}_P\mathbf{x}'(k) + \mathbf{D}_P\mathbf{u}(k) \end{aligned} \tag{2-64}$$

where

$$\begin{aligned} \mathbf{A}_P &= \mathbf{P}^{-1}\mathbf{AP}, \qquad & \mathbf{B}_P &= \mathbf{P}^{-1}\mathbf{B} \\ \mathbf{C}_P &= \mathbf{CP}, & \mathbf{D}_P &= \mathbf{D} \end{aligned} \tag{2-65}$$

Thus for each different $\mathbf{P}$ for which $\mathbf{P}^{-1}$ exists, a different state model of a given system can be found.

The characteristic equation of a matrix $\mathbf{A}$ is defined by the determinant [9]

$$|z\mathbf{I} - \mathbf{A}| = 0 \tag{2-66}$$

and the characteristic values of the matrix are the roots of the characteristic equation; that is, z_i are the characteristic values of $\mathbf{A}$ if

$$|z\mathbf{I} - \mathbf{A}| = (z - z_1)(z - z_2) \cdots (z - z_n) = 0 \tag{2-67}$$

It will be shown in Chapter 7 that (2-67) determines system stability. It is of interest to note that the characteristic equation of the system matrix is unchanged through the linear transformation.

$$\begin{aligned} |z\mathbf{I} - \mathbf{A}_p| &= |z\mathbf{I} - \mathbf{P}^{-1}\mathbf{A}\mathbf{P}| = |z\mathbf{P}^{-1}\mathbf{I}\mathbf{P} - \mathbf{P}^{-1}\mathbf{A}\mathbf{P}| \\ &= |\mathbf{P}^{-1}||z\mathbf{I} - \mathbf{A}||\mathbf{P}| \\ &= |z\mathbf{I} - \mathbf{A}| \end{aligned} \tag{2-68}$$

It is important to recognize that a linear transformation of the form

$$\mathbf{A}_p = \mathbf{P}^{-1}\mathbf{A}\mathbf{P} \tag{2-69}$$

which is called a similarity transformation, has the following properties:

1. As shown above, the characteristic values of the matrix are unchanged under the transformation.
2. The determinant of $\mathbf{A}_p$ is equal to the determinant of $\mathbf{A}$:

$$|\mathbf{A}_p| = |\mathbf{P}^{-1}\mathbf{A}\mathbf{P}| = |\mathbf{P}^{-1}||\mathbf{A}||\mathbf{P}| = |\mathbf{A}|$$

3. The trace of $\mathbf{A}_p$ is equal to the trace of $\mathbf{A}$:

$$\text{tr } \mathbf{A}_p = \text{tr } \mathbf{A}$$

since the trace of a matrix, which is the sum of the diagonal elements, is simply equal to the sum of its characteristic roots. The latter property follows immediately from the first property.

Example 2.20

For the system of Example 2.19, one state-variable model given is

$$\mathbf{x}(k+1) = \begin{bmatrix} 0.8 & 1 \\ 0 & 0.9 \end{bmatrix}\mathbf{x}(k) + \begin{bmatrix} 0 \\ 1 \end{bmatrix}u(k)$$

$$y(k) = [1 \quad 0]\mathbf{x}(k)$$

We arbitrarily choose a linear transformation matrix

$$\mathbf{P} = \begin{bmatrix} 1 & -1 \\ 1 & 1 \end{bmatrix}$$

The inverse of P is given by

$$\mathbf{P}^{-1} = \frac{[\text{Cof}\,[\mathbf{P}]]^T}{|\mathbf{P}|}$$

where $[\cdot]^T$ indicates the transpose, and Cof [**P**] denotes the matrix of cofactors of **P**. Thus

$$\text{Cof}\,[\mathbf{P}] = \begin{bmatrix} 1 & -1 \\ 1 & 1 \end{bmatrix}$$

and

$$|\mathbf{P}| = 2$$

Then

$$\mathbf{P}^{-1} = \begin{bmatrix} 0.5 & 0.5 \\ -0.5 & 0.5 \end{bmatrix}$$

Thus, from (2-65),

$$\mathbf{A}_P = \mathbf{P}^{-1}\mathbf{A}\mathbf{P} = \begin{bmatrix} 0.5 & 0.5 \\ -0.5 & 0.5 \end{bmatrix}\begin{bmatrix} 0.8 & 1 \\ 0 & 0.9 \end{bmatrix}\begin{bmatrix} 1 & -1 \\ 1 & 1 \end{bmatrix}$$

$$= \begin{bmatrix} 0.5 & 0.5 \\ -0.5 & 0.5 \end{bmatrix}\begin{bmatrix} 1.8 & 0.2 \\ 0.9 & 0.9 \end{bmatrix} = \begin{bmatrix} 1.35 & 0.55 \\ -0.45 & 0.35 \end{bmatrix}$$

$$\mathbf{b}_P = \mathbf{P}^{-1}\mathbf{b} = \begin{bmatrix} 0.5 & 0.5 \\ -0.5 & 0.5 \end{bmatrix}\begin{bmatrix} 0 \\ 1 \end{bmatrix} = \begin{bmatrix} 0.5 \\ 0.5 \end{bmatrix}$$

$$\mathbf{c}_P = \mathbf{c}\mathbf{P} = [1 \quad 0]\begin{bmatrix} 1 & -1 \\ 1 & 1 \end{bmatrix} = [1 \quad -1]$$

Then the new state equations are

$$\mathbf{x}'(k+1) = \begin{bmatrix} 1.35 & 0.55 \\ -0.45 & 0.35 \end{bmatrix}\mathbf{x}'(k) + \begin{bmatrix} 0.5 \\ 0.5 \end{bmatrix}u(k)$$

$$y(k) = [1 \quad -1]\mathbf{x}'(k)$$

Note that

$$|z\mathbf{I} - \mathbf{A}| = \begin{vmatrix} z - 0.8 & -1 \\ 0 & z - 0.9 \end{vmatrix} = z^2 - 1.7z + 0.72$$

and

$$|z\mathbf{I} - \mathbf{A}_P| = \begin{vmatrix} z - 1.35 & -0.55 \\ 0.45 & z - 0.35 \end{vmatrix} = z^2 - 1.7z + 0.72$$

Also,

$$|\mathbf{A}| = |\mathbf{A}_P| = 0.72$$

and

$$\operatorname{tr} \mathbf{A} = \operatorname{tr} \mathbf{A}_P = 1.7$$

If a system has distinct characteristic values, we may derive a state-variable model in which the system matrix is diagonal. Consider a vector $\mathbf{m}_i$ and a scalar z_i defined by the equation

$$\mathbf{A}\mathbf{m}_i = z_i\mathbf{m}_i \tag{2-70}$$

or

$$(z_i\mathbf{I} - \mathbf{A})\mathbf{m}_i = \mathbf{0}$$

For a nontrivial solution for this equation to exist, it is required that

$$|z_i\mathbf{I} - \mathbf{A}| = 0$$

Hence, from (2-66), it is seen that z_i is a characteristic value (also called an eigenvalue) of $\mathbf{A}$. The vector $\mathbf{m}_i$ is called a characteristic vector, or eigenvector, of $\mathbf{A}$. From (2-70) we can construct the equation

$$\mathbf{A}[\mathbf{m}_1 \quad \mathbf{m}_2 \quad \cdots \quad \mathbf{m}_n] = [\mathbf{m}_1 \quad \mathbf{m}_2 \quad \cdots \quad \mathbf{m}_n]\begin{bmatrix} z_1 & 0 & \cdot & 0 \\ 0 & z_2 & \cdot & 0 \\ \cdot & \cdot & \cdot & \cdot \\ 0 & 0 & \cdot & z_n \end{bmatrix}$$

or

$$\mathbf{A}\mathbf{M} = \mathbf{M}\mathbf{\Lambda}$$

where $\mathbf{M}$, called the modal matrix, is composed of the characteristic vectors as columns, and $\mathbf{\Lambda}$ is a diagonal matrix with the characteristic values of $\mathbf{A}$ as the diagonal elements.

The characteristic vectors are linearly independent provided the characteristic values are distinct. Hence we can write

$$\mathbf{\Lambda} = \mathbf{M}^{-1}\mathbf{AM} \tag{2-71}$$

Then, in the general similarity transformation, **P** is equal to **M**, the modal matrix. An example will be given to illustrate this procedure.

Example 2.21

For the system of Example 2.20

$$\mathbf{x}(k+1) = \begin{bmatrix} 0.8 & 1 \\ 0 & 0.9 \end{bmatrix} \mathbf{x}(k) + \begin{bmatrix} 0 \\ 1 \end{bmatrix} u(k)$$

$$y(k) = [1 \quad 0]\mathbf{x}(k)$$

From Example 2.20, the characteristic values are $z_1 = 0.8$ and $z_2 = 0.9$. From (2-70),

$$\begin{bmatrix} 0.8 & 1 \\ 0 & 0.9 \end{bmatrix} \begin{bmatrix} m_{11} \\ m_{21} \end{bmatrix} = 0.8 \begin{bmatrix} m_{11} \\ m_{21} \end{bmatrix}$$

Thus

$$0.8m_{11} + m_{21} = 0.8m_{11}$$

$$0.9m_{21} = 0.8m_{21}$$

Hence $m_{21} = 0$ and m_{11} is arbitrary. Let $m_{11} = 1$. Also,

$$\begin{bmatrix} 0.8 & 1 \\ 0 & 0.9 \end{bmatrix} \begin{bmatrix} m_{12} \\ m_{22} \end{bmatrix} = 0.9 \begin{bmatrix} m_{12} \\ m_{22} \end{bmatrix}$$

Then m_{22} is seen to be arbitrary; let $m_{22} = 1$. Then $m_{12} = 10$. The modal matrix and its inverse are

$$\mathbf{M} = \begin{bmatrix} m_{11} & m_{12} \\ m_{21} & m_{22} \end{bmatrix} = \begin{bmatrix} 1 & 10 \\ 0 & 1 \end{bmatrix}$$

and

$$\mathbf{M}^{-1} = \begin{bmatrix} 1 & -10 \\ 0 & 1 \end{bmatrix}$$

From (2-65) and (2-71),

$$\mathbf{\Lambda} = \mathbf{M}^{-1}\mathbf{AM} = \begin{bmatrix} 0.8 & 0 \\ 0 & 0.9 \end{bmatrix}$$

$$\mathbf{b}_P = \mathbf{M}^{-1}\mathbf{b} = \begin{bmatrix} -10 \\ 1 \end{bmatrix}$$

$$\mathbf{c}_P = \mathbf{cM} = [1 \quad 10]$$

Hence the new state model is given by

$$\mathbf{x}'(k+1) = \begin{bmatrix} 0.8 & 0 \\ 0 & 0.9 \end{bmatrix} \mathbf{x}'(k) + \begin{bmatrix} -10 \\ 1 \end{bmatrix} u(k)$$

$$y(k) = [1 \quad 10]\mathbf{x}'(k)$$

2.10 TRANSFER FUNCTIONS

In the techniques described above for obtaining a state-variable formulation, we derived the state model from the transfer function. In this section we demonstrate how one may obtain the transfer function, given the state model.

Given the state equations of a discrete-time system, we can construct a simulation diagram. The transfer function can then be obtained from the simulation diagram, using Mason's gain formula.

Example 2.22

Consider the state model derived in Example 2.20.

$$\mathbf{x}(k+1) = \begin{bmatrix} 1.35 & 0.55 \\ -0.45 & 0.35 \end{bmatrix} \mathbf{x}(k) + \begin{bmatrix} 0.5 \\ 0.5 \end{bmatrix} u(k)$$

$$y(k) = [1 \quad -1]\mathbf{x}(k)$$

A simulation diagram, constructed from these equations, is shown in Figure 2-14. Application of Mason's gain formula to this figure yields

$$\frac{Y(z)}{U(z)} = \frac{0.5z^{-1}(1 - 0.35z^{-1}) - 0.5z^{-1}(1 - 1.35z^{-1}) + (0.5)(0.45)z^{-2} + (0.5)(0.55)z^{-2}}{1 - 1.35z^{-1} - 0.35z^{-1} + (0.45)(0.55)z^{-2} + (1.35z^{-1})(0.35z^{-1})}$$

$$= \frac{z^{-2}}{1 - 1.7z^{-1} + 0.72z^{-2}} = \frac{1}{z^2 - 1.7z + 0.72}$$

Since this is the system transfer function, we see that the derived state model is valid.

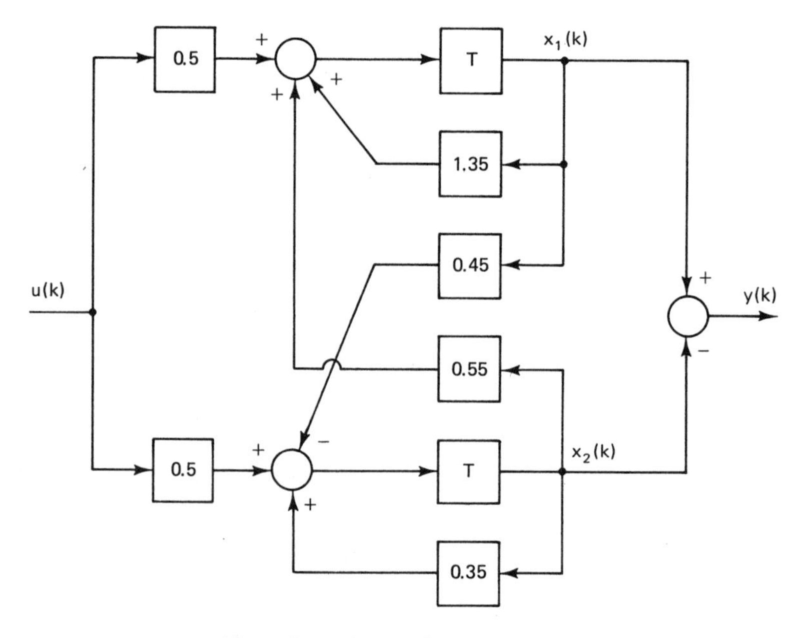

Figure 2-14 System for Example 2.22.

Another approach for obtaining the system transfer function from the state equations is to take the z-transform of the state equations and eliminate $\mathbf{X}(z)$. Since

$$\mathbf{x}(k+1) = \mathbf{A}\mathbf{x}(k) + \mathbf{b}u(k) \tag{2-72}$$

taking the z-transform yields

$$z\mathbf{X}(z) - z\mathbf{x}(0) = \mathbf{A}\mathbf{X}(z) + \mathbf{b}U(z) \tag{2-73}$$

Since, in deriving transfer functions, we ignore initial conditions, then (2-73) can be expressed as

$$[z\mathbf{I} - \mathbf{A}]\mathbf{X}(z) = \mathbf{b}U(z) \tag{2-74}$$

Solving for $\mathbf{X}(z)$, we obtain

$$\mathbf{X}(z) = [z\mathbf{I} - \mathbf{A}]^{-1}\mathbf{b}U(z) \tag{2-75}$$

Also, since

$$y(k) = \mathbf{c}\mathbf{x}(k) + du(k) \tag{2-76}$$

then

$$Y(z) = \mathbf{c}\mathbf{X}(z) + dU(z) \tag{2-77}$$

Substituting (2-75) into (2-77) yields

$$Y(z) = [\mathbf{c}[z\mathbf{I} - \mathbf{A}]^{-1}\mathbf{b} + d]U(z) \tag{2-78}$$

The system transfer function is then seen to be

$$G(z) = \mathbf{c}[z\mathbf{I} - \mathbf{A}]^{-1}\mathbf{b} + d \tag{2-79}$$

This technique will now be illustrated by an example.

Example 2.23

Consider again the state equations of Example 2.22.

$$\mathbf{x}(k+1) = \begin{bmatrix} 1.35 & 0.55 \\ -0.45 & 0.35 \end{bmatrix}\mathbf{x}(k) + \begin{bmatrix} 0.5 \\ 0.5 \end{bmatrix}u(k)$$

$$y(k) = [1 \quad -1]\mathbf{x}(k)$$

Now

$$[z\mathbf{I} - \mathbf{A}] = \begin{bmatrix} z - 1.35 & -0.55 \\ 0.45 & z - 0.35 \end{bmatrix}$$

Thus

$$|z\mathbf{I} - \mathbf{A}| = z^2 - 1.7z + 0.72$$

Also,

$$\text{Cof}\,[z\mathbf{I} - \mathbf{A}] = \begin{bmatrix} z - 0.35 & -0.45 \\ 0.55 & z - 1.35 \end{bmatrix}$$

Then

$$[z\mathbf{I} - \mathbf{A}]^{-1} = \frac{[\text{Cof}\,[z\mathbf{I} - \mathbf{A}]]^T}{|z\mathbf{I} - \mathbf{A}|} = \frac{1}{z^2 - 1.7z + 0.72}\begin{bmatrix} z - 0.35 & 0.55 \\ -0.45 & z - 1.35 \end{bmatrix}$$

From (2-79), since $d = 0$,

$$\begin{aligned} G(z) &= \mathbf{c}[z\mathbf{I} - \mathbf{A}]^{-1}\mathbf{b} \\ &= \frac{1}{z^2 - 1.7z + 0.72}[1 \quad -1]\begin{bmatrix} z - 0.35 & 0.55 \\ -0.45 & z - 1.35 \end{bmatrix}\begin{bmatrix} 0.5 \\ 0.5 \end{bmatrix} \\ &= \frac{1}{z^2 - 1.7z + 0.72}[1 \quad -1]\begin{bmatrix} 0.5z + 0.1 \\ 0.5z - 0.9 \end{bmatrix} \\ &= \frac{1}{z^2 - 1.7z + 0.72} \end{aligned}$$

This, of course, has been shown to be the transfer function.

In this section two techniques are presented for obtaining the transfer function from the state equations. For high-order systems, the second method is more attractive.

The second technique, as given in (2-79), has been implemented as a computer program, to be used for high-order systems. One such program is given in Appendix I.

2.11 SOLUTIONS OF THE STATE EQUATIONS

In this section we develop the general solution of linear time-invariant state equations. It will be seen that the key to the solution of the state equations is the calculation of the state transition matrix. Two related techniques, based on the z-transform, for calculating the state transition matrix are presented. Then the solution of state equations via the digital computer is mentioned.

Recursive Solution

We will first assume that the system is time invariant (fixed) and that $\mathbf{x}(0)$ and $\mathbf{u}(j)$, $j = 0, 1, 2, \ldots$, are known. Now the system equations are

$$\mathbf{x}(k+1) = \mathbf{A}\mathbf{x}(k) + \mathbf{B}\mathbf{u}(k) \tag{2-80}$$

$$\mathbf{y}(k) = \mathbf{C}\mathbf{x}(k) + \mathbf{D}\mathbf{u}(k) \tag{2-81}$$

In a recursive manner it is obvious that

$$\mathbf{x}(1) = \mathbf{A}\mathbf{x}(0) + \mathbf{B}\mathbf{u}(0)$$

and

$$\mathbf{x}(2) = \mathbf{A}\mathbf{x}(1) + \mathbf{B}\mathbf{u}(1)$$

and hence

$$\begin{aligned}\mathbf{x}(2) &= \mathbf{A}(\mathbf{A}\mathbf{x}(0) + \mathbf{B}\mathbf{u}(0)) + \mathbf{B}\mathbf{u}(1)\\ &= \mathbf{A}^2\mathbf{x}(0) + \mathbf{A}\mathbf{B}\mathbf{u}(0) + \mathbf{B}\mathbf{u}(1)\end{aligned}$$

In a similar manner we can show that

$$\mathbf{x}(3) = \mathbf{A}^3\mathbf{x}(0) + \mathbf{A}^2\mathbf{B}\mathbf{u}(0) + \mathbf{A}\mathbf{B}\mathbf{u}(1) + \mathbf{B}\mathbf{u}(2)$$

It is seen, then, that the general solution is given by

$$\mathbf{x}(k) = \mathbf{A}^k\mathbf{x}(0) + \sum_{j=0}^{k-1} \mathbf{A}^{(k-1-j)}\mathbf{B}\mathbf{u}(j) \tag{2-82}$$

If we define

$$\mathbf{\Phi}(k) = \mathbf{A}^k$$

then

$$\mathbf{x}(k) = \mathbf{\Phi}(k)\mathbf{x}(0) + \sum_{j=0}^{k-1} \mathbf{\Phi}(k-1-j)\mathbf{B}\mathbf{u}(j) \tag{2-83}$$

This equation is the general solution to (2-80). From (2-81) and (2-83),

$$\mathbf{y}(k) = \mathbf{C}\mathbf{\Phi}(k)\mathbf{x}(0) + \sum_{j=0}^{k-1} \mathbf{C}\mathbf{\Phi}(k-1-j)\mathbf{B}\mathbf{u}(j) + \mathbf{D}\mathbf{u}(k) \tag{2-84}$$

$\mathbf{\Phi}(k)$ is called the state transition matrix or the fundamental matrix. An example will now be given to illustrate the recursive nature of the solution.

Example 2.24

Consider the transfer function

$$G(z) = \frac{(z+3)}{(z+1)(z+2)}$$

Using the technique of Section 2.8, we write

$$\mathbf{x}(k+1) = \begin{bmatrix} 0 & 1 \\ -2 & -3 \end{bmatrix}\mathbf{x}(k) + \begin{bmatrix} 0 \\ 1 \end{bmatrix}u(k)$$

$$y(k) = [3 \quad 1]\mathbf{x}(k)$$

Assume that the system is initially at rest so that $\mathbf{x}(0) = \mathbf{0}$, and that the input is a unit step; that is,

$$u(k) = 1, \qquad k = 0, 1, 2, \ldots$$

The recursive solution is obtained as follows:

$$\mathbf{x}(1) = \begin{bmatrix} 0 & 1 \\ -2 & -3 \end{bmatrix}\mathbf{x}(0) + \begin{bmatrix} 0 \\ 1 \end{bmatrix}u(0)$$

$$\mathbf{x}(1) = \begin{bmatrix} 0 \\ 1 \end{bmatrix}$$

$$y(1) = [3 \quad 1]\begin{bmatrix} 0 \\ 1 \end{bmatrix}$$

$$y(1) = 1$$

Then

$$\mathbf{x}(2) = \begin{bmatrix} 0 & 1 \\ -2 & -3 \end{bmatrix}\mathbf{x}(1) + \begin{bmatrix} 0 \\ 1 \end{bmatrix}u(1)$$

$$= \begin{bmatrix} 1 \\ -2 \end{bmatrix}$$

$$y(2) = [3 \quad 1]\begin{bmatrix} 1 \\ -2 \end{bmatrix}$$

$$y(2) = 1$$

In a similar manner it can be shown that

$$\mathbf{x}(3) = \begin{bmatrix} -2 \\ 5 \end{bmatrix}, \qquad y(3) = -1$$

and

$$\mathbf{x}(4) = \begin{bmatrix} 5 \\ -10 \end{bmatrix}, \qquad y(4) = 5, \quad \text{etc.}$$

Hence one can recursively determine the states and the output at successive time instants.

z-Transform Method

The general solution to the state equations

$$\mathbf{x}(k+1) = \mathbf{A}\mathbf{x}(k) + \mathbf{B}u(k) \tag{2-85}$$

was developed above, and is given by

$$\mathbf{x}(k) = \mathbf{\Phi}(k)\mathbf{x}(0) + \sum_{j=0}^{k-1} \mathbf{\Phi}(k-1-j)\mathbf{B}\mathbf{u}(j) \tag{2-86}$$

where $\mathbf{\Phi}(k)$, the state transition matrix, is given by

$$\mathbf{\Phi}(k) = \mathbf{A}^k \tag{2-87}$$

One technique for evaluating $\mathbf{\Phi}(k)$ as a function of k is through the use of the z-transform. This technique will now be presented.

In (2-85), let $\mathbf{u}(k) = \mathbf{0}$. Then the z-transform of this equation yields

$$z\mathbf{X}(z) - z\mathbf{x}(0) = \mathbf{A}\mathbf{X}(z) \tag{2-88}$$

Solving for $\mathbf{X}(z)$, we see that

$$\mathbf{X}(z) = z[z\mathbf{I} - \mathbf{A}]^{-1}\mathbf{x}(0) \tag{2-89}$$

Then

$$\mathbf{x}(k) = \mathfrak{z}^{-1}[\mathbf{X}(z)] = \mathfrak{z}^{-1}[z[z\mathbf{I} - \mathbf{A}]^{-1}]\mathbf{x}(0) \tag{2-90}$$

Comparing (2-90) with (2-86), we see that

$$\mathbf{\Phi}(k) = \mathfrak{z}^{-1}[z[z\mathbf{I} - \mathbf{A}]^{-1}] \tag{2-91}$$

To illustrate this method, consider the following example.

Example 2.25

For the state equations of Example 2.24,

$$\mathbf{A} = \begin{bmatrix} 0 & 1 \\ -2 & -3 \end{bmatrix}$$

Then

$$[z\mathbf{I} - \mathbf{A}] = \begin{bmatrix} z & -1 \\ 2 & z+3 \end{bmatrix}$$

and

$$|z\mathbf{I} - \mathbf{A}| = z^2 + 3z + 2 = (z+1)(z+2)$$

Evaluating the inverse matrix in (2-91) and multiplying by z, we obtain

$$z[z\mathbf{I} - \mathbf{A}]^{-1} = \begin{bmatrix} \dfrac{z(z+3)}{(z+1)(z+2)} & \dfrac{z}{(z+1)(z+2)} \\ \dfrac{-2z}{(z+1)(z+2)} & \dfrac{z^2}{(z+1)(z+2)} \end{bmatrix}$$

$$= \begin{bmatrix} \dfrac{2z}{z+1} + \dfrac{-z}{z+2} & \dfrac{z}{z+1} + \dfrac{-z}{z+2} \\ \dfrac{-2z}{z+1} + \dfrac{2z}{z+2} & \dfrac{-z}{z+1} + \dfrac{2z}{z+2} \end{bmatrix} = \mathfrak{z}[\mathbf{\Phi}(k)]$$

Thus

$$\mathbf{\Phi}(k) = \mathfrak{z}^{-1}[z[z\mathbf{I} - \mathbf{A}]^{-1}] = \begin{bmatrix} 2(-1)^k - (-2)^k & (-1)^k - (-2)^k \\ -2(-1)^k + 2(-2)^k & -(-1)^k + 2(-2)^k \end{bmatrix}$$

Transfer-Function Method

As in the preceding method this technique begins with the expression

$$\mathbf{x}(k) = \mathbf{\Phi}(k)\mathbf{x}(0) \tag{2-92}$$

If

$$\mathbf{\Phi}(k) = \begin{bmatrix} \phi_{11}(k) & \cdots & \phi_{1n}(k) \\ \cdot & & \cdot \\ \cdot & & \cdot \\ \cdot & & \cdot \\ \phi_{n1}(k) & \cdots & \phi_{nn}(k) \end{bmatrix}$$

the ith state of the system can be written in the form

$$x_i(k) = \phi_{i1}(k)x_1(0) + \cdots + \phi_{in}(k)x_n(0), \qquad i = 1, \ldots, n \tag{2-93}$$

The z-transform of this expression is then

$$X_i(z) = \mathfrak{z}[\phi_{i1}(k)]x_1(0) + \cdots + \mathfrak{z}[\phi_{in}(k)]x_n(0) \tag{2-94}$$

Hence $X_i(z)$ is known once the $\mathfrak{z}[\phi_{ij}(k)]$ are evaluated. Note that this method is essentially the same as the z-transform method, since $\mathfrak{z}[\mathbf{\Phi}(k)] = z[z\mathbf{I} - \mathbf{A}]^{-1}$. The basic difference between the z-transform method and this approach is that here the $\mathfrak{z}[\phi_{ij}(k)]$ will be evaluated from a simulation diagram or signal flow graph using Mason's gain formula.

Example 2.26

As in the preceding example, we will analyze the system

$$\mathbf{x}(k+1) = \begin{bmatrix} 0 & 1 \\ -2 & -3 \end{bmatrix}\mathbf{x}(k)$$

The signal flow graph for this system is shown in Figure 2-15, where $x_1(0)$ and $x_2(0)$ are the initial conditions. The initial conditions must be entered as shown since, from (2-15),

$$\mathfrak{z}[x(k+1)] = zX(z) - zx(0)$$

we see that

$$X(z) = z^{-1}\mathfrak{z}[x(k+1)] + x(0)$$

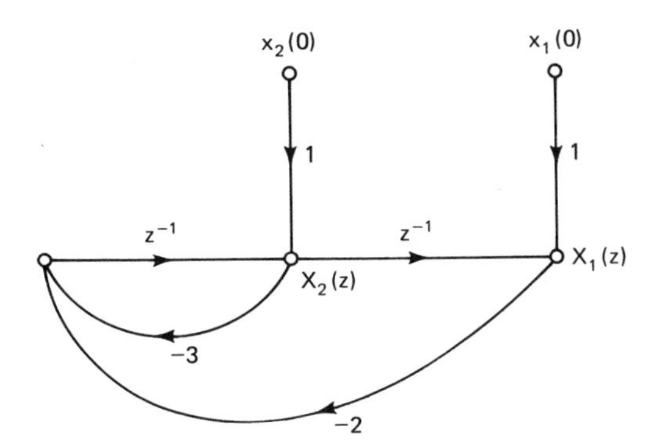

Figure 2-15 Signal flow graph for the system of Example 2.26.

From equation (2-94), we see that $\mathfrak{z}[\phi_{11}(k)]$ is the transfer function from $x_1(0)$ to $X_1(z)$. Applying Mason's gain formula to the signal flow graph, we obtain

$$\mathfrak{z}[\phi_{11}(k)] = \frac{1(1 + 3z^{-1})}{1 + 3z^{-1} + 2z^{-2}}$$

In a similar manner we find that

$$\mathscr{Z}[\phi_{12}(k)] = \frac{X_1(z)}{x_2(0)} = \frac{z^{-1}}{1 + 3z^{-1} + 2z^{-2}}$$

$$\mathscr{Z}[\phi_{21}(k)] = \frac{-2z^{-1}}{1 + 3z^{-1} + 2z^{-2}}$$

$$\mathscr{Z}[\phi_{22}(k)] = \frac{1}{1 + 3z^{-1} + 2z^{-2}}$$

Therefore,

$$\mathbf{X}(z) = \begin{bmatrix} \frac{z(z+3)}{z^2 + 3z + 2} & \frac{z}{z^2 + 3z + 2} \\ \frac{-2z}{z^2 + 3z + 2} & \frac{z^2}{z^2 + 3z + 2} \end{bmatrix} \mathbf{x}(0) = \mathscr{Z}[\mathbf{\Phi}(k)]\mathbf{x}(0)$$

and $\mathbf{\Phi}(k)$ is seen to be the same as that obtained earlier using the z-transform method.

Numerical Method via Digital Computer

The digital computer is ideally suited for evaluating equations of the type

$$\mathbf{x}(k+1) = \mathbf{A}\mathbf{x}(k) + \mathbf{B}\mathbf{u}(k)$$

These equations can be solved recursively by the computer without actually solving for the state transition matrix. A computer program to accomplish this is given in Appendix I.

Another method for finding the state transition matrix is to evaluate the expression

$$\mathbf{\Phi}(k) = \mathbf{A}^k$$

The disadvantage of this procedure is that $\mathbf{\Phi}(k)$ is not found as a general function of k. However, for high-order systems, the evaluation of $\mathbf{\Phi}(k)$ as a function of k is difficult using any method.

Properties of the State Transition Matrix

Three properties of the state transition matrix will now be derived. First, since

$$\mathbf{x}(k) = \mathbf{\Phi}(k)\mathbf{x}(0)$$

then evaluating this expression for $k = 0$ yields

$$\mathbf{\Phi}(0) = \mathbf{I} \tag{2-95}$$

where $\mathbf{I}$ is the identity matrix. Next, since

$$\mathbf{\Phi}(k) = \mathbf{A}^k$$

then

$$\mathbf{\Phi}(k_1 + k_2) = \mathbf{A}^{k_1+k_2} = \mathbf{A}^{k_1}\mathbf{A}^{k_2} = \mathbf{\Phi}(k_1)\mathbf{\Phi}(k_2) \tag{2-96}$$

The third property is seen from the relationships

$$\mathbf{\Phi}(-k) = \mathbf{A}^{-k} = [\mathbf{A}^k]^{-1} = \mathbf{\Phi}^{-1}(k) \tag{2-97}$$

or, taking the inverse of this expression, we obtain an equivalent expression

$$\mathbf{\Phi}(k) = \mathbf{\Phi}^{-1}(-k) \tag{2-98}$$

Example 2.27

We will illustrate the properties of the state transition matrix using the system described by

$$\mathbf{x}(k+1) = \begin{bmatrix} 1 & 0 \\ 0 & 0.5 \end{bmatrix} \mathbf{x}(k)$$

Now

$$\boldsymbol{\Phi}(k) = \mathbf{A}^k = \begin{bmatrix} 1^k & 0 \\ 0 & 0.5^k \end{bmatrix} = \begin{bmatrix} 1 & 0 \\ 0 & 0.5^k \end{bmatrix}$$

Then

$$\boldsymbol{\Phi}(0) = \begin{bmatrix} 1 & 0 \\ 0 & 1 \end{bmatrix} = \mathbf{I}$$

and

$$\boldsymbol{\Phi}(k_1 + k_2) = \begin{bmatrix} 1 & 0 \\ 0 & 0.5^{k_1+k_2} \end{bmatrix} = \begin{bmatrix} 1 & 0 \\ 0 & 0.5^{k_1} \end{bmatrix} \begin{bmatrix} 1 & 0 \\ 0 & 0.5^{k_2} \end{bmatrix} = \boldsymbol{\Phi}(k_1)\boldsymbol{\Phi}(k_2)$$

Also,

$$\boldsymbol{\Phi}^{-1}(-k) = \begin{bmatrix} 1 & 0 \\ 0 & 0.5^{-k} \end{bmatrix}^{-1} = \begin{bmatrix} 1 & 0 \\ 0 & 0.5^k \end{bmatrix} = \boldsymbol{\Phi}(k)$$

2.12 LINEAR TIME-VARYING SYSTEMS

The state equations for a linear time-varying discrete system were given in (2-41) and (2-42), and are repeated here.

$$\mathbf{x}(k+1) = \mathbf{A}(k)\mathbf{x}(k) + \mathbf{B}(k)\mathbf{u}(k) \tag{2-99}$$

$$\mathbf{y}(k) = \mathbf{C}(k)\mathbf{x}(k) + \mathbf{D}(k)\mathbf{u}(k) \tag{2-100}$$

We can find the solution to these equations in a recursive manner. If we denote initial time by k_0 and assume that $\mathbf{x}(k_0)$ is known and that $\mathbf{u}(k)$ is known for $k \geq k_0$, then

$$\mathbf{x}(1 + k_0) = \mathbf{A}(k_0)\mathbf{x}(k_0) + \mathbf{B}(k_0)\mathbf{u}(k_0)$$

Thus

$$\begin{aligned}
\mathbf{x}(2 + k_0) &= \mathbf{A}(1 + k_0)\mathbf{x}(1 + k_0) + \mathbf{B}(1 + k_0)\mathbf{u}(1 + k_0) \\
&= \mathbf{A}(1 + k_0)[\mathbf{A}(k_0)\mathbf{x}(k_0) + \mathbf{B}(k_0)\mathbf{u}(k_0)] + \mathbf{B}(1 + k_0)\mathbf{u}(1 + k_0) \\
&= \mathbf{A}(1 + k_0)\mathbf{A}(k_0)\mathbf{x}(k_0) + \mathbf{A}(1 + k_0)\mathbf{B}(k_0)\mathbf{u}(k_0) + \mathbf{B}(1 + k_0)\mathbf{u}(1 + k_0)
\end{aligned}$$

In a like manner, we see that

$$\begin{aligned}
\mathbf{x}(3 + k_0) &= \mathbf{A}(2 + k_0)\mathbf{x}(2 + k_0) + \mathbf{B}(2 + k_0)\mathbf{u}(2 + k_0) \\
&= \mathbf{A}(2 + k_0)\mathbf{A}(1 + k_0)\mathbf{A}(k_0)\mathbf{x}(k_0) + \mathbf{A}(2 + k_0)\mathbf{A}(1 + k_0)\mathbf{B}(k_0)\mathbf{u}(k_0) \\
&\quad + \mathbf{A}(2 + k_0)\mathbf{B}(1 + k_0)\mathbf{u}(1 + k_0) + \mathbf{B}(2 + k_0)\mathbf{u}(2 + k_0)
\end{aligned}$$

Now, if we define

$$\boldsymbol{\Phi}(k, k_0) = \mathbf{A}(k-1)\mathbf{A}(k-2) \cdots \mathbf{A}(k_0) = \begin{cases} \prod_{j=k_0}^{k-1} \mathbf{A}(j), & k > k_0 \\ \mathbf{I}, & k = k_0 \end{cases} \tag{2-101}$$

where $\mathbf{I}$ is the identity matrix, then the equation above for $\mathbf{x}(3 + k_0)$ can be written as

$$\mathbf{x}(3 + k_0) = \mathbf{\Phi}(3 + k_0, k_0)\mathbf{x}(k_0) + \sum_{j=k_0}^{2+k_0} \mathbf{\Phi}(3 + k_0, j + 1)\mathbf{B}(j)\mathbf{u}(j)$$

Thus it can be shown in general that

$$\mathbf{x}(k) = \mathbf{\Phi}(k, k_0)\mathbf{x}(k_0) + \sum_{j=k_0}^{k-1} \mathbf{\Phi}(k, j + 1)\mathbf{B}(j)\mathbf{u}(j) \tag{2-102}$$

and then

$$\mathbf{y}(k) = \mathbf{C}(k)\mathbf{\Phi}(k, k_0)\mathbf{x}(k_0) + \sum_{j=k_0}^{k-1} \mathbf{C}(k)\mathbf{\Phi}(k, j + 1)\mathbf{B}(j)\mathbf{u}(j) + \mathbf{D}(k)\mathbf{u}(k) \tag{2-103}$$

As before, $\mathbf{\Phi}(k, k_0)$ is called the state transition matrix. Note that since the matrices are time varying, they must be reevaluated at each time instant.

Using (2-101) and (2-102), we can drive the following important properties of the state transition matrix, $\mathbf{\Phi}(k, k_0)$:

$$\begin{aligned} \mathbf{\Phi}(k_0, k_0) &= \mathbf{I} \\ \mathbf{\Phi}(k_2, k_1)\mathbf{\Phi}(k_1, k_0) &= \mathbf{\Phi}(k_2, k_0) \\ \mathbf{\Phi}(k_1, k_2) &= \mathbf{\Phi}^{-1}(k_2, k_1) \end{aligned} \tag{2-104}$$

Note that if the system is time invariant, then $\mathbf{A}$ is not a function k. The equation for the state transition matrix, (2-101), then reduces to that for the time-invariant system, (2-83). It is seen that the time-invariant system is a special case of the time-varying system.

2.13 SUMMARY

In this chapter we have introduced the concepts of discrete-time systems, and the modeling of these systems by difference equations. The z-transform was defined and was shown to be applicable to the solution of difference equations. Next, four methods for determining inverse z-transforms were presented. Finally, the representation of discrete-time systems by simulation diagrams and flow graphs was introduced, which leads naturally to state-variable modeling. Techniques for the solution of linear state equations were then presented.

The foundations for the modeling and the analysis of discrete-time systems was presented in this chapter, and will serve for much of the mathematical basis for the following chapters.

REFERENCES AND FURTHER READING

1. P. G. Brabeck et al., "Final Technical Report—Digital Autopilot Design and Computer Sizing Study," Contract NAS12-2048, Martin Marietta Corporation, Denver, Colo., 1969.
2. A. W. Drake, *Fundamentals of Applied Probability Theory*. New York: McGraw-Hill Book Company, 1967.

3. H. Freeman, *Discrete-Time Systems*. New York: John Wiley & Sons, Inc., 1965.
4. F. Scheid, *Theory and Problems of Numerical Analysis*, (Schaum's Outline Series.) New York, McGraw-Hill Book Company, 1968.
5. C. F. Gerald, *Applied Numerical Analysis*. Reading, Mass.: Addison-Wesley Publishing Company, Inc., 1970.
6. D. F. Lawden, "A General Theory of Sampling Servo Systems," *Proc. IEE*, Lond., Vol. 98, Pt. IV, Oct., 1951.
7. E. A. Guillemin, *The Mathematics of Circuit Analysis*. New York: John Wiley & Sons, Inc., 1949, p. 307.
8. C. R. Wylie, Jr., *Advanced Engineering Mathematics*. New York: McGraw-Hill Book Company, 1951.
9. Paul M. De Russo, R. J. Roy, and C. M. Close, *State Variables for Engineers*. New York, John Wiley and Sons, Inc., 1965.
10. T. R. Benedict and G. W. Bordner, "Synthesis of an Optimal Set of Radar Track-While-Scan Smoothing Equations," *IRE Transactions on Automatic Control*, Vol. AC-7, pp. 27-32; July, 1962.
11. E. I. Jury, *Theory and Application of the z-Transform Method*. New York: John Wiley & Sons, Inc., 1964.
12. B. C. Kuo, *Analysis and Synthesis of Sampled-Data Control Systems*. Englewood Cliffs, N.J.: Prentice-Hall, Inc., 1963.
13. B. C. Kuo, *Discrete-Data Control Systems*. Englewood Cliffs, N.J.: Prentice-Hall, Inc., 1970.
14. J. A. Cadzow and H. R. Martens, *Discrete-Time and Computer Control Systems*. Englewood Cliffs, N.J.: Prentice-Hall, Inc., 1970.
15. R. C. Dorf, *Modern Control Systems*. Reading, Mass.: Addison-Wesley Publishing Company, Inc., 1974.
16. J. L. Melsa, *Computer Programs for Computational Assistance*. New York: McGraw-Hill Book Company, 1970.

PROBLEMS

2-1. Find the z-transform of the number sequence generated by sampling the time function $e(t) = t$ every T seconds, beginning at $t = 0$. Can you express this transform in closed form?

2-2. Find the z-transforms of the number sequences generated by sampling the following time functions every T seconds, beginning at $t = 0$. Express these transforms in closed form.

(a) $e(t) = \epsilon^{-at}$
(b) $e(t) = \epsilon^{-(t-T)}u(t - T)$
(c) $e(t) = \epsilon^{-(t-5T)}u(t - 5T)$

2-3. Find the z-transform, in closed form, of the number sequence generated by sampling the time function $e(t)$ every T seconds beginning at $t = 0$. The function $e(t)$ is specified by its Laplace transform,

$$E(s) = \frac{1 - \epsilon^{-Ts}}{s(s + 1)}$$

2-4. Solve the given difference equation for $x(k)$ using:

(a) the sequential technique

(b) the z-transform

(c) Will the final value theorem give the correct value of $x(k)$ as $k \longrightarrow \infty$?

$$x(k) - 3x(k-1) + 2x(k-2) = e(k)$$

where

$$e(k) = \begin{cases} 1, & k = 0, 1 \\ 0, & k \geq 2 \end{cases}$$

$$x(-2) = x(-1) = 0$$

2-5. Given the difference equation

$$y(k+2) - \tfrac{3}{4}y(k+1) + \tfrac{1}{8}y(k) = e(k)$$

where

$$e(k) = 1, \qquad k = 0, 1, 2, \ldots$$

$$y(0) = y(1) = 0$$

Find the exact expression for $y(100)$.

2-6. Given the difference equation

$$x(k) - x(k-1) + x(k-2) = e(k)$$

where

$$e(k) = \begin{cases} 1, & k = 0 \\ 0, & k = 1, 2, 3, \ldots \end{cases}$$

$$x(-2) = x(-1) = 0$$

(a) Solve for $x(k)$ as a function of k.

(b) Will the final-value theorem give the correct value for $x(k)$ as $k \longrightarrow \infty$? Why?

2-7. Given the difference equation

$$x(k+2) + 3x(k+1) + 2x(k) = e(k)$$

where

$$e(k) = \begin{cases} 1, & k = 0 \\ 0, & \text{otherwise} \end{cases}$$

$$x(0) = 1$$

$$x(1) = -1$$

Solve for $x(k)$ as a function of k.

2-8. Given the difference equation

$$x(k+3) - 2.2x(k+2) + 1.57x(k+1) - 0.36x(k) = e(k)$$

where $e(k) = 1$ for all $k \geq 0$, and $x(0) = x(1) = x(2) = 0$.

(a) Write a digital computer program that will calculate $x(k)$. Run this program, solving for $x(3), x(4), \ldots, x(25)$.

(b) Using the sequential technique, check the first few values obtained in part (a).

2-9. Find $e(0)$ and $e(10)$ for

$$E(z) = \mathfrak{z}\{e(k)\} = \frac{1}{(z-1)(z-0.3)}$$

using the inversion formula. Check the value of $e(0)$ using the initial-value theorem.

2-10. The rectangular rule for numerical integration approximates the integral of a function $x(t)$ by summing rectangular areas as shown in Figure P2-10. Let $y(t)$ be the integral of $x(t)$.

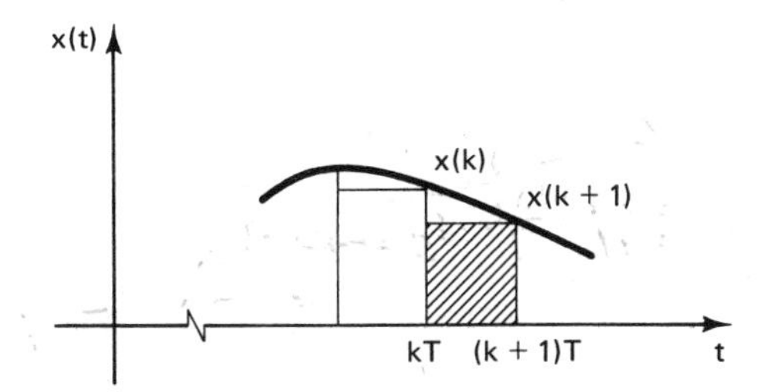

Figure P2-10 Rectangular rule for numerical integration.

(a) Write the difference equation relating $y[(k+1)T]$, $y(kT)$, and $x[(k+1)T]$ for this rule.

(b) Show that the transfer function of this integrator is given by

$$\frac{Y(z)}{X(z)} = \frac{Tz}{z-1}$$

2-11. The trapezoidal rule (modified Euler method) for numerical integration approximates the integral of a function $x(t)$ by summing trapezoid areas as shown in Figure P2-11. Let $y(t)$ be the integral of $x(t)$.

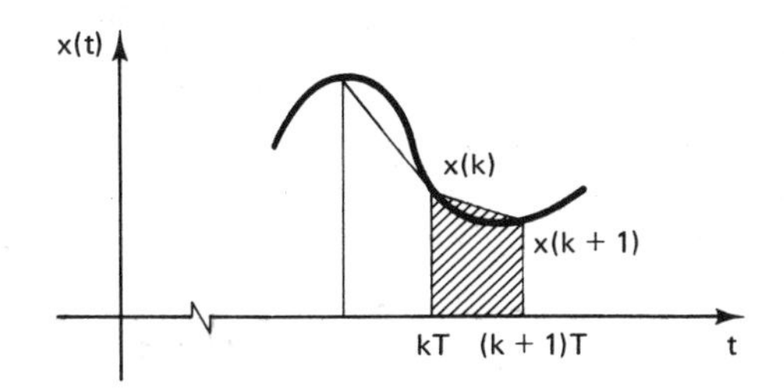

Figure P2-11 Trapezoidal rule for numerical integration.

(a) Write the difference equation relating $y[(k+1)T]$, $y(kT)$, $x[(k+1)T]$, and $x(kT)$ for this rule.

(b) Show that the transfer function for this integrator is given by

$$\frac{Y(z)}{X(z)} = \frac{(T/2)(z+1)}{z-1}$$

2-12. The transfer function for a rectangular-rule integrator is given in Problem 2-10. We would suspect that the reciprocal of this transfer function should yield an approximation to a differentiator. That is, if $w(t)$ is the derivative of $x(t)$,

$$\frac{W(z)}{X(z)} = \frac{z-1}{Tz}$$

(a) Write the difference equation describing this differentiator.

(b) Draw a figure similar to Figure P2-10 illustrating the approximate differentiation.

2-13. Find the inverse z-transform of each $E(z)$ below by the four methods given in the text. Compare the values of $e(k)$, for $k = 0, 1, 2, 3$, obtained by the four methods.

(a) $E(z) = \dfrac{0.5z}{(z-1)(z-0.6)}$ **(b)** $E(z) = \dfrac{0.5}{(z-1)(z-0.6)}$

(c) $E(z) = \dfrac{0.5(z+1)}{(z-1)(z-0.6)}$ **(d)** $E(z) = \dfrac{z(z-0.7)}{(z-1)(z-0.6)}$

2-14. A function $e(t) = 4 \sin 6t$ is sampled every $T = 0.1$ s. Find the z-transform of the resultant number sequence, using the z-transform tables.

2-15. A function $e(t) = A \cos \omega t$ is sampled every $T = 0.2$ s. The z-transform of the resultant number sequence is

$$E(z) = \frac{3z(z-0.6967)}{z^2 - 1.3934z + 1}$$

Solve for A and ω.

2-16. Find $e(2)$ and $e(5)$ if $\mathfrak{z}[e(k)]$ is given by

$$E(z) = \frac{1}{(z-0.9)(z^2 - 0.95z + 0.92)(z^2 - 0.87z + 0.99)}$$

This problem can be solved by inspection. Explain how this can be done.

2-17. For the number sequence $\{e(k)\}$,

$$E(z) = \frac{z}{(z-1)(z+1)}$$

(a) Apply the final-value theorem to $E(z)$.

(b) Check your result in part (a) by finding the inverse z-transform of $E(z)$.

2-18. A function $e(t)$ is sampled, and the resultant number sequence has the z-transform

$$E(z) = \frac{z^3}{z^3 - 3z^2 + 5z - 9}$$

(a) Find the z-transform of $e(t - 3T)u(t - 3T)$.

(b) Find the z-transform of $e(t + T)u(t)$.

2-19. It is desired that the simulation diagram of Figure P2-19 realize the second-order filter described by the transfer function

$$\frac{Y(z)}{E(z)} = \frac{z^2 - 1.96z + 0.99}{z^2 - 1.98z + 0.99}$$

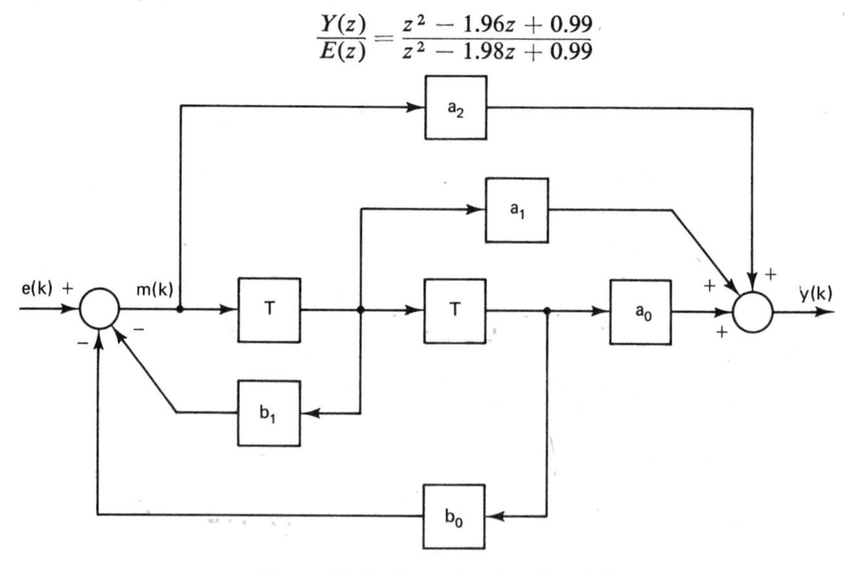

Figure P2-19 Second-order digital filter.

(a) Find the simulation-diagram coefficients b_0, b_1, a_0, a_1, and a_2.

(b) This filter is easiest described by two difference equations. The first expresses $m(k)$ in terms of $e(k)$, $m(k-1)$, and $m(k-2)$, and the second expresses $y(k)$ in terms of $m(k)$, $m(k-1)$, and $m(k-2)$. Write these difference equations.

2-20. Shown in Figure P2-20 are two different realizations of a second-order digital filter.

(a) Write the difference equations describing each realization. For the canonical form, write two equations. The first equation should express $f(k)$ as a function of $e(k)$,

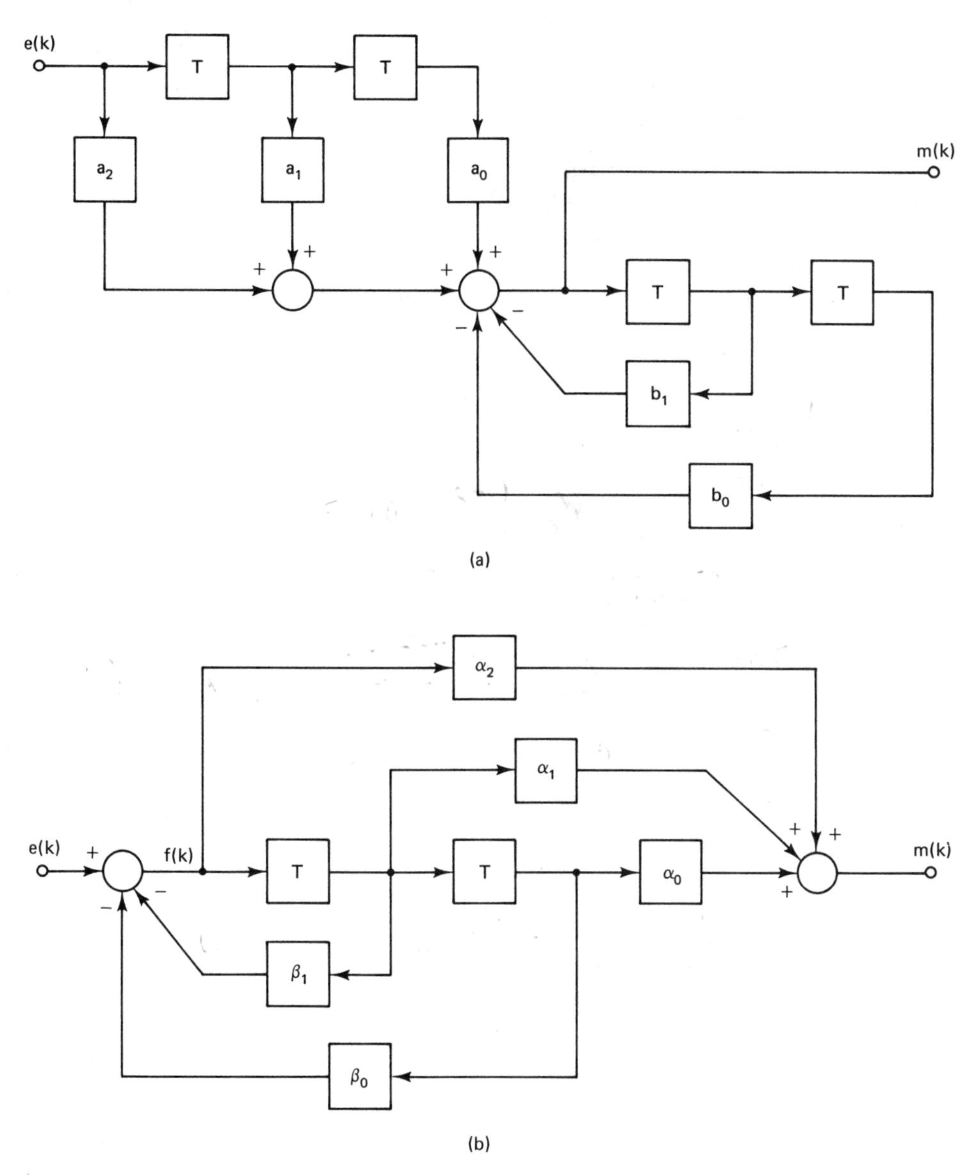

Figure P2-20 Digital filter programming forms: (a) direct form; (b) canonical form.

$f(k-1)$, and $f(k-2)$. The second equation should express $m(k)$ as a function of $f(k)$; and so on.

(b) Using the equations of part (a), derive the transfer functions for the two filter realizations.

(c) Use Mason's gain formula to check the results of part (b).

(d) Given the digital filter transfer function

$$D(z) = \frac{0.99z^2 - 0.976z + 0.983}{z^2 - 0.979z + 1}$$

find the coefficient values in Figure P2-20 such that each filter form realizes $D(z)$.

2-21. The transfer functions of two digital filters are given below. Find difference equations that describe the filters, and draw simulation diagrams for the filters.

(a) $\dfrac{M(z)}{E(z)} = \dfrac{2z^2 - 3.5z + 1.9}{z^3 - 2.7z^2 + z - 0.95}$

(b) $\dfrac{M(z)}{E(z)} = \dfrac{(z-0.95)(z-0.7)}{z(z-0.9)(z-1)}$

2-22. A digital filter designed for both noise reduction and maneuver-following capability of a radar tracking system is the α-β filter [10]. A simulation diagram of this filter is shown in Figure P2-22. In this diagram, $y_R(k)$ is the raw position data of the aircraft under track, obtained from the radar, and $y(k)$ is the filtered position data of the aircraft. In addition, $v_y(k)$ is the filtered velocity data of the aircraft, and T is the system sample period.

(a) Write the difference equations describing this filter.

(b) Find the z-transform of these difference equations, and solve for the transfer functions $Y(z)/Y_R(z)$ and $V_y(z)/Y_R(z)$.

(c) Draw the filter flow graph.

(d) Using Mason's gain formula, solve for the transfer functions of part (b).

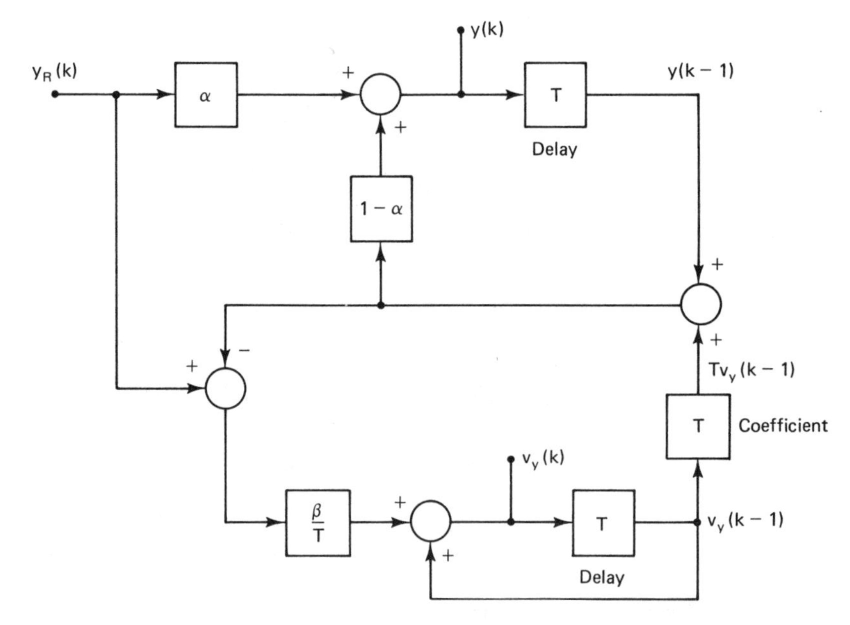

Figure P2-22 α-β tracking filter.

2-23. Given is a BASIC program that solves the difference equation of a digital filter.

```
10   S1=0.
20   E=0.
30   K=0.
40   S2=E-S1
50   M=0.5*S2-S1
60   S1=S2
70   PRINT K,M
80   K=K+1
90   E=E+1
100  GO TO 40
110  END
```

(a) Find the transfer function of the filter.
(b) Find the z-transform of the filter input.
(c) Using the results of parts (a) and (b), find the inverse z-transform of the filter output.
(d) Run the program to check the results of part (c).

2-24. Find two different state-variable formulations that describe the system whose difference equation is:
(a) $y(k+2) + 3y(k+1) + 2y(k) = e(k)$
(b) $y(k+2) + 3y(k+1) + 2y(k) = 2e(k+1) + e(k)$
(c) $y(k+2) + 3y(k+1) + 2y(k) = 2e(k+2) + 2e(k+1) + e(k)$

2-25. Find two different state-variable formulations for the system described by the transfer function

$$\frac{Y(z)}{U(z)} = \frac{3z^2 - 5z - 1}{(z-1)^2(z-2)}$$

2-26. Find a state-variable formulation for each of the following system transfer functions.

(a) $\dfrac{z(z+0.5)}{(z-1)(z-0.5)}$ **(b)** $\dfrac{z}{(z-1)^3}$

(c) $\dfrac{3z^2 - 2z + 2.5}{z^2 - 0.9z + 0.99}$ **(d)** $\dfrac{z^3 + z^2 - 3z - 4}{z^4 - 2z^3 + 2z^2 - 5z + 4}$

2-27. Find a state-variable representation for the system described by the coupled difference equations given below. Consider $y(k+1)$ and $v(k)$ to be the system outputs. *Hint:* Draw a simulation diagram first.

$$y(k+2) + y(k) - v(k) = u(k)$$
$$v(k+1) + v(k) - y(k) = 0$$

2-28. Find a state-variable formulation for the system described by the coupled second-order difference equations given. The system output is $y(k)$, and $e_1(k)$ and $e_2(k)$ are the system inputs. *Hint:* Draw a simulation diagram first.

$$x(k+2) + 2x(k+1) + 3v(k+1) = e_1(k) + e_2(k)$$
$$v(k+2) - v(k) + 3x(k) = e_1(k)$$
$$y(k) = v(k+2) - x(k+1)$$

2-29. Consider the system described by

$$\mathbf{x}(k+1) = \begin{bmatrix} 0 & 1 \\ -2 & -3 \end{bmatrix}\mathbf{x}(k) + \begin{bmatrix} 1 \\ 1 \end{bmatrix}u(k)$$
$$y(k) = [-2 \quad 1]\mathbf{x}(k)$$

(a) Find the transfer function $Y(z)/U(z)$.

(b) Using any similarity transformation, find a different state model for this system. Find the transfer function of the system from the transformed state equations.

(c) Find a similarity transformation that will result in a diagonal system matrix. Check your results by finding the transfer function from the transformed state equations.

2-30. Find the state-transition matrix $\Phi(k)$ for each of the given systems.

(a) $\mathbf{x}(k+1) = \begin{bmatrix} 0 & 0 & 0 \\ 0 & \frac{1}{2} & 0 \\ 0 & 0 & \frac{1}{3} \end{bmatrix} \mathbf{x}(k)$ **(b)** $\mathbf{x}(k+1) = \begin{bmatrix} 0 & 1 \\ -1 & 2 \end{bmatrix} \mathbf{x}(k)$

(c) $\mathbf{x}(k+1) = \begin{bmatrix} 1 & -1 \\ 1 & 3 \end{bmatrix} \mathbf{x}(k)$

2-31. Find the transfer function, $Y(z)/U(z)$, of the system described by

$$\mathbf{x}(k+1) = \begin{bmatrix} 1 & 1 & 0 \\ 0 & 1 & 1 \\ 0 & 0 & 1 \end{bmatrix} \mathbf{x}(k) + \begin{bmatrix} 0 \\ 0 \\ 1 \end{bmatrix} u(k)$$

$$y(k) = [1 \quad 1 \quad 1]\mathbf{x}(k)$$

3

Sampling and Reconstruction

3.1 INTRODUCTION

In Chapter 2 the concept of a discrete system was developed. We found that a discrete system is described (modeled) by a difference equation and that signals within the system are described by number sequences (e.g., $\{e(k)\}$). Some of these number sequences may be generated by sampling a continuous time signal (e.g., in digital control systems). In order to provide a basis for thoroughly understanding the operation of digital control systems, it is necessary to determine the effects of sampling a continuous-time signal. Hence in this chapter these topics will be investigated.

Sections 3.7 and 3.8 are devoted to the internal operation of digital-to-analog (D/A) and analog-to-digital A/D converters. A background in electrical engineering is needed to understand much of this material. However, the nonelectrical engineer will be able to understand the characteristics of different types of A/D and D/A converters by reading these sections.

3.2 SAMPLED-DATA CONTROL SYSTEMS

In this section the type of sampling that generally occurs in sampled-data control systems and in digital control systems will be introduced, and a mathematical model of this sampling will be developed. From this model we may determine the effects of the sampling on the information content of the signal that is sampled.

To introduce sampled-data systems, we will consider the radar tracking system of Figure 3-1a. The closed-loop system is to automatically track the aircraft shown.

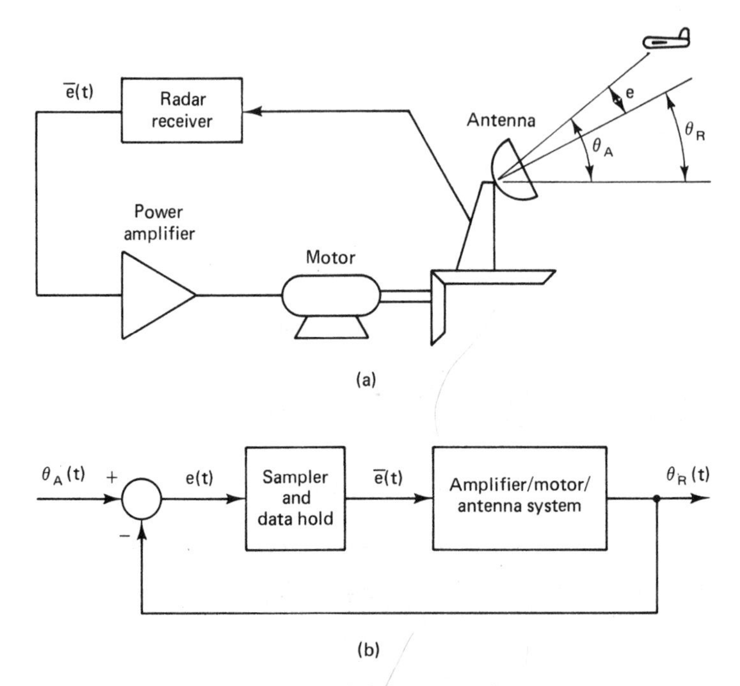

Figure 3-1 Sampled-data control system.

In this figure $\theta_R(t)$ is the pointing angle of the antenna, and $\theta_A(t)$ is the angle to the aircraft. Hence the tracking error is $e(t)$, given by

$$e(t) = \theta_A(t) - \theta_R(t)$$

Assume that the radar transmits every T seconds. Then the error $e(t)$ is known only every T seconds. The block diagram of this system is shown in Figure 3-1b. The radar receiver must output a voltage at every instant of time to the power amplifier. Since only $e(kT)$ is known, a decision must be made as to the form of power amplifier input, $\bar{e}(t)$, for $t \neq kT$.

In general, it is undesirable to apply a signal in sampled form to a plant, because of the high-frequency components inherently present in the sampled signal. Therefore, a data-reconstruction device, called a data hold, is inserted into the system directly following the sampler. The purpose of the data hold is to reconstruct the sampled signal into a form that closely resembles the signal before sampling. The simplest data-reconstruction device, and by far the most common one, is the zero-order hold. The operation of a sampler/zero-order hold combination is described by the signals shown in Figure 3-2. The zero-order hold clamps the output signal to a value equal to that of the input signal at the sampling instant. The sampler and zero-order hold can be represented in block diagram form as shown in Figure 3-3. The signal $\bar{e}(t)$ can be expressed as

$$\begin{aligned}\bar{e}(t) = e(0)[u(t) - u(t-T)] + e(T)[u(t-T) - u(t-2T)] \\ + e(2T)[u(t-2T) - u(t-3T)] + \cdots\end{aligned} \tag{3-1}$$

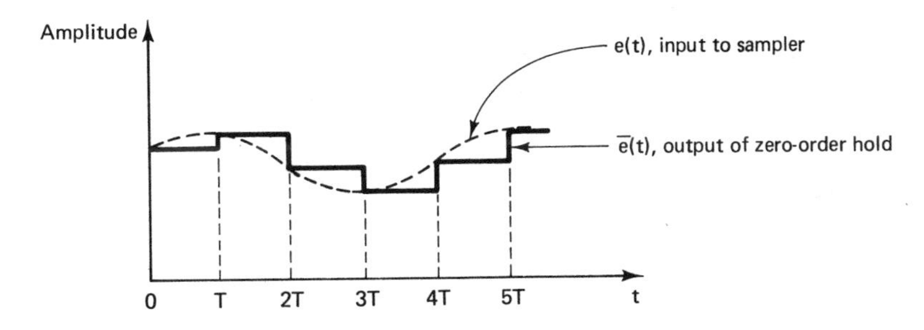

Figure 3-2 Input and output signals of sampler/data hold.

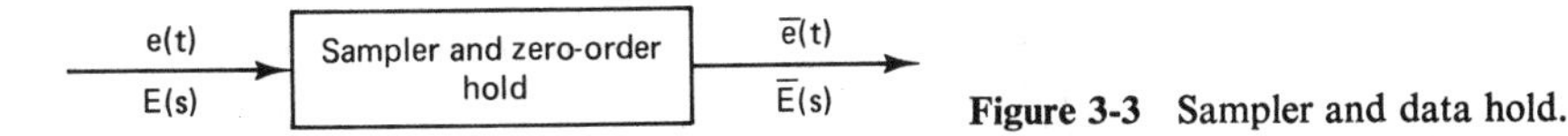

Figure 3-3 Sampler and data hold.

where $u(t)$ is the unit step function. The Laplace transform of $\bar{e}(t)$ is

$$\begin{aligned}\bar{E}(s) &= e(0)\left[\frac{1}{s} - \frac{\epsilon^{-Ts}}{s}\right] + e(T)\left[\frac{\epsilon^{-Ts}}{s} - \frac{\epsilon^{-2Ts}}{s}\right] \\ &\quad + e(2T)\left[\frac{\epsilon^{-2Ts}}{s} - \frac{\epsilon^{-3Ts}}{s}\right] + \cdots \\ &= \left[\frac{1-\epsilon^{-Ts}}{s}\right]\left[e(0) + e(T)\epsilon^{-Ts} + e(2T)^{-2Ts} + \cdots\right] \\ &= \left[\sum_{n=0}^{\infty} e(nT)\epsilon^{-nTs}\right]\left[\frac{1-\epsilon^{-Ts}}{s}\right]\end{aligned} \tag{3-2}$$

The first factor in (3-2) is seen to be a function of the input signal $e(t)$ and the sampling period T. The second factor is seen to be independent of $e(t)$ and therefore can be considered to be a transfer function. Thus the sampler/hold operation can be represented as shown in Figure 3-4. The function $E^*(s)$, called the starred transform, is *defined* as

$$E^*(s) = \sum_{n=0}^{\infty} e(nT)\epsilon^{-nTs} \tag{3-3}$$

and it can be seen that (3-2) is satisfied by the representation in Figure 3-4. It should be emphasized that $E^*(s)$ is not present in the physical system, but appears in the mathematics as a result of factoring (3-2). The sampler in Figure 3-4 does not model a

E(s) — T — E*(s) → $\frac{1-\epsilon^{-Ts}}{s}$ → Ē(s)

Figure 3-4 Representation of sampler and data hold.

physical sampler and the block does not model a physical data hold. However, the combination does accurately model a physical sampler/data-hold device.

The operation symbolized by the sampler in Figure 3-4 cannot be represented by a transfer function. From (3-3) we see that the output of the sampler is a function of $e(t)$ only at $t = kT$, $k = 0, 1, 2, \ldots$. Hence many different input signals can result in the same output signal $E^*(s)$. However, if we could represent the sampler as a transfer function, each different $E(s)$ would result in a different $E^*(s)$. Hence no transfer function exists for the sampler, and this property of the sampler complicates the analysis of systems of the type shown in Figure 3-1.

3.3 THE IDEAL SAMPLER

The inverse Laplace transform of $E^*(s)$ is

$$e^*(t) = \mathcal{L}^{-1}[E^*(s)] = e(0)\delta(t) + e(T)\delta(t - T) + e(2T)\delta(t - 2T) + \cdots \tag{3-4}$$

where $\delta(t)$ is the unit impulse function (Dirac delta function) occurring at $t = 0$. Then $e^*(t)$ is a train of impulse functions whose weights are equal to the values of the signal at the instants of sampling. Thus $e^*(t)$ can be represented as shown in Figure 3-5, since the impulse function has infinite amplitude at the instant it occurs. However, a more common representation of $e^*(t)$ is shown in Figure 3-6.

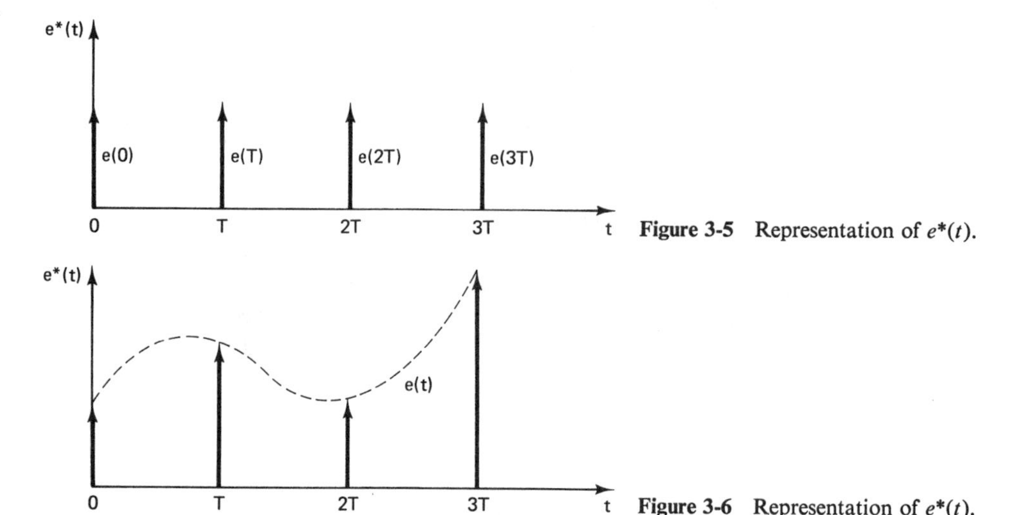

Figure 3-5 Representation of $e^*(t)$.

Figure 3-6 Representation of $e^*(t)$.

The sampler that appears in a sampler/hold model is usually referred to as an *ideal sampler* or an *impulse modulator*. To demonstrate this modulation concept, we define

$$\delta_T(t) = \sum_{n=0}^{\infty} \delta(t - nT) = \delta(t) + \delta(t - T) + \cdots \tag{3-5}$$

Then $e^*(t)$ can be expressed as

$$e^*(t) = e(t)\delta_T(t) = e(t)\delta(t) + e(t)\delta(t - T) + \cdots = e(0)\delta(t) + e(T)\delta(t - T) + \cdots \tag{3-6}$$

In this form it can be readily seen that $\delta_T(t)$ is the carrier in the modulation process, and $e(t)$ is the modulating signal. The function $e(t)$ is assumed to be zero for $t < 0$. For this reason the summation in (3-5) can be taken from $n = -\infty$ to $n = \infty$ with no change in (3-6). Two equivalent representations of the ideal sampler are given in Figure 3-7.

A problem arises in the definition of the ideal sampler output in (3-4) if $e(t)$ has a discontinuity at $t = kT$. For example, if $e(t)$ is a unit step function, what value is used for $e(0)$ in (3-4)? In order to be consistent in the consideration of discontinuous signals, the output signal of an ideal sampler is *defined* as follows:

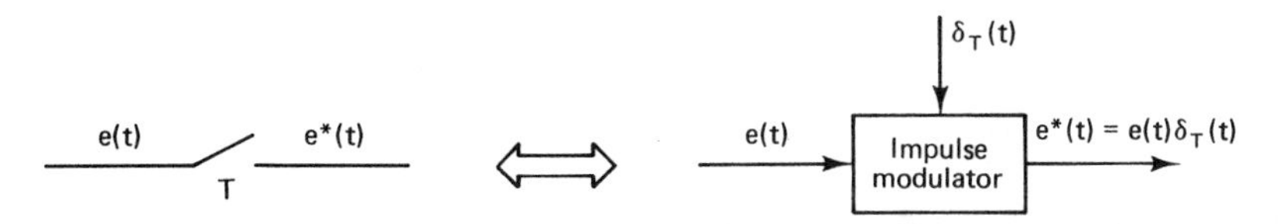

Figure 3-7 Representations of the ideal sampler.

Definition. The output signal of an ideal sampler is defined as the signal whose Laplace transform is

$$E^*(s) = \sum_{n=0}^{\infty} e(nT)\epsilon^{-nTs} \tag{3-7}$$

where $e(t)$ is the input signal to the sampler. If $e(t)$ is discontinuous at $t = kT$, where k is an integer, then $e(kT)$ is taken to be $e(kT^+)$. The notation $e(kT^+)$ indicates the value of $e(t)$ as t approaches kT from the right (i.e., $t = kT + \epsilon$, where ϵ is made arbitrarily small).

It is important to note that the definition of the sampling operation as specified in (3-7) together with the zero-order-hold transfer function defined by

$$G_{ho}(s) = \frac{1 - \epsilon^{-Ts}}{s} \tag{3-8}$$

yield the correct mathematical description of the sampler/hold operation defined by (3-2). It should also be noted that if the signal to be sampled contains an impulse function at a sampling instant, the Laplace transform of the sampled signal does not exist; but since continuous signals in practical situations never contain impulse functions, this limitation is of no practical concern.

Example 3.1

Determine $E^*(s)$ for $e(t) = u(t)$, the unit step. For the unit step, $e(nT) = 1$, $n = 0, 1, 2, \ldots$. Thus from (3-7),

$$E^*(s) = \sum_{n=0}^{\infty} e(nT)\epsilon^{-nTs} = e(0) + e(T)\epsilon^{-Ts} + e(2T)\epsilon^{-2Ts} + \cdots$$

or

$$E^*(s) = 1 + \epsilon^{-Ts} + \epsilon^{-2Ts} + \cdots$$

$E^*(s)$ can be expressed in closed form using the following relationship. For $|x| < 1$,

$$\frac{1}{1-x} = 1 + x + x^2 + \cdots$$

The condition $|x| < 1$ guarantees convergence of the series. Hence the expression for $E^*(s)$ above can be written in closed form as

$$E^*(s) = \frac{1}{1 - \epsilon^{-Ts}}, \qquad |\epsilon^{-Ts}| < 1$$

Example 3.2

Determine $E^*(s)$ for $e(t) = \epsilon^{-t}$. From (3-7),

$$\begin{aligned} E^*(s) &= \sum_{n=0}^{\infty} e(nT)\epsilon^{-nTs} \\ &= 1 + \epsilon^{-T}\epsilon^{-Ts} + \epsilon^{-2T}\epsilon^{-2Ts} + \cdots \\ &= 1 + \epsilon^{-(1+s)T} + (\epsilon^{-(1+s)T})^2 + \cdots \\ &= \frac{1}{1 - \epsilon^{-(1+s)T}}, \qquad |\epsilon^{-(1+s)T}| < 1 \end{aligned}$$

3.4 EVALUATION OF $E^*(s)$

$E^*(s)$, as defined by (3-7), has limited usefulness in analysis because it is expressed as an infinite series. However, for many useful time functions, $E^*(s)$ can be expressed in closed form. In addition, there is a third form of $E^*(s)$ that is useful. These two additional forms of $E^*(s)$ will now be investigated.

As was seen in the preceding section, we can express the inverse Laplace transform of $E^*(s)$ as [see (3-6)]

$$e^*(t) = e(t)\delta_T(t) \tag{3-9}$$

If we then take the Laplace transform of $e^*(t)$ using the complex convolution integral [1], we can derive two additional expressions for $E^*(s)$. These derivations are given in Appendix III. The resultant expressions are

$$E^*(s) = \sum_{\substack{\text{at poles}\\ \text{of } E(\lambda)}} \left[\text{residues of } E(\lambda)\frac{1}{1 - \epsilon^{-T(s-\lambda)}}\right] \tag{3-10}$$

$$E^*(s) = \frac{1}{T}\sum_{n=-\infty}^{\infty} E(s + jn\omega_s) + \frac{e(0)}{2} \tag{3-11}$$

The expression (3-10) is useful in generating tables for the starred transform $E^*(s)$, and expression (3-11) will prove to be useful in analysis. Some examples that illustrate the use of (3-10) will now be given.

Example 3.3

Determine $E^*(s)$ given that

$$E(s) = \frac{1}{(s+1)(s+2)}$$

From (3-10),

$$E(\lambda)\frac{1}{1 - \epsilon^{-T(s-\lambda)}} = \frac{1}{(\lambda+1)(\lambda+2)(1 - \epsilon^{-T(s-\lambda)})}$$

Then

$$\begin{aligned} E^*(s) &= \sum_{\substack{\text{poles of}\\ E(\lambda)}} \left[\text{residues of } E(\lambda)\frac{1}{1 - \epsilon^{-T(s-\lambda)}}\right] \\ &= \frac{1}{(\lambda+2)(1 - \epsilon^{-T(s-\lambda)})}\bigg|_{\lambda=-1} + \frac{1}{(\lambda+1)(1 - \epsilon^{-T(s-\lambda)})}\bigg|_{\lambda=-2} \\ &= \frac{1}{1 - \epsilon^{-T(s+1)}} - \frac{1}{1 - \epsilon^{-T(s+2)}} \end{aligned}$$

Example 3.4

We wish to determine the starred transform of $e(t) = \sin \beta t$. The corresponding $E(s)$ is

$$E(s) = \frac{\beta}{s^2 + \beta^2} = \frac{\beta}{(s - j\beta)(s + j\beta)}$$

Selecting $\lambda_1 = j\beta$ and $\lambda_2 = -j\beta$, then $E^*(s)$ can be evaluated from the expression

$$E^*(s) = \sum_{\substack{\text{poles of} \\ E(\lambda)}} \left[\text{residues of} \frac{\beta}{(\lambda - j\beta)(\lambda + j\beta)(1 - \epsilon^{-T(s-\lambda)})}\right]$$

$$= \frac{\beta}{(\lambda + j\beta)(1 - \epsilon^{-T(s-\lambda)})}\bigg|_{\lambda = j\beta} + \frac{\beta}{(\lambda - j\beta)(1 - \epsilon^{-T(s-\lambda)})}\bigg|_{\lambda = -j\beta}$$

$$= \frac{1}{2j}\left[\frac{1}{1 - \epsilon^{-Ts}\epsilon^{j\beta T}} - \frac{1}{1 - \epsilon^{-Ts}\epsilon^{-j\beta T}}\right]$$

$$= \frac{\epsilon^{-Ts} \sin \beta T}{1 - 2\epsilon^{-Ts} \cos \beta T + \epsilon^{-2Ts}}$$

Example 3.5

Using (3-3) and (3-10), determine $E^*(s)$ given $e(t) = 1 - \epsilon^{-t}$. Now

$$E(s) = \frac{1}{s(s + 1)}$$

and from (3-3),

$$E^*(s) = \sum_{n=0}^{\infty} e(nT)\epsilon^{-nTs}$$

$$= \sum_{n=0}^{\infty} (1 - \epsilon^{-nT})\epsilon^{-nTs}$$

$$= \sum_{n=0}^{\infty} \epsilon^{-nTs} - \sum_{n=0}^{\infty} \epsilon^{-(1+s)nT}$$

$$= \frac{1}{1 - \epsilon^{-Ts}} - \frac{1}{1 - \epsilon^{-(1+s)T}}$$

From (3-10),

$$E^*(s) = \sum_{\substack{\lambda=0 \\ \lambda=-1}} \left[\text{residues of} \frac{1}{\lambda(\lambda + 1)} \frac{1}{1 - \epsilon^{-T(s-\lambda)}}\right]$$

$$= \frac{1}{1 - \epsilon^{-Ts}} - \frac{1}{1 - \epsilon^{-(1+s)T}}$$

It is interesting to consider the case in which the function $e(t)$ contains a time delay. For example, consider a delayed signal of the type

$$e(t) = e_1(t - t_0)u(t - t_0) \tag{3-12}$$

Then

$$E(s) = \epsilon^{-t_0 s}\mathcal{L}[e_1(t)] = \epsilon^{-t_0 s}E_1(s) \tag{3-13}$$

For this case, (3-10) does not apply; instead, special techniques are required to find the starred transform of a delayed signal in closed form. These techniques will be presented in Chapter 4, where the modified z-transform is developed. However, for the special case in which the time signal is delayed a whole number of sampling periods,

(3-10) can be applied in a slightly different form,

$$[\epsilon^{-kTs}E_1(s)]^* = \epsilon^{-kTs} \sum_{\substack{\text{at poles}\\ \text{of } E_1(\lambda)}} \left[\text{residues of } E_1(\lambda)\frac{1}{1-\epsilon^{-T(s-\lambda)}}\right] \tag{3-14}$$

where k is a positive integer (see Problem 3-6).

Example 3.6

The starred transform of $e(t) = [1 - \epsilon^{-(t-1)}]u(t-1)$, with $T = 0.5$s, will now be found. Given $e(t)$ above, $E(s)$ is

$$E(s) = \frac{\epsilon^{-s}}{s} - \frac{\epsilon^{-s}}{s+1} = \frac{\epsilon^{-s}}{s(s+1)}$$

Therefore, in (3-14), k is equal to 2 and

$$E_1(s) = \frac{1}{s(s+1)}$$

Then, from (3-14),

$$\left[\frac{\epsilon^{-s}}{s(s+1)}\right]^* = \epsilon^{-s} \sum_{\lambda=0,-1} \left[\text{residues of } \frac{1}{\lambda(\lambda+1)(1-\epsilon^{-0.5(s-\lambda)})}\right]$$

$$= \epsilon^{-s}\left[\frac{1}{(\lambda+1)(1-\epsilon^{-0.5(s-\lambda)})}\bigg|_{\lambda=0} + \frac{1}{\lambda(1-\epsilon^{-0.5(s-\lambda)})}\bigg|_{\lambda=-1}\right]$$

$$= \epsilon^{-s}\left[\frac{1}{1-\epsilon^{-0.5s}} + \frac{-1}{1-\epsilon^{-0.5(s+1)}}\right] = \frac{(1-\epsilon^{-0.5})\epsilon^{-1.5s}}{(1-\epsilon^{-0.5s})(1-\epsilon^{-0.5(s+1)})}$$

3.5 PROPERTIES OF $E^*(s)$

Two s-plane properties of $E^*(s)$ are given below. These properties are very important and will be used extensively in later derivations.

Property 1. $E^*(s)$ is periodic in s with period $j\omega_s$. This property can be proved using either (3-3), (3-10), or (3-11). From (3-3),

$$E^*(s + jm\omega_s) = \sum_{n=0}^{\infty} e(nT)\epsilon^{-nT(s+jm\omega_s)} \tag{3-15}$$

and since $\omega_s T = 2\pi$,

$$\epsilon^{-jnm\omega_s T} = \epsilon^{-jnm2\pi} = 1, \qquad m \text{ an integer} \tag{3-16}$$

Then

$$E^*(s + jm\omega_s) = \sum_{n=0}^{\infty} e(nT)\epsilon^{-nTs} = E^*(s) \tag{3-17}$$

Property 2. If $E(s)$ has a pole at $s = s_1$, then $E^*(s)$ must have poles at $s = s_1 + jm\omega_s$, $m = 0, \pm 1, \pm 2, \ldots$. This property can be proved from (3-11). Consider $e(t)$ to be continuous at all sampling instants. Then

$$E^*(s) = \frac{1}{T}\sum_{n=-\infty}^{\infty} E(s + jn\omega_s) = \frac{1}{T}[E(s) + E(s + j\omega_s) + E(s + 2j\omega_s) + \cdots + E(s - j\omega_s) + E(s - j2\omega_s) + \cdots] \tag{3-18}$$

If $E(s)$ has a pole at $s = s_1$, then each term of (3-18) will contribute a pole at $s = s_1 + jm\omega_s$, where m is an integer.

It is important to note that no equivalent statement can be made concerning the zeros of $E^*(s)$; that is, the zero locations of $E(s)$ do not uniquely determine the zero locations of $E^*(s)$. However, the zero locations are periodic with period $j\omega_s$, as indicated by the first property of $E^*(s)$.

An example of pole–zero locations of $E^*(s)$ is given in Figure 3-8. The primary strip in the s-plane is defined as the strip for which $-\omega_s/2 \leq \omega \leq \omega_s/2$, as shown in Figure 3-8. Note that if the pole–zero locations are known for $E^*(s)$ in the primary strip, then the pole–zero locations in the entire s-plane are also known.

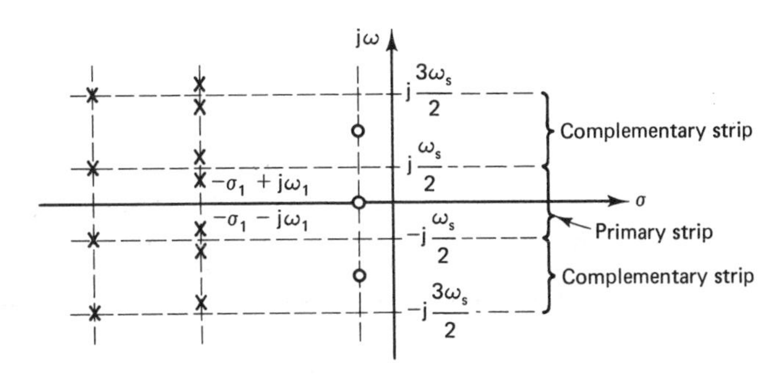

Figure 3-8 Pole–zero locations for $E^*(s)$.

In Figure 3-8, if $E(s)$ has a pole at $-\sigma_1 + j\omega_1$, the sampling operation will generate a pole in $E^*(s)$ at $-\sigma_1 + j(\omega_1 + \omega_s)$. Conversely, if $E(s)$ has a pole at $-\sigma_1 + j(\omega_1 + \omega_s)$, then $E^*(s)$ will have a pole at $-\sigma_1 + j\omega_1$. In fact, a pole location in $E(s)$ at $-\sigma_1 + j(\omega_1 + k\omega_s)$, k an integer, will result in identical pole locations in $E^*(s)$, regardless of the integer value of k. This conclusion can also be seen from the example of Figure 3-9. Note that both signals have the same starred transform, since

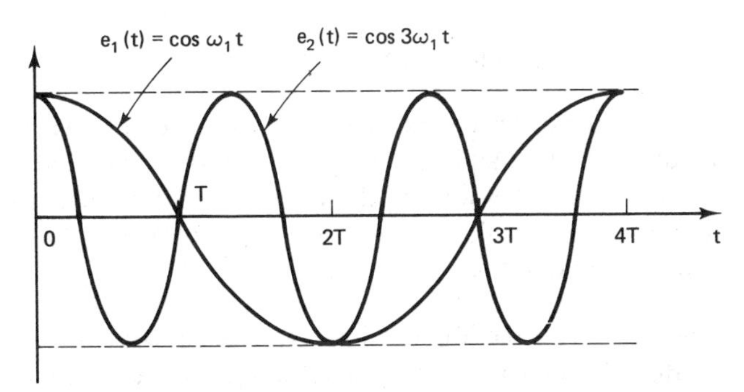

Figure 3-9 Two signals that have the same starred transform.

the two signals have the same value at each sampling instant. One pole of $E_1(s)$ occurs at $s = j\omega_1 = j\omega_s/4$, and one pole of $E_2(s)$ occurs at $s = -j3\omega_1 = j(\omega_1 - \omega_s)$. The other pole of $E_1(s)$ occurs at $s = -j\omega_s/4$, and the other pole of $E_2(s)$ occurs at $s = j3\omega_1 = j(-\omega_1 + \omega_s)$.

Note also that if $E(j\omega)$ has the amplitude spectrum as shown in Figure 3-10a, $E^*(j\omega)$ has the amplitude spectrum as shown in Figure 3-10b. This can be seen by evaluating (3-18) for $s = j\omega$. An ideal filter is a filter with a unity gain in the pass band,

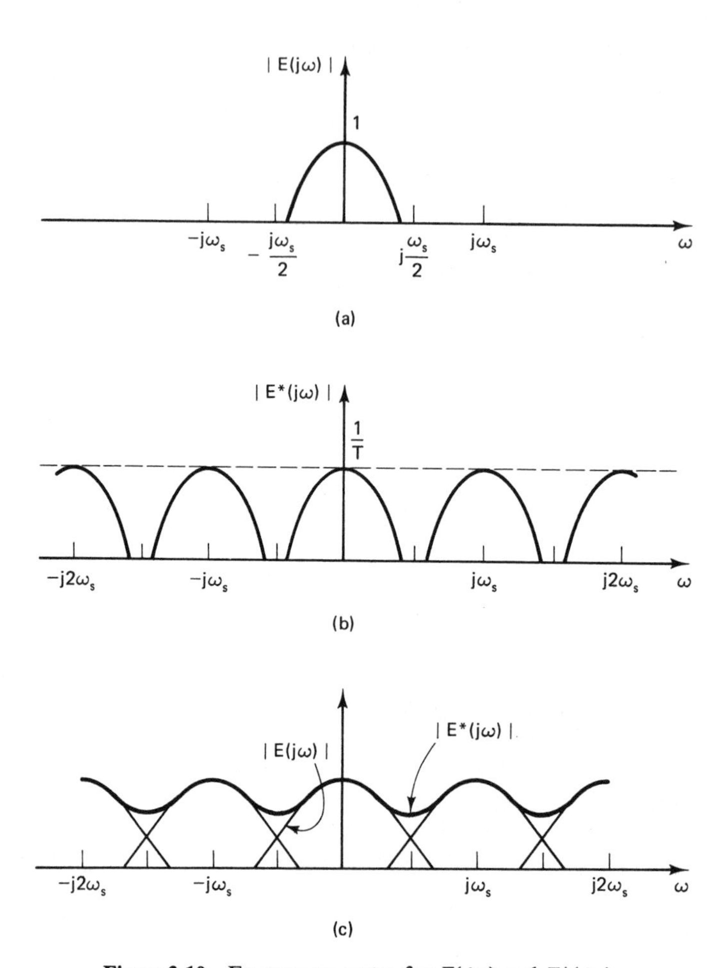

Figure 3-10 Frequency spectra for $E(j\omega)$ and $E^*(j\omega)$.

and a gain of zero outside the pass band. Of course, such a filter is not physically realizable [2]. It is seen from Figure 3-10b that an ideal low-pass filter could completely recover $E(j\omega)[e(t)]$, if the bandwidth of the filter were $\omega_s/2$, for the case that the highest-frequency component present in $E(j\omega)$ is less than $\omega_s/2$. This is, of course, essentially a statement of Shannon's sampling theorem [3].

Shannon's Sampling Theorem. A function of time $e(t)$ which contains no frequency components greater than f_0 hertz is uniquely determined by the values of $e(t)$ at any set of sampling points spaced $\frac{1}{2f_0}$ seconds apart.

Suppose, in Figure 3-10b, that ω_s is decreased until the highest-frequency components present in $E(j\omega)$ are greater than $\omega_s/2$. Then $E^*(j\omega)$ has the amplitude spectrum shown in Figure 3-10c; and for this case, no filtering scheme, ideal or realizable, will recover $e(t)$. Thus, in choosing the sampling rate for a control system, the sampling frequency should be greater than twice the highest-frequency component of significant amplitude of the signal being sampled.

It is seen from the frequency spectrum of $E^*(j\omega)$ that ideal sampling, as defined here, is a form of amplitude modulation. Nonideal sampling (i.e., finite-width pulse amplitude sampling) has the same frequency distribution, but the amplitudes of the generated sidebands have a $(\sin x)/x$ variation, with x proportional to frequency [4].

It is to be recalled that the ideal sampler is not a physical device, and thus the frequency spectrum, as shown in Figure 3-10, is not the spectrum of a signal that appears in a physical system. The ideas above will be extended to signals that do appear in physical systems after an investigation of data holds.

3.6 DATA RECONSTRUCTION

In most feedback control systems employing sampled data, a continuous signal is reconstructed from the sampled signal. The block diagram of a simple sampled-data control system is repeated in Figure 3-11. Suppose that the sampled signal is band-

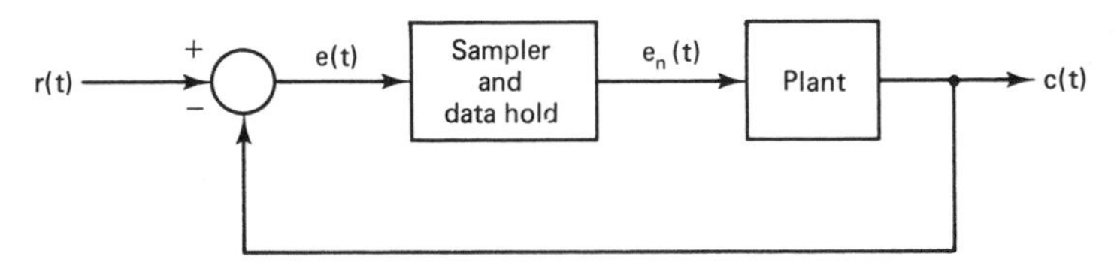

Figure 3-11 Sampled-data control system.

limited in frequency, so that the highest-frequency component of $e(t)$ is less than $\omega_s/2$. Then $E^*(j\omega)$ would have the frequency spectrum shown in Figure 3-10b, and theoretically the signal could be reconstructed exactly by employing an ideal low-pass filter. However, since ideal filters do not exist in physically realizable systems, we must employ approximations. Practical data holds are devices that approximate, in some sense, an ideal low-pass filter.

The reader may ask: Why discuss a sampled-data system as illustrated in Figure 3-11? The system contains a sampler to examine a continuous signal at discrete instants of time. Then a data hold is employed to try to reconstruct the original signal! Our reply is that many existing control systems actually operate in this manner due to hardware implementation techniques. More important, we will later add a compensator block between the sampler and data-hold device in order to improve system performance. In either case, our discussion of data-hold devices in this chapter is prerequisite to closed-loop system analysis and synthesis.

A commonly used method of data reconstruction is polynomial extrapolation. Using a Taylor's series expansion about $t = nT$, we can express $e(t)$ as

$$e(t) = e(nT) + e'(nT)(t - nT) + \frac{e''(nT)}{2!}(t - nT)^2 + \cdots \tag{3-19}$$

where the prime denotes the derivative. $e_n(t)$ is defined as the reconstructed version of

$e(t)$ for the nth sample period; that is,

$$e_n(t) \cong e(t) \qquad \text{for } nT \leq t < (n+1)T \tag{3-20}$$

Thus $e_n(t)$ will be used to represent the output of the data hold. Since $e(t)$ enters the data hold only in sampled form, the values of the derivatives are not known. However, the derivatives may be approximated by the backward difference

$$e'(nT) = \frac{1}{T}\Big[e(nT) - e[(n-1)T]\Big] \tag{3-21}$$

$$e''(nT) = \frac{1}{T}\Big[e'(nT) - e'[(n-1)T]\Big] \tag{3-22}$$

and so forth. Note that by substituting (3-21) into (3-22), we obtain

$$e''(nT) = \frac{1}{T}\Big[\frac{1}{T}\Big[e(nT \; - e[(n-1)T]\Big] - \frac{1}{T}\Big[e[(n-1)T] - e[(n-2)T]\Big]\Big] \tag{3-23}$$

or

$$e''(nT) = \frac{1}{T^2}\Big[e(nT) - 2e[(n-1)T] + e[(n-2)T]\Big] \tag{3-24}$$

Three types of data hold based on the foregoing relationships, the zero-, first-, and fractional-order holds, will now be discussed.

Zero-Order Hold

If only the first term in the expansion of (3-19) is used, the data hold is called a zero-order hold. Here we assume that the function $e(t)$ is approximately constant within the sampling interval at a value equal to that of the function at the preceding sampling instant. Therefore, for the zero-order hold,

$$e_n(t) = e(nT), \qquad nT \leq t < (n+1)T \tag{3-25}$$

Recall that the $e_n(t)$ for a zero-order hold has been defined as $\bar{e}(t)$ in (3-1). Note that no memory is required, and therefore this data hold is the simplest to construct. The transfer function of the zero-order hold was derived in (3-2). However, this transfer function will be derived again using a simpler technique—one that may be easily used in the derivations of transfer functions for other data holds. Using the model of Figure 3-4 for the sampler-data-hold device, we note that the input to the data hold will be only impulse functions. The signal $e_o(t)$ shown in Figure 3-12 then describes the data hold output if the input $e_i(t)$ to the data hold is a unit impulse function. Thus

$$e_o(t) = u(t) - u(t - T)$$

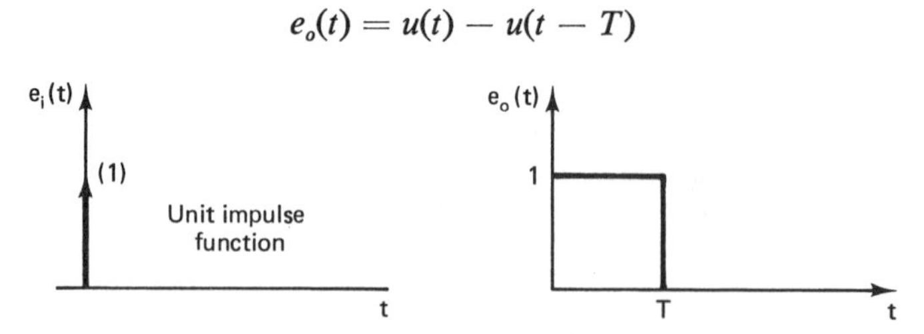

Figure 3-12 Input and output signals for the zero-order hold.

and

$$E_o(s) = \frac{1}{s} - \frac{\epsilon^{-Ts}}{s}$$

Since $E_i(s) = 1$, the transfer function of the zero-order hold is

$$G_{h0}(s) = \frac{E_o(s)}{E_i(s)} = \frac{1 - \epsilon^{-Ts}}{s} \tag{3-26}$$

as has already been shown in (3-2). In addition, recall that (3-26) is not the transfer function of a physical device since this equation was derived assuming that impulse functions occur at the input to the zero-order hold. Thus, as shown earlier in this chapter, if (3-7) is used to describe the sampling operation mathematically, and if (3-26) is used in conjunction with (3-7) to describe the data hold mathematically, a correct mathematical description of the overall sample–hold operation is obtained.

To obtain the frequency response of the zero-order hold, consider the following development:

$$\begin{aligned} G_{h0}(j\omega) &= \frac{1 - \epsilon^{-j\omega T}}{j\omega} \cdot \epsilon^{j(\omega T/2)} \cdot \epsilon^{-j(\omega T/2)} = \frac{2\epsilon^{-j(\omega T/2)}}{\omega}\left[\frac{\epsilon^{j(\omega T/2)} - \epsilon^{-j(\omega T/2)}}{2j}\right] \\ &= T\frac{\sin(\omega T/2)}{\omega T/2} \cdot \epsilon^{-j(\omega T/2)} \end{aligned} \tag{3-27}$$

Since

$$\frac{\omega T}{2} = \frac{\omega}{2}\left(\frac{2\pi}{\omega_s}\right) = \frac{\pi\omega}{\omega_s} \tag{3-28}$$

(3-27) can be expressed as

$$G_{h0}(j\omega) = T \cdot \frac{\sin(\pi\omega/\omega_s)}{\pi\omega/\omega_s} \cdot \epsilon^{-j(\pi\omega/\omega_s)} \tag{3-29}$$

Thus

$$|G_{h0}(j\omega)| = T\left|\frac{\sin(\pi\omega/\omega_s)}{\pi\omega/\omega_s}\right| \tag{3-30}$$

and

$$\underline{/G_{h0}(j\omega)} = -\frac{\pi\omega}{\omega_s} + \theta, \qquad \theta = \begin{cases} 0, & \sin\left(\frac{\pi\omega}{\omega_s}\right) > 0 \\ \pi, & \sin\left(\frac{\pi\omega}{\omega_s}\right) < 0 \end{cases} \tag{3-31}$$

The amplitude and phase plots for $G_{h0}(j\omega)$ are shown in Figure 3-13.

A word is in order concerning the interpretation of the frequency response of the zero-order hold. First, it must be remembered that the data hold must be preceded by an ideal sampler. Now, suppose that a sinusoid of frequency ω_1 is applied to the ideal sampler, where $\omega_1 < \omega_s/2$. Then the output of the sampler contains the frequencies shown in Figure 3-14. Thus the frequency response of the zero-order hold may be used to determine the amplitude spectrum of the data-hold output signal. The output signal components are shown in Figure 3-14c. Note that the output signal amplitude spectrum will be the same as that shown in the figure if the input signal frequency is $\omega = k\omega_s \pm \omega_1$, $k = 0, 1, 2, 3, \ldots$. Note that when $\omega_1 \ll \omega_s/2$, the high-frequency components of $E^*(j\omega)$ will occur near zeros of $G_{h0}(j\omega)$; hence the sampler and zero-order hold will have little effect on system operation.

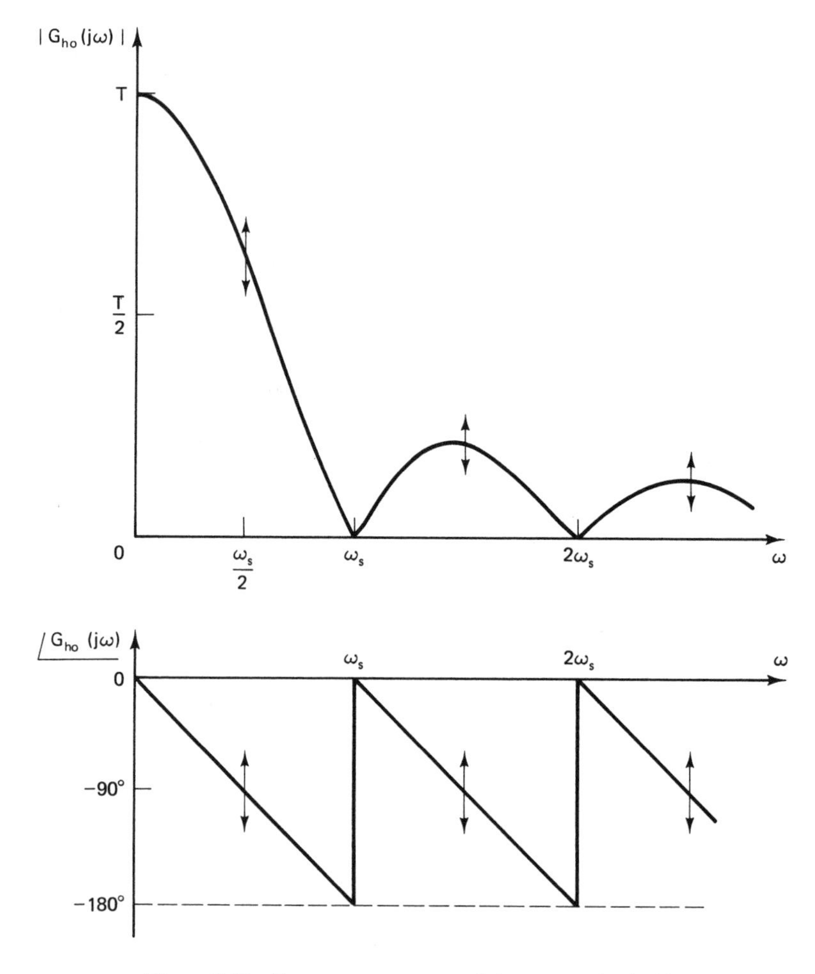

Figure 3-13 Frequency response of the zero-order hold.

Suppose that the frequency ω_1, in Figure 3-14, is equal to $\omega_s/2$. In this case, the frequency components ω_1 and $\omega_s - \omega_1$ are superimposed, and the data-hold output is a function of the phase of the input sinusoid. Note that the amplitude of the data-hold output can range from 0 to a value greater than the amplitude of the input signal. We emphasize this effect by showing arrows at $k\omega_s + \omega_s/2$, $k = 0, 1, 2, \ldots$, in Figure 3-13.

First-Order Hold

The first two terms of (3-19) are used to realize the first-order hold. Therefore,

$$e_n(t) = e(nT) + e'(nT)(t - nT), \qquad nT \leq t < (n + 1)T \tag{3-32}$$

where

$$e'(nT) = \frac{e(nT) - e[(n - 1)T]}{T} \tag{3-33}$$

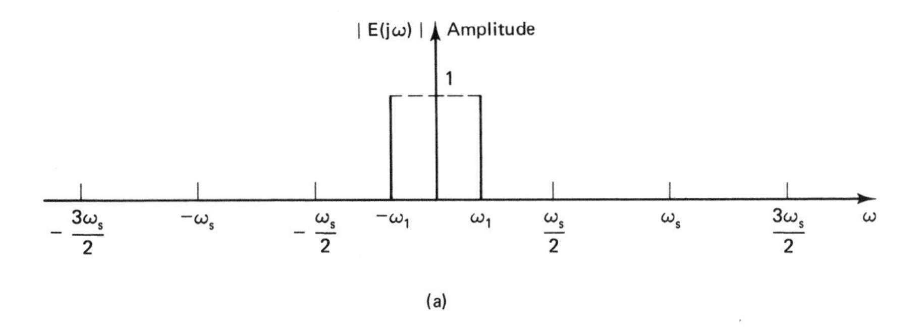

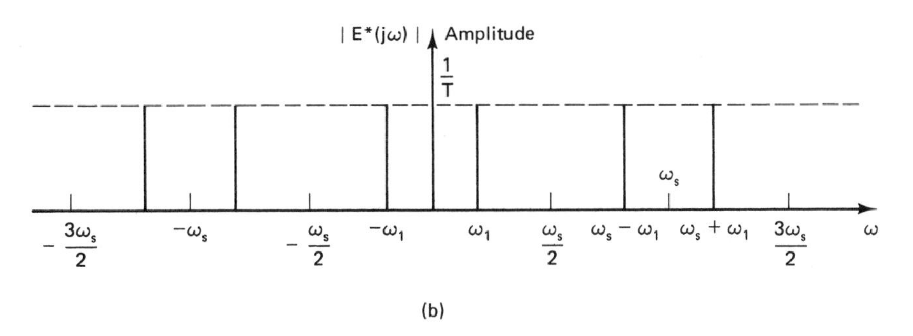

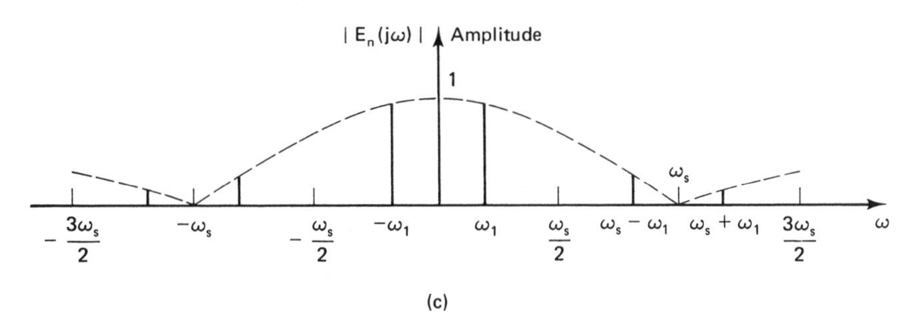

Figure 3-14 Sinusoidal response of the sampler and zero-order hold: (a) input signal to the ideal sampler; (b) output signal of the ideal sampler; (c) output signal from the zero-order hold.

This expression indicates that the extrapolated function within a given interval is linear and that its slope is determined by the values of the function at the sampling instants in the previous interval. Note that memory is required in the realization of this data hold. To determine the transfer function of a first-order hold, assume that the input is a unit impulse function. Then, from (3-32) and (3-33), the output, shown in Figure 3-15, is

$$\begin{aligned} e_o(t) = u(t) + \frac{1}{T}tu(t) - 2u(t-T) - \frac{2}{T}(t-T)u(t-T) + u(t-2T) \\ + \frac{1}{T}(t-2T)u(t-2T) \end{aligned} \tag{3-34}$$

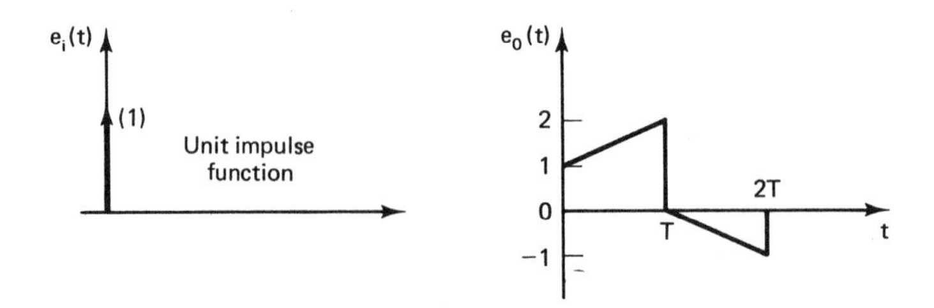

Figure 3-15 Input and output function for the first-order hold.

Since $E_i(s) = 1$,

$$G_{h1}(s) = \frac{E_o(s)}{E_i(s)} = \frac{1}{s} - \frac{2\epsilon^{-Ts}}{s} + \frac{\epsilon^{-2Ts}}{s} + \frac{1}{Ts^2}(1 - 2\epsilon^{-Ts} + \epsilon^{-2Ts}) \tag{3-35}$$

or

$$G_{h1}(s) = \frac{1 + Ts}{T}\left[\frac{1 - \epsilon^{-Ts}}{s}\right]^2$$

An example of the input/output waveforms of a sampler and first-order hold is shown in Figure 3-16.

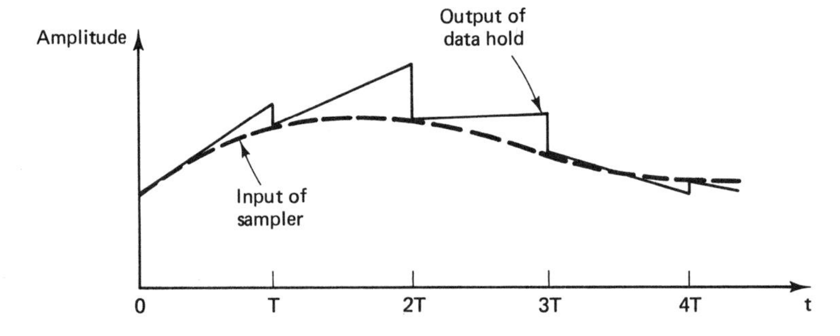

Figure 3-16 Example of the output of a first-order hold.

The frequency response of the first order hold is obtained from (3-35).

$$G_{h1}(j\omega) = \frac{1 + j\omega T}{T}\left[\frac{1 - \epsilon^{-j\omega T}}{j\omega}\right]^2 \tag{3-36}$$

$$|G_{h1}(j\omega)| = \frac{2\pi}{\omega_s}\sqrt{1 + \frac{4\pi^2\omega^2}{\omega_s^2}}\left[\frac{\sin(\pi\omega/\omega_s)}{\pi\omega/\omega_s}\right]^2 \tag{3-37}$$

$$\underline{/G_{h1}(j\omega)} = \tan^{-1}\left(\frac{2\pi\omega}{\omega_s}\right) - \frac{2\pi\omega}{\omega_s} \tag{3-38}$$

The amplitude and phase characteristics of the first-order hold are shown in Figure 3-17. Note that the first-order hold provides a better approximation of the ideal low-pass filter in the vicinity of zero frequency than does the zero-order hold. However, for larger ω, the zero-order hold yields a better approximation. Consider once again the sideband frequencies generated by the sampling process, as illustrated in Figure 3-14b. If $\omega_1 \ll \omega_s/2$, the first-order hold provides better reconstruction of the sampled signal than does the zero-order hold. However, if ω_1 is of the same order of magnitude

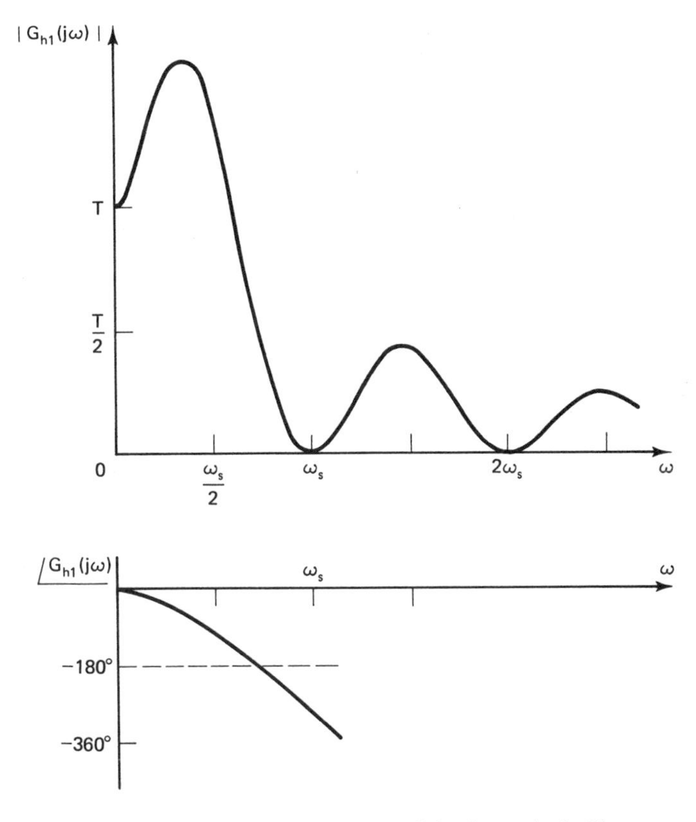

Figure 3-17 Frequency response of the first-order hold.

as $\omega_s/2$, the zero-order hold may yield better results in the reconstruction process. Thus in some applications the zero-order hold is superior to the first-order hold. However, in any case, the zero-order hold is by far the most commonly used device, because of cost considerations.

Fractional-Order Holds

It is important to remember that when using the first-order hold we essentially perform a direct linear extrapolation from one sampling interval to the next; that is, we assume that by using the approximate slope of the signal in the interval from $(n - 1)T$ to nT we can obtain the value of the signal at $(n + 1)T$. The error generated in this process can be reduced by using only a fraction of the slope in the previous interval, as shown in Figure 3-18. In this figure it is assumed that the input is a unit impulse and the value of k ranges from zero to unity. The frequency response of this data hold is shown in Figure 3-19. Note that for $k = 0$, the hold is a zero-order hold; and for $k = 1$, a first-order hold is obtained. Although it is difficult to determine the optimum value of k except in certain specific circumstances, the fractional-order hold may be used to match the data-hold frequency characteristic to the sampled signal's frequency characteristic, thereby generating minimum error extrapolations. By using the techniques

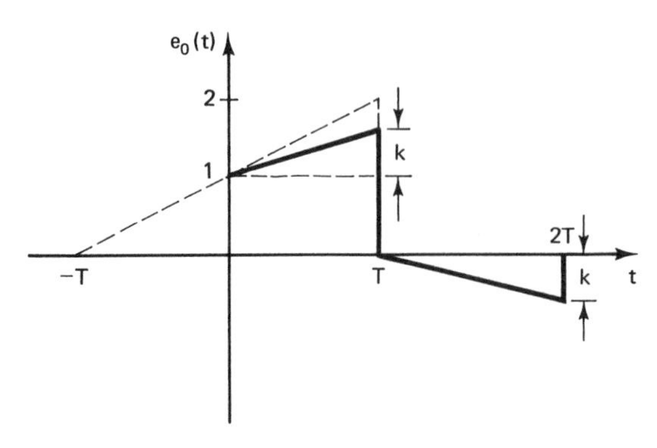

Figure 3-18 Impulse response of the fractional-order hold.

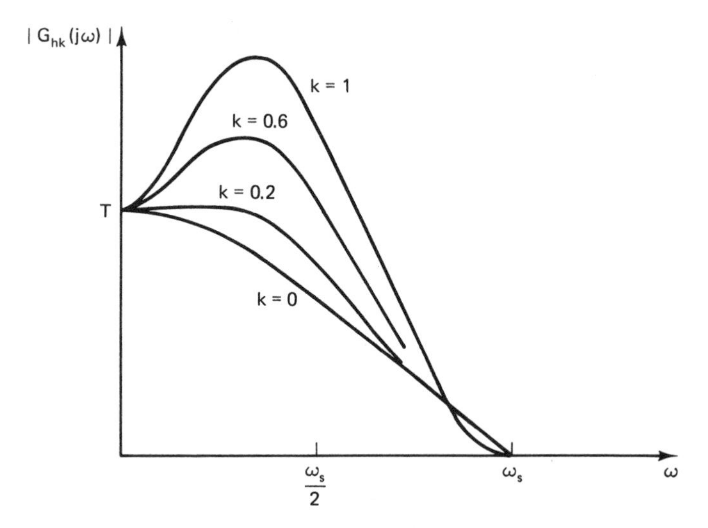

Figure 3-19 Frequency response magnitudes for the fractional-order hold.

given above, the transfer function of the fractional-order hold can be shown to be

$$G_{hk}(s) = (1 - k\epsilon^{-Ts})\frac{1 - \epsilon^{-Ts}}{s} + \frac{k}{Ts^2}(1 - \epsilon^{-Ts})^2 \tag{3-39}$$

The derivation of this transfer function is given as an exercise in Problem 3-15.

3.7 DIGITAL-TO-ANALOG CONVERSION*

This section and Section 3.8 describe the most common methods for digital-to-analog and analog-to-digital conversion. Although the circuit diagrams have been simplified, the functional nature of each method is preserved. Upon completion of the chapter the reader should be able to understand the literature on these subjects.

The basic function of the digital-to-analog converter (DAC) is to convert a

*Section 3.7, contributed by Richard C. Jaeger, is based on his "Data Acquisition Systems," Tutorial Notes, IECI 81, San Francisco, Nov. 9, 1981, and is included here with his permission.

digital representation of a number into its equivalent analog voltage. The output voltage of the D/A converter can be represented as

$$V_0 = V_{fs}[A_1 2^{-1} + A_2 2^{-2} + \cdots + A_n 2^{-n}] \tag{3-40}$$

V_{fs} represents a reference voltage which determines the full-scale output voltage of the converter, and A_1 through A_n represent the binary digits or bits of the input word. A_1 is called the most significant bit (MSB) and corresponds to a voltage of $V_{fs}/2$. A_n is the least significant bit (LSB) and has a weight of $V_{fs}/2^n$. For the sake of our discussion, an "on" bit equals 1 and an "off" bit equals 0. The resolution of the converter is the smallest analog change that can be produced by the converter and is equal to the value of the LSB in volts. However, it is also often specified as a percentage of full scale or just as n-bit resolution.

One of the simplest DAC circuits is given in Figure 3-20a and uses the summing amplifier and weighted resistor network. The input binary word controls the switches with an on bit indicating a closed switch and an off bit corresponding to an open switch. The resistors are weighted progressively by a factor of 2, thereby producing the desired binary weighted contributions to the output. Two problems arise. The first problem is that the DAC requires accurate resistor ratios to be maintained over a very wide range of resistor values (a range of 1000 — 1 for a 10-bit DAC). Also, since the switches are in series with the resistors, this "on" resistance must be very low, and they should have zero offset voltage. These last two requirements can be met using good MOSFETs or JFETs as switches. However, the wide range of resistor values is not suitable for monolithic converters of moderate to high resolution.

The R-$2R$ ladder shown in Figure 3-20b avoids the problem of a wide range of resistor values. It is well suited for integrated circuits since it requires only two resistor values R and $2R$. The value of R typically ranges from 2.5 to 10 kΩ. Taking successive Thévenin equivalent circuits for each bit of the ladder, it is easy to show that the inputs are each reduced by a factor of 2 going from the MSB to the LSB. Again, this network is using the switches in a voltage switching mode and requires low on-resistance, zero offset voltage switches.

Because the currents flowing in the ladder change as the input word changes, power dissipation and heating in the network change, causing nonlinearity in the DAC. Also, the load on the reference voltage depends on the binary input. Because of this, most monolithic versions of this DAC use the configuration shown in Figure 3-20c, known as the inverted R-$2R$ ladder. Here the currents flowing in the ladder are constant with the digital input diverting the current either to ground or to the input of a current-to-voltage converter. This is a popular configuration used with the CMOS process, which provides excellent switching devices. The switches still need to be low-on-resistance devices to minimize errors within the converter. The R-$2R$ ladder must be diffused, implanted, or thin-film, depending on the manufacturer's processing capability and the resolution of the DAC.

Bipolar transistors do not perform well as voltage switches because of their inherent offset voltage in the saturated region of operation. However, they do make excellent current switches, and most DACs realized using bipolar processes use some form of switched current sources.

Figure 3-20d shows conceptually how binary weighted current sources could be

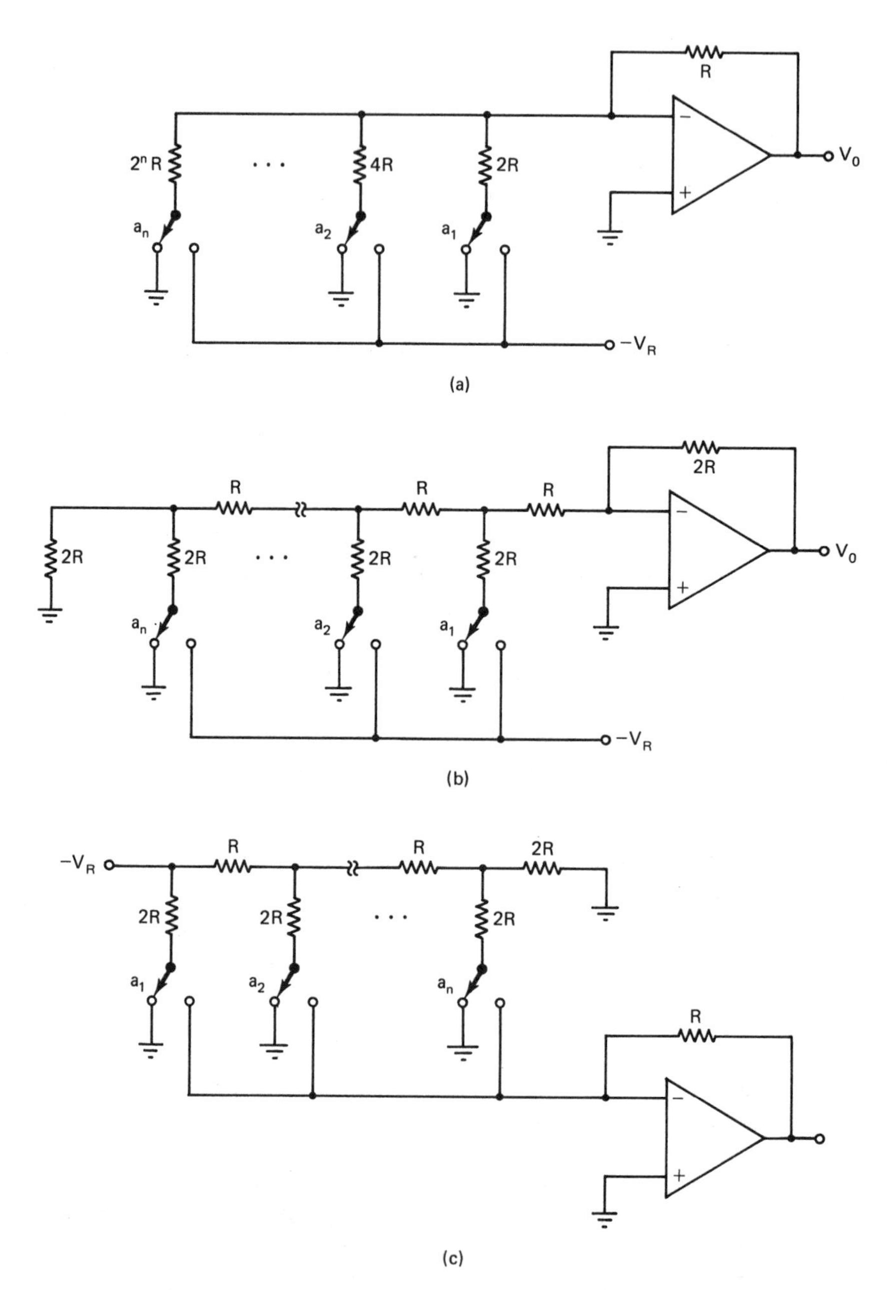

Figure 3-20 Digital-to-analog converters: (a) weighted resistor DAC; (b) R-$2R$ ladder; (c) inverted R-$2R$ ladder; (d) weighted current sources using the R-$2R$ ladder. [From R.C. Jaeger, "Tutorial: Analog Data Acquisition Technology, Part I—Digital-to-Analog Conversion," *IEEE MICRO*, Vol. 2, No. 2, May 1982: (a) Fig. 9, p. 25, (b) Fig. 11, p. 26, (c) Fig. 12, p. 28. ©1982 IEEE.]

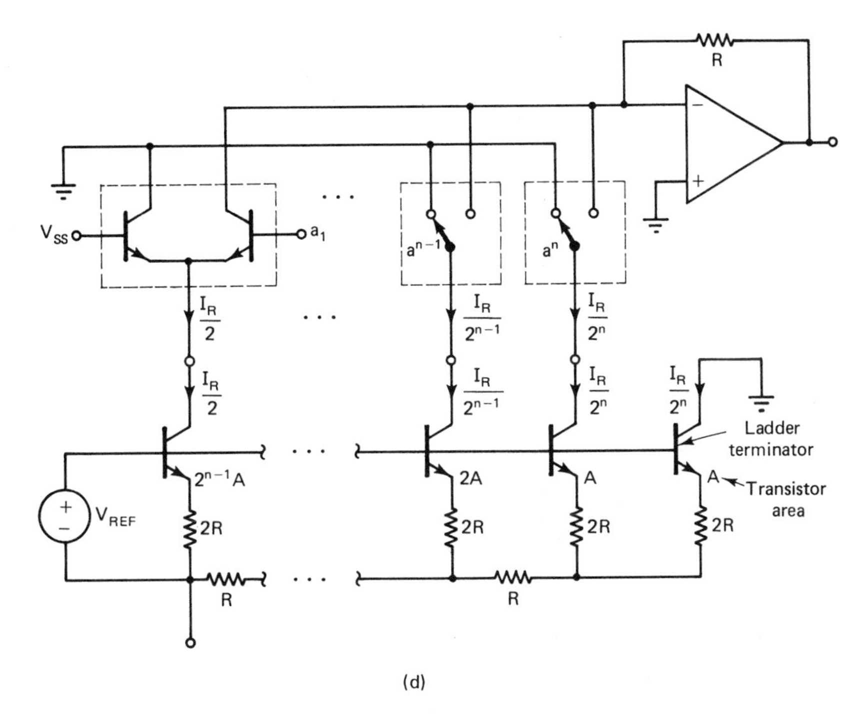

(d)

Figure 3-20 cont.

selectively connected to the input of a current to voltage converter to provide a voltage output DAC. The current sources are easily derived using a standard bipolar current source consisting of a reference, a single transistor, and a resistor. Rather than turning the current sources on and off, the output of the current source is selectively switched to the DAC output or to ground. The switch uses bipolar transistors operating in the same mode as current switching or emitter-coupled logic. Note that the base–emitter voltages must match in order for the weighting of the current sources to be correct. To achieve proper weighting, the transistors must be operated at equal current densities. This is usually achieved by increasing the area of each transistor by a factor of 2 proceeding from the LSB to the MSB.

The 2:1 ratio of the R-$2R$ ladder is employed to generate this weighted current source for the D/A converter. Each transistor carries one-half the current of the preceding device going from left to right down the ladder. This configuration is used in DACs of up to 12-bit resolution.

3.8 ANALOG-TO-DIGITAL CONVERSION*

The basic conversion scheme for most analog-to-digital conversion is shown in Figure 3-21a. The unknown voltage is connected to one input of an analog signal comparator and a time-dependent reference voltage is connected to the second input of the comparator.

*Section 3.8, contributed by Richard C. Jaeger, is based on his "Data Acquisition Systems," Tutorial Notes, IECI 81, San Francisco, Nov. 9, 1981, and is included here with his permission.

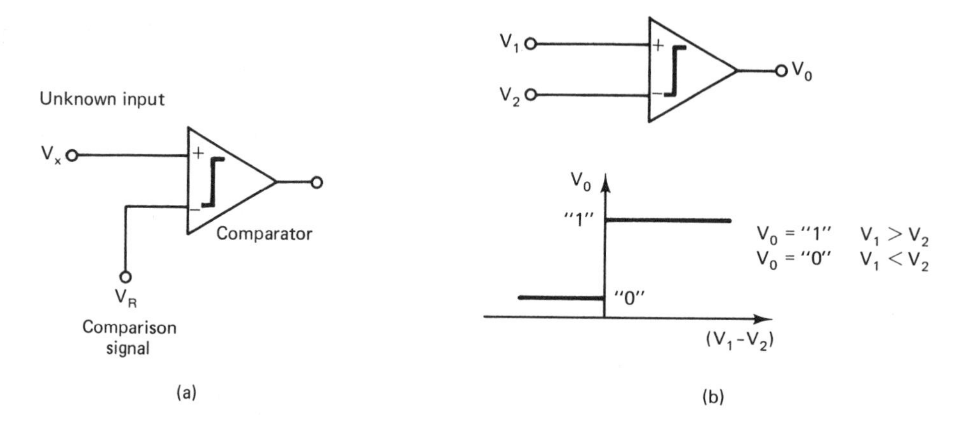

Figure 3-21 Analog-to-digital conversion: (a) general scheme; (b) comparator. [From R.C. Jaeger, "Tutorial: Analog Data Acquisition Technology, Part II—Analog-to-Digital Conversion," *IEEE Micro*, Vol. 2, No. 3, August 1982: (a) Fig. 5, p. 48, (b) Fig. 6, p. 48. ©1982 IEEE.]

The transfer characteristic of the comparator is shown in Figure 3-21b. If the input voltage V_1 is greater than V_2, the output voltage will be at a positive level corresponding to a logic "1." If input V_2 is greater than V_1, the output voltage will be at a low level, corresponding to a logic "0."

To perform a conversion, the reference voltage V_R is varied to determine which of the 2^n possible binary words is closest to the unknown voltage V_x. The reference voltage V_R can assume 2^n different values of the form

$$V_R = V_r \sum_{i=1}^{n} A_i 2^{-i}$$

where V_r is a dc reference voltage and A_i are binary coefficients. The logic of the A/D converter attempts to choose the coefficients A_i so that the difference between the unknown input and the set of possible discrete representations of V_x is a minimum:

Choose A_i such that

$$\text{error} = |V_x - V_R| = |V_x - V_r \sum_{i=1}^{n} A_i 2^{-i}| \tag{3-41}$$

is minimum. The basic difference in converters is in the strategy that is used to vary V_R to determine the binary coefficients A_i.

Counter Ramp Converter

One of the simplest ways of generating the comparison voltage is by using a D/A converter (DAC). An n-bit DAC can be used to generate any one of its possible 2^n outputs by simply applying the appropriate digital inputs. See (3-40) with $V_{fs} = V_r$. The most direct way to determine the unknown voltage V_x is to sequentially compare it to each possible DAC output. By connecting the input of the D/A converter to an n-bit binary counter, a step-by-step comparison with the unknown input can be made, as shown in Figure 3-22. The output of the D/A converter looks like a staircase during the conversion. A reset pulse (start of conversion, SOC) sets the counter output at

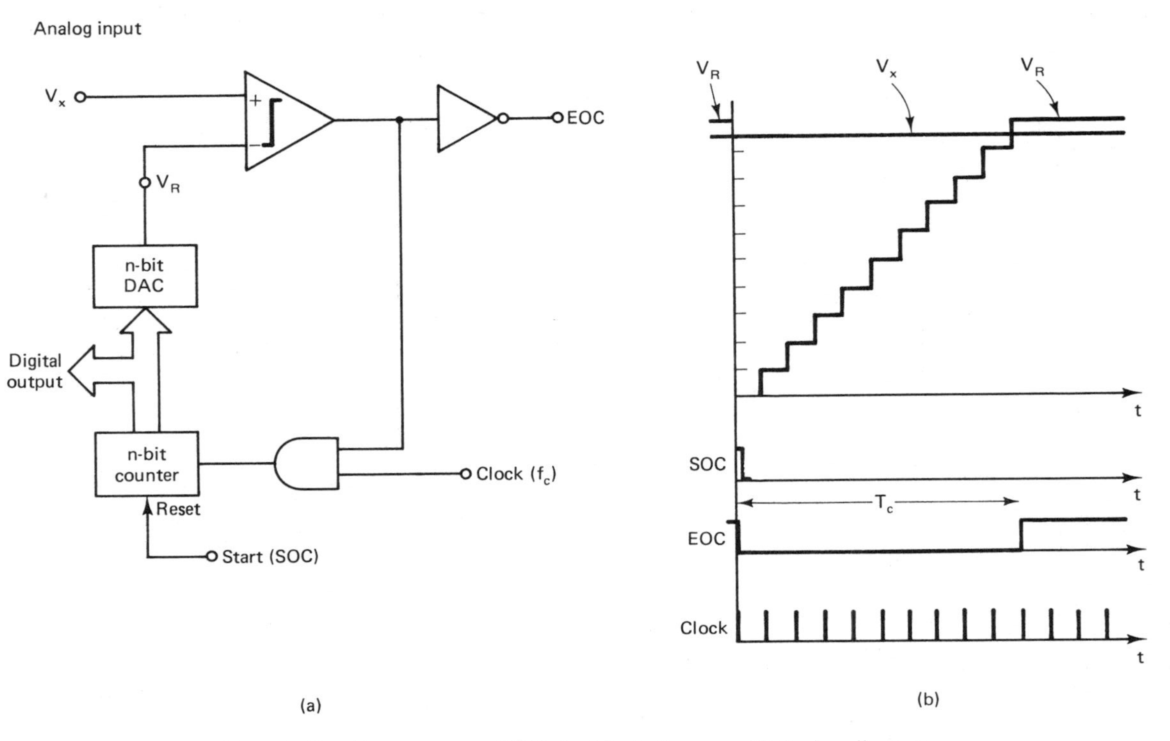

Figure 3-22 Counter ramp ADC: (a) block diagram; (b) timing diagram.

zero. Each successive clock pulse increments the counter until the output of the DAC is larger than the unknown input. At this point the comparator switches state and prevents any further clock pulses from incrementing the counter. The change of state of the comparator output indicates that the conversion cycle is complete and the contents of the binary counter represent the converted value of the input signal. Several features of this converter should be noted. First, the length of the converter cycle is variable and proportional to the unknown input V_x. The maximum conversion period T_c occurs for a full-scale input signal corresponding to 2^n clock periods, or $T_c = 2^n/f_c$, where f_c is the clock pulse frequency. Second, the binary value in the counter represents the smallest DAC output voltage which is larger than the unknown input and is not necessarily the DAC output which is closest to the unknown voltage as we had originally hoped. Finally, if the unknown input should decrease or increase during the conversion period T_c, the binary output of the converter will be a true representation of the input voltage at the instant the comparator stops the counter.

The advantage of this type of converter is that it requires a small amount of hardware and is inexpensive to implement. For this reason, some of the least expensive monolithic and modular converters use this method. The main disadvantage is the relatively low conversion rate for a given converter. An n-bit converter requires 2^n clock periods for its longest conversion. With fast DACs, a conversion rate of 1000 per second can be achieved at 10-bit resolution.

Tracking ADC

One attempt to improve the counting converter's performance is to modify it to track the input signal (see Figure 3-23). An up-down counter is used with slightly more complicated logic to force the output of the DAC to track changes in the unknown input V_x. If the comparator output indicates that the DAC output is less than V_x, the next clock pulse increments the counter. If the DAC output is greater than V_x, the next clock pulse decrements the counter. A staircase DAC output occurs until the converter "acquires" the input signal. If the unknown input is constant, the DAC output will alternate between two output values which differ by one LSB. If the unknown input changes slowly enough, the DAC output will follow the unknown input. Thus the counter contents always contain an accurate representation of the input signal at the time of the last clock pulse, and the converted value may be read from this counter at any time. However, if V_x changes too rapidly, the converter will not be able to follow quickly enough and the counter contents no longer correctly represent the output voltage. Also, when a new input signal is applied, this converter takes exactly the same amount of time to acquire the signal as the counter converter takes to reach conversion. Because of their relatively low conversion rate for a given clock frequency, other types of converters are usually used when the input signal is expected to change by large amounts from one conversion time to the next.

Successive Approximation ADC

The successive approximation converter uses a different strategy in varying the applied reference input to the comparator and results in a converter that requires only n clock periods to complete an n-bit conversion.

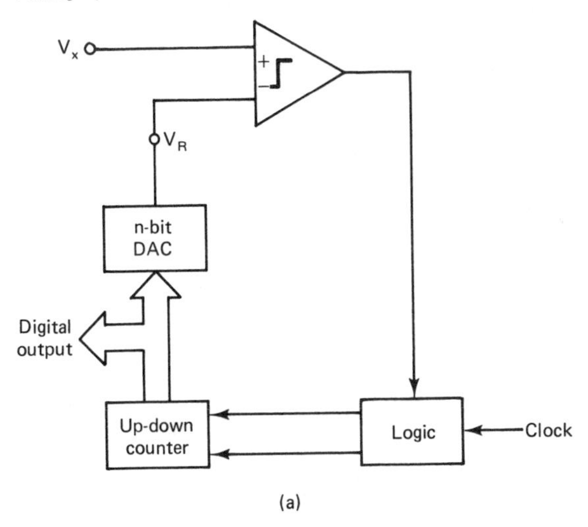

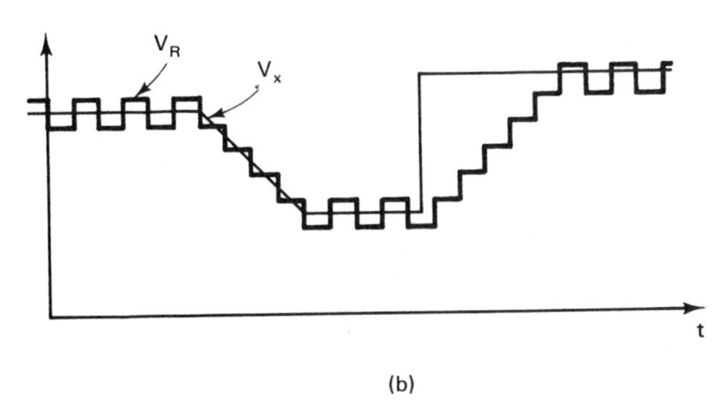

Figure 3-23 Tracking ADC: (a) block diagram; (b) example waveforms. (From R.C. Jaeger, "Tutorial: Analog Data Acquisition Technology, Part II—Analog-to-Digital Conversion," *IEEE MICRO*, Vol. 2, No. 3, August 1982, Fig. 8, p. 50. ©1982 IEEE.)

The operation of a 3-bit converter is shown in Figure 3-24. A "binary search" is used to determine the best approximation to V_x. After reset, the successive approximation logic (SAL) sets the DAC output to $0.5V_{fs}$ and checks the comparator output. At the next clock pulse, the DAC output is set to $0.75V_{fs}$ if the DAC output was less than V_x and to $0.25V_{fs}$ if the DAC output had been greater than V_x. Again the comparator output is tested, and the next clock pulse causes the DAC output to be incremented or decremented by $V_{fs}/8$. A third comparison is made. The final converted binary output is not changed if V_x was larger than the DAC output or is decremented by one LSB if V_x was less than the DAC output. Thus the conversion is obtained at the end of three clock periods for the 3-bit converter or n clock periods for an n-bit converter. The 3-bit DAC code sequence is illustrated in Figure 3-25.

Fast conversion rates are possible with this converter, and it is a very popular type used for 8- to 16-bit converters at speeds up to 10^7 conversions per second. The

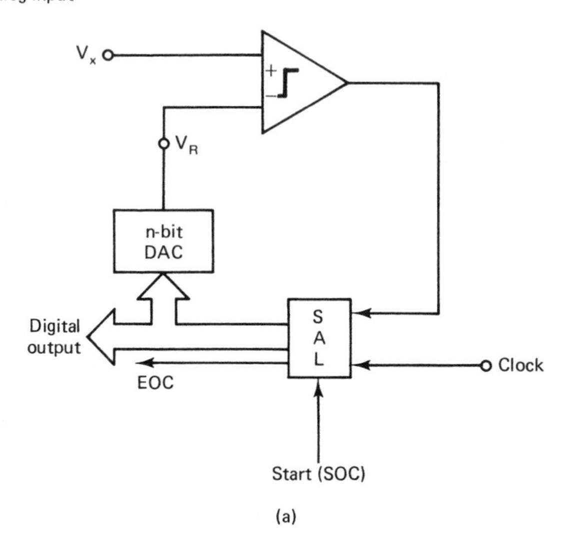

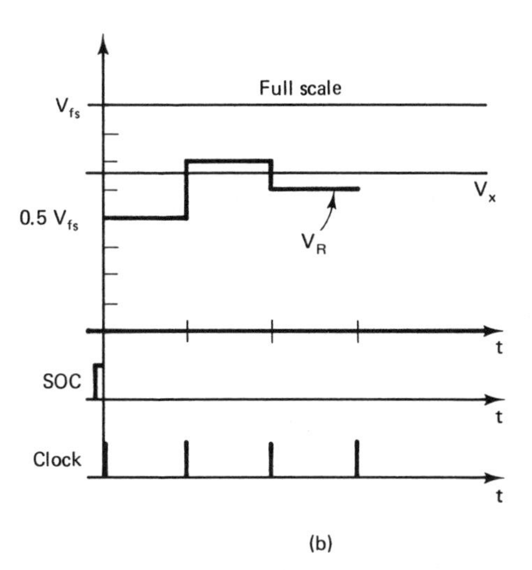

Figure 3-24 Successive approximation ADC: (a) block diagram; (b) timing diagram. (From R.C. Jaeger, "Tutorial: Analog Data Acquisition Technology, Part II—Analog-to-Digital Conversion," *IEEE MICRO*, Vol. 2, No. 3, August 1982, Fig. 9, p. 50. ©1982 IEEE.)

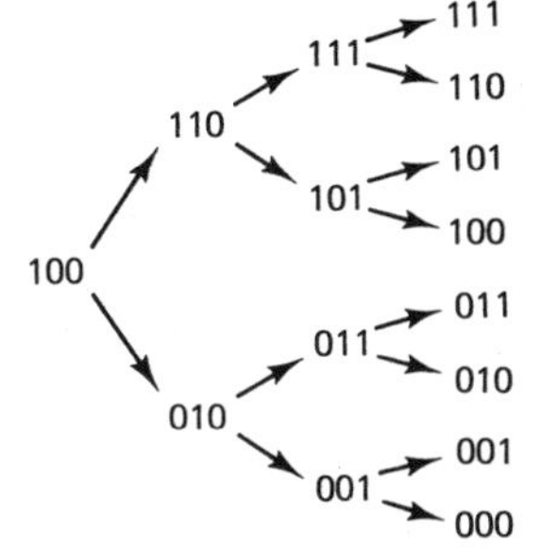

Figure 3-25 Three-bit successive approximation DAC code sequence.

primary factors limiting the speed of this ADC are the time required for the D/A converter output to settle within a fraction of a $V_{fs}/2^n$ (i.e., LSB) and the time for the comparator to respond to input signals which may differ by very small amounts.

A problem in the application of these converters is that the input must remain constant during the full conversion period. Otherwise, the digital output of the converter does not bear any precise relation to the value of the unknown input voltage V_x. Sample-and-hold circuits are usually used ahead of the successive approximation ADC to avoid this problem, as discussed earlier in this chapter.

Single-Ramp Converter

The discrete output of the D/A converter may be replaced with a continuously increasing reference signal as shown in Figure 3-26. In this case the reference is usually called a ramp and varies from slightly below zero to V_{fs}. The length of time required for the ramp signal to become equal to the unknown voltage is then proportional to the unknown voltage. This time period T_c is quantized with a counter to produce the binary representation of the unknown input signal. Referring to the figure, converter operation begins with a start of conversion (SOC) signal, which resets the binary counter and starts the ramp generator at a slightly negative value. As the ramp output crosses through zero, the output of comparator 2 enables the AND-GATE, allowing clock pulses to accumulate in the counter. The counter continues counting until the ramp output equals the unknown voltage V_x. At this time comparator 1 disables the AND-GATE, which stops the counter. The number of clock pulses in the counter is directly proportional to the input voltage since $V_x = KT_c$, where K is the slope of the ramp in volts per second and $T_c = N/f_c$. If the slope of the ramp is chosen to be $V_{fs}f_c/2^n$, the unknown voltage is given by

$$V_x = \frac{V_{fs}N}{2^n}$$

and

$$\frac{V_x}{V_{fs}} = \frac{N}{2^n}$$

The number N in the counter can be directly interpreted as the binary fraction equivalent to V_x/V_{fs}.

The conversion time T_c of the single ramp converter is clearly variable and proportional to the unknown voltage V_x. Maximum conversion time occurs for $V_x = V_{fs}$, or $T_c = 2^n/f_c$. As in the case for the counter converter, the digital output represents the value of V_x at the instant the end of conversion signal occurs.

The ramp voltage is usually generated using an integrator connected to a constant reference voltage as shown in Figure 3-27. When the reset switch is opened, the output increases with a constant slope as given by

$$V_0 = -\frac{1}{RC}\int_0^{T_c} V_r\,dt = -\frac{V_r}{RC}t\Big|_0^{T_c} \tag{3-42}$$

For a constant reference $-V_r$, the slope of the ramp is V_r/RC, which points out one of the major limitations of the converter. The slope is dependent on the absolute values of R and C, which are difficult to maintain constant in the presence of tempera-

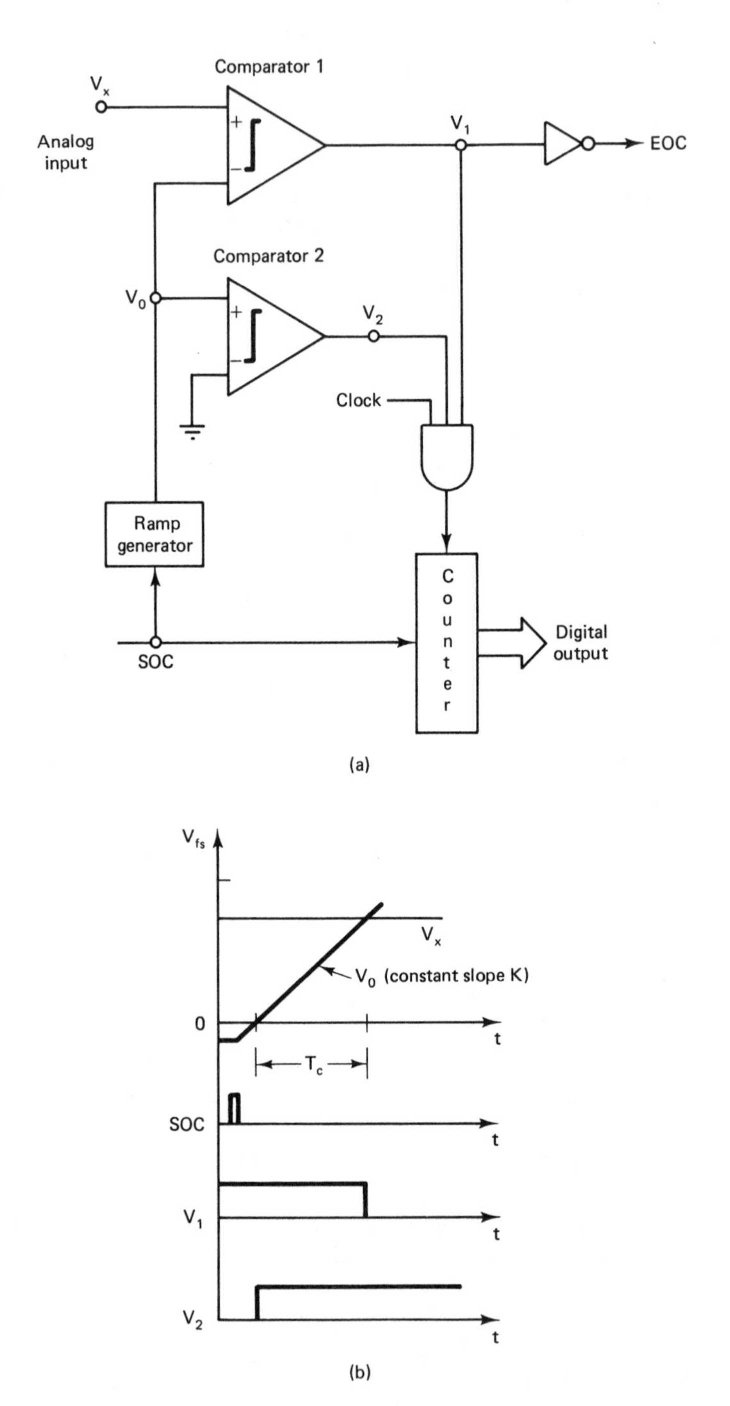

Figure 3-26 Single-ramp ADC: (a) block diagram ($V_x = KT_c$; N, counter output); (b) timing diagrams.

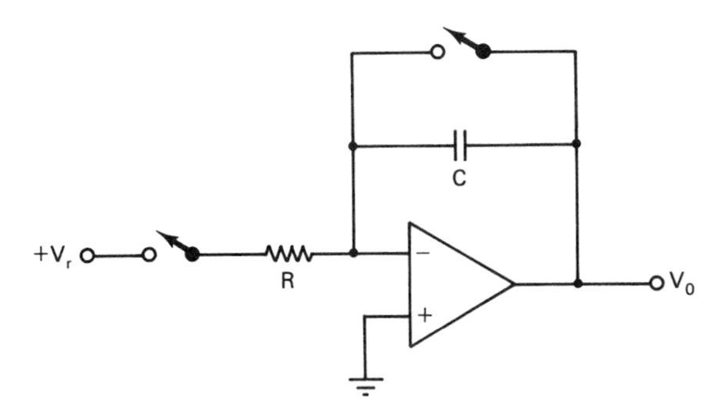

Figure 3-27 Ramp generator.

ture variations and over long periods of time. Because of this problem, the single-ramp converter is seldom used today.

Dual-Ramp Converter

The dual-ramp converter solves the problem associated with the single-ramp converter and is a common converter found in data acquisition and control instrumentation systems. Converter operation may be understood by referring to Figure 3-28. The converter operates with a conversion cycle consisting of two separate integration intervals. First, the unknown V_x is integrated for a known period of time. Then the value of this integral is compared to that of a known reference which is integrated for a variable length of time.

At the start of conversion, the counter is reset, and the integrator is reset to a slightly negative voltage. The unknown input V_x is connected to the integrator input through switch S_1. V_x is integrated for a fixed period of time $T_1 = 2^n/f_c$ which begins when the integrator output crosses through zero. At the end of time T_1 the counter overflows, which causes S_1 to be turned off and the reference input $+V_r$ to be connected to the integrator input through S_2. The integrator output decreases until it crosses through zero, and the comparator changes state, indicating the end of conversion. The number in the counter represents the converted value of the unknown V_x, as demonstrated below.

Circuit operation forces the integrals over the time periods $0+$ to T_1 and T_1 to T_2 to be equal, so that

$$\frac{1}{RC}\int_0^{T_1} V_x\,dt = \frac{1}{RC}\int_{T_1}^{T_1+T_2} V_r\,dt \quad \text{or} \quad \frac{\bar{V}_x T_1}{RC} = \frac{V_R T_2}{RC} \tag{3-43}$$

T_1 is set equal to $2^n/f_c$ since the unknown V_x was integrated for the amount of time needed for the n-bit counter to overflow. $\bar{V}_x$ is the average of V_x. The time period $T_2 - T_1$ is equal to N/f_c where N is the number of counts accumulated in the counter during the second phase of operation. The average value of the input is then given by

$$\bar{V}_x = \frac{V_r N}{2^n} \tag{3-44}$$

assuming that the RC product remained constant throughout the complete conversion cycle. The values of R and C no longer enter directly into the relation between V_x and V_r, and the stability problem associated with the single-ramp converter has been

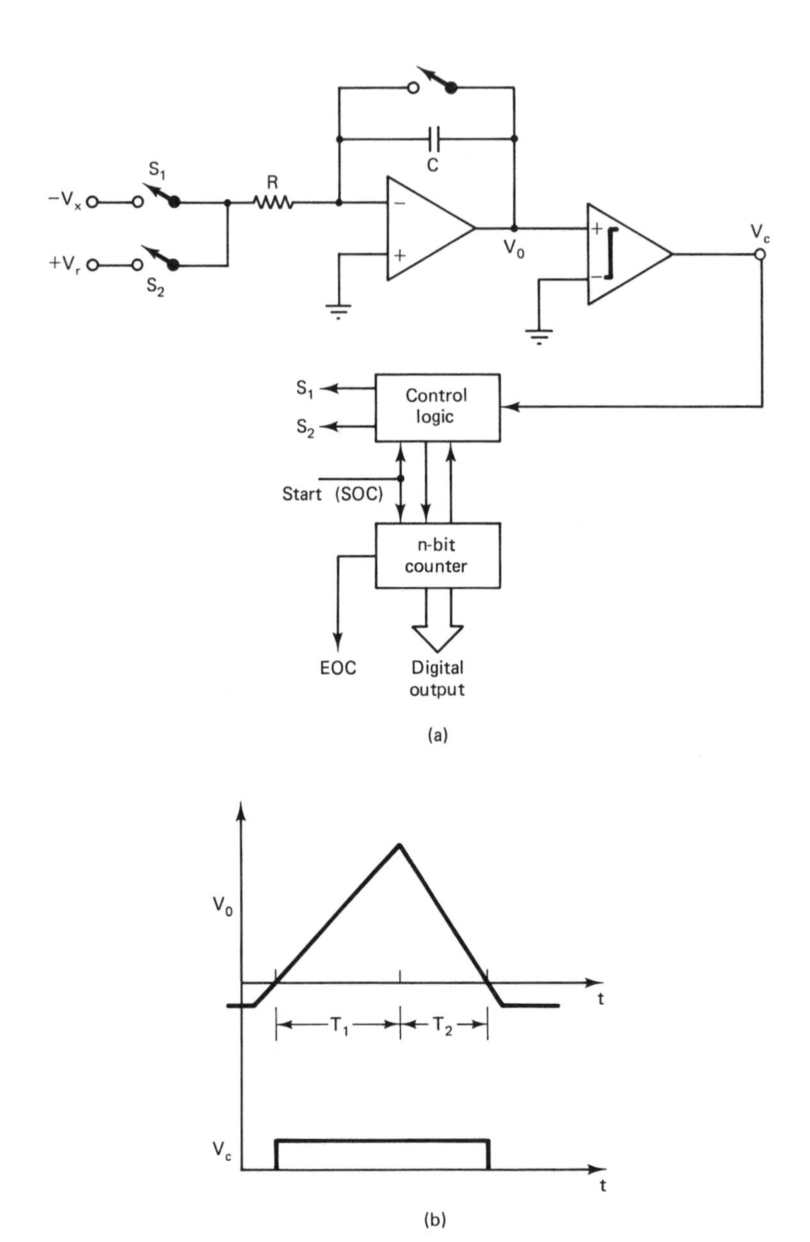

Figure 3-28 Dual-ramp ADC: (a) block diagram; (b) control waveforms. (From R.C. Jaeger, "Tutorial: Analog Data Acquisition Technology, Part II—Analog-to-Digital Conversion," *IEEE MICRO*, Vol. 2, No. 3, August 1982, Fig. 13, p. 53. ©1982 IEEE.)

overcome. Furthermore, the digital output word represents the average value of V_x during the first integration phase. Thus V_x is allowed to change during the conversion cycle of this converter without destroying the validity of the converted output.

The conversion time T_c requires 2^n clock periods for the first integration period and N clock periods for the second integration interval. Thus the conversion time is variable and

$$T_c = \frac{N + 2^n}{f_c} \tag{3-45}$$

As mentioned above, the binary output of the dual-ramp converter represents the average of the input during the first integration phase. The integrator operates as a low-pass filter. Any sinusoidal input signals whose frequencies are exact multiples of the reciprocal of the integration time T_1 will have integrals of zero value and will not disturb the converter output. This property is often used in digital voltmeters which use dual-ramp converters with an integration time which is some fixed multiple of the period of the 50- or 60-Hz power source. Noise sources at multiples of the power-line frequency are removed by the integrating ADC. This property is usually called *normal mode rejection.*

The dual ramp is a widely used converter. Its integrating properties combined with careful design allow accurate conversion at resolutions exceeding 20 bits with low conversion speeds. The basic dual ramp has been modified to include automatic offset elimination phases in a number of monolithic converters.

Parallel Converter (Flash)

The fastest converters use additional hardware to perform a parallel rather than serial conversion. Figure 3-29 schematically shows a 3-bit parallel converter in which the unknown input V_x is simultaneously compared to seven different reference values. The logic network converts the comparator outputs directly to the 3-bit digital value that corresponds to the input. The speed of the converter can be very fast since the conversion speed is limited only by the speed of the comparators and of the logic network. Also, the output continuously represents the input except for the comparator and logic delays. Thus the converter can be thought of as automatically tracking the input signal.

This type of converter is used when maximum speed is needed, and is usually found in relatively low resolution converters since $2^n - 1$ comparators and reference voltages are required for an n-bit converter. Thus the cost of implementing such a converter grows rapidly with resolution. The term "flash" is sometimes used as the name of the parallel converter because of its inherent speed. The flash method has been used successfully in several commerical high-speed converters that perform approximately 30 million 6- or 8-bit conversions per second.

Delta-Sigma ADC

The delta-sigma ADC employs the techniques of delta-sigma modulation demodulation to generate the reference signal necessary for the conversion. The converter continuously generates an output signal whose duty cycle is proportional to the unknown input signal. A simple counter is used to measure the duty cycle to complete the conversion. The duty cycle is the percentage of time a binary signal is in the one state. Time is measured over fixed intervals.

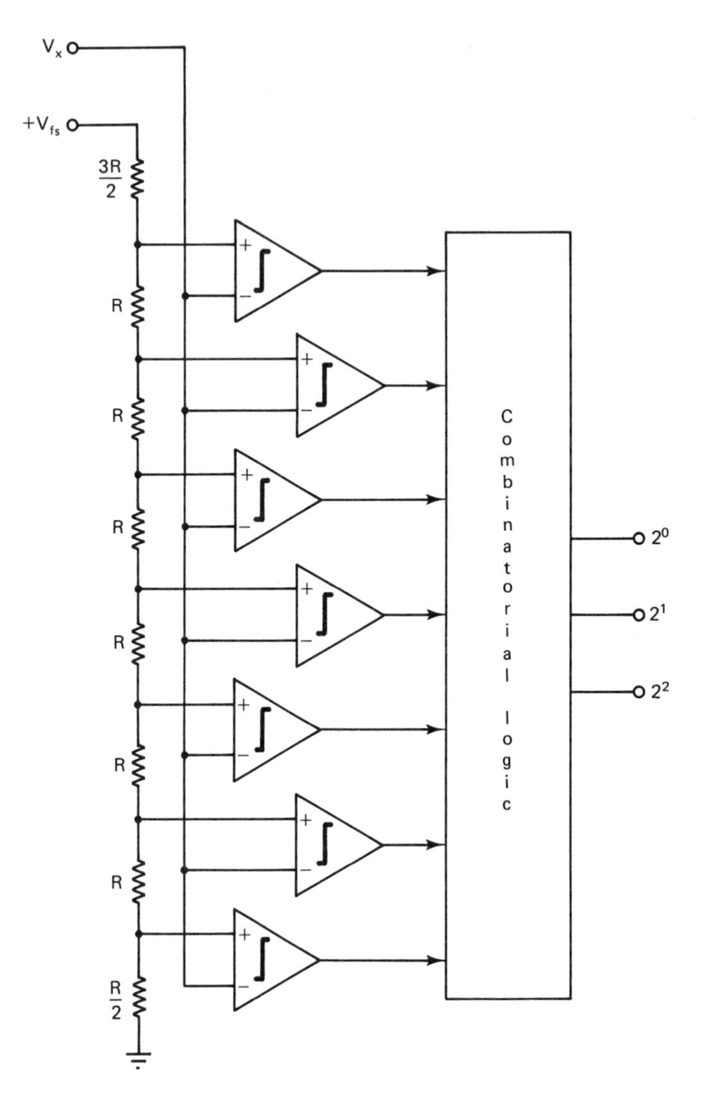

Figure 3-29 Parallel, or "flash" ADC.

As shown in Figure 3-30, the state of the comparator is transferred to the output of the flip-flop (FF) at each clock time. The flip-flop also controls the state of switch *S*, which controls the charging and discharging of capacitor C. For zero input current, the average voltage on C should be zero so that the FF output is in each state 50% of the time, and for negative inputs, the switch will be in position 1 more than 50% of the time.

Basic circuits for realizing such a converter are voltage-to-current converters and current sources. A number of charge-balancing schemes similar to this one have been used in recent monolithic ADC designs. The converter requires $2^n + 1$ clock intervals for an n-bit plus sign conversion. Because of severe timing constraints on the current source switching, the clock frequency is limited to relatively low values in the

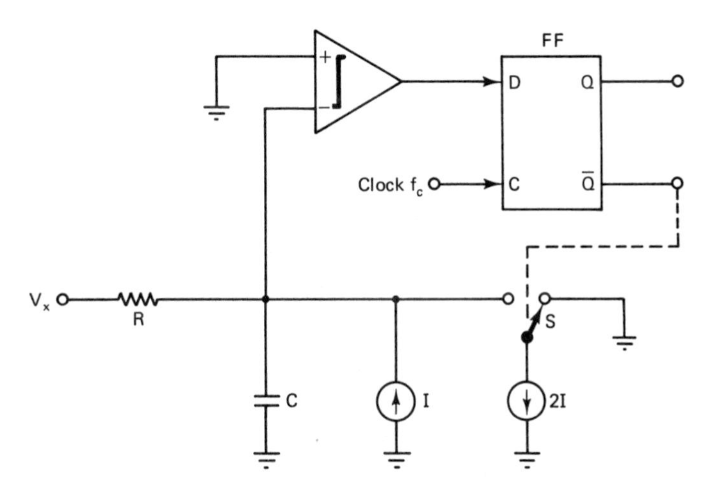

Figure 3-30 Delta-sigma, or charge balancing, ADC.

range 100 to 500 kHz. This results in conversion rates of less than 100 conversions per second, with a few conversions per second being typical at high resolution.

ADC Comparison

A comparison of ADC characteristics is displayed in Table 3-1. The conversion rate and complexity of each scheme is considered. Here we note that speed and cost are

TABLE 3-1 ADC COMPARISON

Converter type	Conversion rate at 12-bit resolution[a]	Cost/complexity	Comments
Counter ramp	Low speed—up to 1000/s	Low	Needs stable input
Tracking		Low	Needs slowly varying input signal; output always available
Successive approximation	Medium to high—up to 10^6/s	Medium	100,000 counts/s at up to 14- to 15-bit resolution; needs stable input
Single ramp	Low—up to 1000/s	Low	Low speed at 12-bit resolution; lacks stability with time and temperature
Dual ramp	Low—up to 1000/s	Medium	Integrates input signal; can be used at high resolution—20 bits or more
Parallel "flash"	Fastest—10^6–10^8/s	High	Output always available

[a]Ramp converters have variable conversion time; successive approximation and parallel converters have fixed conversion time.

directly related. The medium-speed, medium-cost successive approximation ADC is a common choice in digital control systems.

3.9 SUMMARY

The material presented in this chapter is fundamental to the study of sampled-data control systems. The sampling and reconstruction processes are an integral part of these systems, and therefore it is important that the reader possess a thorough understanding of these topics before proceeding to the more advanced topics in analysis and design.

An important result of the topics developed in this chapter is the development of approximate rules for the choice of the sample period T for a given signal. Whereas this chapter emphasized the importance of the frequency contents of a signal in determining its sampling rate, later chapters will show the importance of a system's frequency response in determining the sample rate to be used in that system.

A survey of digital-to-analog and analog-to-digital conversion methods was included in this chapter to provide the reader with some practical information with which to select D/A and A/D converters for particular applications.

REFERENCES AND FURTHER READING

1. M. F. Gardner and J. L. Barnes, *Transients in Linear Systems*, Vol. I. New York: John Wiley & Sons, Inc., 1942.
2. M. E. Van Valkenburg, *Network Analysis*. Englewood Cliffs, N.J.: Prentice-Hall, Inc., 1974.
3. R. M. Oliver, J. R. Pierce, and C. E. Shannon, "The Philosophy of Pulse Code Modulation," *Proc. IRE*, Vol. 36, No. 11, pp. 1324–1331, Nov. 1948.
4. B. C. Kuo, *Analysis and Synthesis of Sampled-Data Control Systems*, Englewood Cliffs, N.J.: Prentice-Hall, Inc., 1963.
5. C. L. Phillips, D. L. Chenoweth, and R. K. Cavin III, "z-Transform Analysis of Sampled-Data Control Systems without Reference to Impulse Functions," *IEEE Trans. Educ.*, Vol. E-11, pp. 141–144, June 1968.
6. C. R. Wylie, Jr., *Advanced Engineering Mathematics*. New York: McGraw-Hill Book Company, 1951.
7. E. A. Guillemin, *The Mathematics of Circuit Analysis*. New York: John Wiley & Sons, Inc., 1949.
8. R. V. Churchill, *Introduction to Complex Variables and Applications*. New York: McGraw-Hill Book Company, 1948.
9. T. Takahashi, *Mathematics of Automatic Control*. New York: Holt, Rinehart and Winston, Inc., 1966.
10. J. T. Tou, *Digital and Sampled-Data Control Systems*. New York: McGraw-Hill Book Company, 1959.
11. G. F. Franklin and J. D. Powell, *Digital Control of Dynamic Systems*. Reading, Mass.: Addison-Wesley Publishing Company, Inc., 1980.

PROBLEMS

3-1. Find $E^*(s)$ for the following functions. Express $E^*(s)$ in closed form.

(a) $e(t) = \epsilon^{-at}$ **(b)** $E(s) = \dfrac{\epsilon^{-Ts}}{s+a}$

(c) $e(t) = \epsilon^{-a(t-5T)}u(t-5T)$ **(d)** $e(t) = \epsilon^{-a(t-T/2)}u\left(t - \dfrac{T}{2}\right)$

3-2. Using the residue method given in (3-10), find $E^*(s)$ for the following functions.

(a) $E(s) = \dfrac{1}{s(s+1)^2}$ **(b)** $E(s) = \dfrac{s}{(s+1)(s+2)}$

(c) $E(s) = \dfrac{s+1}{s(s+2)}$ **(d)** $E(s) = \dfrac{s^2}{(s+1)^2(s+2)}$

(e) $E(s) = \dfrac{s+1}{s^2+25}$ **(f)** $E(s) = \dfrac{s+1}{s^2+2s+26}$

3-3. Using the residue method given in (3-10) and (3-14), find $E^*(s)$ for the following functions.

(a) $E(s) = \dfrac{s+2}{(s-1)(s+1)}$ **(b)** $E(s) = \dfrac{(s+2)\epsilon^{-2Ts}}{(s-1)(s+1)}$

(c) $E(s) = \dfrac{1}{s^2}$ **(d)** $e(t) = (t-T)u(t-T)$

3-4. Find $E^*(s)$ for

$$E(s) = \frac{1 - \epsilon^{-Ts}}{s(s+1)}$$

3-5. Find $E^*(s)$, with $T = 0.5$ s, for

$$E(s) = \frac{(1+0.5s)(1-\epsilon^{-0.5s})^2}{0.5s^2(s+1)}$$

3-6. Express the starred transform of $e(t-kT)u(t-kT)$, k an integer, in terms of the starred transform of $e(t)$.

3-7. Compare the pole–zero locations of $E^*(s)$ in the s-plane with those of $E(s)$, for those functions given in Problem 3-2.

3-8. Find $E^*(s)$, for $T = 0.5$ s, for the two functions below. Explain why the two transforms are equal, first from a time-function approach, and then from a pole–zero approach.

(a) $e_1(t) = \cos(4\pi t)$ **(b)** $e_2(t) = \cos(8\pi t)$

3-9. **(a)** The signal $e_1(t) = \cos(5t)$ is sampled at a frequency of $\omega_s = 20$. List the frequencies present in $E_1^*(s)$.

(b) The signal $e_2(t) = \cos(15t)$ is sampled at a frequency of $\omega_s = 20$. List the frequencies present in $E_2^*(s)$.

(c) Compare the results of parts (a) and (b), and comment on any unusual aspects of these results.

3-10. Find $E^*(s)$ in closed form for the following functions.

(a) $E(s) = \dfrac{1}{(s+1)(s+2)}$ **(b)** $E(s) = \dfrac{\epsilon^{-(T/2)s}}{(s+1)(s+2)}$

(c) $E(s) = \dfrac{\epsilon^{-Ts}}{(s+1)(s+2)}$

3-11. Suppose that the signal $e(t) = \cos[(\omega_s/2)t + \theta]$ is applied to an ideal sampler and zero-order hold.

(a) Show that the amplitude of the time function out of the zero-order hold is a function of the phase angle θ by sketching this time function.

(b) Show that the component of the signal out of the data hold, at the frequency $\omega = \omega_s/2$, is a function of the phase angle θ by finding the Fourier series for the signal.

3-12. A polygonal data hold is a device that reconstructs the sampled signal by the straight-line approximation shown in Figure P3-12. Show that the transfer function of this data hold is

$$G(s) = \frac{\epsilon^{Ts}(1 - \epsilon^{-Ts})^2}{Ts^2}$$

Is this data hold physically realizable?

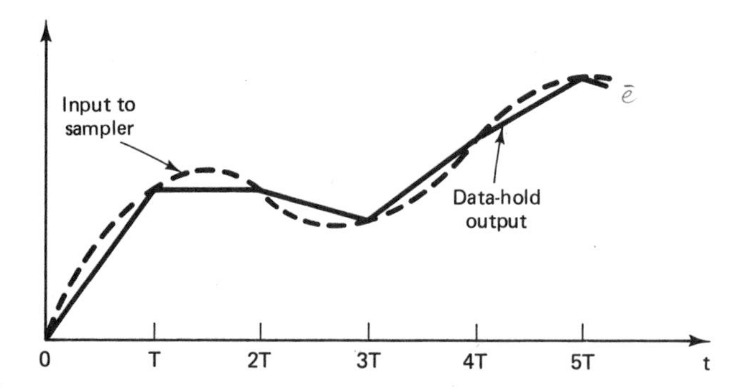

Figure P3-12 Response of a polygonal data hold.

3-13. A data hold is to be constructed that reconstructs the sampled signal by the straight-line approximation shown in Figure P3-13. Note that this device is a polygonal data hold (see Problem 3-12) with a delay of T seconds. *Derive* the transfer function for this data hold. Is this data hold physically realizable?

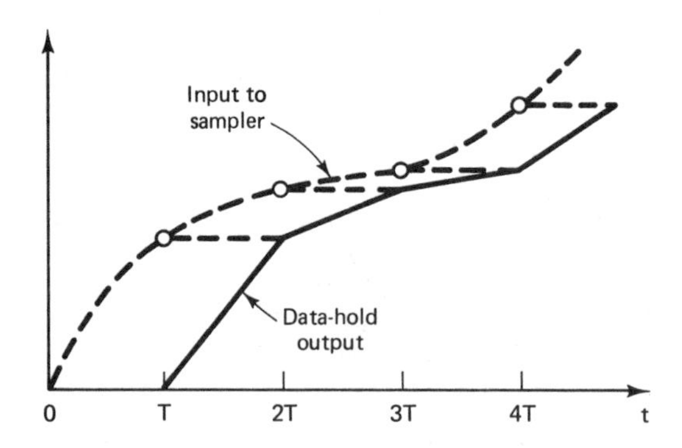

Figure P3-13 Response of a polygonal data hold with time delay.

3-14. Plot the ratio of the frequency responses (in decibels) and phase versus ω for the data holds of problems 3-12 and 3-13. Note the effect on phase of making the data hold realizable.

3-15. Derive the transfer function of the fractional-order hold [see (3-39)].

3-16. Shown in Figure P3-16 is the output of a data hold that clamps the output to the input for the first half of the sampling period, and returns the output to a value of zero for the last half of the sampling period.

(a) Find the transfer function of this data hold.

(b) Plot the frequency response of this data hold.

(c) Comparing this frequency response to that of the zero-order hold, comment on which would be better for data reconstruction.

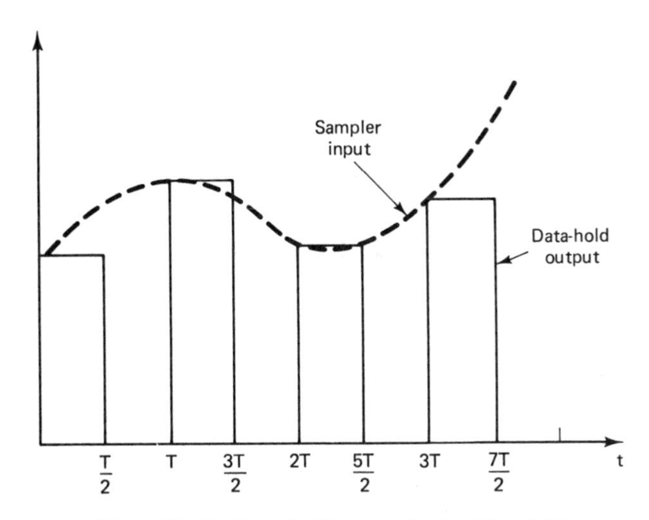

Figure P3-16 Data-hold output for Problem 3-16.

3-17. **(a)** A sinusoid with a frequency of 2 Hz is applied to a sampler/zero-order hold combination. The sampling rate is 10 Hz. List all of the frequencies present in the output that are less than 50 Hz.

(b) Repeat part (a) if the input sinusoid has a frequency of 8 Hz.

3-18. A signal $e(t)$ is sampled by the ideal sampler as specified by (3-3).

(a) List the conditions under which $e(t)$ can be *completely* recovered from $e^*(t)$, that is, the conditions under which *no* loss of information by the sampling process occurs.

(b) State which of the conditions listed in part (a) can occur in a physical system. Recall that the sampling operation itself is not physically realizable.

3-19. The signal $e(t) = 5 \sin 3t$ is applied to a sampler/zero-order hold circuit, with $T = \pi/6$ seconds.

(a) The output of the circuit has a frequency component at $\omega = 3$ rad/s. Find the amplitude and phase of this component.

(b) Repeat part (a) for the frequency component at $\omega = 15$ rad/s and at $\omega = 27$ rad/s.

4

Open-Loop Discrete-Time Systems

4.1 INTRODUCTION

In the preceding chapters the concepts of discrete-time systems, the z-transform, and sampling and data reconstruction were developed. In this chapter these concepts will be utilized to develop methods of analysis for open-loop discrete-time systems. This analysis technique will then be extended to closed-loop systems in the chapters that follow.

4.2 THE RELATIONSHIP BETWEEN $E(z)$ AND $E^*(s)$

In order to provide the proper background for our analysis of open-loop systems, let us establish the relationship that exists between $E(z)$ and $E^*(s)$. Recall that in Chapter 2 the z-transform of the number sequence $\{e(k)\}$ was defined in equation (2-7) to be

$$\mathfrak{z}[\{e(k)\}] = E(z) = e(0) + e(1)z^{-1} + e(2)z^{-2} + \cdots \tag{4-1}$$

In addition, the starred transform for the time function $e(t)$ was defined in equation (3-7) as

$$E^*(s) = e(0) + e(T)\epsilon^{-Ts} + e(2T)\epsilon^{-2Ts} + \cdots \tag{4-2}$$

The similarity between these two transforms is obvious. In fact, if we assume that the number sequence $\{e(k)\}$ is obtained from sampling a time function $e(t)$ [i.e., if $e(k)$ of (4-1) is equal to $e(kT)$ of (4-2)], and if $\epsilon^{sT} = z$ in (4-2), then (4-2) becomes the z-transform. Hence in this case

$$E(z) = E^*(s)\big|_{\epsilon^{sT}=z} \tag{4-3}$$

We will employ this change of variable, and, in general, use the z-transform instead of the starred transform in our analysis of discrete-time systems. One advantage of this approach is illustrated by the following example.

Example 4.1

Let

$$E(s) = \frac{1}{(s+1)(s+2)}$$

Then the starred transform, derived in Example 3.3 via equation (3-10), is

$$E^*(s) = \frac{1}{1-\epsilon^{-Ts}\epsilon^{-T}} - \frac{1}{1-\epsilon^{-Ts}\epsilon^{-2T}}$$

$$= \frac{\epsilon^{-Ts}(\epsilon^{-T}-\epsilon^{-2T})}{(1-\epsilon^{-T}\epsilon^{-Ts})(1-\epsilon^{-2T}\epsilon^{-Ts})}$$

Then, from equation (4-3),

$$E(z) = E^*(s)\big|_{\epsilon^{sT}=z} = \frac{z^{-1}(\epsilon^{-T}-\epsilon^{-2T})}{(1-\epsilon^{-T}z^{-1})(1-\epsilon^{-2T}z^{-1})}$$

$$= \frac{z(\epsilon^{-T}-\epsilon^{-2T})}{(z-\epsilon^{-T})(z-\epsilon^{-2T})}$$

Note that in the example above, $E^*(s)$ has an infinity of poles and zeros in the s-plane. However, $E(z)$ has only a single zero at $z = 0$, and two poles—one at ϵ^{-T} and the other at ϵ^{-2T}. Thus any analysis procedure that utilizes a pole–zero approach is greatly simplified through the use of the z-transform. Other advantages of this approach will become obvious as we proceed through the development of system analysis techniques for sampled-data systems.

$E(z)$ can now be calculated from (3-10) via the substitution in (4-3) as

$$E(z) = \sum_{\substack{\text{at poles}\\ \text{of } E(\lambda)}} \left[\text{residues of } E(\lambda)\frac{1}{1-z^{-1}\epsilon^{T\lambda}}\right] \tag{4-4}$$

Because of the relation between $E(z)$ and $E^*(s)$, the theorems developed in Chapter 2 for the z-transform also apply to the starred transform. In addition, Table 2-2 becomes a table of starred transforms with the substitution $z = \epsilon^{Ts}$. For this reason, a separate table for starred transforms is usually not given.

4.3 THE PULSE TRANSFER FUNCTION

In this section we develop an expression for the z-transform of the output of open-loop systems. This expression will be required later when we form closed-loop systems by feeding back this output signal and sampling it.

Consider the open-loop system of Figure 4-1. Note that $G(s)$ must contain the transfer function of the data hold. It is seen that

$$C(s) = G(s)E^*(s) \tag{4-5}$$

E(s) T E*(s) G(s) C(s)

Figure 4-1 Open-loop sampled-data system.

Assume that $c(t)$ is continuous at all sampling instants. From (3-11),

$$C^*(s) = [G(s)E^*(s)]^* = \frac{1}{T}\sum_{n=-\infty}^{\infty} C(s + jn\omega_s) \tag{4-6}$$

where $[\cdot]^*$ denotes the starred transform of the function in the brackets. Thus, from (4-5) and (4-6),

$$C^*(s) = \frac{1}{T}\sum_{n=-\infty}^{\infty} G(s + jn\omega_s)E^*(s + jn\omega_s) \tag{4-7}$$

It has been shown in equation (3-17) that a property of the starred transform is that

$$E^*(s + jn\omega_s) = E^*(s)$$

Thus (4-7) becomes

$$C^*(s) = E^*(s)\frac{1}{T}\sum_{n=-\infty}^{\infty} G(s + jn\omega_s) = E^*(s)G^*(s) \tag{4-8}$$

and then from (4-3),

$$C(z) = E(z)G(z) \tag{4-9}$$

where $G(z)$ is called the pulse transfer function. It can be shown that (4-9) applies also for the case that $c(t)$ is not continuous at all sampling instants.

Note that the derivation above is completely general. Thus, given any function

$$A(s) = B(s)F^*(s) \tag{4-10}$$

then

$$A^*(s) = B^*(s)F^*(s) \tag{4-11}$$

Hence

$$A(z) = B(z)F(z) \tag{4-12}$$

where $B(s)$ is a function of s and $F^*(s)$ is a function of ϵ^{Ts}; that is, in $F^*(s)$, s appears only in the form ϵ^{Ts}. The following examples illustrate this procedure.

Example 4.2

Suppose that we wish to find the z-transform of

$$A(s) = \frac{1 - \epsilon^{-Ts}}{s(s+1)} = \left[\frac{1}{s(s+1)}\right](1 - \epsilon^{-Ts})$$

From (4-10), we consider

$$B(s) = \frac{1}{s(s+1)}$$

and

$$F^*(s) = 1 - \epsilon^{-Ts}$$

Then, from Table 2-2,

$$B^*(s) = \left[\frac{1}{s(s+1)}\right]^* = \left[\frac{1}{s} - \frac{1}{s+1}\right]^*$$

$$= \frac{1}{1 - \epsilon^{-Ts}} - \frac{1}{1 - \epsilon^{-T}\epsilon^{-Ts}} = \frac{\epsilon^{-Ts}(1 - \epsilon^{-T})}{(1 - \epsilon^{-Ts})(1 - \epsilon^{-T}\epsilon^{-Ts})}$$

and

$$A(z) = B(z)F(z) = \left[\frac{z^{-1}(1 - \epsilon^{-T})}{(1 - z^{-1})(1 - \epsilon^{-T}z^{-1})}\right](1 - z^{-1})$$

$$= \frac{z^{-1}(1 - \epsilon^{-T})}{1 - \epsilon^{-T}z^{-1}} = \frac{1 - \epsilon^{-T}}{z - \epsilon^{-T}}$$

Example 4.3

Given the system shown in Figure 4-2, with input $e(t)$ a unit step, let us determine the output function $C(z)$.

$$C(s) = \left[\frac{1 - \epsilon^{-Ts}}{s(s+1)}\right] E^*(s) = G(s)E^*(s)$$

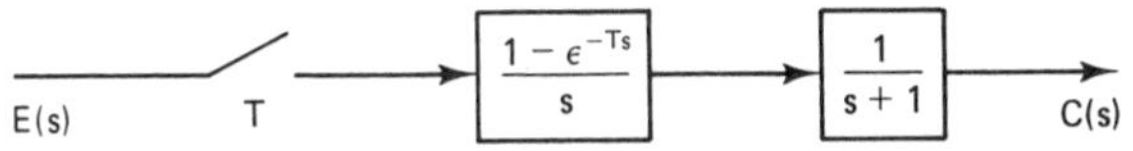

$$E(s) \xrightarrow{\ T\ } \boxed{\frac{1 - \epsilon^{-Ts}}{s}} \longrightarrow \boxed{\frac{1}{s+1}} \longrightarrow C(s)$$

Figure 4-2 Sampled-data system.

In Example 4.2 it was shown that

$$G(z) = \mathfrak{z}\left[\frac{1 - \epsilon^{-Ts}}{s(s+1)}\right] = \frac{1 - \epsilon^{-T}}{z - \epsilon^{-T}}$$

In addition,

$$E(z) = \mathfrak{z}[u(t)] = \frac{z}{z-1}$$

Thus

$$C(z) = \frac{z(1 - \epsilon^{-T})}{(z-1)(z - \epsilon^{-T})} = \frac{z}{z-1} - \frac{z}{z - \epsilon^{-T}}$$

and the inverse z-transform of this function yields

$$c(nT) = 1 - \epsilon^{-nT}$$

Note that if the input to the sampler/zero-order hold is a unit step, the output is also a unit step. Thus the zero-order hold reconstructs the sampled step function exactly. Because of this, the response of the system of Figure 4-2 is simply the step response of a continuous-data system with a transfer function of $1/(s + 1)$. The reader can calculate this response to be

$$c(t) = 1 - \epsilon^{-t}$$

and therefore the results of the z-transform analysis of the sampled-data system are seen to be correct. Note, however, that $c(t)$ cannot be calculated from $C(z)$, except at the sampling instants $t = kT$.

Consider now the system of Figure 4-3a. In this system, there are two plants, and both $G_1(s)$ and $G_2(s)$ contain the transfer functions of the data holds. Now,

$$C(s) = G_2(s)A^*(s) \tag{4-13}$$

and thus

$$C(z) = G_2(z)A(z) \tag{4-14}$$

Also,

$$A(s) = G_1(s)E^*(s) \tag{4-15}$$

and thus

$$A(z) = G_1(z)E(z) \tag{4-16}$$

Then, from (4-14) and (4-16),

$$C(z) = G_1(z)G_2(z)E(z) \tag{4-17}$$

and the total transfer function is the product of the pulse transfer functions.

Consider now the system of Figure 4-3b. Of course, in this case, $G_2(s)$ would not

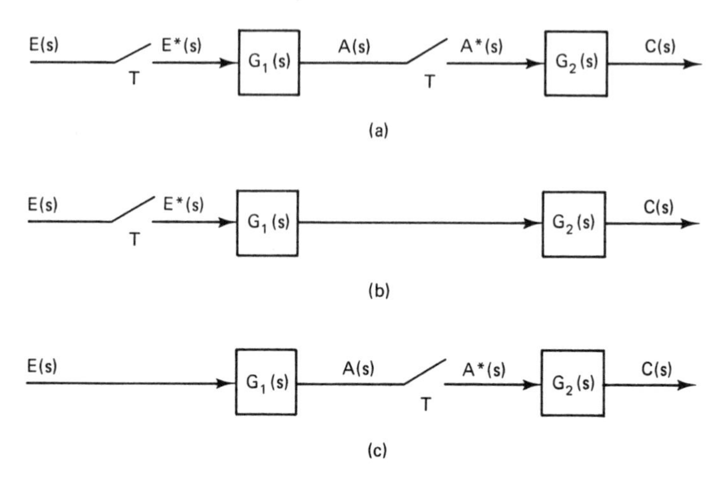

Figure 4-3 Open-loop sampled-data systems.

contain a data-hold transfer function. Then

$$C(s) = G_1(s)G_2(s)E^*(s)$$

and

$$C(z) = \overline{G_1G_2}(z)E(z)$$

where

$$\overline{G_1G_2}(z) = \mathfrak{z}[G_1(s)G_2(s)] \tag{4-18}$$

The bar above a product term indicates that the product must be performed in the s-domain before the z-transform is taken. In addition, one must remember that

$$\overline{G_1G_2}(z) \neq G_1(z)G_2(z) \tag{4-19}$$

For the system of Figure 4-3c,

$$C(s) = G_2(s)A^*(s) = G_2(s)\overline{G_1E}^*(s)$$

Thus

$$C(z) = G_2(z)\overline{G_1E}(z) \tag{4-20}$$

For this system a transfer function cannot be written; that is, we cannot factor $E(z)$ from $\overline{G_1E}(z)$. $E(z)$ contains the values of $e(t)$ only at $t = kT$. But the signal $a(t)$ in Figure 4-3c is a function of all previous values of $e(t)$, not just the values at sampling instants. Since

$$A(s) = G_1(s)E(s)$$

then, from the convolution theorem of the Laplace transform,

$$a(t) = \int_0^t g_1(t - \tau)e(\tau)\, d\tau \tag{4-21}$$

and the dependency of $a(t)$ on all previous value of $e(t)$ is seen. In general, if the input to a sampled-data system is applied directly to a continuous-time part of the system before being sampled, the z-transform of the output of the system cannot be expressed as a function of the z-transform of the input signal. We will see later that this type of system presents no special problems in analysis and design.

4.4 OPEN-LOOP SYSTEMS CONTAINING DIGITAL FILTERS

In the preceding section a transfer-function technique was developed for open-loop sampled-data systems. In this section this technique is extended to cover the case in which the open-loop sampled-data system contains a digital filter.

A manned space platform, such as this one proposed by NASA, will utilize many digital control systems. (Courtesy of NASA.)

In the system shown in Figure 4-4, the A/D converter on the filter input converts the continuous-data signal $e(t)$ to a number sequence $\{e(kT)\}$; the digital filter processes this number sequence $\{e(kT)\}$ and generates the output number sequence $\{m(kT)\}$, which in turn is converted to the continuous-data signal $\bar{m}(t)$ by the D/A converter.

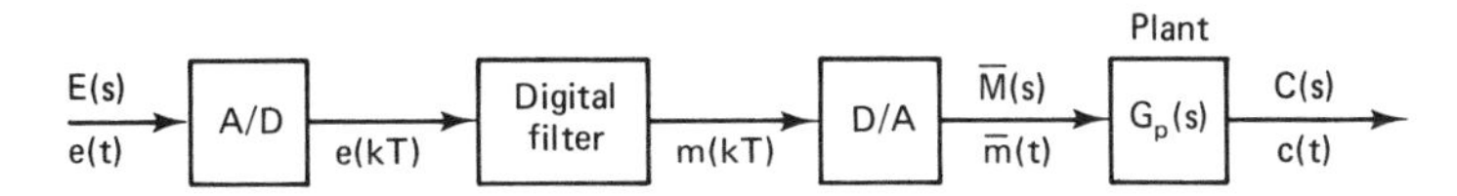

Figure 4-4 Open-loop system with a digital filter.

As was shown in Chapter 2, a digital filter which solves a difference equation can be represented by a transfer function $D(z)$, such that

$$M(z) = D(z)E(z) \tag{4-22}$$

The D/A converter usually has an output data-hold register which gives the D/A the characteristics of a zero-order hold. Thus the Laplace transform of the signal $\bar{m}(t)$ can be expressed as

$$\bar{M}(s) = \left[\frac{1-\epsilon^{-Ts}}{s}\right]M^*(s)$$

Hence

$$C(s) = G_p(s)\bar{M}(s) = G_p(s)\left[\frac{1-\epsilon^{-Ts}}{s}\right]M^*(s)$$

Then, from (4-22),

$$C(s) = G_p(s)\left[\frac{1-\epsilon^{-Ts}}{s}\right]D(z)\big|_{z=\epsilon^{Ts}}E^*(s) \tag{4-23}$$

and we see that the filter and associated A/D and D/A converters can be represented in block-diagram form as shown in Figure 4-5. Hence, from (4-23) or Figure 4-5,

$$C(z) = \mathfrak{z}\left[G_p(s)\left[\frac{1-\epsilon^{-Ts}}{s}\right]\right]D(z)E(z) = G(z)D(z)E(z) \tag{4-24}$$

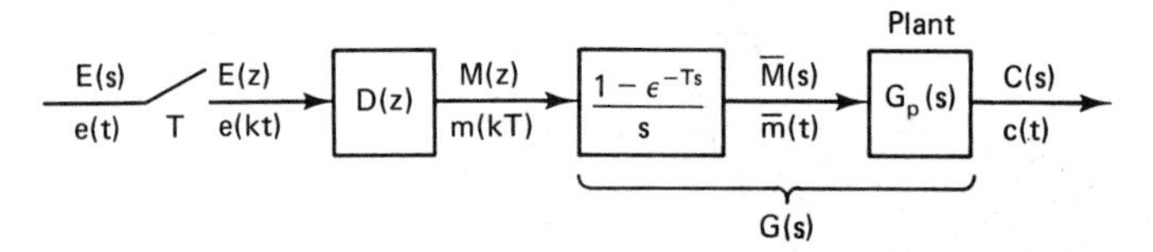

Figure 4-5 Model for the open-loop system.

Perhaps we should emphasize that the digital computing element which implements the digital filter in Figure 4-4 actually processes the values of the input data samples $\{e(kT)\}$. However, our model for the digital filter processes a sequence of *impulse functions* of weight $\{e(kT)\}$. The model must be used as depicted in Figure 4-5; that is, the combination of an ideal sampler, $D(z)$, and a zero-order hold does accurately model the combination of the A/D, digital filter, and D/A.

Example 4.4

Let us determine the step response of the system shown in Figure 4-5. Suppose that the filter is described by the difference equation

$$m(kT) = 2e(kT) - e[(k-1)T]$$

and thus

$$D(z) = \frac{M(z)}{E(z)} = 2 - z^{-1} = \frac{2z-1}{z}$$

In addition, suppose that

$$G(s) = \frac{1}{s+1}$$

Then, as shown in Example 4.3,

$$\mathfrak{z}\left[\frac{1-\epsilon^{-Ts}}{s(s+1)}\right] = \frac{1-\epsilon^{-T}}{z-\epsilon^{-T}}$$

Since $E(z) = z/(z-1)$, from (4-24) we obtain

$$C(z) = \left[\frac{2z-1}{z}\right]\left[\frac{1-\epsilon^{-T}}{z-\epsilon^{-T}}\right]\left[\frac{z}{z-1}\right] = \frac{(2z-1)(1-\epsilon^{-T})}{(z-1)(z-\epsilon^{-T})}$$

The time response can now be obtained using the inversion formula, (2-27). From this expression we obtain, for $k \geq 1$,

$$
\begin{aligned}
c(k) &= (z-1)\left[\frac{(2z-1)(1-\epsilon^{-T})}{(z-1)(z-\epsilon^{-T})}z^{k-1}\right]\Bigg|_{z=1} \\
&\quad + (z-\epsilon^{-T})\left[\frac{(2z-1)(1-\epsilon^{-T})}{(z-1)(z-\epsilon^{-T})}z^{k-1}\right]_{z=\epsilon^{-T}} \\
&= 1^{k-1} + \left[\frac{(2\epsilon^{-T}-1)(1-\epsilon^{-T})}{\epsilon^{-T}-1}\right]\epsilon^{-T(k-1)} \\
&= 1 + \epsilon^{-T(k-1)} - 2\epsilon^{-kT}, \qquad k \geq 1
\end{aligned}
$$

The value of $c(k)$ for $k = 0$ must be calculated separately from $c(k)$ for $k \geq 1$, since the poles of $C(z)z^{k-1}$ for $k = 0$ are different from those of $C(z)z^{k-1}$ for $k \geq 1$. For $k = 0$, $C(z)z^{k-1}$ has a pole at $z = 0$. Therefore, $c(0)$ is obtained from the expression

$$
c(0) = \sum_{\text{at the poles}} \left[\text{residues of } \frac{(2z-1)(1-\epsilon^{-T})}{(z-1)(z-\epsilon^{-T})(z)}\right]
$$

Evaluating the right-hand side of the equation in the manner shown above yields

$$
c(0) = 0
$$

Hence the step response of the system shown in Figure 4-5 is

$$
c(k) = \begin{cases} 0, & k = 0 \\ 1 + \epsilon^{-T(k-1)} - 2\epsilon^{-kT}, & k \geq 1 \end{cases}
$$

Note also that it is obvious from the expression for $C(z)$ that $c(0) = 0$, since the numerator of $C(z)$ is of lower order than the denominator. Use of the power series method for the inverse z-transform shows this.

4.5 THE MODIFIED z-TRANSFORM

The analysis of open-loop systems, including those that contain digital filters, has been presented in preceding sections. However, this technique of analysis does not apply to systems containing ideal time delays. To analyze systems of this type, it is necessary to define the z-transform of a delayed time function. This transform is called the modified z-transform.

The modified z-transform can be developed by considering a time function $e(t)$ that is delayed by an amount ΔT, $0 < \Delta \leq 1$, that is, by considering $e(t - \Delta T)$ $u(t - \Delta T)$. The ordinary z-transform of the delayed time function is

$$
\mathfrak{z}[e(t-\Delta T)u(t-\Delta T)] = \mathfrak{z}[E(s)\epsilon^{-\Delta Ts}] = \sum_{n=1}^{\infty} e(nT-\Delta T)z^{-n} \tag{4-25}
$$

Note that the sampling is not delayed; that is, the sampling instants are $t = 0, T, 2T, \ldots$. The z-transform in (4-25) is called the delayed z-transform, and thus, by definition, the delayed z-transform of $e(t)$ is

$$
E(z, \Delta) = \mathfrak{z}[e(t-\Delta T)u(t-\Delta T)] = \mathfrak{z}[E(s)\epsilon^{-\Delta Ts}] \tag{4-26}
$$

The delayed starred transform is also defined in (4-26), with the substitution $z = \epsilon^{Ts}$. The delayed z-transform will now be illustrated by an example.

Example 4.5

Find $E(z, \Delta)$, if $\Delta = 0.4$, for $e(t) = \epsilon^{-at}u(t)$. From (4-26),

$$E(z, \Delta) = \epsilon^{-0.6aT}z^{-1} + \epsilon^{-1.6aT}z^{-2} + \epsilon^{-2.6aT}z^{-3} + \cdots$$

$$= \epsilon^{-0.6aT}z^{-1}[1 + \epsilon^{-aT}z^{-1} + \epsilon^{-2aT}z^{-2} + \cdots]$$

$$= \frac{\epsilon^{-0.6aT}z^{-1}}{1 - \epsilon^{-aT}z^{-1}} = \frac{\epsilon^{-0.6aT}}{z - \epsilon^{-aT}}$$

The sketches in Figure 4-6 show both $e(t)$ and $e(t - \Delta T)$.

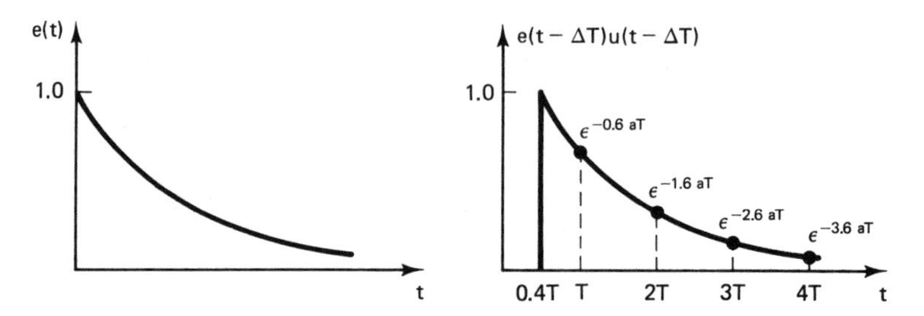

Figure 4-6 Example of the delayed z-transform.

The modified z-transform is defined from the delayed z-transform. By definition, the modified z-transform of a function is equal to the delayed z-transform with Δ replaced by $1 - m$. Thus if we let $E(z, m)$ be the modified z-transform of $E(s)$, then

$$E(z, m) = E(z, \Delta)\,|_{\Delta=1-m} = \mathfrak{z}[E(s)\epsilon^{-\Delta Ts}]\,|_{\Delta=1-m} \tag{4-27}$$

From (4-25) and (4-27),

$$E(z, m) = e(mT)z^{-1} + e[(1 + m)T]z^{-2} + e[(2 + m)T]z^{-3} + \cdots \tag{4-28}$$

Note that

$$E(z, 1) = E(z, m)\,|_{m=1} = E(z) - e(0) \tag{4-29}$$

and

$$E(z, 0) = E(z, m)\,|_{m=0} = z^{-1}E(z) \tag{4-30}$$

Example 4.6

It is desired to find the modified z-transform of $e(t) = \epsilon^{-t}$. From (4-28),

$$E(z, m) = \epsilon^{-mT}z^{-1} + \epsilon^{-(1+m)T}z^{-2} + \epsilon^{-(2+m)T}z^{-3} + \cdots$$

$$= \epsilon^{-mT}z^{-1}[1 + \epsilon^{-T}z^{-1} + \epsilon^{-2T}z^{-2} + \cdots]$$

$$= \frac{\epsilon^{-mT}z^{-1}}{1 - \epsilon^{-T}z^{-1}} = \frac{\epsilon^{-mT}}{z - \epsilon^{-T}}$$

A sketch of the time functions is given in Figure 4-7.

As stated in Chapter 3, the ordinary z-transform tables do not apply to the modified z-transform. Instead, special tables must be derived. These tables are obtained by the following development. One can express the modified z-transform of a function $E(s)$ in the following manner, from the development in Appendix III.

$$E(z, m) = \mathfrak{z}[E(s)\epsilon^{-\Delta Ts}]\,|_{\Delta=1-m} = \mathfrak{z}[E(s)\epsilon^{-(1-m)Ts}] = z^{-1}\mathfrak{z}[E(s)\epsilon^{mTs}] \tag{4-31}$$

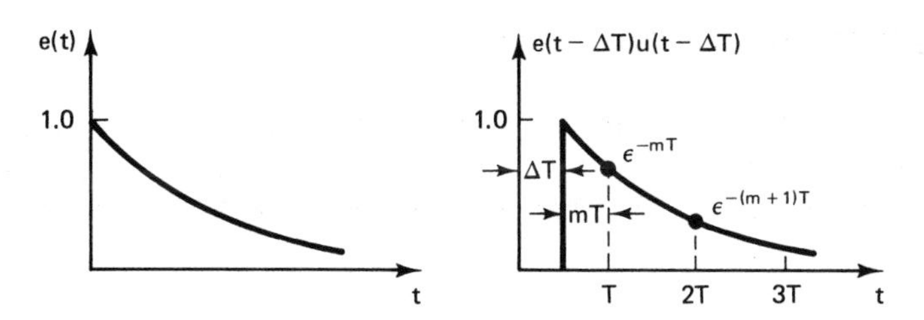

Figure 4-7 Example of the modified z-transform.

Now $\lim_{\lambda\to\infty} \lambda E(\lambda)\epsilon^{mT\lambda}$ is zero in the second integral in (A3-7) (see Figure A3-2), and thus (3-10) applies. Then, from (3-10) and (4-31),

$$E(z, m) = z^{-1} \sum_{\substack{\text{poles of} \\ E(\lambda)}} \text{residues of } E(\lambda)\epsilon^{mT\lambda} \frac{1}{1 - z^{-1}\epsilon^{T\lambda}} \tag{4-32}$$

Also, from (3-11),

$$E^*(s, m) = \frac{1}{T} \sum_{n=-\infty}^{\infty} E(s + jn\omega_s)\epsilon^{-(1-m)(s+jn\omega_s)T} \tag{4-33}$$

provided that $e(t - \Delta T)$ is continuous at all sampling instants. Modified z-transform tables are included with the ordinary z-transform tables in the z-transform tables of Appendix VI.

A useful property of the modified z-transform will now be stated. Since the modified z-transform is the ordinary z-transform of a shifted function, the theorems of the ordinary z-transform derived in Chapter 2 may be applied to the modified z-transform, if care is exercised. In particular, the shifting theorem applies directly. Let $\mathfrak{z}_m[\cdot]$ indicate the modified z-transform; that is,

$$\mathfrak{z}_m[E(s)] = E(z, m) = \mathfrak{z}[\epsilon^{-\Delta Ts}E(s)]\,|_{\Delta=1-m} \tag{4-34}$$

Then, by the shifting theorem, for k a positive integer,

$$\mathfrak{z}_m[\epsilon^{-kTs}E(s)] = z^{-k}\mathfrak{z}_m[E(s)] = z^{-k}E(z, m) \tag{4-35}$$

Example 4.7

We wish to find the modified z-transform of the function $e(t) = t$. It is well known that $E(s) = 1/s^2$. This function has a pole of order 2 at $s = 0$. Therefore, the modified z-transform can be obtained from (4-32) as

$$E(z, m) = z^{-1}\left[\frac{d}{d\lambda}\left[\frac{\epsilon^{mT\lambda}}{1 - z^{-1}\epsilon^{T\lambda}}\right]_{\lambda=0}\right]$$

$$= z^{-1}\left[\frac{(1 - z^{-1}\epsilon^{T\lambda})mT\epsilon^{mT\lambda} - \epsilon^{mT\lambda}(-Tz^{-1}\epsilon^{T\lambda})}{(1 - z^{-1}\epsilon^{T\lambda})^2}\right]_{\lambda=0}$$

$$= z^{-1}\left[\frac{mT(1 - z^{-1}) + Tz^{-1}}{(1 - z^{-1})^2}\right]$$

$$= \frac{mTz^{-1}}{1 - z^{-1}} + \frac{Tz^{-2}}{(1 - z^{-1})^2}$$

$$= \frac{mT}{z - 1} + \frac{T}{(z - 1)^2}$$

4.6 SYSTEMS WITH TIME DELAYS

The modified z-transform may be used to determine the pulse transfer functions of discrete-time systems containing ideal time delays. To illustrate this, consider the system of Figure 4-8, which has an ideal time delay of t_0 seconds. For this system,

$$C(s) = G(s)\epsilon^{-t_0 s}E^*(s) \tag{4-36}$$

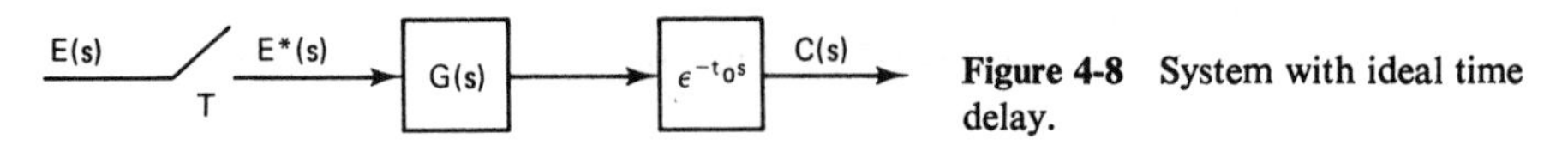

Figure 4-8 System with ideal time delay.

Thus

$$C(z) = \mathfrak{z}[G(s)\epsilon^{-t_0 s}]E(z) \tag{4-37}$$

If we now let

$$t_0 = kT + \Delta T, \qquad 0 < \Delta < 1 \tag{4-38}$$

where k is a positive integer, then from (4-35),

$$C(z) = z^{-k}\mathfrak{z}[G(s)\epsilon^{-\Delta Ts}]E(z) \tag{4-39}$$

or, from (4-27),

$$C(z) = z^{-k}G(z, m)E(z) \tag{4-40}$$

where $m = 1 - \Delta$. The foregoing development will now be illustrated with an example.

Example 4.8

In Figure 4-8, let the input be a unit step, $t_0 = 0.4T$, and

$$G(s) = \frac{1 - \epsilon^{-Ts}}{s(s+1)}$$

Then, from (4-35) and the modified z-transform tables,

$$G(z, m) = \mathfrak{z}_m\left[\frac{1 - e^{-Ts}}{s(s+1)}\right] = (1 - z^{-1})\mathfrak{z}_m\left[\frac{1}{s(s+1)}\right]$$

$$= \frac{z-1}{z}\left[\frac{z(1 - \epsilon^{-mT}) + \epsilon^{-mT} - \epsilon^{-T}}{(z-1)(z - \epsilon^{T})}\right]$$

Thus, since $mT = T - \Delta T = 0.6T$,

$$G(z, m) = \frac{z-1}{z}\left[\frac{z(1 - \epsilon^{-0.6T}) + \epsilon^{-0.6T} - \epsilon^{-T}}{(z-1)(z - \epsilon^{-T})}\right]$$

Since $k = 0$ in (4-40), $C(z)$ is seen to be

$$C(z) = G(z, m)\frac{z}{z-1} = \frac{z(1 - \epsilon^{-0.6T}) + \epsilon^{-0.6T} - \epsilon^{-T}}{(z-1)(z - \epsilon^{-T})}$$

and thus

$$C(z) = (1 - \epsilon^{-0.6T})z^{-1} + (1 - \epsilon^{-1.6T})z^{-2} + (1 - \epsilon^{-2.6T})z^{-3} + \cdots$$

The modified z-transform may also be used to determine the pulse transfer functions of digital control systems in which the computation time of the digital computer cannot be neglected. As given in (2-4), an nth-order linear digital controller solves the difference equation

$$m(k) = \alpha_0 e(k) + \alpha_1 e(k-1) + \cdots + \alpha_n e(k-n) \\ - \beta_1 m(k-1) - \cdots - \beta_n m(k-n) \tag{4-41}$$

every T seconds. Let the time required for the digital controller to compute (4-41) be t_0 seconds. Thus an input at $t = 0$ produces an output at $t = t_0$, an input at $t = T$ produces an output at $t = T + t_0$, and so on. Hence the digital controller may be modeled as a digital controller without time delay, followed by an ideal time delay of t_0 seconds, as shown in Figure 4-9a. An open-loop system containing this controller may be modeled as shown in Figure 4-9b. For this system,

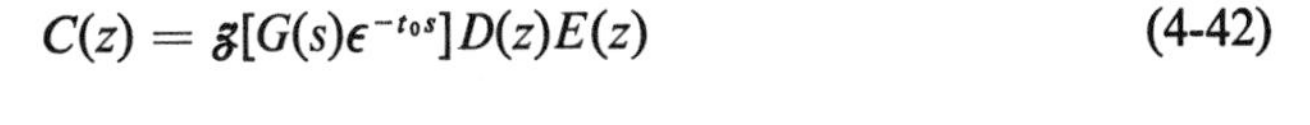

$$C(z) = \mathfrak{z}[G(s)\epsilon^{-t_0 s}]D(z)E(z) \tag{4-42}$$

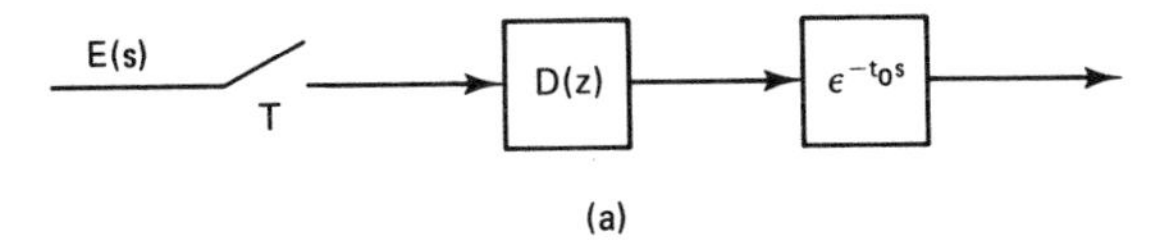

(a)

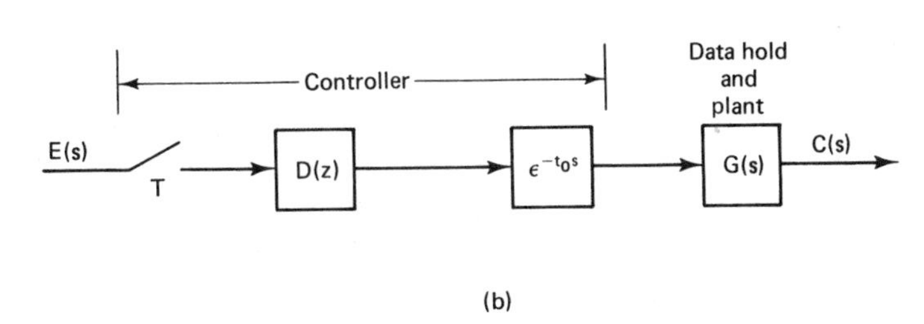

(b)

Figure 4-9 Digital controller with nonzero computation time.

If we let

$$t_0 = kT + \Delta T, \qquad 0 < \Delta < 1 \tag{4-43}$$

with k a positive integer, then from (4-39) and (4-40) we obtain

$$C(z) = z^{-k} G(z, m) D(z) E(z) \tag{4-44}$$

where $m = 1 - \Delta$.

Example 4.9

Consider the system of Figure 4-10. This system is that of Example 4.4, with a computational delay added for the filter. The delay is 1 ms ($t_0 = 10^{-3}$ s) and $T = 0.05$ s. Thus, for this system,

$$D(z) = \frac{2z - 1}{z}$$

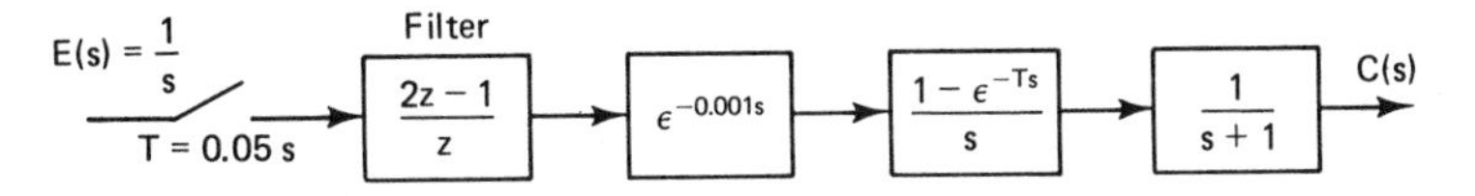

Figure 4-10 System for Example 4.9.

Now

$$mT + \Delta T = T$$

or

$$mT = T - \Delta T = 0.05 - 0.001 = 0.049$$

Then

$$G(z, m) = \mathcal{Z}_m\left[\frac{1 - \epsilon^{-Ts}}{s(s+1)}\right]_{mT=0.049} = \frac{z-1}{z}\mathcal{Z}_m\left[\frac{1}{s(s+1)}\right]_{mT=0.049}$$

From Example 4.8,

$$G(z, m) = \frac{z-1}{z}\left[\frac{z(1 - \epsilon^{-0.049}) + (\epsilon^{-0.049} - \epsilon^{-0.05})}{(z-1)(z - \epsilon^{-0.05})}\right]$$

Since the input is a unit step, then, from (4-44),

$$\begin{aligned} C(z) &= G(z, m)D(z)E(z) \\ &= \frac{z-1}{z}\left[\frac{z(1 - \epsilon^{-0.049}) + (\epsilon^{-0.049} - \epsilon^{-0.05})}{(z-1)(z - \epsilon^{-0.05})}\right]\left[\frac{2z-1}{z}\right]\left[\frac{z}{z-1}\right] \\ &= \frac{(2z-1)[z(1 - \epsilon^{-0.049}) + (\epsilon^{-0.049} - \epsilon^{-0.05})]}{z(z-1)(z - \epsilon^{-0.05})} \end{aligned}$$

4.7 NONSYNCHRONOUS SAMPLING

In the preceding sections, simple open-loop systems and open-loop systems with digital filters and/or ideal time delays were considered. In this section open-loop systems with nonsynchronous sampling are analyzed. Nonsynchronous sampling can be defined by considering the system of Figure 4-11. In this system both samplers operate at the same rate, but are not synchronous.

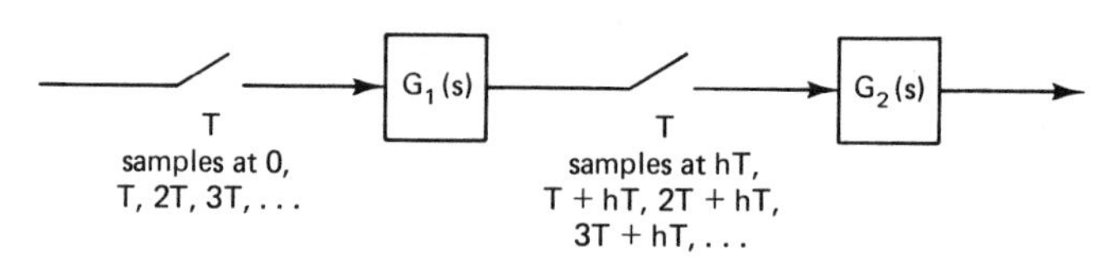

Figure 4-11 System with nonsynchronous sampling.

To develop a method of analysis for systems with nonsynchronous sampling, consider the sampler and data hold in Figure 4-12a. Here the sampler operates at $hT, T + hT, 2T + hT, 3T + hT, \ldots$, where $0 < h < 1$. The data-hold output is as

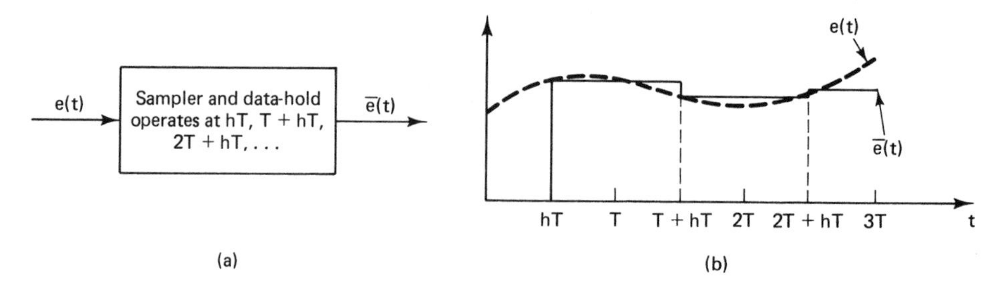

Figure 4-12 Illustration of nonsynchronous sampling.

shown in Figure 4-12b, and may be expressed as

$$\bar{e}(t) = e(hT)[u(t - hT) - u(t - T - hT)] + e(T + hT)[u(t - T - hT) - u(t - 2T - hT)] + e(2T + hT)[u(t - 2T - hT) - u(t - 3T - hT)] + \cdots \tag{4-45}$$

Thus

$$\bar{E}(s) = e(hT)\left[\frac{\epsilon^{-hTs}}{s} - \frac{\epsilon^{-(T+hT)s}}{s}\right] + e(T + hT)\left[\frac{\epsilon^{-(T+hT)s}}{s} - \frac{\epsilon^{-(2T+hT)s}}{s}\right] + e(2T + hT)\left[\frac{\epsilon^{-(2T+hT)s}}{s} - \frac{\epsilon^{-(3T+hT)s}}{s}\right] + \cdots$$

or

$$\bar{E}(s) = \left[\frac{1 - \epsilon^{-Ts}}{s}\right]\epsilon^{-hTs}[e(hT) + e(T + hT)\epsilon^{-Ts} + e(2T + hT)\epsilon^{-2Ts} + \cdots]$$

$$= \left[\frac{1 - \epsilon^{-Ts}}{s}\right]\epsilon^{Ts}\epsilon^{-hTs}[e(hT)\epsilon^{-Ts} + e(T + hT)\epsilon^{-2Ts} + e(2T + hT)\epsilon^{-3Ts} + \cdots]$$

Then, from (4-28),

$$\bar{E}(s) = \left[\frac{1 - \epsilon^{-Ts}}{s}\right]\epsilon^{Ts}\epsilon^{-hTs}E(z, m)\,|_{m=h,\ z=\epsilon^{Ts}} \tag{4-46}$$

Since

$$E(z, m) = \mathfrak{z}[E(s)\epsilon^{-\Delta Ts}]\,|_{\Delta=1-m} \tag{4-47}$$

we see from (4-46) that the sampler and data hold of Figure 4-12 can be modeled as shown in Figure 4-13, where the sampler operates at $t = 0, T, 2T, \ldots$.

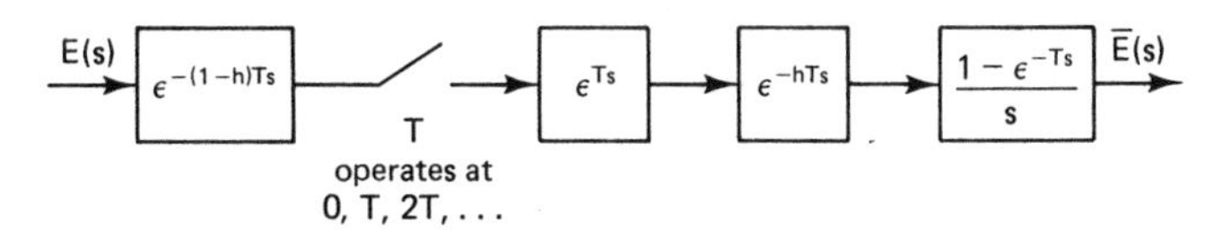

Figure 4-13 Model of a sampler and data hold.

From the development above, we see that the system of Figure 4-11, with nonsynchronous samplers, may be modeled as shown in Figure 4-14. In Figure 4-14, the samplers are synchronous. Now, in Figure 4-14,

$$A(s) = \epsilon^{-(1-h)Ts}G_1(s)E^*(s)$$

and thus

$$A(z) = G_1(z, m)\,|_{m=h}E(z) \tag{4-48}$$

Also,

$$B^*(s) = \epsilon^{Ts}A^*(s)$$

yielding

$$B(z) = zA(z) = zE(z)G_1(z, m)|_{m=h} \tag{4-49}$$

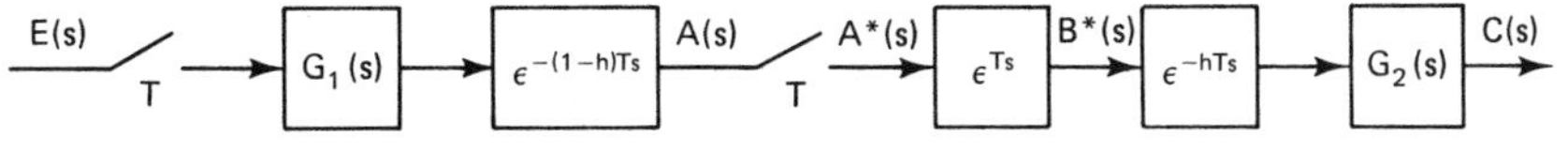

Figure 4-14 Model of the system of Figure 4-11.

Then

$$C(s) = \epsilon^{-hTs} G_2(s) B^*(s)$$

yielding

$$C(z) = G_2(z, m)\big|_{m=1-h} B(z) \tag{4-50}$$

From (4-49) and (4-50), we find $C(z)$ to be given by

$$C(z) = zE(z) G_1(z, m)\big|_{m=h} G_2(z, m)\big|_{m=1-h} \tag{4-51}$$

Example 4.10

We wish to find $C(z)$ for the system of Figure 4-15, which contains nonsynchronous sampling. Now

$$E(z) = \mathfrak{z}\left[\frac{1}{s}\right] = \frac{z}{z-1}$$

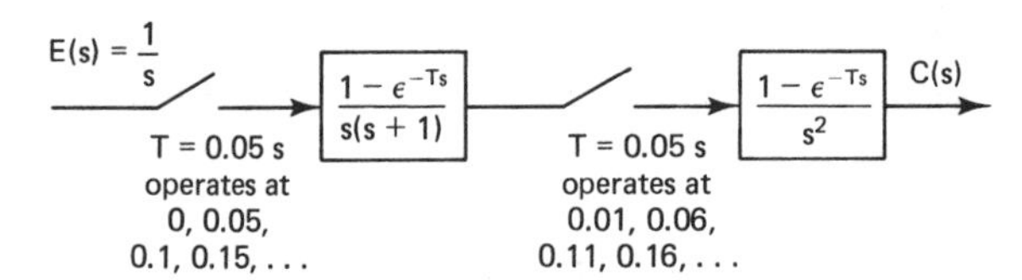

Figure 4-15 System for Example 4.10.

For the system, $T = 0.05$ and $hT = 0.01$. From Example 4.8,

$$G_1(z, m) = \mathfrak{z}_m\left[\frac{1-\epsilon^{-Ts}}{s(s+1)}\right] = \frac{z-1}{z}\left[\frac{z(1-\epsilon^{-mT}) + \epsilon^{-mT} - \epsilon^{-T}}{(z-1)(z-\epsilon^{-T})}\right]$$

Then

$$G_1(z, m)\big|_{mT=0.01} = \frac{z-1}{z}\left[\frac{z(1-\epsilon^{-0.01}) + \epsilon^{-0.01} - \epsilon^{-0.05}}{(z-1)(z-\epsilon^{-0.05})}\right]$$

Also, from the modified z-transform tables,

$$G_2(z, m) = \mathfrak{z}_m\left[\frac{1-\epsilon^{-Ts}}{s^2}\right] = \frac{z-1}{z}\left[\frac{mTz - mT + T}{(z-1)^2}\right]$$

or

$$G_2(z, m)\big|_{m=1-h} = \frac{0.04z + 0.01}{z(z-1)}$$

Then, from the development above and (4-51),

$$\begin{aligned} C(z) &= z\left[\frac{z}{z-1}\right]\left[\frac{z-1}{z}\right]\left[\frac{z(1-\epsilon^{-0.01}) + \epsilon^{-0.01} - \epsilon^{-0.05}}{(z-1)(z-\epsilon^{-0.05})}\right]\left[\frac{0.04z + 0.01}{z(z-1)}\right] \\ &= \frac{(0.04z + 0.01)[z(1-\epsilon^{-0.01}) + \epsilon^{-0.01} - \epsilon^{-0.05}]}{(z-1)^2(z-\epsilon^{-0.05})} \end{aligned}$$

4.8 STATE-VARIABLE MODELS

Thus far in this chapter we have discussed the analysis of open-loop sampled data systems using the transfer-function approach. As was shown in Chapter 2, systems describable by a z-transform transfer function may also be modeled by discrete state equations. The state equations are of the form

$$\begin{aligned} \mathbf{x}(k+1) &= \mathbf{A}\mathbf{x}(k) + \mathbf{B}\mathbf{u}(k) \\ \mathbf{y}(k) &= \mathbf{C}\mathbf{x}(k) + \mathbf{D}\mathbf{u}(k) \end{aligned} \tag{4-52}$$

The techniques used in Chapter 2 may be employed to find the state-variable models of open-loop systems of the type discussed in this chapter. To obtain a state-variable model, first a system flow graph is drawn from the transfer function. Next each time-delay output is labeled as a state variable. Then the state equations are written from the flow graph. This technique will now be illustrated by an example.

Example 4.11

Consider the system of Figure 4-16a, which is the system considered in Example 4.4. Here we are denoting the output as $Y(s)$ instead of $C(s)$, to prevent any notational confusion with the **C** matrix. From Example 4.4,

$$G(z) = \mathcal{Z}\left[\frac{1 - \epsilon^{-Ts}}{s(s+1)}\right] = \frac{1 - \epsilon^{-T}}{z - \epsilon^{-T}}$$

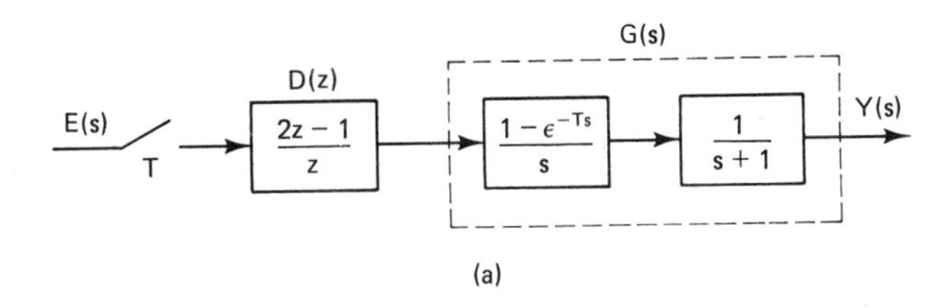

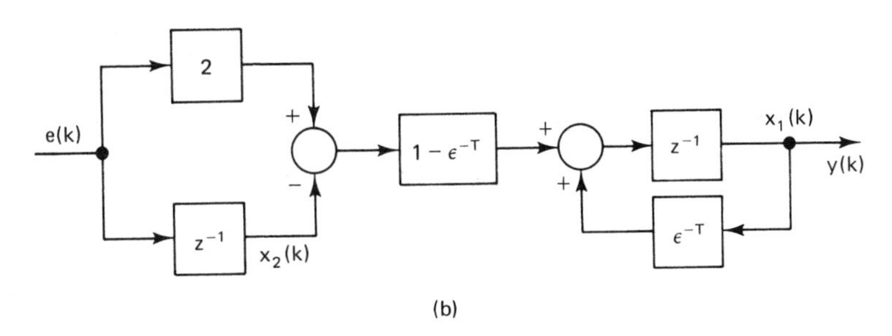

Figure 4-16 System for Example 4.11.

and as is shown in Figure 4-16a,

$$D(z) = \frac{2z - 1}{z}$$

A simulation diagram for this system is given in Figure 4-16b. From this figure we write

$$\mathbf{x}(k+1) = \begin{bmatrix} \epsilon^{-T} & -1 + \epsilon^{-T} \\ 0 & 0 \end{bmatrix} \mathbf{x}(k) + \begin{bmatrix} 2(1 - \epsilon^{-T}) \\ 1 \end{bmatrix} e(k)$$

$$y(k) = [1 \quad 0]\mathbf{x}(k)$$

4.9 REVIEW OF CONTINUOUS STATE VARIABLES

Presented in the preceding section is a technique for obtaining a set of state equations describing a linear time-variant discrete system. This technique is based on transfer functions, and has two major disadvantages. One disadvantage can be illustrated by considering a simple mechanical system, the motion of a unit mass in a frictionless environment. For this system,

$$\frac{d^2x}{dt^2} = \ddot{x} = f(t) \tag{4-53}$$

where x is the displacement, or position, of the mass, and $f(t)$ is the applied force. To obtain a continuous state-variable model, choose as state variables

$$\begin{aligned} v_1(t) &= x(t) = \text{position} \\ v_2(t) &= \dot{v}_1(t) = \dot{x}(t) = \text{velocity} \end{aligned} \tag{4-54}$$

Then the state equations are

$$\begin{bmatrix} \dot{v}_1(t) \\ \dot{v}_2(t) \end{bmatrix} = \begin{bmatrix} 0 & 1 \\ 0 & 0 \end{bmatrix} \begin{bmatrix} v_1(t) \\ v_2(t) \end{bmatrix} + \begin{bmatrix} 0 \\ 1 \end{bmatrix} f(t) \tag{4-55}$$

Here we have chosen as state variables the position and the velocity of the mass. These variables may be considered to be the "natural" states (physical variables) of the system, and would be the desirable states to choose. If this simple system were a part of a sampled-data system, we can easily choose position as one of the states in the discrete state model, by letting position be the output of the simple system. However, in taking the transfer-function approach to discrete state modeling, we would have difficulty in choosing velocity as the second state variable. Thus we lose the natural, and desirable, states of the system. Another disadvantage of the transfer-function approach is the difficulty in deriving the pulse transfer functions for high-order systems.

A different approach for obtaining the discrete state model of a system is presented in the following section. This approach is based on the use of continuous state variables. Hence a brief presentation of the requisite theory of continuous state variables will be made here.

As indicated in the example above of the motion of a mass, continuous state-variable equations for a linear time-invariant system are of the form

$$\begin{aligned} \dot{\mathbf{v}}(t) &= \mathbf{A}_c\mathbf{v}(t) + \mathbf{B}_c\mathbf{u}(t) \\ \mathbf{y}(t) &= \mathbf{C}_c\mathbf{v}(t) + \mathbf{D}_c\mathbf{u}(t) \end{aligned} \tag{4-56}$$

In this equation, $\mathbf{v}(t)$ are the states, $\mathbf{u}(t)$ are the inputs, $\mathbf{y}(t)$ are the outputs, and the matrices are subscripted to indicate continuous state equations. Nonsubscripted matrices will indicate discrete state equations. To illustrate continuous state equations, consider the following example.

Example 4.12

It is desired to find a state model for the mechanical system of Figure 4-17a. Here M is mass, B is the damping factor for linear friction, and K is the stiffness factor for a linear spring. The equations for this system are

$$\ddot{y}_1 + B_1\dot{y}_1 + Ky_1 + B_2(\dot{y}_1 - \dot{y}_2) = 0$$

$$\ddot{y}_2 + B_2(\dot{y}_2 - \dot{y}_1) = u(t)$$

A flow graph for these equations is shown in Figure 4-17b. The transfer functions s^{-1} represent integrators, and the output of each integrator is chosen as a state, as shown.

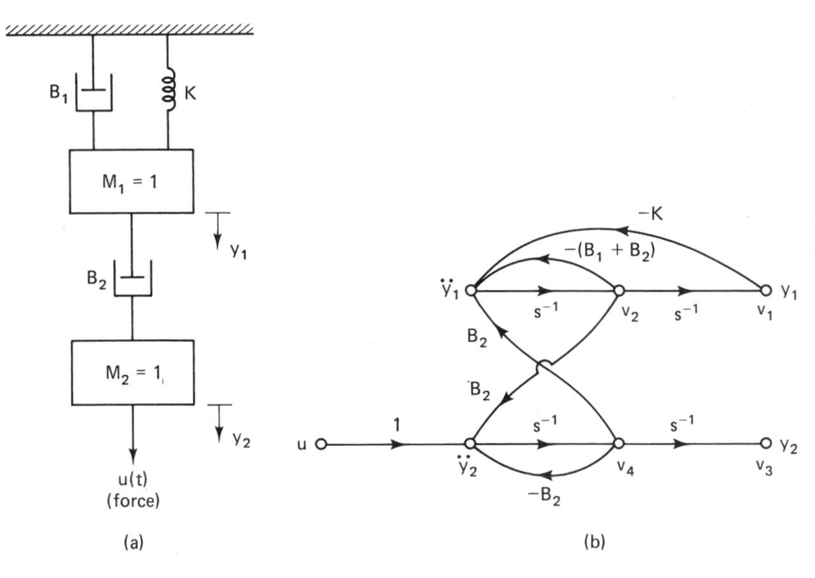

Figure 4-17 System for Example 4.12.

The state-variable equations are written from this flow graph.

$$\dot{\mathbf{v}} = \begin{bmatrix} 0 & 1 & 0 & 0 \\ -K & -(B_1 + B_2) & 0 & B_2 \\ 0 & 0 & 0 & 1 \\ 0 & B_2 & 0 & -B_2 \end{bmatrix} \mathbf{v} + \begin{bmatrix} 0 \\ 0 \\ 0 \\ 1 \end{bmatrix} u$$

$$\mathbf{y} = \begin{bmatrix} 1 & 0 & 0 & 0 \\ 0 & 0 & 1 & 0 \end{bmatrix} \mathbf{v}$$

In the example above, the state variables are positions and velocities. If the technique of drawing a flow graph from the system differential equations (written using physical laws) is used, the states chosen will be the natural states of the system.

Consider now the state equations for a single-input, single-output system.

$$\begin{aligned} \dot{\mathbf{v}}(t) &= \mathbf{A}_c \mathbf{v}(t) + \mathbf{b}_c u(t) \\ y(t) &= \mathbf{c}_c \mathbf{v}(t) + d_c u(t) \end{aligned} \tag{4-57}$$

To obtain the solution of these equations, we will use the Laplace transform. Thus, for (4-57),

$$s\mathbf{V}(s) - \mathbf{v}(0) = \mathbf{A}_c \mathbf{V}(s) + \mathbf{b}_c U(s) \tag{4-58}$$

Solving for $\mathbf{V}(s)$ yields

$$\mathbf{V}(s) = [\mathbf{I}s - \mathbf{A}_c]^{-1}\mathbf{v}(0) + [\mathbf{I}s - \mathbf{A}_c]^{-1}\mathbf{b}_c U(s) \tag{4-59}$$

Define $\mathbf{\Phi}_c(t)$ as

$$\mathbf{\Phi}_c(t) = \mathcal{L}^{-1}\{[\mathbf{I}s - \mathbf{A}_c]^{-1}\} \tag{4-60}$$

Program 5 in Appendix I may be used to calculate $[\mathbf{I}s - \mathbf{A}_c]^{-1}$. The matrix $\mathbf{\Phi}_c(t)$ is called the state transition matrix for (4-57). The inverse Laplace transform of (4-59) is then

$$\mathbf{v}(t) = \mathbf{\Phi}_c(t)\mathbf{v}(0) + \int_0^t \mathbf{\Phi}_c(t-\tau)\mathbf{b}_c u(\tau)\, d\tau \tag{4-61}$$

The state transition matrix $\mathbf{\Phi}_c(t)$ can be calculated as given in (4-60). However, a different expression for $\mathbf{\Phi}_c(t)$ may be derived. This expression is found by assuming $\mathbf{\Phi}_c(t)$ in (4-61), with $u(t) = 0$, to be an infinite series.

$$\mathbf{\Phi}_c(t) = \mathbf{A}_0 + \mathbf{A}_1 t + \mathbf{A}_2 t^2 + \mathbf{A}_3 t^3 + \cdots \tag{4-62}$$

Substitution of (4-61), with $\mathbf{\Phi}_c(t)$ given by (4-62), into (4-57) yields

$$\mathbf{\Phi}_c(t) = \mathbf{I} + \mathbf{A}_c t + \mathbf{A}_c^2 \frac{t^2}{2!} + \mathbf{A}_c^3 \frac{t^3}{3!} + \cdots \tag{4-63}$$

This expression for $\mathbf{\Phi}_c(t)$ will prove to be useful in deriving discrete state models for continuous systems.

A problem that often arises in determining state models of physical systems will now be discussed. This problem is illustrated by the system shown in Figure 4-18a.

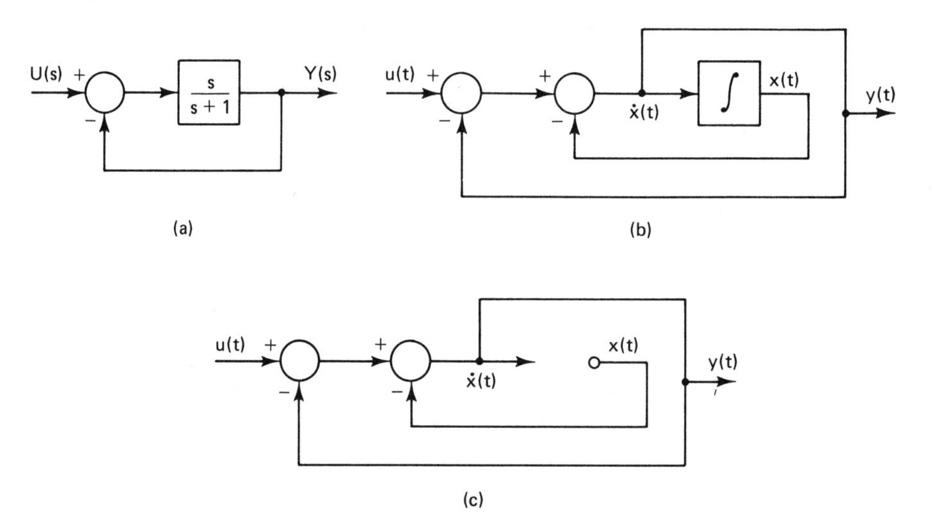

Figure 4-18 System with an algebraic loop.

State equations, written from the simulation diagram of Figure 4-18b, are

$$\dot{x}(t) = -x(t) + [u(t) - \dot{x}(t)]$$

$$y(t) = \dot{x}(t)$$

These equations are not in the standard format for state equations, since $\dot{x}(t)$ appears on the right-hand side of both the $\dot{x}(t)$ equation and the $y(t)$ equation. The $\dot{x}(t)$ equation must be solved for $\dot{x}(t)$, resulting in

$$\dot{x}(t) = -0.5x(t) + 0.5u(t)$$

and hence

$$y(t) = -0.5x(t) + 0.5u(t)$$

If we redraw the simulation diagram of Figure 4-18b with the integrator omitted, Figure 4-18c results. Now we write the equations for $\dot{x}(t)$ and $y(t)$ in terms of the inputs in this diagram, $x(t)$ and $u(t)$. This technique is standard for writing state equations from simulation diagrams, although we usually do not redraw the diagram with the integrators removed. Applying Mason's gain formula to the diagram in Figure 4-18c, we obtain the state equations

$$\dot{x}(t) = \frac{-1}{1+1}x(t) + \frac{1}{1+1}u(t)$$

$$y(t) = \frac{-1}{1+1}x(t) + \frac{1}{1+1}u(t)$$

The loop in Figure 4-18c that does not contain an integrator is called an algebraic loop. For complex systems with algebraic loops, the system equations are initially written as [1]

$$\dot{\mathbf{x}}(t) = \mathbf{A}_1\dot{\mathbf{x}}(t) + \mathbf{A}_2\mathbf{x}(t) + \mathbf{B}_1\mathbf{u}(t)$$
$$\mathbf{y}(t) = \mathbf{C}_1\dot{\mathbf{x}}(t) + \mathbf{C}_2\mathbf{x}(t) + \mathbf{D}_1\mathbf{u}(t)$$

Solving the first equation, we obtain

$$\dot{\mathbf{x}}(t) = [\mathbf{I} - \mathbf{A}_1]^{-1}\mathbf{A}_2\mathbf{x}(t) + [\mathbf{I} - \mathbf{A}_1]^{-1}\mathbf{B}_1\mathbf{u}(t)$$

Then we substitute this equation into the one for $\mathbf{y}(t)$:

$$\mathbf{y}(t) = [\mathbf{C}_1[\mathbf{I} - \mathbf{A}_1]^{-1}\mathbf{A}_2 + \mathbf{C}_2]\mathbf{x}(t) + [\mathbf{C}_1[\mathbf{I} - \mathbf{A}_1]^{-1}\mathbf{B}_1 + \mathbf{D}_1]\mathbf{u}(t)$$

Thus we have the state equations for the system in standard form.

4.10 DISCRETE STATE EQUATIONS

A technique is developed in this section for determining the discrete state equations of a sampled-data system directly from the continuous state equations. In fact, the states of the continuous model become the states of the discrete model. Thus the natural states of the system are preserved.

To develop this technique, consider the state equations for the continuous portion of the system shown in Figure 4-19.

M(s) — T — $\frac{1-\epsilon^{-Ts}}{s}$ — U(s) — G(s) — Y(s)

Figure 4-19 Sampled-data system.

$$\dot{\mathbf{v}}(t) = \mathbf{A}_c\mathbf{v}(t) + \mathbf{b}_c u(t)$$
$$y(t) = \mathbf{c}_c\mathbf{v}(t) + d_c u(t) \qquad (4\text{-}64)$$

As shown in the preceding section, the solution to these equations is

$$\mathbf{v}(t) = \mathbf{\Phi}_c(t - t_0)\mathbf{v}(t_0) + \int_{t_0}^{t} \mathbf{\Phi}_c(t - \tau)\mathbf{b}_c u(\tau)\, d\tau \qquad (4\text{-}65)$$

where the initial time is t_0, and where

$$\mathbf{\Phi}_c(t - t_0) = \sum_{k=0}^{\infty} \frac{\mathbf{A}_c^k(t - t_0)^k}{k!} \qquad (4\text{-}66)$$

To obtain the discrete model we evaluate (4-65) at $t = kT + T$ with $t_0 = kT$, that is,

$$\mathbf{v}(kT + T) = \mathbf{\Phi}_c(T)\mathbf{v}(kT) + m(kT) \int_{kT}^{kT+T} \mathbf{\Phi}_c(kT + T - \tau)\mathbf{b}_c \, d\tau \tag{4-67}$$

It is important to note that we have replaced $u(t)$ with $m(kT)$ since during the time interval $kT \le t < kT + T$, $u(t) = m(kT)$. Compare (4-67) with the discrete state equations [see (4-52)]

$$\mathbf{x}(k + 1) = \mathbf{A}\mathbf{x}(k) + \mathbf{b}m(k) \tag{4-68}$$

Thus, if we let

$$\begin{aligned} \mathbf{x}(kT) &= \mathbf{v}(kT) \\ \mathbf{A} &= \mathbf{\Phi}_c(T) \\ \mathbf{b} &= \int_{kT}^{kT+T} \mathbf{\Phi}_c(kT + T - \tau)\mathbf{b}_c \, d\tau \end{aligned} \tag{4-69}$$

we obtain the discrete state equations for the sampled-data system. The output equation in (4-64), when evaluated at $t = kT$, yields

$$\begin{aligned} y(kT) &= \mathbf{c}_c\mathbf{v}(kT) + d_c u(kT) \\ &= \mathbf{c}_c\mathbf{x}(kT) + d_c m(kT) \end{aligned} \tag{4-70}$$

Thus the discrete $\mathbf{c}$ and d values are equal to the continuous $\mathbf{c}_c$ and d_c values, respectively.

The relationship for $\mathbf{b}$ can be simplified. In (4-69) in the equation for $\mathbf{b}$, let $kT - \tau = -\sigma$. Then this equation becomes

$$\mathbf{b} = \left[\int_0^T \mathbf{\Phi}_c(T - \sigma)\, d\sigma\right]\mathbf{b}_c \tag{4-71}$$

The discrete system matrices $\mathbf{A}$ and $\mathbf{b}$ may be evaluated by finding $\mathbf{\Phi}_c(t)$ using the Laplace transform approach of (4-60). However, in general this approach is cumbersome. A more tractable technique is to use a computer evaluation of $\mathbf{\Phi}_c(T)$ in (4-66); that is, with $t = T$ and $t_0 = 0$, (4-66) becomes

$$\mathbf{\Phi}_c(T) = \mathbf{I} + \mathbf{A}_c T + \mathbf{A}_c^2 \frac{T^2}{2!} + \mathbf{A}_c^3 \frac{T^3}{3!} + \cdots \tag{4-72}$$

Since this is a convergent series, the series can be truncated with adequate resulting accuracy.

The integral of (4-71), necessary for the computation of $\mathbf{b}$, can also be easily evaluated using a series expansion. In (4-71), let $\tau = T - \sigma$. Hence

$$\begin{aligned} \int_0^T \mathbf{\Phi}_c(T - \sigma)\, d\sigma &= \int_T^0 \mathbf{\Phi}_c(\tau)(-d\tau) = \int_0^T \mathbf{\Phi}_c(\tau)\, d\tau \\ &= \int_0^T \left(\mathbf{I} + \mathbf{A}_c\tau + \mathbf{A}_c^2\frac{\tau^2}{2!} + \mathbf{A}_c^3\frac{\tau^3}{3!} + \cdots\right) d\tau \\ &= \mathbf{I}T + \mathbf{A}_c\frac{T^2}{2!} + \mathbf{A}_c^2\frac{T^3}{3!} + \mathbf{A}_c^3\frac{T^4}{4!} + \cdots \end{aligned} \tag{4-73}$$

Comparing (4-72) with (4-73), we see that the computer program used to evaluate $\mathbf{\Phi}_c(T)$ may also be used to evaluate $\int_0^T \mathbf{\Phi}_c(\tau)\,d\tau$, by expanding the program somewhat. Such a program is given in Appendix I.

In the derivations above, only the single-input, single-output case was considered. For the multiple-input, multiple-output case, it is seen that the vectors $\mathbf{b}_c$, $\mathbf{c}_c$, and $\mathbf{d}_c$ are replaced with matrices $\mathbf{B}_c$, $\mathbf{C}_c$, and $\mathbf{D}_c$, respectively.

The derivations above will now be illustrated by an example.

Example 4.13

For the sampled-data system of Figure 4-19, let $T = 0.1$ s and

$$G(s) = \frac{10}{s(s+1)}$$

A continuous state-variable model of this system is

$$\dot{\mathbf{x}}(t) = \begin{bmatrix} 0 & 1 \\ 0 & -1 \end{bmatrix}\mathbf{x}(t) + \begin{bmatrix} 0 \\ 10 \end{bmatrix}u(t)$$

$$y(t) = [1 \quad 0]\mathbf{x}(t)$$

For this example, since the system is second order, $\mathbf{\Phi}_c(t)$ will be found.

$$\mathbf{\Phi}_c(t) = \mathcal{L}^{-1}\{[s\mathbf{I} - \mathbf{A}_c]^{-1}\}$$

$$= \mathcal{L}^{-1}\begin{bmatrix} s & -1 \\ 0 & s+1 \end{bmatrix}^{-1} = \mathcal{L}^{-1}\begin{bmatrix} \dfrac{1}{s} & \dfrac{1}{s(s+1)} \\ 0 & \dfrac{1}{s+1} \end{bmatrix}$$

$$= \begin{bmatrix} 1 & 1 - \epsilon^{-t} \\ 0 & \epsilon^{-t} \end{bmatrix}$$

Also,

$$\int_0^T \mathbf{\Phi}_c(\tau)\,d\tau = \begin{bmatrix} T & T - 1 + \epsilon^{-T} \\ 0 & 1 - \epsilon^{-T} \end{bmatrix}$$

Then

$$\mathbf{A} = \mathbf{\Phi}_c(T)\big|_{T=0.1} = \begin{bmatrix} 1 & 0.0952 \\ 0 & 0.905 \end{bmatrix}$$

and

$$\mathbf{b} = \left[\int_0^T \mathbf{\Phi}_c(\tau)\,d\tau\right]\mathbf{b}_c = \begin{bmatrix} 0.1 & 0.00484 \\ 0 & 0.0952 \end{bmatrix}\begin{bmatrix} 0 \\ 10 \end{bmatrix}$$

$$= \begin{bmatrix} 0.0484 \\ 0.952 \end{bmatrix}$$

Hence the discrete state equations are

$$\mathbf{x}(k+1) = \begin{bmatrix} 1 & 0.0952 \\ 0 & 0.905 \end{bmatrix}\mathbf{x}(k) + \begin{bmatrix} 0.0484 \\ 0.952 \end{bmatrix}m(k)$$

$$y(k) = [1 \quad 0]\mathbf{x}(k)$$

A simulation diagram of this model is shown in Figure 4-20. Computer program 2, given in Appendix I, may also be utilized to obtain the discrete state matrices. For three-significant-figure accuracy, three terms in the series expansion of $\mathbf{\Phi}_c(t)$ [see (4-72)] are required. Five terms in the series expansion yield six-significant-figure accuracy.

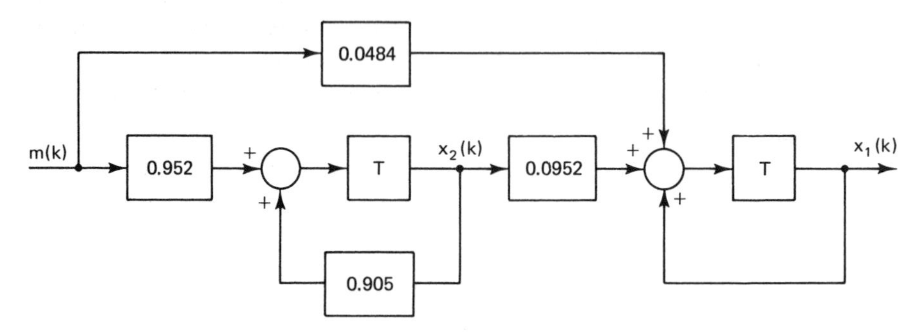

Figure 4-20 Simulation diagram for the system of Example 4.13.

4.11 SUMMARY

In this chapter we have examined various aspects of open-loop discrete-time systems. In particular we discussed the starred transform and showed that it possesses the properties of the z-transform defined in Chapter 2. Next, the starred transform was used to find the pulse transfer function of open-loop systems. The pulse transfer function concepts we extended to analyze open-loop systems containing digital filters. In order to analyze systems containing ideal time delays, the modified z-transform and its properties were derived. Then techniques for finding discrete state-variable models of open-loop systems were presented.

The foundation built in this chapter for open-loop systems will serve as a basis for presenting the analysis of closed-loop discrete-time systems in Chapter 5.

REFERENCES AND FURTHER READING

1. D. M. Look, Direct State Space Formulation of Second-Order Coupled Systems, M. S. thesis, Auburn University, Auburn, Ala., 1971.
2. J. A. Cadzow and H. R. Martens, *Discrete-Time and Computer Control Systems.* Englewood Cliffs, N.J.: Prentice-Hall, Inc., 1970.
3. E. I. Jury, *Sampled-Data Control Systems*, New York: John Wiley & Sons, Inc., 1958.
4. E. I. Jury, *Theory and Application of the z-transform Method.* New York: John Wiley & Sons, Inc., 1964.
5. B. C. Kuo, *Analysis and Synthesis of Sampled-Data Control Systems.* Englewood Cliffs, N.J.: Prentice-Hall, Inc., 1963.

6. B. C. Kuo, *Discrete-Data Control Systems*. Englewood Cliffs, N.J.: Prentice-Hall, Inc., 1970.
7. D. P. Lindorff, *Theory of Sampled-Data Control Systems*. New York: John Wiley & Sons, Inc., 1965.
8. J. R. Ragazzini and G. F. Franklin, *Sampled-Data Control Systems*. New York: McGraw-Hill Book Company, 1958.
9. P. M. DeRusso, R. J. Roy, and C. M. Close, *State Variables for Engineers*. New York: John Wiley & Sons, Inc., 1965.
10. K. Ogata, *State Space Analysis of Control Systems*. Englewood Cliffs, N.J.: Prentice-Hall, Inc., 1967.

PROBLEMS

4-1. With $T = 0.1$ s, find the z-transform of

$$E(s) = \frac{s+1}{(s-1)(s+2)}$$

Compare the pole–zero locations of $E(z)$ in the z-plane of those of $E(s)$ and $E^*(s)$ in the s-plane. Note particularly the locations of the zeros of $E^*(s)$ and the zero of $E(s)$.

4-2. Find the z-transform of the following functions, using z-transform tables. Compare the pole–zero locations of $E(z)$ in the z-plane with those of $E(s)$ and $E^*(s)$ in the s-plane (see Problem 3-2). Let $T = 0.1$ s.

(a) $E(s) = \dfrac{1}{s(s+1)^2}$ **(b)** $E(s) = \dfrac{s}{(s+1)(s+2)}$

(c) $E(s) = \dfrac{s+1}{s(s+2)}$ **(d)** $E(s) = \dfrac{s^2}{(s+1)^2(s+2)}$

(e) $E(s) = \dfrac{s+1}{s^2+25}$ **(f)** $E(s) = \dfrac{s+1}{s^2+2s+26}$

4-3. Shown in Figure P4-3 is a closed-loop temperature control system. For this problem, the microcomputer output $m(kT)$ is specified, and hence the system is open loop. The microcomputer output controls the position of a solenoid value, which in turn controls the amount of steam into the tank coil. Thus the microcomputer controls the temperature of the liquid in the tank.

(a) If $m(kT)$ is constant for a long period of time, the process temperature becomes constant. Then, if $m(kT)$ changes, the process temperature changes. Since the system is linear, the change in temperature can be calculated by considering only the change in $m(kT)$. Let this change in $m(kT)$ be a unit step which occurs at $t = 0$. Solve for the change in process temperature $\theta(kT)$. Let $T = 0.2$ s.

(b) Solve for the final value of temperature in part (a) using two different procedures. One of the procedures must be independent of the z-transform analysis.

(c) Verify your results of part (a) by determining the input to the power amplifier and then calculating $\theta(t)$ by continuous-data techniques.

(d) The gain of the power amplifier is to be changed such that a change in the D/A output of 1 V results in a steady-state change in the process temperature of 30°C. Find the gain required.

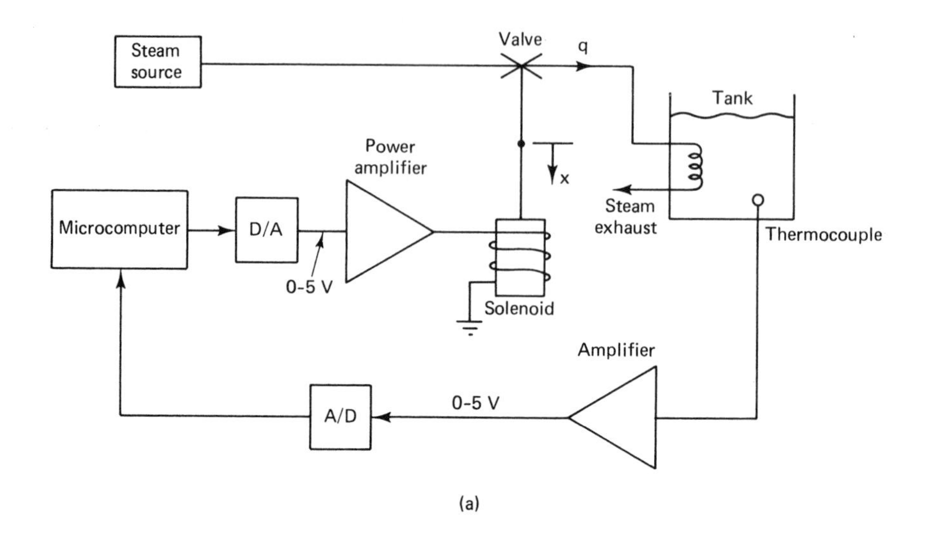

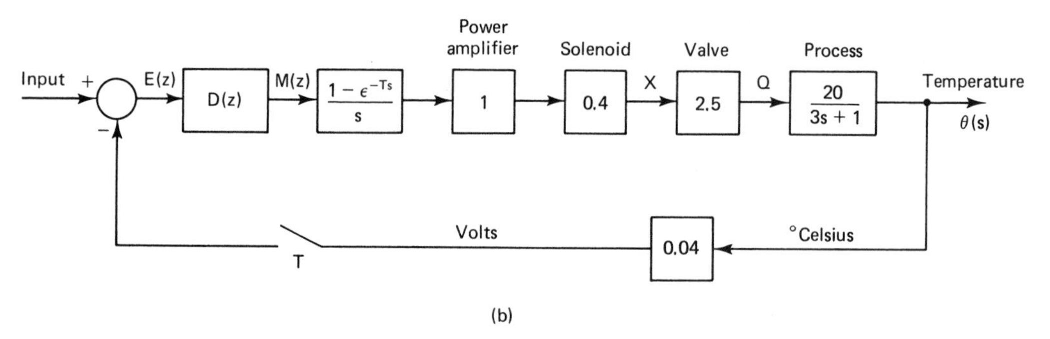

Figure P4-3 Temperature control system.

4-4. **(a)** Find the system response at the sampling instants to a unit step input for the system of Figure P4-4. Plot $c(nT)$ versus time.

(b) Verify your results of part (a) by determining the input to the plant, $m(t)$, and then calculating $c(t)$ by continuous-data techniques.

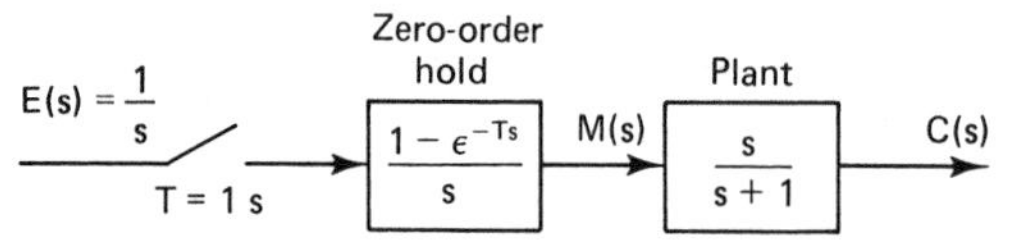

Figure P4-4 System for Problem 4-4.

4-5. Find the system response at the sampling instants to a unit step input for the system of Figure P4-5. Plot $c(nT)$ versus time.

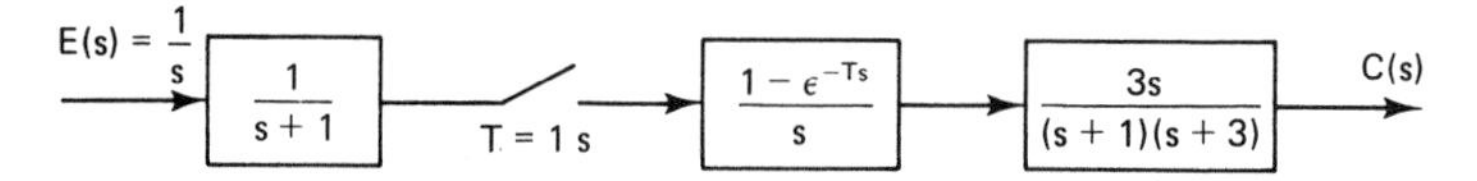

Figure P4-5 System for Problem 4-5.

4-6. Express $C(s)$ and $C(z)$ as a function of the input for the system of Figure P4-6.

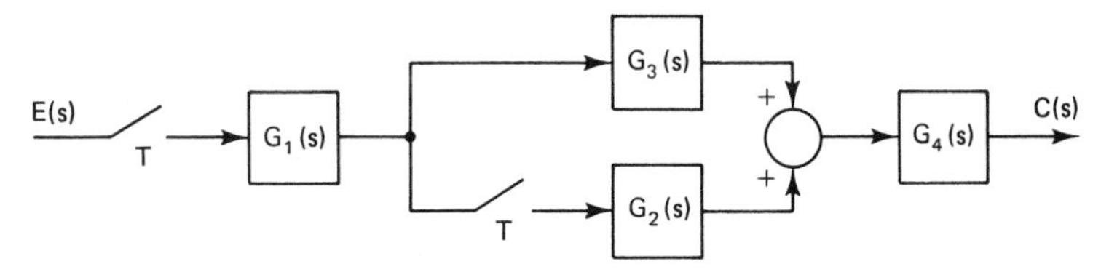

Figure P4-6 System for Problem 4-6.

4-7. Shown in Figure P4-7 is the block diagram of a simplified cruise-control system for an automobile. The actuator controls the throttle position, the carburetor is modeled as a first-order lag with a 1-s time constant, and the total load on the engine is also modeled

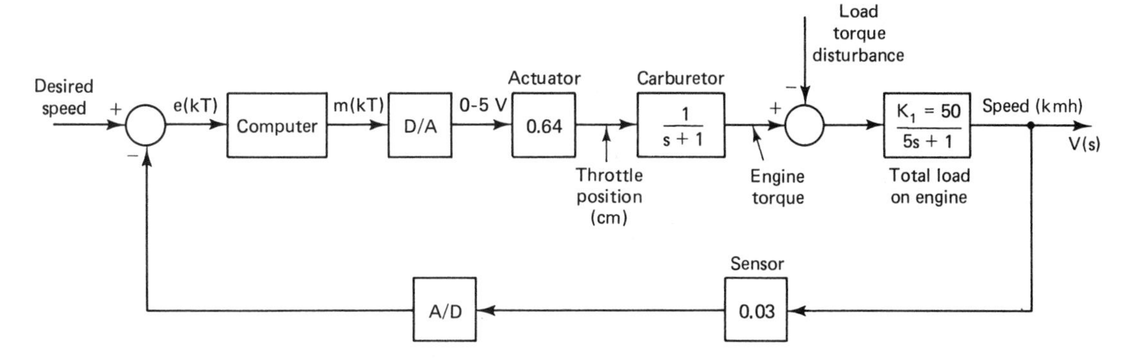

Figure P4-7 Cruise control system.

as a first-order lag with a 5-s time constant. For this problem, the computer output $m(kT)$ is constant; hence the system is open-loop. Also, assume that the disturbance torque is zero. Let $T = 0.2$ s.

(a) Calculate the transfer function

$$G(z) = \frac{V(z)}{M(z)}$$

(b) If the car's speed is constant at 60 kilometers per hour (kmh), find the (constant) value of $m(kT)$.

(c) It is desired to increase the car's speed to 80 kmh. Find the value of $m(kT)$ required.

(d) Parts (b) and (c) can be solved without calculating $G(z)$. Do this.

(e) The computer output $m(kT)$ has been constant at 2.5 for a long time prior to $t = 0$. At $t = 0$, $m(kT)$ is increased to the constant value of 3.0. Plot $v(kT)$ if $T = 0.2$ s. Since the system is linear, the same change in speed occurs if $m(kT)$ is changed from 0 to 0.5 at $t = 0$.

4-8. Repeat Problem 4-7 with the carburetor transfer function set to unity, that is, with the time lag of the carburetor ignored.

4-9. Find the modified z-transform of the output for the system of Problem 4-4 for $m = 0.7$. Plot the response on the same graph as plotted in Problem 4-4.

4-10. Find the modified z-transform of the following functions.

(a) $E(s) = \dfrac{1}{s(s+1)^2}$ **(b)** $E(s) = \dfrac{s}{(s+1)(s+2)}$

(c) $E(s) = \dfrac{s+1}{s(s+2)}$ (d) $E(s) = \dfrac{s^2}{(s+1)^2(s+2)}$

(e) $E(s) = \dfrac{s+1}{s^2+25}$ (f) $E(s) = \dfrac{s+1}{s^2+2s+26}$

4-11. Find the z-transform of the following functions. The results of Problem 4-10 may be useful.

(a) $E(s) = \dfrac{\epsilon^{-0.3Ts}}{s(s+1)^2}$ (b) $E(s) = \dfrac{s\epsilon^{-0.6Ts}}{(s+1)(s+2)}$

(c) $E(s) = \dfrac{(s+1)\epsilon^{-0.75Ts}}{s(s+2)}$ (d) $E(s) = \dfrac{s^2\epsilon^{-0.3Ts}}{(s+1)^2(s+2)}$

(e) $E(s) = \dfrac{(s+1)\epsilon^{-0.1Ts}}{s^2+25}$ (f) $E(s) = \dfrac{(s+1)\epsilon^{-0.2Ts}}{s^2+2s+26}$

4-12. For the system of Figure P4-12, the digital filter solves the difference equation

$$m(k+1) = 0.5e(k+1) - (0.5)(0.99)e(k) + 0.995m(k)$$

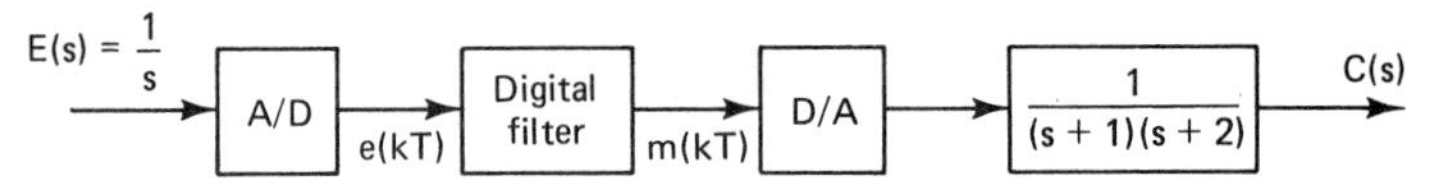

Figure P4-12 System for Problem 4-12.

The sampling rate is 25 Hz.

(a) Find $C(z)$.

(b) Find $c(0)$ and $\lim_{k\to\infty} c(kT)$.

4-13. For the system of Figure P4-12, assume that the digital filter has a computation delay of 5 ms. Repeat Problem 4-12 for this system.

4-14. Note the nonsynchronous sampling in the system of Figure P4-14. Solve for $C(z)$ for this system.

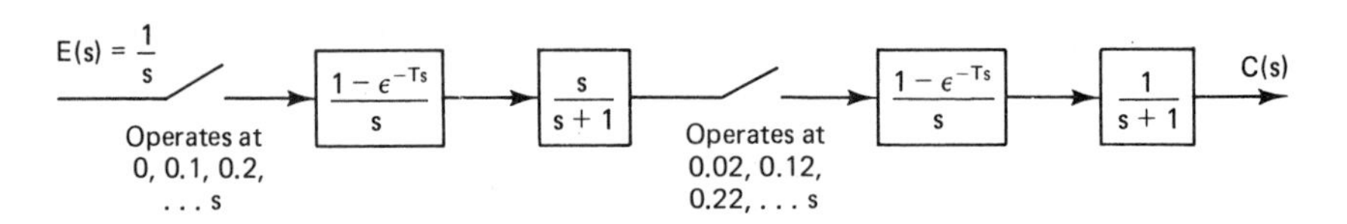

Figure P4-14 System for Problem 4-14.

4-15. Find a discrete-state-variable representation for the system shown in Figure **P4-15**.

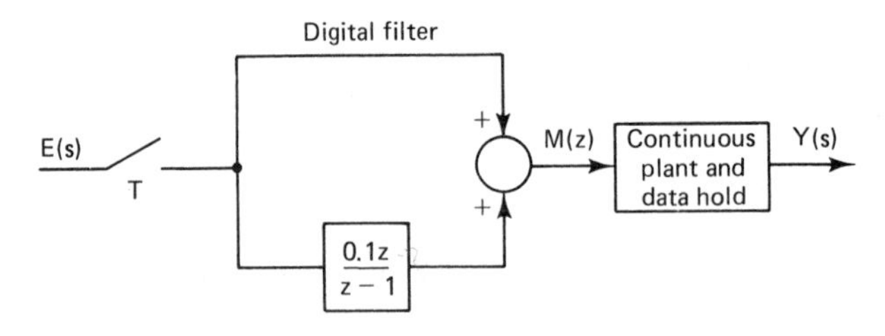

Figure P4-15 System for Problem 4-15.

A discrete-state-variable description of the continuous system is given by

$$\mathbf{x}(k+1) = \begin{bmatrix} 0.9 & 1 & 0 \\ 0 & 0.9 & 0 \\ 0 & 0 & 0.5 \end{bmatrix} \mathbf{x}(k) + \begin{bmatrix} 0 \\ 1 \\ 1 \end{bmatrix} m(k)$$

$$y(k) = [1 \quad 1.5 \quad 1.3]\mathbf{x}(k)$$

4-16. For the temperature control system of Problem 4-3, the digital filter has the transfer function

$$D(z) = 1.2 + \frac{0.1z}{z-1}$$

Find a discrete-state-variable model of the open-loop system with $e(k)$ as the input and $\theta(k)$ as the output, using a transfer-function approach. Let $T = 0.2$ s.

4-17. Consider the system of Problem 4-16.

(a) Find a discrete-state-equation model for the continuous-time part of the system (from the D/A output to the system output), obtaining the discrete state equations from the continuous state equations.

(b) Using the results of part (a), find a discrete-state-equation model of the open-loop system with $e(k)$ as the input and $\theta(k)$ as the output.

4-18. For the automobile cruise-control system of Problem 4-7, the digital filter solves the difference equation

$$m(k) = 2e(k) - 1.9796e(k-1) + 0.99m(k-1)$$

Find a discrete-state-variable model of the open-loop system with $e(k)$ as the input and speed, $v(k)$, as the output. In the state model, choose speed and engine torque as two of the three state variables. In addition, ignore the disturbance torque. Let $T = 0.2$ s.

4-19. For the system shown in Figure P4-19, the filter solves the difference equation

$$m(k) = -m(k-1) + e(k-1)$$

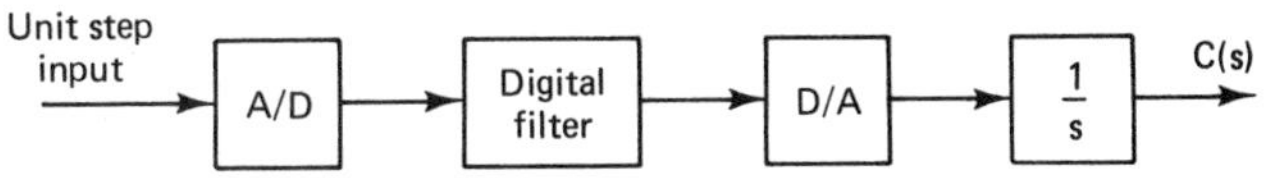

Unit step input
A/D
Digital filter
D/A
$\frac{1}{s}$
C(s)

Figure P4-19 System for Problem 4-19.

The sampling rate is $f_s = 1$ Hz.

(a) Find $C(z)$.

(b) Find $c(kT)$.

4-20. Consider the system described in Problem 4-19.

(a) Using the technique developed in Section 4.10, find a discrete state model for the system.

(b) From the state equations, solve for $c(kT)$, $k = 0, 1, 2, 3, 4, 5$.

4-21. Find a discrete state model for the system of Figure P4-12. The states of the plant are to be the plant output and the derivative of the output. The sampling rate is 25 Hz, and the digital filter is described by the difference equation

$$m(k) = 11.44e(k) - 10.77e(k-1) + 0.33m(k-1)$$

4-22. The plant in Figure P4-22 is described by the differential equation

$$\frac{d^2y}{dt^2} + 0.2\frac{dy}{dt} + 0.02y = \bar{m}(t)$$

Find a discrete state model for this system, with $T = 2$ s.

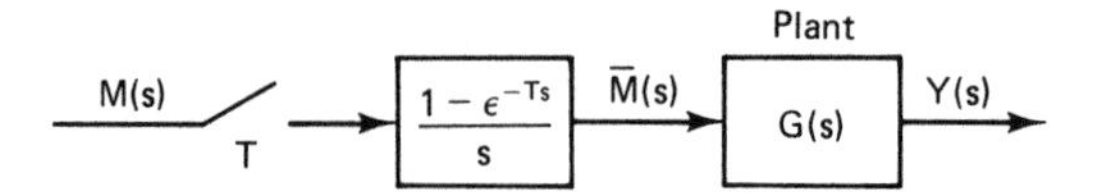

Figure P4-22 System for Problem 4-22.

4-23. Consider the system of Figure P4-23. The filter transfer function is $D(z)$.

(a) Express $C(z)$ as a function of E.

(b) A discrete state model of this system does not exist. Why?

(c) What assumptions concerning $e(t)$ must be made in order to derive an approximate discrete state model.

Figure P4-23 System for Problem 4-23.

4-24. The model of a continuous system with algebraic loops is given as

$$\dot{x}_1(t) = -x_1(t) + 2\dot{x}_2(t) + u_1(t)$$

$$\dot{x}_2(t) = -\dot{x}_2(t) - x_2(t) - \dot{x}_1(t) + x_1(t) + u_2(t)$$

$$y(t) = \dot{x}_2(t)$$

Derive the state equations of this system using the matrix technique of Section 4.9, and check your results using Mason's gain formula.

5

Closed-Loop Systems

5.1 INTRODUCTION

In Chapter 4 a technique was developed for determining the output functions of open-loop discrete-time systems. In this chapter we extend these concepts and employ the signal flow graph to determine the output functions of closed-loop systems. Although methods other than the one given here exist for determining output functions [1, 2], the authors feel that the technique presented in the following material is the simplest, most practical, and easiest to remember. Finally, a technique for developing state-variable models of closed-loop discrete control systems is presented.

5.2 PRELIMINARY CONCEPTS

Before considering simple closed-loop systems, open-loop systems with cascaded plants will be reviewed. As shown in Chapter 4, the transfer function for the system of Figure 5-1a is

$$C(z) = G_1(z)G_2(z)E(z) \tag{5-1}$$

and that of Figure 5-1b is

$$C(z) = \overline{G_1G_2}(z)E(z) \tag{5-2}$$

For the system of Figure 5-1c,

$$C(s) = G_2(s)A^*(s) = G_2(s)\overline{G_1E}^*(s) \tag{5-3}$$

and

$$C(z) = G_2(z)\overline{G_1E}(z) \tag{5-4}$$

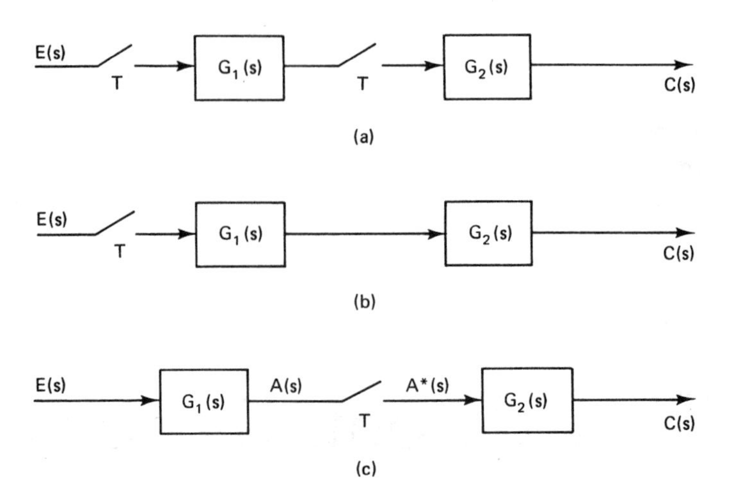

Figure 5-1 Open-loop sampled-data systems.

For this case, no transfer function can be found since $E(z)$ cannot be factored from $\overline{G_1E}(z)$. In general, no transfer function can be written for the system in which the input is applied to a continuous-time element before being sampled. However, the output can always be expressed as a function of the input, and as will be shown later, this type of system presents no particular difficulties in either analysis or design.

Consider next the closed-loop system of Figure 5-2. Now

$$C(s) = G(s)E^*(s) \tag{5-5}$$

and

$$E(s) = R(s) - H(s)C(s) \tag{5-6}$$

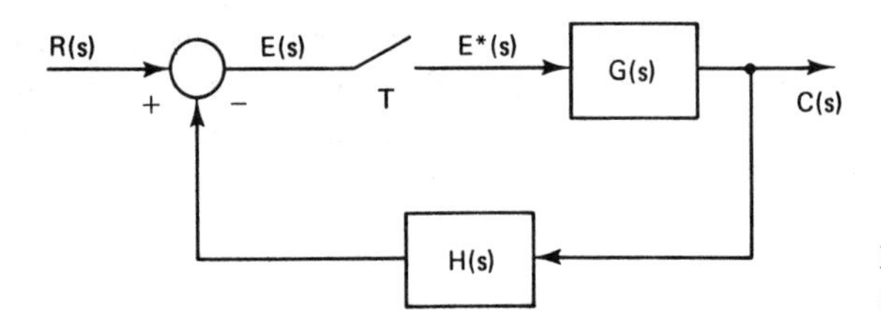

Figure 5-2 Closed-loop sampled-data system.

Substituting (5-5) into (5-6), we obtain

$$E(s) = R(s) - G(s)H(s)E^*(s) \tag{5-7}$$

and, by taking the starred transform (see Section 4.3),

$$E^*(s) = R^*(s) - \overline{GH}^*(s)E^*(s) \tag{5-8}$$

Solving for $E^*(s)$, we obtain

$$E^*(s) = \frac{R^*(s)}{1 + \overline{GH}^*(s)} \tag{5-9}$$

and, from (5-5),

$$C(s) = G(s)\frac{R^*(s)}{1 + \overline{GH}^*(s)} \tag{5-10}$$

which yields an expression for the continuous output. The sampled output is, then,

$$C^*(s) = \frac{G^*(s)R^*(s)}{1 + \overline{GH}^*(s)} \tag{5-11}$$

Trouble can be encountered in deriving the transfer function of a closed-loop system. This can be illustrated for the case above. Suppose that (5-6) had been starred and substituted into (5-5). Then

$$C(s) = G(s)R^*(s) - G(s)\overline{HC}^*(s) \tag{5-12}$$

and $C^*(s)$ is

$$C^*(s) = G^*(s)R^*(s) - G^*(s)\overline{HC}^*(s) \tag{5-13}$$

Now in general $C^*(s)$ cannot be factored from $\overline{HC}^*(s)$. Thus (5-13) cannot be solved for $C^*(s)$.

In general, in analyzing a system, an equation should not be starred if a system variable is lost as a factor, as shown above. However, for systems more complex than the one above, solving the system equations can become quite complex. Therefore, to avert this problem, a simpler method of analysis will be developed.

First, however, consider the system of Figure 5-3. Since the input is not sampled before being applied to a continuous-time element, no transfer function can be derived.

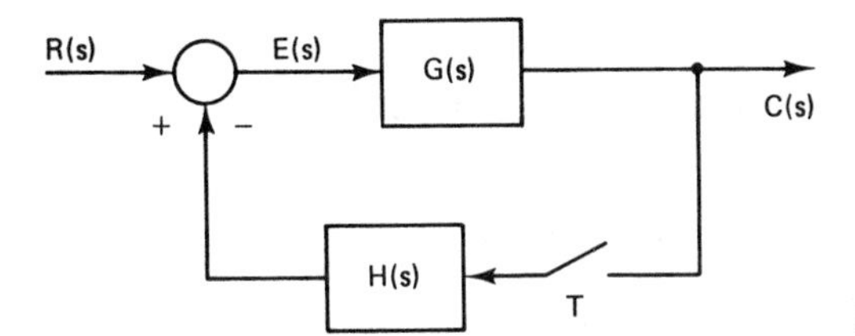

Figure 5-3 Sampled-data system.

Nevertheless, the system can be analyzed. From Figure 5-3 we note that

$$C(s) = G(s)E(s) \tag{5-14}$$

and

$$E(s) = R(s) - H(s)C^*(s) \tag{5-15}$$

Substituting (5-15) into (5-14), we obtain

$$C(s) = G(s)R(s) - G(s)H(s)C^*(s) \tag{5-16}$$

Starring (5-16) yields

$$C^*(s) = \overline{GR}^*(s) - \overline{GH}^*(s)C^*(s) \tag{5-17}$$

In this system, the forcing function $R(s)$ is necessarily lost as a factor in (5-17). Solving (5-17) for $C^*(s)$, we obtain

$$C^*(s) = \frac{\overline{GR}^*(s)}{1 + \overline{GH}^*(s)} \tag{5-18}$$

The continuous output is obtained from (5-16) and (5-18) as

$$C(s) = G(s)R(s) - \frac{G(s)H(s)\overline{GR}^*(s)}{1 + \overline{GH}^*(s)} \tag{5-19}$$

For the system of Figure 5-3, then, no transfer function can be derived. In a practical system, we generally have more than one input. Consider, as an example, an aircraft flying in a closed-loop mode (i.e., flying on autopilot). Suppose also that the autopilot is implemented digitally. Thus this system is a closed-loop digital control system. The commands into this system (e.g., an altitude change) may be inputted into the control computer through an analog-to-digital converter, or may even be generated by the computer itself. Thus this input is sampled, and a transfer function may be developed. However, the vertical component of the wind, which must also be considered as an input into the altitude control system, is not sampled. Hence a transfer function cannot be developed from wind input to aircraft altitude as the output.

5.3 DERIVATION PROCEDURE

The determination of transfer functions for sampled-data systems is difficult because a transfer function for the ideal sampler does not exist. However, for the case in which every signal within the system is sampled, continuous-system procedures are applicable and, using the techniques illustrated in Appendix II, one can usually write the expression for the transfer function by inspection using Mason's gain formula. For example, consider the system shown in Figure 5-4a. The sampled signal flow graph is shown in Figure 5-4b. The simplicity of this flow graph is due to the fact that there is a sampler in each path. Applying Mason's gain formula, the transfer function is

$$\frac{C^*(s)}{R^*(s)} = \frac{G^*(s)}{1 + G^*(s)H^*(s)} \tag{5-20}$$

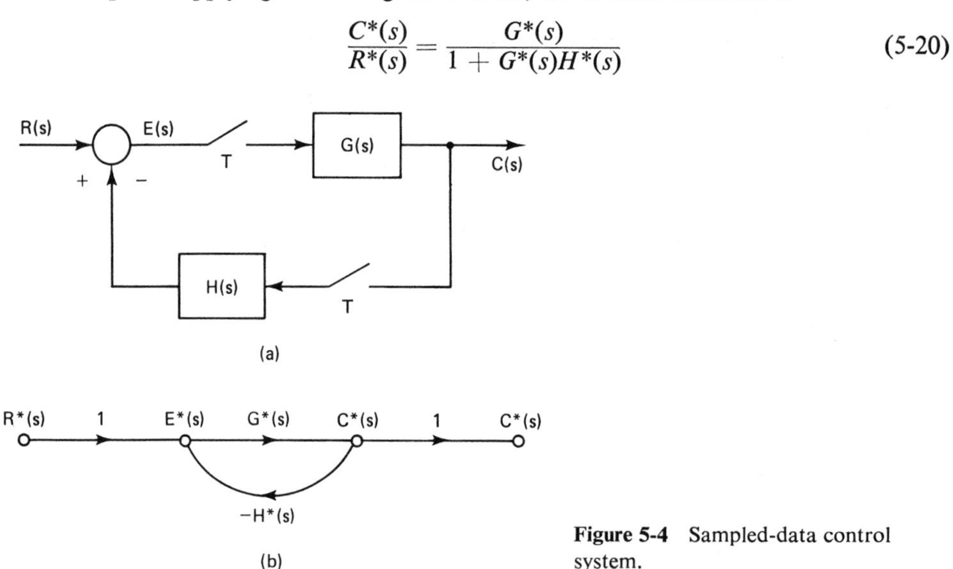

Figure 5-4 Sampled-data control system.

Consider now a system that does not have a sampler in each path, as shown in Figure 5-5. We will omit the sampler from the system signal flow graph, since a transfer function cannot be written for the device. The flow graph is shown in Figure 5-6. Note that the effect of the sampler is included in the flow graph, since the sampler output is shown in starred form, E_1^*, and E_1^* will be treated as an input. This flow graph is referred to as the original signal flow graph. The sampled output $C^*(s)$ can be found for a discrete-time system of this type by employing the following step-by-step procedure.

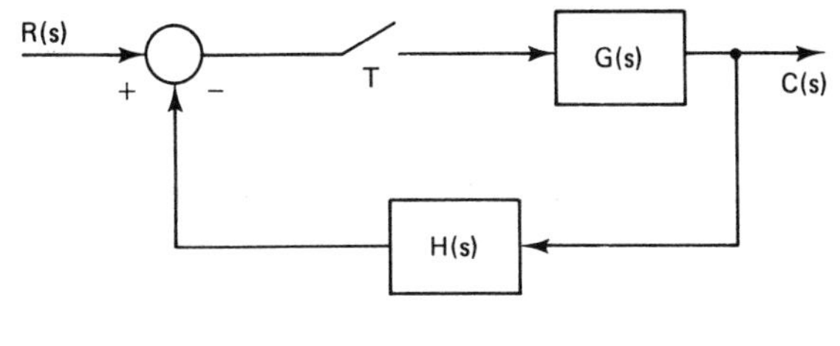

Figure 5-5 Sampled-data control system.

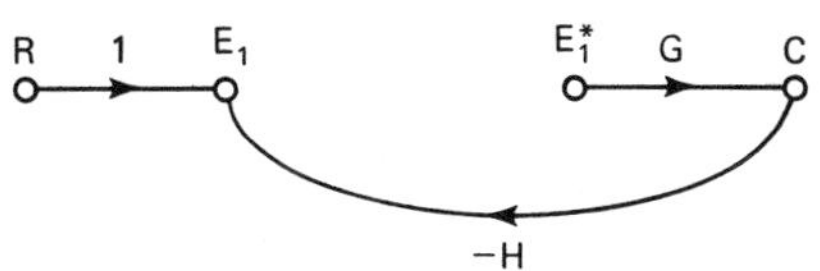

Figure 5-6 Original flow graph for the sampled-data system in Figure 5-5.

1. Construct the original signal flow graph. This has been done for the system and is shown in Figure 5-6.
2. Assign a variable to each sampler input. Then the sampler output is this variable starred. For this example system E_1 is the input to the sampler and E_1^* is the sampler output.
3. Considering each sampler output to be a source, express the sampler inputs and the system output in terms of each sampler output (which is treated as an input in the flow graph), and the system input. For this example,

$$E_1 = R - GHE_1^* \tag{5-21}$$

$$C = GE_1^* \tag{5-22}$$

   where $E_1 = E_1(s)$, and so on. For convenience, the dependency on s will not be shown.
4. Take the starred transform of these equations and solve by any convenient method. For the example,

$$E_1^* = R^* - \overline{GH}^*E_1^* \tag{5-23}$$

$$C^* = G^*E_1^* \tag{5-24}$$

   From (5-23),

$$E_1^* = \frac{R^*}{1 + \overline{GH}^*} \tag{5-25}$$

   and, from (5-24) and (5-25),

$$C^* = \frac{G^*}{1 + \overline{GH}^*} R^* \tag{5-26}$$

If there is more than one sampler in the system, the equations for the starred variables may be solved using Cramer's rule, or any other technique that is applicable to the solution of linear simultaneous equations.

The systems equations can also be solved by constructing a signal flow graph from these equations and applying Mason's gain formula. This method is sometimes superior to Cramer's rule, provided that the sampled signal flow graph is simple enough so that all loops are easily identified. To illustrate this approach, consider again equations (5-22), (5-23), and (5-24). The signal flow graph for these equations is shown in Figure 5-7. From the flow graph,

$$C^* = \frac{G^*}{1 + \overline{GH}^*} R^* \tag{5-27}$$

and this result agrees with (5-26). This method for finding the output function for closed-loop systems will now be illustrated via the following examples.

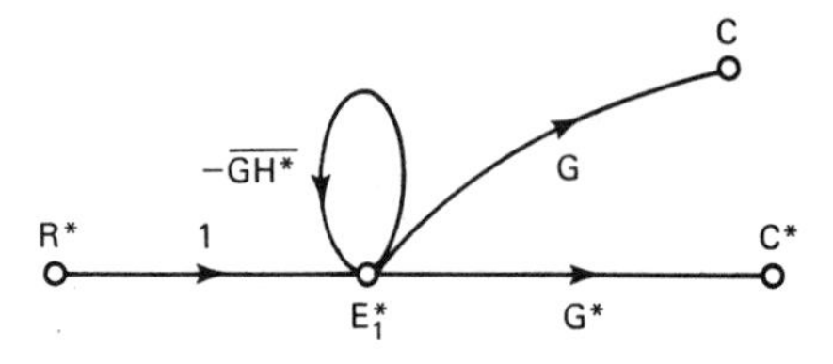

Figure 5-7 Sampled signal flow graph for the system in Figure 5-5.

Example 5.1

Consider the system shown in Figure 5-8. The original signal flow graph is shown in Figure 5-9. The system equations are

$$E_1 = R - G_2 E_2^*$$
$$E_2 = G_1 E_1^* - G_2 H E_2^*$$
$$C = G_2 E_2^*$$

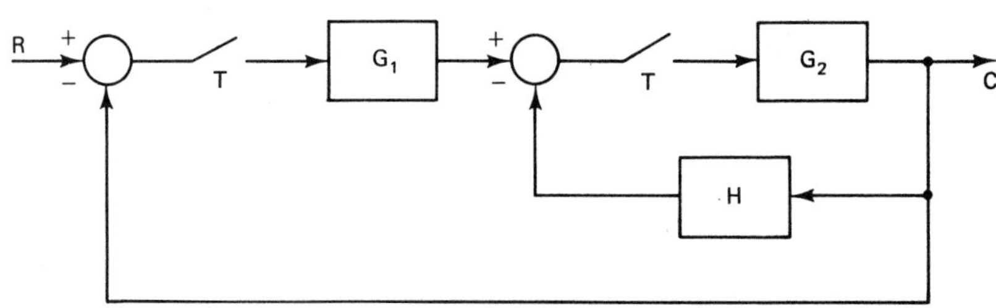

Figure 5-8 Sampled-data system for Example 5.1.

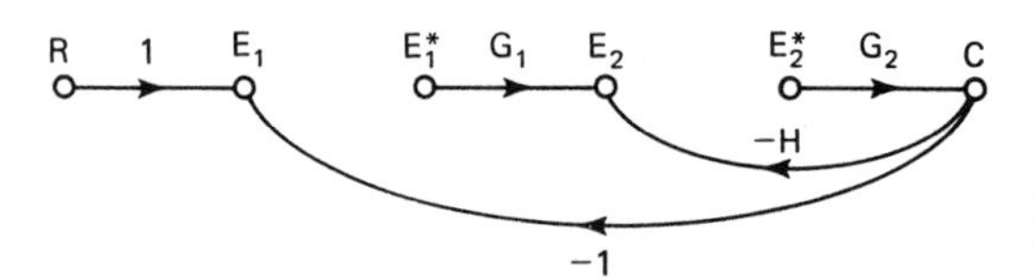

Figure 5-9 Original flow graph for Example 5.1.

Starring these equations, we obtain

$$E_1^* = R^* - G_2^* E_2^*$$
$$E_2^* = G_1^* E_1^* - \overline{G_2 H}^* E_2^*$$
$$C^* = G_2^* E_2^*$$

The sampled flow graph can then be drawn from these equations as shown in Figure 5-10. Then applying Mason's gain formula, we obtain

$$C^* = \frac{G_1^* G_2^*}{1 + G_1^* G_2^* + \overline{G_2 H}^*} R^* \qquad \text{or} \qquad C(z) = \frac{G_1(z) G_2(z) R(z)}{1 + G_1(z) G_2(z) + \overline{G_2 H}(z)}$$

and

$$C = \frac{G_2 G_1^*}{1 + G_1^* G_2^* + \overline{G_2 H}^*} R^*$$

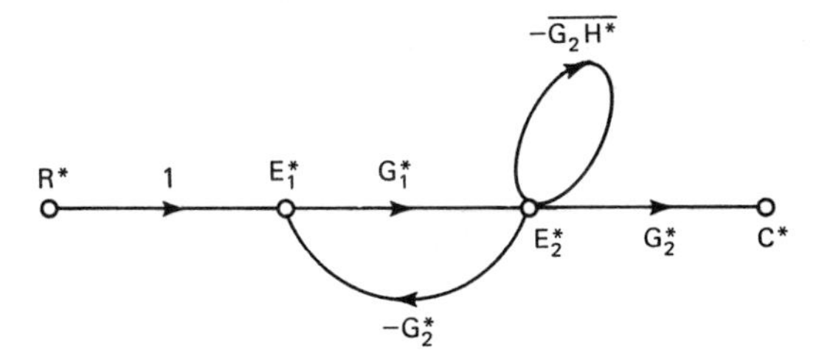

Figure 5-10 Sampled flow graph for Example 5.1.

Example 5.2

As another example, consider the system shown in Figure 5-11. Note that no transfer function may be derived for this system.

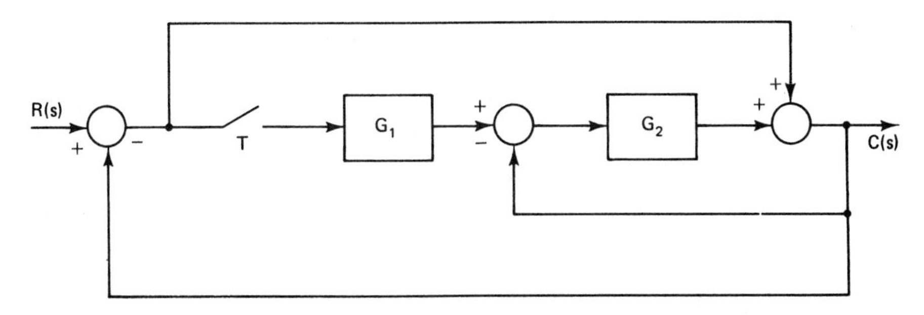

Figure 5-11 Sampled-data system for Example 5.2.

The original signal flow graph is given in Figure 5-12. From this figure note that

$$E_1 = R - C$$

$$C = R - C + G_1G_2E_1^* - G_2C$$

and therefore

$$C = \frac{R}{2 + G_2} + \frac{G_1G_2}{2 + G_2}E_1^*$$

$$E_1 = \frac{(1 + G_2)R}{2 + G_2} - \frac{G_1G_2}{2 + G_2}E_1^*$$

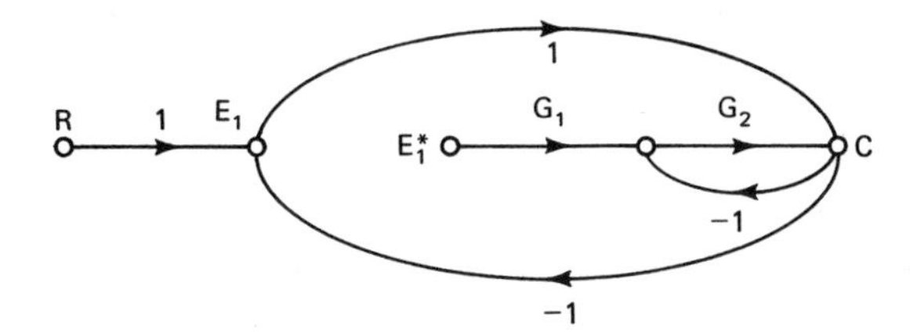

Figure 5-12 Original flow graph for Example 5.2.

Or the equation for E_1 and C can be written directly from Figure 5-12 using Mason's formula. Hence

$$C^* = \left[\frac{R}{2 + G_2}\right]^* + \left[\frac{G_1G_2}{2 + G_2}\right]^* E_1^*$$

$$E_1^* = \left[\frac{(1 + G_2)R}{2 + G_2}\right]^* - \left[\frac{G_1G_2}{2 + G_2}\right]^* E_1^*$$

The sampled flow graph derived from these equations is given in Figure 5-13. Then by employing Mason's gain formula, we obtain

$$C^* = \left[\frac{R}{2+G_2}\right]^* + \frac{\left[\dfrac{G_1G_2}{2+G_2}\right]^*}{1+\left[\dfrac{G_1G_2}{2+G_2}\right]^*}\left[\frac{(1+G_2)R}{2+G_2}\right]^*$$

Note that, as stated above, no transfer function is possible, since the input is fed into a continuous element in the system without first being sampled.

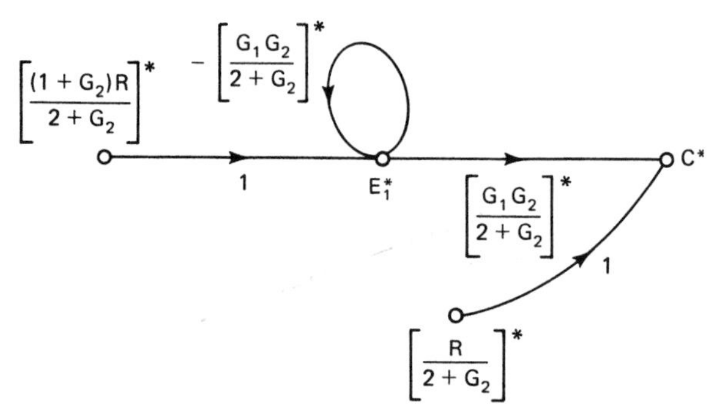

Figure 5-13 Sampled flow graph for Example 5.2.

As mentioned above, matrix methods, or Cramer's rule, may be simpler to use in solving the system equations, if there are many loops in the sampled signal flow graph. This is especially true when all the loops of the flow graph are not easily identified. However, once the reader has gained experience and proficiency in the use of signal flow graphs, these flow graphs can be easily used to obtain quick and accurate solutions to the simpler transfer-function problems.

Example 5.3

Consider the system shown in Figure 5-14, which contains both a feed-forward path and a digital controller. The original flow graph for the system is shown in Figure 5-15.

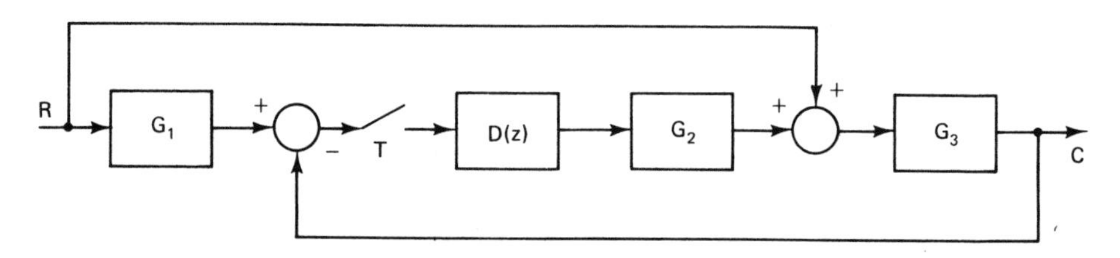

Figure 5-14 Sampled-data system for Example 5.3.

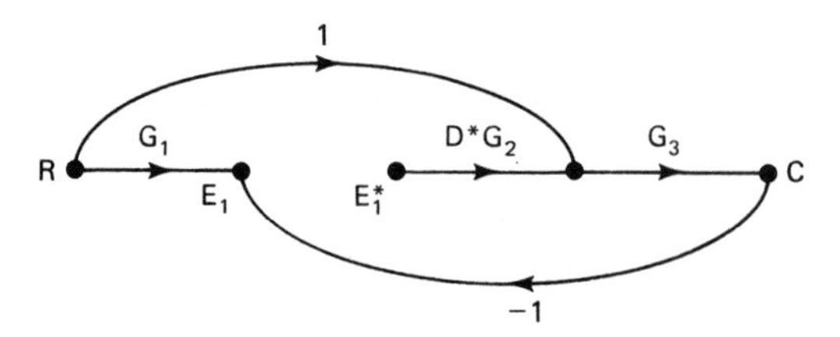

Figure 5-15 Original flow graph for Example 5.3.

The following equations are obtained from this figure.

$$E_1 = RG_1 - RG_3 - D^*G_2G_3E_1^*$$
$$C = RG_3 + D^*G_2G_3E_1^*$$

Hence

$$E_1^* = (RG_1)^* - (RG_3)^* - D^*(G_2G_3)^*E_1^*$$
$$C^* = (RG_3)^* + D^*(G_2G_3)^*E_1^*$$

The equations above can be used to draw the sampled flow graph shown in Figure 5-16, and from this graph via Mason's gain formula we obtain

$$C^* = (RG_3)^* + D^*(G_2G_3)^*\left[\frac{(RG_1)^* - (RG_3)^*}{1 + D^*(G_2G_3)^*}\right] = \frac{D^*(G_2G_3)^*(RG_1)^* + (RG_3)^*}{1 + D^*(G_2G_3)^*}$$

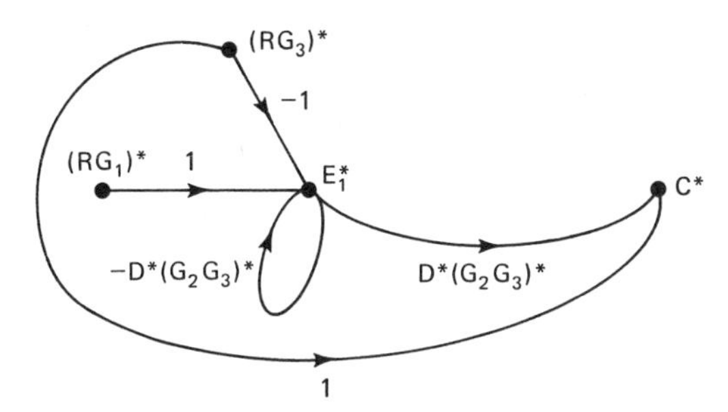

Figure 5-16 Sampled flow graph for Example 5.3.

Example 5.4

As a final example, consider the system of Figure 5-17, which contains a digital controller with a computation time of t_0 seconds, $t_0 < T$. The effect of the computation time is modeled as the ideal time delay (see Chapter 4).

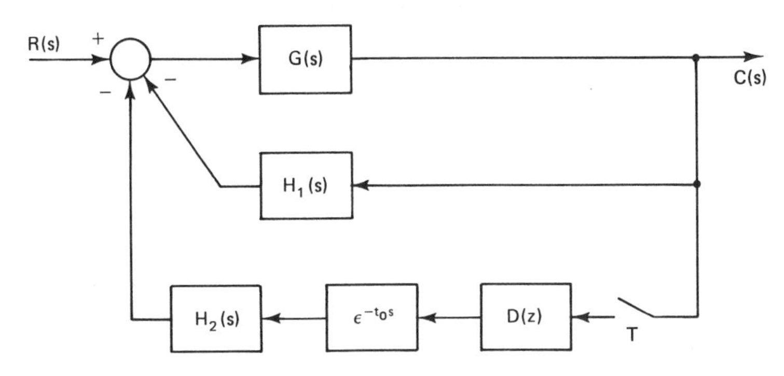

Figure 5-17 System for Example 5.4.

The original signal flow graph is given in Figure 5-18. The system equation is then

$$C = \frac{GR}{1 + GH_1} - \frac{GH_2\epsilon^{-t_0s}}{1 + GH_1}D^*C^*$$

and thus

$$C(z) = \left[\frac{GR}{1 + GH_1}\right](z) - \left[\frac{GH_2}{1 + GH_1}\right](z, m)D(z)C(z)$$

where $mT = T - t_0$. The second equation may be solved directly for $C(z)$.

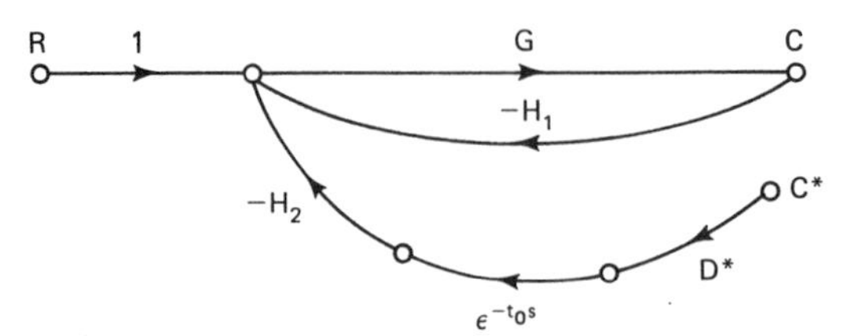

Figure 5-18 Original flow graph for Example 5.4.

$$C(z) = \frac{\left[\dfrac{GR}{1 + GH_1}\right](z)}{1 + \left[\dfrac{GH_2}{1 + GH_1}\right](z, m)D(z)}$$

If the computational delay is greater than one sampling interval (i.e., if $t_0 = kT + \Delta T$, k is a positive integer), the denominator would contain the terms

$$1 + z^{-k}\left[\frac{GH_2}{1 + GH_1}\right](z, m)D(z)$$

where $mT = T - \Delta T$.

5.4 STATE-VARIABLE MODELS

A technique for finding the transfer function of a closed-loop discrete-time system was presented above. As shown in Chapter 4, a discrete-state-variable model may be generated directly from the transfer function. However, this technique has the disadvantage that the system physical variables generally do not appear as discrete state variables.

The technique of converting continuous state equations to discrete state equations, presented in Section 4.10, may also be utilized in determining a discrete-state-variable model for a closed-loop system. The application of this technique will be illustrated by an example. Consider the system of Example 5.1, which is repeated in Figure 5-19. As a first step, the system is redrawn such that zero-order-hold outputs

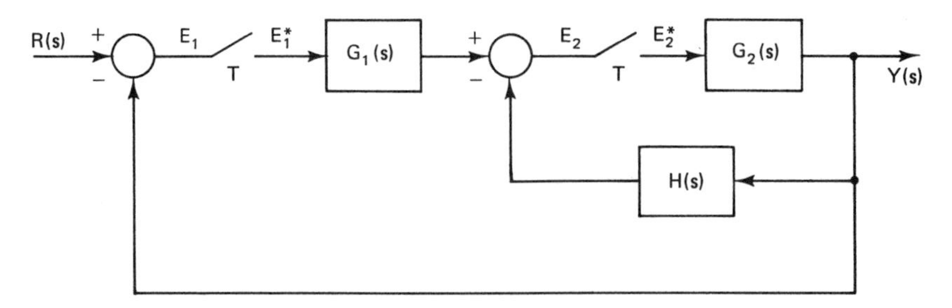

Figure 5-19 Closed-loop system.

are shown as inputs, and sampler inputs and the system output are shown as outputs. The result of performing this operation is shown in Figure 5-20a (e_1 is not shown as an output, since it is determined directly from y). Next the continuous state equations for this system are written, and from these equations, the discrete state equations are

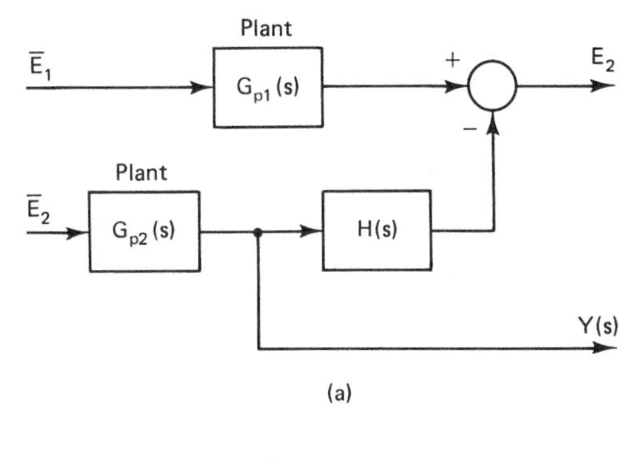

(a)

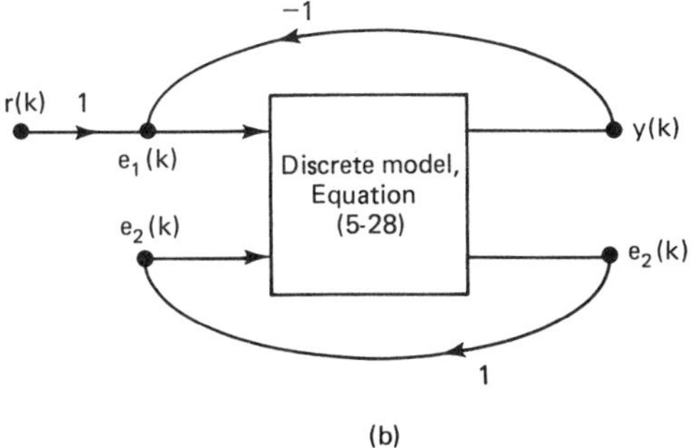

(b)

Figure 5-20 Technique for determining the discrete state model.

generated. For this system, the discrete state equations will be of the form

$$\mathbf{v}(k+1) = \mathbf{A}_1\mathbf{v}(k) + \mathbf{B}_1\begin{bmatrix} e_1(k) \\ e_2(k) \end{bmatrix}$$
$$\begin{bmatrix} y(k) \\ e_2(k) \end{bmatrix} = \mathbf{C}_1\mathbf{v}(k) + \mathbf{D}_1\begin{bmatrix} e_1(k) \\ e_2(k) \end{bmatrix} \tag{5-28}$$

Either a discrete simulation diagram or a flow graph is then constructed from these state equations, and should include all connecting paths of the closed-loop system external to the simulation diagram for (5-28). The result for the system considered is shown in Figure 5-20b. From this simulation diagram the system discrete state equations may be written. An example to illustrate this technique will now be given.

Example 5.5

The discrete state model for the system of Figure 5-19 will be derived, with $T = 0.1$ s, and with

$$G_1(s) = \frac{1-\epsilon^{-Ts}}{s^2(s+1)} = \left[\frac{1-\epsilon^{-Ts}}{s}\right]G_{p1}(s)$$

$$G_2(s) = \frac{2(1-\epsilon^{-Ts})}{s(s+2)}$$

$$H(s) = \frac{10}{s+10}$$

A flow graph of this system is shown in Figure 5-21a. From this flow graph we write the

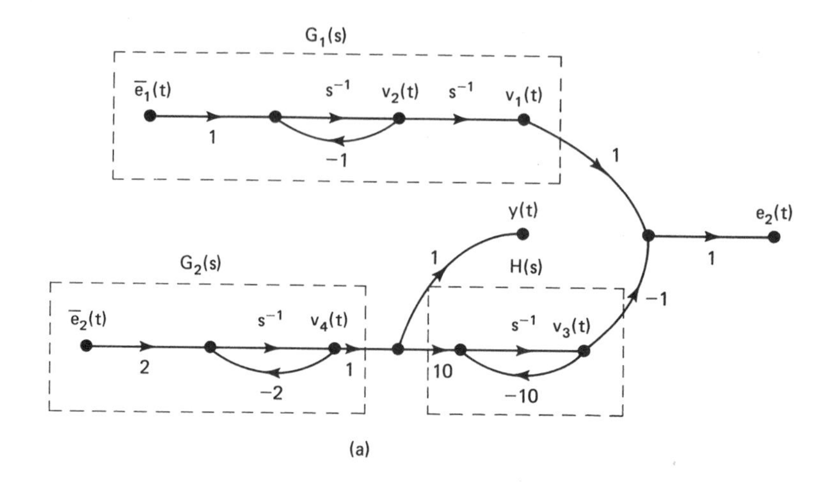

(a)

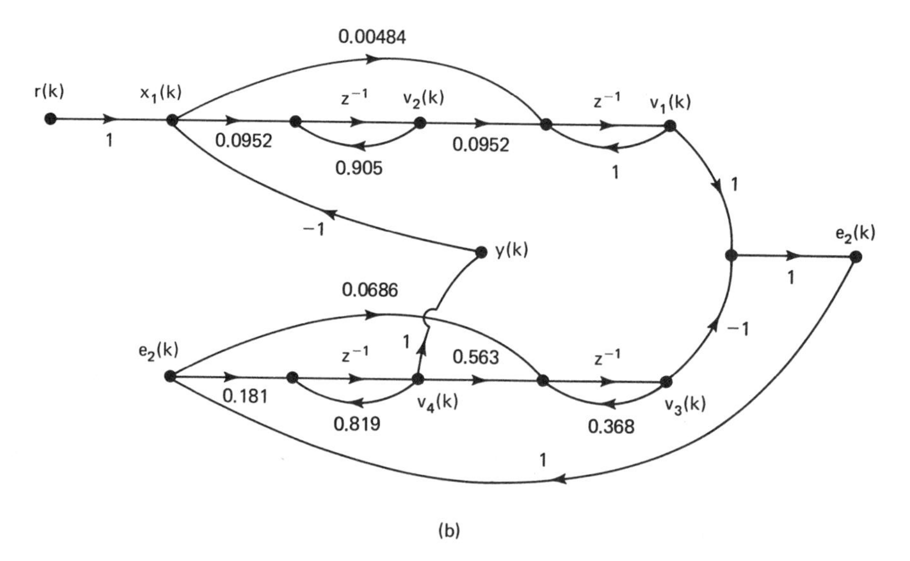

(b)

Figure 5-21 System for Example 5.5.

continuous state equations:

$$\dot{\mathbf{v}}(t) = \begin{bmatrix} 0 & 1 & 0 & 0 \\ 0 & -1 & 0 & 0 \\ 0 & 0 & -10 & 10 \\ 0 & 0 & 0 & -2 \end{bmatrix} \mathbf{v}(t) + \begin{bmatrix} 0 & 0 \\ 1 & 0 \\ 0 & 0 \\ 0 & 2 \end{bmatrix} \begin{bmatrix} \bar{e}_1(t) \\ \bar{e}_2(t) \end{bmatrix}$$

$$\begin{bmatrix} y(t) \\ e_2(t) \end{bmatrix} = \begin{bmatrix} 0 & 0 & 0 & 1 \\ 1 & 0 & -1 & 0 \end{bmatrix} \mathbf{v}(t)$$

In these equations, $\bar{e}_1(t)$ is the zero-order-hold output with $e_1^*(t)$ as its input, and $\bar{e}_2(t)$ is defined in the same manner. To obtain the discrete state matrices, we use the technique of Section 4.10 and the associated computer program given in Appendix I. From the computer program we obtain

$$\mathbf{v}(k+1) = \begin{bmatrix} 1 & 0.0952 & 0 & 0 \\ 0 & 0.905 & 0 & 0 \\ 0 & 0 & 0.368 & 0.563 \\ 0 & 0 & 0 & 0.819 \end{bmatrix} \mathbf{v}(k) + \begin{bmatrix} 0.00484 & 0 \\ 0.0952 & 0 \\ 0 & 0.0686 \\ 0 & 0.181 \end{bmatrix} \begin{bmatrix} e_1(k) \\ e_2(k) \end{bmatrix}$$

$$\begin{bmatrix} y(k) \\ e_2(k) \end{bmatrix} = \begin{bmatrix} 0 & 0 & 0 & 1 \\ 1 & 0 & -1 & 0 \end{bmatrix} \mathbf{v}(k)$$

A simulation diagram of this system, with all external connecting paths, is shown in Figure 5-21b. From this simulation diagram we write the discrete state equation for the closed-loop system.

$$\mathbf{v}(k+1) = \begin{bmatrix} 1 & 0.0952 & 0 & -0.00484 \\ 0 & 0.905 & 0 & -0.0952 \\ 0.0686 & 0 & 0.299 & 0.563 \\ 0.181 & 0 & -0.181 & 0.819 \end{bmatrix} \mathbf{v}(k) + \begin{bmatrix} 0.00484 \\ 0.0952 \\ 0 \\ 0 \end{bmatrix} r(k)$$

$$y(k) = [0 \quad 0 \quad 0 \quad 1]\mathbf{v}(k)$$

Note the similarity of this technique to that developed earlier for finding closed-loop transfer functions. The system is opened at each sampler, each zero-order-hold output is assumed to be an input, and each sampler input is assumed to be an output. Discrete state equations are then written relating these specified inputs and outputs. These state equations are manipulated, through the use of a simulation diagram, to obtain the state equations of the closed-loop system.

The technique above can be applied to low-order systems; however, writing system equations from complex flow graphs is at best tenuous. Instead, a matrix procedure that can be implemented by a computer program is needed, and one will now be developed. Consider Example 5.5. The discrete state equations for the continuous system can be written as

$$\mathbf{v}(k+1) = \mathbf{A}_1\mathbf{v}(k) + \mathbf{B}_1\mathbf{e}(k) \tag{5-29}$$

$$\mathbf{e}(k) = \mathbf{C}_1\mathbf{v}(k) + \mathbf{d}_1 r(k) \tag{5-30}$$

Thus these equations can be combined to yield

$$\mathbf{v}(k+1) = [\mathbf{A}_1 + \mathbf{B}_1\mathbf{C}_1]\mathbf{v}(k) + \mathbf{B}_1\mathbf{d}_1 r(k) \tag{5-31}$$

which is the required equation. An example will now be given.

Example 5.6

Consider the system of Example 5.5. Since $e_2(k) = r(k) - y(k) = r(k) - v_4(k)$, the equations for $\mathbf{e}(k)$ can be written as

$$\begin{bmatrix} e_1(k) \\ e_2(k) \end{bmatrix} = \begin{bmatrix} 0 & 0 & 0 & -1 \\ 1 & 0 & -1 & 0 \end{bmatrix} \mathbf{v}(k) + \begin{bmatrix} 1 \\ 0 \end{bmatrix} r(k)$$

Then, in (5-31),

$$\mathbf{B}_1\mathbf{C}_1 = \begin{bmatrix} 0.00484 & 0 \\ 0.0952 & 0 \\ 0 & 0.0686 \\ 0 & 0.181 \end{bmatrix} \begin{bmatrix} 0 & 0 & 0 & -1 \\ 1 & 0 & -1 & 0 \end{bmatrix}$$

$$= \begin{bmatrix} 0 & 0 & 0 & -0.00484 \\ 0 & 0 & 0 & -0.0952 \\ 0.0686 & 0 & -0.0686 & 0 \\ 0.181 & 0 & -0.181 & 0 \end{bmatrix}$$

and

$$\mathbf{A}_1 + \mathbf{B}_1\mathbf{C}_1 = \begin{bmatrix} 1 & 0.0952 & 0 & 0 \\ 0 & 0.905 & 0 & 0 \\ 0 & 0 & 0.368 & 0.563 \\ 0 & 0 & 0 & 0.819 \end{bmatrix} + \begin{bmatrix} 0 & 0 & 0 & -0.00484 \\ 0 & 0 & 0 & -0.0952 \\ 0.0686 & 0 & -0.0686 & 0 \\ 0.181 & 0 & -0.181 & 0 \end{bmatrix}$$

$$= \begin{bmatrix} 1 & 0.0952 & 0 & -0.00484 \\ 0 & 0.905 & 0 & -0.0952 \\ 0.0686 & 0 & 0.2994 & 0.563 \\ 0.181 & 0 & -0.181 & 0.819 \end{bmatrix}$$

Also, in (5-31),

$$\mathbf{B}_1\mathbf{d}_1 = \begin{bmatrix} 0.00484 & 0 \\ 0.0952 & 0 \\ 0 & 0.0686 \\ 0 & 0.181 \end{bmatrix} \begin{bmatrix} 1 \\ 0 \end{bmatrix} = \begin{bmatrix} 0.00484 \\ 0.0952 \\ 0 \\ 0 \end{bmatrix}$$

These results agree with those obtained in Example 5.5.

The system state equations are more difficult to derive if the system contains a digital controller. A single-loop digital control system is shown in Figure 5-22a. To obtain the state equations, we consider the digital filter and the plant separately and write the state equations for these two parts. We assign states $v_1(k)$ through $v_i(k)$

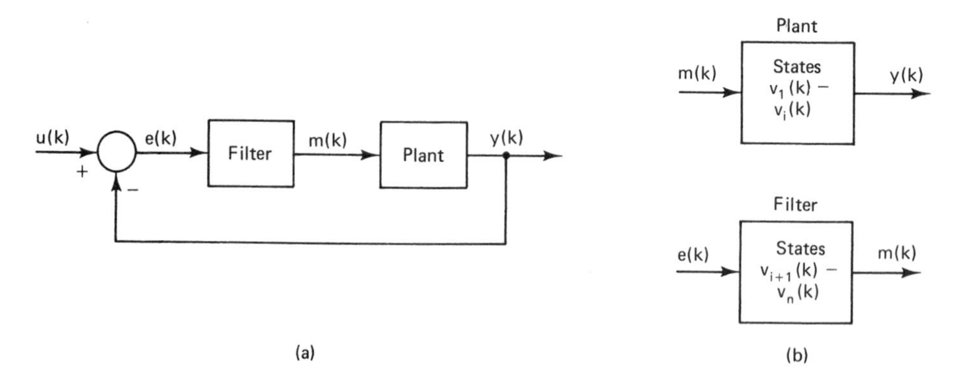

Figure 5-22 Discrete control system.

to the plant, where i is the order of the plant. Then we assign states $v_{i+1}(k)$ through $v_n(k)$ to the filter, where $n - i$ is the order of the filter. Hence we can write the state equations

$$\mathbf{v}(k+1) = \mathbf{A}_1\mathbf{v}(k) + \mathbf{b}_1 m(k) + \mathbf{b}_2 e(k) \tag{5-32}$$

since both $m(k)$ and $e(k)$ are inputs, as shown in Figure 5.22b. Next we write for the filter

$$m(k) = \mathbf{c}_1\mathbf{v}(k) + d_1 e(k) \tag{5-33}$$

and for the plant

$$y(k) = \mathbf{c}\mathbf{v}(k) \tag{5-34}$$

Thus, for the feedback path and from (5-34), we write

$$e(k) = u(k) - y(k) = u(k) - \mathbf{c}\mathbf{v}(k) \tag{5-35}$$

We obtain the system state equations by eliminating $m(k)$ and $e(k)$ from (5-32), (5-33), and (5-35). From (5-33) and (5-35),

$$\begin{aligned} m(k) &= \mathbf{c}_1\mathbf{v}(k) + d_1[u(k) - \mathbf{c}\mathbf{v}(k)] \\ &= [\mathbf{c}_1 - d_1\mathbf{c}]\mathbf{v}(k) + d_1 u(k) \end{aligned} \tag{5-36}$$

Then, substituting (5-35) and (5-36) into (5-32), we obtain

$$\begin{aligned} \mathbf{v}(k+1) &= \mathbf{A}_1\mathbf{v}(k) + \mathbf{b}_1[(\mathbf{c}_1 - d_1\mathbf{c})\mathbf{v}(k) + d_1 u(k)] \\ &\quad + \mathbf{b}_2[u(k) - \mathbf{c}\mathbf{v}(k)] \\ &= [\mathbf{A}_1 + \mathbf{b}_1\mathbf{c}_1 + (-\mathbf{b}_2 - d_1\mathbf{b}_1)\mathbf{c}]\mathbf{v}(k) + [d_1\mathbf{b}_1 + \mathbf{b}_2]u(k) \end{aligned} \tag{5-37}$$

which is the desired relationship. An example will now be given.

Example 5.7

The state equations for the system of Example 4.13 (shown in Figure 5-23a) will be developed. From Example 4.13, the state equations for the plant are

$$\begin{bmatrix} v_1(k+1) \\ v_2(k+1) \end{bmatrix} = \begin{bmatrix} 1 & 0.0952 \\ 0 & 0.905 \end{bmatrix} \begin{bmatrix} v_1(k) \\ v_2(k) \end{bmatrix} + \begin{bmatrix} 0.0484 \\ 0.952 \end{bmatrix} m(k)$$

$$y(k) = [1 \quad 0] \begin{bmatrix} v_1(k) \\ v_2(k) \end{bmatrix}$$

The filter is modeled as shown in Figure 5-23b. The state equations for the filter are

$$\begin{aligned} v_3(k+1) &= 0.9v_3(k) + e(k) \\ m(k) &= (0.81 - 0.8)v_3(k) + 0.9e(k) \\ &= 0.01v_3(k) + 0.9e(k) \end{aligned}$$

Combining the state equations for $\mathbf{v}(k)$, we obtain

$$\begin{bmatrix} v_1(k+1) \\ v_2(k+1) \\ v_3(k+1) \end{bmatrix} = \begin{bmatrix} 1 & 0.0952 & 0 \\ 0 & 0.905 & 0 \\ 0 & 0 & 0.9 \end{bmatrix} \begin{bmatrix} v_1(k) \\ v_2(k) \\ v_3(k) \end{bmatrix} + \begin{bmatrix} 0.0484 \\ 0.952 \\ 0 \end{bmatrix} m(k)$$

$$+ \begin{bmatrix} 0 \\ 0 \\ 1 \end{bmatrix} e(k)$$

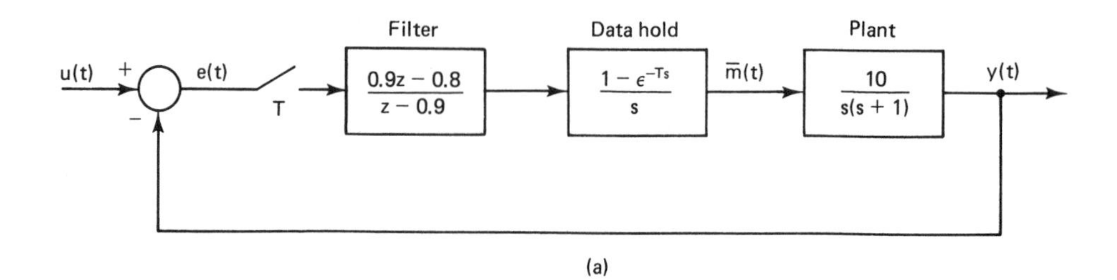

(a)

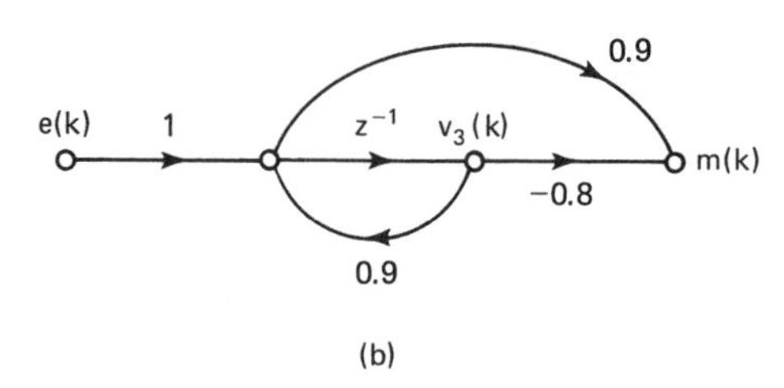

(b)

Figure 5-23 System for Example 5.7.

Also, from Figure 5-23,

$$e(k) = u(k) - y(k) = u(k) - [1 \quad 0 \quad 0]\mathbf{v}(k)$$

and, from above,

$$m(k) = [0 \quad 0 \quad 0.01]\mathbf{v}(k) + 0.9e(k)$$

Comparing the equation above for $\mathbf{v}(k)$ with (5-32), we see that

$$\mathbf{A}_1 = \begin{bmatrix} 1 & 0.0952 & 0 \\ 0 & 0.905 & 0 \\ 0 & 0 & 0.9 \end{bmatrix}, \qquad \mathbf{b}_1 = \begin{bmatrix} 0.0484 \\ 0.952 \\ 0 \end{bmatrix}, \qquad \mathbf{b}_2 = \begin{bmatrix} 0 \\ 0 \\ 1 \end{bmatrix}$$

From the equation above for $m(k)$ and (5-33),

$$\mathbf{c}_1 = [0 \quad 0 \quad 0.01], \qquad d_1 = 0.9$$

and from the equation above for $e(k)$ and (5-35),

$$\mathbf{c} = [1 \quad 0 \quad 0]$$

Then, in (5-37),

$$\mathbf{b}_1\mathbf{c}_1 = \begin{bmatrix} 0.0484 \\ 0.952 \\ 0 \end{bmatrix} [0 \quad 0 \quad 0.01] = \begin{bmatrix} 0 & 0 & 0.000484 \\ 0 & 0 & 0.00952 \\ 0 & 0 & 0 \end{bmatrix}$$

$$(-\mathbf{b}_2 - d_1\mathbf{b}_1)\mathbf{c} = \begin{bmatrix} 0 - 0.04356 \\ 0 - 0.8568 \\ -1 - 0 \end{bmatrix} [1 \quad 0 \quad 0]$$

$$= \begin{bmatrix} -0.04356 & 0 & 0 \\ -0.8568 & 0 & 0 \\ -1 & 0 & 0 \end{bmatrix}$$

Thus

$$\mathbf{A}_1 + \mathbf{b}_1\mathbf{c}_1 + (-\mathbf{b}_2 - d_1\mathbf{b}_1)\mathbf{c} = \begin{bmatrix} 0.9564 & 0.0952 & 0.000484 \\ 0.8568 & 0.905 & 0.00952 \\ -1 & 0 & 0.9 \end{bmatrix}$$

$$d_1\mathbf{b}_1 + \mathbf{b}_2 = \begin{bmatrix} 0.04356 + 0 \\ 0.8568 \quad + 0 \\ 0 \quad + 1 \end{bmatrix}$$

Thus the state equations for this system are, from (5-37) and (5-34),

$$\mathbf{v}(k+1) = \begin{bmatrix} 0.9564 & 0.0952 & 0.000484 \\ -0.8568 & 0.905 & 0.00952 \\ -1 & 0 & 0.9 \end{bmatrix}\mathbf{v}(k) + \begin{bmatrix} 0.04356 \\ 0.8568 \\ 1 \end{bmatrix} u(k)$$

$$y(k) = [1 \quad 0 \quad 0]\mathbf{v}(k)$$

Examples 5.6 and 5.7 illustrate the derivation of discrete state models for digital control systems. Of course, some systems are more complex than these in the examples above. However, the technique used to derive the state equations of (5-37) may be employed for more complex systems. For example, if $\mathbf{y}(k)$ in (5-34) is also a function of $m(k)$, the derivation is somewhat more complicated (see Problem 5-16).

5.5 SUMMARY

In this chapter we have described a technique for finding the Laplace and z-transforms of the output of a closed-loop discrete-time system. The technique assumes the availability of a block diagram or signal flow graph of the discrete-time system. From this specification, a sampled signal flow graph is derived which can be used to determine the Laplace transform or the z-transform of the output of the closed-loop system. In addition, a technique is developed for determining the state-variable model of a closed-loop discrete-time system. In the following chapter these techniques will be utilized in analyzing closed-loop discrete-time systems.

REFERENCES AND FURTHER READING

1. B. C. Kuo, *Analysis and Synthesis of Sampled-Data Control Systems.* Englewood Cliffs, N.J.: Prentice-Hall, Inc., 1963.
2. M. Sedlar and G. A. Bekey, "Signal Flow Graphs of Sampled Data Systems: A New Formulation," *IEEE Trans. Autom. Control,* Vol. AC-12, No. 2, pp. 606–608. Oct., 1967.
3. C. L. Phillips and S. M. Seltzer, "Design of Advanced Sampled-Data Control Systems," Contract DAAHO1-72-C-0901, Auburn University, Auburn, Ala., July 1973.
4. C. L. Phillips, "Digital Compensation of the Thrust Vector Control System," Contract NAS8-11274, Auburn University, Auburn, Ala., Nov. 1964.
5. J. A. Cadzow and H. R. Martens, *Discrete-Time and Computer Control Systems.* Englewood Cliffs, N.J.: Prentice-Hall, Inc., 1970.

6. E. I. Jury, *Sampled-Data Control Systems.* New York: John Wiley & Sons, Inc., 1958.
7. E. I. Jury, *Theory and Application of the z-Transform Method.* New York: John Wiley & Sons, Inc., 1964.
8. B. C. Kuo, *Discrete-Data Control Systems*, Englewood Cliffs, N.J.: Prentice-Hall, Inc., 1970.
9. D. P. Lindorff, *Theory of Sampled-Data Control Systems.* New York: John Wiley & Sons, Inc., 1965.
10. J. R. Ragazzini and G. F. Franklin, *Sampled-Data Control Systems.* New York: McGraw-Hill Book Company, 1958.

PROBLEMS

5-1. For each of the systems of Figure P5-1, express $C(z)$ as a function of the input and the transfer functions shown.

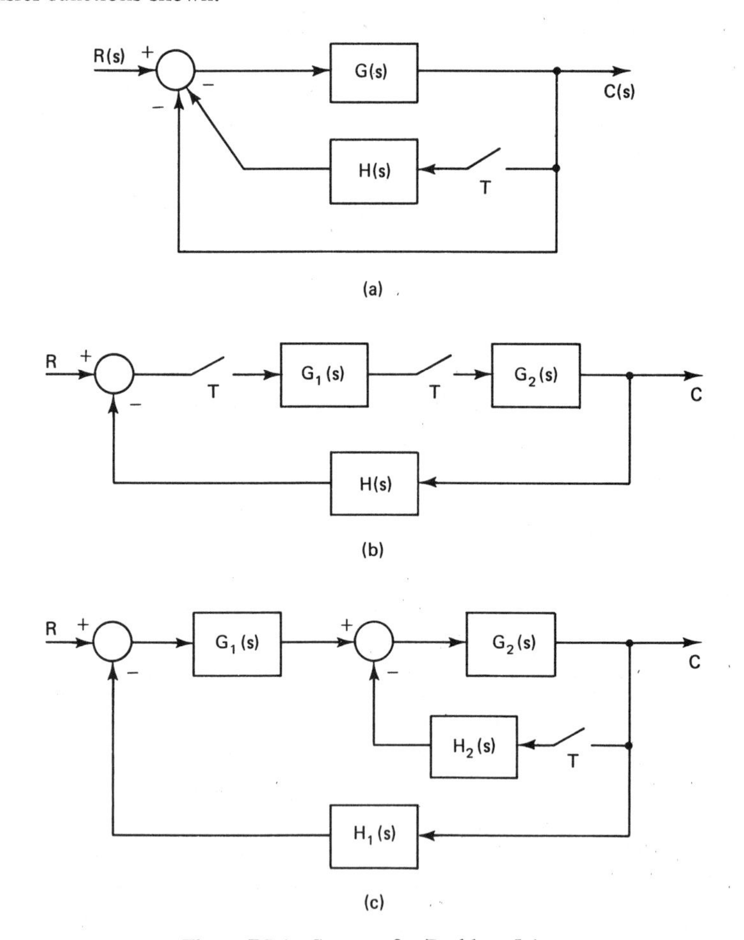

Figure P5-1 Systems for Problem 5-1.

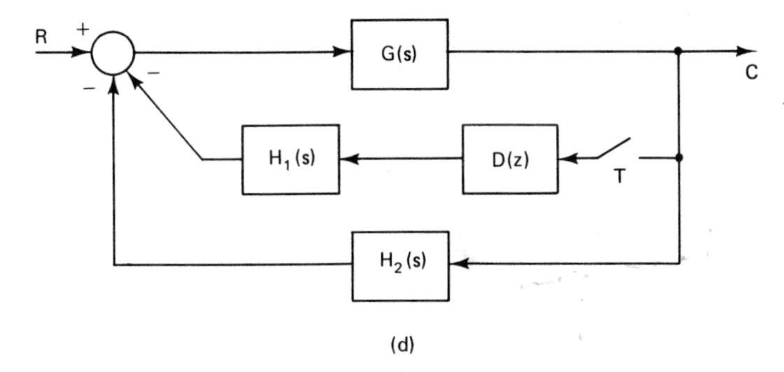

Figure P5-1 cont.

5-2. For the system of Figure P5-2, express $C(z)$ as a function of the input and the transfer functions shown.

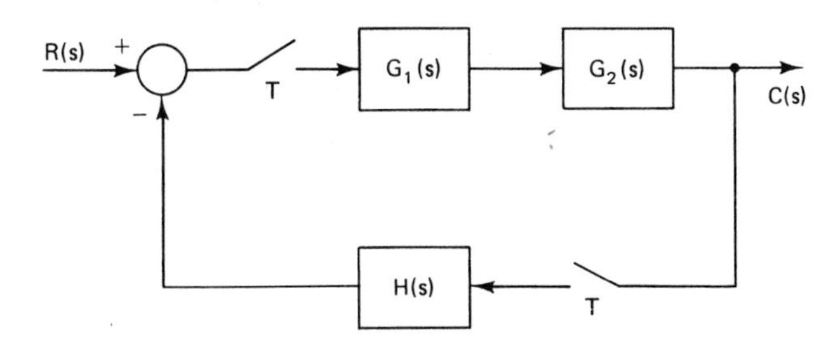

Figure P5-2 System for Problem 5-2.

5-3. Repeat Problem 5-2 for the case that the sampler in the forward path samples at $t = 0, T, 2T, \ldots$, and the sampler in the feedback path operates at $t = T/2, 3T/2, 5T/2, \ldots$.

5-4. The system of Figure P5-4 contains a digital filter with the transfer function $D(z)$. Express $\phi_m(z)$ as a function of the input. The roll-axis control system of the Pershing missile is of this configuration [3].

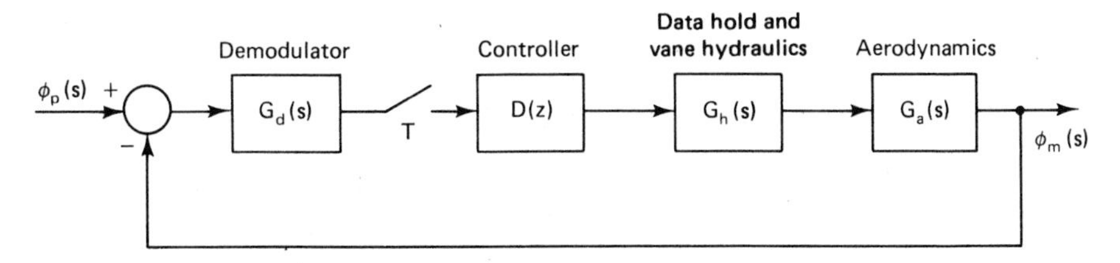

Figure P5-4 Roll-axis control system for a Pershing missile.

5-5. Shown in Figure P5-5 is a closed-loop temperature control system, which was also considered in Problem 4-3. Let $G_p(s)$ be defined as the transfer function from the input of the power amplifier to the temperature $\theta(t)$, and let $\theta_R(kT)$ be the input signal.

(a) Solve for the transfer function $\Theta(z)/\Theta_R(z)$.
(b) Let $D(z) = 1$. Evaluate the transfer function $\Theta(z)/\Theta_R(z)$, with $T = 0.5$ s.

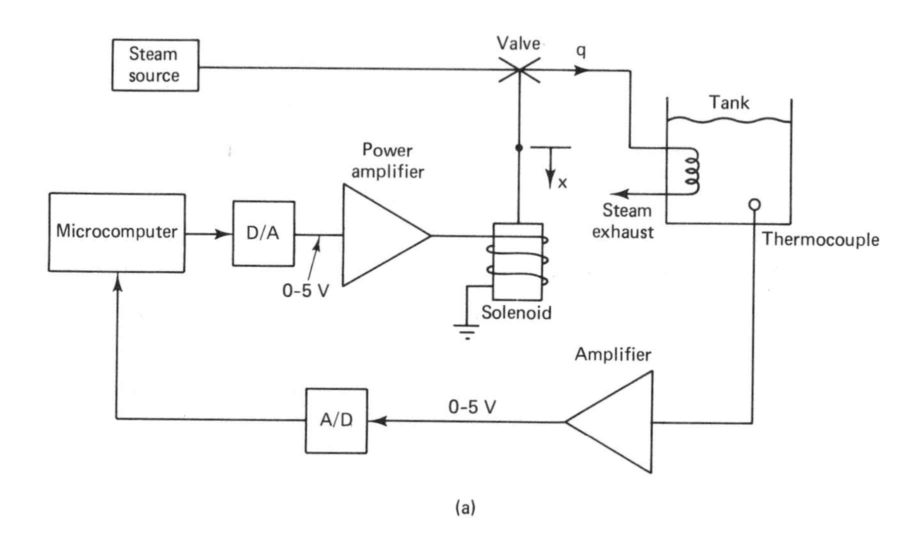

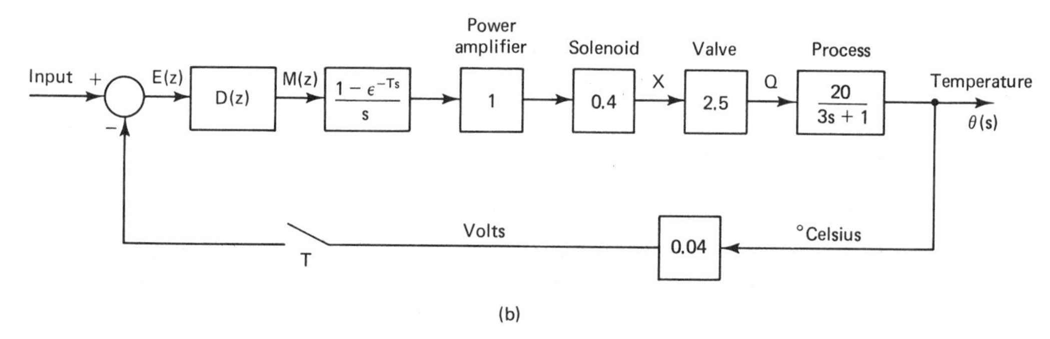

Figure P5-5 Temperature control system.

5-6. Shown in Figure P5-6 is a simplified Saturn V booster pitch-axis attitude control system [4]. For this system $G_2(s) = sG_1(s)$. Even though positive feedback is normally shown, $G_1(s)$ [and thus $G_2(s)$] contains a minus sign, and the system does use negative feedback. The sampler appears in the system because the onboard computer is required to calculate the attitude ϕ_c. This calculation is performed 25 times per second.
(a) Express $\phi_c(z)$ as a function of the system input.
(b) Suppose that each calculation of attitude position requires t_0 seconds to perform. Express $\phi_c(z)$ as a function of the system input with the effect of this time delay included.

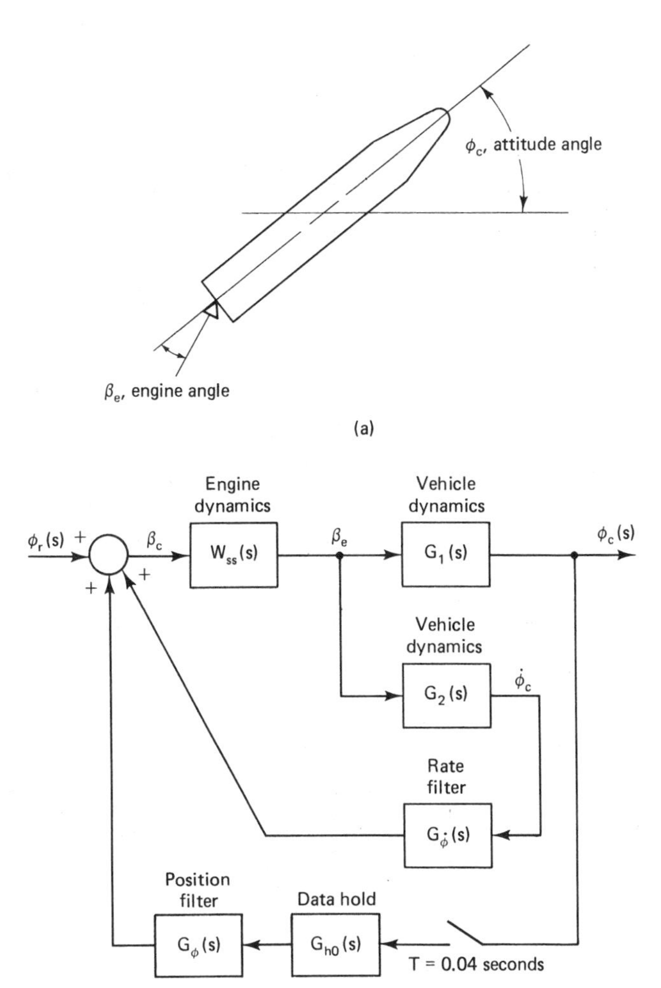

Figure P5-6 Simplified Saturn V attitude control system.

5-7. Figure P5-7 illustrates an environmental plant growth chamber, designed to allow scientists to study the effects of certain parameters on plant growth. A temperature control system is shown. Air conditioning is utilized to cool the chamber and controlled electric heaters then maintain the desired temperature. The heaters are rated at a maximum of 240 V, rms. Electronic switches called triacs are utilized to vary the effective voltage into the heaters from 0 V to 240 V, rms. The temperature is measured by a thermistor in a bridge circuit, and the bridge circuit output voltage is amplified to give a voltage in the range of ± 5 V, which is linearly related to temperature in the range of ± 64°C. A simplified block diagram of the system is given in the figure. Let

$$G_p(s) = \frac{1}{60s + 1}$$

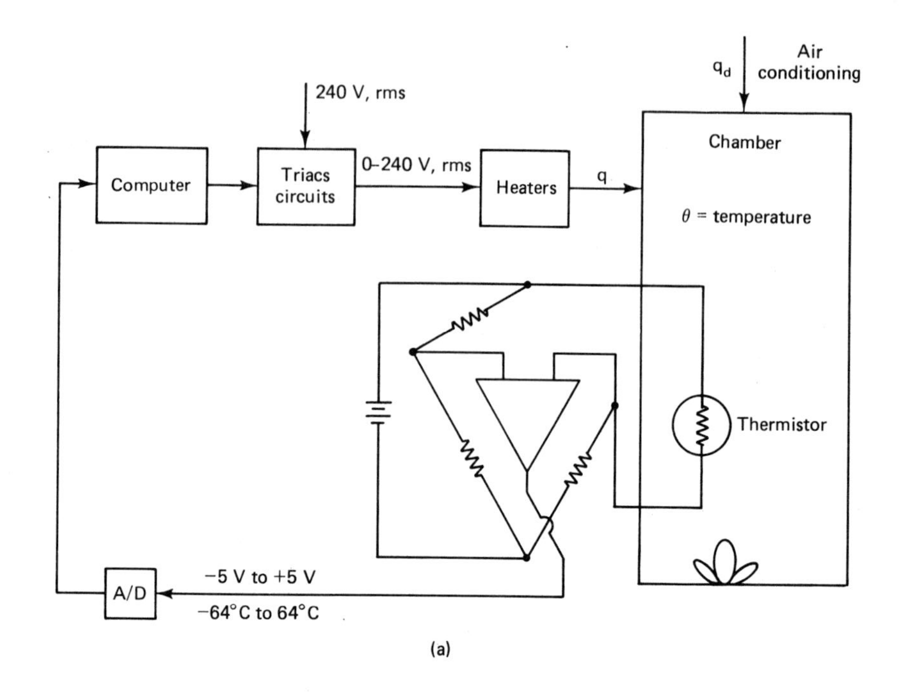

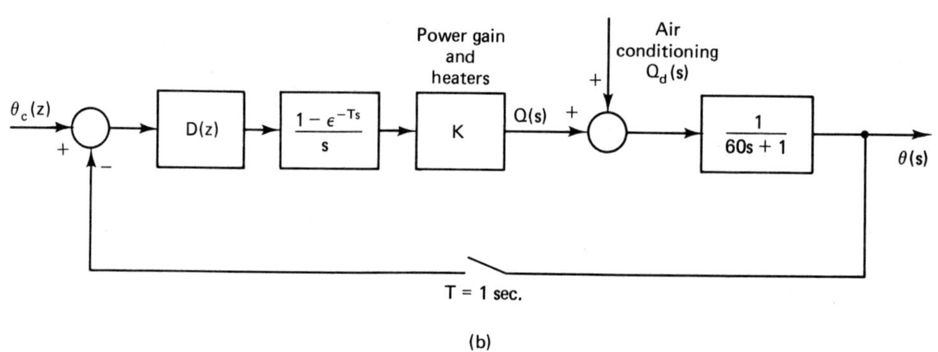

Figure P5-7 Plant growth chamber control system.

(a) Express $\Theta(z)$ as a function of $\Theta_c(z)$, with $Q_d(s) = 0$.

(b) Express $\Theta(z)$ as a function of $\Theta_c(z)$ and $Q_d(s)$.

(c) Assume that $q_d(t)$ varies so slowly that this signal can be considered to be the output of a zero-order hold. Repeat part (b).

5-8. Figure P5-8 gives the simplified block diagram of an automobile cruise-control system, which was also considered in Problem 4-7. Let

$$M(z) = D(z)E(z)$$

$$G_1(s) = \frac{0.64}{s+1}, \qquad G_2(s) = \frac{50}{5s+1}$$

Express the speed $V(z)$ as a function of the reference input (desired speed) and the load torque disturbance.

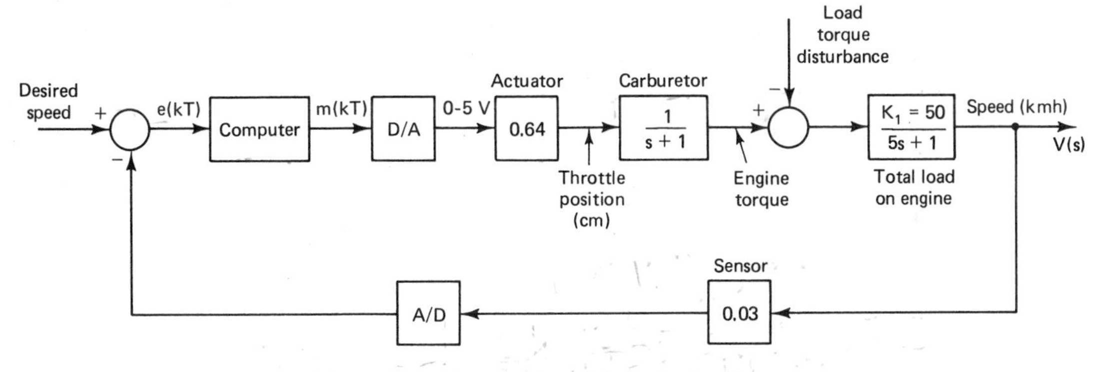

Figure P5-8 Automobile cruise-control system.

5-9. The system of Figure P5-9 contains a disturbance input (undesired input) $U(s)$, in addition to the reference input $R(s)$.

(a) Express $C(z)$ as a function of the two inputs.

(b) Suppose that $D_2(z)$ and $D_3(z)$ are chosen such that $D_3(z) = D_2(z)\overline{G_1G_2}(z)$. Find $C(z)$ as a function of the two inputs.

(c) What is the advantage of the choice in part (b) if it is desired to minimize the response, $C(z)$, to the disturbance, $U(s)$?

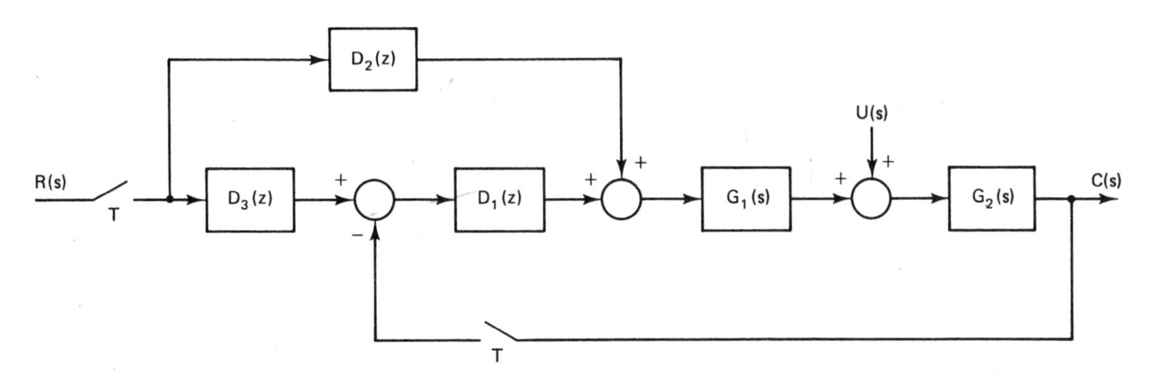

Figure P5-9 System for Problem 5-9.

5-10. For the system of Figure P5-1b, let

$$G_1(s) = G_2(s) = \frac{1 - \epsilon^{-Ts}}{s(s+1)}, \qquad H(s) = 1$$

with $T = 0.1$ s. Find a discrete state-variable model for the system.

5-11. For the system of Figure P5-1d, let

$$G(s) = \frac{1}{s^2}, \qquad H_1(s) = \frac{1 - \epsilon^{-Ts}}{s}$$

$$H_2(s) = 1, \qquad D(z) = 2 + \frac{0.5z}{z-1}$$

with $T = 0.2$ s. Find a discrete-state-variable model for the system. Assume that $r(t)$

changes slowly with time, such that it can be considered to be constant over any sample interval. What is the effect on the model derivation if this assumption is not made?

5-12. The plant in Figure P5-12 is described by the differential equation

$$\frac{d^2y}{dt^2} + 0.2\frac{dy}{dt} + 0.02y = \bar{m}(t)$$

Derive a set of discrete state equations for the closed-loop system such that the first state is $y(kT)$ and the second state is $dy(kT)/dt$. Let $T = 2$ s.

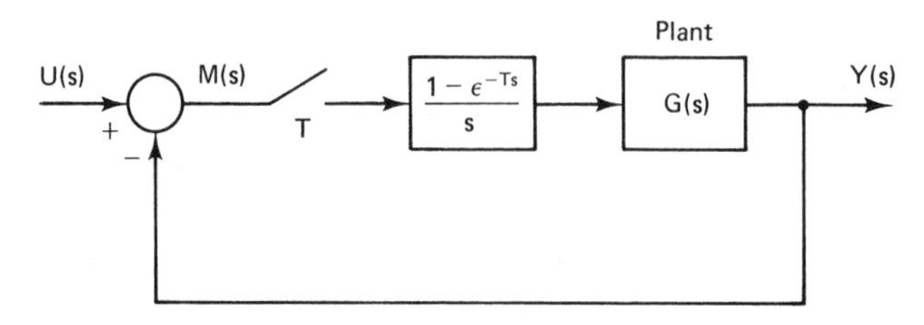

Figure P5-12 System for Problem 5-12.

5-13. Find a discrete-state-variable model of the closed-loop system shown in Figure P5-12 if the discrete state model of the plant is given by

$$\mathbf{x}(k+1) = \begin{bmatrix} 0.9 & 1 & 0 \\ 0 & 0.9 & 0 \\ 0 & 0 & 0.5 \end{bmatrix}\mathbf{x}(k) + \begin{bmatrix} 0 \\ 1 \\ 1 \end{bmatrix} m(k)$$

$$y(k) = [1 \quad 1.5 \quad 1.3]\mathbf{x}(k)$$

5-14. Consider the temperature control system of Problem 5-5 and Figure P5-5. Suppose that the digital filter transfer function is given by

$$D(z) = 1.2 + \frac{0.1z}{z-1}$$

with $T = 0.5$ s.

(a) Using the closed-loop transfer function, derive a state model for the system.

(b) Derive a discrete state model for the plant, as in Problem 4-17. Then derive the state model of the closed-loop system by adding the filter and the feedback path to the flow graph of the plant.

(c) Repeat part (b), but use the matrix method of equation (5-37) rather than the flow-graph method.

5-15. Consider the automobile cruise-control system of Problem 5-8 and Figure P5-8. Let the digital filter transfer function be unity; [i.e., $D(z) = 1$]. In addition, let the disturbance torque be zero, and $T = 0.2$ s.

(a) Derive a state model for the system, using the closed-loop transfer function.

(b) Derive a discrete state model for the plant, as in Problem 4-18. Then derive the state model of the closed-loop system by adding the feedback path to the flow graph of the plant.

(c) Assume that the disturbance torque is nonzero and varies so slowly that it can be assumed to be transmitted through a zero-order hold. Repeat part (b).

(d) Repeat part (b), but with the digital filter solving the difference equation

$$m(k) = 2e(k) - 1.9796e(k - 1) + 0.99m(k - 1)$$

5-16. Suppose that, for the system of Figure 5-22, equation (5-34) is

$$y(k) = \mathbf{c}^T\mathbf{v}(k) + d_2 m(k)$$

(a) Derive the state model of equation (5-37) for this case.
(b) This system has an algebraic loop. Identify this loop.
(c) The gain of the algebraic loop is $-d_1 d_2$. What is the effect on the system equations if $d_1 d_2 = -1$?
(d) We can argue that algebraic loops as in this case cannot occur in physical systems, since time delay is always present in signal transmission. Quite often we can ignore this delay. Under what conditions can we obviously *not* ignore delay in this system?

5-17. Shown in Figure P5-17 is a control rod positioning system for a nuclear reactor. The system positions control rods in a nuclear reactor to obtain a desired radiation level.

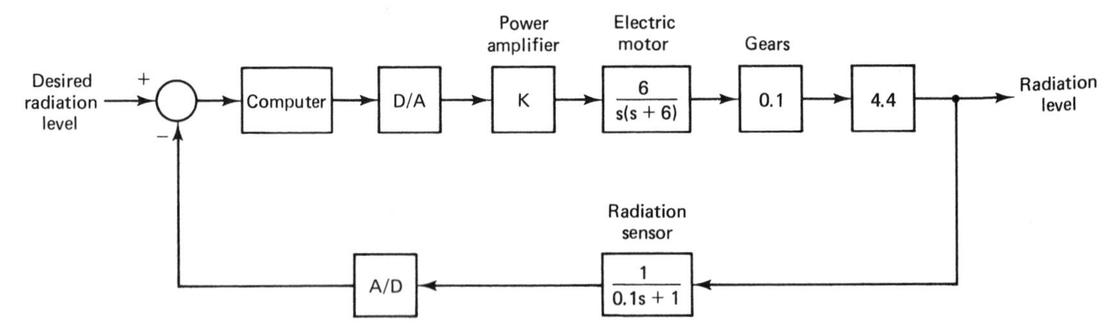

Figure P5-17 Nuclear reactor control rod positioning system.

The gain of 4.4 is the conversion factor from rod position to radiation level. The radiation sensor has a dc gain of unity and a time constant of 0.1 s. Let $D(z)$ be the digital filter transfer function, and

$$G(s) = \frac{2.64\,(1 - \epsilon^{-Ts})}{s^2(s + 6)}$$

$$H(s) = \frac{1}{0.1s + 1}$$

(a) Derive the closed-loop transfer function as a function of $D(z)$, $G(s)$, and $H(s)$.
(b) If $T = 0.02$ s and $D(z) = 1$, evaluate this transfer function.
(c) For part (b), derive a state model of the system.

5-18. Problem 5-7 describes a temperature control system for an environmental plant growth chamber. An additional parameter to be controlled in the chamber is carbon dioxide (CO_2) content. A carbon dioxide control system is shown in Figure P5-18a. A simplified block diagram is given in Figure P5-18b. The system sensor is a gas analyzer, which requires 45 s to analyze a gas sample. The analyzer output is 0 to 5 V, which is linearly related to CO_2 content of 0 to 400 parts per million (ppm). The sample period T is 45 s.
(a) Derive and evaluate the closed-loop transfer function for $D(z) = 1$ and $K = 1$.
(b) Repeat part (a) for $T = 60$ s.

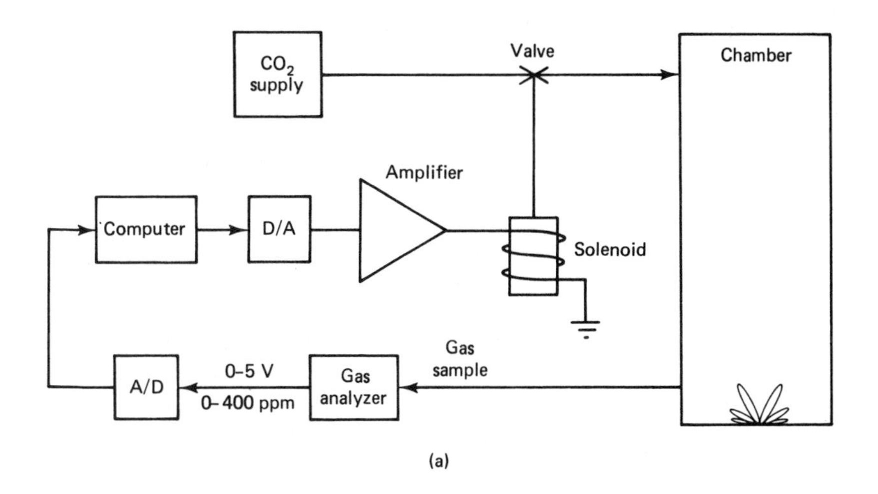

(a)

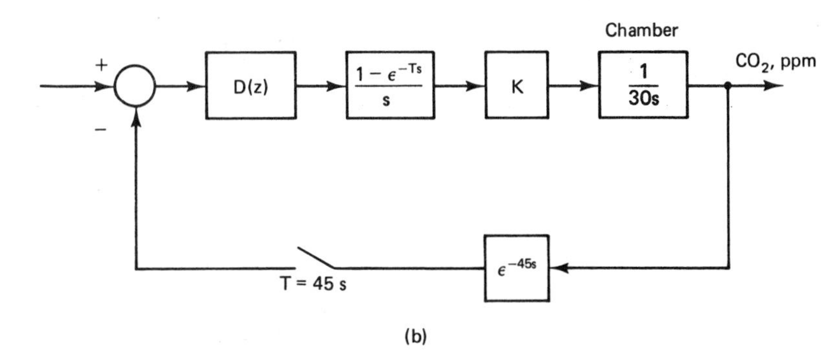

(b)

Figure P5-18 Carbon dioxide control system.

6

System Time-Response Characteristics

6.1 INTRODUCTION

In this chapter we consider five important topics. First, the time response of a discrete-time system is investigated. Next, regions in the s-plane are mapped into regions in the z-plane. Then by using the correlation between regions in the two planes, the effect of the closed-loop poles on the system transient response is discussed. Next, the effects of the system transfer characteristics on the steady-state system error are considered. Finally, the simulation of analog and discrete-time systems is introduced.

6.2 SYSTEM TIME RESPONSE

In this section the time response of discrete-time systems is introduced via examples. In these examples some of the salient features and ramifications of determining the system time response are illustrated.

Example 6.1

The step response will be found for the system in Figure 6-1a. Using the technique developed in Chapter 5, we can express the system output as

$$C(z) = \frac{\left[G_{h0}\dfrac{G_1}{1 + G_1}\right](z)}{1 + \left[G_{h0}\dfrac{G_1}{1 + G_1}\right](z)} R(z)$$

For convenience, we will define $G(z)$ as

$$G(z) = \left[G_{h0}\frac{G_1}{1 + G_1}\right](z)$$

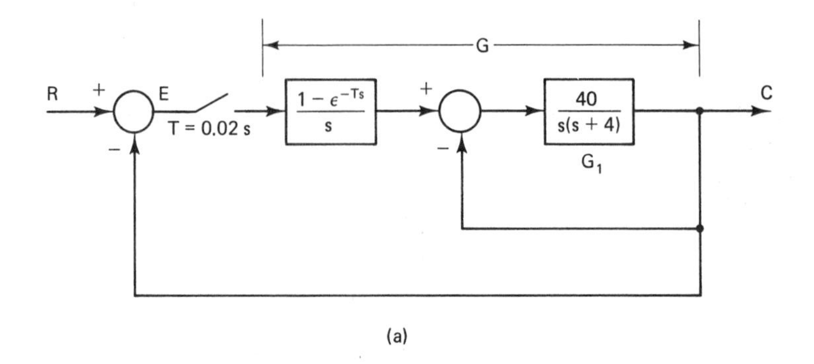

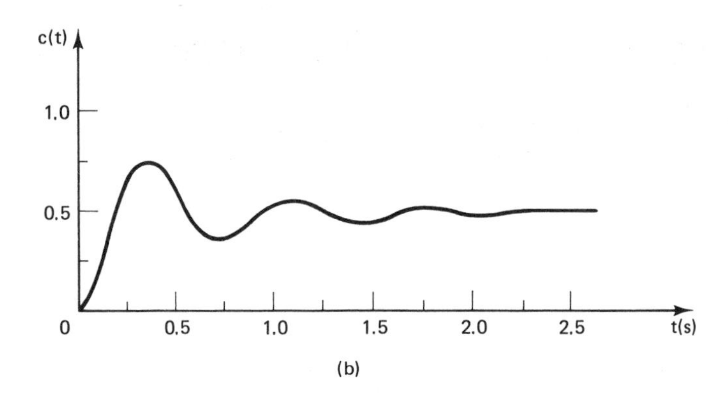

Figure 6-1 System and response for Example 6.1.

Then

$$G(z) = \mathfrak{z}\left[\frac{40(1 - \epsilon^{-Ts})}{s(s^2 + 4s + 40)}\right]$$

$$= \left(\frac{z-1}{z}\right)\mathfrak{z}\left[\frac{1}{s} - \frac{s+4}{s^2 + 4s + 40}\right]$$

$$= \left(\frac{z-1}{z}\right)\mathfrak{z}\left[\frac{1}{s} - \frac{s+2}{(s+2)^2 + 6^2} - \frac{1}{3}\frac{6}{(s+2)^2 + 6^2}\right]$$

With $G(z)$ in this form, we can obtain the z-transform from the tables.

$$G(z) = \left(\frac{z-1}{z}\right)\left[\frac{z}{z-1} - \frac{z^2 - z\epsilon^{-2T}\cos 6T + 1/3z\epsilon^{-2T}\sin 6T}{z^2 - 2z\epsilon^{-2T}\cos 6T + \epsilon^{-4T}}\right]$$

With $T = 0.02$ s, we evaluate the terms in $G(z)$.

$$2\epsilon^{-0.04}\cos 0.12 = 1.907760$$

$$\tfrac{1}{3}\epsilon^{-0.04}\sin 0.12 = 0.038339$$

$$\epsilon^{-0.08} = 0.923116$$

Therefore,

$$G(z) = \left(\frac{z-1}{z}\right)\left[\frac{z}{z-1} - \frac{z^2 - 0.91554z}{z^2 - 1.90776z + 0.92312}\right]$$

$$= \frac{0.00778z + 0.00758}{z^2 - 1.90776z + 0.92312}$$

The closed-loop transfer function is then

$$\frac{G(z)}{1 + G(z)} = \frac{0.00778z + 0.00758}{z^2 - 1.90z + 0.9307}$$

For the input a unit step, $R(z) = z/(z - 1)$, and

$$\begin{aligned}
C(z) &= \frac{G(z)}{1 + G(z)} R(z) \\
&= \frac{0.00778z^2 + 0.00758z}{z^3 - 2.90z^2 + 2.83z - 0.9307} \\
&= 0.00778z^{-1} + 0.0301z^{-2} + 0.0654z^{-3} + 0.1112z^{-4} + \cdots
\end{aligned}$$

Note that many terms must be calculated to determine the response. If more terms are calculated, the output is seen to approach a final value. Thus we may calculate the final value of $c(kT)$ using the final-value theorem.

$$\begin{aligned}
\lim_{n\to\infty} c(nT) &= (z - 1)C(z)\big|_{z=1} = (z - 1)\frac{G(z)}{1 + G(z)} R(z)\big|_{z=1} \\
&= (z - 1)\frac{G(z)}{1 + G(z)} \cdot \frac{z}{z - 1}\bigg|_{z=1} = \frac{G(z)}{1 + G(z)}\bigg|_{z=1} \\
&= \frac{G(1)}{1 + G(1)} = 0.50
\end{aligned}$$

Since the system input is a constant value of unity, we see from this derivation that the dc gain of a stable system is simply the closed-loop transfer function evaluated at $z = 1$.

We may also calculate the continuous response. By the technique developed in Chapter 5, we find that, with $z = \epsilon^{Ts}$,

$$\begin{aligned}
C(s) &= \frac{R(z)}{1 + G(z)} G(s) \\
&= \left[\frac{z}{z - 1}\right]\left[\frac{z^2 - 1.90776z + 0.92312}{z^2 - 1.89998z + 0.9307}\right]\left[\frac{(1 - z^{-1})40}{s(s^2 + 4s + 40)}\right] \\
&= \left[\frac{z^2 - 1.90776z + 0.92312}{z^2 - 1.89998z + 0.9307}\right]\left[\frac{40}{s(s^2 + 4s + 40)}\right]
\end{aligned}$$

As shown above,

$$\frac{40}{s(s^2 + 4s + 40)} = \frac{1}{s} - \frac{s + 2}{(s + 2)^2 + 6^2} - \frac{1}{3}\frac{6}{(s + 2)^2 + 6^2}$$

The inverse Laplace transform of this function is

$$\mathcal{L}^{-1}\left[\frac{40}{s(s^2 + 4s + 40)}\right] = 1 - \epsilon^{-2t}(\cos 6t + \tfrac{1}{3}\sin 6t)$$

In addition, by dividing the denominator into the numerator, we may express the first term in $C(s)$ above as

$$\frac{z^2 - 1.90776z + 0.92312}{z^2 - 1.89998z + 0.9307} = 1 - 0.00778z^{-1} - 0.02236z^{-2} + \cdots$$

where $z = \epsilon^{Ts}$. Then

$$C(s) = [1 - 0.00778\epsilon^{-Ts} - 0.02236\epsilon^{-2Ts} + \cdots]\left[\frac{40}{s(s^2 + 4s + 40)}\right]$$

Hence, taking the inverse Laplace transform, we obtain

$$\begin{aligned} c(t) = 1 &- \epsilon^{-2t}(\cos 6t + \tfrac{1}{3}\sin 6t) \\ &- 0.00778[1 - \epsilon^{-2(t-T)}[\cos 6(t-T) + \tfrac{1}{3}\sin 6(t-T)]]u(t-T) \\ &- 0.02236[1 - \epsilon^{-2(t-2T)}[\cos 6(t-2T) + \tfrac{1}{3}\sin 6(t-2T)]]u(t-2T) \\ &+ \cdots \end{aligned}$$

where $T = 0.02$ s. This expression gives the value of $c(t)$ at any time, but because of its complexity, it is of little practical value. A plot of $c(t)$, obtained from a digital simulation of the system, is given in Figure 6-1b.

It is instructive to compare this system response with that of the analog (continuous-time) system response, that is, the response of the system of Figure 6-1a with the sampler and zero-order hold removed. For the analog system, the open-loop transfer function is

$$G_{OL}(s) = \frac{G_1(s)}{1 + G_1(s)} = \frac{40}{s^2 + 4s + 40}$$

Then the closed-loop transfer function is given by

$$G_{CL}(s) = \frac{G_{OL}(s)}{1 + G_{OL}(s)} = \frac{40}{s^2 + 4s + 80} = \frac{0.5\omega_n^2}{s^2 + 2\zeta\omega_n s + \omega_n^2}$$

This transfer function is the standard form for second-order systems, with $\omega_n = 8.944$ and $\zeta = 0.2236$. Hence the percent overshoot to a step input, given by [8]

$$\text{percent overshoot} = 100\epsilon^{-\zeta\pi/\sqrt{1-\zeta^2}}$$

is 49%. An inspection of Figure 6-1b shows that the sampled-data step response has the same overshoot. In fact, a simulation of the analog system results in a step response that is essentially the same as that of Figure 6-1b.

We will investigate this point further. The closed-loop transfer function of the analog system, derived above, can be expressed as

$$G_{CL}(s) = \frac{40}{s^2 + 4s + 80} = \frac{40}{(s+2)^2 + (8.72)^2}$$

Thus the frequency $\omega = 8.72$ rad/s is excited by the system input. This frequency has a period of $2\pi/\omega$, or 0.72 s. Hence this frequency is sampled 36 times per cycle ($T = 0.02$ s), which results in a very good description of the signal. Also, the time constant of the poles of the closed-loop transfer function, given by $1/\zeta\omega_n$, is 0.5 s. Thus we are sampling 25 times per time constant. A rule of thumb often given for selecting sample rates is that a rate of at least five times per time constant is a good choice. Hence, for this system, we would expect very little degradation in system response because of the sampling. In fact, we could increase T with very little effect.

A final point will be made concerning this example. For a continuous-time system whose output is

$$C(s) = G(s)E(s)$$

the final value of $c(t)$ is obtained using the final-value theorem of the Laplace transform,

$$\lim_{t\to\infty} c(t) = sC(s)\big|_{s=0} = sG(s)E(s)\big|_{s=0}$$

provided that the system is stable. If the input is a unit step, $E(s) = 1/s$ and

$$\lim_{t\to\infty} c(t) = sG(s)\frac{1}{s}\bigg|_{s=0} = G(s)\bigg|_{s=0}$$

Thus the dc gain is equal to the transfer function evaluated at $s = 0$. In the system of Figure 6-1a, if the input to the sampler is constant, the output of the zero-order hold is also constant and equal to the sampler input. Thus the sampler and data hold have no effect and may be removed, resulting in a continuous-time system. Thus, for a stable sampled-data system, the system dc gain may be found by removing the sampler and data hold, and evaluating the resulting system transfer function at $s = 0$. For this example, the open-loop dc gain is

$$\left.\frac{G_1(s)}{1 + G_1(s)}\right|_{s=0} = \left.\frac{40}{s^2 + 4s + 40}\right|_{s=0} = 1$$

Thus the closed-loop dc gain is 0.5, which agrees with the dc gain calculated above via the z-transform.

Two important points were made in this example concerning the calculation of the steady-state gain of a sampled-data system with a constant input applied (i.e., the calculation of the dc gain). For a stable system, the system output approaches a constant value as time increases for a constant input. The dc gain may be calculated by evaluating the transfer function with $z = 1$. In addition, the same value of dc gain is obtained by evaluating the transfer function of the analog system (sampler and zero-order hold removed) with $s = 0$. This second calculation applies in any case that the input to the sampler is constant.

Example 6.2

The system for this example is shown in Figure 6-2. As in the first example, we will calculate the unit step response. This system will also appear in many of the following examples. As was demonstrated in Chapter 5, the system output can be expressed as

$$C(z) = \frac{G(z)}{1 + G(z)} R(z)$$

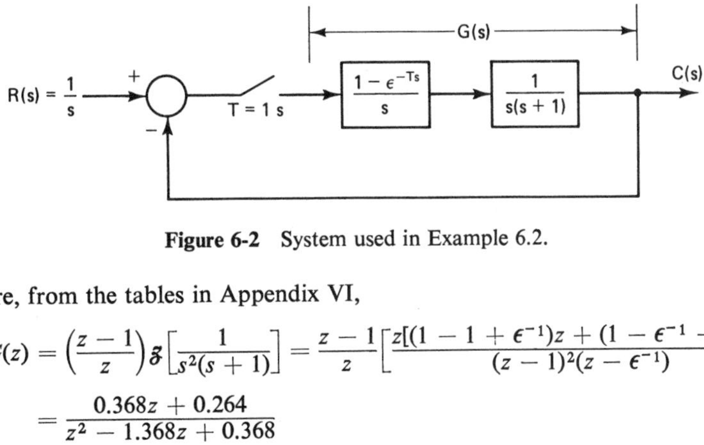

Figure 6-2 System used in Example 6.2.

where, from the tables in Appendix VI,

$$G(z) = \left(\frac{z-1}{z}\right)\mathcal{Z}\left[\frac{1}{s^2(s+1)}\right] = \frac{z-1}{z}\left[\frac{z[(1 - 1 + \epsilon^{-1})z + (1 - \epsilon^{-1} - \epsilon^{-1})]}{(z-1)^2(z-\epsilon^{-1})}\right]$$

$$= \frac{0.368z + 0.264}{z^2 - 1.368z + 0.368}$$

Then

$$\frac{G(z)}{1 + G(z)} = \frac{0.368z + 0.264}{z^2 - z + 0.632}$$

Since

$$R(z) = \frac{z}{z-1}$$

then

$$C(z) = \frac{z(0.368z + 0.264)}{(z-1)(z^2 - z + 0.632)} = 0.368z^{-1} + 1.00z^{-2} + 1.40z^{-3}$$
$$+ 1.40z^{-4} + 1.15z^{-5} + 0.90z^{-6} + 0.80z^{-7} + 0.87z^{-8} \tag{6-1}$$
$$+ 0.99z^{-9} + 1.08z^{-10} + 1.08z^{-11} + 1.00z^{-12} + 0.98z^{-13}$$
$$+ \cdots$$

The final value of $c(nT)$, obtained using the final-value theorem, is

$$\lim_{n\to\infty} c(nT) = \lim_{z\to 1} (z-1)C(z) = \frac{0.632}{0.632} = 1$$

The step response for this system is plotted in Figure 6-3. The response between sampling instants was obtained from a simulation of the system, and, of course, the response at the sampling instants is given in (6-1). Also plotted in Figure 6-3 is the

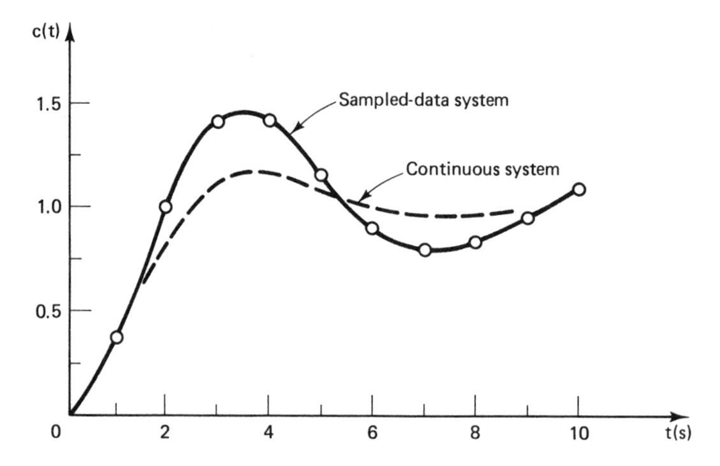

Figure 6-3 Step response of the systems analyzed in Example 6.2.

response of the system with the sampler and data hold removed. For this continuous-time system, the transfer function is, in standard notation [8],

$$\frac{C(s)}{R(s)} = \frac{\omega_n^2}{s^2 + 2\zeta\omega_n s + \omega_n^2} = \frac{1}{s^2 + s + 1}$$

Therefore, $\omega_n = 1$, $\zeta = 0.5$, and hence the overshoot is approximately 18%. Thus, as shown in Figure 6-3, the sampling has a destabilizing effect on the system. In general, it is desirable that the effects of sampling be negligible, that is, that the continuous system response and the discrete system response be approximately equal. For this system the sampling period and the plant time constant are equal. Hence the sampling frequency is too low, and should be increased, if allowed by hardware constraints. These effects will be discussed in detail later.

The system time response can also be calculated using a difference-equations approach. From the expression for the closed-loop transfer function

$$\frac{C(z)}{R(z)} = \frac{0.368z^{-1} + 0.264z^{-2}}{1 - z^{-1} + 0.632z^{-2}} \tag{6-2}$$

or

$$C(z)[1 - z^{-1} + 0.632z^{-2}] = R(z)[0.368z^{-1} + 0.264z^{-2}] \tag{6-3}$$

Taking the inverse z-transform of (6-3), we obtain the difference equation

$$\begin{aligned} c(kT) &= 0.368r[(k-1)T] + 0.264r[(k-2)T] \\ &\quad + c[(k-1)T] - 0.632c[(k-2)T] \end{aligned} \tag{6-4}$$

Both $c(kT)$ and $r(kT)$ are zero for $k < 0$. Thus, from (6-4), $c(0) = 0$ and $c(1) = 0.368$. For $k \geq 2$, (6-4) becomes

$$c(kT) = 0.632 + c[(k-1)T] - 0.632c[(k-2)T]$$

Solving (6-4) for $c(kT)$ yields the same values as found in (6-1).

In this section the time responses of two different sampled-data systems were calculated using a z-transform approach. In addition, the continuous output was calculated for the first example, using the Laplace transform. Because of the complexities involved, it should be evident to the reader that in general we do not calculate the continuous output. In fact, system response is normally determined from either a digital simulation or a hybrid simulation of the system, and not from a transform approach. For high-order systems, simulation is the only practical technique for calculating the time response.

In many of the examples presented, the sampling frequency has been purposely chosen low, for two reasons. First, a sample period of 1 s is often selected to make the numerical calculations simpler. Second, if the sample period is large, only a few terms in the series expansion of $C(z)$ are required to give a good indication of the character of the response. In Example 6.2, if $T = 0.1$ s, the response of the sampled-date system is approximately the same as that of the continuous system. However, for $T = 0.1$ s, 21 terms in the series for $C(z)$ are required to obtain the system response from $t = 0$ to $t = 2$ s. Hence the only practical technique of calculating the system response for this system is by simulation (which is usually the case).

6.3 SYSTEM CHARACTERISTIC EQUATION

Consider a sampled-data system of the type shown in Figure 6-2. For this system,

$$C(z) = \frac{G(z)R(z)}{1 + G(z)} = \frac{K \prod^{m} (z - z_i)}{\prod^{n} (z - p_i)} R(z)$$

Using the partial-fraction expansion, we can express $C(z)$ as

$$C(z) = \frac{k_1 z}{z - p_1} + \cdots + \frac{k_n z}{z - p_n} + C_R(z) \tag{6-5}$$

where $C_R(z)$ contains the terms of $C(z)$ which originate in the poles of $R(z)$. The first n terms of (6-5) are the transient-response terms of $C(z)$ [i.e., the terms that are always present in $C(z)$]. The inverse z-transform of the ith term yields

$$\mathcal{Z}^{-1}\left[\frac{k_i z}{z - p_i}\right] = k_i (p_i)^k$$

It is seen that these terms determine the nature, or character, of the system transient response. Since the p_i originate in the roots of the equation

$$1 + G(z) = 0$$

this equation is then the system characteristic equation. The roots of the characteristic equation are the poles of the closed-loop transfer function. If a transfer function cannot be written, the roots of the characteristic equation are the poles of $C(z)$ that are independent of the input function.

6.4 MAPPING THE s-PLANE INTO THE z-PLANE

It is of interest to be able to assign response characteristics to transfer-function pole locations (characteristic-equation zero locations) in the z-plane in the same manner as is done in the s-plane for continuous-time systems. To gain insight into the characteristics of pole locations in the z-plane, several mappings will be considered.

Consider first the mapping of the left-half-plane portion of the primary strip of the s-plane into the z-plane as shown in Figure 6-4. Along the $j\omega$-axis,

$$z = \epsilon^{sT} = \epsilon^{\sigma T}\epsilon^{j\omega T} = \epsilon^{j\omega T} = \cos \omega T + j \sin \omega T = 1\underline{/\omega T}$$

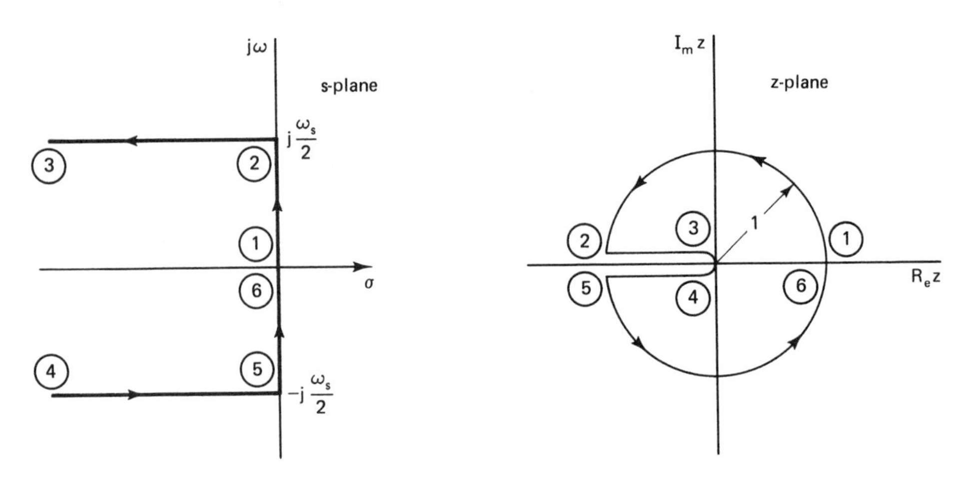

Figure 6-4 Mapping the primary strip into the z-plane.

Since for $\omega = \omega_s/2$, $\omega T = \pi$, then the $j\omega$-axis between $-j(\omega_s/2)$ and $+j(\omega_s/2)$ maps into the unit circle in the z-plane. In fact, any portion of the $j\omega$ axis of length ω_s maps into the unit circle in the z-plane. The right-half-plane portion of the primary strip maps into the exterior of the unit circle, and the left-half-plane portion of the primary strip maps into the interior of the unit circle. Thus, since the stable region of the s-plane is the left half-plane, the stable region of the z-plane is the interior of the unit circle. Stability will be discussed in detail in Chapter 7.

Constant damping loci in the s-plane (i.e., straight lines with σ constant) map into circles in the z-plane as shown in Figure 6-5. This can be seen using the relationship

$$z = \epsilon^{\sigma_1 T}\epsilon^{j\omega T} = \epsilon^{\sigma_1 T}\underline{/\omega T}$$

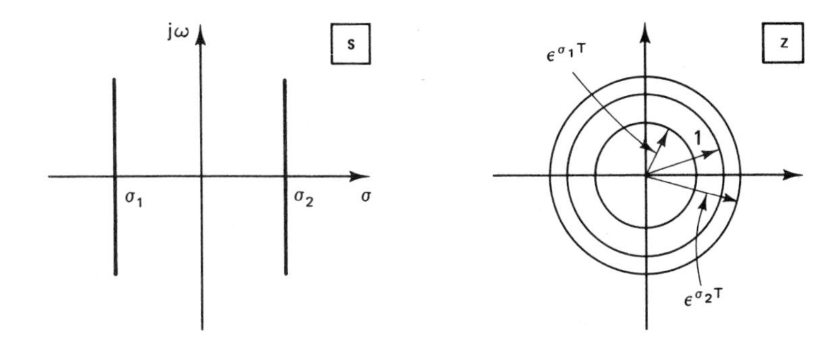

Figure 6-5 Mapping constant damping loci into the z-plane.

Constant frequency loci in the s-plane map into rays as shown in Figure 6-6. For constant damping ratio loci, σ and ω are related by

$$\frac{\omega}{\sigma} = \tan \beta$$

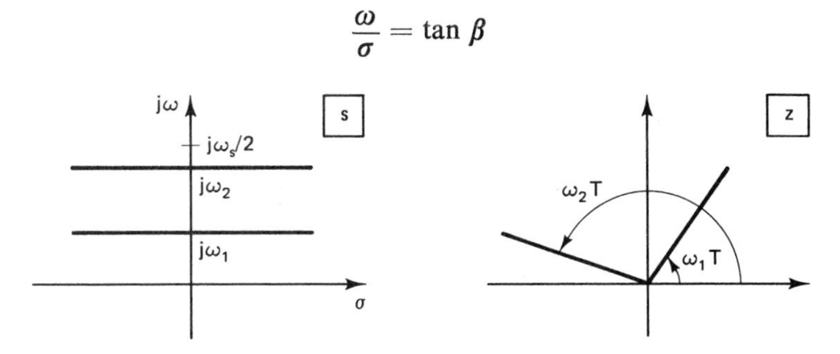

Figure 6-6 Mapping constant frequency loci into the z-plane.

where β is constant. Then

$$z = \epsilon^{sT} = \epsilon^{\sigma T} \underline{/\sigma T \tan \beta} \tag{6-6}$$

Since σ is negative in the second and third quadrants of the s-plane, (6-6) describes a logarithmic spiral whose amplitude decreases with σ increasing in magnitude. This is illustrated in Figure 6-7.

The characteristics of a sampled time function at the sampling instants are the same as those of the time function before sampling. That is, a sampled exponential function is, of course, exponential in nature at the sampling instants. Thus, using the

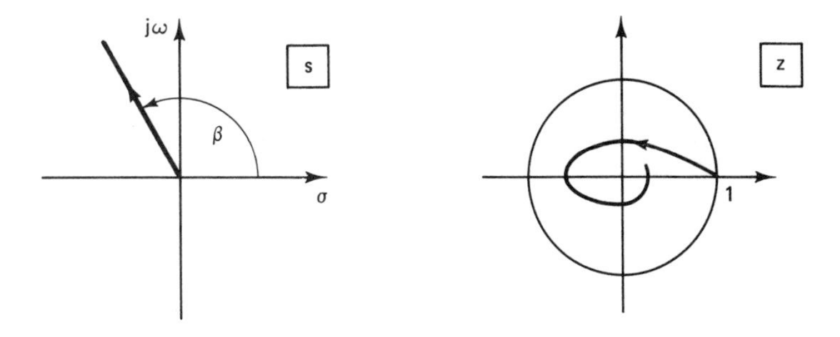

Figure 6-7 Mapping constant damping-ratio loci into the z-plane.

mappings illustrated in Figures 6-4 through 6-7, we may assign time-response characteristics to characteristic-equation root locations in the z-plane.

Corresponding pole locations between the s-plane and the z-plane are shown in Figure 6-8. The time-response characteristics of the z-plane pole locations are illus-

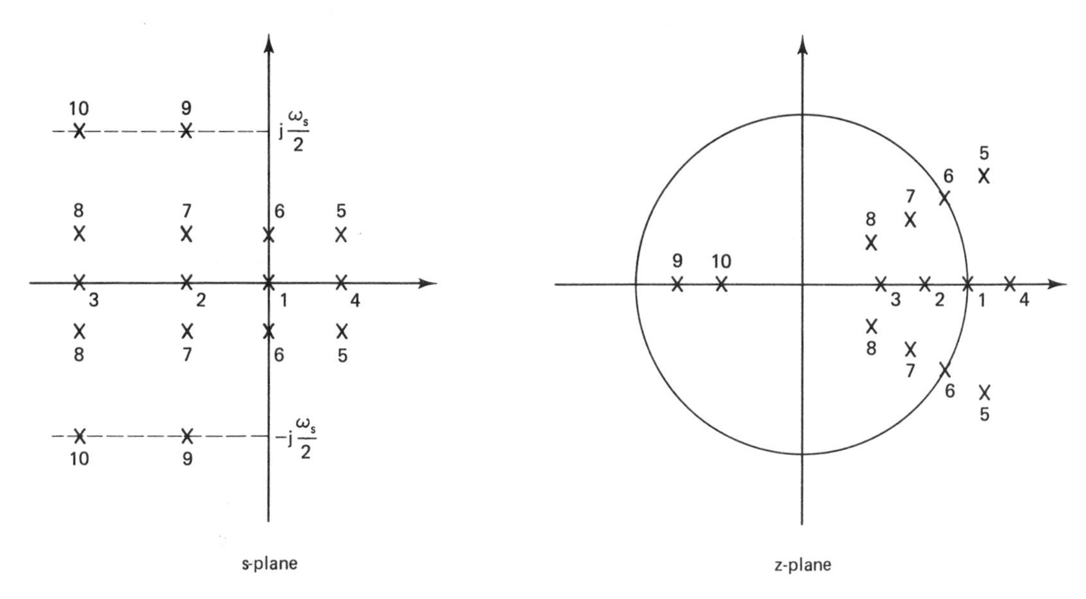

Figure 6-8 Corresponding pole locations between the s-plane (a) and the z-plane (b).

trated in Figure 6-9. The reader should note that since $z = \epsilon^{sT}$, the response characteristics are a function of both s and T. In addition, the response characteristics shown in Figure 6-9 apply not only to systems with closed-loop transfer functions but also to systems for which no closed-loop transfer function exists, since in both cases the response is determined by the roots of the system's characteristic equation.

Consider the case in Figure 6-8 that s-plane poles occur at $s = \sigma \pm j\omega$. These poles result in a system transient-response term of the form $k_1 \epsilon^{\sigma t} \cos(\omega t + \phi)$. When sampling occurs, these s-plane poles result in z-plane poles at

$$z = \epsilon^{sT}\big|_{s=\sigma\pm j\omega} = \epsilon^{\sigma T}\epsilon^{\pm j\omega T} = \epsilon^{\sigma T}\underline{/\pm\omega T} = r\underline{/\pm\theta} \tag{6-7}$$

Thus roots of the characteristic equation that appear at $z = r\underline{/\pm\theta}$ result in a transient-response term of the form

$$k_1 \epsilon^{\sigma kT} \cos(\omega kT + \phi) = k_1 (r)^k \cos(\theta k + \phi)$$

Example 6.3

As an example, the frequency that appears in the transient response of the system of Example 6.1 will be calculated. The step response for this system is shown in Figure 6-1. The closed-loop transfer function for this system was found to be

$$\frac{G(z)}{1 + G(z)} = \frac{0.00778z + 0.00758}{z^2 - 1.90z + 0.9307}$$

Hence the characteristic equation is

$$z^2 - 1.90z + 0.9307 = 0$$

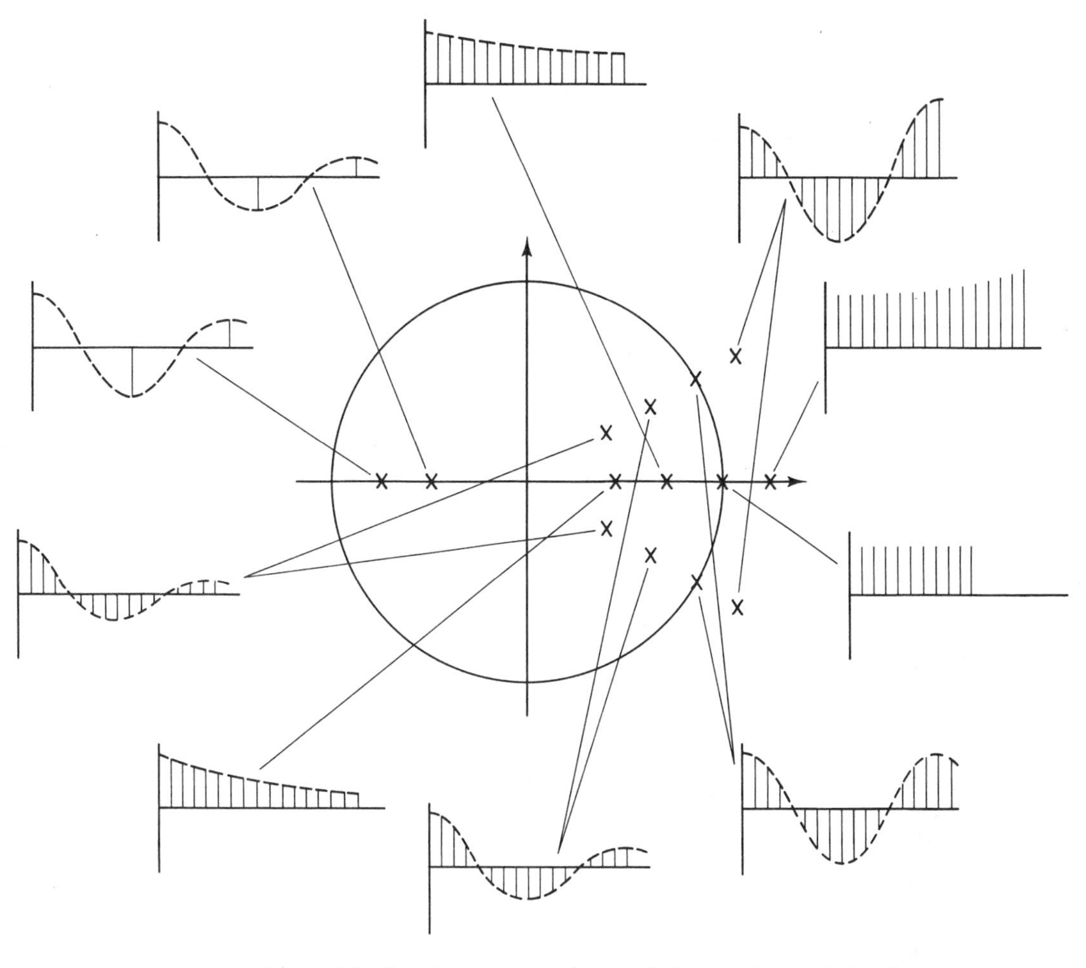

Figure 6-9 Transient response characteristics of z-plane pole locations.

which has the roots

$$z = 0.95 \pm j0.168 = 0.965\underline{/\pm 10.03^\circ} = 0.965\underline{/\pm 0.175 \text{ rad}}$$

We compare these root locations to the general relationship in (6-7):

$$z = \epsilon^{\sigma T}\underline{/\pm \omega T} = 0.965\underline{/\pm 0.175} \text{ rad}$$

Then, since $T = 0.02$ s, $\sigma = -1.80$ (and the time constant is $1/1.80 = 0.56$ s) and $\omega = 8.75$ rad/s.

We will now mathematically relate the s-plane pole locations and the z-plane pole locations. We express in standard form the s-plane second-order transfer function

$$G(s) = \frac{\omega_n^2}{s^2 + 2\zeta\omega_n s + \omega_n^2}$$

which has the poles

$$s_{1,2} = -\zeta\omega_n \pm j\omega_n\sqrt{1 - \zeta^2}$$

where ζ is the damping ratio and ω_n is the natural frequency. The equivalent z-plane poles occur at

$$z = \epsilon^{sT}\big|_{s_{1,2}} = \epsilon^{-\zeta\omega_n T}\underline{/\pm\omega_n T\sqrt{1-\zeta^2}} = r\underline{/\pm\theta}$$

Hence

$$\epsilon^{-\zeta\omega_n T} = r$$

or

$$\zeta\omega_n T = -\ln r \tag{6-8}$$

Also,

$$\omega_n T\sqrt{1-\zeta^2} = \theta \tag{6-9}$$

Taking the ratio of (6-8) and (6-9), we obtain

$$\frac{\zeta}{\sqrt{1-\zeta^2}} = \frac{-\ln r}{\theta}$$

Solving this equation for ζ yields

$$\zeta = \frac{-\ln r}{\sqrt{\ln^2 r + \theta^2}} \tag{6-10}$$

We find ω_n by substituting (6-10) in (6-8).

$$\omega_n = \frac{1}{T}\sqrt{\ln^2 r + \theta^2} \tag{6-11}$$

The time constant, τ, of the poles is then given by

$$\tau = \frac{1}{\zeta\omega_n} = \frac{-T}{\ln r} \tag{6-12}$$

Thus, given the complex pole location in the z-plane, we find the damping ratio, the natural frequency, and the time constant of the pole from (6-10), (6-11), and (6-12), respectively.

Example 6.4

We will consider the system of Example 6.1. We found the closed-loop system transfer function to be

$$\frac{G(z)}{1+G(z)} = \frac{0.00778z + 0.00758}{z^2 - 1.90z + 0.9307}$$

Thus the pole locations are, from Example 6.3,

$$z_{1,2} = 0.95 \pm j0.168 = 0.965\underline{/\pm 0.175} \text{ rad}$$

Hence, from (6-10), (6-11), and (6-12), respectively,

$$\zeta = \frac{-\ln(0.965)}{[\ln^2(0.965) + (0.175)^2]^{1/2}} = 0.199$$

$$\omega_n = \frac{1}{0.02}[\ln^2(0.965) + (0.175)^2]^{1/2} = 8.93$$

$$\tau = \frac{-0.02}{\ln(0.965)} = 0.56 \text{ s}$$

The equivalent values for the closed-loop analog system (with sampler and data hold removed) were found in Example 6.1 to be $\zeta = 0.22$, $\omega_n = 8.94$, and $\tau = 0.5$ s. Thus there is little effect from the sampling. The interested reader is encouraged to derive the same parameters for Example 6.2 (see Problem 6-16).

A word is in order concerning the probable location of characteristic-equation zeros. Transfer-function pole locations in the s-plane transform change into z-plane pole locations as

$$s + a \longrightarrow z - \epsilon^{-aT}$$

$$(s + a)^2 + b^2 \longrightarrow z^2 - 2z\epsilon^{-aT} \cos bT + \epsilon^{-2aT} = (z - z_1)(z - \bar{z}_1)$$

where $z_1 = \epsilon^{-aT}\epsilon^{jbT} = \epsilon^{-aT}\underline{/bT}$.
The s-plane time constant for the real pole is $1/a$. In order for the sampling to have negligible effect, T must be much less than $1/a$. Thus for the simple pole, the z-plane pole location will be in the vicinity of $z = 1$, since aT is much less than 1. For the complex pole, an additional requirement is that several samples be taken per cycle of the sinusoid, or that $bT < 1$. Thus, once again, $T \ll 1/a$, and since $z_1\bar{z}_1 = \epsilon^{-2aT}$, then $|z_1| = \epsilon^{-aT}$ and the z-plane pole locations again will be in the vicinity of $z = 1$. In general, for a discrete-time control system, the transfer-function pole locations (characteristic-equation zero locations) are placed in the vicinity of $z = 1$, if system constraints allow a sufficiently high sample rate to be chosen.

6.5 STEADY-STATE ACCURACY

An important characteristic of a control system is its ability to follow, or track, certain inputs with a minimum of error. The control system designer attempts to minimize the system error to certain anticipated inputs. In this section the effects of the system transfer characteristics on the steady-state system errors are considered.

Consider the system of Figure 6-10. For this system,

$$\frac{C(z)}{R(z)} = \frac{G(z)}{1 + G(z)} \tag{6-13}$$

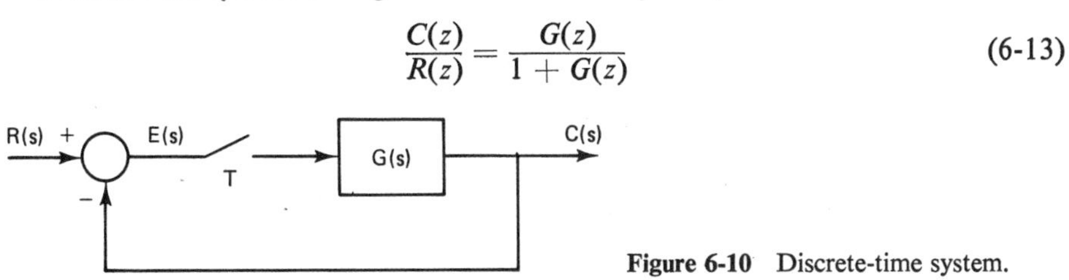

Figure 6-10 Discrete-time system.

where $G(z) = \mathfrak{z}[G(s)]$. The plant transfer function can always be expressed as

$$G(z) = \frac{K\prod^{m}(z - z_i)}{(z - 1)^N \prod^{p}(z - z_j)}, \qquad z_j \neq 1 \tag{6-14}$$

For convenience in the following development, we define

$$K_{\text{dc}} = \left. \frac{K\prod^{m}(z - z_i)}{\prod^{p}(z - z_j)} \right|_{z=1} \tag{6-15}$$

Note that K_{dc} is the open-loop system dc gain with all poles at $z = 1$ removed.

The system error, $e(t)$, is defined as the difference between the system input and the system output. Or

$$E(z) = \mathfrak{z}[e(t)] = R(z) - C(z) \tag{6-16}$$

Then, from (6-13) and (6-16),

$$E(z) = \frac{R(z)}{1 + G(z)}$$

Then, from the final-value theorem, the steady-state error is seen to be

$$e_{ss}(kT) = \lim_{z \to 1} (z - 1)E(z) = \lim_{z \to 1} \frac{(z - 1)R(z)}{1 + G(z)} \tag{6-17}$$

provided that $e_{ss}(kT)$ has a final value.

The steady-state errors will now be derived for two common inputs—a position (step) input and a velocity (ramp) input. First, for the step input,

$$R(z) = \frac{z}{z - 1}$$

The steady-state error, from (6-17), is seen to be

$$e_{ss}(kT) = \lim_{z \to 1} \frac{z}{1 + G(z)} = \frac{1}{1 + \lim_{z \to 1} G(z)}$$

We now define the position error constant as

$$K_p = \lim_{z \to 1} G(z) \tag{6-18}$$

Then in (6-14), if $N = 0$ [i.e., no poles in $G(z)$ at $z = 1$], $K_p = K_{dc}$ and

$$e_{ss}(kT) = \frac{1}{1 + K_p} = \frac{1}{1 + K_{dc}} \tag{6-19}$$

For $N \geq 1$, $K_p = \infty$ and the steady-state error is zero.

Consider next the ramp input. In this case $r(t) = t$, and

$$R(z) = \frac{Tz}{(z - 1)^2}$$

Then, from (6-17),

$$e_{ss}(kT) = \lim_{z \to 1} \frac{Tz}{(z - 1) + (z - 1)G(z)} = \frac{T}{\lim_{z \to 1} (z - 1)G(z)}$$

We now define the velocity error constant as

$$K_v = \lim_{z \to 1} \frac{1}{T}(z - 1)G(z) \tag{6-20}$$

Then, if $N = 0$, $K_v = 0$ and $e_{ss}(kT) = \infty$. For $N = 1$, $K_v = K_{dc}/T$ and

$$e_{ss}(kT) = \frac{1}{K_v} = \frac{T}{K_{dc}} \tag{6-21}$$

For $N \geq 2$, $K_v = \infty$ and $e_{ss}(kT)$ is zero.

The development above illustrates that, in general, increased system gain and/or the addition of poles at $z = 1$ to the open-loop transfer function tend to decrease steady-state errors. However, as will be demonstrated in Chapter 7, both large system gain and poles of $G(z)$ at $z = 1$ have destabilizing effects on the system. Generally, trade-offs exist between small steady-state errors and adequate system stability (or acceptable system transient response).

Example 6.5

The steady-state errors will be calculated for the system of Figure 6-10, in which the open-loop transfer function is given as

$$G(s) = \left[\frac{1 - \epsilon^{-Ts}}{s}\right]\left[\frac{K}{s(s+1)}\right]$$

Thus

$$G(z) = K\mathfrak{z}\left[\frac{1 - \epsilon^{-Ts}}{s^2(s+1)}\right] = \frac{K(z-1)}{z}\mathfrak{z}\left[\frac{1}{s^2(s+1)}\right]$$

$$= \frac{K(z-1)}{z}\mathfrak{z}\left[\frac{1}{s^2} + \frac{-1}{s} + \frac{1}{s+1}\right]$$

$$= \frac{K(z-1)}{z}\left[\frac{Tz}{(z-1)^2} + \frac{-z}{z-1} + \frac{z}{z-\epsilon^{-T}}\right]$$

$$= \frac{K[(\epsilon^{-T} + T - 1)z + (1 - \epsilon^{-T} - T\epsilon^{-T})]}{(z-1)(z-\epsilon^{-T})}$$

Then, from (6-14) and (6-20),

$$K_v = \frac{K_{dc}}{T} = \frac{K[(\epsilon^{-T} + T - 1) + (1 - \epsilon^{-T} - T\epsilon^{-T})]}{T(1-\epsilon^{-T})} = K$$

Since $G(z)$ has one pole at $z = 1$, the steady-state error to a step input is zero, and to a ramp input is, from (6-21),

$$e_{ss}(kT) = \frac{1}{K_v} = \frac{1}{K}$$

provided that the system is stable. The question of stability is considered in Chapter 7.

Example 6.6

As a second example, consider again the system of Figure 6-10, where

$$G(z) = \mathfrak{z}\left[\frac{1 - \epsilon^{-Ts}}{s(s+1)}\right] = \frac{1 - \epsilon^{-T}}{z - \epsilon^{-T}}$$

Suppose that the design specification for this system requires that the steady-state error to a unit ramp input be less than 0.01. Thus, from (6-20), it is necessary that the open-loop transfer function have a pole at $z = 1$. Since the plant does not contain a pole at $z = 1$, a digital compensator of the form

$$D(z) = \frac{K_I z}{z-1} + K_p$$

will be added, to produce the resultant system shown in Figure 6-11. The compensator, called a PI or proportional-plus-integral compensator, is of a form commonly used to reduce steady-state errors. This compensator is discussed in Chapter 8. For this system (6-20) becomes

$$K_v = \lim_{z \to 1} \frac{1}{T}(z-1)D(z)G(z)$$

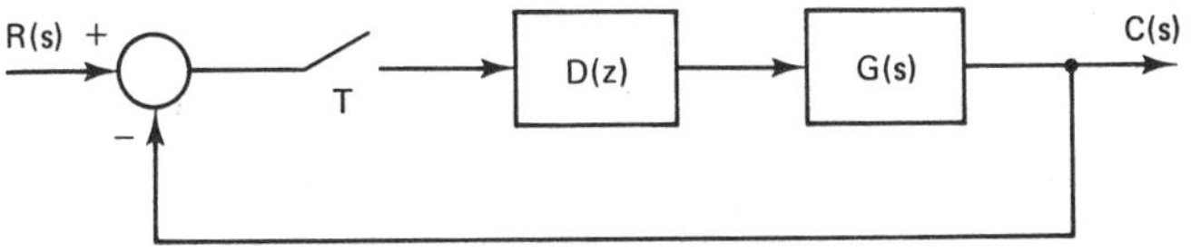

Figure 6-11 System for Example 6.5.

Employing the expressions above for $D(z)$ and $G(z)$, we see that

$$K_v = \lim_{z \to 1} (z-1) \left[\frac{(K_I + K_p)z - K_p}{T(z-1)}\right] \left[\frac{1-\epsilon^{-T}}{z-\epsilon^{-T}}\right] = \frac{K_I}{T}$$

Thus K_I equals $100T$ for the required steady-state error, provided that the system is stable. The latter point is indeed an important consideration since the error analysis is meaningless unless stability of the system is guaranteed. Chapter 7 illustrates a number of techniques for analyzing system stability.

6.6 SIMULATION

Thus far we have determined the time response of a system via a transform approach. In Example 6.1 we determined the system response at the sampling instants using the z-transform. A Laplace transform technique for calculating the system response for all time was also illustrated, but was discarded for being too unwieldy.

A different approach for determining a system's response is through simulation. The simulation of a continuous-time (analog) system may be via the integration of the system's differential equations using electronic circuits. The analog computer is designed to perform this function. The interconnection of a digital computer with the analog computer is called a hybrid computer and is useful in simulating digital control systems. The analog parts of the control system are simulated on the analog computer with the digital controller simulated on the digital computer.

If a numerical algorithm is used to integrate a system's differential equations on a digital computer, we then have a digital simulation of the system. This is the type of simulation that we consider here.

The problem of numerical integration of a time function can be illustrated using Figure 6-12. We wish to numerically integrate $y(t)$; that is, we wish to find $x(t)$, where

$$x(t) = \int_0^t y(t)\, dt + x(0) \tag{6-22}$$

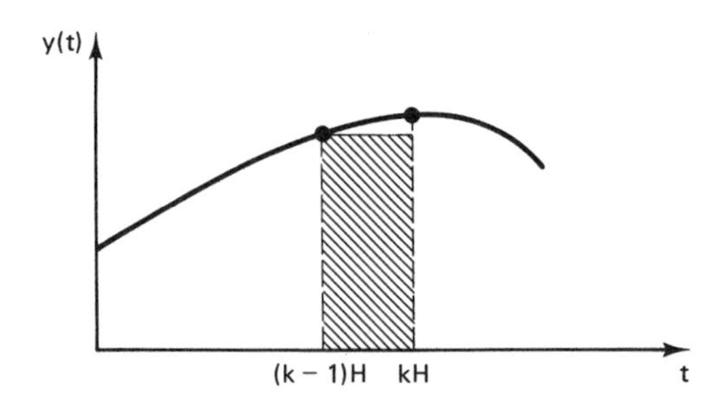

Figure 6-12 Rectangular rule for numerical integration.

Suppose that we know $x[(k-1)H]$ and we want to calculate $x(kH)$, where H is called the numerical integration increment and is the step size in the algorithm. Perhaps the simplest numerical integration algorithm is obtained by assuming $y(t)$ constant at the value $y[(k-1)H]$ for $(k-1)H \leq t < kH$. Then

$$x(kH) = x[(k-1)H] + Hy[(k-1)H] \tag{6-23}$$

Of course, $x(kH)$ in (6-23) is only an approximation to $x(t)$ in (6-22) evaluated at $t = kH$. Note in Figure 6-12 that we are approximating the area under the $y(t)$ curve for $(k - 1)H \leq t < kH$ with the area of the rectangle shown. For this reason, this numerical integration algorithm is called the rectangular rule, and is also known as the Euler method.

We are interested in the integration of differential equations [i.e., in (6-22), $y(t) = \dot{x}(t)$]. We will illustrate this case with a simple example. Suppose that we wish to solve the differential equation

$$\dot{x}(t) + x(t) = 0, \qquad x(0) = 1 \tag{6-24}$$

The solution is obviously

$$x(t) = \epsilon^{-t}$$

However, by numerical integration in (6-23),

$$x(kH) = x[(k - 1)H] + H[-x(k - 1)H] \tag{6-25}$$

since, from (6-24), for this differential equation

$$y(t) = \dot{x}(t) = -x(t)$$

Suppose that we choose $H = 0.1$ s. Then, solving (6-25) iteratively starting with $k = 1$ [we know $x(0)$],

$$x(0.1) = x(0) - Hx(0) = 1.0 - (0.1)(1.0) = 0.9$$

$$x(0.2) = x(0.1) - Hx(0.1) = 0.9 - 0.09 = 0.81$$

$$\cdots$$

$$x(1.0) = x(0.9) - Hx(0.9) = 0.3487$$

Since, for this example, we know the solution, we calculate $x(t)$ at $t = 1.0$ s as

$$x(1.0) = \epsilon^{-1.0} = 0.3678$$

and we can see the error due to the numerical integration.

In the example above, if we choose H larger, the error is larger. If we decrease H, the error decreases to a minimum value. Then a further decrease in H will result in an increase in the error, due to round-off in the computer. With H equal to 0.1 s, 10 iterations are required to calculate $x(1)$. With $H = 0.001$ s, 1000 iterations are required to calculate $x(1)$. The round-off errors in the computer are larger for the latter case, since more calculations are made. If H is made sufficiently smaller, this round-off error becomes appreciable.

Next we will consider the simulation of an analog system. Assume that the system's state equations are given by

$$\dot{\mathbf{x}}(t) = \mathbf{A}_c\mathbf{x}(t) + \mathbf{B}_c\mathbf{u}(t) \tag{6-26}$$

Then (6-22) becomes

$$\mathbf{x}(t) = \int_0^T \dot{\mathbf{x}}(t)\, dt + \mathbf{x}(0) \tag{6-27}$$

and the rectangular rule for the vector case becomes

$$\mathbf{x}(kH) = \mathbf{x}[(k - 1)H] + H\dot{\mathbf{x}}[(k - 1)H] \tag{6-28}$$

where, from (6-26),

$$\dot{\mathbf{x}}[(k - 1)H] = \mathbf{A}_c\mathbf{x}[(k - 1)H] + \mathbf{B}_c\mathbf{u}[(k - 1)H] \tag{6-29}$$

The numerical integration algorithm becomes:

1. Let $k = 1$.
2. Evaluate $\dot{\mathbf{x}}[(k-1)H]$ in (6-29).
3. Evaluate $\mathbf{x}(kH)$ in (6-28).
4. Let $k = k + 1$.
5. Go to step 2.

Refer again to Figure 6-12. This figure illustrates the rectangular rule for numerical integration. Other numerical integration algorithms differ from this rule in the manner that $\mathbf{x}[kH]$ is calculated, based on the knowledge of $\dot{\mathbf{x}}[(k-1)H]$. As an example of a different algorithm, consider the trapezoidal rule illustrated in Figure 6-13. The integral for $[(k-1)H] \leq t < kH$ is approximated by the area of the trapezoid. For this rule, with $x(t)$ equal to the integral of $y(t)$,

$$x(kH) = x[(k-1)H] + H\left[\frac{y[(k-1)H] + y(kH)}{2}\right] \tag{6-30}$$

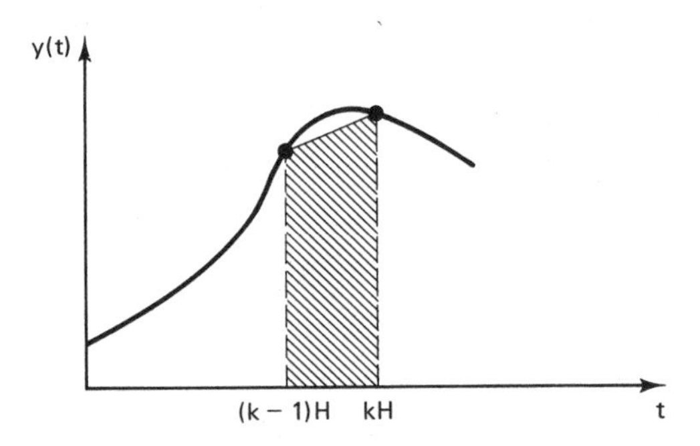

Figure 6-13 Trapezoidal rule for numerical integration.

For $y(t)$ equal to $\dot{x}(t)$, then

$$x(kH) = x[(k-1)H] + \frac{H}{2}[\dot{x}[(k-1)H] + \dot{x}(kH)] \tag{6-31}$$

Since $x[(k-1)H]$ is known, $\dot{x}[(k-1)H]$ can be calculated from the differential equation that is to be solved. However, we cannot calculate $\dot{x}(kH)$ without a knowledge of $x(kH)$. A method for solving this problem will now be presented.

We use a different rule to "predict" the value of $x(kH)$. Using this predicted value, we calculate the predicted value of $\dot{x}(kH)$, and substitute this value in the trapezoidal rule to "correct" the value of $x(kH)$. A method of this type is called a predictor–corrector algorithm. A commonly used rule for prediction is the rectangular rule.

The state model of an analog system given in (6-26) will be used to illustrate the trapezoidal rule. It is assumed that $\mathbf{x}[kH]$ is known for $k = 0$, and thus $\mathbf{x}[kH]$ is to be calculated, for $k = 1, 2, 3, \ldots$. The predictor algorithm, which is the rectangular rule, is given by

$$\dot{\mathbf{x}}[(k-1)H] = \mathbf{A}_c\mathbf{x}[(k-1)H] + \mathbf{B}_c\mathbf{u}[(k-1)H] \tag{6-32}$$

$$\mathbf{x}(kH) = \mathbf{x}[(k-1)H] + H\dot{\mathbf{x}}[(k-1)H] \tag{6-33}$$

The corrector algorithm is given by

$$\dot{\mathbf{x}}(kH) = \mathbf{A}_c\mathbf{x}(kH) + \mathbf{B}_c\mathbf{u}(kH) \tag{6-34}$$

$$\mathbf{x}(kH) = \mathbf{x}[(k-1)H] + \frac{H}{2}[\dot{\mathbf{x}}[(k-1)H] + \dot{\mathbf{x}}(kH)] \tag{6-35}$$

The predicted value of $\mathbf{x}(kH)$ is given by (6-33), and the corrected, and final, value of $\mathbf{x}(kH)$ is given by (6-35).

Thus the numerical integration algorithm is given by:

1. Let $k = 1$.
2. Evaluate $\dot{\mathbf{x}}[(k-1)H]$ in (6-32).
3. Evaluate $\mathbf{x}(kH)$ in (6-33).
4. Evaluate $\dot{\mathbf{x}}(kH)$ in (6-34), using the result of step 3.
5. Evaluate the final value of $\mathbf{x}(kH)$ in (6-35).
6. Let $k = k + 1$.
7. Go to step 2.

This algorithm obviously requires more calculations per iteration than does the rectangular rule. However, for a specified accuracy of the solution, a much larger increment H may be used such that the total computer execution time for the trapezoidal rule is less than that for the rectangular rule. The trapezoidal rule will now be illustrated by an example.

Example 6.7

Consider once again the differential equation

$$\dot{x}(t) + x(t) = 0, \qquad x(0) = 1$$

which has the solution

$$x(t) = \epsilon^{-t}$$

Since

$$\dot{x}(t) = -x(t)$$

implementing the trapezoidal rule for $H = 0.1$ s yields

$K = 1$:

$$\dot{x}(0) = -x(0) = -1$$
$$x(0.1) = x(0) + H\dot{x}(0) = 1 + 0.1(-1) = 0.9$$
$$\dot{x}(0.1) = -x(0.1) = -0.9$$
$$x(0.1) = x(0) + H/2[\dot{x}(0) + \dot{x}(0.1)]$$
$$= 1 + 0.05(-1 - 0.9) = 0.905$$

$K = 2$:

$$\dot{x}(0.1) = -x(0.1) = -0.905$$
$$x(0.2) = x(0.1) + H\dot{x}(0.1) = 0.905 - 0.0905 = 0.8145$$
$$\dot{x}(0.2) = -x(0.2) = -0.8145$$
$$x(0.2) = x(0.1) + H/2[\dot{x}(0.1) + \dot{x}(0.2)]$$
$$= 0.905 + 0.05(-0.905 - 0.8145) = 0.8190$$

for the first two iterations. If these calculations are continued, we find $x(1.0)$ to be 0.3685. Recall that the exact value of $x(1)$ is 0.3678, and the value of $x(1)$ calculated above using the rectangular rule is 0.3487. Hence much greater accuracy results from use of the trapezoidal rule as compared to the rectangular rule, but at the expense of more calculations. If H is increased to 0.333 s, $x(1)$ is calculated to be 0.3767 with the trapezoidal rule. Here the accuracy is still significantly greater than that of the rectangular rule with $H = 0.1$ s, although the amount of calculations required is approximately the same.

In the discussion above, only two of many numerical integration rules were discussed. One of the most commonly used rules for digital simulation is the fourth-order Runge–Kutta rule [1–4]. The interested reader is referred to the many texts available in this area.

Two additional points concerning simulation should be made. First, in digital simulations we approximate differential equations by difference equations and solve the difference equations. Thus we are replacing a continuous-time system with a discrete-time system that has approximately the same response.

The second point is that, in the simulation of nonlinear systems, the nonlinearities appear only in the calculation of $\dot{\mathbf{x}}[(k-1)H]$, given $\mathbf{x}[(k-1)H]$. If the nonlinearities are of a form that can be easily expressed in a mathematical form, the simulation is not appreciably more difficult to write.

In the discussion above, we considered the simulation of analog systems only. The addition of a sampler and zero-order hold to an analog system requires that logic be added to the system simulation. The logic will hold the zero-order-hold output constant over any sample period, and equal to the value of sampler input at the beginning of that sample period. The addition of a digital controller requires that the controller difference equation be solved only at the beginning of each sample period, and that the controller output then remain constant over the sample period.

6.7 SIMULATION PROGRAMMING LANGUAGES

The basic ideas of numerical integration and digital simulation were presented in the preceding section. These ideas may be utilized to develop a digital simulation of a digital control system, written in a high-level programming language such as FORTRAN or BASIC. For a high-order control system with nonlinearities, the development of such a digital simulation is difficult and time consuming. To aid in the development of digital simulations, many application-oriented simulation languages have been written. One of these languages, the Continuous System Modeling Program (CSMP), is introduced in this section.

CSMP was developed by IBM, and although the name implies the simulation of only continuous systems, discrete-time systems are easily simulated using the language. Only a small number of the capabilities of CSMP are covered here; those readers interested in a complete description are referred to Ref. 4.

In any digital simulation, the structure and the parameters of the system must be specified. CSMP allows the system structure to be given as block diagrams and/or equations. In this development, only the equations approach will be given. As was seen in the preceding section, the system's state equations must be given. The basic statement required for simulation is one that implements integration, and for CSMP the programming statement is

```
Y = INTGRL (IC,X)
```

which implements the integral

$$Y = \int_0^t X\,dt + \mathrm{IC}$$

where IC is the initial condition on X.

To implement the timing required for sampling, the statement

```
F = IMPULS (t0, T)
```

is available. In the execution of this statement, F is equal to unity at times t_0, $t_0 + T$, $t_0 + 2T, \ldots$. At all other times F is zero. The zero-order hold is implemented through the statement

```
Y = ZHOLD (F, X1)
```

For this statement, Y is equal to X1 for $F > 0$, and does not change value from the last iteration for $F \leq 0$. For example, the two statements

```
F = IMPULS (0., 0.1)
Y = ZHOLD (F, X1)
```

implement a sampler/zero-order hold, with X1 the input, Y the output, and the sample period T equal to 0.1 s.

The data statement

```
PARAMETER A = 1.0 , B = (2.0, 2.5)
```

is utilized to set parameter values. In this statement the parameter A is equal to 1.0. The use of parentheses in a PARAMETER statement is a facility that allows multiple simulations to be run. For the statement above, two complete simulations will be run, first with B equal to 2.0, and then with B equal to 2.5. The data statement

```
INCON   S1 = 0. , S2 = -3.1
```

allows initial conditions to be set for variables other than those that are the outputs of integrators.

The variables of the simulation may be both printed and plotted. The statement

```
PRTPLT   X1, X1DOT
```

both prints and plots the variables X1 and X1DOT.

The execution control is achieved through the statement

```
TIMER   FINTIM = 10.0 , OUTDEL = 0.1 , DELT = 0.01
```

FINTIM specifies the simulation time of the run in seconds, OUTDEL specifies the print increment for the print-plot output in seconds, and DELT specifies the numerical integration increment.

The final statement needed is for the specification of the numerical integration algorithm. Eight algorithms are available; a fixed-increment method should be specified if digital filtering is present in the system. Otherwise, problems can occur at the sampling instants, when the filter equations must be solved. The statement required for specifying the numerical integration algorithm is

```
METHOD    RECT
```

which, in this case, specifies the rectangular rule.

Most FORTRAN statements may be used in CSMP programs. For example, if we wish to specify an input R1 to be cos $2t$, we use the statement

```
R1 = COS (2.*TIME)
```

The independent variable time is available in CSMP; the variable name is TIME. The FORTRAN statements in a CSMP program in general will not be executed in the order that they appear in the program. If a certain sequence of execution is required, this sequence must be preceded by a NOSORT statement, and terminated by a SORT statement. Variable names generally follow the same rules as FORTRAN; that is, the symbolic name of a variable contains one to six alphameric characters. The first character must be alphabetic. However, all variables that are represented by symbolic names are treated as being real.

As an example of a CSMP simulation, we will write a program to simulate the system of Figure 6-14. This is a design example taken from Section 8.11. The simulation is written directly from the simulation diagrams given in Figure 6-15, with the variable names as shown. The number stored in the unit delay in the filter in Figure 6-15 is called S1; this variable is initially set to zero. The simulation will calculate the system step response. The complete simulation is as follows:

```
  INCON    S1 = 0.
  R = 1.
  E = R-X1
  X2DOT = 0.814*Y-X2
  X2 = INTGRL (0., X2DOT)
  X1 = INTGRL (0., X2)
  F = IMPULS (0., 0.1)
  NOSORT
  IF (F.NE.1.) GO TO 1
  M = E + 0.7*S1
  Y = 3.15*M - 2.85*S1
  S1 = M
1 CONTINUE
  SORT
  METHOD RECT
  TIMER FINTIM = 10., OUTDEL = 0.1, DELT = 0.01
  PRTPLT X1,Y
  END
  STOP
```

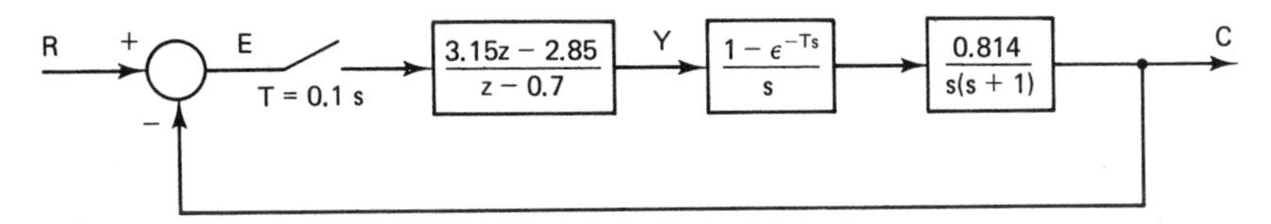

Figure 6-14 System for CSMP simulation.

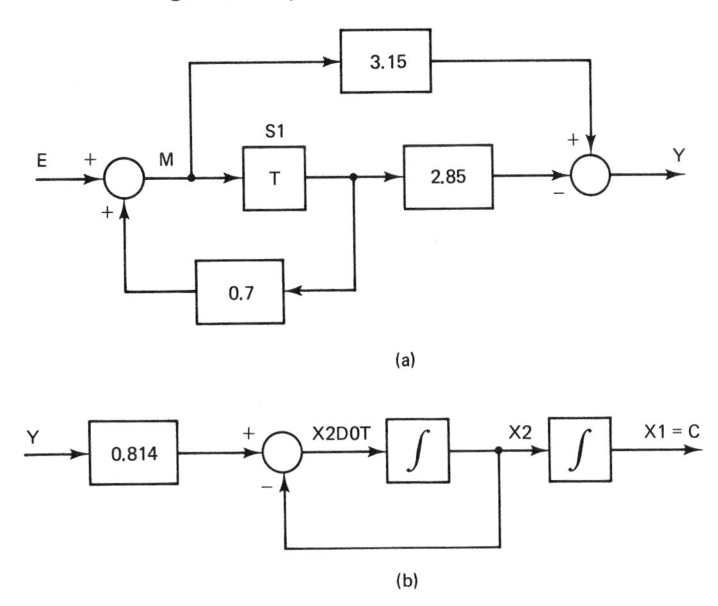

Figure 6-15 Simulation diagrams for CSMP simulation: (a) digital filter; (b) plant.

This section presents a very brief introduction to a simulation language. The capabilities of this language are far greater than those given here. The interested reader is referred to Ref. 4 for a complete description of CSMP.

6.8 SUMMARY

The time response of discrete-time closed-loop systems has been discussed. Both steady-state and transient responses have been considered. The correlation between the s-plane and the z-plane which has been presented provides a mechanism for the transfer of many of the continuous-system tools needed for both the analysis and design of closed-loop discrete-time systems. Finally, a brief introduction into the digital simulation of systems is presented.

REFERENCES AND FURTHER READING

1. M. L. Dertouzos et al., *Systems, Networks, and Computation: Basic Concepts*. New York: McGraw-Hill Book Company, 1972.
2. J. L. Melsa, *Computer Programs for Computational Assistance*. New York: McGraw-Hill Book Company, 1970.

3. A. W. Bennett, *Introduction to Computer Simulation*. St. Paul Minn.: West Publishing Company, 1974.
4. F. H. Speckhart and W. L. Green, *A Guide to Using CSMP—The Continuous System Modeling Program*. Englewood Cliffs, N.J.; Prentice-Hall, Inc., 1976.
5. R. Saucedo and E. E. Schiring, *Introduction to Continuous and Digital Control Systems*. New York: Macmillan Publishing Co., Inc., 1968.
6. B. C. Kuo, *Discrete-Data Control Systems*. Englewood Cliffs, N.J.: Prentice-Hall, Inc., 1970.
7. J. A. Cadzow and H. R. Martens, *Discrete-Time and Computer Control Systems*. Englewood Cliffs, N.J.: Prentice-Hall, Inc., 1970.
8. E. I. Jury, *Sampled-Data Control Systems*. New York: John Wiley & Sons, Inc., 1958.
9. E. I. Jury, *Theory and Application of the z-Transform Method*. New York: John Wiley & Sons, Inc., 1964.
10. B. C. Kuo, *Analysis and Synthesis of Sampled-Data Control Systems*. Englewood Cliffs, N.J.: Prentice-Hall, Inc., 1963.
11. J. R. Ragazzini and G. F. Franklin, *Sampled-Data Control Systems*. New York: McGraw-Hill Book Company, 1958.
12. R. C. Dorf, *Modern Control Systems*, 3rd ed. Reading, Mass.: Addison-Wesley Publishing Company, Inc., 1980.

PROBLEMS

6-1. Shown in Figure P6-1 is a closed-loop temperature control system, which was described in Problem 4-3.

(a) With $D(z) = 1$ and $T = 0.5$ s, evaluate and plot the system response if the input is a 10°C step function. Note that the magnitude of the step at the indicated input will not be equal to 10.

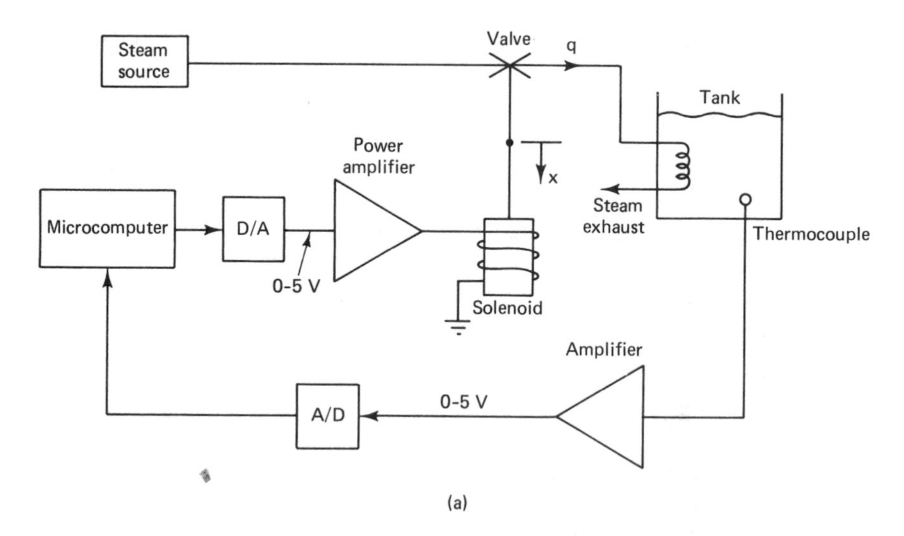

Figure P6-1 Temperature control system.

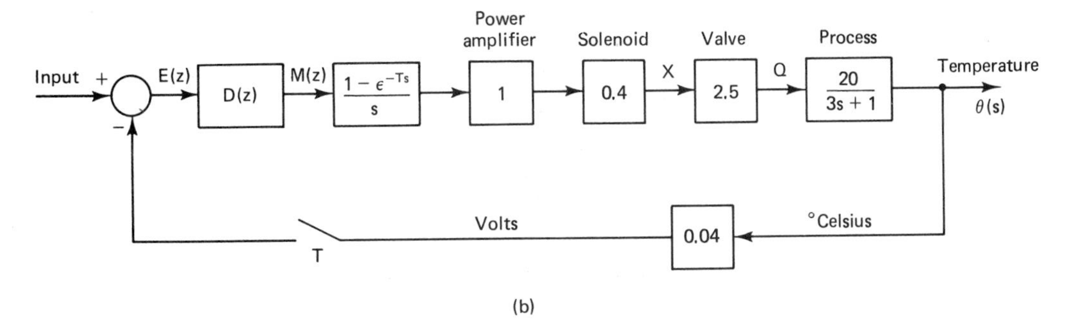

Figure P6-1 cont.

(b) Repeat part (a) with the sampling and compensation removed. Both the sampler and the zero-order hold must be omitted. Plot the response on the same graph as that of part (a).

6-2. Repeat Problem 6-1, with $T = 3$ s. Compare this step response with that obtained in Problem 6-1. Note the effects of increasing T from 0.5 s to 3 s, when the plant has a time constant of 3 s.

6-3. Shown in Figure P6-3 is a closed-loop automobile cruise-control system, which was described in Problem 4-7. Let $T = 0.2$ s, and the disturbance torque be zero.

(a) With the digital controller transfer function equal to unity, evaluate the response to a step input of 5 kilometers per hour (kmh). Since the system is linear, this is also the change in response to any step change of 5 kmh, provided that the car is operating at a constant speed at the time the step is applied. Note that the magnitude of the step at the indicated input will not be equal to 5.

(b) Repeat part (a) with the carburetor transfer function equal to unity. Plot the results of parts (a) and (b) on the same graph for $0 \leq t \leq 20$ s, and note the effects of ignoring the carburetor time constant.

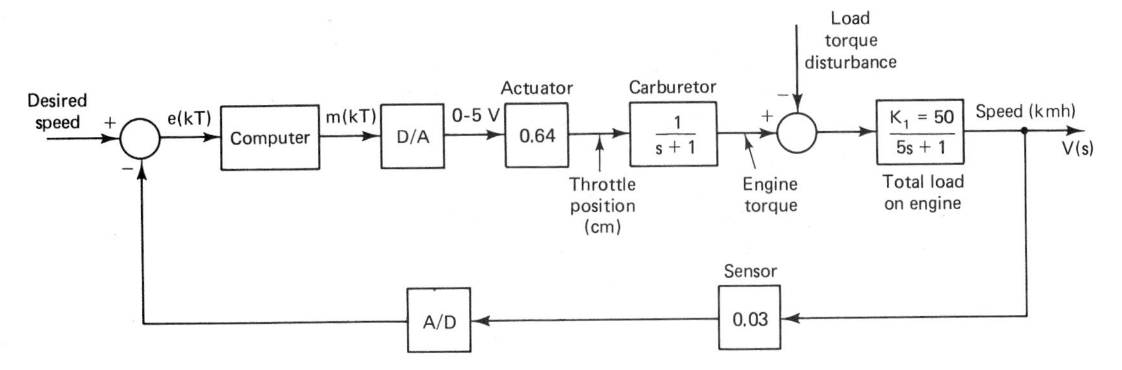

Figure P6-3 Cruise control system.

6-4. Shown in Figure P6-4 is a carbon dioxide control system for an environmental plant growth chamber, which was also considered in Problem 5-18.

(a) The magnitude of maximum signal out of the digital controller in the simplified block diagram is unity. It is desired that the carbon dioxide content change no more than 10 parts per million (ppm) over a sample period. Find the maximum value of K that satisfies this constraint. A z-transform analysis is not necessary.

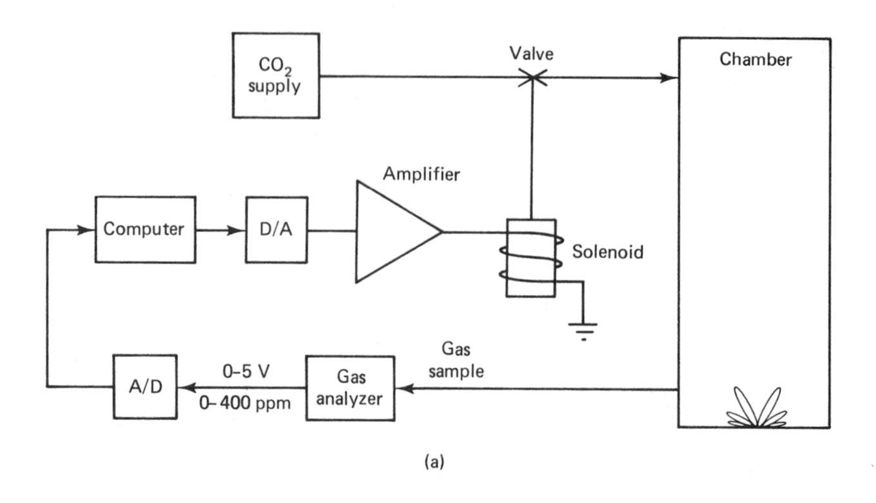

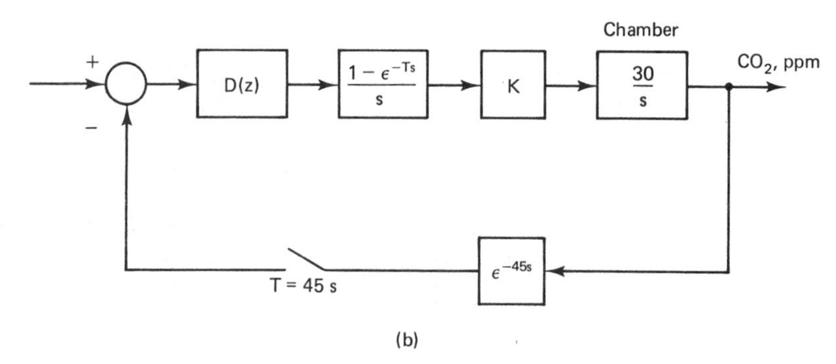

Figure P6-4 Carbon dioxide control system.

(b) With $K = 0.005$, find the response to a step input of 20 ppm. What is the form of the response between sampling instants? Let $D(z) = 1$.

6-5. Consider the system of Figure P6-5. This system is called a regulator control system, in which it is desired to maintain the output, $c(t)$, at a value of zero in the presence of a disturbance, $f(t)$. In this problem the disturbance is a unit step.

(a) With $D(z) = 1$ (i.e., no compensation) find the steady-state value of $c(t)$.

(b) For $f(t)$ to have no effect on the steady-state value of $c(kT)$, $D(z)$ should have a

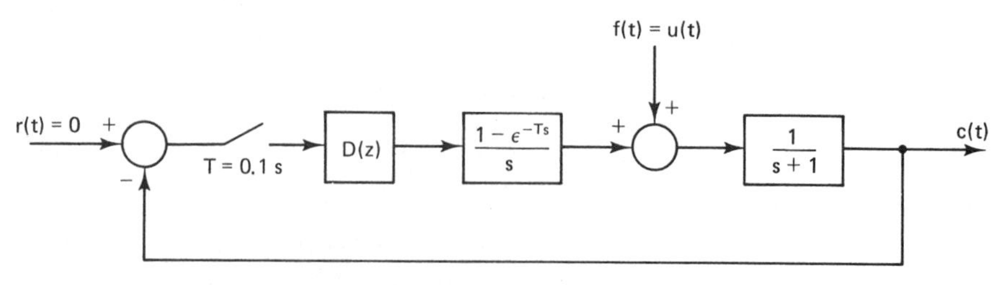

Figure P6-5 System for Problem 6-5.

pole at $z = 1$. Let

$$D(z) = 1 + \frac{0.1z}{z - 1}$$

Determine the steady-state value of $c(kT)$.

6-6. Predict the nature of the transient response of a discrete-time system whose characteristic equation is given by:

(a) $z^2 - z + 0.5 = 0$ **(b)** $z^2 - 1.7z + 0.72 = 0$

(c) $z^2 - z + 1 = 0$ **(d)** $z^2 - z + 2 = 0$

(e) $z^2 - 1 = 0$ **(f)** $z^2 - 0.64 = 0$

6-7. Shown in Figure P6-7 is a solar collector utilized in generating electricity from solar

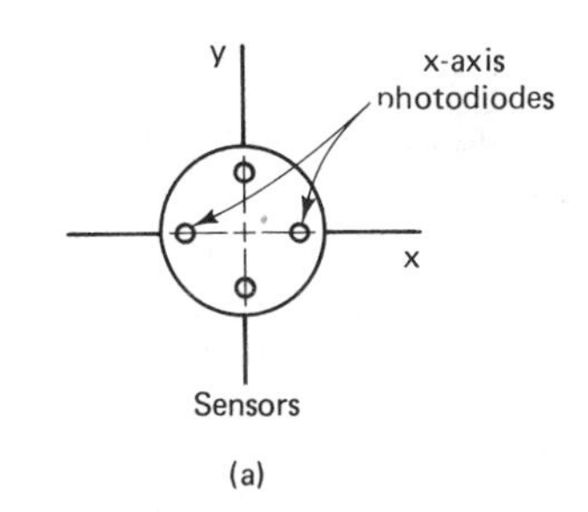

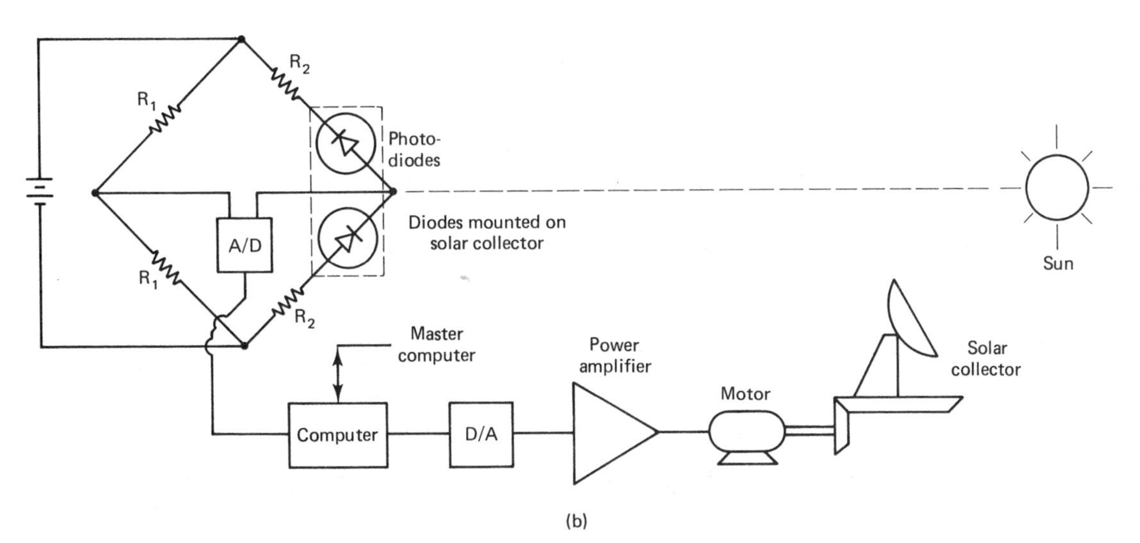

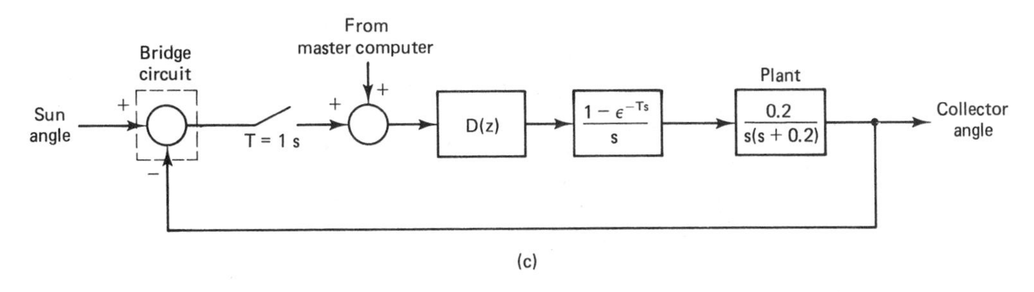

Figure P6-7 Solar collector control system.

energy. The solar energy heats a heat-transfer liquid that circulates through the solar collector. This heat energy is used to generate steam, which drives a steam turbine/alternator to generate the electricity. The closed-loop control system shown results in the solar collector tracking the sun. Two photodiodes, placed as shown in Figure P6-7a, form the sensor when placed in the bridge circuit shown. If the incidence of sun light is greater on one of the diodes, the bridge is unbalanced and an error signal of the correct polarity is developed. The signal from the master computer aligns the collector with the sun each morning, and after any temporary loss of track because of clouds. A tracking error of 0.5° results in an energy loss of approximately 5%.

(a) With $D(z) = 1$, calculate the response to a step input of 10° from the master computer on a cloudy day. You must consider the sensor output if the sun is hidden, and make a reasonable assumption.

(b) Find the steady-state error for $D(z) = 1$, the sun angle input equal to $0.004t$, and a signal of zero from the master computer.

(c) Where would you place a pole of $D(z)$ to reduce the error in part (b) to zero, assuming that the resulting system is stable?

6-8. For the system of Problem 6-5, let $f(t) = 0$. Find the steady-state errors for $r(t)$ a step input and then a ramp input, for $D(z)$ as given in both the (a) and (b) parts of Problem 6-5.

6-9. For the system of Figure P6-9, find $c(t)$ for $t = 10$ s. The input is as shown and the digital filter is described by

$$m(kT) = e(kT) - 0.5e[(k-1)T] + m[(k-1)T]$$

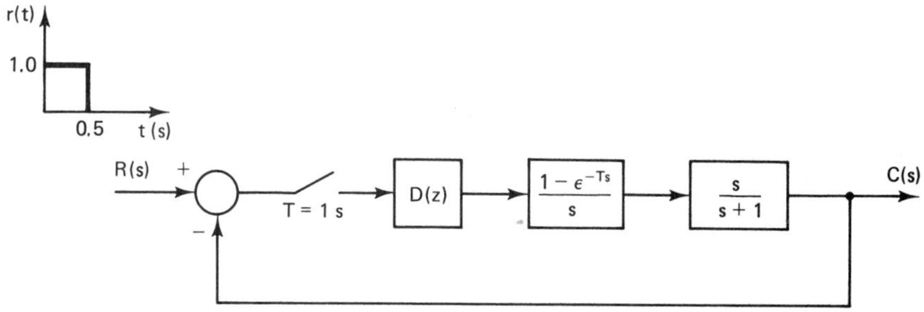

r(t)
1.0
0.5
t (s)

R(s)
+
−
T = 1 s
D(z)
$\frac{1-\epsilon^{-Ts}}{s}$
$\frac{s}{s+1}$
C(s)

Figure P6-9 System for Problem 6-9.

6-10. Consider the system of Figure P6-1, with $T = 0.5$ s and $D(z) = 1$.

(a) With the sampler and zero-order hold removed, the closed-loop system is first order. The resultant transient response is of the form $k_1\epsilon^{-t/\tau}$, where τ is the time constant. Find τ.

(b) With the sampler and zero-order hold present as shown, the closed-loop system is first order. The resultant transient response is of the form $k_1'\epsilon^{-kT/\tau}$, where τ is the time constant. Find τ for both $T = 0.5$ s and $T = 3.0$ s.

6-11. Consider the system of Figure P6-9, with the plant transfer function equal to $1/(s^2 + 1)$ rather than that given, and with $D(z) = 1$.

(a) Find the frequency present in the closed-loop transient response with the sampler and zero-order hold removed.

(b) Find the frequency present in the sampled output of the closed-loop transient response with the sampler and zero-order hold in the system as shown, and with $T = 0.1$ s.

(c) Repeat part (b) with $T = 1.0$ s.

6-12. Consider the system of Figure P6-1, with $D(z) = 1$.

(a) With the sampler and zero-order hold removed, write the system differential equation.

(b) Using the rectangular rule for numerical integration with the numerical integration increment equal to 0.25 s, evaluate the unit step response for $0 \leq t \leq 1.5$ s.

(c) Repeat part (b) with the sampler and zero-order hold in the system, and with a sample period $T = 0.5$ s.

(d) Solve for the exact unit step responses for parts (b) and (c), and compare results.

6-13. Repeat Problem 6-12, but with the process transfer function given by

$$G(s) = \frac{20}{s(3s + 1)}$$

Solve for the unit step responses for $0 \leq t \leq 1.0$ s.

6-14. Write a CSMP simulation for the system of part (b) of Problem 6-5. If facilities are available, run this simulation.

6-15. Write CSMP simulations for the systems of Problem 6-13, both with sampling and without sampling. If facilities are available, run these simulations.

6-16. For the closed-loop system of Example 6.2, calculate the damping ratio ζ, the natural frequency ω_n, and the time constant τ of the poles for both the sampled-data system and the analog system (sampler and data hold removed).

7

Stability Analysis Techniques

7.1 INTRODUCTION

In this chapter stability analysis techniques for discrete-time systems are emphasized. In general, the stability analysis techniques applicable to linear continuous-time systems may also be applied to the analysis of discrete-time systems, if certain modifications are made. These techniques include the Routh–Hurwitz criterion, root-locus procedures, and frequency-response techniques. The Jury stability test, a technique developed for discrete systems, is also presented.

In the design of discrete control systems, generally the very minimum specification is that the system must be stable. However, frequency-response techniques not only determine stability, but also suggest the designs necessary to achieve stability if the system proves to be unstable. In addition, these techniques suggest the designs necessary to achieve acceptable transient and steady-state responses. Thus frequency-response techniques are emphasized.

7.2 STABILITY

To introduce the concepts of stability, consider the system shown in Figure 7-1. For this system,

$$C(z) = \frac{G(z)R(z)}{1 + \overline{GH}(z)} = \frac{K \prod\limits^{m} (z - z_i)}{\prod\limits^{n} (z - p_i)} R(z)$$

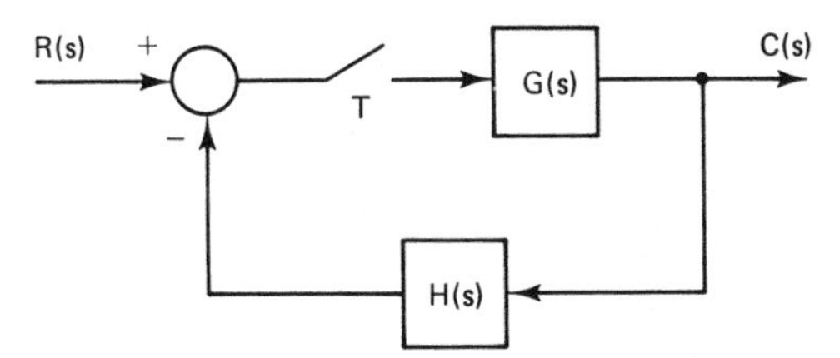

Figure 7-1 Sampled-data system.

Using the partial-fraction expansion method, we can express $C(z)$ as

$$C(z) = \frac{k_1 z}{z - p_1} + \cdots + \frac{k_n z}{z - p_n} + C_R(z) \tag{7-1}$$

where $C_R(z)$ contains the terms of $C(z)$ which originate in the poles of $R(z)$. The first n terms of (7-1) are the transient-response terms of $C(z)$. If the inverse z-transform of these terms tend to zero as time increases, the system is stable. The inverse z-transform of the ith term is

$$\mathfrak{z}^{-1}\left[\frac{k_i z}{z - p_i}\right] = k_i(p_i)^k \tag{7-2}$$

Thus, if the magnitude of p_i is less than 1, this term approaches zero as k approaches infinity. Note that the factors $(z - p_i)$ originate in the characteristic equation of the system, that is, in

$$1 + \overline{GH}(z) = 0 \tag{7-3}$$

The system is stable provided that all the roots of (7-3) lie inside the unit circle in the z-plane. Of course, (7-3) can also be expressed as

$$1 + \overline{GH}^*(s) = 0 \tag{7-4}$$

and since the area within the unit circle of the z-plane corresponds to the left half of the s-plane, the roots in (7-4) must lie in the left half of the s-plane for stability. The system characteristic equation may be calculated by either (7-3) or (7-4).

We have demonstrated previously that for certain discrete-time control systems, transfer functions cannot be derived. A method for finding the characteristic equation for control systems of this type will now be developed. To illustrate this method, consider the system of Figure 7-2a. This system was considered in Example 5.2, and the output expression developed there is

$$C(z) = \left[\frac{R}{2 + G_2}\right](z) + \frac{\left[\dfrac{G_1 G_2}{2 + G_2}\right](z)}{1 + \left[\dfrac{G_1 G_2}{2 + G_2}\right](z)} \left[\frac{(1 + G_2)R}{2 + G_2}\right](z)$$

Hence that part of the denominator of $C(z)$ that is independent of the input R is

$$1 + \left[\frac{G_1 G_2}{2 + G_2}\right](z)$$

and this function set equal to zero is then the characteristic equation.

This characteristic equation can be developed by a different procedure. Since stability is independent of the input, we set $R = 0$ in the system. In addition, we open

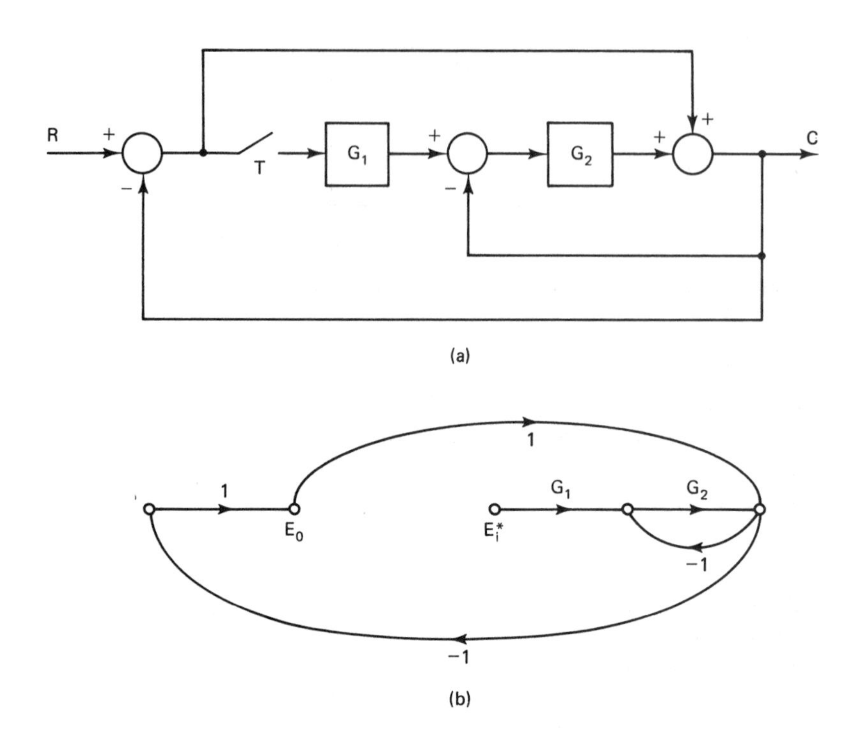

Figure 7-2 Discrete-time system.

the system in front of the sampler, since we can always write a transfer function if an input signal is sampled prior to being applied to a continuous-time part of a system. We denote the input signal at this open as $E_i(s)$, and the output signal as $E_o(s)$. Thus the system flow graph with no input and an open at the sampler appears as shown in Figure 7-2b. By Mason's gain formula (or any other applicable technique), we write

$$E_o(s) = \left[\frac{-G_1G_2}{2 + G_2}\right]E_i^*(s)$$

Taking the z-transform of this equation, we obtain

$$E_o(z) = -\left[\frac{G_1G_2}{2 + G_2}\right](z)E_i(z) \tag{7-5}$$

We will denote this open-loop transfer function as

$$G_{op}(z) = -\left[\frac{G_1G_2}{2 + G_2}\right](z) \tag{7-6}$$

For the closed-loop system, $E_i(z) = E_o(z)$, and (7-5) and (7-6) become

$$[1 - G_{op}(z)]E_o(z) = 0$$

Since we can set initial conditions on the system such that $E_o(z) \neq 0$, then

$$1 - G_{op}(z) = 0$$

and this relationship must be the system characteristic equation. Hence the characteristic equation for this system is

$$1 + \left[\frac{G_1G_2}{2 + G_2}\right](z) = 0$$

which checks the results of Example 5.2. Another example will now be given.

Example 7.1

Consider the system of Example 5.1, which is repeated in Figure 7-3a. The flow graph of the system is opened at the first sampler as shown in Figure 7-3b. From this flow graph we write

$$E_1 = G_1 E_i^* - G_2 H E_1^*$$

$$E_o = -G_2 E_1^*$$

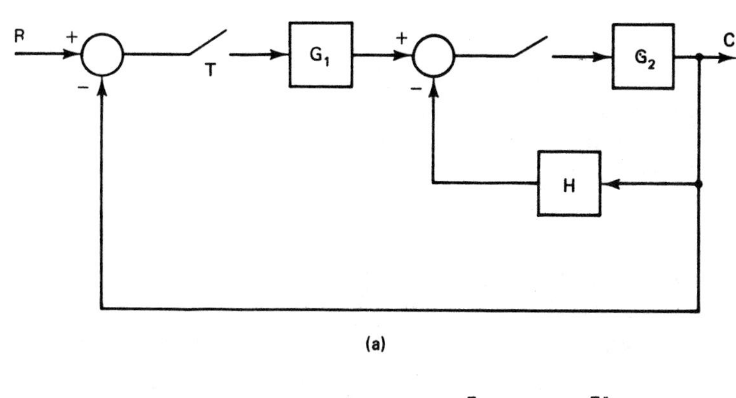

(a)

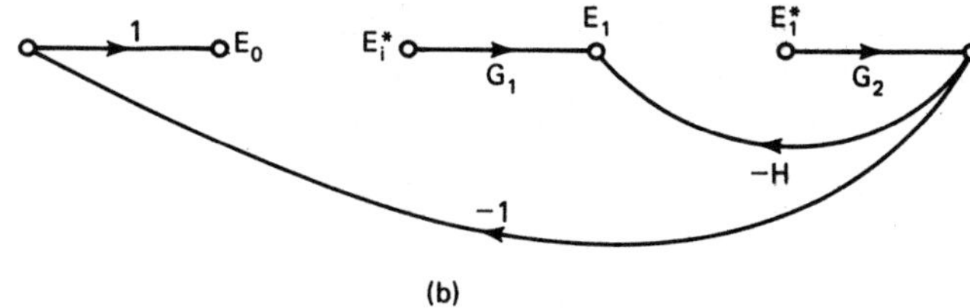

(b)

Figure 7-3 System for Example 7.1.

Starring the first equation and solving for E_1^*, we obtain

$$E_1^* = \frac{G_1^* E_i^*}{1 + \overline{G_2H}^*}$$

Starring the second equation and substituting in the value for E_1^*, we obtain

$$E_o^* = \frac{-G_1^* G_2^*}{1 + \overline{G_2H}^*} E_i^*$$

Since $E_i(z) = E_o(z)$ in the closed-loop system,

$$\left[1 + \frac{G_1(z)G_2(z)}{1 + \overline{G_2H}(z)}\right] E_o(z) = 0$$

Thus we can write the characteristic equation

$$1 + G_1(z)G_2(z) + \overline{G_2H}(z) = 0$$

We leave the derivation of the characteristic equation by opening the system at the second sampler as an exercise for the reader.

The characteristic equation of a system can also be calculated from a state-variable approach. Suppose that the state-variable model of the system of Figure 7-1 is

$$\mathbf{x}(k+1) = \mathbf{A}\mathbf{x}(k) + \mathbf{b}r(k)$$
$$y(k) = \mathbf{c}\mathbf{x}(k) + dr(k)$$

where the output is now denoted as $y(k)$ rather than $c(k)$. By taking the z-transform of these state equations and eliminating $\mathbf{X}(z)$, it was shown in Chapter 2 [see (2-79)] that the system transfer function is given by

$$\frac{Y(z)}{R(z)} = \mathbf{c}[z\mathbf{I} - \mathbf{A}]^{-1}\mathbf{b} + d$$

The denominator of the transfer function is seen to be $|z\mathbf{I} - \mathbf{A}|$, and thus the characteristic equation for the system is given by

$$|z\mathbf{I} - \mathbf{A}| = 0 \tag{7-7}$$

7.3 BILINEAR TRANSFORMATION

Many analysis and design techniques for continuous-time systems, such as the Routh–Hurwitz criterion and Bode techniques, are based on the property that in the s-plane the stability boundary is the imaginary axis. Thus these techniques cannot be applied to discrete-time systems in the z-plane, since the stability boundary is the unit circle. However, through the use of the transformation

$$z = \frac{1 + (T/2)w}{1 - (T/2)w} \tag{7-8}$$

or solving for w,

$$w = \frac{2}{T}\frac{z-1}{z+1} \tag{7-9}$$

the unit circle of the z-plane transforms into the imaginary axis of the w-plane. This can be seen through the following development. On the unit circle in the z-plane, $z = \epsilon^{j\omega T}$ and

$$w = \frac{2}{T}\frac{z-1}{z+1}\bigg|_{\epsilon^{j\omega T}} = \frac{2}{T}\frac{\epsilon^{j\omega T} - 1}{\epsilon^{j\omega T} + 1} = \frac{2}{T}\frac{\epsilon^{j\omega T/2} - \epsilon^{-j\omega T/2}}{\epsilon^{j\omega T/2} + \epsilon^{-j\omega T/2}}$$
$$= j\frac{2}{T}\tan\left(\frac{\omega T}{2}\right) \tag{7-10}$$

Using trigonometric identities, we can also express (7-10) as

$$w = j\frac{2}{T}\frac{\sin\omega T}{1 + \cos\omega T}$$

Thus it is seen that the unit circle of the z-plane transforms into the imaginary axis of the w-plane. The mappings of the primary strip of the s-plane into both the z-plane and the w-plane are shown in Figure 7-4, and it is noted that the stable region of the w-plane is the left half-plane.

Let $j\omega_w$ be the imaginary part of w. Then (7-10) can be expressed as

$$\omega_w = \frac{2}{T} \tan\left(\frac{\omega T}{2}\right) \tag{7-11}$$

and this expression gives the relationship between frequencies in the s-plane and frequencies in the w-plane.

A different definition of the w-transform is used in many textbooks. The definition is

$$w = \frac{z-1}{z+1}$$

However, it is felt that the definition given here is better. For the frequency range such that $\omega T/2$ is small, (7-11) becomes

$$\omega_w = \frac{2}{T} \tan\left(\frac{\omega T}{2}\right) \approx \frac{2}{T}\left(\frac{\omega T}{2}\right) = \omega$$

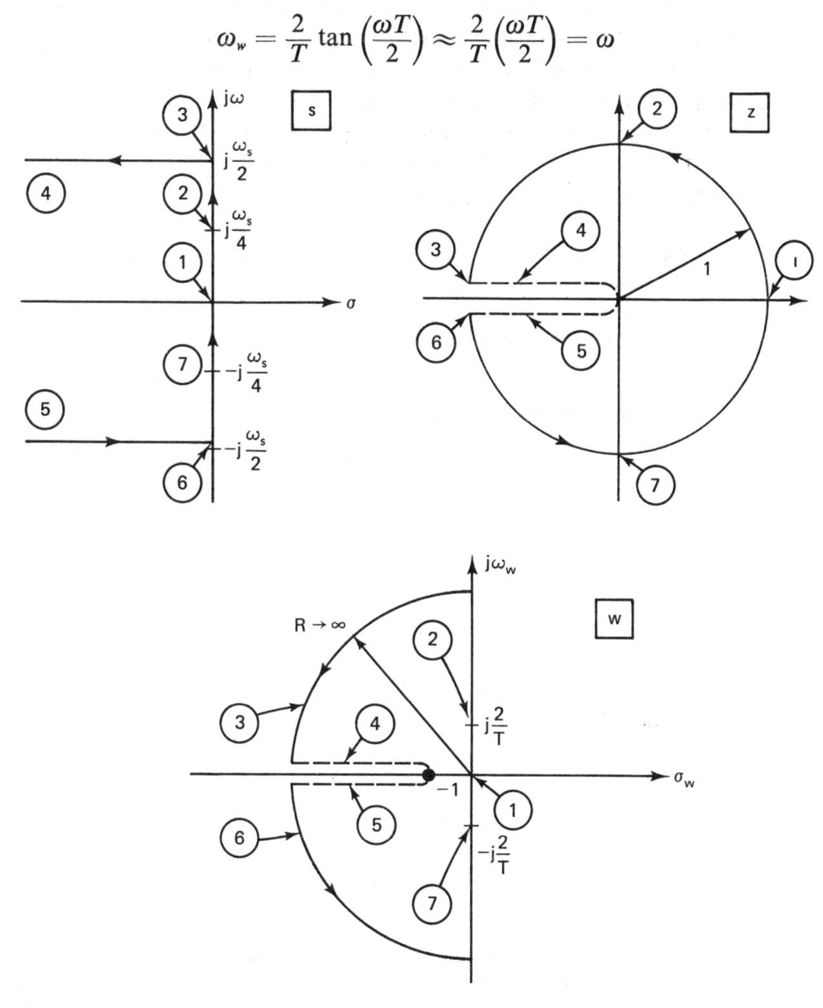

Figure 7-4 Mapping from s-plane to z-plane to w-plane.

Thus w-plane frequency is approximately equal to s-plane frequency. This approximation is valid for

$$\frac{\omega T}{2} \ll 1$$

or for

$$\omega \ll \frac{2}{T} = \frac{\omega_s}{\pi}$$

7.4 THE ROUTH–HURWITZ CRITERION

The Routh–Hurwitz criterion [1] may be used in the analysis of continuous-time systems to determine if any roots of a given equation are in the right half of the s-plane. If this criterion is applied to the characteristic equation of a discrete-time system when expressed as a function of z, no useful information on stability is obtained. However, if the characteristic equation is expressed as a function of the bilinear transform variable w, then the stability of the system may be determined by directly applying the Routh–Hurwitz criterion.

We assume that the reader is familiar with the procedures for applying the Routh–Hurwitz criterion. The procedure is summarized in Table 7-1. The technique will now be illustrated via examples.

TABLE 7-1 BASIC PROCEDURE FOR APPLYING THE ROUTH–HURWITZ CRITERION

1. Given a characteristic equation of the form

$$F(w) = b_n w^n + b_{n-1} w^{n-1} + \cdots + b_1 w + b_0 = 0$$

form the Routhian array as

$$\begin{array}{c|cccc} w^n & b_n & b_{n-2} & b_{n-4} & \cdots \\ w^{n-1} & b_{n-1} & b_{n-3} & b_{n-5} & \cdots \\ w^{n-2} & c_1 & c_2 & c_3 & \cdots \\ \cdot & d_1 & d_2 & d_3 & \cdots \\ \cdot & & & & \\ \cdot & & & & \\ w^1 & j_1 & & & \\ w^0 & k_1 & & & \end{array}$$

2. Only the first two rows of the array are obtained from the characteristic equation. The remaining rows are calculated as follows.

$$c_1 = \frac{b_{n-1}b_{n-2} - b_n b_{n-3}}{b_{n-1}} \qquad d_1 = \frac{c_1 b_{n-3} - b_{n-1} c_2}{c_1}$$

$$c_2 = \frac{b_{n-1}b_{n-4} - b_n b_{n-5}}{b_{n-1}} \qquad d_2 = \frac{c_1 b_{n-5} - b_{n-1} c_3}{c_1}$$

$$c_3 = \frac{b_{n-1}b_{n-6} - b_n b_{n-7}}{b_{n-1}} \qquad \vdots$$

3. Once the array has been formed, the Routh–Hurwitz criterion states that the number of roots of the characteristic equation with positive real parts is equal to the number of sign changes of the coefficients in the first column of the array.

Example 7.2

Consider the system shown in Figure 7-5, with $T = 0.1$ s. The open-loop transfer function is

$$G(s) = \left[\frac{1 - \epsilon^{-Ts}}{s}\right]\left[\frac{1}{s(s+1)}\right]$$

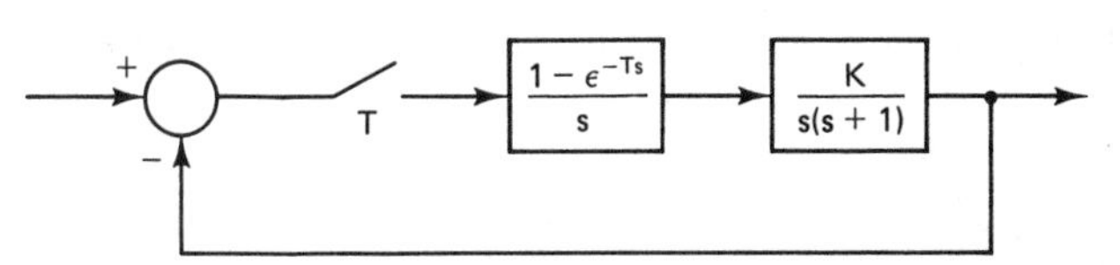

Figure 7-5 System for Examples 7.2 and 7.3.

Hence, from the z-transform tables we obtain

$$G(z) = \frac{z-1}{z}\left[\frac{(\epsilon^{-T} + T - 1)z^2 + (1 - \epsilon^{-T} - T\epsilon^{-T})z}{(z-1)^2(z - \epsilon^{-T})}\right]$$

$$= \frac{0.00484z + 0.00468}{(z-1)(z-0.905)}$$

Then $G(w)$ is given by

$$G(w) = G(z)\big|_{z=[1+(T/2)w]/[1-(T/2)w]} = G(z)\big|_{z=(1+0.05w)/(1-0.05w)}$$

or

$$G(w) = \frac{-0.00016w^2 - 0.1872w + 3.81}{3.81w^2 + 3.80w}$$

Then the characteristic equation is given by

$$1 + KG(w) = (3.81 - 0.00016K)w^2 + (3.80 - 0.1872K)w + 3.81K = 0$$

The Routhian array derived from this equation is

$$\begin{array}{c|ll} w^2 & 3.81 - 0.00016K & 3.81K \\ w^1 & 3.80 - 0.1872K & \\ w^0 & 3.81K & \end{array}$$

Hence, for no sign changes to occur in the first column, it is necessary that K be in the range $0 < K < 20.3$, and this is the range of K for stability.

Example 7.3

Consider again the system of Example 6.2 (shown again in Figure 7-5 with $T = 1$ s) with a gain factor K added to the plant. The characteristic equation is given by

$$1 + KG(w) = 1 + KG(z)\big|_{z=(1+0.5w)/(1-0.5w)}$$

$$= 1 + \frac{K\left[0.368\left[\dfrac{1+0.5w}{1-0.5w}\right] + 0.264\right]}{\left[\dfrac{1+0.5w}{1-0.5w}\right]^2 - 1.368\left[\dfrac{1+0.5w}{1-0.5w}\right] + 0.368}$$

or

$$1 + KG(w) = 1 + \frac{-0.0381K(w-2)(w+12.14)}{w(w+0.924)}$$

$$= \frac{(1 - 0.0381K)w^2 + (0.924 - 0.386K)w + 0.924K}{w(w+0.924)}$$

Thus the characteristic equation may be expressed as

$$(1 - 0.0381K)w^2 + (0.924 - 0.386K)w + 0.924K = 0$$

The Routhian array is then

$$\begin{array}{c|ll} w^2 & 1 - 0.0381K & 0.924K \\ w^1 & 0.924 - 0.386K & \\ w^0 & 0.924K & \end{array}$$

An examination of the first column of the array indicates that the system is stable for

$$0 < K < \frac{0.924}{0.386} = 2.39$$

From our knowledge of continuous-time systems we know that the Routh–Hurwitz criterion can be used to determine the value of K at which the root locus crosses into the right half-plane (i.e., the value of K at which the system becomes unstable). That value of K is the gain at which the system is marginally stable, and thus can also be used to determine the resultant frequency of oscillation. Therefore, the value $K = 2.39$ determined above is the value of K for which the root locus crosses the unit circle in the z-plane.

In a manner similar to that employed in continuous-time systems, the frequency of oscillation at $K = 2.39$ can be found from the w^2 row of the array. Recalling that ω_w is the imaginary part of w, we obtain the auxiliary equation

$$(1 - 0.0381K)w^2 + 0.924K|_{K=2.39} = 0$$

Solving this expression for ω_w yields

$$\omega_w = 1.559$$

Then from (7-11),

$$\omega = \frac{2}{T}\tan^{-1}\left(\frac{\omega_w T}{2}\right)$$
$$= 1.32 \text{ rad/s}$$

and is the s-plane (real) frequency at which this system will oscillate if K is set equal to 2.39.

7.5 JURY'S STABILITY TEST

For continuous-data systems, the Routh–Hurwitz criterion offers a simple and convenient technique for determining the stability of low-ordered systems. However, since the stability boundary in the z-plane is different from that in the s-plane, the Routh–Hurwitz criterion cannot be directly applied to discrete-time systems if the system characteristic equation is expressed as a function of z. A stability criterion for discrete-time systems that is similar to the Routh–Hurwitz criterion and can be applied to the characteristic equation written as a function of z is the Jury stability test [2]. Other stability criteria, such as the Schur–Cohn criterion [3], are available, but only Jury's technique is presented here.

Jury's test will now be presented. Let the characteristic equation of a discrete-time system be expressed as

$$F(z) = a_n z^n + a_{n-1}z^{n-1} + \cdots + a_1 z + a_0 = 0, \qquad a_n > 0 \qquad (7\text{-}12)$$

Then form the array as shown in Table 7-2. Note that the elements of each of the even-numbered rows are the elements of the preceding row in reverse order. The ele-

TABLE 7-2 ARRAY FOR JURY'S STABILITY TEST

z^0	z^1	z^2	$\cdots$	z^{n-k}	$\cdots$	z^{n-1}	z^n
a_0	a_1	a_2	$\cdots$	a_{n-k}	$\cdots$	a_{n-1}	a_n
a_n	a_{n-1}	a_{n-2}	$\cdots$	a_k	$\cdots$	a_1	a_0
b_0	b_1	b_2	$\cdots$	b_{n-k}	$\cdots$	b_{n-1}	
b_{n-1}	b_{n-2}	b_{n-3}	$\cdots$	b_{k-1}	$\cdots$	b_0	
c_0	c_1	c_2	$\cdots$	c_{n-k}	$\cdots$		
c_{n-2}	c_{n-3}	c_{n-4}	$\cdots$	c_{k-2}	$\cdots$		
$\vdots$	$\vdots$	$\vdots$	$\vdots$	$\vdots$			
l_0	l_1	l_2	l_3				
l_3	l_2	l_1	l_0				
m_0	m_1	m_2					

ments of the odd-numbered rows are defined as

$$b_k = \begin{vmatrix} a_0 & a_{n-k} \\ a_n & a_k \end{vmatrix}, \qquad c_k = \begin{vmatrix} b_0 & b_{n-1-k} \\ b_{n-1} & b_k \end{vmatrix}$$
$$d_k = \begin{vmatrix} c_0 & c_{n-2-k} \\ c_{n-2} & c_k \end{vmatrix} \qquad \cdots \tag{7-13}$$

The necessary and sufficient conditions for the polynomial $F(z)$ to have no roots outside or on the unit circle, with $a_n > 0$, are as follows:

$$\begin{aligned} F(1) &> 0 \\ (-1)^n F(-1) &> 0 \\ |a_0| &< a_n \\ |b_0| &> |b_{n-1}| \\ |c_0| &> |c_{n-2}| \\ |d_0| &> |d_{n-3}| \\ &\vdots \\ |m_0| &> |m_2| \end{aligned} \tag{7-14}$$

Note that for a second-order system, the array contains only one row. For each additional order, two additional rows are added to the array. Note also that, for an nth-order system, there are a total of $n + 1$ constraints.

Example 7.4

Consider again the system of Example 6.2 (and Example 7.3). Suppose that a gain factor K is added to the plant, and it is desired to determine the range of K for which the system is stable. Now, from Example 6.2, the system characteristic equation is

$$1 + KG(z) = 1 + \frac{(0.368z + 0.264)K}{z^2 - 1.368z + 0.368} = 0$$

or

$$z^2 + (0.368K - 1.368)z + (0.368 + 0.264K) = 0$$

The Jury array is

z^0	z^1	z^2
$0.368 + 0.264K$	$0.368K - 1.368$	1

The constraint $F(1) > 0$ yields

$$1 + (0.368K - 1.368) + (0.368 + 0.264K) > 0$$

or

$$K > 0$$

The constraint $(-1)^2F(-1) > 0$ yields

$$K < 26.3$$

and the constraint $|a_0| < a_2$ yields

$$K < 2.39$$

Thus the system is stable for

$$0 < K < 2.39$$

The system is marginally stable for $K = 2.39$. For this value of K, the characteristic equation is

$$z^2 + (0.368K - 1.368)z + (0.368 + 0.264K)|_{K=2.39}$$
$$= z^2 - 0.49z + 1 = 0$$

The roots of this equation are

$$z = 0.245 \pm j0.97 = 1\underline{/\pm 75.8^\circ} = 1\underline{/\pm 1.32}\text{ rad} = 1\underline{/\pm \omega T}$$

Since $T = 1$ s, the system will oscillate at a frequency of 1.32 rad/s.

Example 7.5

Suppose that the characteristic equation for a closed-loop discrete-time system is given by the expression

$$F(z) = z^3 - 3z^2 + 2.25z - 0.5$$

and we wish to determine via the Jury stability test if the system is stable. In this case the Jury array is

z^0	z^1	z^2	z^3
-0.5	2.25	-3	1
1	-3	2.25	-0.5
-0.75	1.875	-0.75	

where the last row has been calculated as follows:

$$b_0 = \begin{vmatrix} -0.5 & 1 \\ 1 & -0.5 \end{vmatrix} = -0.75, \qquad b_1 = \begin{vmatrix} -0.5 & -3 \\ 1 & 2.25 \end{vmatrix} = 1.875$$

$$b_2 = \begin{vmatrix} -0.5 & 2.25 \\ 1 & -3 \end{vmatrix} = -0.75$$

The constraint $F(1) > 0$ is not satisfied since

$$F(1) = -0.25 \not> 0$$

The constraint $(-1)^3F(-1) > 0$ is satisfied since $(-1)^3F(-1) = 6.75$ and $|a_0| < a_3$ is obviously satisfied. However, the constraint $|b_0| > |b_2|$ implies that $0.75 > 0.75$, which is not true. Therefore, this system is unstable.

The characteristic equation can be factored as

$$F(z) = (z - 0.5)^2(z - 2)$$

This form of the equation clearly indicates the system's instability.

Example 7.6

In this example we wish to determine the range of values of K_p in Example 6.6 so that the system is stable. The characteristic equation for the system is given by the expression

$$1 + D(z)G(z) = 0$$

Substituting into this expression the proper functions for the system components yields

$$z^2 - [(1 + \epsilon^{-T}) - (1 - \epsilon^{-T})(K_I + K_p)]z + \epsilon^{-T} - (1 - \epsilon^{-T})K_p = 0$$

Suppose that $T = 0.1$ s and $K_I = 100T = 10$. Then the characteristic equation becomes

$$z^2 - [0.953 - 0.0952K_p]z + 0.905 - 0.0952K_p = 0$$

For this system, the Jury array is

z^0	z^1	z^2
$0.905 - 0.0952K_p$	$0.0952K_p - 0.953$	1

The constraint $F(1) > 0$ yields

$$1 + 0.0952K_p - 0.953 + 0.905 - 0.0952K_p > 0$$

Thus this constraint is satisfied independent of K_p. The constraint $(-1)^2F(-1) > 0$ yields

$$1 - 0.0952K_p + 0.953 + 0.905 - 0.0952K_p > 0$$

or

$$K_p < 15.01$$

The constraint $|a_0| < a_2$ yields

$$|0.905 - 0.0952K_p| < 1$$

or

$$K_p < 20.0$$

Thus the stability constraint on K_p is $K_p < 15.01$.

7.6 ROOT LOCUS

For the sampled-data system of Figure 7-6,

$$\frac{C(z)}{R(z)} = \frac{KG(z)}{1 + K\overline{GH}(z)} \tag{7-15}$$

The system characteristic equation is, then,

$$1 + K\overline{GH}(z) = 0 \tag{7-16}$$

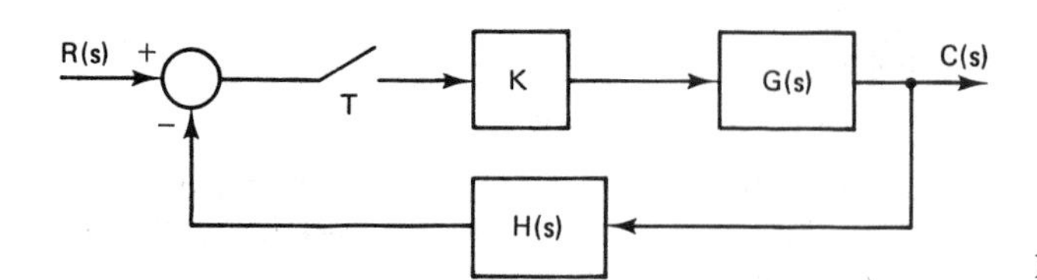

Figure 7-6 Sampled-data system.

The root locus for this system is a plot of the locus of roots of (7-16) in the z-plane as a function of K. Thus the rules of root-locus construction for discrete-time systems are identical to those for continuous-time systems, since the roots of any equation $F(z) = 0$ in the z-plane are the same as those of $F(s) = 0$ in the s-plane. Since the rules for root-locus construction are numerous, and appear in any standard text for continuous-time control systems [1], only the most important rules will be repeated here in abbreviated form. These rules are given in Table 7-3.

TABLE 7-3 RULES FOR ROOT-LOCUS CONSTRUCTION

1. Loci originate on poles of $\overline{GH}(z)$ and terminate on the zeros of $\overline{GH}(z)$.
2. The root locus on the real axis always lies in a section of the real axis to the left of an odd number of poles and zeros on the real axis.
3. The root locus is symmetrical with respect to the real axis.
4. The number of asymptotes is equal to the number of poles of $\overline{GH}(z)$. n_p, minus the number of zeros of $\overline{GH}(z)$, n_z, with angles given by $(2k + 1)\pi/(n_p - n_z)$.
5. The asymptotes intersect the real axis at σ, where

$$\sigma = \frac{\sum \text{poles of } \overline{GH}(z) - \sum \text{zeros of } \overline{GH}(z)}{n_p - n_z}$$

6. The breakaway points are given by the roots of

$$\frac{d[1/\overline{GH}(z)]}{dz} = 0$$

However, while the rules for construction of both s-plane and z-plane root loci are the same, there are important differences in the interpretation of the root loci. For example, in the z-plane, the stable region is the interior of the unit circle. In addition, root locations in the z-plane have different meanings from those in the s-plane from the standpoint of the system time response, as was seen in Figure 6-9.

In (7-16) the root locus was described as the loci of the closed-loop poles as the gain K varies, with T the sampling period, held constant. The root locus may also be plotted as a function of T, with K constant [3]. However, the rules are more complex, and the construction is more difficult except in the case of simple low-order systems.

The two examples that follow review not only the root-locus technique, but also the part it plays in stability analyses.

Example 7.7

Consider again the system of Example 6.2. For this system

$$KG(z) = \frac{0.368K(z + 0.717)}{(z - 1)(z - 0.368)}$$

Thus the loci originate at $z = 1$ and $z = 0.368$, and terminate at $z = -0.717$ and $z = \infty$. There is one asymptote, at 180°. The breakaway points, obtained from

$$\frac{d}{dz}\left[\frac{1}{G(z)}\right] = 0$$

occur at $z = 0.65$ for $K = 0.196$, and at $z = -2.08$ for $K = 15.0$. The root locus is then as shown in Figure 7-7. The points of intersection of the root loci with the unit

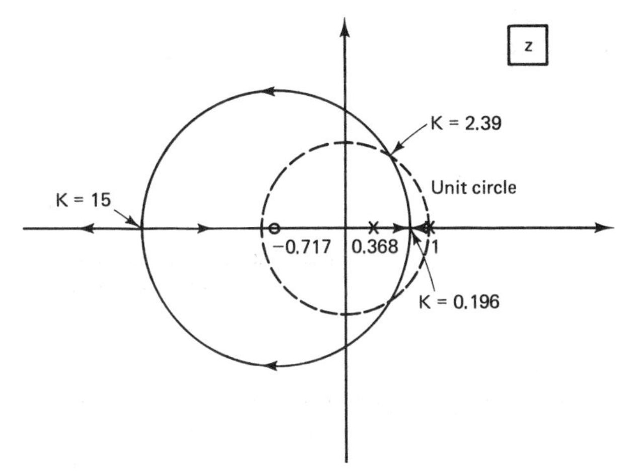

Figure 7-7 Root locus for Example 7.7.

circle may be found by graphical construction, the Jury stability test, or the Routh–Hurwitz criterion. To illustrate the use of the Jury stability test, consider the results of Example 7.4. The value of gain for marginal stability (i.e., for the roots to appear on the unit circle) is $K = 2.39$. For this value of gain, the characteristic equation is

$$z^2 - 0.49z + 1 = 0$$

The roots of this equation are

$$z = 0.245 \pm j0.97 = 1\underline{/\pm 75.8^\circ} = 1\underline{/\pm 1.32}\text{ rad} = 1\underline{/\pm \omega T}$$

and thus these are the points at which the root locus crosses the unit circle. Note that the frequency of oscillation for this case is $\omega = 1.32$, since $T = 1$ s. This was also calculated in Example 7.3 using the Routh–Hurwitz criterion, and in Example 7.4 using the Jury test.

The value of the gain at points where the root locus crosses the unit circle can also be determined using the root-locus condition that at any point along the locus the magnitude of the open-loop function must be equal to 1 (i.e., $K|GH| = 1$). Using the condition and Figure 7-8, we note that

$$\frac{0.368K(Z_1)}{(P_1)(P_2)} = 1$$

From Figure 7-8 the following values can be calculated: $Z_1 = 1.364$, $P_1 = 1.229$, and $P_2 = 0.978$. Using these values in the equation above yields $K = 2.39$.

Example 7.8

Consider the system shown in Figure 7-9. We wish to determine the form of the root locus and the range of K for stability.

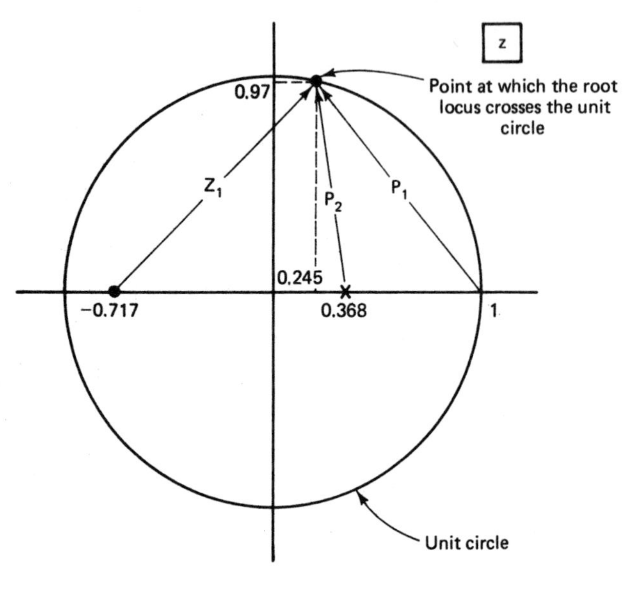

Figure 7-8 Determination of the system gain at the crossover point on the unit circle.

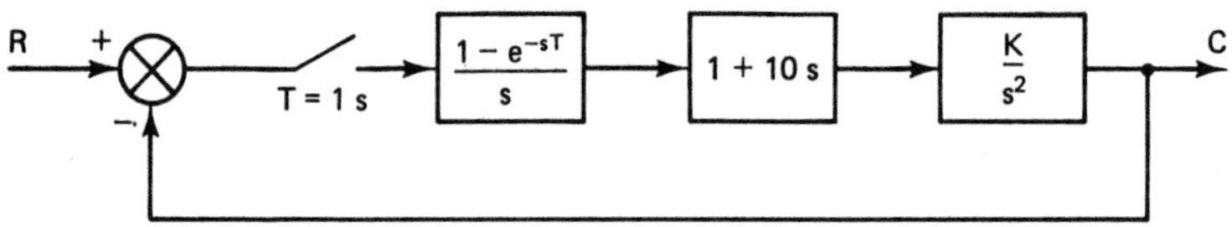

Figure 7-9 System for Example 7.8.

The open-loop function is

$$KG(s) = \left[\frac{1 - e^{-sT}}{s}\right]\left[\frac{K(1 + 10s)}{s^2}\right]$$

Applying the z-transform, we obtain

$$KG(z) = \frac{10.5K(z - 0.9048)}{(z - 1)^2}$$

The loci originate at $z = 1$ and terminate at $z = 0.9048$ and $z = -\infty$. There is one asymptote at 180°. The root locus is shown in Figure 7-10. The system becomes unstable when the closed-loop pole leaves the interior of the unit circle at point A shown in Figure 7-10. The value of K at this point can be determined from the condition $K|G| = 1$. Therefore,

$$\frac{10.5K(Z_1)}{(P_1)(P_2)} = 1$$

where Z_1 is the distance from the open-loop zero to point A, and P_1 and P_2 represent the distances from the open-loop poles to point A. Hence

$$\frac{(10.5)(K)(1.9048)}{(2)(2)} = 1$$

or

$$K = 0.2$$

An examination of the root locus indicates that the system is stable for $0 < K < 0.2$.

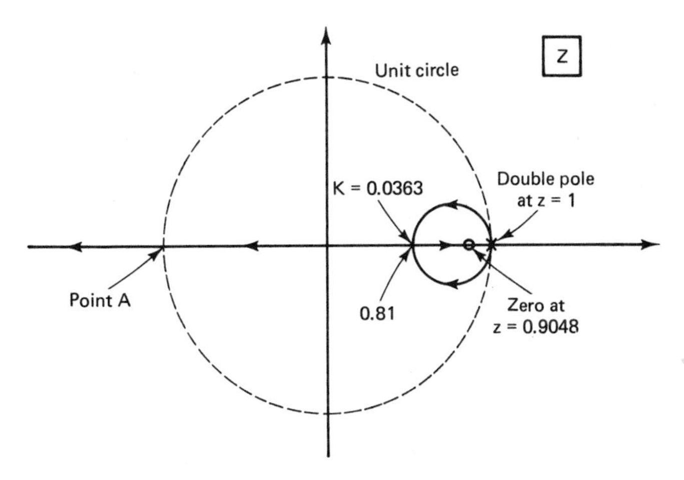

Figure 7-10 Root locus for the system of Example 7.8.

7.7 THE NYQUIST CRITERION

In order to develop the Nyquist criterion for discrete-time systems, consider the two systems shown in Figure 7-11. In Figure 7-11b, $G(s)$ contains the transfer function of a data hold. It has been shown that the transfer function for the continuous-time system of Figure 7-11a is

$$\frac{C(s)}{R(s)} = \frac{G(s)}{1 + G(s)H(s)} \tag{7-17}$$

and for the sampled-data system of Figure 7-11b the transfer function is

$$\frac{C^*(s)}{R^*(s)} = \frac{G^*(s)}{1 + \overline{GH}^*(s)} \tag{7-18}$$

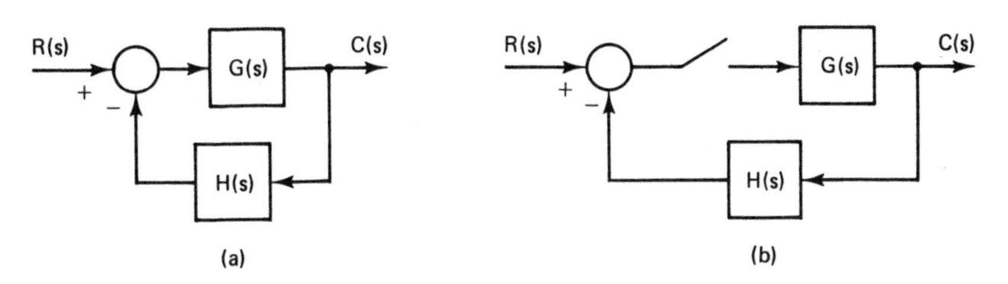

Figure 7-11 Continuous-time and sampled-data systems.

Thus the characteristic equation for the continuous-time system is

$$1 + G(s)H(s) = 0 \tag{7-19}$$

and for the sampled-data system, the characteristic equation is

$$1 + \overline{GH}^*(s) = 0 \tag{7-20}$$

The characteristic equation of the sampled-data system can also be written as

$$1 + \overline{GH}(z) = 0 \tag{7-21}$$

Recall that the continuous-time system is stable if the roots of (7-19) are all contained in the left half-plane. Similarly, the sampled-data system is stable if the roots of (7-20) all lie in the left half-plane, or if the roots of (7-21) all lie within the unit circle.

The Nyquist criterion is based on Cauchy's Principle of Argument [4].

Theorem. Let $f(z)$ be the ratio of two polynomials in z. Let the closed curve C in the z-plane be mapped into the complex plane through the mapping $f(z)$. If $f(z)$ is analytic within and on C, except at a finite number of poles, and if $f(z)$ has neither poles nor zeros on C, then

$$N = Z - P$$

where Z is the number of zeros of $f(z)$ in C, P is the number of poles of $f(z)$ in C, and N is the number of encirclements of the origin, taken in the same sense as C.

In order to determine stability via the Nyquist criterion, a complex-plane plot of the open-loop transfer function, $G(s)H(s)$, is made. This plot is referred to as the Nyquist diagram. The closed curve C of Cauchy's Principle of Argument is called the Nyquist path, and for continuous-time systems encloses the right half-plane as shown in Figure 7-12. The Nyquist criterion states that if the complex-plane plot of $G(s)H(s)$, for values for values of s on the Nyquist path, is made, then

$$N = Z - P \tag{7-22}$$

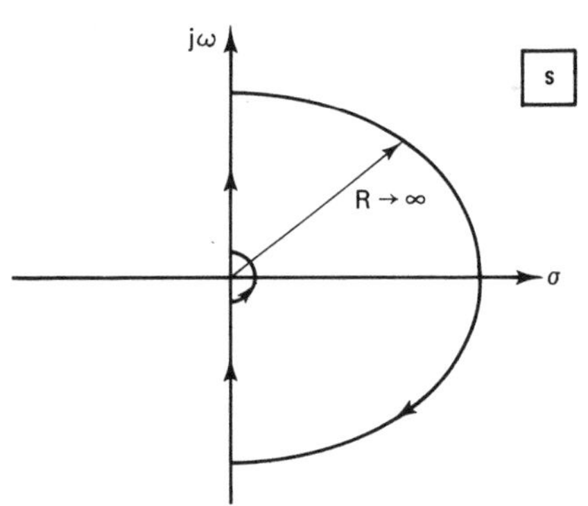

Figure 7-12 Nyquist's paths in the s-plane.

where N is the number of clockwise encirclements of the -1 point made by the plot of $G(s)H(s)$, Z is the number of zeros of the characteristic equation enclosed by the Nyquist path, and P is the number of poles of the open-loop transfer function enclosed by the Nyquist path. Since the Nyquist path encloses the right half-plane, Z in (7-22) must be zero for stability.

Example 7.9

As an example of the Nyquist criterion applied to a continuous-time system, consider the system of Figure 7-13. The Nyquist path and the resultant Nyquist diagram are shown in Figure 7-14. The small detour taken by the Nyquist path around the origin is necessary, since $G(s)$ has a pole at the origin. Along this detour, let

$$s = \rho\epsilon^{j\theta}, \qquad \rho \ll 1$$

Then

$$G(s)\big|_{s=\rho\epsilon^{j\theta}} \approx \frac{K}{\rho\epsilon^{j\theta}} = \frac{K}{\rho}\underline{/-\theta}$$

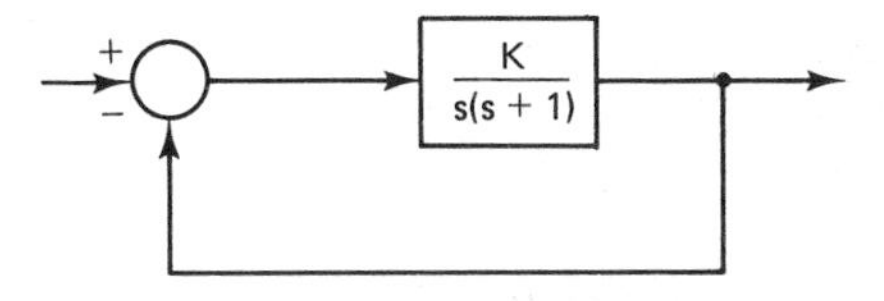

Figure 7-13 System for Example 7.9.

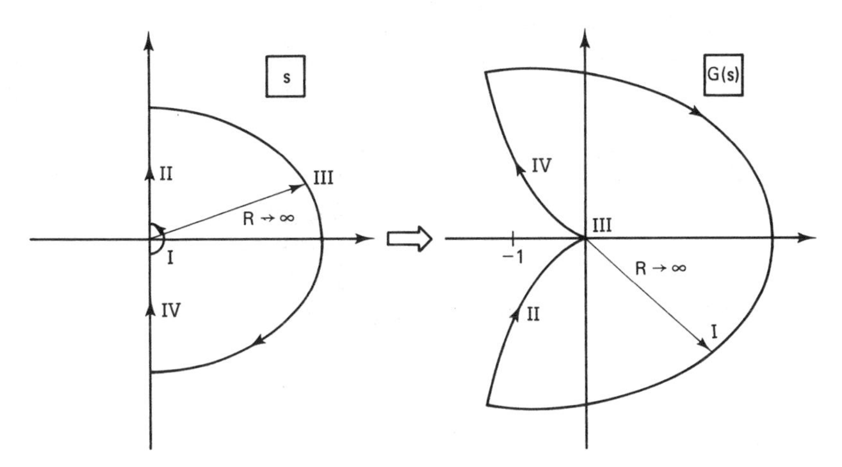

Figure 7-14 Nyquist diagram for Example 7.9.

Thus the small detour generates the large arc on the Nyquist diagram. For the Nyquist path along the $j\omega$-axis (II),

$$G(s)|_{s=j\omega} = \frac{K}{j\omega(1+j\omega)} = \frac{K}{\omega\sqrt{1+\omega^2}}\underline{/-90^\circ - \tan^{-1}\omega}$$

This portion of the Nyquist diagram is simply the frequency response of the system. Along the large arc of the Nyquist path, $G(s) \approx 0$. Now let us apply the Nyquist criterion. We can see from the Nyquist diagram in Figure 7-14 that $N = 0$. Since the open-loop transfer function has no poles in the right half-plane, $P = 0$, and therefore from the equation

$$N = Z - P$$

we find that $Z = 0$. Thus the system is stable. In addition, from the Nyquist diagram in Figure 7-14 it can be seen that the system is stable for all $K > 0$.

Consider now the sampled-data system of Figure 7-11b. The characteristic equation is given in (7-20), and thus the Nyquist diagram for the system may be generated by using the s-plane Nyquist path of Figure 7-12 (i.e., the same path as that used for continuous-time systems). However, recall from Chapter 3 that $\overline{GH}^*(s)$ is periodic in s with period $j\omega_s$. Thus it is necessary only that $\overline{GH}^*(j\omega)$ be plotted for $-\omega_s/2 \leq \omega \leq \omega_s/2$, in order to obtain the frequency response. Consider also the relationship, from (3-11),

$$\begin{aligned}\overline{GH}^*(j\omega) &= \frac{1}{T}\sum_{n=-\infty}^{\infty} GH(j\omega + jn\omega_s) \\ &= \frac{1}{T}[GH(j\omega) + GH(j\omega + j\omega_s) + GH(j\omega - j\omega_s) + \cdots]\end{aligned} \tag{7-23}$$

Since physical systems are generally low pass, $\overline{GH}^*(j\omega)$ may be approximated by only a few terms of (7-23). In general, however, the approximation will not apply for $\omega > \omega_s/2$. A digital computer program may easily be written for (7-23) (see Appendix I), and thus the Nyquist diagram of a sampled-data system may be obtained without calculating the z-form of the transfer function.

In (7-23), if the system is sufficiently low pass with respect to $\omega_s/2$, $\overline{GH}^*(j\omega)$ may be approximated by the first term $(1/T)GH(j\omega)$. Recall, from Figure 3-13, that the gain of a data hold for $\omega \ll \omega_s/2$ is approximately equal to T. Since G in (7-23) must contain a data hold, then for this case the Nyquist diagram becomes a plot of the continuous-time system frequency response; that is, the sampler and zero-order hold may be ignored with very little resulting error. If this approximation is made, generally the error in phase has more effect than the error in magnitude.

The Nyquist diagram may, however, be generated directly from the z-plane. The Nyquist path for the z-plane is the unit circle, and the path direction is counterclockwise, as shown in Figure 7-15. To apply Cauchy's Principle of Argument, let

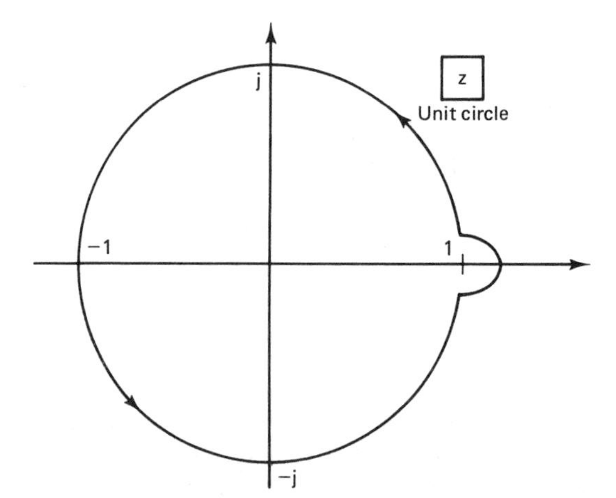

Figure 7-15 Nyquist path in the z-plane.

Z_i and P_i be the zeros of the characteristic equation and the poles of the open-loop transfer function, respectively, inside the unit circle. And let Z_o and P_o be the zeros of the characteristic equation and the poles of the open-loop transfer function, respectively, outside the unit circle. Then, from (7-22),

$$N = -(Z_i - P_i) \tag{7-24}$$

where N is the number of clockwise encirclements of the -1 point made by the Nyquist diagram of $\overline{GH}(z)$. The minus sign appears in (7-24) since the Nyquist path is counterclockwise. Now, in general, the order of the numerator and the order of the denominator of $1 + \overline{GH}(z)$ are the same. Let this order be n. Then

$$Z_o + Z_i = n \tag{7-25}$$

$$P_o + P_i = n \tag{7-26}$$

Solving (7-25) and (7-26) for Z_i and P_i, respectively, and substituting the results in (7-24), we obtain

$$N = Z_o - P_o \tag{7-27}$$

Thus the Nyquist criterion is given by (7-27), with the Nyquist path shown in Figure 7-15. The Nyquist diagram for the system of Figure 7-11b is obtained by plotting $\overline{GH}(z)$ for values of z on the unit circle. Then

$$N = Z - P \tag{7-28}$$

where N = number of clockwise encirclements of the -1 point
Z = number of zeros of the characteristic equation outside the unit circle
P = number of poles of the open-loop transfer function, or equivalently poles of the characteristic equation, that are outside the unit circle

Example 7.10

The Nyquist criterion will now be illustrated using the system of Figure 7-16. From Example 6.2,

$$G(z) = \frac{0.368z + 0.264}{(z - 1)(z - 0.368)}$$

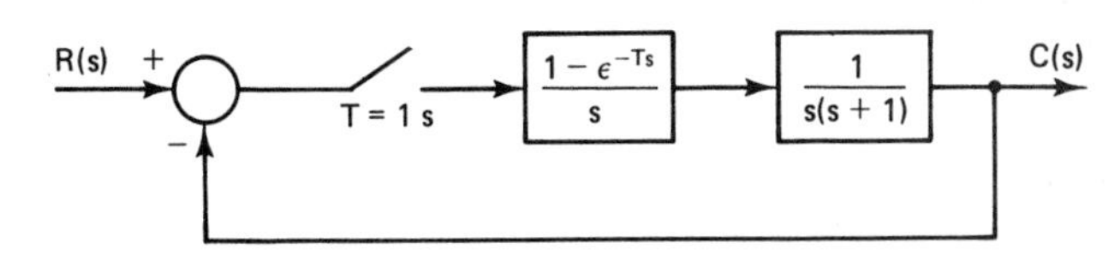

Figure 7-16 System for Example 7.10.

The Nyquist path and the Nyquist diagram are both shown in Figure 7-17. The detour around the $z = 1$ point on the Nyquist path (Figure 7-17a) is necessary since $G(z)$ has a pole at this point. On this detour,

$$z = 1 + \rho\epsilon^{j\theta}, \qquad \rho \ll 1$$

and

$$G(z)\big|_{z=1+\rho\epsilon^{j\theta}} \approx \frac{0.632}{\rho\epsilon^{j\theta}(0.632)} = \frac{1}{\rho}\underline{/-\theta}$$

Thus this detour generates the large arc on the Nyquist diagram (Figure 7-17b). For z on the unit circle,

$$G(z)\big|_{z=\epsilon^{j\omega T}} = \frac{0.368\epsilon^{j\omega T} + 0.264}{(\epsilon^{j\omega T} - 1)(\epsilon^{j\omega T} - 0.368)}$$

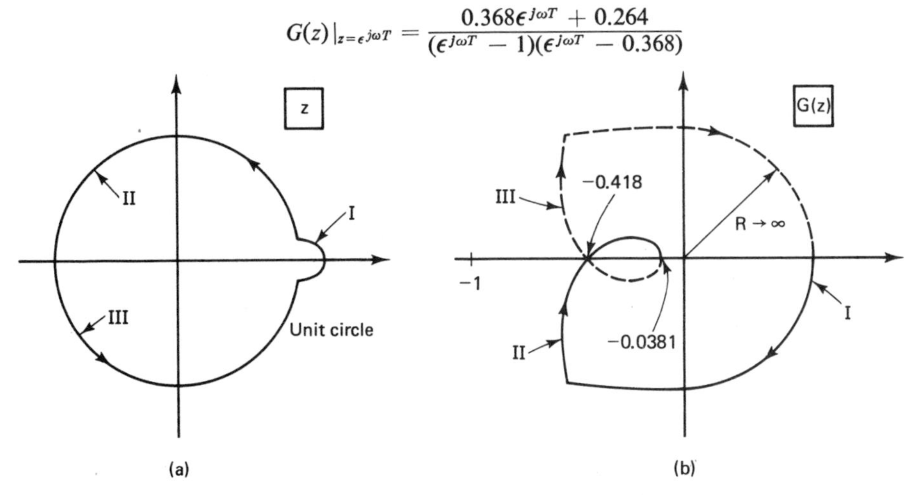

Figure 7-17 Nyquist path and Nyquist diagram for Example 7.10.

In this equation ω varies from $-\omega_s/2$ to $\omega_s/2$. Since $G(\epsilon^{j\omega T})$ for $0 > \omega > -\omega_s/2$ is the complex conjugate of $G(\epsilon^{j\omega T})$ for $0 < \omega < \omega_s/2$, it is necessary to calculate $G(\epsilon^{j\omega T})$ only for $0 < \omega < \omega_s/2$. This calculation results in the frequency response for $G(z)$. Note that $G(-1) = -0.0381$.

The calculation of the Nyquist diagram for the portion of the diagram labeled II may be accomplished in several ways. We may use

$$G^*(s)\,|_{s=j\omega} \qquad \text{or} \qquad G(z)\,|_{z=\epsilon^{j\omega T}}$$

as the basis for calculation. The actual computations may be done graphically or numerically. The recommended technique is numerical evaluation using a higher-level programming language such as FORTRAN or BASIC. FORTRAN programs for calculating the Nyquist diagram are given in Appendix I.

It is to be noted that this system can be forced into instability by increasing the system gain by a factor of 1/0.418, or 2.39. However, the same system without sampling, as shown in Example 7.9, is stable for all positive values of gain. Thus, for this system, the sampling effect is destabilizing. This destabilizing effect can be seen from the frequency response of the zero-order hold. One should recall that the zero-order hold introduces phase lag into the system, and in general, phase lag is a destabilizing effect.

Example 7.11

Consider again the system of Example 7.10. For this system

$$G(z) = \frac{0.368z + 0.264}{z^2 - 1.368z + 0.368}$$

Then $G(w)$ is given by

$$G(w) = \frac{0.368\left[\dfrac{1 + 0.5w}{1 - 0.5w}\right] + 0.264}{\left[\dfrac{1 + 0.5w}{1 - 0.5w}\right]^2 - 1.368\left[\dfrac{1 + 0.5w}{1 - 0.5w}\right] + 0.368} = \frac{-0.0381(w - 2)(w + 12.14)}{w(w + 0.924)}$$

The Nyquist diagram can be obtained from this equation by allowing w to assume values from $j0$ to $j\infty$. Since $G(w)$ has a pole at the origin, the Nyquist path must detour around this point. The Nyquist diagram generated by $G(w)$ is identical to that generated using $G(z)$, as shown in Figure 7-17.

Three different transfer functions may be used to generate the Nyquist diagram: $G^*(s)$, $G(z)$, and $G(w)$. Of course, the Nyquist diagrams generated will be identical in each case.

For a continuous-time system, the *gain margin* is defined as the factor by which the gain must change to force the system to marginal stability. The *phase margin* is defined as the angle through which the Nyquist diagram must be rotated such that the diagram intersects the -1 point. The gain and phase margin definitions for discrete-time systems are exactly the same; that is, in the Nyquist diagram shown in Figure 7-18, the gain margin is $1/a$, and the phase margin is ϕ_m.

At this point a discussion of certain properties of pulse transfer functions is in order. For the system of Figure 7-19a,

$$C(z) = G(z)E(z)$$

where we can express $G(z)$ as

$$G(z) = \frac{a_m z^m + a_{m-1}z^{m-1} + \cdots + a_0}{z^n + b_{n-1}z^{n-1} + \cdots + b_0} \tag{7-29}$$

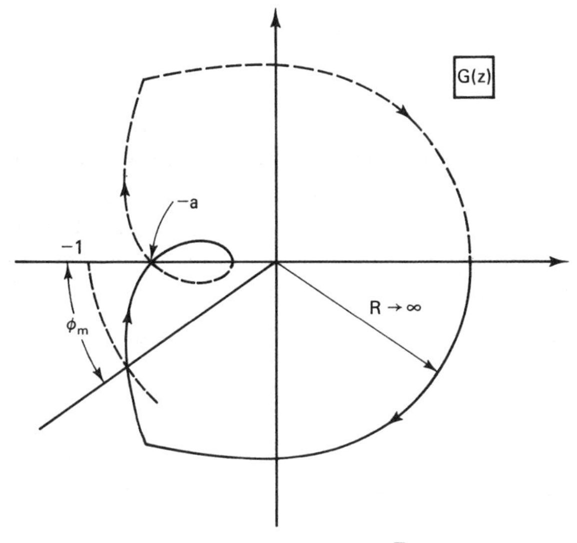

Figure 7-18 Nyquist diagram illustrating stability margins.

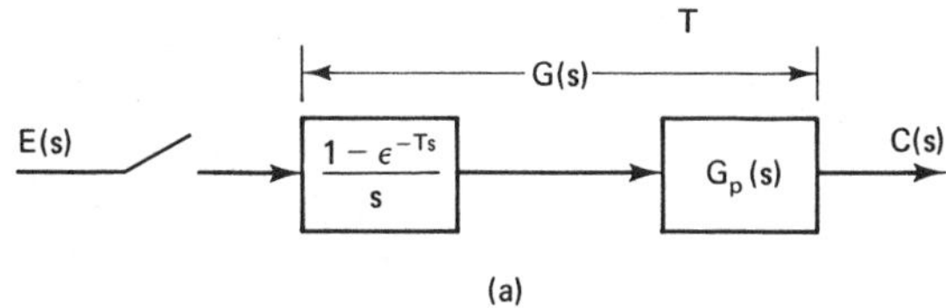

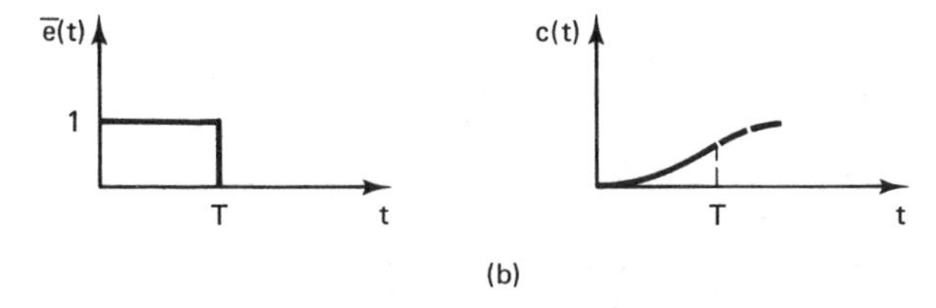

Figure 7-19 Discrete unit pulse response.

By dividing the denominator of $G(z)$ into its numerator, we can express $G(z)$ as

$$G(z) = g(n - m)z^{-(n-m)} + g(n - m + 1)z^{-(n-m+1)} + \cdots \tag{7-30}$$

A discrete unit pulse function is defined by the transform

$$E(z) = 1 \tag{7-31}$$

Thus a discrete unit pulse function is a number sequence $\{e(k)\}$ that has a value of unity for $k = 0$, and a value of zero for $k = 1, 2, 3, \ldots$. Hence, if in Figure 7-19, $E(z)$ is the discrete unit pulse function, then

$$C(z) = G(z)E(z) = G(z) \tag{7-32}$$

Thus (7-30) is the system pulse response.

Figure 7-19b shows $\bar{e}(t)$ for the case that $E(z)$ is the discrete unit pulse function. For most physical systems we would expect a response $c(t)$ as shown in the figure. The plant does not respond instantaneously to the input, but it will have responded to some nonzero value at $t = T$. Hence we may express $C(z)$ as

$$C(z) = c(1)z^{-1} + c(2)z^{-2} + \cdots \tag{7-33}$$

Comparing (7-33) with (7-30), we see that, in general,

$$n - m = 1 \tag{7-34}$$

Then, for a physical system, we generally expect the order of the numerator of its pulse transfer function to be one less than the order of the denominator. If the plant can respond instantaneously, the order of the numerator will be equal to the order of the denominator. If the order of the numerator is greater than the order of the denominator, the plant will respond prior to the application of an input. This case, of course, is not physically realizable.

7.8 THE BODE DIAGRAM

The convenience of frequency-response plots for continuous-time systems in the form of the Bode diagram in both analysis and design stems from the straight-line approximations that are made, and these straight-line approximations are based on the independent variable, $j\omega$, being completely imaginary. Thus Bode diagrams of discrete-time systems may be plotted, using straight-line approximations, provided that the w-plane form of the transfer function is used. For convenience, a summary of the first-order terms used in the construction of a Bode diagram is given in Table 7-4. Since terms with complex zeros or poles are not amenable to straight-line approximations, these terms have been omitted.

TABLE 7-4 SUMMARY OF TERMS EMPLOYED IN A BODE DIAGRAM

1. *A constant term k.* When this term is present, the log magnitude plot is shifted up or down by the amount $20 \log_{10} k$.
2. *The term $j\omega_w$ or $1/j\omega_w$.* If the term $j\omega_w$ is present, the log magnitude is $20 \log_{10} \omega_w$, which is a straight line with a slope of $+20$ dB/decade, and the phase is constant at $+90°$. If the term $1/j\omega_w$ is present, the log magnitude is $-20 \log \omega_w$, which is a straight line with a slope of -20 dB/decade, and the phase is constant at $-90°$. The Bode plots for these terms are shown in Figure 7-20.
3. *The term $(1 + j\omega_w\tau)$ or $[1/(1 + j\omega_w\tau)]$.* The term $(1 + j\omega_w\tau)$ has a log magnitude of $20 \log_{10} \sqrt{1 + \omega_w^2\tau^2}$ which can be approximated as $20 \log_{10} 1 = 0$ when $\omega_w\tau \ll 1$ and as $20 \log_{10} \omega_w\tau$, $\omega_w\tau \gg 1$. The corner or "break" frequency is $\omega_w = 1/\tau$. The phase is given by the expression $\tan^{-1} \omega_w\tau$. The term $[1/(1 + j\omega_w\tau)]$ is handled in a similar manner. The Bode plots for these functions are shown in Figure 7-20.

Let us now employ the Bode diagram in the analysis of a familiar example.

Example 7.12

Consider again the system of Example 7.3. For this system,

$$G(w) = -\frac{0.0381(w - 2)(w + 12.14)}{w(w + 0.924)}$$

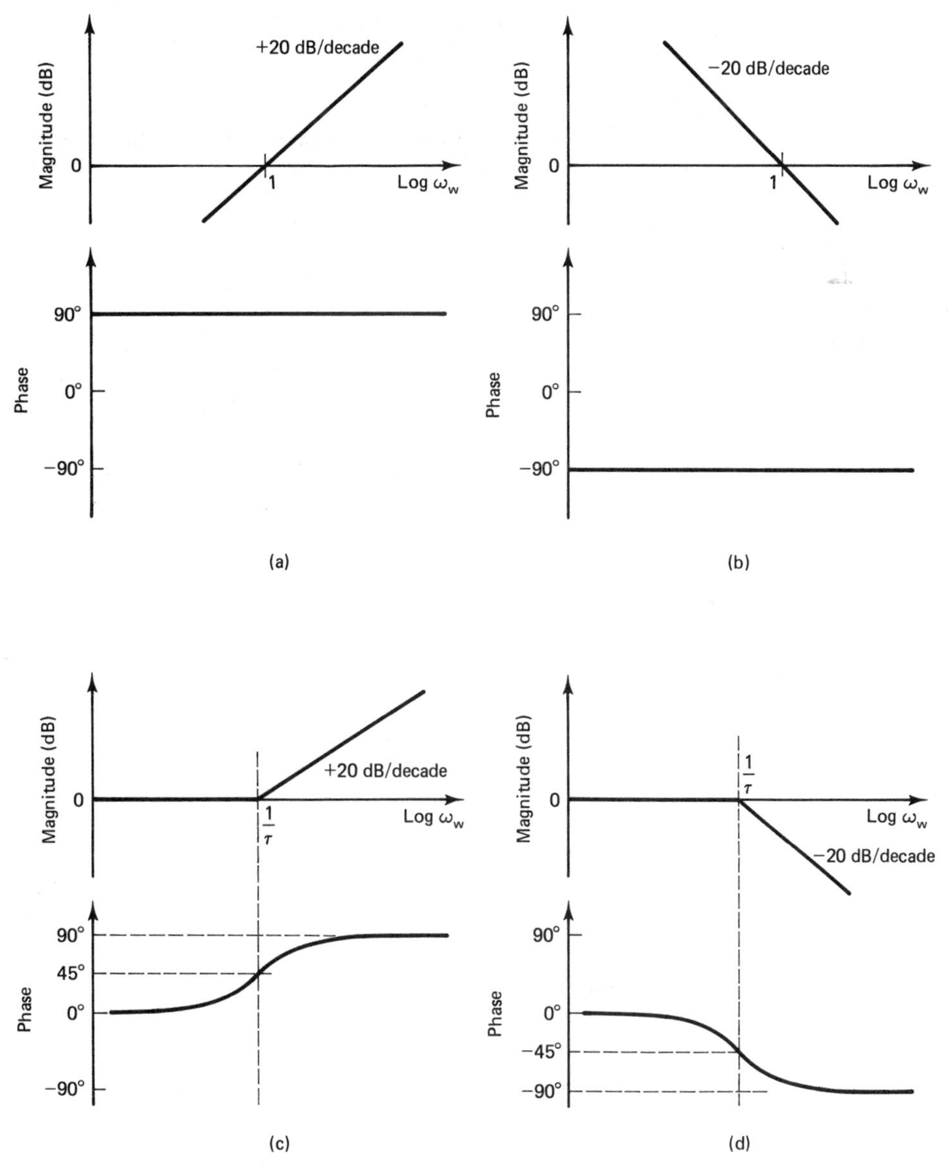

Figure 7-20 Short summary of terms employed in Bode diagrams: (a) Bode plot for $j\omega_w$; (b) Bode plot for $1/j\omega_w$; (c) Bode plot for $1 + j\omega_w\tau$; (d) Bode plot for $1/1 + j\omega_w\tau$.

and

$$G(j\omega_w) = -\frac{0.0381(j\omega_w - 2)(j\omega_w + 12.14)}{j\omega_w(j\omega_w + 0.924)}$$

$$= \frac{-\left(j\frac{\omega_w}{2} - 1\right)\left(\frac{j\omega_w}{12.14} + 1\right)}{j\omega_w\left(\frac{j\omega_w}{0.924} + 1\right)}$$

Note that the numerator break frequencies are $\omega_w = 2$ and $\omega_w = 12.14$ and the denominator break frequencies are $\omega_w = 0$ and $\omega_w = 0.924$. The Bode diagram for this system, using straight-line approximations, is shown in Figure 7-21. Both the gain and phase margins of the system are shown on the diagram.

As stated in Example 7.10, this system can be made unstable by increasing the gain. This can also be seen in Figure 7-21. Increasing the gain is equivalent to shifting the entire magnitude curve vertically upward. Therefore, if the gain is increased by an amount equal to the gain margin, a condition of marginal stability will exist.

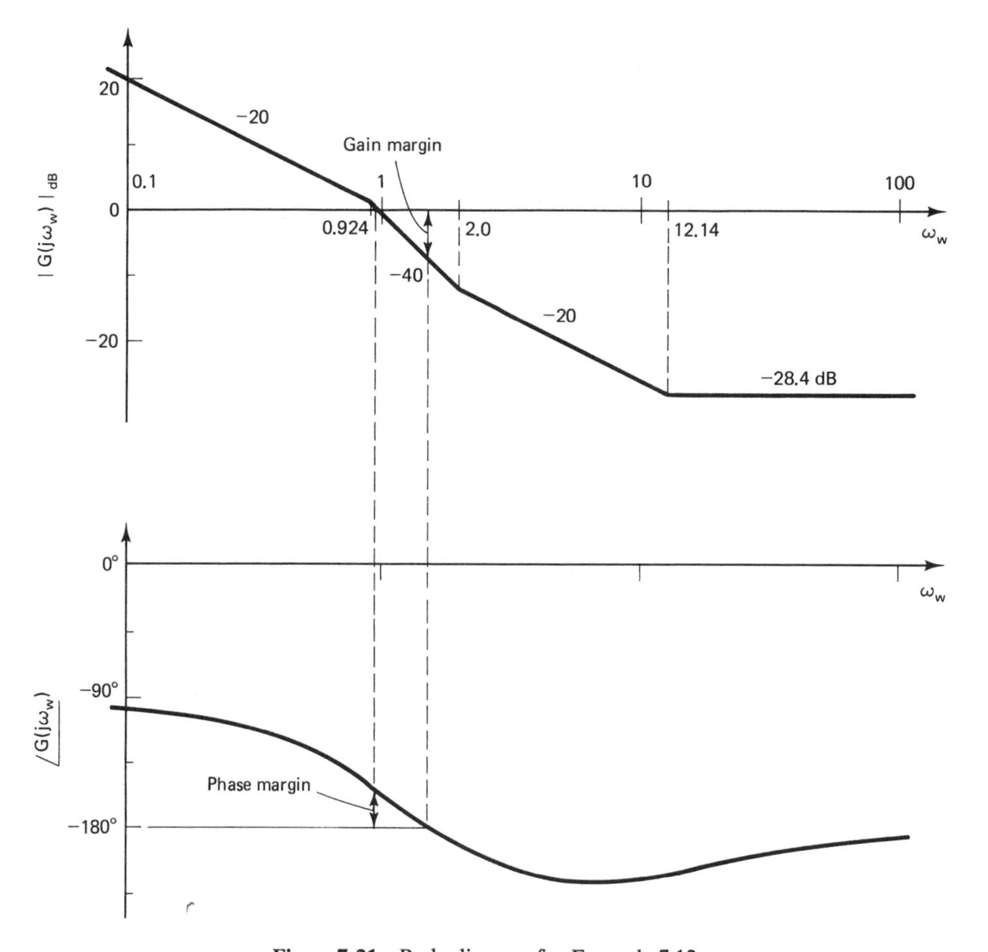

Figure 7-21 Bode diagram for Example 7.12.

While this section presents an approximate technique for generating Bode diagrams, in practice these diagrams are usually plotted from computer-calculated data. The frequency response may be calculated from either $G^*(s)$ or $G(z)$, and for each s-plane frequency ω, the w-plane frequency ω_w is obtained from (7-11); that is,

$$\omega_w = \frac{2}{T} \tan\left(\frac{\omega T}{2}\right)$$

Thus it is unnecessary to construct the function $G(w)$. Computer programs for calculating frequency responses are given in Appendix I.

The gain–phase plot of the frequency response of a system is a plot of the same information shown by a Bode diagram, plotted on different axes. For the gain–phase plot, the frequency response is plotted as gain versus phase on rectangular axes, with frequency as a parameter. Thus the frequency response of a discrete-time system may be plotted on gain–phase axes, as well as on a Bode diagram. As an example, the gain–phase plot of the system of Example 7.12 is shown in Figure 7-22.

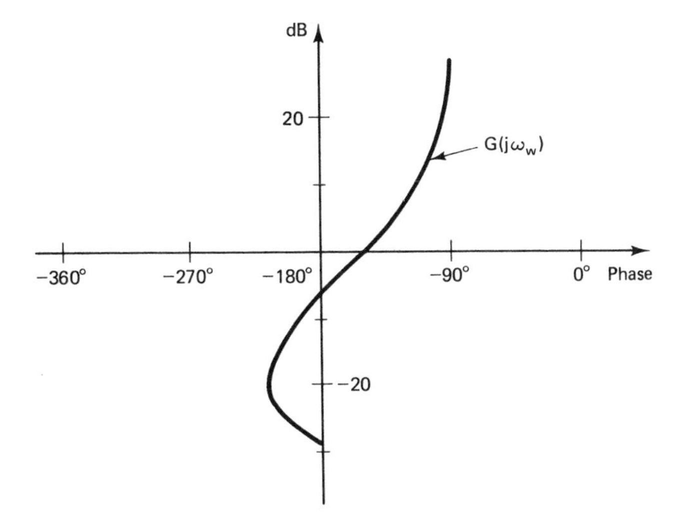

Figure 7-22 Gain–phase plot for Example 7.12.

7.9 INTERPRETATION OF THE FREQUENCY RESPONSE

Throughout this chapter the term "frequency response" has been used in relation to discrete-time systems. The physical meaning of a system frequency response in relation to continuous-time systems is well known. In this section the physical meaning of the frequency response of a discrete-time system is developed.

Consider a discrete-time system described by

$$C(z) = G(z)E(z)$$

and suppose that the system input is a sampled sine wave; that is,

$$E(z) = \mathfrak{z}[\sin \omega t] = \frac{z \sin \omega T}{(z - \epsilon^{j\omega T})(z - \epsilon^{-j\omega T})}$$

Then

$$C(z) = \frac{G(z)z \sin \omega T}{(z - \epsilon^{j\omega T})(z - \epsilon^{-j\omega T})} = \frac{k_1 z}{z - \epsilon^{j\omega T}} + \frac{k_2 z}{z - \epsilon^{-j\omega T}} + C_g(z) \qquad (7\text{-}35)$$

where $C_g(z)$ are those components of $C(z)$ that originate in the poles of $G(z)$. If the system is stable, these components of $c(nT)$ will tend to zero with increasing time, and the system steady-state response is given by

$$C_{ss}(z) = \frac{k_1 z}{z - \epsilon^{j\omega T}} + \frac{k_2 z}{z - \epsilon^{-j\omega T}}$$

Now, from (7-35),

$$k_1 = \frac{G(\epsilon^{j\omega T}) \sin \omega T}{\epsilon^{j\omega T} - \epsilon^{-j\omega T}} = \frac{G(\epsilon^{j\omega T})}{2j} \tag{7-36}$$

Expressing $G(\epsilon^{j\omega T})$ as

$$G(\epsilon^{j\omega T}) = |G(\epsilon^{j\omega T})| \epsilon^{j\theta}$$

we see that (7-36) becomes

$$k_1 = \frac{|G(\epsilon^{j\omega T})| \epsilon^{j\theta}}{2j} \tag{7-37}$$

Since k_2 is the complex conjugate of k_1, then

$$k_2 = \frac{-|G(\epsilon^{j\omega T})| \epsilon^{-j\theta}}{2j} \tag{7-38}$$

Thus, from (7-36), (7-37), and (7-38),

$$\begin{aligned} c_{ss}(kT) &= k_1(\epsilon^{j\omega T})^k + k_2(\epsilon^{-j\omega T})^k \\ &= |G(\epsilon^{j\omega T})| \left[\frac{\epsilon^{j(\omega kT + \theta)} - \epsilon^{-j(\omega kT + \theta)}}{2j} \right] \\ &= |G(\epsilon^{j\omega T})| \sin(\omega kT + \theta) \end{aligned} \tag{7-39}$$

From the development above, it is seen that if the input to a stable discrete-time system is a sinusoid of frequency ω, the steady-state system response is also sinusoidal at the same frequency. The amplitude of the response is equal to the amplitude of the input multiplied by $|G(\epsilon^{j\omega T})|$, and the phase of the response is equal to the phase of the input plus the angle of $G(\epsilon^{j\omega T})$. Thus it is seen that $G(\epsilon^{j\omega T})$ is the true system frequency response at the sampling instants.

7.10 CLOSED-LOOP FREQUENCY RESPONSE

In this chapter many of the techniques presented for closed-loop stability analysis are based on the system open-loop frequency response. Of course, the system transfer characteristics are determined by the closed-loop frequency response. In this section a simple technique is given for finding the closed-loop frequency response, given the open-loop frequency response, provided that the system is of a standard form.

Consider the discrete-time system shown in Figure 7-23a. For this system,

$$\frac{C(z)}{R(z)} = \frac{G(z)}{1 + G(z)}$$

The closed-loop frequency response is given by

$$\frac{C(\epsilon^{j\omega T})}{R(\epsilon^{j\omega T})} = \frac{G(\epsilon^{j\omega T})}{1 + G(\epsilon^{j\omega T})} \tag{7-40}$$

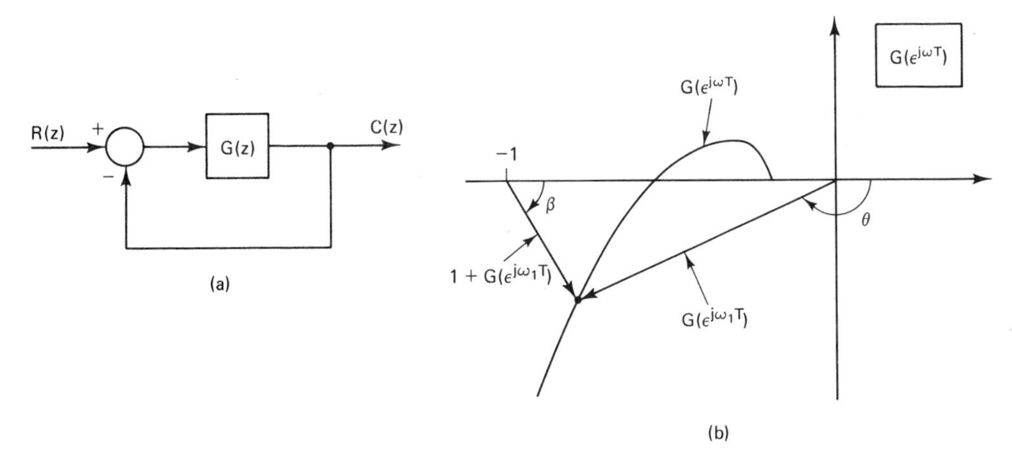

Figure 7-23 Determining closed-loop frequency response from open-loop frequency response.

Suppose that $G(\epsilon^{j\omega T})$ is as shown in the polar plot in Figure 7-23b. Then the numerator and the denominator of (7-40) are the vectors shown for the frequency $\omega = \omega_1$, and the frequency response at ω_1 is the ratio of these vectors. Let the ratio of these vectors be denoted as

$$\frac{C(\epsilon^{j\omega_1 T})}{R(\epsilon^{j\omega_1 T})} = \frac{|G(\epsilon^{j\omega_1 T})|\epsilon^{j\theta}}{|1 + G(\epsilon^{j\omega_1 T})|\epsilon^{j\beta}} = M\epsilon^{j(\theta-\beta)} = M\epsilon^{j\phi} \tag{7-41}$$

The locus of points in the $G(\epsilon^{j\omega T})$ plane for which the magnitude of the closed-loop frequency response, M, is a constant is called a constant magnitude locus, or a constant M circle. To see that these loci are in fact circles, consider the following development. Let

$$G(\epsilon^{j\omega T}) = X + jY \tag{7-42}$$

Then, from (7-41),

$$M^2 = \frac{X^2 + Y^2}{(1 + X)^2 + Y^2}$$

Hence

$$X^2(1 - M^2) - 2M^2 X - M^2 + (1 - M^2)Y^2 = 0 \tag{7-43}$$

For $M \neq 1$, we can express this relationship as

$$\left[X + \frac{M^2}{M^2 - 1}\right]^2 + Y^2 = \frac{M^2}{(M^2 - 1)^2} \tag{7-44}$$

This relationship is the equation of a circle of radius $|M/(M^2 - 1)|$ with center at $X = -M^2/(M^2 - 1)$ and $Y = 0$. For $M = 1$, (7-43) yields $X = -\frac{1}{2}$, which is a straight line. Figure 7-24 illustrates the constant M circles.

The loci of points of constant phase are also circles. It is seen from (7-41) and (7-42) that

$$\phi = \theta - \beta = \tan^{-1}\left(\frac{Y}{X}\right) - \tan^{-1}\left(\frac{Y}{1 + X}\right)$$

Then, letting $N = \tan \phi$,

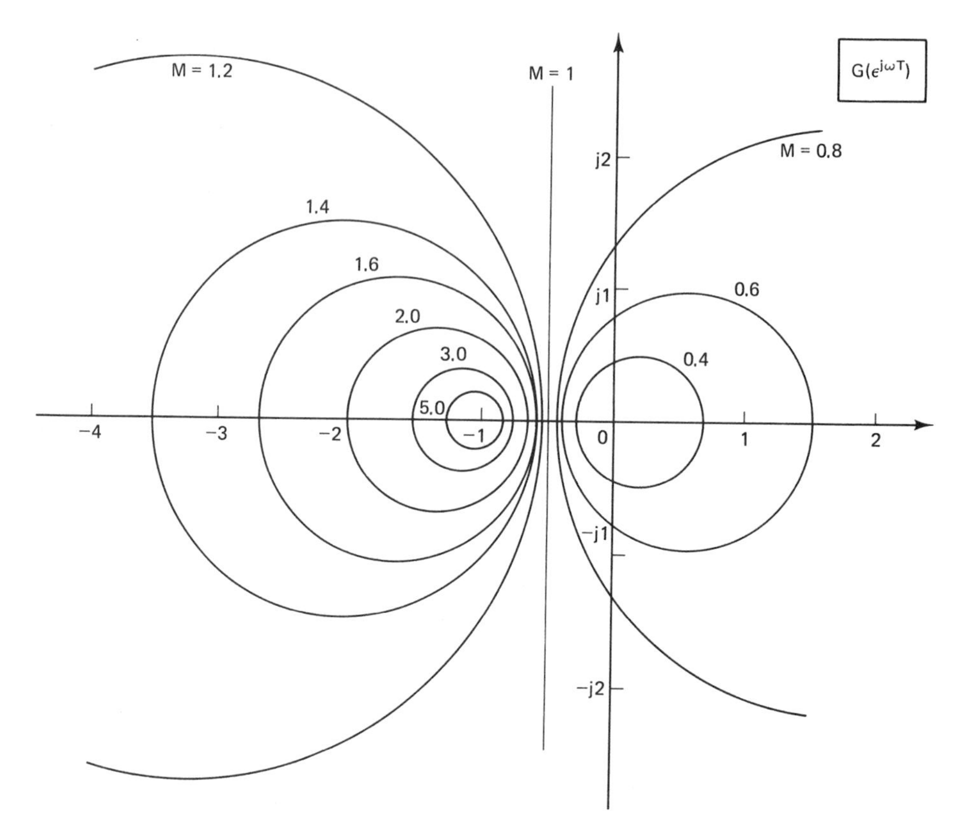

Figure 7-24 Constant magnitude circles.

$$N = \tan(\theta - \beta) = \frac{\tan\theta - \tan\beta}{1 + \tan\theta\tan\beta} = \frac{\dfrac{Y}{X} - \dfrac{Y}{1+X}}{1 + \dfrac{Y}{X}\left[\dfrac{Y}{1+X}\right]}$$

Hence

$$X^2 + X + Y^2 - \frac{1}{N}Y = 0$$

This equation can then be expressed as

$$\left(X + \frac{1}{2}\right)^2 + \left(Y - \frac{1}{2N}\right)^2 = \frac{1}{4} + \left(\frac{1}{2N}\right)^2 \tag{7-45}$$

This is the equation of a circle with radius of $\sqrt{1/4 + (1/2N)^2}$ with center at $X = -1/2$ and $Y = 1/(2N)$. Figure 7-25 illustrates the constant N (phase) circles. Note that accuracy is very limited in the region of the -1 point. For this reason, the Nichols chart is often used, instead of the constant M and N circles. The Nichols chart will now be discussed.

The gain–phase plane was introduced in Section 7.8, and was shown to be useful in graphically presenting a system frequency response. For example, the frequency response $G(\epsilon^{j\omega T})$ presented as a polar plot in Figure 7-23 could also be plotted in the gain–phase plane, as shown in Figure 7-22. Also the constant M circles of Figure

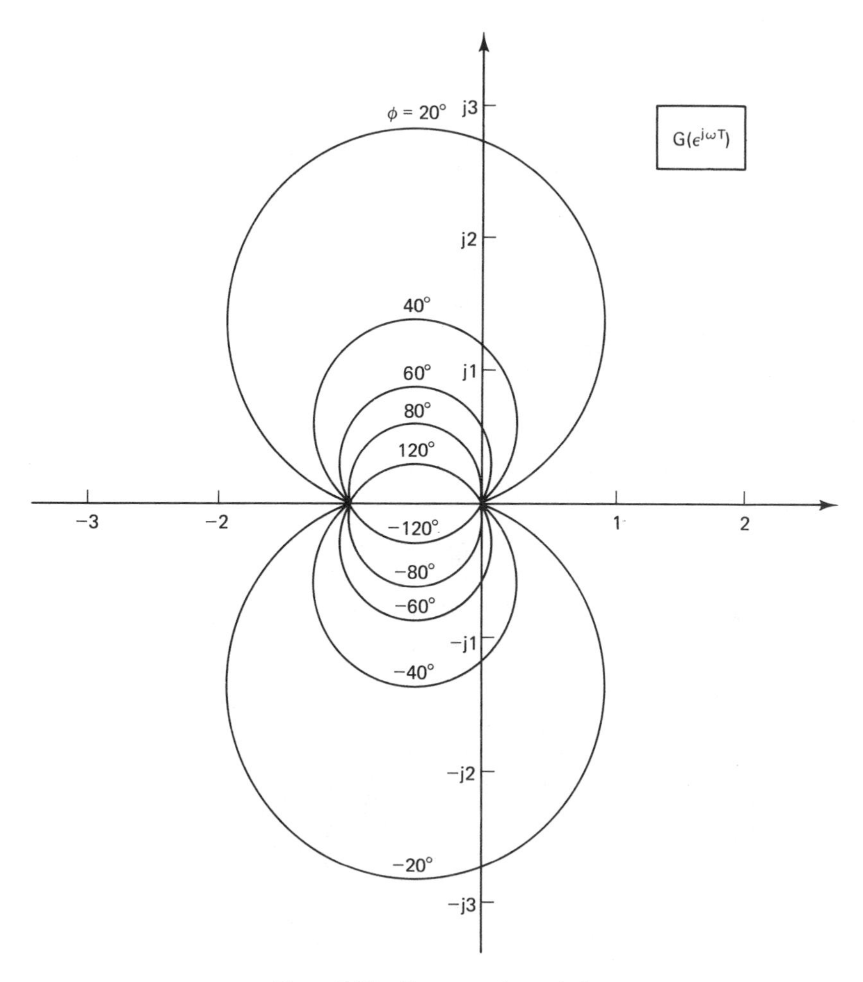

Figure 7-25 Constant phase circles.

7-24 and the constant N circles of Figure 7-25 can be plotted in the gain–phase plane. Such a plot is called the Nichols chart [11], and is often used to determine closed-loop frequency response. Figure 7-26 illustrates the Nichols chart.

The frequency response $G(\epsilon^{j\omega T})$ is usually computed via computer programs, of the type given in Appendix I. Thus it is logical to add statements to these programs that will, at the same time, calculate the closed-loop frequency response, rather than going to a graphical procedure as described above. The computer procedure is more accurate, saves time, and is not limited to the system of Figure 7-23a.

An example will now be given which illustrates closed-loop frequency response.

Example 7.13

Again the system of Example 6.2 will be considered. For this example, $T = 1$ s and

$$G^*(s) = \left[\frac{1 - \epsilon^{-Ts}}{s^2(s + 1)}\right]^*$$

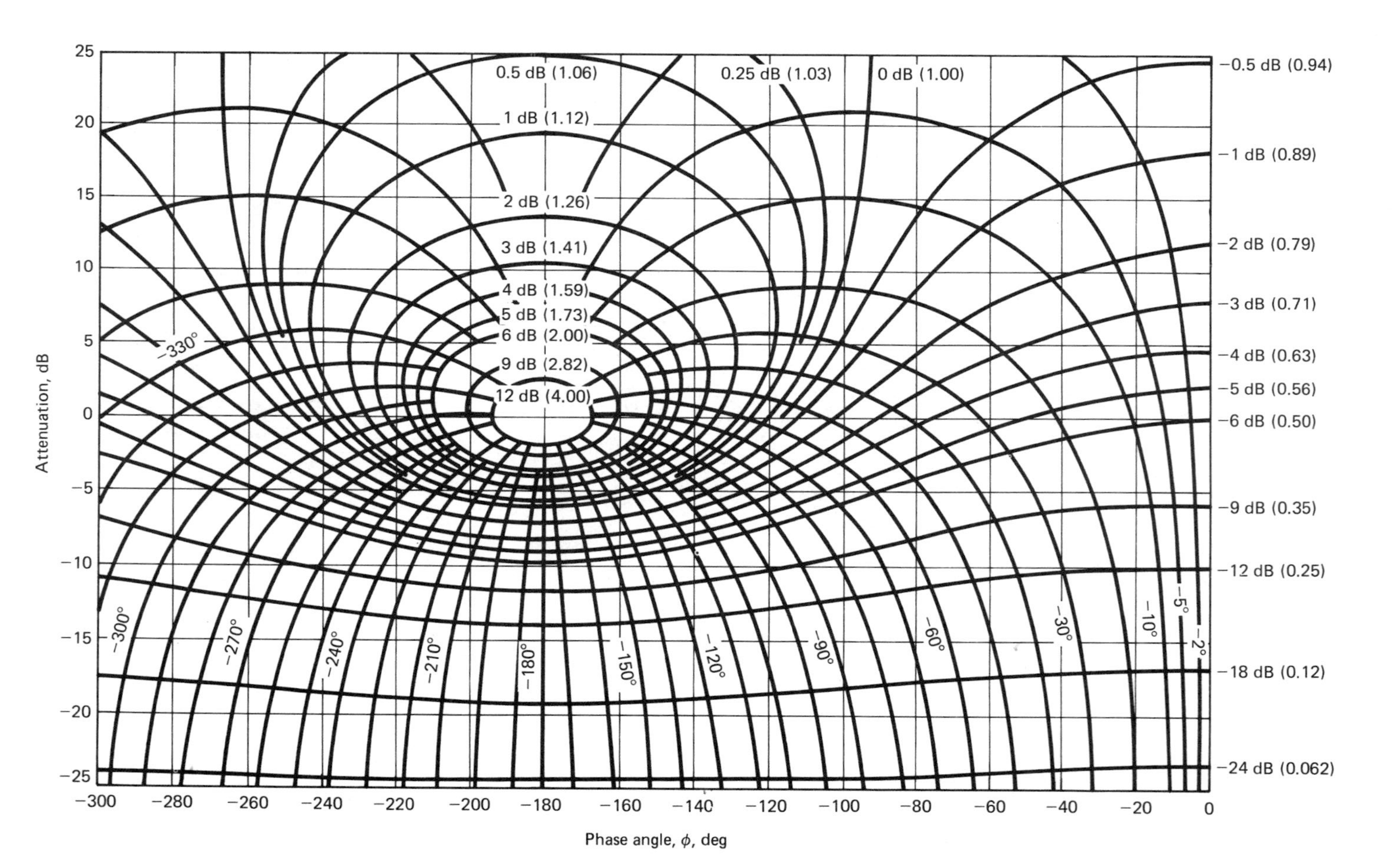

Figure 7-26 Nichols chart.

The open-loop frequency response for this system was calculated using the computer program given in Appendix I for $G^*(j\omega)$. In this case, three terms in the expansion

$$G^*(j\omega) = \frac{1}{T}[G(j\omega) + G(j\omega + j\omega_s) + G(j\omega - j\omega_s) + G(j\omega + j2\omega_s) + \cdots]$$

are required to yield four-significant-figure accuracy in the calculations. The closed-loop frequency response was calculated by altering the computer programs in Appendix I, and is plotted in Figure 7-27 versus real frequency ω. To illustrate the effects on the step response of the closed-loop frequency response, the closed-loop frequency response was then calculated for $T = 0.1$ s and is also shown in Figure 7-27. Note that a resonant, or peaking, effect is more pronounced at $T = 1$ s. A more pronounced resonance in

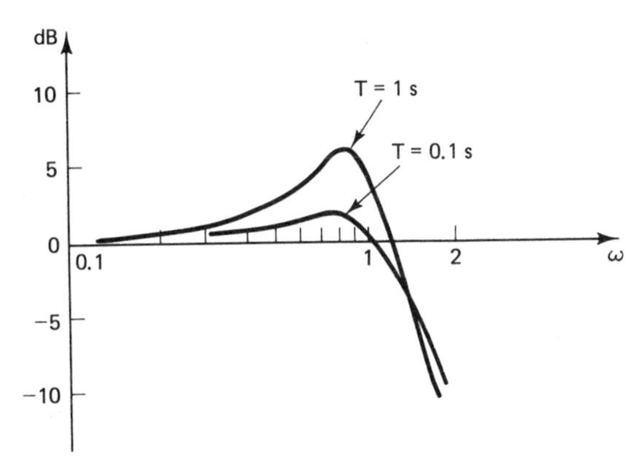

Figure 7-27 Closed-loop frequency responses for Example 7.13.

the closed-loop frequency response generally indicates more overshoot in the step response. The step responses for the system were obtained by simulation, and are shown in Figure 7-28. The peak overshoot for $T = 1$ s is 45%, and that for $T = 0.1$ s is 18%. The correlation between the closed-loop frequency response and the time response of a system is investigated in greater detail in Chapter 8.

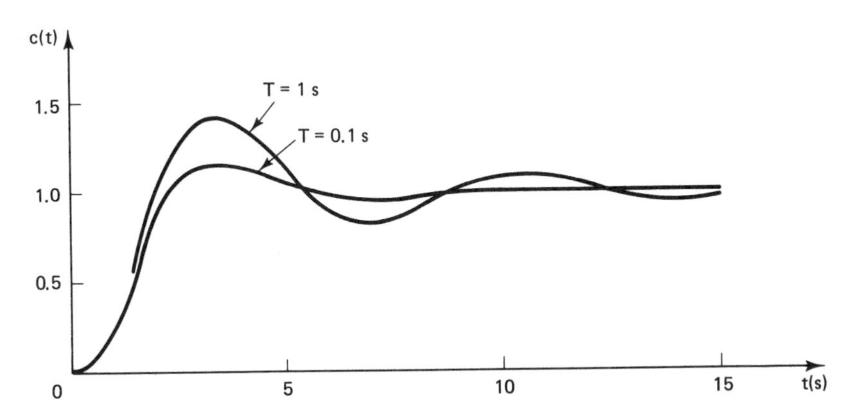

Figure 7-28 Step responses for Example 7.13.

7.11 SUMMARY

In this chapter a number of techniques for analyzing the stability of discrete-time systems have been presented. It has been shown that many of the methods used in the analysis of continuous-time systems are applicable to sampled-data systems also. The chapter contains a number of examples, and the use of one of them throughout the chapter provides a common thread and basis for comparison among the various stability analysis techniques.

REFERENCES AND FURTHER READING

1. R. C. Dorf, *Modern Control Systems.* Reading, Mass.: Addison-Wesley Publishing Company, Inc., 1974.
2. E. I. Jury, *Theory and Application of the z-Transform Method.* New York: John Wiley & Sons, Inc., 1964.
3. B. C. Kuo, *Analysis and Synthesis of Sampled-Data Control Systems.* Englewood Cliffs, N.J.: Prentice-Hall, Inc., 1963.
4. B. C. Kuo, *Automatic Control Systems.* Englewood Cliffs, N.J.: Prentice-Hall, Inc., 1962.
5. H. M. James, N. B. Nichols, and R. S. Phillips, *Theory of Servomechanisms.* New York: McGraw-Hill Book Company, 1947.
6. J. A. Cadzow and H. R. Martens, *Discrete-Time and Computer Control Systems.* Englewood Cliffs, N.J.: Prentice-Hall, Inc., 1970.
7. E. I. Jury, *Sampled-Data Control Systems.* New York: John Wiley & Sons, Inc., 1958.
8. B. C. Kuo, *Discrete-Data Control Systems.* Englewood Cliffs, N.J.: Prentice-Hall, Inc., 1970.
9. D. P. Lindorff, *Theory of Sampled-Data Control Systems.* New York: John Wiley & Sons, Inc., 1965.
10. J. R. Ragazzini and G. F. Franklin, *Sampled-Data Control Systems.* New York: McGraw-Hill Book Company, 1958.
11. W. R. Evans, *Control System Dynamics.* New York: McGraw-Hill Book Company, 1954.

PROBLEMS

7-1. A temperature control system is shown in Figure P7-1. This system was described in Problem 4-3. Let the digital filter transfer function be unity, the power amplifier gain be K, and $T = 0.5$ s. Derive the characteristic equations of the system, both with and without sampling. Determine the range of K for stability in each case.

7-2. Suppose that, in Problem 7-1, the sample period is increased to $T = 3$ s. Note that the process time constant is also 3 s. Determine the range of K for stability, and compare the results to those obtained in Problem 7-1.

7-3. An automobile cruise-control system is shown in Figure P7-3. This system was described in Problem 4-7. Let $T = 0.2$ s and $D(z) = K$ (i.e., the digital controller is simply a gain).

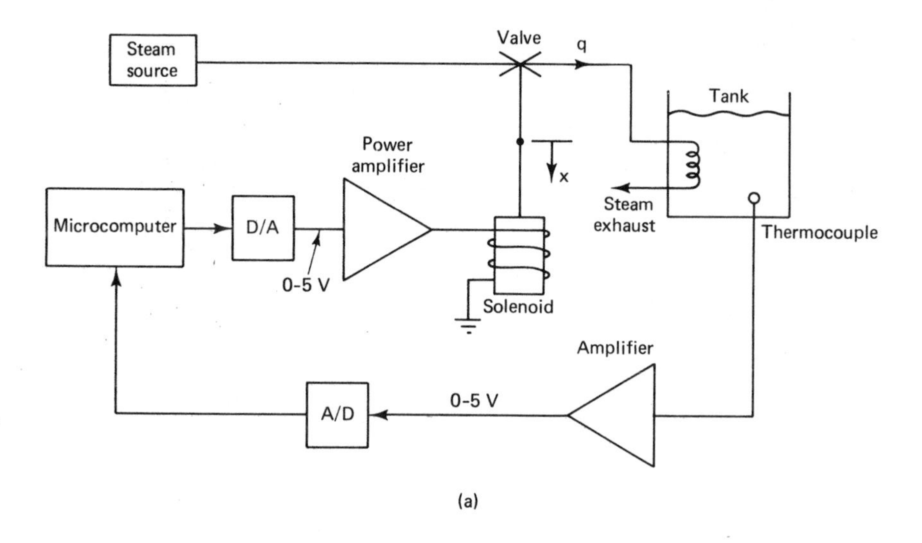

(a)

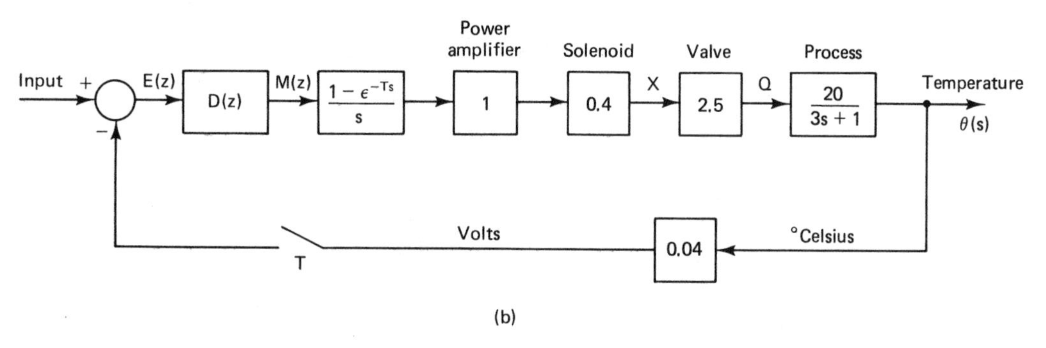

(b)

Figure P7-1 Temperature control system.

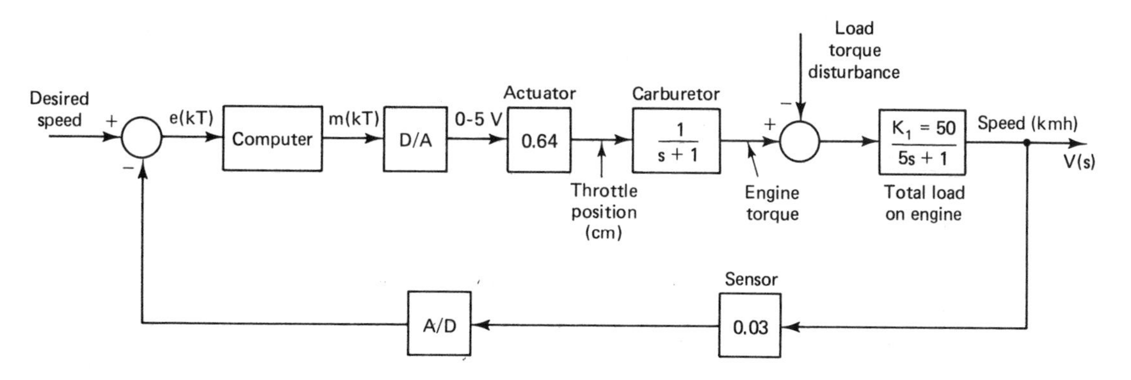

Figure P7-3 Cruise control system.

(a) Derive the system characteristic equation, and determine the range of K for which the system is stable.

(b) Repeat part (a) with the carburetor time constant ignored, that is, with the carburetor transfer function equal to unity.

7-4. The unit pulse response of a discrete system with input $R(z)$ is defined as the system response $c(kT)$, with $r(kT) = 1$, $k = 0$, and $r(kT) = 0$, $k \geq 1$ [i.e., the input $R(z) = 1$]. Show that if the discrete system is stable, the system unit pulse response, $c(kT)$, approaches zero as $k \longrightarrow \infty$.

7-5. Shown in Figure P7-5 is the attitude control system of a space booster. The aerodynamic center of pressure is labeled CP, the center of gravity is CG, F_α is the normal aerodynamic force, and v is the vehicle velocity. The angle of the engine is δ, and the engine develops a thrust F_T. The vehicle transfer function is $G_P(s)$, and is open-loop unstable if the center of pressure CP is forward of the center of gravity. For this case the vehicle will tend to tumble. The block diagram shows analog rate feedback and sampled position feedback, as was the case for the Saturn V booster stage. For the Saturn V, the rigid-body transfer function was

$$G_P(s) = \frac{0.9407}{s^2 - 0.0297}$$

This transfer function does not include the effects of vehicle bending and liquid fuel

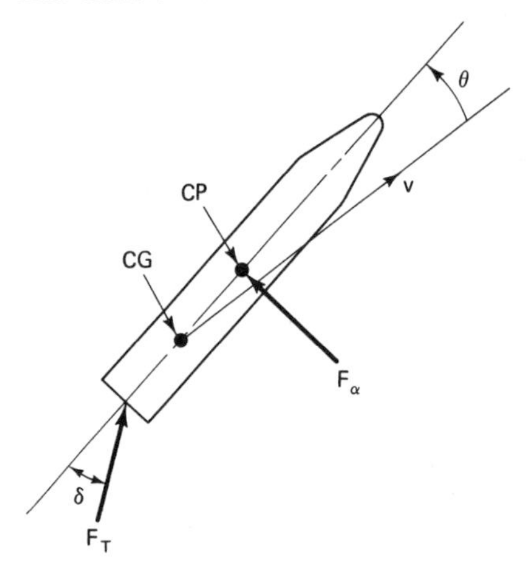

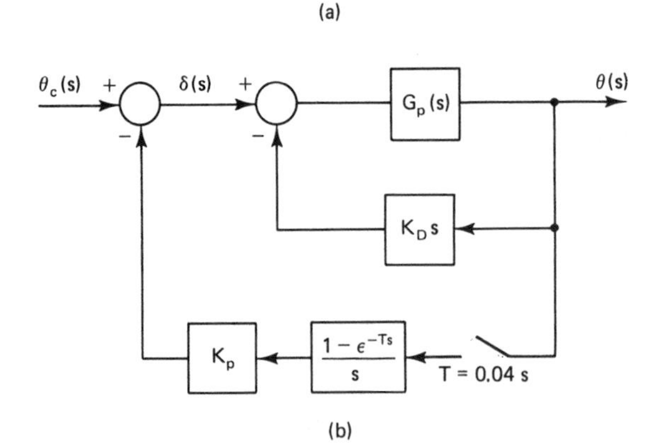

Figure P7-5 Booster attitude control system.

slosh, which add 22 orders to the transfer function. In addition, the dynamics of the actuator that controls δ has been omitted.

(a) With $T = 0.04$ s (the value for the Saturn V) and $K_D = 1$, find the range of K_P for stability.

(b) Repeat part (a) with $K_D = 0$ (i.e., with no rate feedback).

7-6. Given below are the characteristic equations of certain discrete-time systems. Using Jury's test, determine if the systems are stable.

(a) $z^2 - 1.5z + 0.5 = 0$ **(b)** $z^2 - 1.7z + 0.66 = 0$

(c) $z^3 - 2.3z^2 + 1.7z - 0.4 = 0$ **(d)** $z^3 - 2.2z^2 + 1.51z - 0.33 = 0$

7-7. Consider the temperature control system of Problem 7-1, with $D(z) = 1$, the power amplifier gain equal to K, and $T = 0.5$ s. Plot the root locus for this system, both with sampling and without sampling. From the root-locus plots, determine the range of K for stability.

7-8. Repeat Problem 7-7, with $T = 3.0$ s. (See Problem 7-2.)

7-9. Consider the automobile cruise-control system of Problem 7-3. Let $T = 0.2$ s and $D(z) = K$. Plot the root locus, and from the root locus determine the range of K for stability.

7-10. Shown in Figure P7-10 is a carbon dioxide control system for an environmental plant growth chamber. This system was described in Problem 5-18. Let $D(z) = 1$.

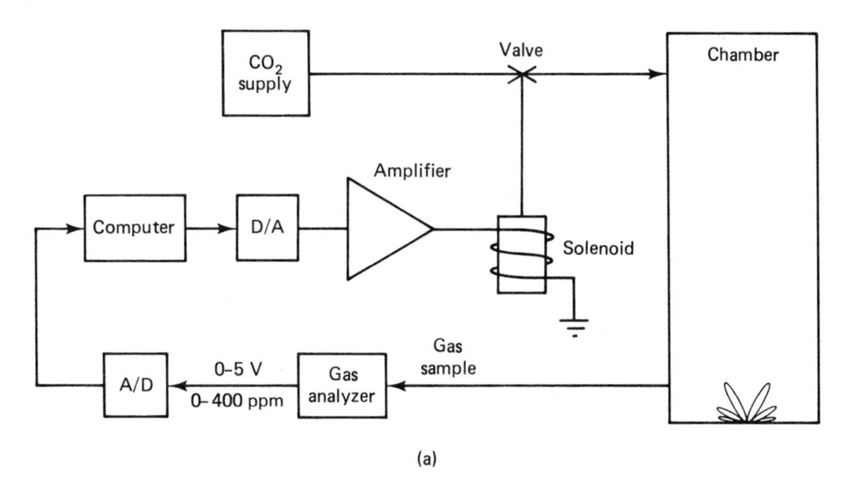

(a)

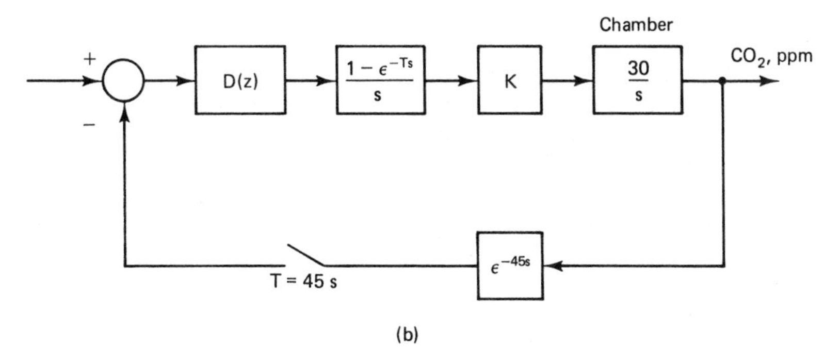

(b)

Figure P7-10 Carbon dioxide control system.

(a) Plot the root locus for the system. Determine the range of K for stability from the root locus.

(b) Repeat part (a) for $T = 60$ s.

7-11. In the temperature control system of Problem 7-1, let $D(z) = 1$ and $T = 0.5$ s. Plot the Nyquist diagram for the system and determine the gain and phase margins. Check the gain margin using the system characteristic equation.

7-12. Plot the Nyquist diagram for the cruise-control system of Problem 7-3, with $D(z) = 1$ and $T = 0.2$ s. Determine the gain and phase margins. Check the gain margin using the system characteristic equation.

7-13. **(a)** Plot the Nyquist diagram for the carbon dioxide control system of Problem 7-10, with $D(z) = 1$. Determine the gain and phase margins, and check the gain margin using the system characteristic equation.

(b) Repeat part (a) for $T = 60$ s.

7-14. Consider the system of Figure P7-14. A sketch (not to scale) of the system Nyquist diagram is shown for a certain gain K. Determine if the system is stable for this value of K.

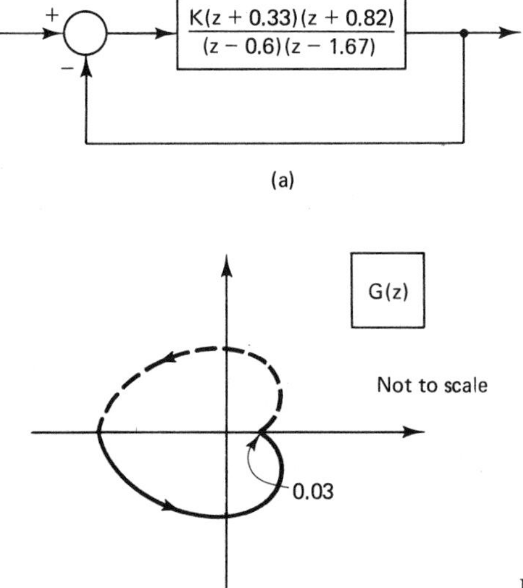

Figure P7-14 System for Problem 7-14: (a) system; (b) Nyquist diagram (not to scale).

7-15. Shown in Figure P7-15 is the sun-tracking control system for a solar collector. This system is described in Problem 6-7. Let $D(z) = 1$ and $T = 1$ s.

(a) Sketch the Bode diagram for this system.

(b) Determine the gain and phase margins.

(c) Check the gain margin using the Routh–Hurwitz criterion.

(d) Let $D(z) = K$, pure gain. This gain K is increased until the system is marginally stable. Find the real (s-plane) frequency at which the system will oscillate.

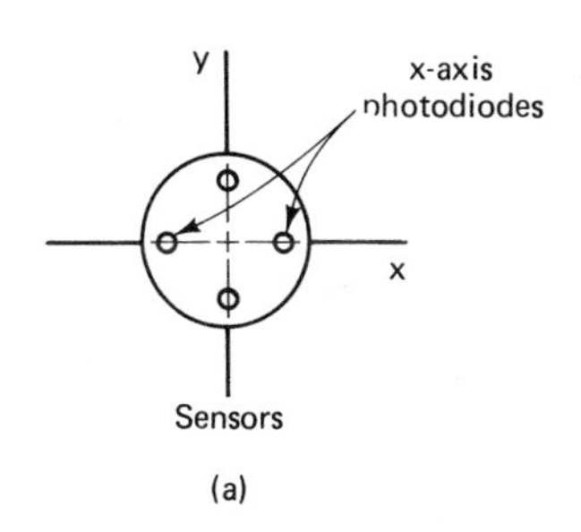

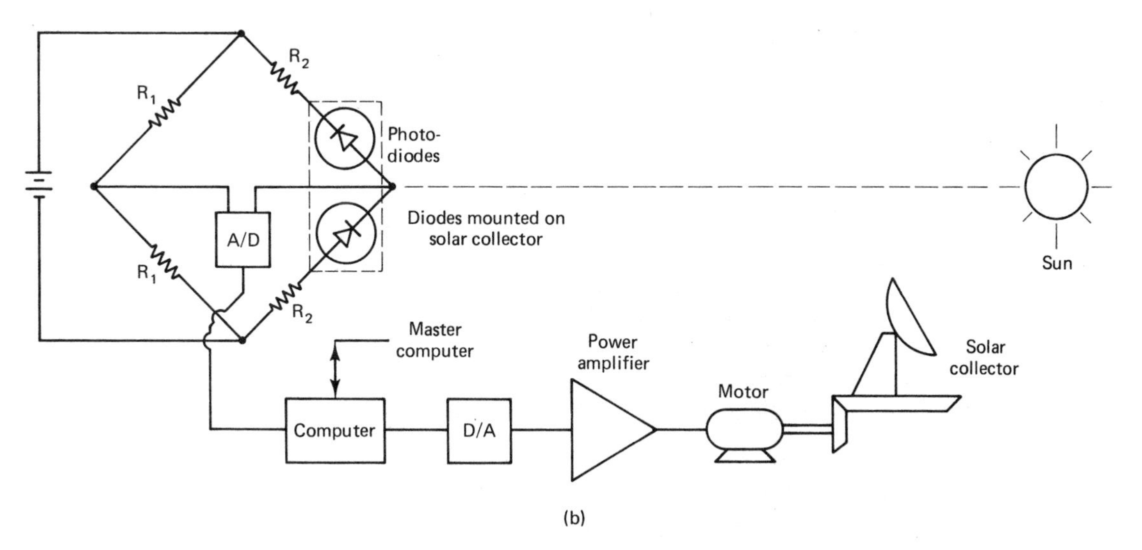

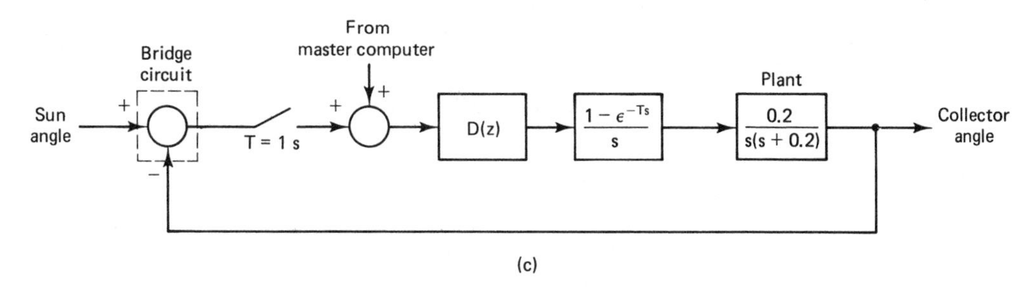

Figure P7-15 Solar collector control system.

7-16. Consider the temperature control system of Problem 7-1. Let $D(z) = 1$ and $T = 0.5$ s.

(a) Sketch the Bode diagram for this system.

(b) Determine the gain and phase margins.

(c) The gain of the power amplifier is increased until the system is marginally stable. Find the real (s-plane) frequency at which the system will oscillate.

7-17. Consider the cruise-control system of Problem 7-3. Let $D(z) = 1$ and $T = 0.2$ s.

(a) Sketch the Bode diagram for this system.

(b) Determine the gain and phase margins.

(c) Let $D(z) = K$, a pure gain. This gain K is increased until the system response is periodic at a constant amplitude. Find the real (s-plane) frequency of the system response.

(d) Check the results of part (c) using the Routh–Hurwitz criterion.

7-18. For the system of Figure P7-18,

$$G(z) = \frac{K(z + 0.9)}{(z - 1)(z - 0.7)}$$

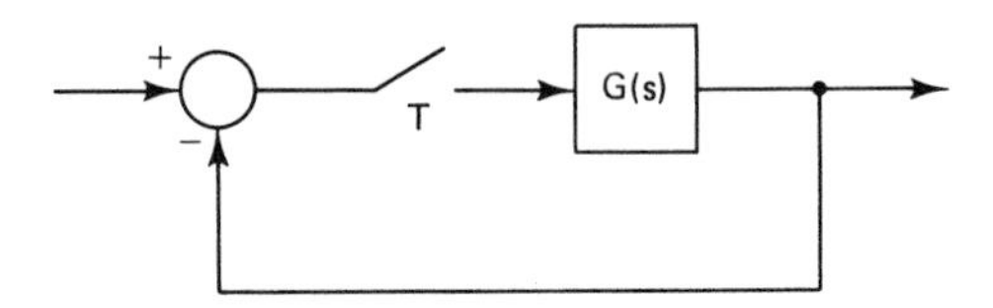

Figure P7-18 System for Problem 7-18.

Determine the range of K for stability by:

(a) Finding the roots of the characteristic equation

(b) The Routh–Hurwitz criterion

(c) Jury's stability test

7-19. For a discrete-time system with $T = 0.1$ s, the characteristic equation is given by

$$(z - 0.99)(z^2 - 0.5z + 1.0) = 0$$

The system is marginally stable. Find the frequency of oscillation.

7-20. For the system of Figure P7-20, the discrete model of the continuous system is given by

$$y(k) - 1.3y(k - 1) + 0.42y(k - 2) = K[m(k - 1) + 0.95m(k - 2)]$$

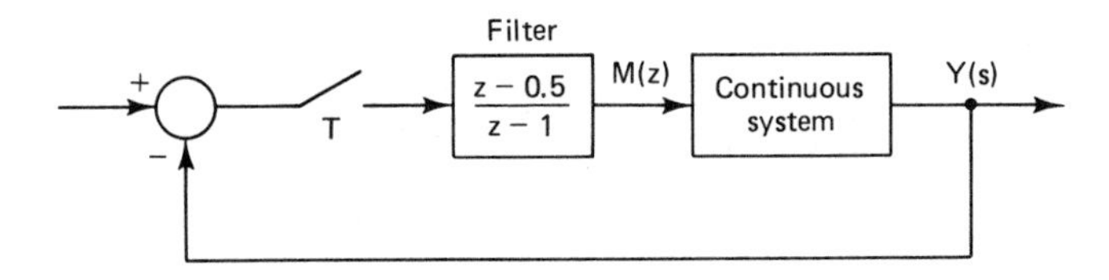

Figure P7-20 System for Problem 7-20.

(a) Determine the stability of the system if $K = 1$.

(b) Calculate the frequency response by computer, and plot the Nyquist diagram, for $K = 1$.

(c) Sketch the root locus.

(d) For $T = 0.04$ s, find the frequency of oscillation if the system gain K is varied until the system is marginally stable.

7-21. For the system shown in Figure P7-21,

$$G(z) = \mathfrak{z}\left[\frac{1 - \epsilon^{-Ts}}{s} G_p(s)\right] = \frac{K(z - 0.8)}{(z - 0.9)(z - 0.6)}$$

(a) Plot the Bode diagram by calculating the frequency response by computer, with $K = 1$.

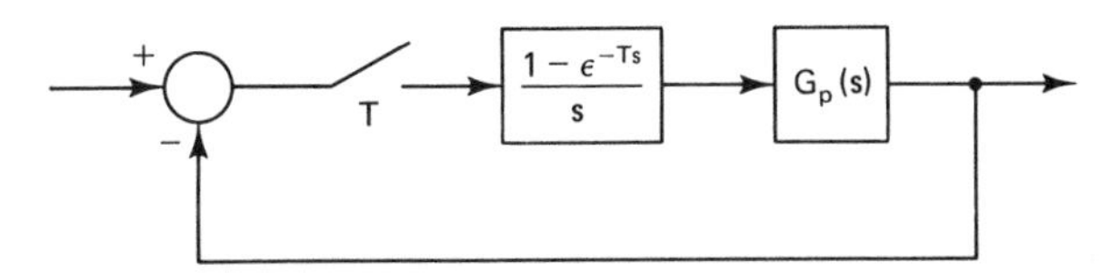

Figure P7-21 System for Problem 7-21.

(b) From the Bode diagram, determine the gain and phase margin.

(c) Find the value of K for marginal stability.

(d) If K is set equal to the value found in part (c), determine the frequency of oscillation.

7-22. List four different relationships that may be used to calculate points for a Nyquist diagram. For each relationship, give the required range of the independent variable.

7-23. Give a technique for calculating the frequency response of a plant that does not require determining the plant's pulse transfer function $G(z)$.

7-24. List four different techniques for determining the stability of a discrete-time system.

7-25. Suppose that you have developed a digital simulation of a discrete-time system. How would you verify that the simulation is correct?

7-26. Suppose that you are given the pulse transfer function, $G(z)$, of a plant. How can you calculate the frequency response required for plotting the Bode diagram without first calculating $G(w)$?

7-27. Given the pulse transfer function $G(z)$ of a plant. For $\omega = 2$ rad/s, $G(\epsilon^{j\omega T})$ is equal to the complex number $1.3\underline{/-25^\circ}$. The signal $5 \cos 2t$ is applied to the input (sampler and data hold) of the plant. Find the steady-state sampled output.

7-28. A Nyquist diagram for a discrete-time system is calculated using $G(\epsilon^{j\omega T})$, $0 \leq \omega \leq \omega_s/2$. A Nyquist diagram for the same system is also calculated using $G(j\omega_w)$, $0 \leq \omega_w < \infty$. In what ways will these two Nyquist diagrams differ?

8

Digital Controller Design

8.1 INTRODUCTION

In the preceding chapters we have been concerned primarily with analysis. We have assumed that the control system was given, and we analyzed the system to determine stability, stability margins, time response, frequency response, and so forth. Some simple design problems were considered: for example, the determination of gains required to meet steady-state error specifications.

In this chapter we consider the total design problem: How do we design a digital controller transfer function (or difference equation) that will satisfy design specifications for a given control system? In this chapter the major emphasis is on classical design techniques (frequency-response techniques). First, phase-lag and phase-lead controllers are considered. Then a particular type of lag–lead controller, called a proportional-plus-integral-plus-derivative (PID) controller, is developed. Further design techniques, which are based on the state-variable model of the plant, are developed in Chapters 9 and 10.

8.2 CONTROL SYSTEM SPECIFICATIONS

The design of a control system involves the changing of system parameters and/or the addition of subsystems (called compensators) to achieve certain desired system characteristics. The desired characteristics, or performance specifications, generally relate to steady-state accuracy, transient response, relative stability, sensitivity to change in system parameters, and disturbance rejection. These performance specifications will now be discussed.

Modern military aircraft utilize many digital control systems. (Courtesy of McDonnell-Douglas Corporation.)

Steady-State Accuracy

Since steady-state accuracy was discussed in detail in Section 6.5, only a brief review will be given here. In Section 6.5, it was shown that steady-state accuracy is increased if poles at $z = 1$ are added to the open-loop transfer function, and/or if the open-loop gain is increased. However, added poles at $z = 1$ in the open-loop transfer function introduce phase lag into the open-loop frequency response, resulting in reduced stability margins. Thus stability problems may ensue. In addition, an increase in the the open-loop gain generally results in stability problems, as was seen in Chapter 7. Thus a control system design is usually a trade-off between steady-state accuracy and acceptable relative stability.

Transient Response

Figure 8-1 illustrates a typical step response for a system that has two dominant complex poles. Typical performance criteria are rise-time t_r, peak overshoot M_p, time-to-peak overshoot t_p, and settling time t_s. Rise time in this figure is the time required for the response to rise from 10% to 90% of the final value. However, other definitions are also used for rise time, but all are similar. Settling time t_s is defined as the time required for the response to settle to within a certain per cent of the final value. Typical percentage values used are 2% and 5%.

For a given system, the time response is uniquely related to the closed-loop frequency response. However, except for first- and second-order systems, the exact relationship is complex and is generally not used. As indicated in Section 7.10, a typical closed-loop frequency response is as shown in Figure 8-2. In this figure, M_r is the resonant peak value of the frequency response, and as was implied in Example 7.13,

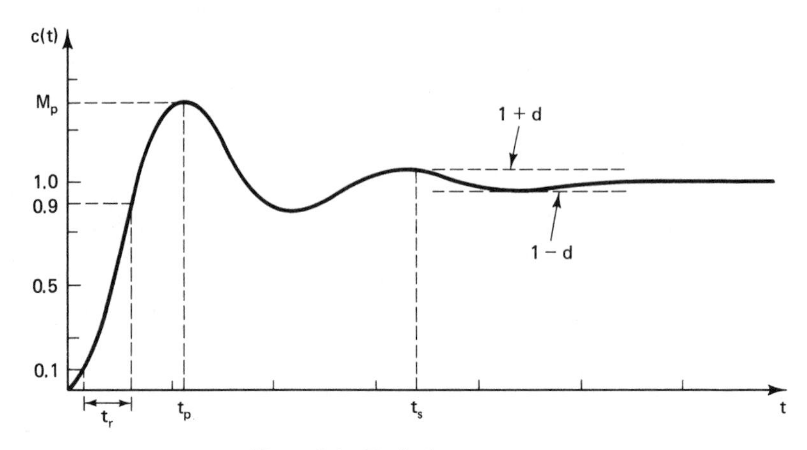

Figure 8-1 Typical step response.

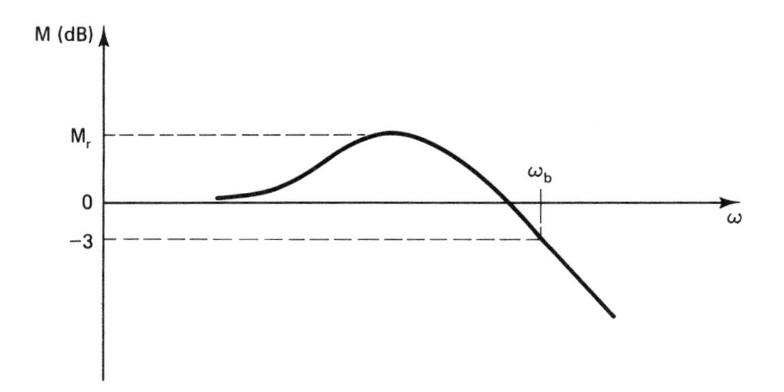

Figure 8-2 Typical closed-loop frequency response.

a larger resonant peak value indicates a larger peak overshoot M_p in the step response. For example, a control-system specification sometimes used is to limit M_r to 2 dB in order to limit M_p to a reasonable value.

The transient response is also related to the system bandwidth, shown as ω_b in Figure 8-2. For a system the product of rise time and bandwidth (i.e., the product $t_r\omega_b$), is approximately constant [1]. Thus, in order to decrease rise time and increase speed of response, it is necessary that the system bandwidth be increased. However, if significant high-frequency noise sources are present in the system, a larger bandwidth will increase the system response to these noise sources. In this case, a trade-off must be made between a fast rise time and an acceptable noise response.

Relative Stability

In Chapter 7 the relative stability measurements, gain margin and phase margin, were introduced. These margins are an approximate indication of the closeness of the Nyquist diagram (open-loop frequency response) to the -1 point. As was shown in Section 7.10, the closeness of the open-loop frequency response to the -1 point in the complex plane determines the resonant peak value M_r of the closed-loop frequency response (see Figure 8-2). And M_r is related, in an approximate sense, to the peak overshoot M_p (Figure 8-1) in the step response. Thus, in an approximate sense, the stability margins are related to peak overshoot M_p. It can be shown that, for a continuous-time second-order system, phase margin and peak overshoot M_p are directly related [2]. For example, for a phase margin of 45°, the peak overshoot is 21%. However, this relationship between phase margin and peak overshoot is only approximate for higher-order systems.

Sensitivity

Generally, any control system will contain parameters that change with temperature, humidity, altitude, age, and so on. However, we prefer that the control system characteristics not vary as these parameters vary. Of course, the system characteristics are a function of the system parameters, but in some cases the sensitivity of system characteristics to parameter variations can be reduced. A simple case will now be discussed.

Consider the system of Figure 8-3a. For this system the closed-loop transfer function $T(z)$ is given by

$$T(z) = \frac{G(z)}{1 + G(z)} \tag{8-1}$$

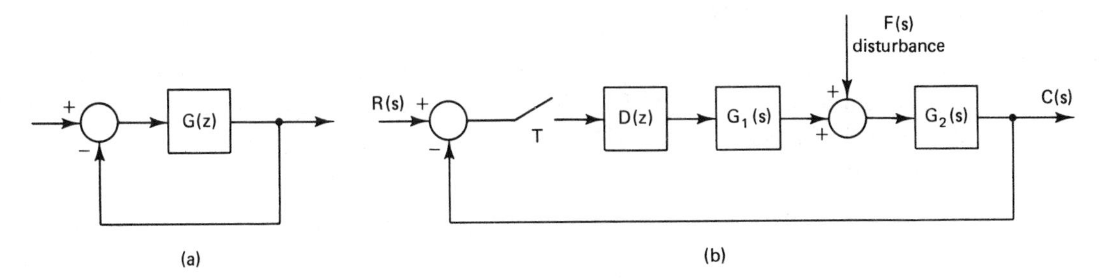

Figure 8-3 Discrete control systems.

Sensitivity to a parameter, a, is normally defined as a measure of the percentage change in $T(z)$ to a percentage change in the parameter, a. One such definition is

$$\text{sensitivity} \approx \frac{\Delta T/T}{\Delta a/a} = \frac{\Delta T}{\Delta a}\frac{a}{T} \tag{8-2}$$

where ΔT is the variation in T caused by Δa, the variation in parameter a. If the limit of (8-2) is taken as $\Delta a \longrightarrow 0$, we get the usual definition of sensitivity; that is,

$$S_a^T = \frac{\partial T}{\partial a}\frac{a}{T} \tag{8-3}$$

We will now find the sensitivity of T with respect to G.

$$S_G^T = \frac{\partial T}{\partial G} \cdot \frac{G}{T} = \frac{1 + G(z) - G(z)}{[1 + G(z)]^2} \cdot \frac{G(z)}{G(z)/[1 + G(z)]}$$

$$= \frac{1}{1 + G(z)}$$

At the frequency ω,

$$S_G^T = \frac{1}{1 + G(\epsilon^{j\omega T})} \tag{8-4}$$

For this sensitivity to be small within the system bandwidth, we require that $G(\epsilon^{j\omega T})$ be large. Thus we can reduce the sensitivity of T to G by increasing the open-loop gain. But, as noted before, increasing the open-loop gain can cause stability problems. Thus, once again, in design we are faced with trade-offs.

Consider now that $G(z)$ is a function of the parameter a. Then we can express (8-3) as

$$S_a^T = \frac{\partial T}{\partial a} \cdot \frac{a}{T} = \frac{\partial T}{\partial G} \cdot \frac{\partial G}{\partial a} \cdot \frac{a}{T} \tag{8-5}$$

Thus

$$\begin{aligned} S_a^T &= \frac{1 + G(z) - G(z)}{[1 + G(z)]^2} \cdot \frac{a}{G(z)/[1 + G(z)]} \cdot \frac{\partial G(z)}{\partial a} \\ &= \frac{a \dfrac{\partial G(z)}{\partial a}}{G(z)[1 + G(z)]} \end{aligned} \tag{8-6}$$

Then, as in (8-4), to reduce the sensitivity we must increase loop gain.

Disturbance Rejection

A control system will generally have inputs other than the one to be used to control the system output. An example is shown in Figure 8-3b. In this system $F(s)$ is a disturbance. Since $R(s)$ is the control input, we design the system such that $c(t)$ is approximately equal to $r(t)$. If $F(s)$ is zero, then

$$C(z) = \frac{D(z)\overline{G_1 G_2}(z)}{1 + D(z)\overline{G_1 G_2}(z)} R(z)$$

Hence, in terms of the frequency response, we require that

$$D(\epsilon^{j\omega T})\overline{G_1 G_2}(\epsilon^{j\omega T}) \gg 1$$

over the desired system bandwidth. Then

$$C(\epsilon^{j\omega T}) \approx R(\epsilon^{j\omega T})$$

If we consider only the disturbance input in Figure 8-3b, then

$$C(z) = \frac{\overline{G_2 F}(z)}{1 + D(z)\overline{G_1 G_2}(z)}$$

Hence, over the desired system bandwidth,

$$C(\epsilon^{j\omega T}) \approx \frac{\overline{G_2 F}(\epsilon^{j\omega T})}{D(\epsilon^{j\omega T})\overline{G_1 G_2}(\epsilon^{j\omega T})}$$

Since the denominator of this expression is large, the disturbance response will be small, provided that the numerator is not large. Therefore, we generally have good disturbance rejection in a system provided that we have a high loop gain, and provided that the high loop gain does not occur in the direct path between the disturbance input and the system output [$G_2(s)$ in Figure 8-3b].

8.3 COMPENSATION

In this chapter we will, for the most part, limit the discussion to the design of compensators for single-input, single-output systems. A simple system of this type is shown in Figure 8-4a. For this system,

$$\frac{C(z)}{R(z)} = \frac{D(z)G(z)}{1 + D(z)\overline{GH}(z)} \tag{8-7}$$

and hence the characteristic equation is

$$1 + D(z)\overline{GH}(z) = 0 \tag{8-8}$$

We call compensation of the type shown in Figure 8-4a cascade, or series, compensation. The effects of this compensation on system characteristics are given by the characteristic equation (8-8).

It is sometimes more feasible to place the compensator within a loop internal to the system. Such a system is illustrated in Figure 8-4b. For this system,

$$C(z) = \frac{\mathfrak{z}\left[\dfrac{G_1G_2R}{1 + G_1G_2H_1}\right]}{1 + D(z)\mathfrak{z}\left[\dfrac{G_2H_2}{1 + G_1G_2H_1}\right]} \tag{8-9}$$

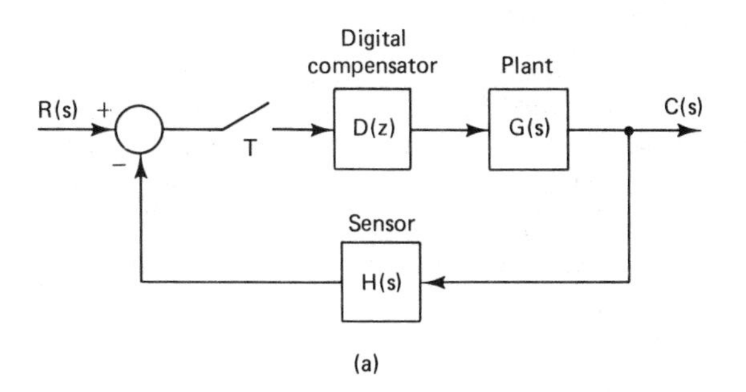

(a)

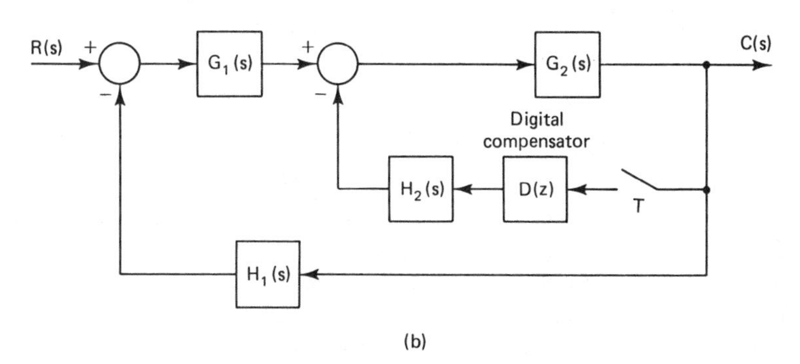

(b)

Figure 8-4 Digital control systems.

The system characteristic equation is then

$$1 + D(z)\mathfrak{z}\left[\frac{G_2H_2}{1 + G_1G_2H_1}\right] = 0 \tag{8-10}$$

This type of compensation is termed feedback, parallel, or minor-loop compensation. For this system, the effects of compensation on system characteristics are given by (8-10).

In the following three sections, basically we will consider compensation by a first-order device. Thus the transfer function is of the form

$$D(z) = \frac{K_d(z - z_0)}{(z - z_p)} \tag{8-11}$$

The design of the compensator will be performed in the frequency domain using Bode techniques; thus we will be working in the w-plane. The transformation of $D(z)$ to the w-plane yields $D(w)$; that is,

$$D(w) = D(z)\Big|_{z=[1+(T/2)w]/[1-(T/2)w]} \tag{8-12}$$

Thus $D(w)$ is also first order, and we will assume it to be of the form

$$D(w) = \frac{1 + w/\omega_{w0}}{1 + w/\omega_{wp}} \tag{8-13}$$

where ω_{w0} is the zero location and ω_{wp} is the pole location, in the w-plane. The dc gain of the compensator is found in (8-11) by letting $z = 1$, or in (8-13) by letting $w = 0$. Thus, in (8-13), we are assuming a unity dc gain for the compensator. A nonunity dc gain is obtained by multiplying the right side of (8-13) by a constant equal to the value of the desired dc gain.

Some designs require a compensator with a nonunity dc gain, to improve steady-state response, increase disturbance rejection, etc. Then, for design, we add the required increase in gain to the plant transfer function, and design the unity dc gain compensator of (8-13) based on the new plant transfer function. Then the compensator is realized as the transfer function of (8-13) multiplied by the required gain factor.

To realize the compensator, the transfer function must be expressed in z, as in (8-11). Then, from (8-13),

$$D(z) = \frac{1 + \dfrac{w}{\omega_{w0}}}{1 + \dfrac{w}{\omega_{wp}}}\Bigg|_{w=(2/T)[(z-1)/(z+1)]} = \frac{\omega_{wp}(\omega_{w0} + 2/T)}{\omega_{w0}(\omega_{wp} + 2/T)}\left[\frac{z + \dfrac{\omega_{w0} - 2/T}{\omega_{w0} + 2/T}}{z + \dfrac{\omega_{wp} - 2/T}{\omega_{wp} + 2/T}}\right] \tag{8-14}$$

Hence, in (8-11),

$$K_d = \frac{\omega_{wp}(\omega_{w0} + 2/T)}{\omega_{w0}(\omega_{wp} + 2/T)}, \qquad z_0 = \frac{2/T - \omega_{w0}}{2/T + \omega_{w0}}, \qquad z_p = \frac{2/T - \omega_{wp}}{2/T + \omega_{wp}} \tag{8-15}$$

The compensator of (8-13) is classified by the location of the zero, ω_{w0}, relative to that of the pole, ω_{wp}. If $\omega_{w0} < \omega_{wp}$, the compensation is called phase lead. If $\omega_{w0} > \omega_{wp}$, the compensation is called phase lag. The phase-lag compensator will be discussed first, because phase-lag design techniques are simpler than those for phase lead. It is not implied that phase-lag compensation is more commonly used than is phase-lead compensation.

8.4 PHASE-LAG COMPENSATION

In (8-13), for $\omega_{w0} > \omega_{wp}$, the frequency response of $D(w)$ exhibits a negative phase angle, or phase lag. The frequency response of $D(w)$, as given by a Bode plot, is shown in Figure 8-5. The dc gain is unity, and the high-frequency gain is

$$(\text{high-frequency gain})_{\text{dB}} = 20 \log \frac{\omega_{wp}}{\omega_{w0}} \tag{8-16}$$

The phase characteristic is also shown in Figure 8-5. The maximum phase shift is denoted as ϕ_M, and has a value between 0 and $-90°$, depending on the ratio ω_{w0}/ω_{wp}.

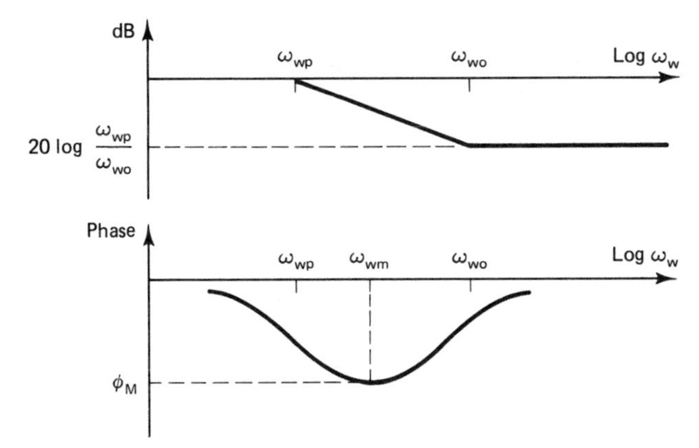

Figure 8-5 Phase-lag digital filter frequency-response characteristics.

Design using phase-lag digital compensators will be discussed relative to the system of Figure 8-6. For this system, the characteristic equation is given by

$$1 + D(z)G(z) = 0 \tag{8-17}$$

where

$$G(z) = \mathcal{Z}\left[\frac{1 - \epsilon^{-Ts}}{s} G_p(s)\right] \tag{8-18}$$

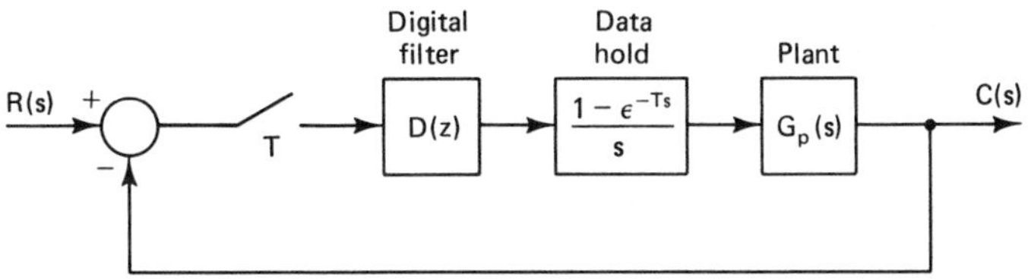

Figure 8-6 Digital control system.

For a system configuration differing from that of Figure 8-6, the characteristic equation is formed as in (8-17), and the frequency response of the transfer function that multiplies $D(z)$, as in (8-17), is calculated. From that point, the design procedure follows that given below.

As is seen from Figure 8-5, phase-lag filters introduce both reduced gain and phase lag. Since, in general, phase lag tends to destabilize a system (rotates the Nyquist diagram toward the -1 point) the break frequencies, ω_{wp} and ω_{w0}, must be chosen such that the phase lag does not occur in the vicinity of the 180° crossover point of the plant frequency response $G(j\omega_w)$, where

$$G(w) = \mathfrak{z}\left[\frac{1-\epsilon^{-Ts}}{s}G_p(s)\right]_{z=[1+(T/2)w]/[1-(T/2)w]} \tag{8-19}$$

for the system of Figure 8-6. However, for stability purposes, it is necessary that the filter introduce the reduced gain in the vicinity of 180° crossover. Thus, both ω_{wp} and ω_{w0} must be much smaller than the 180° crossover frequency. Figure 8-7 illustrates design by phase-lag compensation.

Note that, in Figure 8-7, both the system gain margin and the system phase margin have been increased, increasing relative stability. In addition, the low-frequency gain has not been reduced, and thus steady-state errors and low-frequency sensitivity have not been increased to attain the improved relative stability. The bandwidth has been decreased, which will generally result in a slower system time response.

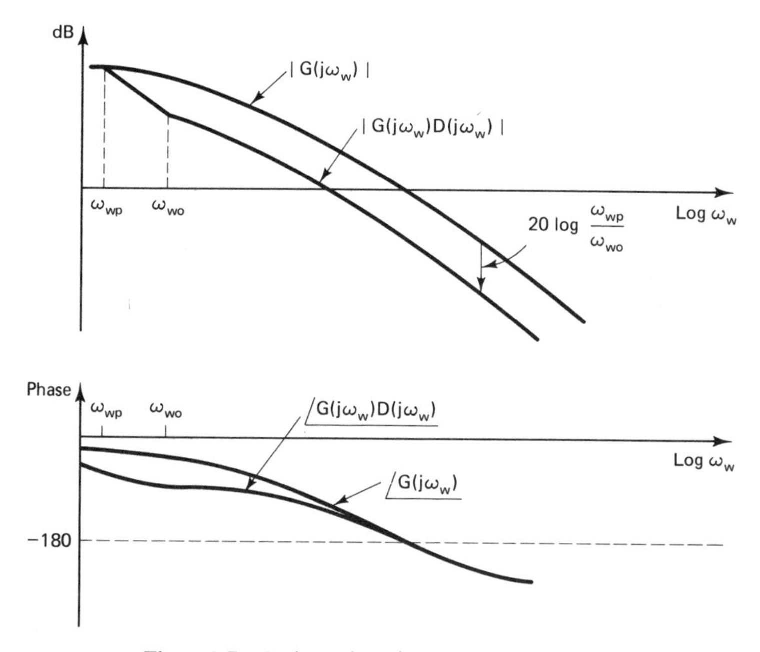

Figure 8-7 Design using phase-lag compensation.

In Figure 8-7, if we keep ω_{w0} constant and decrease ω_{wp}, the dc gain of the compensator can be made greater than unity while maintaining the same stability margins. For this case, the steady-state errors and the low-frequency sensitivity are improved. However, for a given system, the increase in phase lag added to system phase characteristics may push the phase characteristic below the $-180°$ line (see Figure 8-7). Then we have a conditionally stable system, that is, one that can be forced unstable by reducing gain. If the system contains a saturation nonlinearity, large signals into this nonlinearity reduce its effective gain [3]. Thus a phase-lag compensated system may exhibit instabilities for large signals (nonlinear operation).

A technique for determining ω_{wp} and ω_{w0} to yield a desired phase margin will now be given. It is assumed that the system gain has been adjusted to give satisfactory steady-state errors. Suppose that a phase margin of ϕ_m is desired. Refer to Figure 8-7. First determine the frequency, ω_{w1}, at which the phase angle $(-180° + \phi_m + 5°)$ occurs. (The 5° term will be explained later.) It is desired that the 0 dB crossover occur at this frequency. Let $(\text{dB})_1$ be the magnitude of $G(j\omega_w)$ at the frequency ω_{w1}. Then $D(j\omega_w)$ must reduce the gain by an amount $(\text{db})_1$ at this frequency. Or

$$-(\text{dB})_1 = 20 \log_{10} \frac{\omega_{wp}}{\omega_{w0}} \tag{8-20}$$

Now, choose

$$\omega_{w0} = 0.1\omega_{w1} \tag{8-21}$$

to ensure that little phase lag is introduced at the 180° crossover point. Substitution of (8-21) in (8-20) yields an equation with only one unknown. Thus $D(w)$ may be determined. The 5° term was added to compensate for the angle of $D(w)$ at ω_{w1}, since this angle will be negative and approximately 5°. Once $D(w)$ is known, then $D(z)$ is obtained from (8-14).

Example 8.1

We will consider the design of a servomotor system as described in Section 1.5. Suppose that the servo is to control the horizontal (azimuth) angle for pointing a radar antenna. Then, in the closed-loop system of Figure 8-6, $c(t)$ is the azimuth angle of the antenna, and $r(t)$ is the commanded, or desired, azimuth angle. The plant transfer function derived in Section 1.5 is second order; however, we will assume that the armature inductance cannot be neglected, resulting in a third-order transfer function. Then, suppose that the parameters of the system are such that

$$G_p(s) = \frac{1}{s(s+1)(0.5s+1)}$$

Since the fastest time constant is 0.5 s, we will choose T to be one-tenth that value (a rule of thumb often used), or $T = 0.05$ s. Then

$$\begin{aligned} G(z) &= \frac{z-1}{z} \mathfrak{z}\left[\frac{1}{s^2(s+1)(0.5s+1)}\right] \\ &= \frac{z-1}{z} \mathfrak{z}\left[\frac{1}{s^2} + \frac{-1.5}{s} + \frac{2}{s+1} + \frac{-0.5}{s+2}\right] \\ &= \frac{z-1}{z}\left[\frac{0.05z}{(z-1)^2} - \frac{1.5z}{z-1} + \frac{2z}{z-0.9512} - \frac{0.5z}{z-0.9048}\right] \end{aligned}$$

The frequency response of this system was calculated using program 4 of Appendix I, and is given in Table 8-1 and plotted in Figure 8-8. Suppose that it is desired to design a phase-lag compensator to achieve a phase margin of 55°. Then, using the foregoing procedure, we see that the frequency ω_{w1} occurs where the phase of $G(j\omega_w)$ is $-120°$, or $\omega_{w1} \approx 0.36$. At this frequency, $|G(j\omega_w)|_{\text{dB}} = 8.0$ dB. Then, from (8.21),

$$\omega_{w0} = 0.1\omega_{w1} = 0.036$$

and from (8-20),

$$-8.0 = 20 \log_{10} \frac{\omega_{wp}}{\omega_{w0}} = 20 \log_{10} \frac{\omega_{wp}}{0.036}$$

TABLE 8-1 FREQUENCY RESPONSE OF THE PLANT IN EXAMPLE 8.1

ω	$\lvert G(\epsilon^{j\omega T})\rvert_{dB}$	$\underline{/G(\epsilon^{j\omega T})}$	ω_w
0.010	40.0	−90.9	0.010
0.050	26.0	−94.4	0.050
0.100	19.9	−98.7	0.100
0.200	13.8	−107.3	0.200
0.300	9.99	−115.6	0.300
0.400	7.15	−123.7	0.400
0.500	4.79	−131.3	0.500
0.600	2.73	−138.5	0.600
0.700	0.87	−145.3	0.700
0.800	−0.85	−151.6	0.800
0.900	−2.46	−157.5	0.900
1.000	−3.97	−163.0	1.000
1.200	−6.79	−172.9	1.200
1.370	−8.99	−180.3	1.371
1.500	−10.6	−185.4	1.501
2.000	−16.0	−201.4	2.001
3.000	−24.6	−222.3	3.006
5.000	−36.7	−244.3	5.026

Thus

$$\omega_{wp} = 0.0143$$

Then, from (8-14),

$$D(z) = \frac{0.3974(z - 0.998202)}{(z - 0.999285)} = \frac{0.3974z - 0.39669}{z - 0.999285}$$

This filter results in a gain margin of approximately 16 dB and a phase margin of approximately 55°.

With phase-lag compensation numerical problems may occur in the realization of the filter coefficients. To illustrate this point, suppose that a microprocessor is used to implement the digital controller. Suppose, in addition, that filter coefficients are realized by a binary word that employs 8 bits to the right of the binary point. Then the fractional part of the coefficient can be represented as [4]

$$\text{fraction} = b_7 * \frac{1}{2} + b_6 * \frac{1}{4} + b_5 * \frac{1}{8} + \cdots + b_0 * \frac{1}{2^8}$$

where b_i is the ith bit, and has a value of either zero or one. For example, the binary number

$$(0.11000001)_2 = \left(\frac{1}{2} + \frac{1}{2^2} + \frac{1}{2^8}\right)_{10} = (0.75390625)_{10}$$

The maximum value that the fraction can assume is $[1 - 1/(2)^8]$, or 0.99609375. Note, in Example 8.1, that a denominator coefficient of 0.999285 is required, but a value of 0.99609375 will be implemented (b_7 to b_0 are all equal to 1). The numerator coefficients,

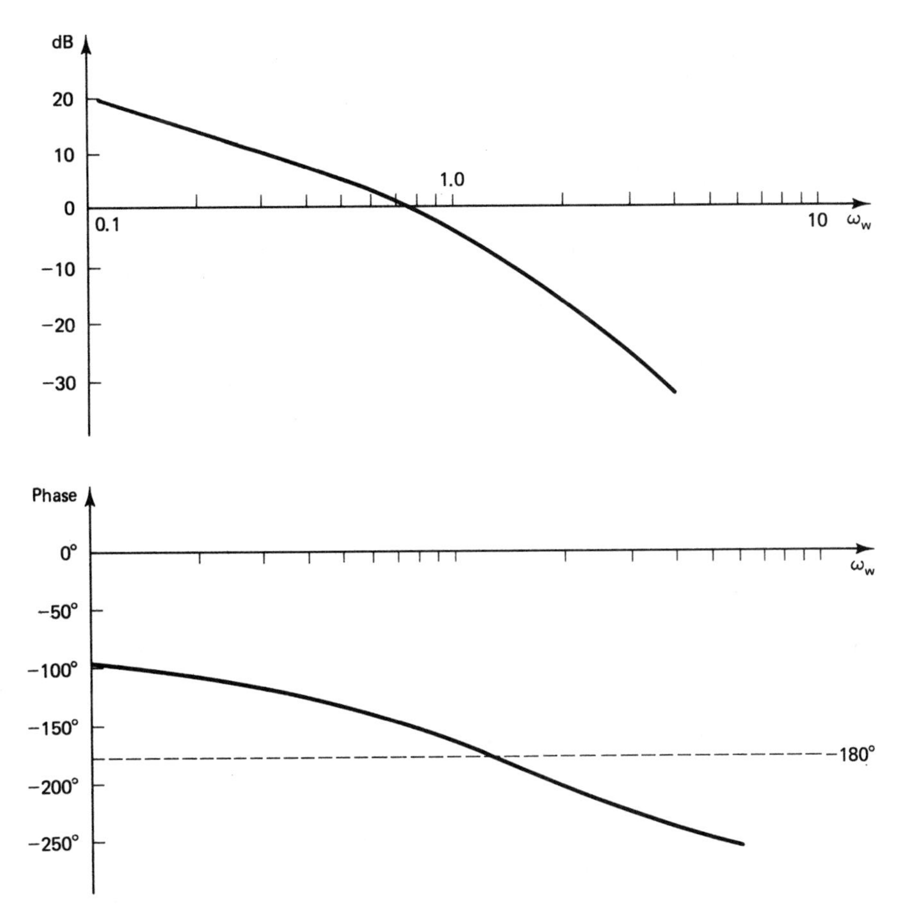

Figure 8-8 Frequency response for the system of Example 8.1.

when converted by standard decimal-to-binary conversion algorithms [4], become

$$(0.3974)_{10} \Longrightarrow (0.01100101)_2 = 0.39453125$$

$$(0.39669)_{10} \Longrightarrow (0.01100101)_2 = 0.39453125$$

Thus the zero has been shifted to $z = 1$, and the digital filter that is implemented has the transfer function

$$D(z) = \frac{0.39453125z - 0.39453125}{z - 0.99609375}$$

Shown in Figure 8-9 are the frequency responses of the designed filter and the implemented filter, and the effects of coefficient quantization are evident. The resultant system stability margins, when the implemented filter is used, are: phase margin 70° (designed value 55°), and gain margin 18 dB (designed value 16 dB). However, the implemented filter has a dc gain of zero; thus the system will not respond correctly to a constant input. Hence more bits must be used to represent the filter coefficients. Coefficient quantization is described in detail in Chapter 14.

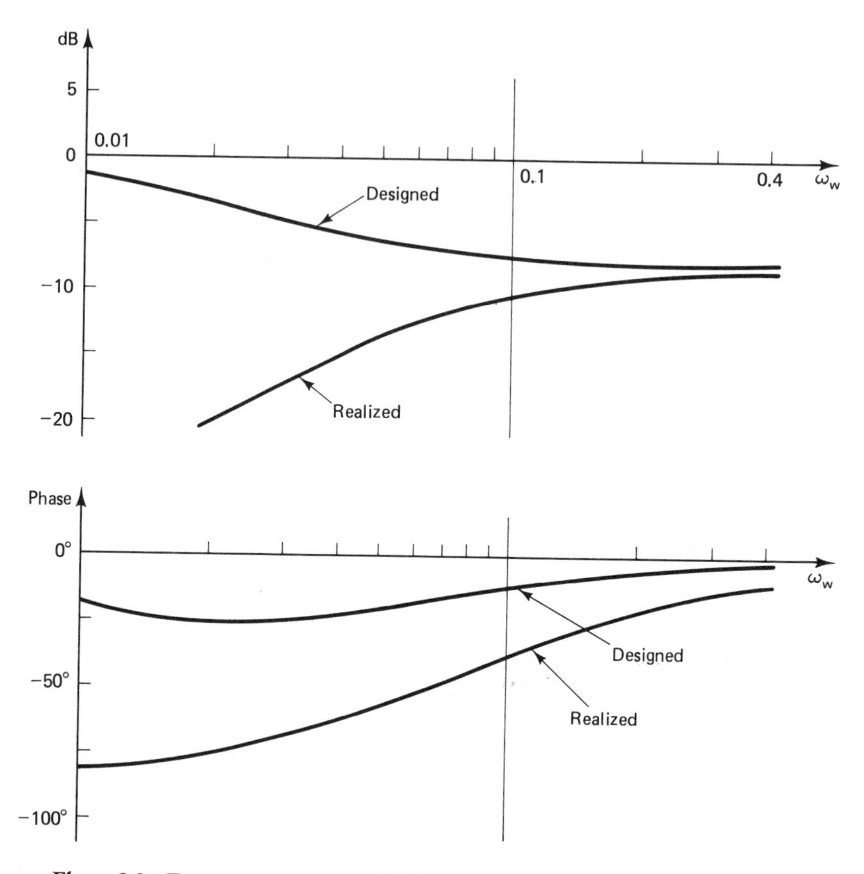

Figure 8-9 Frequency responses of designed and of realized digital controllers.

We can view the coefficient quantization problem as one that results from the choice of the sample period T. We place digital filters in a physical system in order to change its real (s-plane) frequency response, and we want this change to occur over a certain frequency (ω) range. The choice of T places this frequency range on a certain part of the unit circle in the z-plane, since $z = \epsilon^{j\omega T} = 1\underline{/\omega T}$. Thus since phase-lag filtering occurs for ω small, the choice of T small requires that the filtering occur in the vicinity of the $z = 1$ point. Thus the phase-lag pole and zero will occur close to $z = 1$, and thus close to each other. If T is chosen to be a larger value, the phase-lag pole and zero will move away for the $z = 1$ point, and the numerical accuracy required for the filter coefficients will not be as great.

8.5 PHASE-LEAD COMPENSATION

Phase-lead compensation will now be discussed. For a phase-lead compensator, in (8-13) $\omega_{w0} < \omega_{wp}$, and the compensator frequency response is as shown in Figure 8-10. The maximum phase shift, θ_M, occurs at a frequency ω_{wm}, where ω_{wm} is the geometric

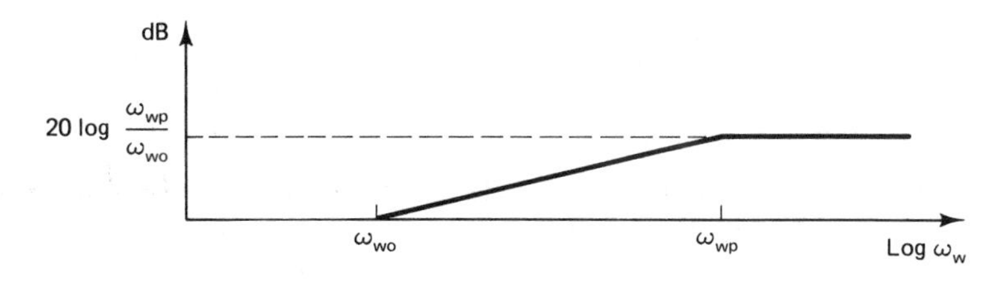

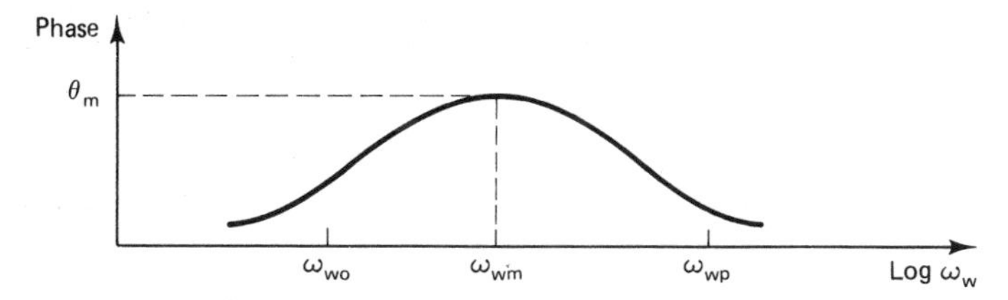

Figure 8-10 Phase-lead digital filter frequency-response characteristics.

mean of ω_{w0} and ω_{wp}, that is,

$$\omega_{wm} = \sqrt{\omega_{w0}\omega_{wp}} \tag{8-22}$$

A plot of θ_M versus the ratio ω_{wp}/ω_{w0} is given in Figure 8-11. This plot was obtained through the following development. We can express (8-13) as

$$D(j\omega_w) = |D(j\omega_w)|\epsilon^{j\theta} = \frac{1 + j(\omega_w/\omega_{w0})}{1 + j(\omega_w/\omega_{wp})} \tag{8-23}$$

Then

$$\tan\theta = \tan\left[\tan^{-1}\frac{\omega_w}{\omega_{w0}} - \tan^{-1}\frac{\omega_w}{\omega_{wp}}\right] = \tan(\alpha - \beta) \tag{8-24}$$

Thus

$$\tan\theta = \frac{\tan\alpha - \tan\beta}{1 + \tan\alpha\tan\beta} = \frac{\omega_w/\omega_{w0} - \omega_w/\omega_{wp}}{1 + \omega_w^2/\omega_{w0}\omega_{wp}} \tag{8-25}$$

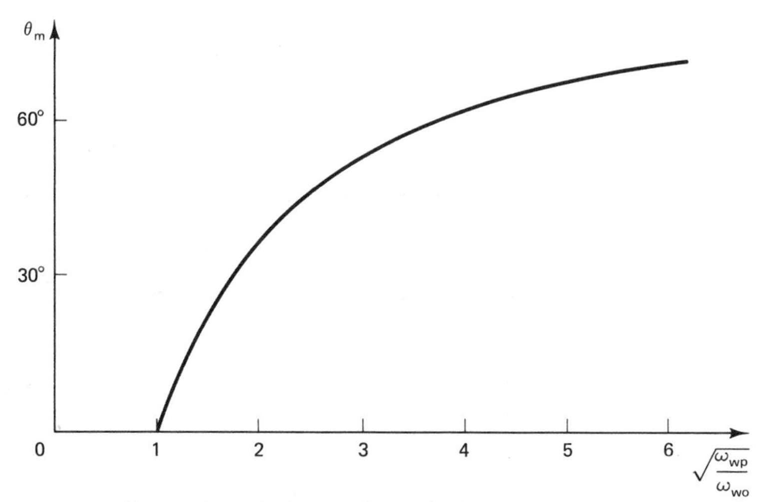

Figure 8-11 Maximum phase shift for a phase-lead filter.

Then, from (8-22) and (8-25),

$$\tan \theta_M = \frac{1}{2}\left[\sqrt{\frac{\omega_{wp}}{\omega_{w0}}} - \sqrt{\frac{\omega_{w0}}{\omega_{wp}}}\right] \tag{8-26}$$

From this equation, θ_M is seen to be a function only of the ratio ω_{wp}/ω_{w0}. Figure 8-11 is a plot of (8-26). Note also that

$$|D(j\omega_{wm})| = \left.\frac{\sqrt{1 + (\omega_w/\omega_{w0})^2}}{\sqrt{1 + (\omega_w/\omega_{wp})^2}}\right|_{\omega_{wm}} = \sqrt{\frac{1 + \omega_{wp}/\omega_{w0}}{1 + \omega_{w0}/\omega_{wp}}} = \sqrt{\frac{\omega_{wp}}{\omega_{w0}}} \tag{8-27}$$

It is seen in Figure 8-10 that phase-lead compensation introduces phase lead, which is a stabilizing effect, but also increases gain, which is a destabilizing effect. Design using phase-lead compensation is illustrated in Figure 8-12. The phase lead is

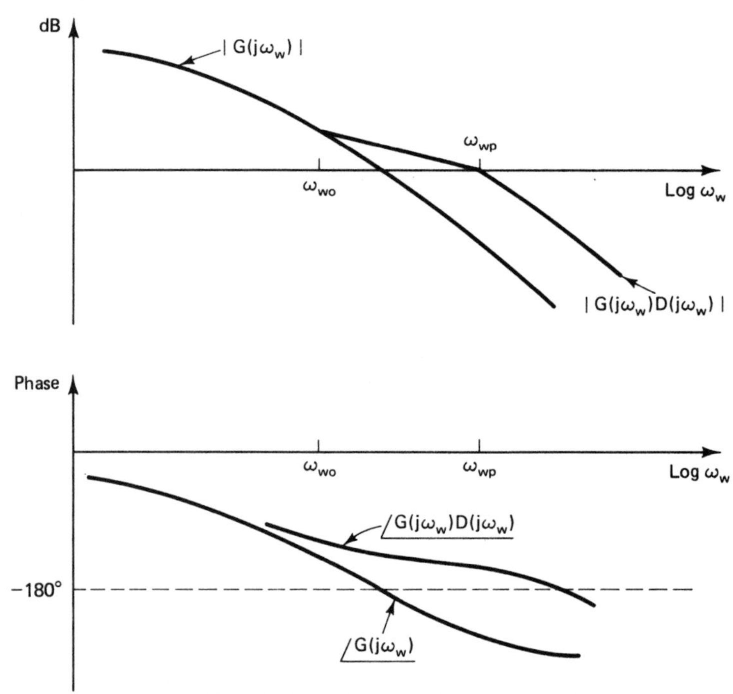

Figure 8-12 Design using phase-lead compensation.

introduced in the vicinity of the plant's 180° crossover frequency, in order to increase the system's stability margins. Note that system bandwidth is also increased, resulting in a faster time response.

The design of phase-lead compensation tends to be more of a trial-and-error procedure since, in the frequency range that the stabilizing phase lead is added, destabilizing gain is also added. Given in the next section is a procedure that will yield a specified phase margin, but has no control on the gain margin.

8.6 PHASE-LEAD DESIGN PROCEDURE

Presented in this section is a design procedure that will yield a specified phase margin in a discrete control system, provided that the designed system is stable. That is, the procedure will set the gain and phase of the open-loop transfer function to specified values at a given frequency, and we choose the specified gain to be 0 dB and the specified phase to be $(180^\circ + \phi_m)$, where ϕ_m is the desired phase margin. Thus the procedure does not determine the gain margin, and may in fact result in an unstable system. Then, as a later step in the design procedure, it will be necessary to check the gain margin to ensure that it is adequate.

The characteristic equation for the system of Figure 8-6 is

$$1 + D(w)G(w) = 0 \tag{8-28}$$

where $D(w)$ is given by (8-13). Our design problem may then be stated with respect to this system:

> Determine $D(w)$ in (8-28) such that, at some frequency ω_{w1},
>
> $$D(j\omega_{w1})G(j\omega_{w1}) = 1\underline{/180^\circ + \phi_m} \tag{8-29}$$
>
> and, in addition, the system possesses an adequate gain margin.

The design equations will now be developed. From (8-29) we see that $D(j\omega_{w1})$ must satisfy the relationships

$$|D(j\omega_{w1})| = \frac{1}{|G(j\omega_{w1})|} \tag{8-30}$$

and

$$\underline{/D(j\omega_{w1})} = 180^\circ + \phi_m - \underline{/G(j\omega_{w1})}$$

where the symbol $\underline{/D}$ denotes the angle associated with the complex number D. Let the angle associated with $D(j\omega_{w1})$ be denoted as θ; that is,

$$\theta = \underline{/D(j\omega_{w1})} = 180^\circ + \phi_m - \underline{/G(j\omega_{w1})} \tag{8-31}$$

In the equation for $D(w)$, (8-13), we see that there are two unknowns, namely ω_{w0} and ω_{wp}. Equations (8-30) and (8-31) give two constraints on $D(j\omega_{w1})$. Solving these two equations for ω_{w0} and ω_{wp} yields

$$\omega_{w0} = \left[\frac{\sin\theta}{\dfrac{1}{|G(j\omega_{w1})|} - \cos\theta}\right]\omega_{w1} \tag{8-32}$$

$$\omega_{wp} = \left[\frac{\sin\theta}{\cos\theta - |G(j\omega_{w1})|}\right]\omega_{w1} \tag{8-33}$$

The derivation of these relationships is algebraically long and complicated [5,6]. However, they may be verified by substitution into (8-30) and (8-31), as shown in Appendix V.

For phase-lead compensation, $D(w)$ of (8-13) must satisfy three constraints:

1. $\theta > 0°$
2. $|D(j\omega_{w1})| > 1$
3. $\cos\theta > \dfrac{1}{|D(j\omega_{w1})|}$

Constraints 1 and 2 can be seen from Figure 8-10, and constraint 3 in required for the filter pole, ω_{wp}, to be positive [see (8-33)].

It should be noted that (8-32) and (8-33) may also be utilized in phase-lag filter design, since nothing in the development of these equations is based on the filter being phase-lead. For phase-lag design, the three constraints become:

1. $\theta < 0°$
2. $|D(j\omega_{w1})| < 1$
3. $\cos\theta < |G(j\omega_{w1})|$

The proof of these constraints is left as an exercise for the interested reader.

Example 8.2

Consider again the system of Example 8.1, whose frequency response is given in Figure 8-8 and Table 8-1. A phase margin of 55° is to be achieved, and a phase-lead compensator will be employed. Consider Table 8-1. We must choose a frequency ω_{w1} such that the $|G(j\omega_{w1})|$ is less than 1 [see (8-30) and constraint 2], or, when expressed in dB, is negative. We then choose $\omega_{w1} = 1.200$. Then

$$|G(j\omega_{w1})| = -6.79 \text{ dB} = 0.4576$$

From (8-31),

$$\theta = 180° + 55° - (-172.9°) = 407.9° = 47.9°$$

Hence, from (8-32),

$$\omega_{w0} = \left[\frac{\sin(47.9°)}{(1/0.4576) - \cos(47.9°)}\right] 1.2 = 0.5878$$

and, from (8-33),

$$\omega_{wp} = \left[\frac{\sin(47.9°)}{\cos(47.9°) - 0.4576}\right] 1.2 = 4.184$$

We then obtain the filter transfer function from (8-14):

$$D(z) = \frac{6.539(z - 0.9710)}{(z - 0.8106)}$$

This filter results in a phase margin of 55° and a gain margin of 12.3 dB. If we choose ω_{w1} as a value different from 1.2 above, the phase margin will remain at 55°, but the gain margin will be different. If ω_{w1} is chosen larger, then θ, the angle of $D(j\omega_{w1})$, is larger (see Table 8-1), and the ratio ω_{wp}/ω_{w0} is larger. Thus, from Figure 8-10, the high-frequency gain increases, which increases the system bandwidth. Choosing ω_{w1} less than 1.2 will reduce the system bandwidth.

The coefficient quantization problem observed for the phase-lag filter generally does not occur in phase-lead filters. For the phase-lag filter, the pole and zero were almost coincident, making their placement critical. For the phase-lead filter, the pole and zero are well separated, and any small shift in either one has little effect on the filter frequency response.

Examples 8.1 and 8.2 illustrate simple phase-lag and phase-lead compensation. The effect of phase-lag compensation is to reduce the system gain at higher frequencies, which in turn reduces system bandwidth. The system gain at lower frequencies is not reduced, and thus steady-state errors are not increased. The effect of phase-lead compensation is to increase system gain at higher frequencies, and thus increase system bandwidth. The frequency responses of the system of the foregoing two examples without compensation, with phase-lag compensation, and with phase-lead compensation are shown in Figure 8-13. Note the reduced bandwidth for phase-lag compensa-

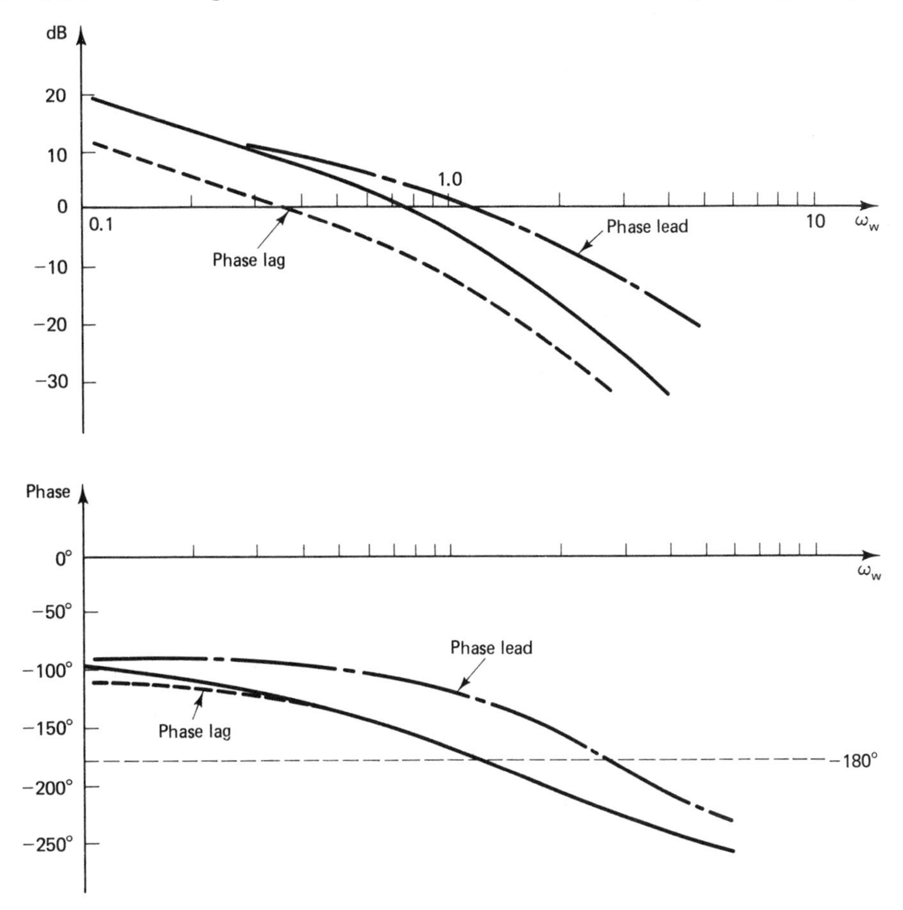

Figure 8-13 Frequency responses for systems of Examples 8.1 and 8.2.

tion and increased bandwidth for phase-lead compensation. Shown in Figure 8-14 are the step responses of the system with phase-lag compensation and with phase-lead compensation, obtained by digital simulation. Note that the system step response is much faster for the phase-lead case, because of the increased system bandwidth. Table 8-2 gives several step-response characteristics (see Figure 8-1) for the two cases. Note

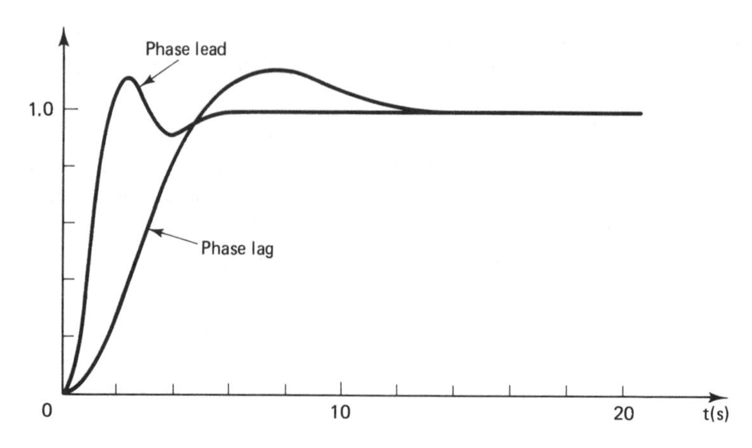

Figure 8-14 Step responses for Examples 8.1 and 8.2.

TABLE 8-2 STEP-RESPONSE CHARACTERISTICS

	Phase lag	Phase lead
Steady-state error	0	0
Percent overshoot	15	13
Rise time, t_r (s)	3.2	1.0
t_p (s)	7.3	2.2
t_s, for $d = 0.05$ (s)	11.7	4.7

that, while the phase margins are equal, the peak overshoots are not. Thus phase margin alone does not determine peak overshoot.

An additional point should be made concerning phase-lead compensation. From Figure 8-10 we see that the high-frequency gain of the digital filter can be quite large. Hence, if the control system is burdened with high-frequency noise, the phase-lead compensation may lead to noise problems. If this is the case, some compromise in design may be required. One solution involves the use of a phase-lag compensator cascaded with a phase-lead compensator. The lag compensation is employed to realize a part of the required stability margins, thus reducing the amount of phase-lead compensation required.

A second possible solution to this noise problem would be to add one or more poles to the filter transfer function. The pole (or poles) are placed at high frequencies such that the phase lag introduced by the poles does not decrease the system stability margins. The required transfer function for a single pole is of the form

$$D_h(w) = \frac{1}{1 + w/\omega_{wp1}}$$

and the total filter transfer function is then

$$D(w) = \left[\frac{1 + w/\omega_{w0}}{1 + w/\omega_{wp}}\right]\left[\frac{1}{1 + w/\omega_{wp1}}\right]$$

The resultant filter frequency response is as shown in Figure 8-15.

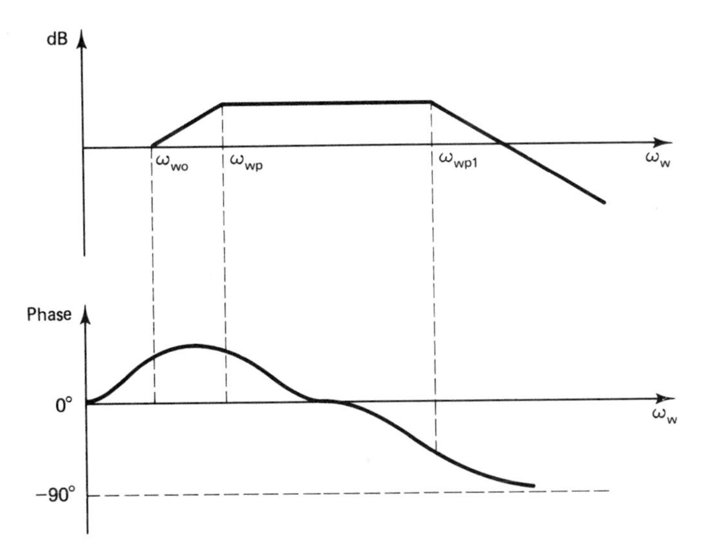

Figure 8-15 Phase-lead filter with an added pole.

Another problem originates in the large high-frequency gain of the phase-lead compensator. This problem is evident in the filter transfer function of Example 8.2. For this example,

$$D(z) = \frac{6.539(z - 0.9710)}{z - 0.8106} = 6.539 - 1.048z^{-1} + \cdots$$

Thus a step change of 1 unit in the filter input results in a step change of 6.539 units in the filter output. This large signal out of the filter may force the plant into a nonlinear region of operation (e.g., an amplifier may saturate), or the digital-to-analog converter may saturate. Since the design is based on a linear plant model, the effects of forcing the system into nonlinear regions of operation will not, in general, be obvious.

In summary, some possible advantages of phase-lag compensation are:

1. Improvement of stability margins (or steady-state performance)
2. Attenuation of high-frequency noise

Some possible disadvantages are:

1. Slow response
2. Numerical problems with filter coefficients

For phase-lead compensation, some possible advantages are:

1. Improvement of stability margins
2. Fast response
3. Required for some systems

Some possible disadvantages are:

1. Accentuation of high-frequency noise problems
2. Operation of the system in nonlinear regions

8.7 LAG–LEAD COMPENSATION

In the preceding sections, only simple first-order compensators were considered. In many system design projects, however, the system specifications cannot be satisfied by a first-order compensator. In these cases higher-order filters must be used. To illustrate this point, suppose that smaller steady-state errors to ramp inputs are required for the system of Example 8.2. Then the low-frequency gain of the system must be increased. If phase-lead compensation is employed, this increase in gain must be reflected at all frequencies. It is then unlikely that one first-order section of phase-lead compensation can be designed to give adequate stability margins. One solution to this problem would be to cascade two first-order phase-lead filters. However, if noise in the control system is a problem, the increased gain at high frequencies may lead to noise problems. A different approach is to cascade a phase-lag filter with a phase-lead filter. This filter is usually referred to as a lag–lead compensator.

A lag–lead filter has the characteristics shown in Figure 8-16. The purpose of the lag section is to increase the low-frequency gain, and the lead section increases the bandwidth and the stability margins. The lag–lead filter will now be illustrated by an example.

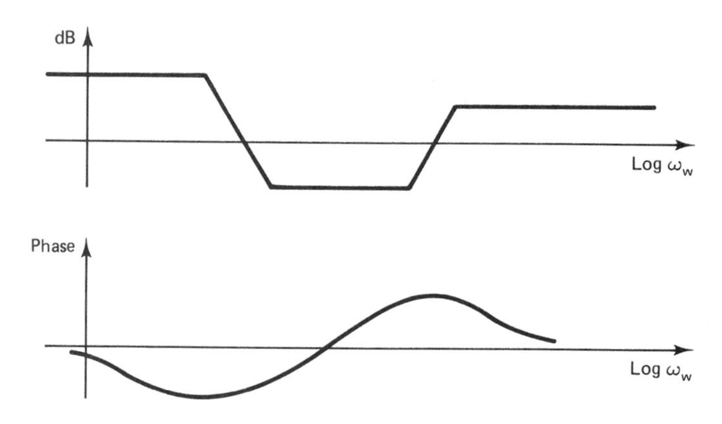

Figure 8-16 Frequency response of a lag–lead filter.

Example 8.3

In this example, the system of Examples 8.1 and 8.2 will again be considered. The steady-state error to a unit ramp input is given by

$$e_{ss}(kT) = \frac{T}{\lim_{z \to 1} (z - 1)G(z)} \tag{6-19}$$

From Example 8.1,

$$\lim_{z\to 1}(z-1)G(z) = \lim_{z\to 1}\frac{(z-1)^2}{z}\left[\frac{0.05z}{(z-1)^2} - \frac{1.5z}{z-1} + \frac{2z}{z-0.9512} - \frac{0.5z}{z-0.9048}\right]$$
$$= 0.05$$

Hence, from (6-19),

$$e_{ss}(kT) = \frac{0.05}{0.05} = 1$$

Suppose that the design specifications require a steady-state error to a unit ramp input of 0.50 and a phase margin of 55°. We will use a phase-lag filter, $D_1(z)$, to increase the low-frequency gain by a factor of 2, and then design a phase-lead filter, $D_2(z)$, to yield the 55° phase margin. We will choose the pole–zero locations of $D_1(z)$ to be the same as those of the phase-lag filter in Example 8.1. Then

$$\lim_{z\to 1} D_1(z) = \lim_{z\to 1}\frac{K(z-0.998202)}{z-0.999285} = 2$$

From this expression, we see that $K = 0.7948$, or

$$D_1(z) = \frac{0.7948(z-0.998202)}{z-0.999285}$$

In order to design the phase-lead filter, we must calculate the frequency response $D_1(z)G(z)$. The equations of Section 8.6 can then be utilized to find the phase-lead filter transfer function $D_2(z)$. Calculation of the required frequency response yields the point

$$D_1(w)G(w)\Big|_{w=j1.20} = 0.365\underline{/-173.9^\circ}$$

We now substitute these values into (8-31), (8-32), and (8-33), with G in these equations replaced with D_1G above. Hence, from (8-31),

$$\theta = 180^\circ + 55^\circ - (-173.9^\circ) = 408.9^\circ = 48.9^\circ$$

From (8-32),

$$\omega_{w0} = \left[\frac{\sin(48.9^\circ)}{(1/0.365) - \cos(48.9^\circ)}\right]1.2 = 0.434$$

and, from (8-33), $\omega_{wp} = 3.097$. Thus, from (8-14),

$$D_2(z) = \frac{6.68(z-0.9785)}{z-0.857}$$

The compensated system has a phase margin of 55° and a gain margin of 11.2 dB. The step response for this system is plotted in Figure 8-17, together with those from Examples 8.1 and 8.2. Note that the step responses of the phase-lead system and the lag–lead system are approximately the same; however, the steady-state error for a ramp input for the lag–lead system is only one-half that for the phase-lead system. The total filter transfer function is

$$D(z) = D_1(z)D_2(z) = \frac{5.31(z-0.998202)(z-0.9785)}{(z-0.999285)(z-0.857)}$$

A sketch of the Bode diagram for this filter is shown in Figure 8-18.

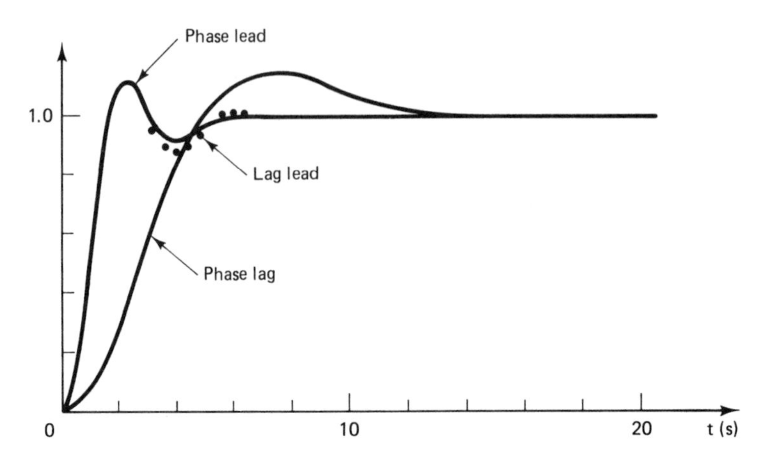

Figure 8-17 Step responses for Examples 8.1, 8.2, and 8.3.

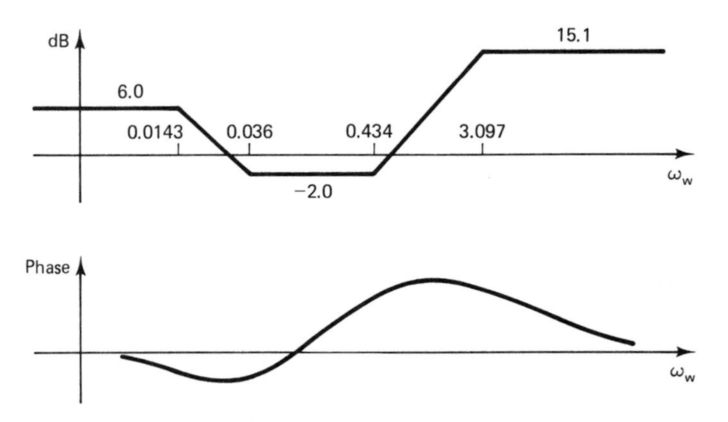

Figure 8-18 Frequency response for the lag–lead filter for Example 8.3.

8.8 INTEGRATION AND DIFFERENTIATION FILTERS

A somewhat different technique of design by classical frequency-response methods is presented in this and the following section. This technique is used extensively in the chemical processing industry and to a lesser degree in the aerospace industry.

To introduce this technique, we will first consider a technique for digital-filter integration. Suppose that we desire to digitally integrate a signal $e(t)$, and to accomplish this, we will use the trapezoidal technique [7]. The trapezoidal rule is illustrated in Figure 8-19. Let $m(kT)$ be the integral of $e(t)$. Then, from Figure 8-19, the value of the integral at $t = (k+1)T$ is equal to the value at kT plus the area added from kT to $(k+1)T$. Then

$$m[(k+1)T] = m(kT) + \frac{T}{2}\{e[(k+1)T] + e(kT)\} \tag{8-34}$$

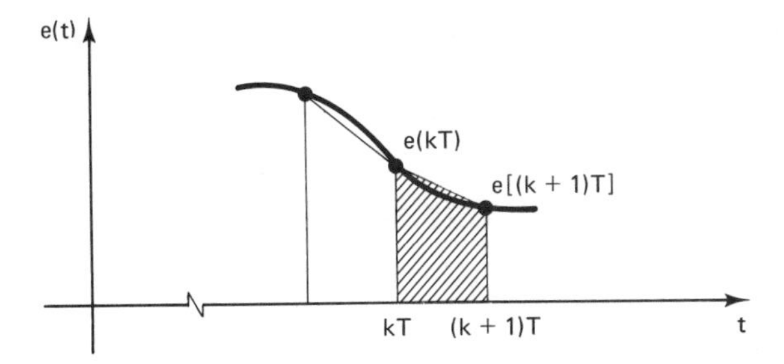

Figure 8-19 Trapezoidal rule for numerical integration.

Taking the z-transform, we obtain

$$zM(z) = M(z) + \frac{T}{2}[zE(z) + E(z)] \tag{8-35}$$

Thus

$$\frac{M(z)}{E(z)} = \frac{T}{2}\left[\frac{z+1}{z-1}\right] \tag{8-36}$$

Hence (8-36) is the transfer function of a discrete integrator. Of course, there are many other discrete transfer functions that may be used to integrate a number sequence (see Chapter 11).

Now consider a technique for the digital-filter differentiation a function $e(t)$. Figure 8-20 illustrates one method. The slope of $e(t)$ at $t = kT$ is approximated to be

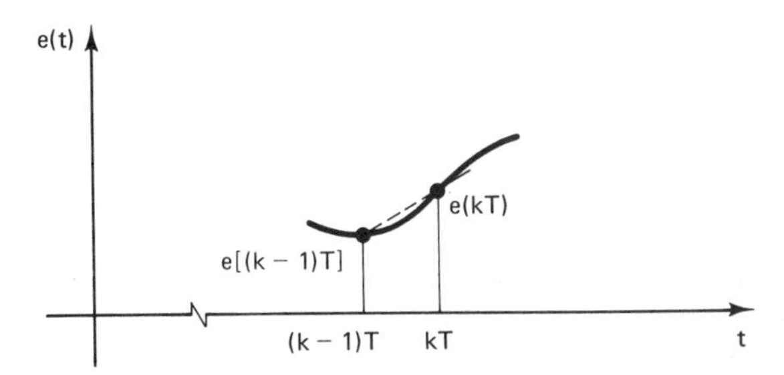

Figure 8-20 Illustration of numerical differentiation.

the slope of the straight line connecting $e[(k-1)T]$ with $e(kT)$. Then, from Figure 8-20, letting the derivative of $e(t)$ at $t = kT$ be $m(kT)$, we can write

$$m(kT) = \frac{e(kT) - e[(k-1)T]}{T} \tag{8-37}$$

Thus

$$\frac{M(z)}{E(z)} = \frac{z-1}{Tz} \tag{8-38}$$

Note that this differentiator is the reciprocal of the transfer function for the rectangular-rule integrator (see Problem 2-10). Another transfer function of a discrete differentiator is the reciprocal of the transfer function for trapezoidal integration, given in (8-36). Numerical differentiation and integration are covered in detail in Chapter 11.

It is evident from the discussion above that may different discrete transfer functions may be used for numerical integration or numerical differentiation. For the development in this section, (8-36) and the reciprocal of this transfer function will be used for integration and differentiation, respectively. Then, from (8-36), letting $D_I(z)$ be the integrator transfer function, we obtain

$$D_I(w) = \frac{T}{2}\left[\frac{z+1}{z-1}\right]_{z=[1+(T/2)w]/[1-(T/2)w]} = \frac{1}{w} \tag{8-39}$$

Letting $D_D(z)$, the differentiator transfer function, be the reciprocal of (8-39),

$$D_D(w) = D_D(z)\Big|_{z=[1+(T/2)w/[1-(T/2)w]} = w \tag{8-40}$$

The frequency responses of these transfer functions will now be investigated. Recall that, for continuous systems, a differentiator has a transfer function of s, and an integrator a transfer function of $1/s$. The frequency responses are obtained by replacing s with $j\omega$. Now, from (7-11),

$$\omega_w = \frac{2}{T}\tan\left(\frac{\omega T}{2}\right) \tag{8-41}$$

For $\omega T/2$ small compared to 1, we see that

$$\omega_w \approx \omega \tag{8-42}$$

Then, for the integrator transfer function (8-39),

$$D_I(j\omega_w) = \frac{1}{j\omega_w} \tag{8-43}$$

Substituting (8-42) into (8-43), we obtain

$$D_I(j\omega) \approx \frac{1}{j\omega} \tag{8-44}$$

This approximation is good provided that

$$\tan\left(\omega\frac{T}{2}\right) = \tan\left(\pi\frac{\omega}{\omega_s}\right) \approx \pi\frac{\omega}{\omega_s}$$

If this expression is satisfied, we would expect to obtain accurate differentiation and integration.

8.9 PID CONTROLLERS

We will now discuss a frequency-response design technique that considers phase-lead phase-lag controllers from a somewhat different viewpoint. The resultant controller is called a PID (Proportional-plus-Integral-plus-Derivative) controller. This controller is a special type of lag–lead controller.

The transfer function of a PID controller, using integrator and differentiator transfer functions developed in the preceding section, is given by

$$D(w) = K_p + \frac{K_I}{w} + K_D w \tag{8-45}$$

In this expression, K_p is the gain in the proportional path, K_I the gain in the integral path, and K_D the gain in the derivative path.

Consider first a proportional-plus-integral (PI) controller. The filter transfer function is

$$D(w) = K_P + \frac{K_I}{w} = \frac{K_P w + K_I}{w} = K_I \frac{1 + w/\omega_{w0}}{w} \tag{8-46}$$

where $\omega_{w0} = K_I/K_P$. Note that this is a phase-lag filter of the type given in (8-13), but the pole is placed at $\omega_{wp} = 0$. The filter frequency response is illustrated in Figure 8-21. Note that the PI controller increases the low-frequency gain (infinite at $\omega_w = 0$),

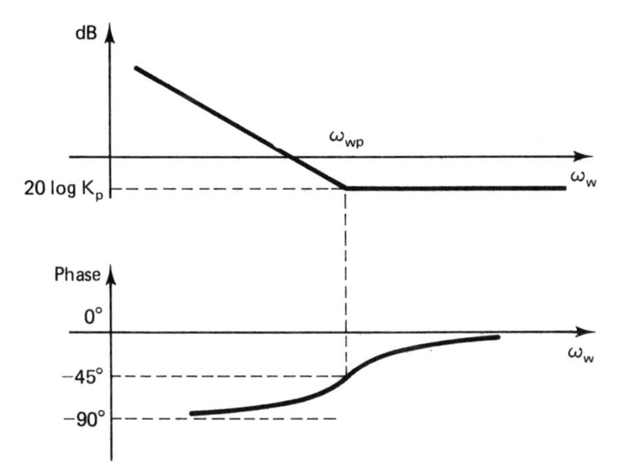

Figure 8-21 Frequency response for a PI controller.

and thus reduces steady-state errors. The purpose is then the same as that of the phase-lag controller of (8-13): to increase stability margins and/or reduce steady-state errors.

Consider next the PD (proportional-plus-derivative) controller. The filter transfer function is

$$D(w) = K_p + K_D w = K_p \left(1 + \frac{w}{\omega_{w0}}\right) \tag{8-47}$$

where $\omega_{w0} = K_p/K_D$. Note that this is a phase-lead controller of the type given in (8-13), with the pole placed at $\omega_{wp} = \infty$. The filter frequency response is illustrated in Figure 8-22. The purposes of the PD controller are to add positive phase angles to the

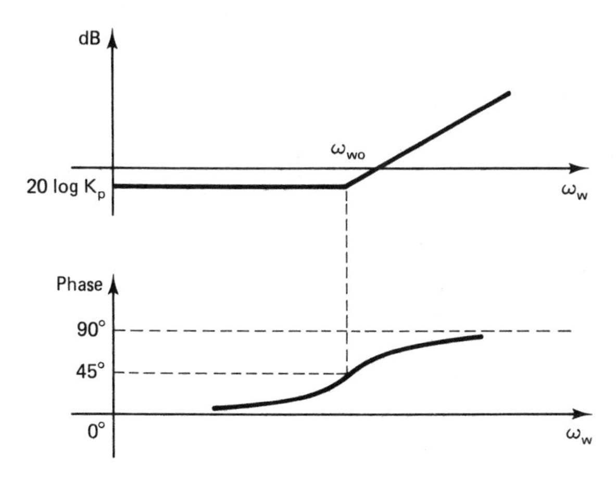

Figure 8-22 Frequency response for a PD controller.

system frequency response so as to improve system stability, and to increase system bandwidth so as to increase the speed of response. The effects of the PD controller appear at high frequencies, as opposed to the low-frequency effects of the PI controller.

The PID filter is a composite of the two filters discussed above, and has the transfer function

$$D(w) = K_p + \frac{K_I}{w} + K_D w \tag{8-48}$$

The frequency response of the PID controller is illustrated in Figure 8-23. The design techniques can follow those given for phase-lag (the PI part) and phase-lead (the PD part) filters. However, a different design procedure will be developed in the following section. Since the K_p term is common to both the PI and PD parts, the design of the PD part affects the PI part, and vice versa.

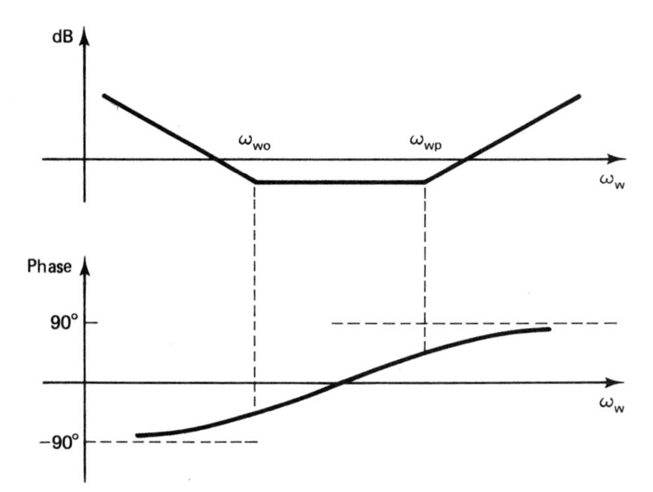

Figure 8-23 Frequency response of a PID controller.

A problem is obvious in the use of the PID filter. As is seen in Figure 8-23, the filter gain continues to increase without limit as frequency increases. As ω_w approaches infinity, z approaches -1. In general, $G(z)$ does not have a zero at $z = -1$. Thus $|G(z)D(z)|$ will approach infinity as z approaches -1, which will probably lead to encirclements of the -1 point on the system Nyquist diagram. Hence the system will be unstable. To avert this problem, a pole (or poles) must be added to the derivative term, resulting in the filter transfer function

$$D(w) = K_p + \frac{K_I}{w} + \frac{K_D w}{\prod (1 + w/\omega_{wpi})} \tag{8-49}$$

However, the usual solution to this problem is to add a single pole using the differentiator of Figure 8-20, which has the transfer function given in (8-38).

$$D_D(w) = \left.\frac{z-1}{Tz}\right|_{z=[1+(T/2)w]/[1-(T/2)w]} = \frac{w}{1+(T/2)w} = \frac{w}{1+(w/\omega_{wp})} \tag{8-50}$$

Hence this algorithm for differentiation adds a pole at $2/T$. And since $\omega_s = 2\pi/T$, then

$$\omega_{wp} = \frac{2}{T} = \frac{\omega_s}{\pi} \tag{8-51}$$

This pole position is normally well outside the bandwidth of the system, and will contribute very little phase lag at the frequency that the phase margin occurs. Hence in many cases this pole can be ignored (at least initially) in the design process. If high-frequency noise is a problem in the control system, additional poles may be required in the derivative term.

The PID filter transfer function as a function of z is, from (8-45), (8-36), and (8-38),

$$D(z) = K_p + K_I \frac{T}{2}\left[\frac{z+1}{z-1}\right] + K_D\left[\frac{z-1}{Tz}\right] \tag{8-52}$$

The filter difference equations can be written directly from this equation. In some cases it is possible to determine K_P, K_I, and K_D experimentally using the physical control system, when only a rudimentary knowledge of the plant characteristics is available. An educated guess of K_P, K_I, and K_D is made, which must result in a stable closed-loop system. Then K_P, K_I, and K_D are varied in some systematic manner until an acceptable response is obtained.

The PID controller offers an advantage if, for some reason, the sample period T must be changed after the system design has been completed. Suppose that, for a given design, T is chosen such that accurate differentiation and integration occur. If, for a different value of T, accurate differention and integration still occur, then the gains K_P, K_I, and K_D will not change. Hence a new design is not required. A design procedure for PID controllers will be developed in the following section.

8.10 PID CONTROLLER DESIGN

In this section a design process is developed for PID controllers [8]. The development will closely parallel that of Section 8.6 for phase-lead filters (controllers).

We assume initially that the PID controller has a transfer function given by

$$D(w) = K_P + \frac{K_I}{w} + K_D w \tag{8-45}$$

Hence we are ignoring the effects of the required pole of (8-50) in the derivative path; these effects will be considered later. The controller frequency response is given by

$$D(j\omega_w) = K_P + j\left(K_D\omega_w - \frac{K_I}{\omega_w}\right) = |D(j\omega_w)|\underline{/\theta} \tag{8-53}$$

Here, as in Section 8.6, the design problem is to choose $D(w)$ (i.e., choose K_P, K_I, and K_D) such that

$$D(j\omega_{w1})G(j\omega_{w1}) = 1\underline{/180^\circ + \phi_m} \tag{8-54}$$

at a chosen frequency ω_{w1}. Now, from (8-53),

$$K_P + j\left(K_D\omega_{w1} - \frac{K_I}{\omega_{w1}}\right) = |D(j\omega_{w1})|(\cos\theta + j\sin\theta) \tag{8-55}$$

where

$$\theta = 180^\circ + \phi_m - \underline{/G(j\omega_{w1})} \tag{8-56}$$

Therefore, from (8-55),

$$K_P = |D(j\omega_{w1})| \cos\theta = \frac{\cos\theta}{|G(j\omega_{w1})|} \tag{8-57}$$

$$K_D\omega_{w1} - \frac{K_I}{\omega_{w1}} = \frac{\sin\theta}{|G(j\omega_{w1})|} \tag{8-58}$$

The design equations are then (8-57) and (8-58). For a given plant $[G(w)]$, the choice of ω_{w1} and ϕ_m uniquely determines K_p, from (8-57). However, K_D and K_I are not uniquely determined, as is evident in (8-58). This results from (8-55) yielding two equations, but with the three unknowns K_P, K_I, and K_D. In satisfying (8-58), in general increasing K_D will increase the bandwidth, while increasing K_I will decrease steady-state errors, *provided* that the system retains an acceptable gain margin. Note that if (8-57) and (8-58) are satisfied, varying K_D and K_I will change the gain margin, while the phase margin remains constant.

Equations (8-57) and (8-58) also apply to the design of PI and PD controllers, with the appropriate gain (K_I or K_D) set to zero. For this case all gains are uniquely determined.

As shown in Section 8.9, in general a pole is required in the derivative term. A commonly used transfer function was given in (8-50), which results in a PID controller transfer function

$$D(w) = K_P + \frac{K_I}{w} + \frac{K_D w}{1 + (T/2)w} \tag{8-59}$$

Then

$$\begin{aligned} D(j\omega_w) &= K_P - j\frac{K_I}{\omega_w} + \frac{K_D j\omega_w}{1 + j\omega_w T/2} \\ &= \left(K_P + \frac{K_D\omega_w^2(2/T)}{(2/T)^2 + \omega_w^2}\right) + j\left(\frac{K_D\omega_w(2/T)^2}{(2/T)^2 + \omega_w^2} - \frac{K_I}{\omega_w}\right) \end{aligned} \tag{8-60}$$

Hence (8-57) and (8-58) become

$$K_P + \frac{K_D\omega_{w1}^2(2/T)}{(2/T)^2 + \omega_{w1}^2} = \frac{\cos\theta}{|G(j\omega_{w1})|} \tag{8-61}$$

and

$$\frac{K_D\omega_{w1}(2/T)^2}{(2/T)^2 + \omega_{w1}^2} - \frac{K_I}{\omega_{w1}} = \frac{\sin\theta}{|G(j\omega_{w1})|} \tag{8-62}$$

Note that if $\omega_{w1} \ll 2/T$, these equations reduce to (8-57) and (8-58); otherwise, no simple procedure has been found for calculating K_P, K_I, and K_D. For the PD controller, (8-61) and (8-62) contain only two unknowns and may be solved directly for

$$K_D = \left[\frac{\sin\theta}{|G(j\omega_{w1})|}\right]\left[\frac{(2/T)^2 + \omega_{w1}^2}{(2/T)^2\omega_{w1}}\right] \tag{8-63}$$

and

$$K_P = \frac{\cos\theta}{|G(j\omega_{w1})|} - \frac{K_D\omega_{w1}^2(2/T)}{(2/T)^2 + \omega_{w1}^2} \tag{8-64}$$

Two examples of PID design will now be given.

Example 8.4

First the design problem of Example 8.3 will be repeated, but a PI filter will be utilized. In Example 8.3 we required a 55° phase margin and a steady-state error to a unit ramp input of 0.5. Since $D(z)$ adds a pole at $z = 1$ to the one already present in $G(z)$, then $D(z)G(z)$ has two poles at $z = 1$. Thus the steady-state error to a ramp input is zero (see Section 6.5), satisfying the steady-state error design criteria. From Table 8.1, the frequency response of $G(z)$, we choose $\omega_{w1} = 0.400$. Then,

$$G(j\omega_{w1}) = G(j0.4) = 7.15 \text{ dB}\underline{/-123.7^\circ}$$
$$= 2.278\underline{/-123.7^\circ}$$

From (8-56),

$$\theta = 180^\circ + 55^\circ - (-123.7^\circ) = 358.7^\circ = -1.3^\circ$$

Then, from (8-57) and (8-58), respectively,

$$K_p = \frac{\cos(-1.3^\circ)}{2.278} = 0.439$$

$$K_I = \left[\frac{-\sin(-1.3^\circ)}{2.278}\right] 0.4 = 0.00398$$

From (8-46), the zero of the PI controller is placed at

$$\omega_{w0} = \frac{K_I}{K_P} = 0.00907$$

which is quite low. Hence we have reduced the system bandwidth considerably. Even though we get good steady-state error response, a relatively long time will be required to achieve it. In fact, a simulation of the system shows that the error to a ramp input is 2.0 after 20 s, and has descreased to only 1.5 at the end of 50 s. If we choose $\omega_{w1} = 0.3$, then the PI gains calculated are $K_p = 0.313$ and $K_I = 0.01556$. Then

$$\omega_{w0} = \frac{K_I}{K_P} = 0.0497$$

and the system bandwidth has been increased. For this PI controller, the error to a ramp input is 1.3 after 20 s, and has descreased to 0.11 at the end of 50 s.

Example 8.5

The design problem of Example 8.4 will be repeated, except that in this example a PID controller will be utilized. The design equations (8-57) and (8-58) will be used (i.e., initially the pole in the derivative term will be ignored). From (8-57),

$$K_p = \frac{\cos\theta}{|G(j\omega_{w1})|}$$

Thus the larger we choose ω_{w1}, the smaller is $|G(j\omega_{w1})|$ (see Table 8-1), and hence the larger is K_p. This increases the open-loop gain, which is desirable for many reasons. We choose $\omega_{w1} = 1.2$. Then

$$G(j1.2) = -6.79 \text{ dB}\underline{/-172.9^\circ} = 0.4576\underline{/-172.9^\circ}$$

From (8-56),

$$\theta = 180^\circ + 55^\circ - (-172.9^\circ) = 407.9^\circ = 47.9^\circ$$

and from (8-57),

$$K_p = \frac{\cos(47.9^\circ)}{0.4576} = 1.465$$

From (8-58),

$$K_D\omega_{w1} - \frac{K_I}{\omega_{w1}} = \frac{\sin\theta}{|G(j\omega_{w1})|}$$

Hence, with $G(j\omega_{w1})$ from above,

$$K_D = \left[\frac{\sin(47.9°)}{0.4576}\right]\frac{1}{1.2} + K_I\left(\frac{1}{1.2}\right)^2$$

$$= 1.351 + 0.694K_I$$

After some trial and error using a simulation of the system, K_I was chosen as 0.004, and K_D is then 1.354. This choice of gains results in a phase margin of 53.5° and a gain margin of 16 dB, when the pole of (8-50) is added to the derivative path. Thus this pole has little effect on the phase margin. The step response of this system is shown in Figure 8-24, together with the step response of the phase-lead system of Example 8.2. An additional simulation of the system shows that the error to a ramp input has decreased to 0.65 at $t = 20$ s.

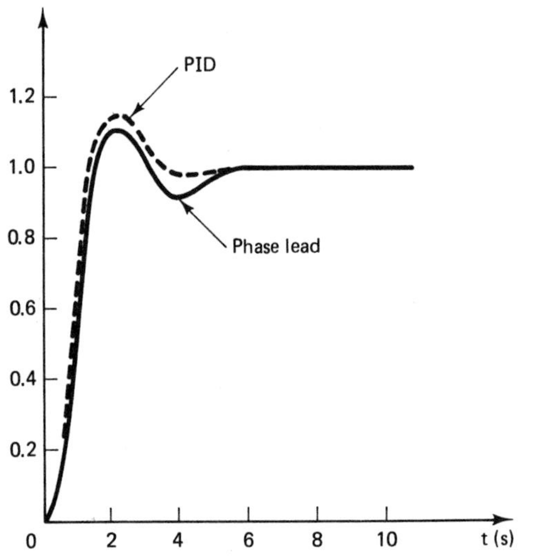

Figure 8-24 Responses of Examples 8.2 and 8.5.

8.11 DESIGN BY ROOT LOCUS

In the frequency-response design procedures described above, we attempted to reshape the system open-loop frequency response to achieve certain stability margins, transient-response characteristics, steady-state response characteristics, and so on. Even though design equations were developed, the techniques are still largely trial and error.

A different design technique is presented in this section: the root-locus procedure. Recall that the root locus for a system is a plot of the roots of the system's characteristic equation as gain is varied. Hence the character of the transient response of a system is evident from the root locus. The design procedure is to add poles and zeros via a digital controller so as to shift the roots of the characteristic equation to more appropriate locations in the z-plane.

Consider again the system of Figure 8-6, which has the characteristic equation

$$1 + KD(z)G(z) = 0 \tag{8-65}$$

where K is the gain that is to be varied to generate the root locus. Then a point z_a is on the root locus provided that (8-65) is satisfied for $z = z_a$, or that

$$K = \frac{1}{|D(z_a)G(z_a)|} \tag{8-66}$$

$$\underline{/D(z_a)G(z_a)} = \pm 180° \tag{8-67}$$

Since we allow K to vary from zero to infinity, a value of K will always exist such that (8-66) is satisfied. Then the condition for a point z_a to be on the root locus is simply (8-67).

Suppose, as an example, that $D(z) = 1$ and $KG(z)$ is given by

$$KG(z) = \frac{K(z - z_1)}{(z - z_2)(z - z_3)} \tag{8-68}$$

Figure 8-25 illustrates the testing of point z to determine if it is on the root locus. If z is on the root locus, then, from (8-67),

$$\theta_1 - \theta_2 - \theta_3 = \pm 180° \tag{8-69}$$

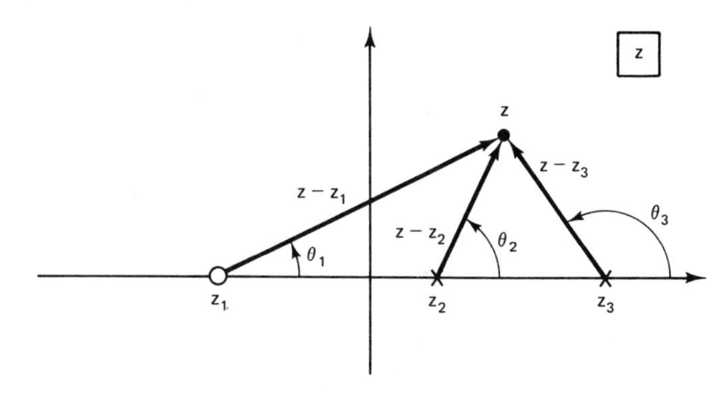

Figure 8-25 Point on the root locus.

The value of K that places a root of the characteristic equation at this point is, from (8-66),

$$K = \frac{|z - z_2||z - z_3|}{|z - z_1|} \tag{8-70}$$

With this brief discussion of the root locus, we will now consider design procedures. Consider the first-order controller

$$D(z) = \frac{K_1(z - z_0)}{z - z_p} \tag{8-71}$$

We will require that $D(1) = 1$ so as not to affect the steady-state response; hence

$$K_1 = \frac{1 - z_p}{1 - z_0} \tag{8-72}$$

The controller pole is restricted to real values inside the unit circle. For a phase-lead controller $z_0 > z_p$, and thus $K_1 > 1$. For a phase-lag controller $z_0 < z_p$ and $K_1 < 1$.

We will first consider phase-lag design. We will illustrate the design for the plant, $G(z)$, of (8-68). The uncompensated root locus is sketched in Figure 8-26a. Suppose that the root locations z_a and $\bar{z}_a$ give a satisfactory transient response, but that the loop gain must be increased to produce smaller steady-state errors, improved disturbance rejection, and so on. We add the controller pole and zero as shown in Figure 8-26b, assuming that $z_3 = 1$. Since the pole and zero are very close to $z = 1$ (recall Example 8.1), the scale in the vicinity of this point is greatly expanded. Hence these two poles and one zero will appear as a single pole as in Figure 8-26a, when determining the root location at z_a. We see, then, that the compensator pole and zero cause the root at z_a to shift only slightly to z_a'. However,

$$KD(z)G(z) = \frac{KK_1(z - z_0)(z - z_1)}{(z - z_p)(z - z_2)(z - z_3)}$$

Hence, for a root to appear at z_a', from (8-66),

$$K = \frac{|z_a' - z_p||z_a' - z_2||z_a' - z_3|}{K_1|z_a' - z_0||z_a' - z_1|} \approx \frac{|z_a - z_2||z_a - z_3|}{K_1|z_a - z_1|} \tag{8-73}$$

Let K_0 be the gain required to place the root at z_a in the uncompensated system, Figure 8-26a. Then

$$K_0 = \frac{|z_a - z_2||z_a - z_3|}{|z_a - z_1|} \tag{8-74}$$

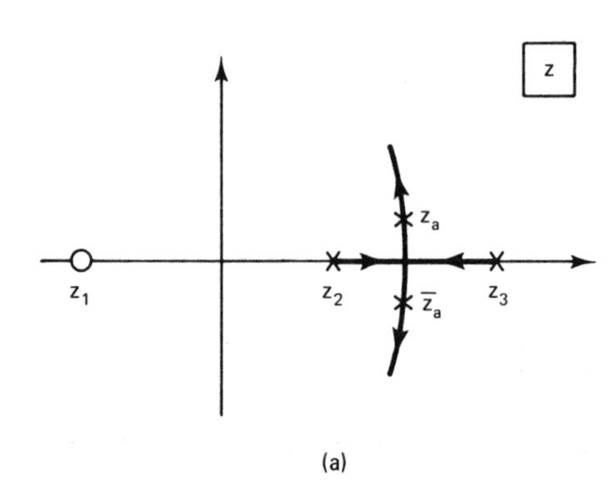

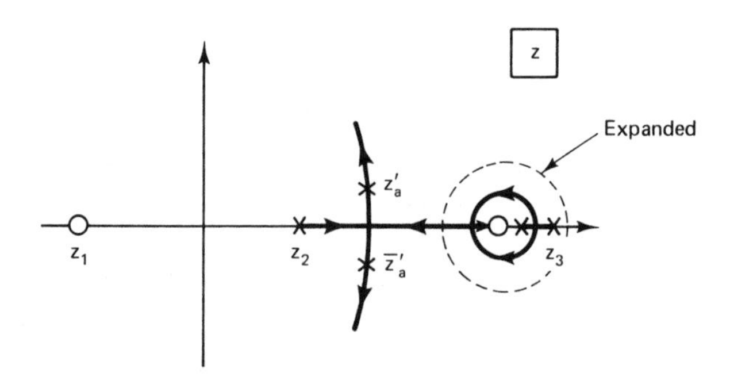

Figure 8-26 Phase-lag design.

Hence, from (8-73) and (8-74),

$$K \approx \frac{K_0}{K_1}$$

and $K_1 < 1$. So we see that phase-lag compensation allows us to increase the open-loop gain while maintaining the approximate same roots in the characteristic equation. Of course, as seen from Figure 8-26b, we have added a root close to $z = 1$, which has a long time constant. Thus the system will have an increased settling time.

Phase-lead design is illustrated in Figure 8-27. Here, to simplify the discussion, we place the controller zero coincident with the plant pole at $z = z_2$. The controller pole is placed to the left of the zero, which yields the phase-lead controller. Thus the root locus is shifted to the left. The root at z_b has a smaller time constant than that at z_a; thus the system responds faster (larger bandwidth).

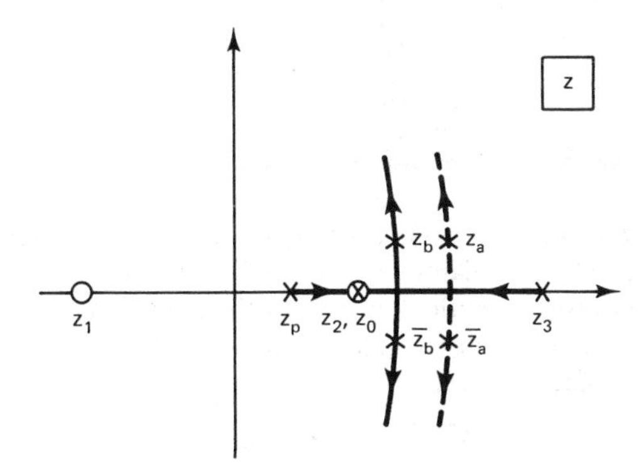

Figure 8-27 Phase-lead design.

In summary, the phase-lag controller shifts the root locus very little, but allows a higher open-loop gain to be used. Or if the same open-loop gain is used, the system is more stable. The phase-lead controller shifts the root locus to the left, resulting in a system that responds faster. A phase-lead example will now be given.

Example 8.6

Phase-lead design will be illustrated in this example. Suppose that the plant transfer function is given by

$$G_p(s) = \frac{K}{s(s+1)}$$

Since the time constant of the pole at $s = -1$ is 1 s, we choose $T = 0.1$ s. Then

$$G(z) = \frac{z-1}{z} \mathfrak{z}\left[\frac{K}{s^2(s+1)}\right]$$

$$= \frac{0.004837K(z+0.9672)}{(z-1)(z-0.9048)}$$

The root locus for $G(z)$ is shown in Figure 8-28. Note that K is equal to 0.244 for critical damping, with roots at $z = 0.952$. We will choose a phase-lead controller with the zero at 0.9048, in order to cancel one of the plant poles. We will place the controller pole at $z = 0.7$, which should increase the system speed of response. Then

$$D(z) = \frac{3.15(z-0.9048)}{z-0.7}$$

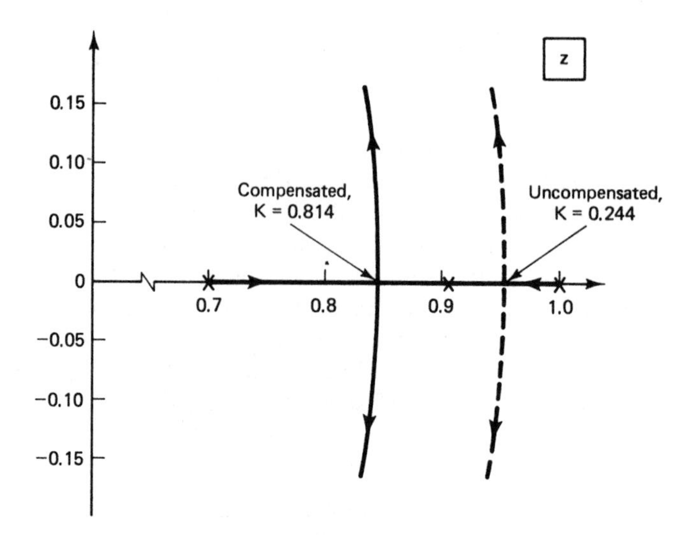

Figure 8-28 Root locus for Example 8.6.

The root locus of the compensated system is also shown in Figure 8-28. Here a value of $K = 0.814$ results in a critically damped system, with roots at $z = 0.844$. We choose critical damping as our design criterion. A pole in the s-plane at $s = -a$ has a time constant of $\tau = 1/a$ and an equivalent z-plane location of $\epsilon^{-aT} = \epsilon^{-T/\tau}$. Thus, for the uncompensated critically damped case,

$$\epsilon^{-0.1/\tau} = 0.952$$

or $\tau = 2.03$ s. For the compensated critically damped case,

$$\epsilon^{-0.1/\tau} = 0.844$$

or $\tau = 0.59$ s. Thus the compensated system responds much faster. A plot giving the step responses for both the uncompensated system and the compensated system is shown in Figure 8-29.

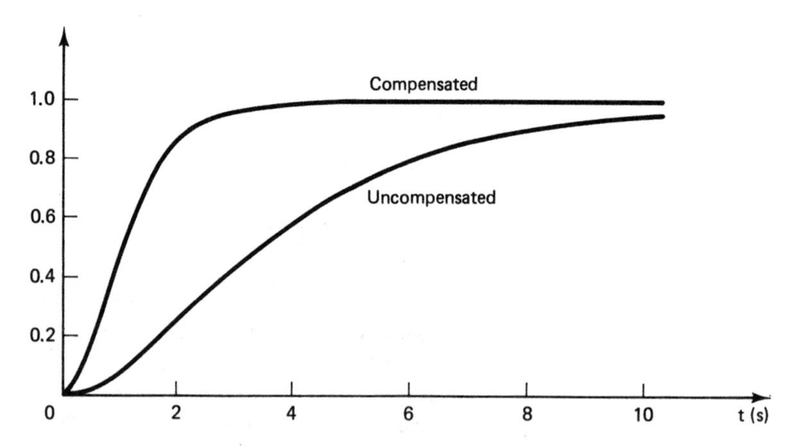

Figure 8-29 Step responses for Example 8.6.

A point must be made concerning the last example. It appears that we can increase the speed of response of this system by a very large factor, simply by moving the pole of the controller to the left. However, this movement of the pole increases the high-frequency gain of the controller, and this increase may not be acceptable. To illustrate this, suppose that the plant is a servomotor, and that the gain K is realized in a power amplifier that precedes the motor. For the uncompensated system, with a unit step input, the maximum error is 1, which occurs at $t = 0$. Thus the maximum amplifier output is 0.244. For the compensated system, the maximum error is also 1, at $t = 0$. At this instant, the controller output is 3.15 (the interested reader can prove this). Hence, since the amplifier gain in 0.814, the amplifier output is 2.56, which is an increase by more than a factor of 10 over that of the uncompensated system. This increase in motor voltage accounts for the faster response. However, the larger voltages generated by the phase-lead compensation may force the system into nonlinear regions of operation. A discussion of the possible effects of this operation is beyond the scope of this book; however, the results of the linear analysis given above will no longer be applicable.

8.12 SUMMARY

Various criteria used in the specification of control systems are presented in this chapter. Next, digital controller design techniques using phase-lead and phase-lag compensation are developed. These techniques, based on frequency responses, tend to be largely trial and error, but are among the most commonly used techniques in compensator design. Then the three-term controller (PID) is developed, and is seen to be a special type of lag–lead design. Finally, design by root-locus techniques is developed.

REFERENCES AND FURTHER READING

1. M. E. Van Valkenberg, *Network Analysis*. Englewood Cliffs, N.J.: Prentice-Hall, Inc., 1974.
2. C. J. Savant, Jr., *Control System Design*. New York: McGraw-Hill Book Company, 1964.
3. K. Ogata, *Modern Control Engineering*. New York: McGraw-Hill Book Company, 1970.
4. H. T. Nagle, B. D. Carroll, and J. D. Irwin, *An Introduction to Computer Logic*. Englewood Cliffs, N.J.: Prentice-Hall, Inc., 1975.
5. W. R. Wakeland, "Bode Compensator Design," *IEEE Trans. Autom. Control*, Vol. AC-21, p. 771, Oct. 1976.
6. J. R. Mitchell, "Comments on Bode Compensator Design," *IEEE Trans. Autom. Control*, Vol. AC-22, p. 869, Oct. 1977.
7. M. L. Dertouzos, M. Athans, R. N. Spann, and S. J. Mason, *Systems, Networks, and Computation: Basic Concepts*. New York: McGraw-Hill Book Company, 1972.
8. C. I. Huang, "Computer Aided Design of Digital Controllers," M.S. thesis, Auburn University, Auburn, Ala., 1981.

9. B. C. Kuo, *Analysis and Synthesis of Sampled-Data Control Systems.* Englewood Cliffs, N.J.: Prentice-Hall, Inc., 1963.

10. J. A. Cadzow and H. R. Martens, *Discrete-Time and Computer Control Systems.* Englewood Cliffs, N.J.: Prentice-Hall, Inc., 1970.

11. R. C. Dorf, *Modern Control Systems.* Reading Mass.: Addison-Wesley Publishing Company, Inc., 1974.

PROBLEMS

8-1. For the uncompensated system of Figure P8-1, the phase margin is approximately 30°.

(a) Design a phase-lag compensator that will increase the phase margin to approximately 50°.

(b) Obtain the step response by simulation. Compare this step response with that of the uncompensated system, given in Figure 6-3.

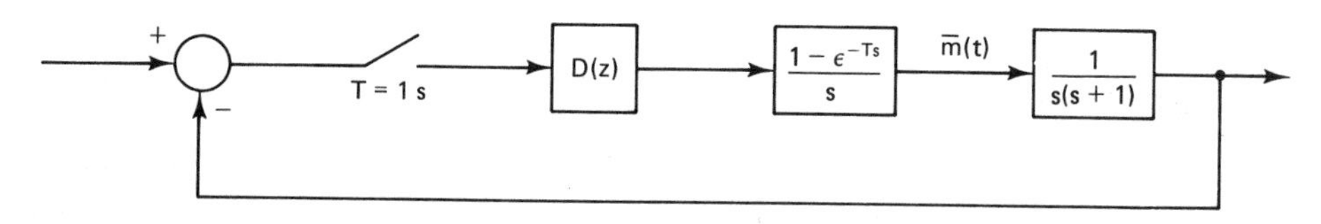

Figure P8-1 System for Problem 8-1.

8-2. For the uncompensated system of Figure P8-1, the phase margin is approximately 30°.

(a) Design a phase-lead compensator that will increase the phase margin to approximately 50°.

(b) Obtain the step response by simulation. Compare this step response with that of the uncompensated system, given in Figure 6-3, and with that obtained in Problem 8-1, if available.

8-3. For the uncompensated system of Figure P8-1, with $T = 0.1$ s the phase margin is approximately 50°. It is desired that the steady-state error constant for a unit ramp input, K_v of (6-20), be 4, resulting in a steady-state error of 0.25 unit.

(a) Increase the plant gain to that value which results in $K_v = 4$. Then find the new phase margin.

(b) Design a phase-lead compensation that results in a 50° phase margin with $K_v = 4$.

8-4. The temperature control system shown in Figure P8-4 is described in Problem 4-3. With $T = 0.5$ s, the steady-state error is to be less than 2% for a constant input.

(a) Find the percent steady-state error for a constant input, for $D(z) = 1$. Recall that, to command the output to change by 1°C, the input step amplitude must be 0.04.

(b) Let $D(z) = K$, a pure gain. Find the value of K that gives a 2% error to a step input. Is the system stable for this value of K?

(c) Design a phase-lag compensator for the system of part (b), such that the 2% steady-state error is realized, the phase margin is greater than 40°, and the gain margin greater than 6 dB. Give the total transfer function $D(z)$ of the compensator.

(d) Can the design of part (c) be achieved using a phase-lead compensator? Justify your answer.

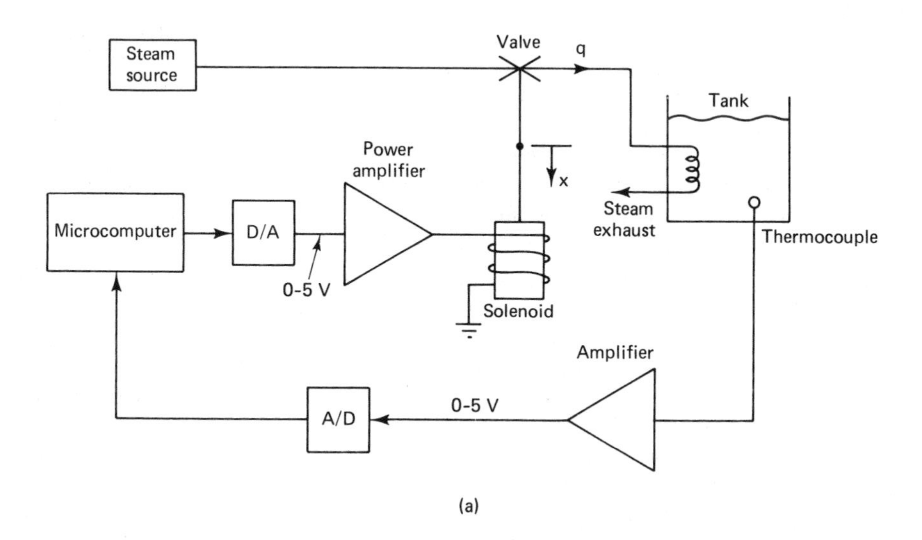

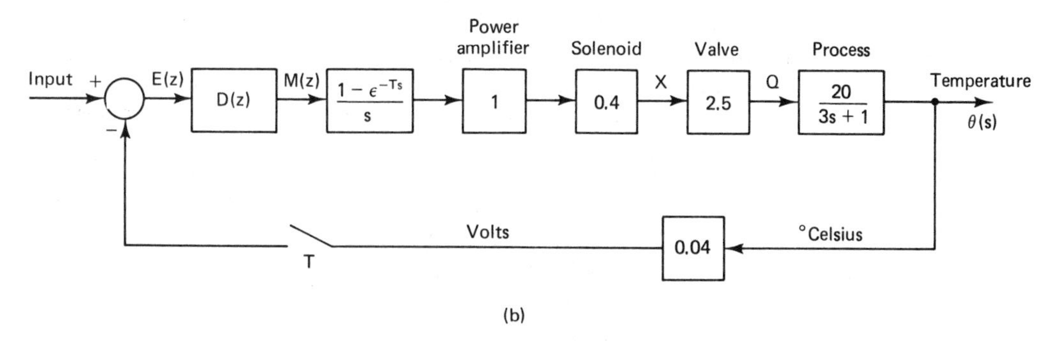

Figure P8-4 Temperature control system.

8-5. The automobile cruise-control system shown in Figure P8-5 is described in Problem 4-7. With $T = 0.2$ s, the steady-state error is to be less than 3% for a constant input. Assume that the disturbance torque is zero.

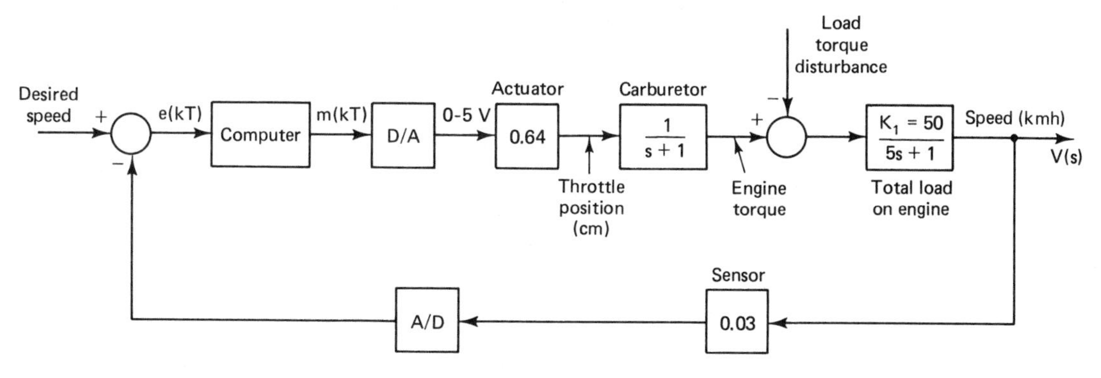

Figure P8-5 Cruise-control system.

(a) Find the percent steady-state error for a constant input, with $D(z) = 1$. Recall that, to command the output to change by 1 kilometer per hour (kmh), the input step amplitude must be 0.03.

(b) Let $D(z) = K$, a pure gain. Find the value of K that gives a 3% error to a step input. Is the system stable for this value of K?

(c) Design a phase-lag compensator for the system of part (b), such that the 3% steady-state error is realized, the phase margin is greater than 45°, and the gain margin is greater than 10 dB. Give the total transfer function of the compensator.

8-6. The automobile cruise-control system shown in Figure P8-5 is described in Problem 4-7. With $T = 0.2$ s, the steady-state error is to be less than 5% for a constant input. Assume that the disturbance torque is zero.

(a) Find the percent steady-state error for a constant input, with $D(z) = 1$. Recall that, to command the output to change by 1 kilometer per hour (kmh), the input step amplitude must be 0.03.

(b) Let $D(z) = K$, a pure gain. Find the value of K that gives a 5% error to a step input. Is the system stable for this value of K?

(c) Design a phase-lead compensator for the system of part (b), such that the 5% steady-state error is realized, the phase margin is greater than 45°, and the gain margin is greater than 8 dB. Give the total transfer function of the compensator.

8-7. (a) Consider the temperature control system of Problem 8-4. With $T = 0.5$ s, design a PI compensator that yields a phase margin of 45° and a gain margin greater than 8 dB. Find the percent steady-state error to a constant input for the compensated system.

(b) Consider the cruise-control system of Problem 8-5. With $T = 0.2$ s, design a PI compensator that yields a phase margin of 45° and a gain margin greater than 8 dB. Find the percent steady-state error to a constant input for the compensated system.

8-8. Repeat Problem 8-1, using a proportional-plus-integral digital filter.

8-9. Repeat Problem 8-2, using a proportional-plus-derivative filter.

8-10. Consider the system of Example 8.1.

(a) Design a PI filter to achieve a phase margin of 60°.

(b) Obtain the system step response by simulation. Compare this response to that of the system of Example 8.1, which is plotted in Figure 8-14.

8-11. Consider the system of Example 8.1. Design a PID controller to achieve a phase margin of 60°.

8-12. Shown in Figure P8-12 is a solar collector control system, which results in the collector tracking the sun. This system is described in Problem 6-7.

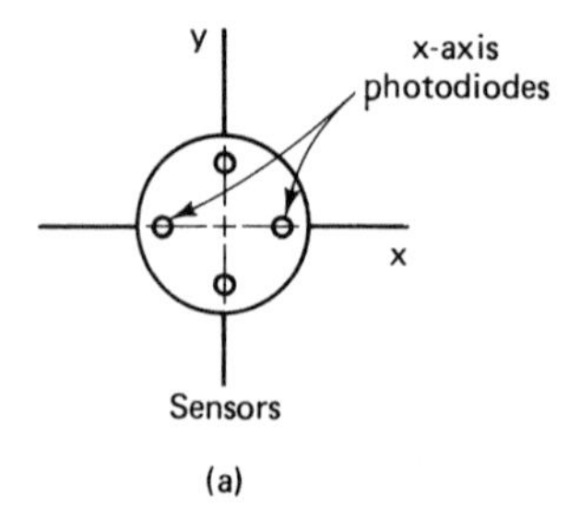

Figure P8-12 Solar collector control system.

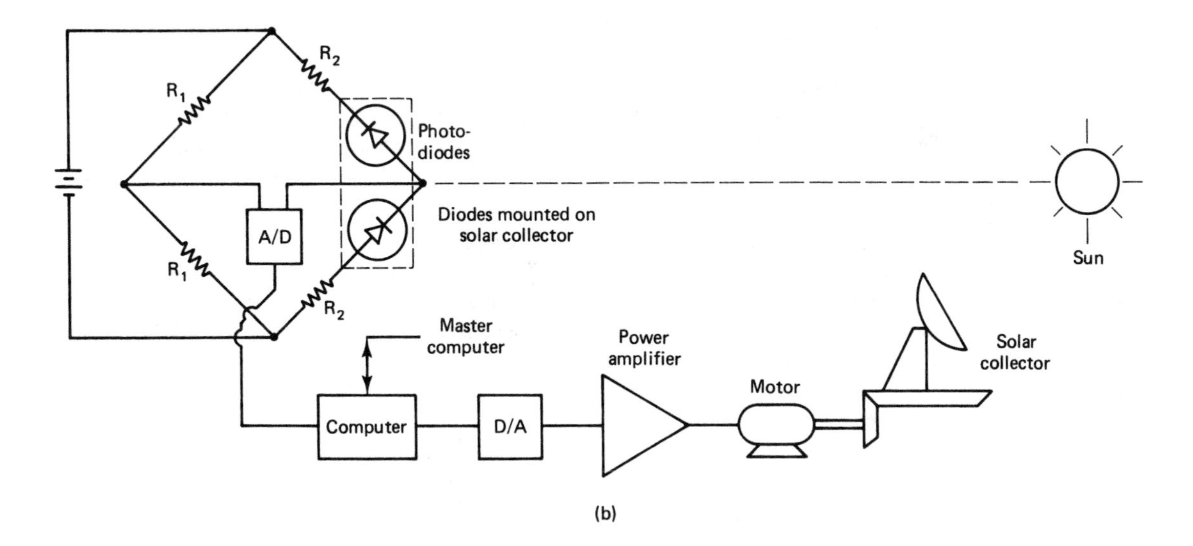

(b)

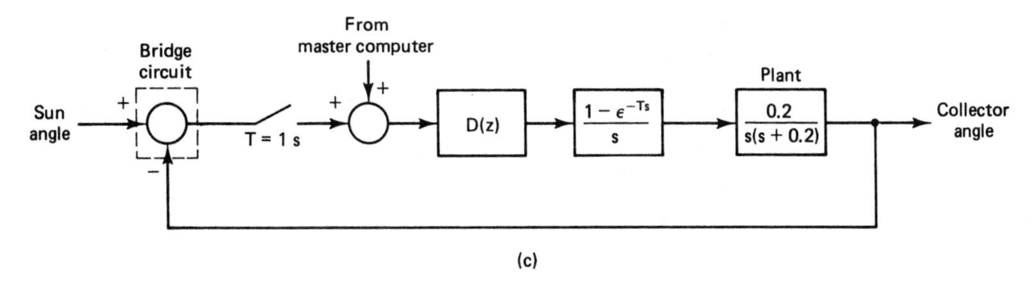

(c)

Figure P8-12 cont.

(a) With $D(z) = 1$, find the steady-state error to the ramp input of $0.004t$, which approximates the sun angle in degrees.

(b) Design a PI controller such that the system phase margin is 50°. Find the steady-state error to a ramp input of $0.004t$, for the compensated system.

8-13. Consider the solar collector tracking system of Problem 8-12.

(a) Let $D(z) = K$. Plot the system root locus, and find the value of K such that the system poles are equal (critical damping).

(b) Find the time constants of the poles found in part (a).

(c) Using root-locus techniques, design a phase-lead compensator which results in critical damping, with poles whose time constant is less than or equal to one-third the value found in part (b).

8-14. Consider the system of Figure P8-1, with a gain factor K added in the numerator.

(a) Using root-locus techniques, find the value of K that results in critical damping for the system.

(b) Find the time constant of the roots of part (a).

(c) Design a phase-lag compensator such that twice the value of gain found in part (a) results in critical damping, with roots having approximately the same time constant as in part (b).

(d) Design a phase-lead compensator which results in critical damping, with roots whose time constant is less than or equal to one-half the value found in part (b).

8-15. Consider the system of Example 8.6. Design a phase-lag compensator such that, with $K = 1.0$, the system is critically damped with roots having a time constant of 2.03 s. This time constant is the same as that of the uncompensated system with $K = 0.244$.

8-16. For the design in Example 8.6, suppose that K is increased to 1.5. Find ζ, the damping factor, and τ, the time constant, of the resulting roots.

8-17. Shown in Figure P8-17 is a carbon dioxide control system for a plant growth chamber. This system is described in Problem 5-18. This system has been implemented in a slightly different form, and operates satisfactorily with $D(z) = K_p$, a pure gain. For $K = 0.005$, find the maximum value of K_p such that the gain margin is at least 10 dB and the phase margin at least 60°.

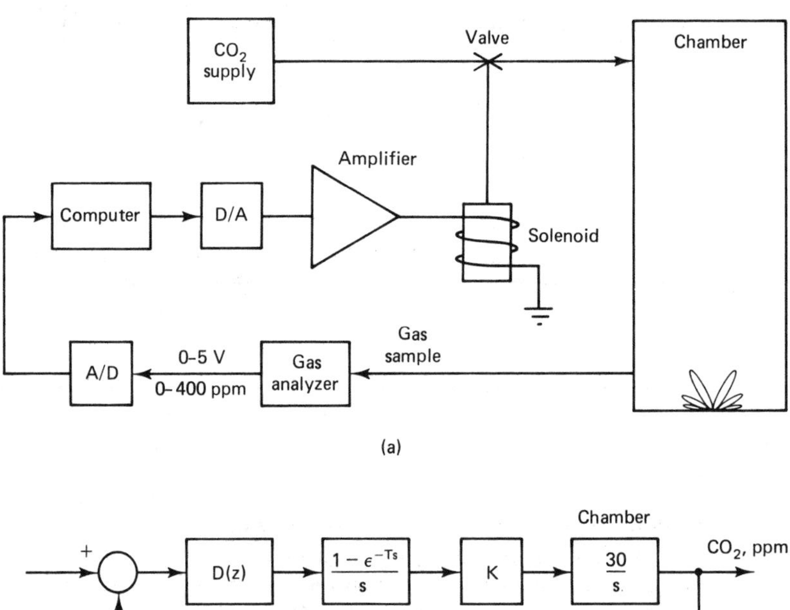

(a)

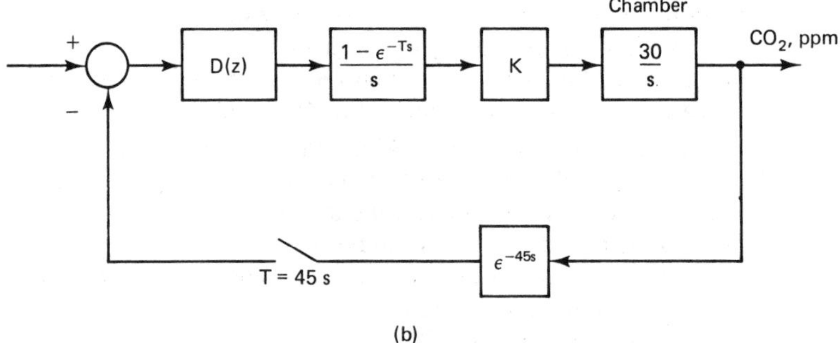

(b)

Figure P8-17 Carbon dioxide control system.

9

Pole-Assignment Design and State Estimation

9.1 INTRODUCTION

In Chapter 8 we considered discrete control design both from a frequency-response point of view and from a root-locus point of view. In every case we considered that only one signal was to be fed back. It seems reasonable that if we determine more about the present condition of a system, and use this additional information to generate the control input, then we should be able to control the system in a manner that is, in some sense, better. Of course, we have a complete description of the condition of a system if we measure its state vector. Hence we might conclude that if we are able to specify mathematically what defines the very best control system, the design of this control system might require that we have all the states of the system available for feedback. Generally, the design of systems of this type does require that we assume that the full state vector is available.

For most control systems the measurement of the full state vector is impractical. In order to implement a design based on full state feedback, we must then estimate the states of a system using measurements that are practical. Fortunately, we can separate the design into two phases. During the first phase, we design the system as though all states of the system will be measured. The second phase is then the design of the state estimator. In this chapter we consider both phases of the design process, and the effects that the state estimator has on closed-loop system operation.

9.2 POLE ASSIGNMENT

In this section a design procedure generally known as pole assignment, or pole placement, is developed. The design results in the assignment of the poles of the closed-loop transfer function (zeros of the characteristic equation) to any desired locations. There are, of course, practical implications that will be discussed as the technique is developed.

To introduce the pole-assignment technique, we will consider the model of a servomotor, developed in Chapter 1. This model is second-order, and an example is shown in Figure 9-1. The state model for this system was calculated in Section 4.10, and is given by

$$\begin{aligned} \mathbf{x}(k+1) &= \begin{bmatrix} 1 & 0.0952 \\ 0 & 0.905 \end{bmatrix} \mathbf{x}(k) + \begin{bmatrix} 0.00484 \\ 0.0952 \end{bmatrix} u(k) \\ y(k) &= [1 \quad 0]\mathbf{x}(k) \end{aligned} \tag{9-1}$$

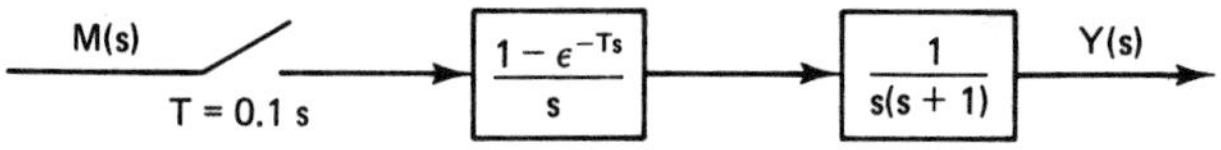

Figure 9-1 Servomotor system.

In this model $x_1(k)$ is the position of the motor shaft, which can easily be measured. The state $x_2(k)$ is shaft velocity, which can be measured using a tachometer or some other suitable sensor. Thus, for this system the full state vector can be measured.

We choose to generate the control input $u(k)$ to be a linear combination of the states; that is,

$$u(k) = -K_1 x_1(k) - K_2 x_2(k) = -\mathbf{K}\mathbf{x}(k) \tag{9-2}$$

where the gain matrix $\mathbf{K}$ is

$$\mathbf{K} = [K_1 \quad K_2] \tag{9-3}$$

Then (9-1) can be written as

$$\begin{aligned} \mathbf{x}(k+1) &= \begin{bmatrix} 1 & 0.0952 \\ 0 & 0.905 \end{bmatrix} \mathbf{x}(k) - \begin{bmatrix} 0.00484 \\ 0.0952 \end{bmatrix} [K_1 x_1(k) + K_2 x_2(k)] \\ &= \begin{bmatrix} 1 - 0.00484K_1 & 0.0952 - 0.00484K_2 \\ -0.0952K_1 & 0.905 - 0.0952K_2 \end{bmatrix} \begin{bmatrix} x_1(k) \\ x_2(k) \end{bmatrix} \end{aligned} \tag{9-4}$$

The matrix of (9-4) is the closed-loop system matrix, which we will call $\mathbf{A}_f$. The system characteristic equation is then

$$|z\mathbf{I} - \mathbf{A}_f| = 0 \tag{9-5}$$

Substitution of (9-4) into (9-5) yields, after some calculations, the characteristic equation

$$\begin{aligned} z^2 &+ (0.00484K_1 + 0.0952K_2 - 1.905)z \\ &+ 0.00468K_1 - 0.0952K_2 + 0.905 = 0 \end{aligned} \tag{9-6}$$

Suppose that, by some process, we choose the desired characteristic-equation zero locations to be λ_1 and λ_2. Then the desired characteristic equation, denoted $\alpha_c(z)$, is given by

$$\alpha_c(z) = (z - \lambda_1)(z - \lambda_2) = z^2 - (\lambda_1 + \lambda_2)z + \lambda_1\lambda_2 \tag{9-7}$$

Equating coefficients in (9-6) and (9-7) yields the equations

$$\begin{aligned} 0.00484K_1 + 0.0952K_2 &= -(\lambda_1 + \lambda_2) + 1.905 \\ 0.00468K_1 - 0.0952K_2 &= \lambda_1\lambda_2 - 0.905 \end{aligned} \tag{9-8}$$

These equations are linear in K_1 and K_2, and upon solving yield

$$\begin{aligned} K_1 &= 105[\lambda_1\lambda_2 - (\lambda_1 + \lambda_2) + 1.0] \\ K_2 &= 14.67 - 5.34\lambda_1\lambda_2 - 5.17(\lambda_1 + \lambda_2) \end{aligned} \tag{9-9}$$

Thus we can find the gain matrix **K** that will realize any desired characteristic equation.

We wish now to make the following points. By some process we choose the root locations so as to satisfy design criteria such as speed of response, overshoot in the transient response, and the like. However, at this time no input has been shown; hence we cannot speak of overshoot to a step input. Once we have chosen λ_1 and λ_2, (9-9) will give the gain matrix needed to realize these characteristic-equation zeros. Then, to the extent that (9-1) is an accurate model of the physical servomotor, we will realize these zeros in the physical system. Obviously, we cannot increase the speed of response of a servo system without limit, even though (9-9) seems to indicate that we can. If we attempt to force the system to respond too fastly, large signals will be generated and the plant will enter a nonlinear mode of operation; then (9-1) will no longer accurately model the plant. Hence, in choosing λ_1 and λ_2, we must consider only those root locations that can be attained by the physical system.

We will now develop a general procedure of pole assignment. Our nth-order plant is modeled by

$$\mathbf{x}(k + 1) = \mathbf{A}\mathbf{x}(k) + \mathbf{B}u(k) \tag{9-10}$$

We generate the control input $u(k)$ by the relationship

$$u(k) = -\mathbf{K}\mathbf{x}(k) \tag{9-11}$$

where

$$\mathbf{K} = [K_1 \quad K_2 \quad \cdots \quad K_n] \tag{9-12}$$

Then (9-10) can be written as

$$\mathbf{x}(k + 1) = (\mathbf{A} - \mathbf{B}\mathbf{K})\mathbf{x}(k) \tag{9-13}$$

We choose the desired pole locations

$$z = \lambda_1, \lambda_2, \ldots, \lambda_n \tag{9-14}$$

Then the system characteristic equation is

$$\alpha_c(z) = |z\mathbf{I} - \mathbf{A} + \mathbf{B}\mathbf{K}| = (z - \lambda_1)(z - \lambda_2) \cdots (z - \lambda_n) \tag{9-15}$$

In this equation there are n unknowns $K_1, K_2, \ldots, K_n$, and n known coefficients in the right-hand-side polynomial. We can solve for the unknown gains by equating coefficients in (9-15), as illustrated by the servomotor example above. An example will now be given to illustrate this procedure.

Example 9.1

In this example the servomotor system of Figure 9-1, with the state equations given in (9-1), will be considered. With full state feedback, the characteristic equation is given in (9-6). Consider first that the closed-loop system is implemented with the usual unity feedback; then $K_1 = 1$ and $K_2 = 0$. From (9-6), the characteristic equation is given by

$$z^2 - 1.9z + 0.91 = 0$$

This equation has roots at

$$z_{1,2} = 0.954\underline{/\pm 0.091 \text{ rad}} = r\underline{/\pm\theta}$$

From (6-10), we calculate the damping factor of these roots to be

$$\zeta = \frac{-\ln r}{\sqrt{\ln^2 r + \theta^2}} = \frac{-\ln(0.954)}{\sqrt{\ln^2(0.954) + (0.091)^2}} = 0.46$$

and from (6-12), the time constant is

$$\tau = \frac{-T}{\ln r} = \frac{-0.1}{\ln(0.954)} = 2.12 \text{ s}$$

Suppose that we decide that this value of ζ is satisfactory, but that a time constant of 1.0 s is required. Then

$$\ln r = -\frac{T}{\tau} = -0.1$$

or $r = 0.905$. Solving (6-10) for θ, we have

$$\theta^2 = \frac{\ln^2 r}{\zeta^2} - \ln^2 r = \frac{\ln^2(0.905)}{(0.46)^2} - \ln^2(0.905)$$

or θ is equal to 0.193 rad, or 11.04°. Then the desired root locations are

$$\lambda_{1,2} = 0.905\underline{/\pm 11.04^\circ} = 0.888 \pm j0.173$$

Hence the desired characteristic equation is given by

$$(z - 0.888 - j0.173)(z - 0.888 + j0.173) = z^2 - 1.776z + 0.819$$

From (9-9),

$$\begin{aligned} K_1 &= 105[\lambda_1\lambda_2 - (\lambda_1 + \lambda_2) + 1.0] \\ &= 105[0.819 - (1.776) + 1.0] = 4.52 \\ K_2 &= 14.67 - 5.34\lambda_1\lambda_2 - 5.17(\lambda_1 + \lambda_2) = 1.12 \end{aligned}$$

The initial-condition response of this system is given in Figure 9-2, with $x_1(0) = 1.0$ and $x_2(0) = 0$. Both $y(t)$, the output (Figure 9-2a), and $u(t)$, the sampler input (Figure 9-2b), are shown.

Suppose that we decide that the foregoing value of $\zeta = 0.46$ is satisfactory, but that a time constant of $\tau = 0.5$ s is required. This design is given as Problem 9-1. The gains obtained from this design are $K_1 = 16.0$ and $K_2 = 3.26$, and the initial condition response is also given in Figure 9-2. Compare the control inputs $u(t)$ for the two designs in Figure 9-2. The maximum amplitude of $u(t)$ for $\tau = 1.0$ s is 4.52, while that for $\tau = 0.5$ s is 15.0. Thus we see that an attempt to increase the speed of response of the system results in larger signals at the plant input. These larger signals may force the system into a nonlinear region of operation, which in some cases may be undesirable.

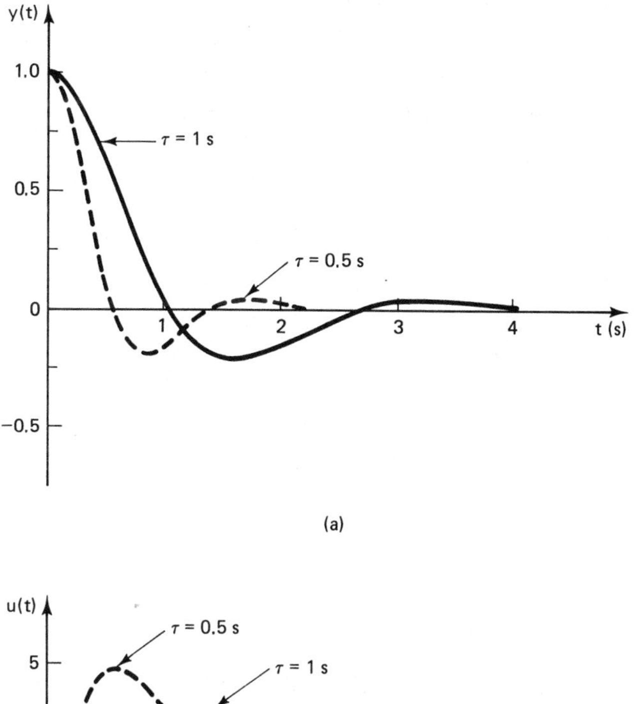

(a)

(b)

Figure 9-2 Initial condition responses for the systems of Example 9.1.

In the procedure above the gain matrix **K** was calculated by equating coefficients in the characteristic equation of (9-15). This calculation is greatly simplified if the state model of the plant is in the control canonical form, which is the form

$$\mathbf{x}(k+1) = \begin{bmatrix} 0 & 1 & 0 & \cdots & 0 \\ 0 & 0 & 1 & \cdots & 0 \\ & & & \vdots & \\ & & & \vdots & \\ -b_0 & -b_1 & -b_2 & \cdots & -b_{n-1} \end{bmatrix} \mathbf{x}(k) + \begin{bmatrix} 0 \\ 0 \\ \vdots \\ \vdots \\ 1 \end{bmatrix} u(k) \qquad (9\text{-}16)$$

This plant has the characteristic equation (see Section 2.8)

$$\alpha(z) = |z\mathbf{I} - \mathbf{A}| = z^n + b_{n-1}z^{n-1} + \cdots + b_1 z + b_0 = 0 \qquad (9\text{-}17)$$

For this state model, in (9-15) **BK** is equal to

$$\mathbf{BK} = \begin{bmatrix} 0 \\ 0 \\ \cdot \\ \cdot \\ \cdot \\ 1 \end{bmatrix} [K_1 \quad K_2 \quad \cdots \quad K_n] = \begin{bmatrix} 0 & 0 & \cdots & 0 \\ 0 & 0 & \cdots & 0 \\ & & \vdots & \\ K_1 & K_2 & \cdots & K_n \end{bmatrix} \tag{9-18}$$

Hence the closed-loop system matrix is

$$\mathbf{A} - \mathbf{BK} = \begin{bmatrix} 0 & 1 & \cdots & 0 \\ 0 & 0 & \cdots & 0 \\ & & \vdots & \\ -(b_0 + K_1) & -(b_1 + K_2) & \cdots & -(b_{n-1} + K_n) \end{bmatrix} \tag{9-19}$$

and the characteristic equation becomes

$$\begin{aligned} |z\mathbf{I} - \mathbf{A} + \mathbf{BK}| = z^n + (b_{n-1} + K_n)z^{n-1} + \cdots + (b_1 + K_2)z \\ + (b_0 + K_1) = 0 \end{aligned} \tag{9-20}$$

If we write the desired characteristic equation as

$$\alpha_c(z) = z^n + \alpha_{n-1}z^{n-1} + \cdots + \alpha_1 z + \alpha_0 = 0 \tag{9-21}$$

we can calculate the required gains by equating coefficients in (9-20) and (9-21).

$$K_{i+1} = \alpha_i - b_i, \qquad i = 0, 1, \ldots, n-1 \tag{9-22}$$

In general, the techniques for developing state models for plants do not result in the control canonical form (see Section 4.10). A more practical procedure for calculating the gain matrix **K** is the use of Ackermann's formula [1]. The proof of Ackermann's formula is given in Section 9.5, and is based on transformations from a general system matrix **A** to that of the control canonical form. Ackermann's formula will now be given.

We begin with the plant model

$$\mathbf{x}(k+1) = \mathbf{Ax}(k) + \mathbf{B}u(k) \tag{9-23}$$

The matrix polynomial $\alpha_c(\mathbf{A})$ is formed using the coefficients of the desired characteristic equation (9-21).

$$\alpha_c(\mathbf{A}) = \mathbf{A}^n + \alpha_{n-1}\mathbf{A}^{n-1} + \cdots + \alpha_1\mathbf{A} + \alpha_0\mathbf{I} \tag{9-24}$$

Then Ackermann's formula for the gain matrix **K** is given by

$$\mathbf{K} = [0 \quad 0 \quad \cdots \quad 0 \quad 1][\mathbf{B} \quad \mathbf{AB} \quad \cdots \quad \mathbf{A}^{n-2}\mathbf{B} \quad \mathbf{A}^{n-1}\mathbf{B}]^{-1}\alpha_c(\mathbf{A}) \tag{9-25}$$

The problems with the existence of the inverse matrix in (9-25) are discussed in Section 9.7. Example 9.1 will now be solved using Ackermann's formula.

Example 9.2

We will solve the design in Example 9.1 via Ackermann's formula. The plant model is given by

$$\mathbf{x}(k+1) = \mathbf{Ax}(k) + \mathbf{B}u(k) = \begin{bmatrix} 1 & 0.0952 \\ 0 & 0.905 \end{bmatrix}\mathbf{x}(k) + \begin{bmatrix} 0.00484 \\ 0.0952 \end{bmatrix}u(k)$$

and the desired characteristic equation is

$$\alpha_c(z) = z^2 - 1.776z + 0.819$$

Hence

$$\alpha_c(\mathbf{A}) = \begin{bmatrix} 1 & 0.0952 \\ 0 & 0.905 \end{bmatrix}^2 - 1.776\begin{bmatrix} 1 & 0.0952 \\ 0 & 0.905 \end{bmatrix} + 0.819\begin{bmatrix} 1 & 0 \\ 0 & 1 \end{bmatrix}$$

or

$$\alpha_c(\mathbf{A}) = \begin{bmatrix} 0.043 & 0.01228 \\ 0 & 0.03075 \end{bmatrix}$$

Also,

$$\mathbf{AB} = \begin{bmatrix} 1 & 0.0952 \\ 0 & 0.905 \end{bmatrix}\begin{bmatrix} 0.00484 \\ 0.0952 \end{bmatrix} = \begin{bmatrix} 0.0139 \\ 0.0862 \end{bmatrix}$$

Thus

$$[\mathbf{B}\ \mathbf{AB}]^{-1} = \begin{bmatrix} 0.00484 & 0.0139 \\ 0.0952 & 0.0862 \end{bmatrix}^{-1} = \begin{bmatrix} -95.13 & 15.34 \\ 105.1 & -5.342 \end{bmatrix}$$

Then the gain matrix is, from (9-25),

$$\mathbf{K} = [0 \quad 1]\begin{bmatrix} -95.13 & 15.34 \\ 105.1 & -5.342 \end{bmatrix}\begin{bmatrix} 0.043 & 0.01228 \\ 0 & 0.03075 \end{bmatrix}$$
$$= [4.52 \quad 1.12]$$

These results are the same as those obtained in Example 9.1.

9.3 STATE ESTIMATION

In the preceding section a design technique was developed which requires that all the plant states be measured. In general, the measurement of all the plant states is impractical, if not impossible, in all but the simplest of systems. In this section we develop a technique for estimating the states of a plant from the information that is available concerning the plant. The system that estimates the states of another system is generally called an *observer* [2], or a *state estimator*. Thus in this section we consider the design of observers.

Suppose that the plant is described by the following equations:

$$\begin{aligned} \mathbf{x}(k+1) &= \mathbf{Ax}(k) + \mathbf{Bu}(k) \\ \mathbf{y}(k) &= \mathbf{Cx}(k) \end{aligned} \tag{9-26}$$

where $\mathbf{y}(k)$ are the plant signals that will be measured. Hence, in (9-26), we know the matrices $\mathbf{A}$, $\mathbf{B}$, and $\mathbf{C}$, and the signals $\mathbf{y}(k)$ and $\mathbf{u}(k)$. The control inputs $\mathbf{u}(k)$ are known since we generate them. The problem of observer design can be depicted as shown in Figure 9-3. The states of the system to be observed are $\mathbf{x}(k)$; the states of the observer

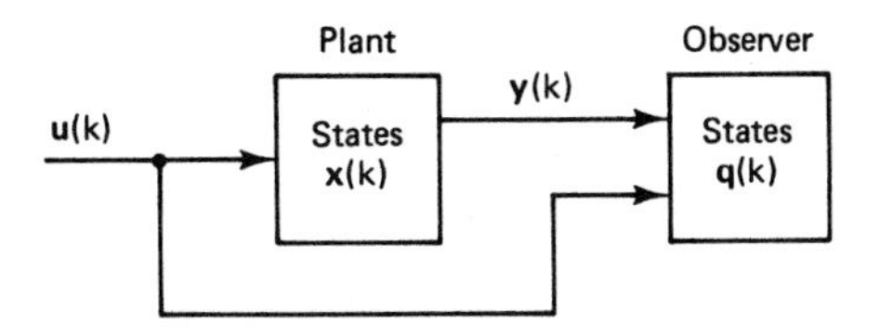

Figure 9-3 System illustrating state estimation.

are $\mathbf{q}(k)$ and we desire that $\mathbf{q}(k)$ be approximately equal to $\mathbf{x}(k)$. Since the observer will be implemented on a computer, the signals $\mathbf{q}(k)$ are then available for feedback calculations.

Equations describing observers can be developed in several different ways; we choose here to take a transfer-function approach. The observer design criterion to be used is that the transfer-function matrix from the input $\mathbf{u}(k)$ to the observer state $q_i(k)$ be equal to that from the input $\mathbf{u}(k)$ to the system state $x_i(k)$, for $i = 1, 2, \ldots, n$. Equations describing the observer of Figure 9-3 will now be developed, based on this criterion.

First we solve (9-26) for $\mathbf{X}(z)$:

$$z\mathbf{X}(z) = \mathbf{AX}(z) + \mathbf{BU}(z) \tag{9-27}$$

or

$$\mathbf{X}(z) = (z\mathbf{I} - \mathbf{A})^{-1}\mathbf{BU}(z) \tag{9-28}$$

Now, since the observer has two inputs, $\mathbf{y}(k)$ and $\mathbf{u}(k)$, we write the observer state equations as

$$\mathbf{q}(k + 1) = \mathbf{Fq}(k) + \mathbf{Gy}(k) + \mathbf{Hu}(k) \tag{9-29}$$

Solving this equation for $\mathbf{Q}(z)$ yields

$$\mathbf{Q}(z) = (z\mathbf{I} - \mathbf{F})^{-1}[\mathbf{GY}(z) + \mathbf{HU}(z)] \tag{9-30}$$

From (9-26),

$$\mathbf{Y}(z) = \mathbf{CX}(z) \tag{9-31}$$

Next we substitute (9-31) and (9-28) in (9-30):

$$\begin{aligned} \mathbf{Q}(z) &= (z\mathbf{I} - \mathbf{F})^{-1}[\mathbf{GCX}(z) + \mathbf{HU}(z)] \\ &= (z\mathbf{I} - \mathbf{F})^{-1}[\mathbf{GC}(z\mathbf{I} - \mathbf{A})^{-1}\mathbf{B} + \mathbf{H}]\mathbf{U}(z) \end{aligned} \tag{9-32}$$

Recall that the design criterion is that the transfer-function matrix from $\mathbf{U}(z)$ to $\mathbf{Q}(z)$ be the same as that from $\mathbf{U}(z)$ to $\mathbf{X}(z)$. Thus, from (9-28),

$$\mathbf{Q}(z) = (z\mathbf{I} - \mathbf{A})^{-1}\mathbf{BU}(z) \tag{9-33}$$

must be satisfied. Then, from (9-32) and (9-33),

$$(z\mathbf{I} - \mathbf{A})^{-1}\mathbf{B} = (z\mathbf{I} - \mathbf{F})^{-1}\mathbf{GC}(z\mathbf{I} - \mathbf{A})^{-1}\mathbf{B} + (z\mathbf{I} - \mathbf{F})^{-1}\mathbf{H} \tag{9-34}$$

or

$$[\mathbf{I} - (z\mathbf{I} - \mathbf{F})^{-1}\mathbf{GC}](z\mathbf{I} - \mathbf{A})^{-1}\mathbf{B} = (z\mathbf{I} - \mathbf{F})^{-1}\mathbf{H} \tag{9-35}$$

This equation can be expressed as

$$(z\mathbf{I} - \mathbf{F})^{-1}[z\mathbf{I} - (\mathbf{F} + \mathbf{GC})](z\mathbf{I} - \mathbf{A})^{-1}\mathbf{B} = (z\mathbf{I} - \mathbf{F})^{-1}\mathbf{H} \tag{9-36}$$

or

$$(z\mathbf{I} - \mathbf{A})^{-1}\mathbf{B} = [z\mathbf{I} - (\mathbf{F} + \mathbf{GC})]^{-1}\mathbf{H} \tag{9-37}$$

Hence we choose $\mathbf{H}$ equal to $\mathbf{B}$, and

$$\mathbf{A} = \mathbf{F} + \mathbf{GC} \tag{9-38}$$

Therefore, from (9-29) and (9-38), we write the observer state equations as

$$\mathbf{q}(k + 1) = (\mathbf{A} - \mathbf{GC})\mathbf{q}(k) + \mathbf{Gy}(k) + \mathbf{Bu}(k) \tag{9-39}$$

and the design criterion (9-33) is satisfied. Note that $\mathbf{G}$ is unspecified.

We now consider the errors in the state-estimation process. Define the error vector $\mathbf{e}(k)$ as

$$\mathbf{e}(k) = \mathbf{x}(k) - \mathbf{q}(k) \tag{9-40}$$

Then, from (9-40), (9-26), and (9-36),

$$\begin{aligned}\mathbf{e}(k+1) &= \mathbf{x}(k+1) - \mathbf{q}(k+1) \\ &= \mathbf{Ax}(k) + \mathbf{Bu}(k) - (\mathbf{A} - \mathbf{GC})\mathbf{q}(k) - \mathbf{GCx}(k) - \mathbf{Bu}(k)\end{aligned} \tag{9-41}$$

or

$$\mathbf{e}(k+1) = (\mathbf{A} - \mathbf{GC})[\mathbf{x}(k) - \mathbf{q}(k)] = (\mathbf{A} - \mathbf{GC})\mathbf{e}(k) \tag{9-42}$$

Hence the error dynamics have a characteristic equation given by

$$|z\mathbf{I} - (\mathbf{A} - \mathbf{GC})| = 0 \tag{9-43}$$

which is the same as that of the observer [see (9-39)].

We look next at the sources of the errors in state estimation. First, we have assumed that we have an exact model of the physical system involved. For example, in (9-41) we assumed that the $\mathbf{A}$ matrix in the plant's state equations is identical to that in the observer's equations. In a practical case these two matrices are not identical. However, more effort expended in obtaining an accurate plant model will lead to improved state estimation.

A second source of errors in our state estimation is in the choice of initial conditions for the observer. Since the initial states of the plant will not be known, the initial error $\mathbf{e}(0)$ will not be zero. However, provided that the observer is stable, this error will tend to zero with increasing time, with the dynamics given by (9-43).

The third source of errors is the plant disturbances and the sensor errors. The complete state equations of the plant, (9-26), become

$$\begin{aligned}\mathbf{x}(k+1) &= \mathbf{Ax}(k) + \mathbf{Bu}(k) + \mathbf{B}_1\mathbf{w}(k) \\ \mathbf{y}(k) &= \mathbf{Cx}(k) + \mathbf{v}(k)\end{aligned} \tag{9-44}$$

when the plant disturbance vector $\mathbf{w}(k)$ and the measurement inaccuracies $\mathbf{v}(k)$ are considered. Hence, if (9-44) is used in deriving the error state equations (9-42), we obtain

$$\mathbf{e}(k+1) = (\mathbf{A} - \mathbf{GC})\mathbf{e}(k) + \mathbf{B}_1\mathbf{w}(k) - \mathbf{Gv}(k) \tag{9-45}$$

For the practical case, the errors in the state estimation will tend to zero if the estimator is stable (we will certainly require this); however, the excitation terms in (9-45) will prevent the errors from reaching a value of zero.

Consider now the observer state equations

$$\mathbf{q}(k+1) = (\mathbf{A} - \mathbf{GC})\mathbf{q}(k) + \mathbf{Gy}(k) + \mathbf{Bu}(k) \tag{9-39}$$

All matrices in this equation are determined by the plant equations (9-26) except $\mathbf{G}$; however, the observer design criterion (9-33) is satisfied independent of the choice of $\mathbf{G}$. Note that $\mathbf{G}$ determines the error dynamics in (9-43). We generally use this fact in choosing $\mathbf{G}$. A characteristic equation $\alpha_e(z)$ is chosen for the error dynamics. Note from (9-39) and (9-43) that this is also the characteristic equation of the observer. Then

$$\alpha_e(z) = |z\mathbf{I} - (\mathbf{A} - \mathbf{GC})| = z^n + \alpha_{n-1}z^{n-1} + \cdots + \alpha_1 z + \alpha_0 \tag{9-46}$$

For the case of a single-output system [i.e., $\mathbf{y}(k)$ is a scalar],

$$\mathbf{G} = \begin{bmatrix} g_1 \\ g_2 \\ \cdot \\ \cdot \\ \cdot \\ g_n \end{bmatrix} \tag{9-47}$$

and (9-46) yields n equations by equating coefficients. Hence the solution of these equations will yield the **G** matrix of (9-47).

For this case Ackermann's equation may also be employed. First we compare (9-46) to (9-15).

$$\alpha_c(z) = |z\mathbf{I} - (\mathbf{A} - \mathbf{BK})| \tag{9-15}$$

Ackermann's equation for the solution of this equation is given in (9-25).

$$\mathbf{K} = [0 \quad 0 \quad \cdots \quad 0 \quad 1][\mathbf{B} \quad \mathbf{AB} \quad \cdots \quad \mathbf{A}^{n-2}\mathbf{B} \quad \mathbf{A}^{n-1}\mathbf{B}]^{-1}\alpha_c(\mathbf{A}) \tag{9-25}$$

Thus Ackermann's equation for **G** in (9-46) is seen to be

$$\mathbf{G} = \alpha_e(\mathbf{A}) \begin{bmatrix} \mathbf{C} \\ \mathbf{CA} \\ \cdot \\ \cdot \\ \cdot \\ \mathbf{CA}^{n-1} \end{bmatrix}^{-1} \begin{bmatrix} 0 \\ 0 \\ \cdot \\ \cdot \\ \cdot \\ 1 \end{bmatrix} \tag{9-48}$$

(The derivation of this result is given as Problem 9-2.)

Thus we see that we can design the observer, once we decide on an appropriate characteristic equation in (9-46). An approach usually suggested for choosing $\alpha_e(z)$ is to make the observer two to four times faster than the closed-loop control system. The fastest time constant of the closed-loop system, determined from the system characteristic equation,

$$|z\mathbf{I} - (\mathbf{A} - \mathbf{BK})| = 0 \tag{9-49}$$

is calculated first. The time constants of the observer are then set at a value equal to from one-fourth to one-half this fastest time constant.

The choice of $\alpha_e(z)$ can be approached from a different viewpoint. The observer state equation, from (9-39), can be written as

$$\mathbf{q}(k+1) = \mathbf{Aq}(k) + \mathbf{G}[\mathbf{y}(k) - \mathbf{Cq}(k)] + \mathbf{Bu}(k) \tag{9-50}$$

Thus the plant–observer system can be modeled as in Figure 9-4.

The choice of **G** can be viewed in the following manner. In Figure 9-4, if the output of the observer plant model and of the plant itself are approximately equal, there is little effect from the feedback through **G**. Hence $\mathbf{q}(k)$ is determined principally from $\mathbf{u}(k)$. However, if the effects of disturbances on the plant cause $\mathbf{q}(k)$ to differ significantly from $\mathbf{x}(k)$, the effect of **G** is much more important. Here the measurement of $\mathbf{y}(k)$ is more important in determining $\mathbf{q}(k)$ than in the first case. Thus we can view the choice of **G** as relating to the relative importance that we attach to the effects of $\mathbf{u}(k)$ and to the effects of the disturbances on $\mathbf{x}(k)$. And, of course, we will have the

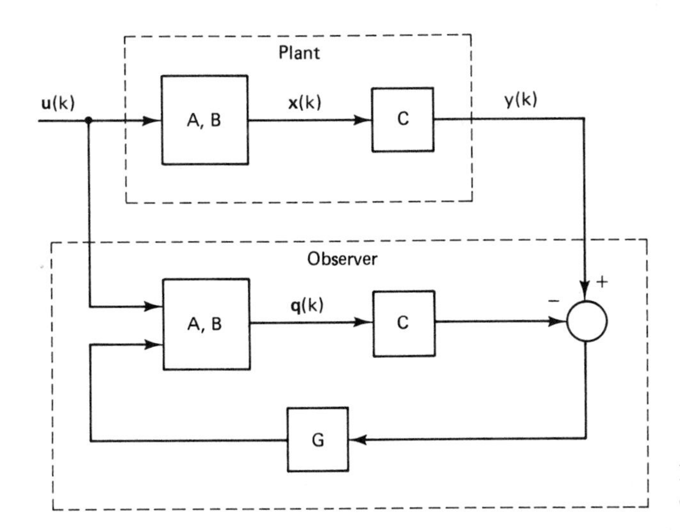

Figure 9-4 Different view of state estimation.

effects of measurement noise (inaccuracies) on $\mathbf{y}(k)$. If these effects are great, the observer design cannot allow $\mathbf{y}(k)$ to have a large weight in determining $\mathbf{q}(k)$.

We will now summarize the paragraph above. We can view **G** as furnishing a correction input to the plant model in Figure 9-4, to account for unknowns in the plant. If the unknowns are significant, **G** should be relatively large. However, if measurement noise in $\mathbf{y}(k)$ is significant, we cannot rely heavily on the measurement $\mathbf{y}(k)$, and **G** should be relatively small. Hence, for practical systems, perhaps the best method for choosing **G** is through the use of an accurate simulation of the system of Figure 9-4 that includes both the disturbances and the measurement noise. Simulations should be run for different choices of **G**, with the final choice of **G** resulting from the best system response.

An example of the design of an observer will now be given.

Example 9.3

We will design an observer for the system of Example 9.1, which has the plant state equations

$$\mathbf{x}(k+1) = \begin{bmatrix} 1 & 0.0952 \\ 0 & 0.905 \end{bmatrix}\mathbf{x}(k) + \begin{bmatrix} 0.00484 \\ 0.0952 \end{bmatrix}u(k)$$

$$y(k) = [1 \quad 0]\mathbf{x}(k)$$

With the gain matrix

$$\mathbf{K} = [4.52 \quad 1.12]$$

the closed-loop system characteristic equation is given by

$$\alpha_c(z) = z^2 - 1.776z + 0.819 = 0$$

The time constant of the roots of this equation, from Example 9.1, is 1.0 s. Thus we will choose the time constant of the observer to be 0.5 s. We also choose the observer roots to be real. Hence the root locations are

$$z = \epsilon^{-T/\tau} = \epsilon^{-0.1/0.5} = 0.819$$

The observer characteristic equation is then

$$\alpha_e(z) = (z - 0.819)^2 = z^2 - 1.638z + 0.671 = 0$$

The matrix $\mathbf{G}$ is given by (9-48):

$$\mathbf{G} = \alpha_e(\mathbf{A})\begin{bmatrix}\mathbf{C}\\ \mathbf{CA}\end{bmatrix}^{-1}\begin{bmatrix}0\\1\end{bmatrix}$$

Now

$$\alpha_e(\mathbf{A}) = \begin{bmatrix}1 & 0.0952\\ 0 & 0.905\end{bmatrix}^2 - 1.638\begin{bmatrix}1 & 0.0952\\ 0 & 0.905\end{bmatrix} + 0.671\begin{bmatrix}1 & 0\\ 0 & 1\end{bmatrix}$$

$$= \begin{bmatrix}0.033 & 0.0254\\ 0 & 0.00763\end{bmatrix}$$

$$\begin{bmatrix}\mathbf{C}\\ \mathbf{CA}\end{bmatrix}^{-1} = \begin{bmatrix}1 & 0\\ 1 & 0.0952\end{bmatrix}^{-1} = \begin{bmatrix}1 & 0\\ -10.51 & 10.51\end{bmatrix}$$

Then

$$\mathbf{G} = \begin{bmatrix}0.033 & 0.0254\\ 0 & 0.00763\end{bmatrix}\begin{bmatrix}1 & 0\\ -10.51 & 10.51\end{bmatrix}\begin{bmatrix}0\\1\end{bmatrix}$$

$$= \begin{bmatrix}0.267\\ 0.0802\end{bmatrix}$$

The system matrix of the estimator is given by

$$\mathbf{F} = \mathbf{A} - \mathbf{GC} = \begin{bmatrix}0.733 & 0.0952\\ -0.0802 & 0.905\end{bmatrix}$$

From (9-39), the estimator's state equations are

$$\mathbf{q}(k+1) = \begin{bmatrix}0.733 & 0.0952\\ -0.0802 & 0.905\end{bmatrix}\mathbf{q}(k) + \begin{bmatrix}0.267\\ 0.0802\end{bmatrix}y(k) + \begin{bmatrix}0.00484\\ 0.0952\end{bmatrix}u(k)$$

But

$$u(k) = -\mathbf{Kq}(k) = -4.52q_1(k) - 1.12q_2(k)$$

Then the estimator's state equations become

$$\mathbf{q}(k+1) = \begin{bmatrix}0.711 & 0.0898\\ -0.510 & 0.798\end{bmatrix}\mathbf{q}(k) + \begin{bmatrix}0.267\\ 0.0802\end{bmatrix}y(k)$$

Initial-condition responses for the observer-based control system are given in Figure 9-5. Two responses are given: one with

$$\mathbf{q}(0) = \mathbf{x}(0) = \begin{bmatrix}1\\0\end{bmatrix}$$

and the other with

$$\mathbf{q}(0) = \begin{bmatrix}0\\0\end{bmatrix}, \quad \mathbf{x}(0) = \begin{bmatrix}1\\0\end{bmatrix}$$

Thus we can see an effect of not knowing the initial state of the system. It should be noted that, with $\mathbf{q}(0)$ and $\mathbf{x}(0)$ equal, the response of the observer-based control system is identical to that of the full-state feedback system of Example 9.1. Shown in Figure 9-6 are the initial-condition responses of the observer-based control system and the full-state feedback system with a unit-step disturbance. The disturbance enters as given in (9-44), with

$$\mathbf{B}_1 = \begin{bmatrix}0\\1\end{bmatrix}$$

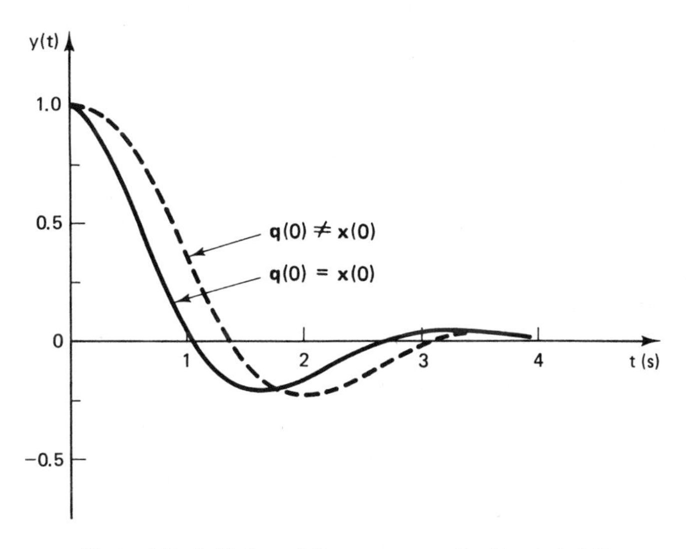

Figure 9-5 Initial conditions responses for Example 9.3.

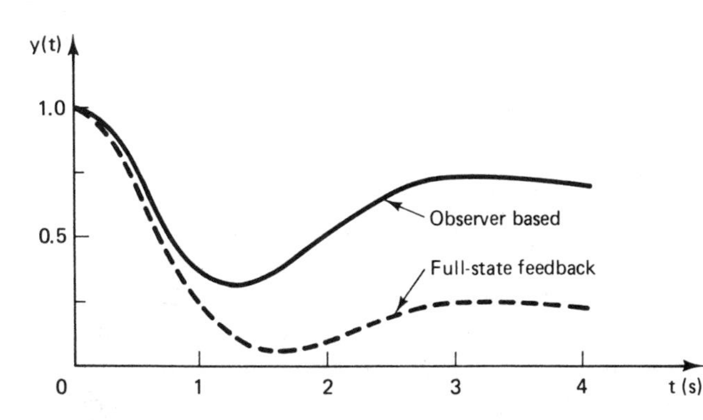

Figure 9-6 Disturbance responses for Example 9.3.

Since the observer does not take disturbances into account, the response of the observer-based control system differs significantly from that of the full-state feedback system.

As a final point, we must investigate the effects on the closed-loop system characteristic equation of the addition of the observer. For full-state feedback, the characteristic equation is given by

$$\alpha_c(z) = |z\mathbf{I} - \mathbf{A} + \mathbf{BK}| = 0 \tag{9-51}$$

We will now derive the system characteristic equation for the observer-based control system.

To derive this equation, we will use the error variables of (9-40).

$$\mathbf{e}(k) = \mathbf{x}(k) - \mathbf{q}(k) \tag{9-40}$$

The plant state equations of (9-26) can then be expressed as

$$\mathbf{x}(k+1) = \mathbf{Ax}(k) - \mathbf{BKq}(k) = (\mathbf{A} - \mathbf{BK})\mathbf{x}(k) + \mathbf{BKe}(k) \tag{9-52}$$

The state equations for the error variables are

$$\mathbf{e}(k+1) = (\mathbf{A} - \mathbf{GC})\mathbf{e}(k) \tag{9-42}$$

We can adjoin the variables of (9-52) and (9-42) into a single state vector, with the resulting equations

$$\begin{bmatrix} \mathbf{x}(k+1) \\ \mathbf{e}(k+1) \end{bmatrix} = \begin{bmatrix} \mathbf{A} - \mathbf{BK} & \mathbf{BK} \\ 0 & \mathbf{A} - \mathbf{GC} \end{bmatrix} \begin{bmatrix} \mathbf{x}(k) \\ \mathbf{e}(k) \end{bmatrix} \tag{9-53}$$

with the states of the observer a linear combination of $\mathbf{x}(k)$ and $\mathbf{e}(k)$, given in (9-40). Thus the characteristic equation of the state equations of (9-53) is also the closed-loop system characteristic equation. This equation is seen to be

$$|z\mathbf{I} - \mathbf{A} + \mathbf{BK}||z\mathbf{I} - \mathbf{A} + \mathbf{GC}| = \alpha_c(z)\alpha_e(z) = 0 \tag{9-54}$$

We see then that the roots of the characteristic equation of the closed-loop system are the roots obtained by the pole-placement design plus those of the observer. Hence the pole-placement design is independent of the observer design.

This section has presented only a brief introduction to observers. The interested reader is referred to Ref. 3, which contains a bibliography of current papers on state estimation. Different approaches to find solutions to some of the shortcomings of observers mentioned above will be found in this literature.

9.4 REDUCED-ORDER OBSERVERS

In the example in the preceding section, we estimated $x_1(k)$ [position] and $x_2(k)$ [velocity], given the measurement of position. However, if an accurate measurement of position is available, it is not reasonable to try to estimate position; we already know it. In general, if accurate measurements of certain states are made, one needs to estimate only the remaining states, and the accurately measured signals are then used directly for feedback. The resulting observer is called a reduced-order observer. However, if the measurements are relatively inaccurate (noisy), the full-order observer may yield better results.

To develop the design equations for the reduced-order observer, we first partition the state vector as

$$\mathbf{x}(k) = \begin{bmatrix} \mathbf{x}_a(k) \\ \mathbf{x}_b(k) \end{bmatrix} \tag{9-55}$$

where $\mathbf{x}_a(k)$ are the states to be measured and $\mathbf{x}_b(k)$ are the states to be estimated. Then the plant state equation of (9-26) can be partitioned as

$$\begin{bmatrix} \mathbf{x}_a(k+1) \\ \mathbf{x}_b(k+1) \end{bmatrix} = \begin{bmatrix} \mathbf{A}_{aa} & \mathbf{A}_{ab} \\ \mathbf{A}_{ba} & \mathbf{A}_{bb} \end{bmatrix} \begin{bmatrix} \mathbf{x}_a(k) \\ \mathbf{x}_b(k) \end{bmatrix} + \begin{bmatrix} \mathbf{B}_a \\ \mathbf{B}_b \end{bmatrix} \mathbf{u}(k)$$

$$\mathbf{y}(k) = [\mathbf{I} \quad \mathbf{0}] \begin{bmatrix} \mathbf{x}_a(k) \\ \mathbf{x}_b(k) \end{bmatrix} \tag{9-56}$$

Note that in this case we are considering both multiple inputs and multiple outputs. The equations for the measured states can be written as

$$\mathbf{x}_a(k+1) = \mathbf{A}_{aa}\mathbf{x}_a(k) + \mathbf{A}_{ab}\mathbf{x}_b(k) + \mathbf{B}_a\mathbf{u}(k) \tag{9-57}$$

Collecting all the known terms on the left side of the equation, we write

$$\mathbf{x}_a(k+1) - \mathbf{A}_{aa}\mathbf{x}_a(k) - \mathbf{B}_a\mathbf{u}(k) = \mathbf{A}_{ab}\mathbf{x}_b(k) \tag{9-58}$$

For the reduced-order observer we consider the left side to be the "known measurements." From (9-56), the equations for the estimated states are

$$\mathbf{x}_b(k+1) = \mathbf{A}_{ba}\mathbf{x}_a(k) + \mathbf{A}_{bb}\mathbf{x}_b(k) + \mathbf{B}_b\mathbf{u}(k) \tag{9-59}$$

The term

$$\mathbf{A}_{ba}\mathbf{x}_a(k) + \mathbf{B}_b\mathbf{u}(k)$$

is then considered to be the "known inputs." We now compare the state equations for the full-order observer to those for the reduced-order observer.

$$\mathbf{x}(k+1) = \mathbf{A}\mathbf{x}(k) + \mathbf{B}\mathbf{u}(k) \tag{9-26}$$

$$\mathbf{x}_b(k+1) = \mathbf{A}_{bb}\mathbf{x}_b(k) + [\mathbf{A}_{ba}\mathbf{x}_a(k) + \mathbf{B}_b\mathbf{u}(k)] \tag{9-59}$$

and

$$\mathbf{y}(k) = \mathbf{C}\mathbf{x}(k) \tag{9-26}$$

$$\mathbf{x}_a(k+1) - \mathbf{A}_{aa}\mathbf{x}_a(k) - \mathbf{B}_a\mathbf{u}(k) = \mathbf{A}_{ab}\mathbf{x}_b(k) \tag{9-58}$$

We then obtain the reduced-order observer equations by making the following substitutions into the full-order observer equations, (9-39):

$$\begin{aligned}
\mathbf{x}(k) &\longleftarrow \mathbf{x}_b(k)\\
\mathbf{A} &\longleftarrow \mathbf{A}_{bb}\\
\mathbf{B}\mathbf{u}(k) &\longleftarrow \mathbf{A}_{ba}\mathbf{x}_a(k) + \mathbf{B}_b\mathbf{u}(k)\\
\mathbf{y}(k) &\longleftarrow \mathbf{x}_a(k+1) - \mathbf{A}_{aa}\mathbf{x}_a(k) - \mathbf{B}_a\mathbf{u}(k)\\
\mathbf{C} &\longleftarrow \mathbf{A}_{ab}
\end{aligned}$$

If we make these substitutions into (9-39),

$$\mathbf{q}(k+1) = (\mathbf{A} - \mathbf{G}\mathbf{C})\mathbf{q}(k) + \mathbf{G}\mathbf{y}(k) + \mathbf{B}\mathbf{u}(k) \tag{9-39}$$

we obtain the equations

$$\begin{aligned}
\mathbf{q}_b(k+1) = (\mathbf{A}_{bb} - \mathbf{G}\mathbf{A}_{ab})\mathbf{q}_b(k) &+ \mathbf{G}[\mathbf{x}_a(k+1) - \mathbf{A}_{aa}\mathbf{x}_a(k)\\
&- \mathbf{B}_a\mathbf{u}(k)] + \mathbf{A}_{ba}\mathbf{x}_a(k) + \mathbf{B}_b\mathbf{u}(k)
\end{aligned} \tag{9-60}$$

From (9-56),

$$\mathbf{y}(k) = \mathbf{x}_a(k) \tag{9-61}$$

Then (9-60) can be written as

$$\begin{aligned}
\mathbf{q}_b(k+1) = (\mathbf{A}_{bb} - \mathbf{G}\mathbf{A}_{ab})\mathbf{q}_b(k) &+ \mathbf{G}\mathbf{y}(k+1) + (\mathbf{A}_{ba} - \mathbf{G}\mathbf{A}_{aa})\mathbf{y}(k)\\
&+ (\mathbf{B}_b - \mathbf{G}\mathbf{B}_a)\mathbf{u}(k)
\end{aligned} \tag{9-62}$$

The following points should be made about the reduced-order observer. The observer characteristic equation is

$$\alpha_e(z) = |z\mathbf{I} - \mathbf{A}_{bb} + \mathbf{G}\mathbf{A}_{ab}| = 0 \tag{9-63}$$

and $\mathbf{G}$ is determined from (9-63) in the same manner that the $\mathbf{G}$ matrix is obtained from (9-46) for the full-order observer. For the case of a single measurement [i.e., $y(k)$ is

$x_1(k)]$ Ackermann's formula is given by

$$\mathbf{G} = \alpha_e(\mathbf{A}_{bb}) \begin{bmatrix} \mathbf{A}_{ab} \\ \mathbf{A}_{ab}\mathbf{A}_{bb} \\ \cdot \\ \cdot \\ \cdot \\ \mathbf{A}_{ab}\mathbf{A}_{bb}^{n-2} \end{bmatrix}^{-1} \begin{bmatrix} 0 \\ 0 \\ \cdot \\ \cdot \\ \cdot \\ 1 \end{bmatrix} \tag{9-64}$$

Note also that, in (9-62), the measurements $\mathbf{y}(k+1)$ is required to estimate $\mathbf{q}_b(k+1)$. However, for the full-order observer, in (9-39), $\mathbf{q}(k+1)$ is estimated using only the measurements $\mathbf{y}(k)$.

An example of the reduced-order estimator will now be given.

Example 9.4

We will again consider the design of the system of Example 9.1. The plant model is given by

$$\mathbf{x}(k+1) = \begin{bmatrix} 1 & 0.0952 \\ 0 & 0.905 \end{bmatrix} \mathbf{x}(k) + \begin{bmatrix} 0.00484 \\ 0.0952 \end{bmatrix} u(k)$$

$$y(k) = [1 \quad 0]\mathbf{x}(k)$$

Thus we are measuring position, $x_1(k)$, and will estimate velocity, $x_2(k)$. The closed-loop system characteristic equation was chosen to be

$$\alpha_c(z) = z^2 - 1.776z + 0.819 = 0$$

In Example 9.3 we chose the estimator characteristic-equation roots be at $z = 0.819$; we will make the same choice here. However, the reduced-order observer is first order; hence

$$\alpha_e(z) = z - 0.819 = 0$$

From the plant state equations above, and (9-56), the partitioned matrices are seen to be

$$\mathbf{A}_{aa} = 1 \qquad \mathbf{A}_{ab} = 0.0952 \qquad \mathbf{B}_a = 0.00484$$

$$\mathbf{A}_{ba} = 0 \qquad \mathbf{A}_{bb} = 0.905 \qquad \mathbf{B}_b = 0.0952$$

Thus we have all the terms required for Ackermann's formula in (9-64).

$$\begin{aligned} \mathbf{G} &= \alpha_e(\mathbf{A}_{bb})[\mathbf{A}_{ab}]^{-1}[1] = [0.905 - 0.819][0.0952]^{-1}[1] \\ &= 0.903 \end{aligned}$$

The observer equation is given by (9-62):

$$\begin{aligned} q(k+1) = {} & [0.905 - (0.903)(0.0952)]q(k) + 0.903y(k+1) \\ & + [0 - (0.903)(1)]y(k) + [0.0952 - (0.903)(0.00484)]u(k) \end{aligned}$$

or

$$q(k+1) = 0.819q(k) + 0.903y(k+1) - 0.903y(k) + 0.0908u(k)$$

Here $q(k)$ is the estimate of velocity, $x_2(k)$. Since we have considered the measurement $y(k)$ to be the measurement at the present time, the implementation of the observer is more obvious and we write the observer equation above as

$$q(k) = 0.819q(k-1) + 0.903y(k) - 0.903y(k-1) + 0.0908u(k-1)$$

and $q(k)$ is the estimate at the present time. From Example 9.1, the control law is given by

$$u(k) = -4.52x_1(k) - 1.12x_2(k)$$

which is implemented as

$$u(k) = -4.52y(k) - 1.12q(k)$$

Hence we can write the observer equation as

$$q(k+1) = 0.819q(k) + 0.903y(k+1) - 0.903y(k) + 0.0908[-4.52y(k) - 1.12q(k)]$$

or

$$q(k+1) = 0.717q(k) + 0.903y(k+1) - 1.313y(k)$$

The control system is then implemented as follows. A measurement $y(k)$ is made at $t = kT$. The observer state is calculated from

$$q(k) = 0.717q(k-1) + 0.903y(k) - 1.313y(k-1)$$

Then the control input is calculated, using

$$u(k) = -4.52y(k) - 1.12q(k)$$

The initial-condition response, obtained by simulation, is approximately the same as that obtained using full-state feedback. In addition, the effects of the disturbance input are less than those for the full-order observer system.

For the case of the single-input, single-output plant, the control system can be represented in block-diagram form as shown in Figure 9-7. This block diagram applies to both the full-order and the reduced-order observer. From Figure 9-7 we see that

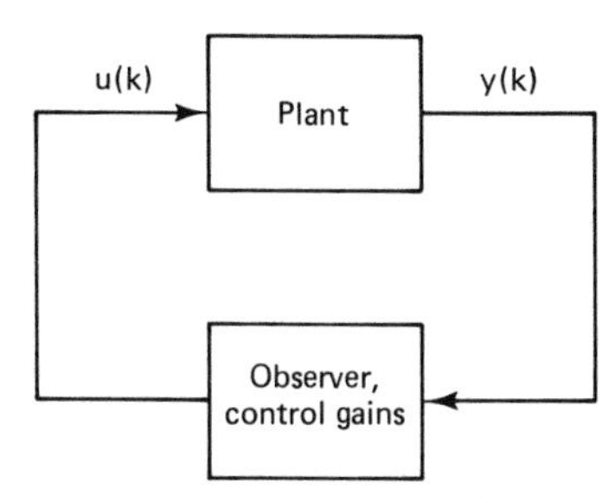

Figure 9-7 Observer-based control system.

the observer and control gains part of the system, which is implemented in a digital computer, can be considered to be a digital filter with the transfer function

$$-D(z) = \frac{U(z)}{Y(z)} \tag{9-65}$$

We will denote the plant transfer function as $G(z)$. Then the open-loop transfer function, with the system opened at either $u(k)$ or $y(k)$, is

$$\text{open-loop transfer function} = -D(z)G(z) \tag{9-66}$$

and hence the system characteristic equation is given by

$$1 + D(z)G(z) = 0 \tag{9-67}$$

We see then that, for this case, pole-placement and observer design yields a digital filter, as do the classical design techniques of Chapter 8. To illustrate this point, we will derive the filter transfer function for the system of Example 9.4.

Example 9.5

Consider the system of Example 9.4. The design resulted in the control equation

$$u(k) = -4.52y(k) - 1.12q(k)$$

and the observer equation

$$q(k+1) = 0.819q(k) + 0.903y(k+1) - 0.903y(k) + 0.0908u(k)$$

Substituting the first equation into the second yields

$$q(k+1) = 0.717q(k) + 0.903y(k+1) - 1.313y(k)$$

Thus

$$Q(z) = \left[\frac{0.903z - 1.313}{z - 0.717}\right] Y(z)$$

Substituting this into the control equation

$$U(z) = -4.52\,Y(z) - 1.12Q(z)$$

yields

$$U(z) = -\left[4.52 + \frac{1.12(0.903z - 1.313)}{z - 0.717}\right] Y(z)$$

$$= -\left[\frac{5.53z - 4.71}{(z - 0.717)}\right] Y(z)$$

Thus, from (9-65), the digital filter transfer function is given by

$$D(z) = \frac{5.53z - 4.71}{z - 0.717} = \frac{5.53(z - 0.852)}{z - 0.717}$$

We see then that the design has resulted in a phase-lead filter with the dc gain

$$D(z)\big|_{z=1} = 2.90$$

If desired, we can determine the gain margin and the phase margin by finding the plant transfer function $G(z)$, and calculating the frequency response for $D(z)G(z)$.

9.5 ACKERMANN'S FORMULA

In the preceding sections, Ackermann's formula proved to be useful in calculating the gain matrices required for both pole-placement design and for observer design. In this section a derivation of Ackermann's formula is given.

We start with the state equations of the plant:

$$\mathbf{x}(k+1) = \mathbf{A}\mathbf{x}(k) + \mathbf{B}u(k)$$
$$y(k) = \mathbf{C}\mathbf{x}(k) \tag{9-68}$$

First we consider a particular form of the state equations which is called the observer canonical form. The form is illustrated in Figure 9-8, in which the state vector is $\mathbf{w}(k)$ and all the feedback signals come from the observed (output) signal. Both the

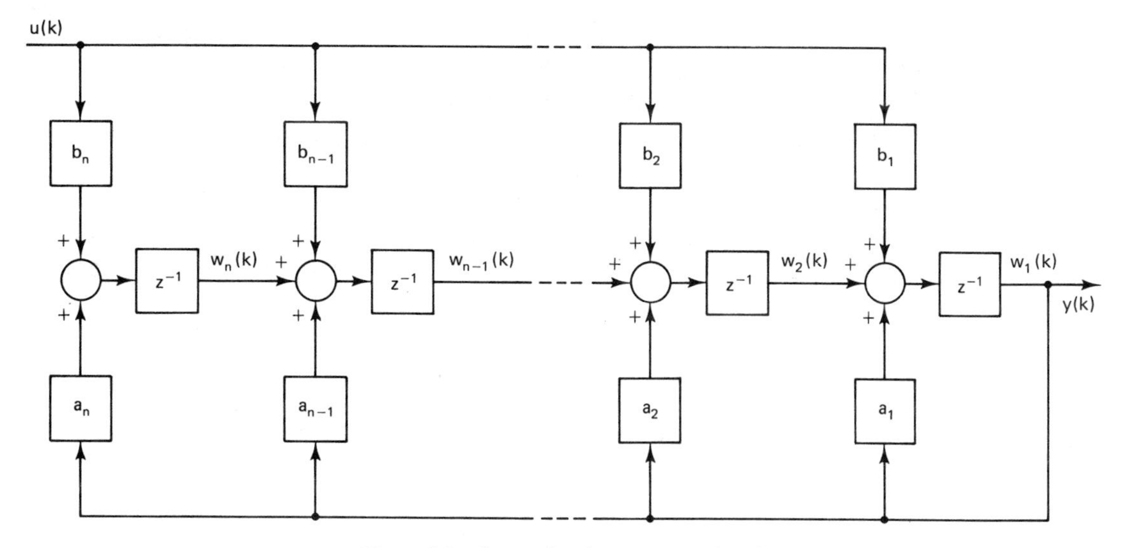

Figure 9-8 System in observer canonical form.

plants' state equations and transfer function can be determined by inspection from Figure 9-7. The state equations are

$$\begin{bmatrix} w_1(k+1) \\ w_2(k+1) \\ w_3(k+1) \\ \cdot \\ \cdot \\ \cdot \\ w_{n-1}(k+1) \\ w_n(k+1) \end{bmatrix} = \begin{bmatrix} a_1 & 1 & 0 & \cdots & 0 \\ a_2 & 0 & 1 & \cdots & 0 \\ a_3 & 0 & 0 & \cdots & 0 \\ \cdot & \cdot & \cdot & \cdot & \\ \cdot & \cdot & \cdot & \cdot & \\ \cdot & \cdot & \cdot & \cdot & \\ a_{n-1} & 0 & 0 & \cdots & 1 \\ a_n & 0 & 0 & \cdots & 0 \end{bmatrix} \begin{bmatrix} w_1(k) \\ w_2(k) \\ w_3(k) \\ \cdot \\ \cdot \\ \cdot \\ w_{n-1}(k) \\ w_n(k) \end{bmatrix} + \begin{bmatrix} b_1 \\ b_2 \\ b_3 \\ \cdot \\ \cdot \\ \cdot \\ b_{n-1} \\ b_n \end{bmatrix} u(k) \tag{9-69}$$

$$y(k) = [1 \quad 0 \quad 0 \quad \cdots \quad 0 \quad 0] \begin{bmatrix} w_1(k) \\ w_2(k) \\ w_3(k) \\ \cdot \\ \cdot \\ \cdot \\ w_{n-1}(k) \\ w_n(k) \end{bmatrix}$$

and the transfer function is

$$G(z) = \frac{b_1 z^{n-1} + b_2 z^{n-2} + \cdots + b_{n-1} z + b_n}{z^n - a_1 z^{n-1} - a_2 z^{n-2} - \cdots - a_{n-1} z - a_n} \tag{9-70}$$

It should be noted that the transfer-function coefficients appear directly in the state equations; thus we can write the transfer function directly from the state equations, with no calculations required. Filter structures in which the coefficients appear explicitly are called direct forms. Several other direct forms are presented in Chapter 12.

Consider now the effects of an arbitrary transformation of the state vector $\mathbf{x}(k)$ to $\mathbf{w}(k)$ (see Section 2.9),

$$\mathbf{x}(k) = \mathbf{T}\mathbf{w}(k) \tag{9-71}$$

where the nonsingular matrix $\mathbf{T}$ is $n \times n$, and the new state vector $\mathbf{w}(k)$ and the old state vector $\mathbf{x}(k)$ are both $n \times 1$. From (9-71) and (9-68) a new set of system state equations can be formed.

$$\mathbf{x}(k+1) = \mathbf{A}\mathbf{x}(k) + \mathbf{B}u(k) \tag{9-68}$$

$$\mathbf{T}\mathbf{w}(k+1) = \mathbf{A}\mathbf{T}\mathbf{w}(k) + \mathbf{B}u(k)$$

$$\mathbf{w}(k+1) = \mathbf{T}^{-1}\mathbf{A}\mathbf{T}\mathbf{w}(k) + \mathbf{T}^{-1}\mathbf{B}u(k) \tag{9-72}$$

and

$$y(k) = \mathbf{C}\mathbf{x}(k) \tag{9-68}$$

$$y(k) = \mathbf{C}\mathbf{T}\mathbf{w}(k) \tag{9-73}$$

The system described by (9-72) and (9-73) can be made to fit the observer canonical form of (9-69) if the correct choice of the matrix $\mathbf{T}$ is made. For notational purposes, if (9-72) and (9-73) do in fact describe a system that is in observer canonical form, these equations will be written as

$$\begin{aligned} \mathbf{w}(k+1) &= \mathbf{A}_o\mathbf{w}(k) + \mathbf{B}_o u(k) \\ y(k) &= \mathbf{C}_o\mathbf{w}(k) \end{aligned} \tag{9-74}$$

where

$$\begin{aligned} \mathbf{A}_o &= \mathbf{T}^{-1}\mathbf{A}\mathbf{T} \\ \mathbf{B}_o &= \mathbf{T}^{-1}\mathbf{B} \\ \mathbf{C}_o &= \mathbf{C}\mathbf{T} \end{aligned} \tag{9-75}$$

The subscript o signifies that the matrix is in the observer canonical form. Note that $\mathbf{T}$ is a matrix and *not* the sample interval.

Before continuing further with the derivation, it is necessary to define a matrix $\boldsymbol{\Theta}$, which is called the observability matrix of a system. The reasons for this name will become evident in the following section. The observability matrix for the plant of (9-68) is defined as

$$\boldsymbol{\Theta} = \begin{bmatrix} \mathbf{C} \\ \mathbf{C}\mathbf{A} \\ \mathbf{C}\mathbf{A}^2 \\ \cdot \\ \cdot \\ \cdot \\ \mathbf{C}\mathbf{A}^{n-1} \end{bmatrix} \tag{9-76}$$

and that of the transformed state equations of (9-74) is

$$\boldsymbol{\Theta}_o = \begin{bmatrix} \mathbf{C}_o \\ \mathbf{C}_o\mathbf{A}_o \\ \mathbf{C}_o\mathbf{A}_o^2 \\ \cdot \\ \cdot \\ \cdot \\ \mathbf{C}_o\mathbf{A}_o^{n-1} \end{bmatrix} \tag{9-77}$$

Substituting (9-75) into (9-77) gives

$$\boldsymbol{\Theta}_o = \begin{bmatrix} \mathbf{CT} \\ \mathbf{CT}(\mathbf{T}^{-1}\mathbf{AT}) \\ \mathbf{CT}(\mathbf{T}^{-1}\mathbf{AT})^2 \\ \cdot \\ \cdot \\ \cdot \\ \mathbf{CT}(\mathbf{T}^{-1}\mathbf{AT})^{n-1} \end{bmatrix} = \begin{bmatrix} \mathbf{CT} \\ \mathbf{CAT} \\ \mathbf{CA}^2\mathbf{T} \\ \cdot \\ \cdot \\ \cdot \\ \mathbf{CA}^{n-1}\mathbf{T} \end{bmatrix} \tag{9-78}$$

Comparing (9-78) and (9-76) gives the relationship

$$\boldsymbol{\Theta}_o = \boldsymbol{\Theta}\mathbf{T} \tag{9-79}$$

and solving for $\mathbf{T}$ yields

$$\mathbf{T} = \boldsymbol{\Theta}^{-1}\boldsymbol{\Theta}_o \tag{9-80}$$

Later this will prove to be a useful expression for the matrix $\mathbf{T}$.

Next, we will derive Ackermann's equation for the full-order observer. The observer state equations are given by

$$\mathbf{q}(k+1) = \mathbf{Aq}(k) + \mathbf{B}u(k) + \mathbf{G}y(k) - \mathbf{GCq}(k) \tag{9-39}$$

Consider now the transformation

$$\mathbf{q}(k) = \mathbf{Tw}(k) \tag{9-81}$$

Substitution of this transformation in the observer state equation yields

$$\mathbf{w}(k+1) = \mathbf{T}^{-1}\mathbf{ATw}(k) + \mathbf{T}^{-1}\mathbf{B}u(k) + \mathbf{T}^{-1}\mathbf{G}y(k) - \mathbf{T}^{-1}\mathbf{GCTw}(k) \tag{9-82}$$

We assume that $\mathbf{T}$ is chosen such that (9-82) is in the observer canonical form, with respect to the $\mathbf{A}$ matrix. Then, using the notation of (9-75), (9-82) becomes

$$\mathbf{w}(k+1) = \mathbf{A}_o\mathbf{w}(k) + \mathbf{B}_o u(k) + \mathbf{G}_o y(k) - \mathbf{G}_o\mathbf{C}_o\mathbf{w}(k) \tag{9-83}$$

where

$$\mathbf{G}_o = \mathbf{T}^{-1}\mathbf{G} \tag{9-84}$$

The problem now is to develop an expression for the gain matrix $\mathbf{G}_o$. [We have an expression for $\mathbf{T}$ in (9-80).] From the expression for $\mathbf{G}_o$, we solve (9-84) for $\mathbf{G}$.

$$\mathbf{G} = \mathbf{TG}_o \tag{9-85}$$

An expression for the matrix $\mathbf{G}_o$ can be found by matching the coefficients of the observer's characteristic polynomial to the coefficients of the desired characteristic polynomial. The characteristic polynomial of the observer, in terms of the matrix $\mathbf{G}_o$, is given by

$$\alpha_e(z) = |z\mathbf{I} - [\mathbf{A}_o - \mathbf{G}_o\mathbf{C}_o]| \tag{9-86}$$

But the matrix $[\mathbf{A}_o - \mathbf{G}_o\mathbf{C}_o]$ will be in observer canonical form; therefore, the characteristic polynomial can immediately be written

$$\begin{aligned}\alpha_e(z) = z^n - (a_1 - g_{o1})\, z^{n-1} - (a_2 - g_{o2})z^{n-2} - \cdots - (a_{n-1} - g_{o(n-1)})z \\ - (a_n - g_{on})\end{aligned} \tag{9-87}$$

If the desired characteristic polynomial of the observer is

$$\alpha_e(z) = z^n + \alpha_1 z^{n-1} + \alpha_2 z^{n-2} + \cdots + \alpha_{n-1}z + \alpha_n \tag{9-88}$$

then the matching of the coefficients of (9-87) and (9-88) gives

$$-\alpha_i = a_i - g_{oi}, \qquad i = 1, 2, \ldots, n$$

In general terms,

$$-\boldsymbol{\alpha} = \mathbf{a} - \mathbf{G}_o \tag{9-89}$$

or

$$\mathbf{G}_o = \mathbf{a} + \boldsymbol{\alpha} \tag{9-90}$$

where $\mathbf{a}$ is an $n \times 1$ column vector of the coefficients for the system's characteristic polynomial, and $\boldsymbol{\alpha}$ is an $n \times 1$ column vector of coefficients for the observer's desired characteristic polynomial.

A relationship between these polynomial coefficients and the system matrix $\mathbf{A}_o$ is now necessary. This is done through the use of the Cayley–Hamilton theorem [4]. This theorem states that a matrix satisfies its own characteristic equation. For the system matrix $\mathbf{A}_o$, this theorem gives

$$\mathbf{A}_o^n - a_1\mathbf{A}_o^{n-1} - a_2\mathbf{A}_o^{n-2} - \cdots - a_{n-1}\mathbf{A}_o - a_n\mathbf{I} = 0 \tag{9-91}$$

Next the polynomial $\alpha(\mathbf{A}_o)$ is formed, which is the observer's desired characteristic polynomial with the matrix $\mathbf{A}_o$ substituted for the variable z.

$$\alpha(\mathbf{A}_o) = \mathbf{A}_o^n + \alpha_1\mathbf{A}_o^{n-1} + \alpha_2\mathbf{A}_o^{n-2} + \cdots + \alpha_{n-1}\mathbf{A}_o + \alpha_n\mathbf{I} \tag{9-92}$$

Solving (9-91) for $\mathbf{A}_o^n$ and substituting it into (9-92) will give the relationship required:

$$\begin{aligned}\alpha(\mathbf{A}_o) = (a_1 + \alpha_1)\mathbf{A}_o^{n-1} + (a_2 + \alpha_2)\mathbf{A}_o^{n-2} + \cdots + (a_{n-1} + \alpha_{n-1})\mathbf{A}_o \\ + (a_n + \alpha_n)\mathbf{I}\end{aligned} \tag{9-93}$$

At this point we must define the unit vector, $\mathbf{e}_i^n$. Let $\mathbf{e}_i^n$ be an $n \times 1$ column vector which is equal to the ith column of the $n \times n$ identity matrix. Since the matrix $\mathbf{A}_o$ is in observer canonical form, an interesting thing happens when the unit vector $\mathbf{e}_n^n$ is premultiplied by the matrix $\mathbf{A}_o$,

$$\mathbf{A}_o\mathbf{e}_n^n = \begin{bmatrix} 0 \\ 0 \\ \cdot \\ \cdot \\ \cdot \\ 1 \\ 0 \end{bmatrix} = \mathbf{e}_{n-1}^n \tag{9-94}$$

and premultiplying (9-94) by the matrix $\mathbf{A}_o$ again gives

$$\mathbf{A}_o(\mathbf{A}_o\mathbf{e}_n^n) = \mathbf{A}_o^2\mathbf{e}_n^n = \begin{bmatrix} 0 \\ 0 \\ \cdot \\ \cdot \\ \cdot \\ 1 \\ 0 \\ 0 \end{bmatrix} = \mathbf{e}_{n-2}^n \tag{9-95}$$

Continuing in this manner will generate successive unit vectors, until

$$\mathbf{A}_o^{n-1}\mathbf{e}_n^n = \begin{bmatrix} 1 \\ 0 \\ 0 \\ \cdot \\ \cdot \\ \cdot \\ 0 \end{bmatrix} = \mathbf{e}_1^n \tag{9-96}$$

Therefore, if (9-93) is postmultiplied by $\mathbf{e}_n^n$, the following polynomial is obtained:

$$\begin{aligned} \alpha(\mathbf{A}_o)\mathbf{e}_n^n = (a_1 + \alpha_1)\mathbf{e}_1^n + (a_2 + \alpha_2)\mathbf{e}_2^n + \cdots + (a_{n-1} + \alpha_{n-1})\mathbf{e}_{n-1}^n \\ + (a_n + \alpha_n)\mathbf{e}_n^n \end{aligned} \tag{9-97}$$

By examining (9-97), with the help of the relationship shown in (9-90), the following becomes apparent:

$$\alpha(\mathbf{A}_o)\mathbf{e}_n^n = \mathbf{G}_o \tag{9-98}$$

which is the needed expression of the gain matrix $\mathbf{G}_o$.

With (9-98) and the relationship given in (9-85), an expression for the gain matrix $\mathbf{G}$ can now be written.

$$\begin{aligned} \mathbf{G} &= \mathbf{T}\mathbf{G}_o \\ &= \mathbf{T}\alpha(\mathbf{A}_o)\mathbf{e}_n^n \\ &= \mathbf{T}\alpha(\mathbf{T}^{-1}\mathbf{A}\mathbf{T})\mathbf{e}_n^n \\ &= \mathbf{T}\mathbf{T}^{-1}\alpha(\mathbf{A})\mathbf{T}\mathbf{e}_n^n \end{aligned} \tag{9-99}$$

or

$$\mathbf{G} = \alpha(\mathbf{A})\mathbf{T}\mathbf{e}_n^n \tag{9-100}$$

The expression shown in (9-100) can be used to calculate the gain matrix $\mathbf{G}$ if the transformation matrix $\mathbf{T}$ has been found. But calculating the $\mathbf{T}$ matrix is not necessary if the expression developed for the $\mathbf{T}$ matrix in (9-80) is substituted into (9-100). Doing this gives

$$\mathbf{G} = \alpha(\mathbf{A})\mathbf{\Theta}^{-1}\mathbf{\Theta}_o\mathbf{e}_n^n \tag{9-101}$$

It should now be noted that the product of any $n \times n$ matrix, $\mathbf{R}$, and the unit vector, $\mathbf{e}_i^n$, will give the ith column of the matrix $\mathbf{R}$. Therefore, the last two terms of (9-101) will give the nth column of the observability matrix $\mathbf{\Theta}_o$. But if the matrix $\mathbf{\Theta}_o$ is completed in (9-77), it will be found that its nth column is again the unit vector $\mathbf{e}_n^n$; therefore,

$$\mathbf{\Theta}_o \mathbf{e}_n^n = \mathbf{e}_n^n \tag{9-102}$$

Substituting (9-102) into (9-101) gives

$$\mathbf{G} = \alpha(\mathbf{A})\mathbf{\Theta}^{-1}\mathbf{e}_n^n \tag{9-103}$$

which is recognized to be Ackermann's formula, as stated in (9-48).

9.6 CONTROLLABILITY AND OBSERVABILITY

In the preceding sections Ackermann's formula was useful in both pole-assignment design, (9-25), and in observer design, (9-48). In pole-assignment design, it was necessary that the inverse of the matrix

$$[\mathbf{B} \quad \mathbf{AB} \quad \mathbf{A}^2\mathbf{B} \quad \cdots \quad \mathbf{A}^{n-1}\mathbf{B}] \tag{9-104}$$

exist, and in observer design it was necessary that the inverse of the matrix

$$\begin{bmatrix} \mathbf{C} \\ \mathbf{CA} \\ \mathbf{CA}^2 \\ \cdot \\ \cdot \\ \cdot \\ \mathbf{CA}^{n-1} \end{bmatrix} \tag{9-105}$$

exist. We will now relate the existence of the inverses of (9-104) and (9-105) to the important concepts of controllability and observability, respectively.

We will introduce the concept of *controllability* with respect to the system of Figure 9-9. The system characteristic equation is given by

$$(z - 0.9)(z - 0.8) = 0 \tag{9-106}$$

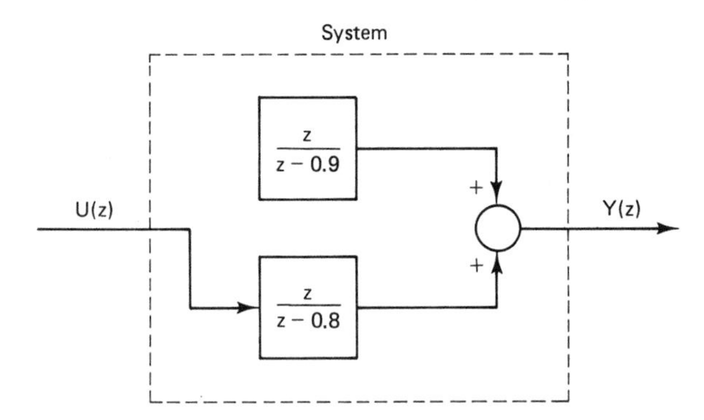

Figure 9-9 Uncontrollable system.

However, the mode of the transient response $(0.9)^k$ is not excited by the input $u(k)$. This system is said to be *uncontrollable*.

Definition 1. A system

$$\mathbf{x}(k+1) = \mathbf{Ax}(k) + \mathbf{Bu}(k) \tag{9-107}$$

is controllable provided that there exists a sequence of input $\mathbf{u}(0), \mathbf{u}(1), \mathbf{u}(2), \ldots, \mathbf{u}(N)$ that will translate the system from any initial state $\mathbf{x}(0)$ to any final state $\mathbf{x}(N)$, with N finite.

Note that, in the system of Figure 9-9, the input $u(k)$ has no influence on the state of the upper block; thus the system is uncontrollable.

We will now derive the conditions for the system of (9-107) to be controllable. Now

$$\begin{aligned}\mathbf{x}(1) &= \mathbf{A}\mathbf{x}(0) + \mathbf{B}\mathbf{u}(0)\\ \mathbf{x}(2) &= \mathbf{A}\mathbf{x}(1) + \mathbf{B}\mathbf{u}(1)\\ &= \mathbf{A}^2\mathbf{x}(0) + \mathbf{A}\mathbf{B}\mathbf{u}(0) + \mathbf{B}\mathbf{u}(1)\end{aligned} \tag{9-108}$$

$$\cdots$$

$$\begin{aligned}\mathbf{x}(N) &= \mathbf{A}^N\mathbf{x}(0) + \mathbf{A}^{N-1}\mathbf{B}\mathbf{u}(0) + \cdots + \mathbf{A}\mathbf{B}\mathbf{u}(N-2) + \mathbf{B}\mathbf{u}(N-1)\\ &= \mathbf{A}^N\mathbf{x}(0) + [\mathbf{B} \quad \mathbf{A}\mathbf{B} \quad \cdots \quad \mathbf{A}^{N-1}\mathbf{B}]\begin{bmatrix}\mathbf{u}(N-1)\\ \mathbf{u}(N-2)\\ \cdots\\ \mathbf{u}(0)\end{bmatrix}\end{aligned} \tag{9-109}$$

Hence, with $\mathbf{x}(N)$ and $\mathbf{x}(0)$ known, this equation can be written as

$$[\mathbf{B} \quad \mathbf{A}\mathbf{B} \quad \cdots \quad \mathbf{A}^{N-1}\mathbf{B}]\begin{bmatrix}\mathbf{u}(N-1)\\ \mathbf{u}(N-2)\\ \cdots\\ \mathbf{u}(0)\end{bmatrix} = \mathbf{x}(N) - \mathbf{A}^N\mathbf{x}(0) \tag{9-110}$$

Since the order of the state vector $\mathbf{x}(k)$ is n, then (9-110) yields n linear simultaneous equations. Hence, for a solution to exist, the rank of the coefficient matrix [which is (9-104) for N equal to n]

$$[\mathbf{B} \quad \mathbf{A}\mathbf{B} \quad \cdots \quad \mathbf{A}^{N-1}\mathbf{B}] \tag{9-111}$$

must be n [7]. For the case of a single input, $\mathbf{B}$ is a column matrix and (9-111) is $n \times n$ for N equal to n. Then the inverse of (9-111) must exist; this is also the condition for the existence of the solution to Ackermann's formula, in (9-25).

We consider next the concept of *observability*.

Definition 2. A system

$$\begin{aligned}\mathbf{x}(k+1) &= \mathbf{A}\mathbf{x}(k) + \mathbf{B}\mathbf{u}(k)\\ \mathbf{y}(k) &= \mathbf{C}\mathbf{x}(k)\end{aligned} \tag{9-112}$$

is observable provided that the initial state $\mathbf{x}(0)$, for any $\mathbf{x}(0)$, can be calculated from the N measurements $\mathbf{y}(0), \mathbf{y}(1), \ldots, \mathbf{y}(N-1)$, with N finite.

A system that is obviously unobservable is shown in Figure 9-10, since the state of the upper block does not contribute to the output $y(k)$.

The derivation of the criteria for observability closely parallels that for controllability. To simplify the derivation, we will assume that $\mathbf{u}(k)$ is zero; the derivation for $\mathbf{u}(k)$ not equal to zero is similar, and is given as Problem 9-10. Now, from (9-112),

$$\begin{aligned} \mathbf{y}(0) &= \mathbf{Cx}(0) \\ \mathbf{y}(1) &= \mathbf{Cx}(1) = \mathbf{CAx}(0) \\ &\cdots \\ \mathbf{y}(N-1) &= \mathbf{Cx}(N-1) = \mathbf{CA}^{N-1}\mathbf{x}(0) \end{aligned} \tag{9-113}$$

or

$$\begin{bmatrix} \mathbf{y}(0) \\ \mathbf{y}(1) \\ \cdots \\ \mathbf{y}(N-1) \end{bmatrix} = \begin{bmatrix} \mathbf{C} \\ \mathbf{CA} \\ \cdots \\ \mathbf{CA}^{N-1} \end{bmatrix} \mathbf{x}(0) \tag{9-114}$$

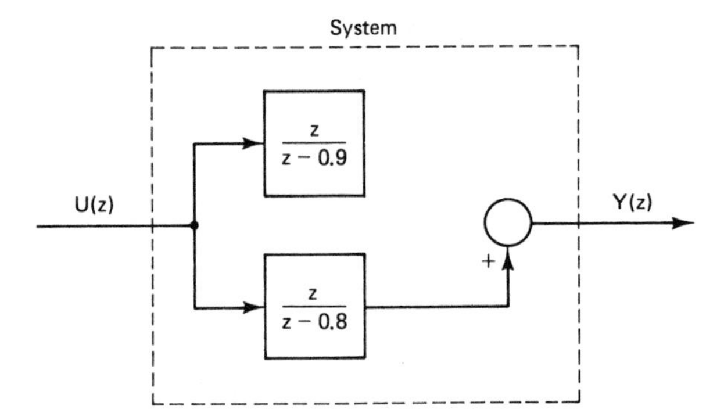

Figure 9-10 Unobservable system.

Hence, by the same arguments as for controllability, the rank of the coefficient matrix [which is (9-105) for N equal to n]

$$\begin{bmatrix} \mathbf{C} \\ \mathbf{CA} \\ \cdots \\ \mathbf{CA}^{N-1} \end{bmatrix} \tag{9-115}$$

must be n. For the case of a single output, $\mathbf{C}$ is a row matrix and (9-115) is $n \times n$ for N equal to n. Then the inverse of (9-115) must exist; this is the condition for the existence of the solution to Ackermann's formula, (9-48).

Example 9.6

We will test the system of Figure 9-11 for both controllability and observability. This system is composed of two first-order subsystems, each of which is controllable and observable, which is seen by inspection. The system state equations are given by

$$\mathbf{x}(k+1) = \begin{bmatrix} -0.2 & 0 \\ -1 & 0.8 \end{bmatrix} \mathbf{x}(k) + \begin{bmatrix} 1 \\ 1 \end{bmatrix} u(k)$$

$$y(k) = [-1 \quad 1]\mathbf{x}(k) + u(k)$$

First we will test for controllability, using (9-111).

$$\mathbf{AB} = \begin{bmatrix} -0.2 & 0 \\ -1 & 0.8 \end{bmatrix} \begin{bmatrix} 1 \\ 1 \end{bmatrix} = \begin{bmatrix} -0.2 \\ -0.2 \end{bmatrix}$$

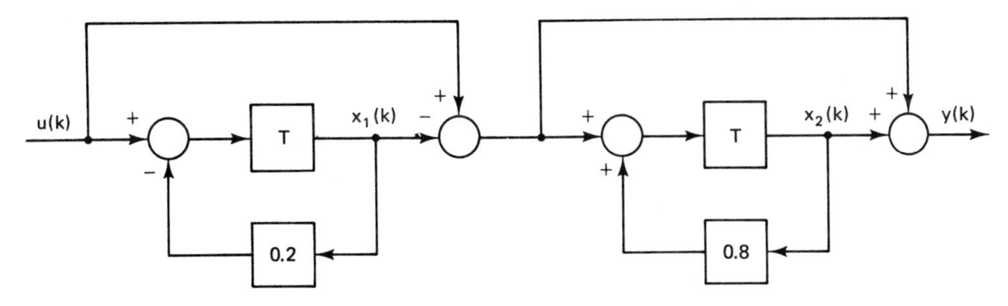

Figure 9-11 System for Example 9.6.

and thus

$$[\mathbf{B} \quad \mathbf{AB}] = \begin{bmatrix} 1 & -0.2 \\ 1 & -0.2 \end{bmatrix}$$

This matrix is of rank 1 and its inverse does not exist; hence the system is uncontrollable. Next we will test for observability, using (9-115).

$$\mathbf{CA} = [-1 \quad 1]\begin{bmatrix} -0.2 & 0 \\ -1 & 0.8 \end{bmatrix} = [-0.8 \quad 0.8]$$

and thus

$$\begin{bmatrix} \mathbf{C} \\ \mathbf{CA} \end{bmatrix} = \begin{bmatrix} -1 & 1 \\ -0.8 & 0.8 \end{bmatrix}$$

This matrix is also of rank 1; hence the system is unobservable. It is instructive to calculate some transfer functions. By Mason's gain formula and from Figure 9-11,

$$\frac{Y(z)}{U(z)} = \frac{1(1 - 0.6z^{-1} - 0.16z^{-2}) - z^{-1}(1 - 0.8z^{-1}) + z^{-1}(1 + 0.2z^{-1}) - z^{-2}}{1 + 0.2z^{-1} - 0.8z^{-1} - 0.16z^{-2}}$$

$$= 1$$

$$\frac{X_2(z)}{U(z)} = \frac{z^{-1}(1 + 0.2z^{-1}) - z^{-2}}{1 - 0.6z^{-1} - 1.6z^{-2}} = \frac{z^{-1}}{1 + 0.2z^{-1}}$$

$$\frac{X_1(z)}{U(z)} = \frac{z^{-1}}{1 + 0.2z^{-1}}$$

$$\frac{Y(z)}{X_1(z)} = \frac{Y(z)}{U(z)}\frac{U(z)}{X_1(z)} = \frac{1 + 0.2z^{-1}}{z^{-1}}$$

From the transfer function $X_2(z)/U(z)$, we see that the mode $(0.8)^k$ is not excited by the input. From the transfer function $Y(z)/X_1(z)$, we see that the mode $(-0.2)^k$ does not appear in the output.

9.7 SYSTEMS WITH INPUTS

In the pole-assignment design procedures presented in this chapter, the design resulted in a system in which the initial conditions were driven to zero in some prescribed manner; however, the system had no input. This type of system is called a *regulator control system.* However, in many control systems, it is necessary that the system output $y(t)$ follow, or track, a system input $r(t)$. An example of this type of system is the vertical (or altitude) control system of an aircraft automatic landing system.

The aircraft follows a prescribed glide slope, which is a ramp function. In this section some techniques will be presented for modifying the pole assignment design method for systems that must track an input.

We will first consider the case of low-order systems in which all states are measured; thus no observer is required. The system equations are given by

$$\begin{aligned} \mathbf{x}(k+1) &= \mathbf{A}\mathbf{x}(k) + \mathbf{B}u(k) \\ y(k) &= \mathbf{C}\mathbf{x}(k) \end{aligned} \tag{9-116}$$

where, for a regulator control system,

$$u(k) = -\mathbf{K}\mathbf{x}(k) \tag{9-117}$$

Since $u(k)$ is the control input available for the plant, this input must then be a function of the system input $r(k)$. Thus we modify $u(k)$ to be a linear function of $r(k)$,

$$u(k) = -\mathbf{K}\mathbf{x}(k) + Nr(k) \tag{9-118}$$

where N is a constant to be determined. Then, from (9-116) and (9-118),

$$\begin{aligned} \mathbf{x}(k+1) &= (\mathbf{A} - \mathbf{B}\mathbf{K})\mathbf{x}(k) + \mathbf{B}Nr(k) \\ y(k) &= \mathbf{C}\mathbf{x}(k) \end{aligned} \tag{9-119}$$

Taking the z-transform of these equation and solving for the transfer function yields

$$\frac{Y(z)}{R(z)} = \mathbf{C}(z\mathbf{I} - \mathbf{A} + \mathbf{B}\mathbf{K})^{-1}\mathbf{B}N \tag{9-120}$$

Note that the choice of N does not affect the system's relative frequency response, but only the amplitude of the frequency response. Thus N is an open-loop influence, and is not a part of the closed loop.

It is informative to determine the zeros of the system transfer function, (9-120). To locate the zeros, consider the z-transform of (9-119).

$$\begin{aligned} (z\mathbf{I} - \mathbf{A} + \mathbf{B}\mathbf{K})\mathbf{X}(z) - \mathbf{B}NR(z) &= 0 \\ \mathbf{C}\mathbf{X}(z) &= Y(z) \end{aligned} \tag{9-121}$$

If $z = z_0$ is a zero of the transfer function, (9-120), then $Y(z_0)$ is zero with $R(z_0)$ and $\mathbf{X}(z_0)$ nonzero. Hence we can express (9-121) as

$$\begin{bmatrix} z_0\mathbf{I} - \mathbf{A} + \mathbf{B}\mathbf{K} & -\mathbf{B}N \\ \mathbf{C} & 0 \end{bmatrix} \begin{bmatrix} \mathbf{X}(z_0) \\ R(z_0) \end{bmatrix} = \mathbf{0} \tag{9-122}$$

The solution to this equation is nontrivial, since $R(z_0)$ and $\mathbf{X}(z_0)$ are not zero. Hence the determinant of the coefficient matrix, which is a polynomial in z_0, must be zero [4]. The roots of this polynomial are then the zeros of the system transfer function.

The locations of the zeros of the transfer function are not evident from (9-122). However, we can rewrite (9-122) as

$$\begin{bmatrix} z_0\mathbf{I} - \mathbf{A} & -\mathbf{B} \\ \mathbf{C} & 0 \end{bmatrix} \begin{bmatrix} \mathbf{X}(z_0) \\ NR(z_0) - \mathbf{K}\mathbf{X}(z_0) \end{bmatrix} = \mathbf{0} \tag{9-123}$$

Here the coefficient matrix is independent of $\mathbf{K}$, and is in fact that of the plant alone. Hence the zeros of the closed-loop transfer function are the same as those of the transfer function of the plant, and cannot be changed by the design procedure. This design procedure is then of limited value. An example illustrating this design procedure will now be given.

Example 9.7

The system of Example 9.1 will again be considered, but with an input added as in (9-118). For this system, the plant state equations are given by

$$\mathbf{x}(k+1) = \begin{bmatrix} 1 & 0.0952 \\ 0 & 0.905 \end{bmatrix} \mathbf{x}(k) + \begin{bmatrix} 0.00484 \\ 0.0952 \end{bmatrix} u(k)$$

$$y(k) = [1 \quad 0]\mathbf{x}(k)$$

The design in Example 9.1 resulted in the gain matrix

$$\mathbf{K} = [4.52 \quad 1.12]$$

Hence

$$\mathbf{BK} = \begin{bmatrix} 0.00484 \\ 0.0952 \end{bmatrix} [4.52 \quad 1.12] = \begin{bmatrix} 0.0219 & 0.00542 \\ 0.430 & 0.107 \end{bmatrix}$$

Then

$$z\mathbf{I} - \mathbf{A} + \mathbf{BK} = \begin{bmatrix} z - 0.978 & -0.0898 \\ 0.430 & z - 0.798 \end{bmatrix}$$

and

$$(z\mathbf{I} - \mathbf{A} + \mathbf{BK})^{-1} = \frac{1}{z^2 - 1.776z + 0.819} \begin{bmatrix} z - 0.798 & 0.0898 \\ -0.430 & z - 0.978 \end{bmatrix}$$

The system transfer function is given by (9-120):

$$\frac{Y(z)}{R(z)} = \mathbf{C}(z\mathbf{I} - \mathbf{A} + \mathbf{BK})^{-1}\mathbf{B}N = \frac{(0.00484z + 0.00468)N}{z^2 - 1.776z + 0.819}$$

Note that the system dc gain is given by

$$\left.\frac{Y(z)}{R(z)}\right|_{z=1} = 0.221N$$

The dc gain can be increased by the choice of N; however, N is a gain on the input signal and is an open-loop parameter. Thus this method of increasing the dc gain is generally unacceptable. The unit step response, for $N = 1$, is shown in Figure 9-12. The dynamics of the response appear satisfactory, but the steady-state error is very large.

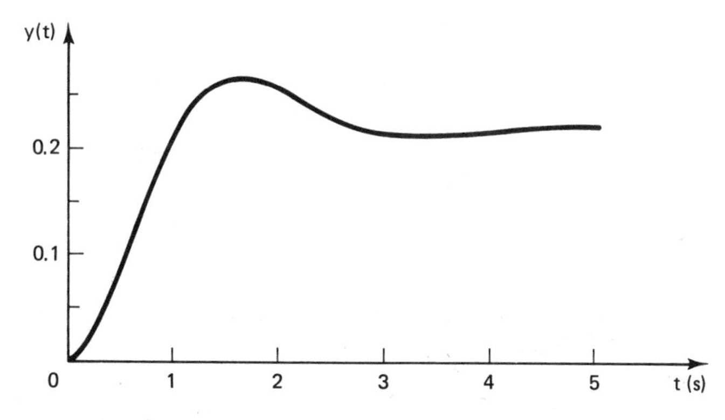

Figure 9-12 Step response for Example 9.7.

We will now consider the case that an observer is used to implement the control system. The plant is described by

$$\mathbf{x}(k+1) = \mathbf{A}\mathbf{x}(k) + \mathbf{B}u(k)$$
$$y(k) = \mathbf{C}\mathbf{x}(k) \tag{9-124}$$

and the observer, from (9-39), is described by

$$\mathbf{q}(k+1) = (\mathbf{A} - \mathbf{G}\mathbf{C})\mathbf{q}(k) + \mathbf{G}y(k) + \mathbf{B}u(k) \tag{9-39}$$

The control is implemented by

$$u(k) = -\mathbf{K}\mathbf{q}(k) \tag{9-125}$$

To be general, we can allow the input to enter both (9-39) and (9-125). These equations become

$$\mathbf{q}(k+1) = (\mathbf{A} - \mathbf{G}\mathbf{C})\mathbf{q}(k) + \mathbf{G}y(k) + \mathbf{B}u(k) + \mathbf{M}r(k) \tag{9-126}$$

$$u(k) = -\mathbf{K}\mathbf{q}(k) + Nr(k) \tag{9-127}$$

The design problem then is to choose $\mathbf{M}$ and N so as to achieve certain design objectives, such as zeros of the system transfer function, steady-state errors, and so on.

We will consider the case that only the error

$$e(t) = r(t) - y(t) \tag{9-128}$$

is measured, and thus only this signal is available for control. For example, in radar tracking systems, the radar return indicates only the error $e(t)$ between the antenna pointing direction and the direction to the target. The error-control condition can be satisfied in (9-126) and (9-127) by choosing N to be zero and $\mathbf{M}$ equal to $-\mathbf{G}$. Then (9-126) and (9-127) become, respectively,

$$\mathbf{q}(k+1) = (\mathbf{A} - \mathbf{G}\mathbf{C})\mathbf{q}(k) + \mathbf{G}[y(k) - r(k)] + \mathbf{B}u(k) \tag{9-129}$$

$$u(k) = -\mathbf{K}\mathbf{q}(k) \tag{9-130}$$

Note that this design procedure also gives no choice in the selection of the system transfer-function zeros. An example will now be given to illustrate this design technique.

Example 9.8

The system of Example 9.7 will be considered, with the observer designed in Example 9.3 to be employed. The plant state equations are given in Example 9.7. The gain matrices for the controller and for the observer are given in Example 9.3. The required matrices are then

$$\mathbf{A} = \begin{bmatrix} 1 & 0.0952 \\ 0 & 0.905 \end{bmatrix}, \quad \mathbf{B} = \begin{bmatrix} 0.00484 \\ 0.0952 \end{bmatrix}, \quad \mathbf{C} = [1 \quad 0]$$

$$\mathbf{K} = [4.52 \quad 1.12], \quad \mathbf{G} = \begin{bmatrix} 0.267 \\ 0.0802 \end{bmatrix}$$

Substitution of these matrices into (9-129) and (9-130), and then substitution of (9-130) into (9-129) yields the observer equations

$$\mathbf{q}(k+1) = \begin{bmatrix} 0.711 & 0.0898 \\ -0.510 & 0.798 \end{bmatrix}\mathbf{q}(k) + \begin{bmatrix} 0.267 \\ 0.0802 \end{bmatrix}[y(k) - r(k)]$$

Note that no design is required beyond that performed in Example 9.3. Step responses for this design are given in Figure 9-13. For the response shown by the solid line, the initial conditions of both the plant and the observer are all zero. For the response shown by the dashed line, the initial conditions are given by

$$\mathbf{x}(0) = \begin{bmatrix} -0.5 \\ 0 \end{bmatrix}, \qquad \mathbf{q}(0) = \begin{bmatrix} 0 \\ 0 \end{bmatrix}$$

For this design procedure, the transient response appears satisfactory; in addition, the steady-state error is zero.

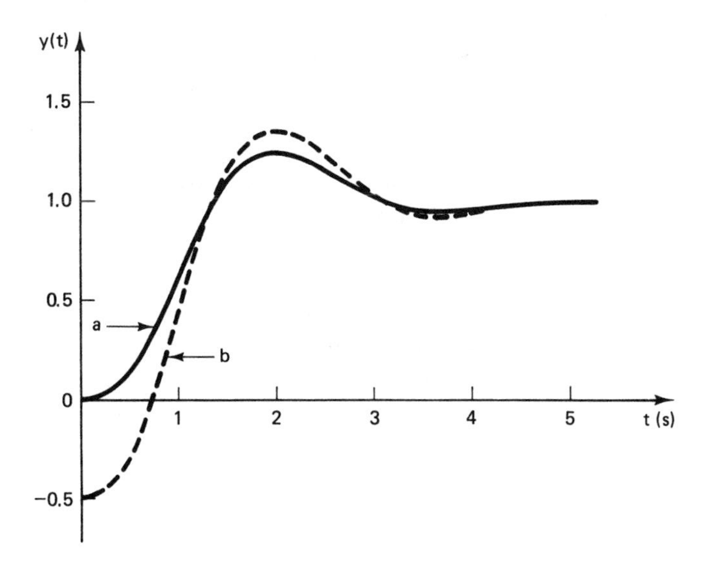

Figure 9-13 Responses for the system of Example 9.8.

9.8 SUMMARY

In this chapter a design technique has been presented which is based on the state model of the system, compared to the design methods of Chapter 8, which are based on transfer functions. The design technique presented is the pole-assignment, or pole-placement, technique. This technique allows us to place all the zeros of the characteristic equation at desirable locations; however, the technique requires that all the system states be known, a condition that at least is generally impractical. To circumvent this condition, an observer is constructed. An observer estimates the system states, based on the known system inputs and the outputs that can be measured. Then, for the pole-placement design, the estimated states, rather than the actual states, are used to calculate feedback signals.

The design above results in a regulator control system, that is, a control system with no reference input. A method was presented which modifies the pole-placement design for systems that have reference inputs. For the case that the system error is to be used in the control of the system, the modifications required are simple and straightforward.

REFERENCES AND FURTHER READING

1. J. E. Ackermann, "Der Entwulf Linearer Regelungs Systems in Zustandsraum," *Regelungstech. und Prozessdatenverarb.*, Vol. 7, pp. 297–300, 1972.
2. D. G. Luenberger, "An Introduction to Observers," *IEEE Trans. Autom. Control*, Vol. AC-16, pp. 596–602, Dec. 1971.
3. Part II, *IEEE Trans. Autom. Control*, Vol. AC-26, Aug. 1981.
4. P. M. DeRusso, R. J. Roy, and C. M. Close, *State Variable for Engineers*, New York: John Wiley & Sons, Inc., 1965.
5. J. C. Williams and S. K. Mitter, "Controllability, Observability, Pole Allocation, and State Reconstruction," *IEEE Trans. Autom. Control*, Vol. AC-16, pp. 582–602, Dec. 1971.
6. W. M. Wonham, "On Pole Assignment in Multi-input Controllable Linear Systems," *IEEE Trans. Autom. Control*, Vol. AC-12, pp. 660–665, Dec. 1967.
7. C. T. Leondes and L. M. Novak, "Reduced-Order Observers for Linear Discrete-Time Systems," *IEEE Trans. Autom. Control*, Vol. AC-19, pp. 42–46, Feb. 1974.
8. R. F. Wilson, "An Observer Based Aircraft Automatic Landing System," M. S. thesis, Auburn University, Auburn, Ala., 1981.
9. G. F. Franklin and J. D. Powell, *Digital Control of Dynamic Systems*. Reading Mass.: Addison-Wesley Publishing Company, 1980.
10. B. C. Kuo, *Digital Control Systems*. New York: Holt, Rinehart and Winston, Inc., 1980.

PROBLEMS

9-1. The plant of Example 9.1 has the state equations

$$\mathbf{x}(k+1) = \begin{bmatrix} 1 & 0.0952 \\ 0 & 0.905 \end{bmatrix} \mathbf{x}(k) + \begin{bmatrix} 0.00484 \\ 0.0952 \end{bmatrix} u(k)$$

Find the gain matrix **K** required to realize the closed-loop characteristic equation with zeros which have a damping ratio, ζ, of 0.46 and a time constant τ of 0.5 s. Use pole-assignment design.

9-2. Assume that equation (9-25), Ackermann's formula for pole-assignment design, is the solution of (9-15). Based on this result, show that (9-48), Ackermann's formula for observer design, is the solution of (9-46).

9-3. Consider the plant of Problem 9-1. Using pole-assignment design, find the gain matrix **K** required to realize the closed-loop characteristic equation with zeros which have a damping ratio $\zeta = 1$ (critical damping) and a time constant $\tau = 1.0$ s.

9-4. A satellite may be modeled as shown in Figure P9-4 (see Chapter 1).

(a) Develop a discrete model for this system.

(b) A closed-loop control system is to be designed using the pole-placement technique. The desired characteristic equation is given by

$$\alpha_c(z) = z^2 - 1.62z + 0.665$$

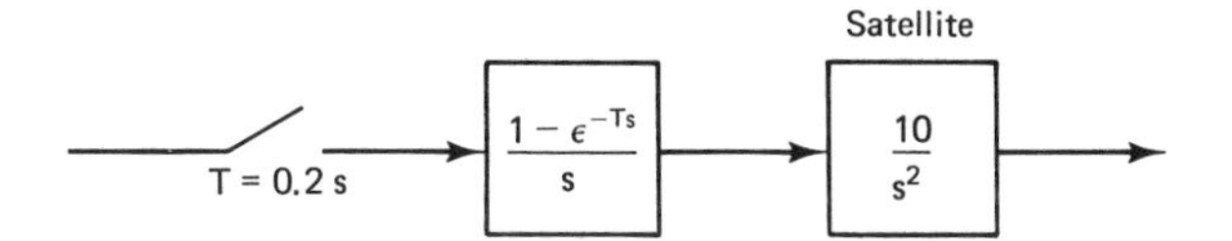

Figure P9-4 System for Problem 9-4.

Find the damping ratio and the time constant of the roots of this characteristic equation.

(c) Find the gain matrix that will realize the characteristic equation of part (b).

9-5. The system designed in Problem 9-1 is to be realized using a full-order observer. Choosing the roots of the characteristic equation of the observer to be real, equal, and with a time constant of 0.25 s, design the observer. For this system, $y(k) = x_1(k)$.

9-6. The system designed in Problem 9-4 is to be realized using a full-order observer. Choosing the roots of the characteristic equation of the observer to be real, equal, and with a time constant equal to one-fourth that of the roots of the control-system characteristic equation, design the observer.

9-7. For the plant in Problem 9-1, assume that the output $y(t)$ is equal to $x_1(t)$. The design in Problem 9-1 is to be realized using a reduced-order observer. Let the root of the characteristic equation of the observer be real with a time constant equal to 0.25 s.

(a) Design the observer.

(b) The implementation of the control system will be as shown in Figure P9-7. Find the transfer function, $D(z)$, of the digital filter.

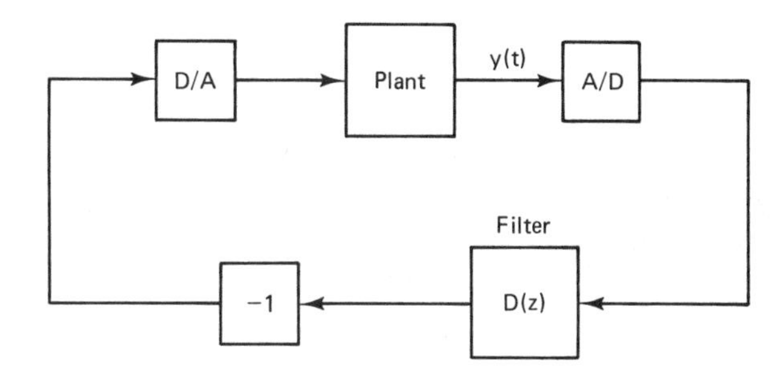

Figure P9-7 System for Problem 9-7.

9-8. For the plant in Problem 9-3, assume that the output $y(t)$ is equal to $x_1(t)$. The design in Problem 9-3 is to be realized using a reduced-order observer. Let the root of the characteristic equation of the observer be real with a time constant equal to 0.75 s.

(a) Design the observer.

(b) The implementation of the control system will be as shown in Figure P9-7. Find the transfer function, $D(z)$, of the digital filter.

9-9. For the plant in Problem 9-4, assume that $y(t)$ is equal to $x_1(t)$. The design in Problem 9-4 is to be realized using a reduced-order observer. Let the root of the characteristic equation of the observer be real with a time constant equal to one-half that of the roots of the control-system characteristic equation.

(a) Design the observer.

(b) The implementation of the control system will be as shown in Figure P9-7. Find the transfer function, $D(z)$, of the digital filter.

9-10. Consider a system described by (9-112).

$$\begin{aligned}\mathbf{x}(k+1) &= \mathbf{A}\mathbf{x}(k) + \mathbf{B}\mathbf{u}(k) \\ \mathbf{y}(k) &= \mathbf{C}\mathbf{x}(k)\end{aligned} \tag{9-112}$$

For the case that $\mathbf{u}(k)$ is not zero, derive the conditions for observability.

9-11. Consider the plant of Problem 9-5, which is

$$\mathbf{x}(k+1) = \begin{bmatrix} 1 & 0.0952 \\ 0 & 0.905 \end{bmatrix} \mathbf{x}(k) + \begin{bmatrix} 0.00484 \\ 0.0952 \end{bmatrix} u(k)$$

Suppose that the output is given by

$$y(k) = [0 \quad 1]\mathbf{x}(k)$$

(a) Is this system observable?

(b) Explain the reason for your answer in part (a) in terms of the physical aspects of the system.

9-12. Consider the reduced-order observer design of Example 9.4. The plant is described by

$$\mathbf{x}(k+1) = \begin{bmatrix} 1 & 0.0952 \\ 0 & 0.905 \end{bmatrix} \mathbf{x}(k) + \begin{bmatrix} 0.00484 \\ 0.0952 \end{bmatrix} u(k)$$

$$y(k) = [1 \quad 0]\mathbf{x}(k)$$

The observer is described by

$$q(k+1) = 0.819q(k) + 0.903y(k+1) - 0.903y(k) + 0.0908u(k)$$

(a) Construct a single set of state equations for the plant–observer system, where the state vector is given by

$$\begin{bmatrix} x_1(k) \\ x_2(k) \\ q(k) \end{bmatrix}$$

and the input is $u(k)$ (i.e., there is no feedback).

(b) Determine if this plant–observer system is controllable.

9-13. For the design of Problem 9-5, an input $r(k)$ is to be added to the closed-loop system in the form $[y(k) - r(k)]$. Write the resultant observer equations in the form

$$\mathbf{q}(k+1) = \mathbf{A}_1\mathbf{q}(k) + \mathbf{G}[y(k) - r(k)]$$

9-14. For the design of Problem 9-6, an input $r(k)$ is to be added to the closed-loop system in the form $[y(k) - r(k)]$. Write the resultant observer equations in the form

$$\mathbf{q}(k+1) = \mathbf{A}_1\mathbf{q}(k) + \mathbf{G}[y(k) - r(k)]$$

9-15. Shown in Figure P9-15 is the simplified design of the attitude control system of a large space booster such as the Saturn V. Rate feedback is required to stabilize this system, which is described in Problem 7-5. For the Saturn V, the rigid-body transfer function is given by

$$G_p(s) = \frac{0.9407}{s^2 - 0.0297}$$

(a) Develop a discrete model of this system, such that the two state variables are $\theta(kT)$ and $\dot{\theta}(kT)$.

(b) The closed-loop control system is to be designed to replace the feedback elements shown, using the pole-placement technique. The desired characteristic equation is given by

$$\alpha_c(z) = z^2 - 1.98z + 0.9802$$

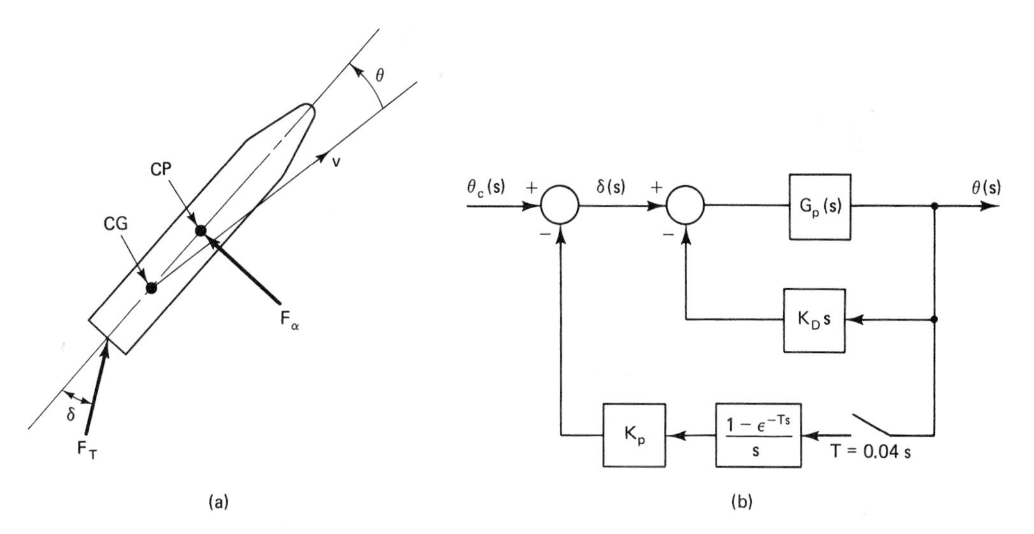

Figure P9-15 Space-booster control system.

Find the damping ratio and the time constant of the roots of the characteristic equation.

(c) Find the gain matrix that will realize the characteristic equation of part (b).

(d) Suppose that $\theta(t)$ is the only signal measured. Design a reduced-order observer for the system.

(e) Determine if the compensator is phase-lead or phase-lag.

(f) For this system $\dot{\theta}(t)$ can be measured more easily than can $\theta(t)$. If $\dot{\theta}(t)$ is the only signal measured, can a reduced-order observer be designed?

10

Linear Quadratic Optimal Control

10.1 INTRODUCTION

Chapter 8 presented design techniques that are termed classical, or traditional. The two techniques developed are the frequency-response techniques and the root-locus technique. Both methods are very effective, but are largely trial and error, with experience very useful. Even when an acceptable design is completed, the question remains as to whether a "better" design could be found. The pole-assignment design technique of Chapter 9 is termed a modern technique, and is based on the state-variable model of the plant, rather than on the transfer function as required by the classical methods. In this procedure we assumed that we know the exact locations required for the closed-loop transfer-function poles, and we can realize these locations, at least in the linear model. Of course, for the physical system, there are boundaries to the regions in which the pole locations can be placed.

In the pole-assignment technique, we assume that we know the pole locations that yield the "best" control system. In this chapter we develop a different technique that yields the "best" control system. This technique is an optimal design technique, and assumes that we can write a mathematical function which is called the *cost function.* The optimal design procedure minimizes this cost function: hence the term "optimal." However, in most cases the choice of the cost function involves some trial and error; that is, we are not sure of the exact form that the cost function should take.

For discrete systems, the cost function (also called the *performance index*) is generally of the form

$$J_N = \sum_{k=0}^{N} L[\mathbf{y}(k), \mathbf{r}(k), \mathbf{u}(k)] \tag{10-1}$$

In this relationship k is the sample instant and N is the terminal sample instant. The control system outputs are $\mathbf{y}(k)$, the inputs are $\mathbf{r}(k)$, and $\mathbf{u}(k)$ are the control inputs to the plant.

For a physical system, the control inputs are always constrained. For example, the amplitude of each component of the control vector may be limited, such that

$$|u_i(k)| \leq U_i$$

where each U_i is a given constant, and the subscript i denotes the vector component. For the case of limited control energy, we have

$$u_i^2(k) \leq M_i$$

where each M_i is a given constant. The availability of finite control energy may be represented by a term in the cost function (10-1) as

$$\sum_{k=0}^{N} \mathbf{u}^T(k)\mathbf{R}(k)\mathbf{u}(k)$$

where $\mathbf{R}(k)$ is a weighting matrix and is not related to $\mathbf{r}(k)$. This function is called a *quadratic form*, and will be considered in detail in the following sections. In any case, the control must satisfy certain constraints; any control that satisfies these constraints is called an admissible control.

In this chapter the design technique for optimal linear regulator control systems with quadratic cost functions is developed. This design results in a control law of the form

$$\mathbf{u}(k) = -\mathbf{K}(k)\mathbf{x}(k) \tag{10-2}$$

Hence we have linear, time-varying, full-state feedback. The same limitations of full-state feedback apply here, as in Chapter 9. It will generally be necessary to utilize an observer to implement the control law. The observer, of course, is designed by the techniques presented in Chapter 9.

10.2 THE QUADRATIC COST FUNCTION

We will begin the development of optimal control design by considering the case of a quadratic cost function of the states and the control, that is,

$$J_N = \sum_{k=0}^{N} \mathbf{x}^T(k)\mathbf{Q}(k)\mathbf{x}(k) + \mathbf{u}^T(k)\mathbf{R}(k)\mathbf{u}(k) \tag{10-3}$$

where N is finite and $\mathbf{Q}(k)$ and $\mathbf{R}(k)$ are symmetric. The system is described by

$$\begin{aligned} \mathbf{x}(k+1) &= \mathbf{A}(k)\mathbf{x}(k) + \mathbf{B}(k)\mathbf{u}(k) \\ \mathbf{y}(k) &= \mathbf{C}(k)\mathbf{x}(k) \end{aligned} \tag{10-4}$$

Note that both the system and the cost-function matrices are allowed to be time varying.

The quadratic cost function is considered because the development is simple and the cost function is logical. For example, suppose that the system order is 2. Then, let

$$\begin{aligned} F = \mathbf{x}^T\mathbf{Q}\mathbf{x} &= [x_1 \quad x_2]\begin{bmatrix} q_{11} & q_{12} \\ q_{21} & q_{22} \end{bmatrix}\begin{bmatrix} x_1 \\ x_2 \end{bmatrix} \\ &= q_{11}x_1^2 + (q_{12} + q_{21})x_1x_2 + q_{22}x_2^2 \end{aligned} \tag{10-5}$$

Note first that F is a scalar and that there is no loss of generality in assuming $\mathbf{Q}$ to be symmetric. Now, if the quadratic form in (10-5) is positive semidefinite [1], which we will require, then

$$\begin{aligned} F &\geq 0, \qquad \mathbf{x} \neq \mathbf{0} \\ F &= 0, \qquad \mathbf{x} = \mathbf{0} \end{aligned}$$

Then minimizing F will minimize the magnitudes of states that contribute to F, in some sense. For example, if

$$F = x_1^2 + x_2^2$$

then minimizing F will tend to minimize the magnitudes of x_1 and x_2. However, if

$$F = 100x_1^2 + x_2^2$$

then minimizing F will also minimize $|x_1|$ and $|x_2|$, but $|x_1|$ should be much smaller than $|x_2|$. As a third case, suppose that

$$F = x_1^2$$

(i.e., F is not a function of x_2). Then in minimizing F only x_1 is minimized, and x_2 is determined by its relationship to x_1.

Consider now the contribution of the control input $\mathbf{u}(k)$. Suppose that

$$\begin{aligned} G = \mathbf{u}^T\mathbf{R}\mathbf{u} &= [u_1 \quad u_2]\begin{bmatrix} r_{11} & r_{12} \\ r_{21} & r_{22} \end{bmatrix}\begin{bmatrix} u_1 \\ u_2 \end{bmatrix} \\ &= r_{11}u_1^2 + (r_{12} + r_{21})u_1u_2 + r_{22}u_2^2 \end{aligned} \tag{10-6}$$

Note again that there is no loss in generality in assuming the matrix in a quadratic form is symmetric. If G in (10-6) is positive definite [1], which we will require, then

$$\begin{aligned} G &> 0, \qquad \mathbf{u} \neq \mathbf{0} \\ G &= 0, \qquad \mathbf{u} = \mathbf{0} \end{aligned}$$

Thus if we minimize G, we are minimizing the control functions. If G were allowed to be only positive semidefinite, some components of the control vector could be quite large when G is minimized.

Consider the total cost function of (10-3):

$$J_N = \sum_{k=0}^{N} \mathbf{x}^T(k)\mathbf{Q}(k)\mathbf{x}(k) + \mathbf{u}^T(k)\mathbf{R}(k)\mathbf{u}(k) \tag{10-3}$$

If $\mathbf{Q}(k)$ were positive semidefinite and $\mathbf{R}(k)$ were the null matrix, minimization of (10-3) would force the $\mathbf{x}(k)$ vector toward zero very quickly, which would require a large $\mathbf{u}(k)$. For a physical system, $\mathbf{u}(k)$ is always bounded, and in general the large $\mathbf{u}(k)$ would not be realizable. Hence the positive definite $\mathbf{R}(k)$ is added to the cost function to limit $\mathbf{u}(k)$ to realizable values. Then, in using the cost function (10-3), we include the term involving $\mathbf{x}(k)$ so that, in some sense, the magnitudes of the states are driven toward zero. The term involving $\mathbf{u}(k)$ is included in order that the components of the control vector will be limited in magnitude such that the design is physically realizable.

10.3 THE PRINCIPLE OF OPTIMALITY

The optimal control design problem posed in Section 10.2 may be stated as follows:

For a linear discrete system described by (10-4),

$$\begin{aligned}\mathbf{x}(k+1) &= \mathbf{A}(k)\mathbf{x}(k) + \mathbf{B}(k)\mathbf{u}(k) \\ \mathbf{y}(k) &= \mathbf{C}(k)\mathbf{x}(k)\end{aligned} \tag{10-4}$$

determine the control

$$\mathbf{u}^o(k) = \mathbf{f}[\mathbf{x}(k)] \tag{10-7}$$

that minimizes the quadratic cost function of (10-3):

$$J_N = \sum_{k=0}^{N} \mathbf{x}^T(k)\mathbf{Q}(k)\mathbf{x}(k) + \mathbf{u}^T(k)\mathbf{R}(k)\mathbf{u}(k) \tag{10-3}$$

where N is finite, $\mathbf{Q}(k)$ is positive semidefinite, and $\mathbf{R}(k)$ is positive definite. In (10-7), the superscript o denotes that the control is optimal.

The solution (10-7) can be obtained by several different approaches. The approach to be used here will be through the principle of optimality, developed by Richard Bellman [2,3]. For our purposes, the principle may be stated as follows:

If a closed-loop control $\mathbf{u}^o(k) = \mathbf{f}[\mathbf{x}(k)]$ is optimal over the interval $0 \le k \le N$, it is also optimal over *any* subinterval $m \le k \le N$, where $0 \le m \le N$.

We can show the rationale for this principle by the following argument. Let J_N be the cost function that is minimized. (It is not necessary that J_N be quadratic.) Also, let S_m be the cost from $k = m$ to $k = N$; that is,

$$S_m = J_N - J_{N-m} \tag{10-8}$$

First assume that, although $\mathbf{u}^o(k) = \mathbf{f}[\mathbf{x}(k)]$ is optimal over $0 \le k \le N$, a different control law $\mathbf{u}(k) = \mathbf{g}[\mathbf{x}(k)]$ yields a smaller value of S_m of (10-8) over the interval $m \le k \le N$, $m > 0$. Then we construct a second control law composed of $\mathbf{f}[\mathbf{x}(k)]$ over $0 \le k \le m-1$, and of $\mathbf{g}[\mathbf{x}(k)]$ over $m \le k \le N$. The cost of either control law over the interval $0 \le k \le m-1$ is the same, since the control laws over this interval are identical. Since the cost of the second control law is assumed less than that of the first law over the interval $m \le k \le N$, the total cost over the interval $0 \le k \le N$ is less for the second control law. But this is a contradiction. Hence if $\mathbf{u}^o(k) = \mathbf{f}[\mathbf{x}(k)]$ is optimal over $0 \le k \le N$, it is also optimal over $m \le k \le N$, $m > 0$. Note that this argument does not apply for a control law over the interval $p_1 \le k \le p_2$, where $p_1 \ge 0$ and $p_2 < N$. That is, $\mathbf{u}^o(k) = \mathbf{f}[\mathbf{x}(k)]$ is not necessarily optimal over the interval $p_1 \le k \le p_2$, $p_2 < N$. Design using the principal of optimality is also known as dynamic programming.

A simple example will now be given to illustrate optimal control design using the principal of optimality. Then the general design procedure will be developed in the following section. Consider the first-order system described by

$$x(k+1) = 2x(k) + u(k) \tag{10-9}$$

Note that this system is unstable. We wish to determine the control law $u(k)$ which minimizes

$$J_2 = \sum_{k=0}^{2} x^2(k) + u^2(k)$$

Hence we wish to choose $u(k)$ to minimize

$$J_2 = x^2(0) + u^2(0) + x^2(1) + u^2(1) + x^2(2) + u^2(2) \tag{10-10}$$

subject to the relationship, or constraint, (10-9). Note first that $u(2)$ does not affect any of the other terms in (10-10); thus we must choose $u(2) = 0$. This requirement will also appear in the following development.

First, from (10-8),

$$S_1 = J_N - J_{N-1} = J_2 - J_1 = x^2(2) + u^2(2)$$

By the principal of optimality, $u(2)$ must minimize this function. Hence

$$\frac{\partial S_1}{\partial u(2)} = 2u(2) = 0$$

since $x(2)$ is independent of $u(2)$. Thus $u(2) = 0$, and

$$S_1^o = x^2(2) \tag{10-11}$$

Next, we calculate S_2.

$$\begin{aligned} S_2 &= J_2 - J_0 = J_2 - J_1 + J_1 - J_0 \\ &= S_1^o + x^2(1) + u^2(1) \end{aligned}$$

From (10-9) and (10-11), S_2 may be expressed as

$$S_2 = [2x(1) + u(1)]^2 + x^2(1) + u^2(1)$$

By the principle of optimality,

$$\frac{\partial S_2}{\partial u(1)} = 2[2x(1) + u(1)] + 2u(1) = 0$$

since $x(1)$ is independent of $u(1)$. Thus

$$u(1) = -x(1)$$

and

$$S_2^o = 3x^2(1)$$

By the same procedure,

$$S_3 = J_2 = J_2 - J_0 + J_0 = S_2^o + x^2(0) + u^2(0)$$

Hence

$$\begin{aligned} S_3 &= 3x^2(1) + x^2(0) + u^2(0) \\ &= 3[2x(0) + u(0)]^2 + x^2(0) + u^2(0) \end{aligned}$$

and

$$\frac{\partial S_3}{\partial u(0)} = 6[2x(0) + u(0)] + 2u(0) = 0$$

Thus

$$u(0) = -1.5x(0)$$

Therefore, the optimal control sequence is given by

$$\begin{aligned} u(0) &= -1.5x(0) \\ u(1) &= -x(1) \\ u(2) &= 0 \end{aligned} \tag{10-12}$$

The minimum-cost function is then

$$\begin{aligned} S_3^o = J_2^o &= 3[0.5x(0)]^2 + x^2(0) + [-1.5x(0)]^2 \\ &= 4x^2(0) \end{aligned}$$

Thus *no* choice of $\{u(0), u(1), u(2)\}$ will yield a smaller value of J_2 than will that of (10-12).

The following points should be made about the example above. First, while no assumption was made concerning the form of the control law, the optimal control law is of the form

$$u(k) = -K(k)x(k)$$

Next, even though the plant is time-invariant, the feedback gains required are time varying. The third point is that the control law is linear. In the following section a general solution to this optimal control design problem will be derived, and the points above will be shown to also apply to the general case.

Note also that the optimal control design is solved in reverse time. The optimal gain $K(i)$ cannot be calculated until all the remaining $K(j)$, $i < j \leq N$ are known.

10.4 LINEAR QUADRATIC OPTIMAL CONTROL

We will now solve the linear quadratic optimal design problem posed in Section 10.4, which is as follows:

For a linear discrete system described by (10-4),

$$\begin{aligned} \mathbf{x}(k+1) &= \mathbf{A}(k)\mathbf{x}(k) + \mathbf{B}(k)\mathbf{u}(k) \\ \mathbf{y}(k) &= \mathbf{C}(k)\mathbf{x}(k) \end{aligned} \tag{10-4}$$

determine the control (10-7)

$$\mathbf{u}^o(k) = \mathbf{f}[\mathbf{x}(k)] \tag{10-7}$$

that minimizes the quadratic cost function of (10-3):

$$J_N = \sum_{k=0}^{N} \mathbf{x}^T(k)\mathbf{Q}(k)\mathbf{x}(k) + \mathbf{u}^T(k)\mathbf{R}(k)\mathbf{u}(k) \tag{10-3}$$

where N is finite, $\mathbf{Q}(k)$ is positive semidefinite, and $\mathbf{R}(k)$ is positive definite.

First a short review of the differentiation of quadratic functions is in order (see Appendix IV). Given the quadratic function $\mathbf{x}^T\mathbf{Q}\mathbf{x}$, then

$$\frac{\partial}{\partial \mathbf{x}}[\mathbf{x}^T\mathbf{Q}\mathbf{x}] = 2\mathbf{Q}\mathbf{x} \tag{10-13}$$

Given the bilinear form $\mathbf{x}^T\mathbf{Q}\mathbf{y}$, then

$$\frac{\partial}{\partial \mathbf{x}}[\mathbf{x}^T\mathbf{Q}\mathbf{y}] = \mathbf{Q}\mathbf{y} \tag{10-14}$$

and

$$\frac{\partial}{\partial \mathbf{y}}[\mathbf{x}^T\mathbf{Q}\mathbf{y}] = \mathbf{Q}^T\mathbf{x} \tag{10-15}$$

In (10-13) we can assume that $\mathbf{Q}$ is symmetric. However, in (10-14) and (10-15), $\mathbf{Q}$ is not necessarily symmetric.

In (10-8) we defined S_m to be the cost from the $(N - m)$ sample instant to the N(terminal) sample instant, that is,

$$S_m = J_N - J_{N-m} \tag{10-8}$$

Hence

$$\begin{aligned} S_{m+1} &= J_N - J_{N-m-1} = J_N - J_{N-m} + J_{N-m} - J_{N-m-1} \\ &= S_m + J_{N-m} - J_{N-m-1} \end{aligned} \tag{10-16}$$

From the cost function (10-3), we can then write

$$\begin{aligned} S_{m+1} = S_m &+ \mathbf{x}^T(N-m)\mathbf{Q}(N-m)\mathbf{x}(N-m) \\ &+ \mathbf{u}^T(N-m)\mathbf{R}(N-m)\mathbf{u}(N-m) \end{aligned} \tag{10-17}$$

Thus, from the principle of optimality,

$$\frac{\partial S_{m+1}}{\partial \mathbf{u}(N-m)} = \mathbf{0} \tag{10-18}$$

if S_m in (10-17) is replaced with S_m^o.

Consider first S_1:

$$S_1 = J_N - J_{N-1} = \mathbf{x}^T(N)\mathbf{Q}(N)\mathbf{x}(N) + \mathbf{u}^T(N)\mathbf{R}(N)\mathbf{u}(N) \tag{10-19}$$

Since $\mathbf{x}(N)$ is independent of $\mathbf{u}(N)$, $\mathbf{u}^o(N) = \mathbf{0}$ and

$$S_1^o = \mathbf{x}^T(N)\mathbf{Q}(N)\mathbf{x}(N) \tag{10-20}$$

We see then that S_1^o is quadratic in $\mathbf{x}(N)$.

For S_2, we get the expression, from (10-17), that

$$S_2 = S_1^o + \mathbf{x}^T(N-1)\mathbf{Q}(N-1)\mathbf{x}(N-1) + \mathbf{u}^T(N-1)\mathbf{R}(N-1)\mathbf{u}(N-1) \tag{10-21}$$

where, from (10-4) and (10-20),

$$S_1^o = [\mathbf{A}\mathbf{x}(N-1) + \mathbf{B}\mathbf{u}(N-1)]^T\mathbf{Q}(N)[\mathbf{A}\mathbf{x}(N-1) + \mathbf{B}\mathbf{u}(N-1)]\,|_{\mathbf{u}^o(N-1)} \tag{10-22}$$

With (10-22) substituted into (10-21), we solve

$$\frac{\partial S_2}{\partial \mathbf{u}(N-1)} = \mathbf{0}$$

for $\mathbf{u}(N-1)$, with the result of the form

$$\mathbf{u}^o(N-1) = -\mathbf{K}(N-1)\mathbf{x}(N-1) \tag{10-23}$$

The derivation of this result is straightforward and is given as Problem 10-1. We will not at this time solve for $\mathbf{K}(N-1)$. However, when (10-23) is substituted into (10-21), the result is seen to be quadratic in $\mathbf{x}(N-1)$. Hence we can write

$$S_2^o = \mathbf{x}^T(N-1)\mathbf{P}(N-1)\mathbf{x}(N-1) \tag{10-24}$$

where $\mathbf{P}(N-1)$ is symmetric.

Following the development in the paragraph above in solving for S_3, we see that S_3^o is quadratic. Hence we may write

$$S_3^o = \mathbf{x}^T(N-2)\mathbf{P}(N-2)\mathbf{x}(N-2) \tag{10-25}$$

Thus the general relationship of S_m^o is seen to be

$$S_m^o = \mathbf{x}^T(N-m+1)\mathbf{P}(N-m+1)\mathbf{x}(N-m+1) \tag{10-26}$$

From (10-4), we can then express S_m^o as

$$S_m^o = [\mathbf{A}\mathbf{x}(N-m) + \mathbf{B}\mathbf{u}(N-m)]^T\mathbf{P}(N-m+1)[\mathbf{A}\mathbf{x}(N-m) + \mathbf{B}\mathbf{u}(N-m)] \tag{10-27}$$

Next we substitute (10-27) into (10-17) to get the expression for S_{m+1}. Then we differentiate S_{m+1} with respect to $\mathbf{u}(N-m)$, set this to zero, and solve for $\mathbf{u}^o(N-m)$. The algebra becomes unwieldy, and to simplify this, we will drop the notational dependence of $(N-m)$. Then, in the final solution, we will reinsert this dependence.

From (10-27) and (10-17),

$$S_{m+1} = [\mathbf{A}\mathbf{x} + \mathbf{B}\mathbf{u}]^T\mathbf{P}(N-m+1)[\mathbf{A}\mathbf{x} + \mathbf{B}\mathbf{u}] + \mathbf{x}^T\mathbf{Q}\mathbf{x} + \mathbf{u}^T\mathbf{R}\mathbf{u} \tag{10-28}$$

Then, from (10-13), (10-14), and (10-15),

$$\frac{\partial S_{m+1}}{\partial \mathbf{u}} = \mathbf{B}^T\mathbf{P}(N-m+1)[\mathbf{A}\mathbf{x} + \mathbf{B}\mathbf{u}] + \mathbf{B}^T\mathbf{P}(N-m+1)[\mathbf{A}\mathbf{x} + \mathbf{B}\mathbf{u}] + 2\mathbf{R}\mathbf{u} = \mathbf{0}$$

or

$$2\mathbf{B}^T\mathbf{P}(N-m+1)\mathbf{A}\mathbf{x} + 2[\mathbf{B}^T\mathbf{P}(N-m+1)\mathbf{B} + \mathbf{R}]\mathbf{u} = \mathbf{0}$$

Thus the desired solution is

$$\mathbf{u}^o = -[\mathbf{B}^T\mathbf{P}(N-m+1)\mathbf{B} + \mathbf{R}]^{-1}\mathbf{B}^T\mathbf{P}(N-m+1)\mathbf{A}\mathbf{x} \tag{10-29}$$

Hence, from (10-23) and (10-29), the optimal gain matrix is

$$\begin{aligned}\mathbf{K}(N-m) = {} & [\mathbf{B}^T(N-m)\mathbf{P}(N-m+1)\mathbf{B}(N-m) + \mathbf{R}(N-m)]^{-1} \\ & \times \mathbf{B}^T(N-m)\mathbf{P}(N-m+1)\mathbf{A}(N-m)\end{aligned} \tag{10-30}$$

and

$$\mathbf{u}^o(N-m) = -\mathbf{K}(N-m)\mathbf{x}(N-m) \tag{10-31}$$

Next we develop the expression for S_{m+1}^o. In (10-28), from (10-31),

$$\mathbf{A}\mathbf{x} + \mathbf{B}\mathbf{u} = [\mathbf{A} - \mathbf{B}\mathbf{K}]\mathbf{x}$$

and

$$S_{m+1}^o = \{[\mathbf{A} - \mathbf{B}\mathbf{K}]\mathbf{x}\}^T\mathbf{P}(N-m+1)[\mathbf{A} - \mathbf{B}\mathbf{K}]\mathbf{x} + \mathbf{x}^T\mathbf{Q}\mathbf{x} + [\mathbf{K}\mathbf{x}]^T\mathbf{R}\mathbf{K}\mathbf{x}$$

or

$$S_{m+1}^o = \mathbf{x}^T\{[\mathbf{A} - \mathbf{B}\mathbf{K}]^T\mathbf{P}(N-m+1)[\mathbf{A} - \mathbf{B}\mathbf{K}] + \mathbf{Q} + \mathbf{K}^T\mathbf{R}\mathbf{K}\}\mathbf{x} \tag{10-32}$$

But, from (10-26),

$$S_{m+1}^o = \mathbf{x}^T\mathbf{P}(N-m)\mathbf{x} \tag{10-33}$$

Hence

$$\begin{aligned}\mathbf{P}(N-m) = [\mathbf{A}(N-m) - \mathbf{B}(N-m)\mathbf{K}(N-m)]^T\mathbf{P}(N-m+1) \\ \times [\mathbf{A}(N-m) - \mathbf{B}(N-m)\mathbf{K}(N-m)] \\ + \mathbf{Q}(N-m) + \mathbf{K}^T(N-m)\mathbf{R}(N-m)\mathbf{K}(N-m)\end{aligned} \tag{10-34}$$

The final design equations will now be summarized. The design progresses backward in time from $k = N$. We know the final value of the $\mathbf{P}$ matrix from (10-20) and (10-26).

$$\mathbf{P}(N) = \mathbf{Q}(N) \tag{10-35}$$

For a time-invariant system and cost function, from (10-30) we obtain the optimal gain-matrix expression

$$\mathbf{K}(N-m) = [\mathbf{B}^T\mathbf{P}(N-m+1)\mathbf{B} + \mathbf{R}]^{-1}\mathbf{B}^T\mathbf{P}(N-m+1)\mathbf{A} \tag{10-36}$$

Hence we can solve for $\mathbf{K}(N-1)$. And from (10-34),

$$\begin{aligned}\mathbf{P}(N-m) = [\mathbf{A} - \mathbf{B}\mathbf{K}(N-m)]^T\mathbf{P}(N-m+1)[\mathbf{A} - \mathbf{B}\mathbf{K}(N-m)] \\ + \mathbf{Q} + \mathbf{K}^T(N-m)\mathbf{R}\mathbf{K}(N-m)\end{aligned} \tag{10-37}$$

This equation can be expressed in other forms (see Problems 10-2 and 10-3). One of the simpler forms is

$$\mathbf{P}(N-m) = \mathbf{A}^T\mathbf{P}(N-m+1)[\mathbf{A} - \mathbf{B}\mathbf{K}(N-m)] + \mathbf{Q} \tag{10-38}$$

Thus we may solve either (10-37) or (10-38) for $\mathbf{P}(N-1)$. We see then that (10-36) and (10-37) or (10-38) form a set of nonlinear difference equations which may be solved recursively for $\mathbf{K}(N-m)$ and $\mathbf{P}(N-m)$. Thus (10-35), (10-36), and (10-38) form the design equations. Note from (10-16) that

$$\begin{aligned}J_N &= J_N - J_0 + J_0 = S_N + J_0 \\ &= S_N + \mathbf{x}^T(0)\mathbf{Q}(0)\mathbf{x}(0) + \mathbf{u}^T(0)\mathbf{R}(0)\mathbf{u}(0)\end{aligned} \tag{10-39}$$

But from (10-28), this expression is simply S_{N+1}. Thus from (10-26), the minimum cost is given by

$$J_N^o = S_{N+1}^o = \mathbf{x}^T(0)\mathbf{P}(0)\mathbf{x}(0) \tag{10-40}$$

For the case that the system and/or the cost function are time varying, the following substitutions must be made in (10-36) and (10-38):

$$\begin{array}{ccc} \textit{Replace} & & \textit{with} \\ \mathbf{A} & \longrightarrow & \mathbf{A}(N-m) \\ \mathbf{B} & \longrightarrow & \mathbf{B}(N-m) \\ \mathbf{Q} & \longrightarrow & \mathbf{Q}(N-m) \\ \mathbf{R} & \longrightarrow & \mathbf{R}(N-m) \end{array} \tag{10-41}$$

Two examples illustrating this design procedure will now be given.

Example 10.1

We wish to design an optimal control law for the system

$$x(k+1) = 2x(k) + u(k)$$

with the cost function

$$J_2 = \sum_{k=0}^{2} x^2(k) + u^2(k)$$

Note that this is the same design problem considered in Section 10.3. Now, the required parameters are

$$A = 2 \qquad Q = 1$$
$$B = 1 \qquad R = 1$$

From (10-35),

$$P(2) = Q(2) = Q = 1$$

From (10-36), the gain required is

$$K(2-1) = K(1) = [B^T P(2) B + R]^{-1} B^T P(2) A$$
$$= [1 + 1]^{-1}(1)(1)(2) = 1$$

From (10-38),

$$P(1) = A^T P(2)\{A - BK(1)\} + Q$$
$$= 2(1)\{2 - (1)(1)\} + 1 = 3$$

Then, from (10-36),

$$K(0) = [B^T P(1) B + R]^{-1} B^T P(1) A$$
$$= [3 + 1]^{-1}(1)(3)(2) = 1.5$$

and from (10-38),

$$P(0) = 2(3)\{2 - 1.5\} + 1 = 4$$

Hence the optimal gain schedule is

$$\{K(0), K(1)\} = \{1.5, 1\}$$

and, from (10-39), the minimum cost is

$$J_2 = x^T(0)P(0)x(0) = 4x^2(0)$$

The results check those obtained in Section 10.3.

Example 10.2

As a second example, we consider the servo system utilized in several examples of pole-assignment design in Chapter 9. The system is shown in Figure 10-1, and has the state model (see Example 9.1)

$$\mathbf{x}(k+1) = \begin{bmatrix} 1 & 0.0952 \\ 0 & 0.905 \end{bmatrix} \mathbf{x}(k) + \begin{bmatrix} 0.00484 \\ 0.0952 \end{bmatrix} u(k)$$

$$y(k) = [1 \quad 0]\mathbf{x}(k)$$

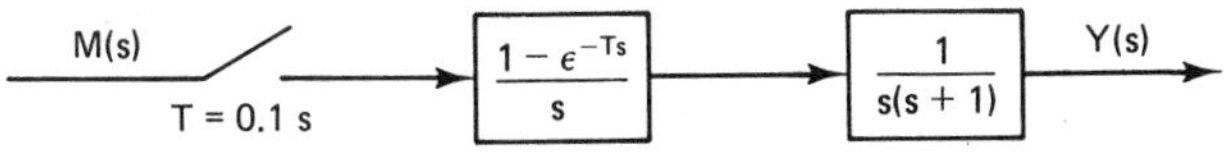

Figure 10-1 Servomotor system.

We will choose the cost function to be

$$J_2 = \sum_{k=0}^{2} \mathbf{x}^T(k)\mathbf{Q}\mathbf{x}(k) + Ru^2(k)$$

with

$$\mathbf{Q} = \begin{bmatrix} 1 & 0 \\ 0 & 0 \end{bmatrix}, \qquad R = 1$$

Thus velocity is ignored in the cost function. We are attempting to minimize the magnitude of the position $x_1(k)$ without regard to the velocity $x_2(k)$ required. The weight of the control, R, in the cost function is chosen arbitrarily. N is chosen equal to 2 so that the solution can be calculated by hand. Thus, from (10-35),

$$\mathbf{P}(2) = \mathbf{Q} = \begin{bmatrix} 1 & 0 \\ 0 & 0 \end{bmatrix}$$

In (10-36), for $m = 1$,

$$\mathbf{B}^T\mathbf{P}(2)\mathbf{B} + R = [0.00484 \quad 0.0952]\begin{bmatrix} 1 & 0 \\ 0 & 0 \end{bmatrix}\begin{bmatrix} 0.00484 \\ 0.0952 \end{bmatrix} + 1$$

$$= [0.00484 \quad 0]\begin{bmatrix} 0.00484 \\ 0.0952 \end{bmatrix} + 1 = 1.0000234$$

Then, in (10-36),

$$\mathbf{K}(1) = [\mathbf{B}^T\mathbf{P}(2)\mathbf{B} + R]^{-1}\mathbf{B}^T\mathbf{P}(2)\mathbf{A}$$

$$= \frac{1}{1.0000234}[0.00484 \quad 0.0952]\begin{bmatrix} 1 & 0 \\ 0 & 0 \end{bmatrix}\begin{bmatrix} 1 & 0.0952 \\ 0 & 0.905 \end{bmatrix}$$

$$= [0.00484 \quad 0.000461]$$

To calculate $\mathbf{P}(1)$ from (10-38), we need

$$\mathbf{BK}(1) = \begin{bmatrix} 0.00484 \\ 0.0952 \end{bmatrix}[0.00484 \quad 0.000461]$$

$$= \begin{bmatrix} 2.34 \times 10^{-5} & 2.23 \times 10^{-6} \\ 4.61 \times 10^{-4} & 4.39 \times 10^{-5} \end{bmatrix}$$

Then

$$[\mathbf{A} - \mathbf{BK}(1)] = \begin{bmatrix} 1.00 & 0.0952 \\ -4.61 \times 10^{-4} & 0.905 \end{bmatrix}$$

and

$$\mathbf{A}^T\mathbf{P}(2)[\mathbf{A} - \mathbf{BK}(1)] = \begin{bmatrix} 1 & 0.0952 \\ 0.0952 & 0.00906 \end{bmatrix}$$

Then, from (10-38),

$$\mathbf{P}(1) = \mathbf{A}^T\mathbf{P}(2)[\mathbf{A} - \mathbf{BK}(1)] + \mathbf{Q}$$

$$= \begin{bmatrix} 2.0 & 0.0952 \\ 0.0952 & 0.00906 \end{bmatrix}$$

The calculations above are carried through in detail to illustrate the calculations required for one step of the solution of the difference equations (10-36) and (10-38) for a second-order system with a single input. Obviously, a computer solution is required. If these difference equations are solved again, the results are

$$\mathbf{K}(0) = [0.0187 \quad 0.00298]$$

and

$$\mathbf{P}(0) = \begin{bmatrix} 3.0 & 0.276 \\ 0.276 & 0.0419 \end{bmatrix}$$

Example 10.3

Next a more practical design will be performed, using a computer solution; that is, (10-36) and (10-38) are solved recursively via the computer. We will compare the results to those of Example 9.1. In Example 9.1 the plant equations are (see Figure 10-1)

$$\mathbf{x}(k+1) = \begin{bmatrix} 1 & 0.0952 \\ 0 & 0.905 \end{bmatrix} \mathbf{x}(k) + \begin{bmatrix} 0.00484 \\ 0.0952 \end{bmatrix} u(k)$$

and the closed-loop characteristic equation zeros had a time constant of 1.0 s. We generally consider the response to have settled out in five time constants, or 5 s in this case. Hence we will choose N of the cost function of (10-3) to be 51 and compare responses for the first 5 s. The **Q** and **R** matrices of (10-3) are chosen to be the same as in Example 10.2, that is,

$$\mathbf{Q} = \begin{bmatrix} 1 & 0 \\ 0 & 0 \end{bmatrix}, \qquad R = 1$$

The optimal gains obtained for this design are plotted in Figure 10-2. Also shown are the optimal gains for values of R of 0.1 and 0.03, with **Q** unchanged. As the value of R is reduced, the contribution of the control effort to the cost function is reduced. Hence the control effort will be increased in order to reduce the magnitudes of the states. This effect is seen in Figure 10-2.

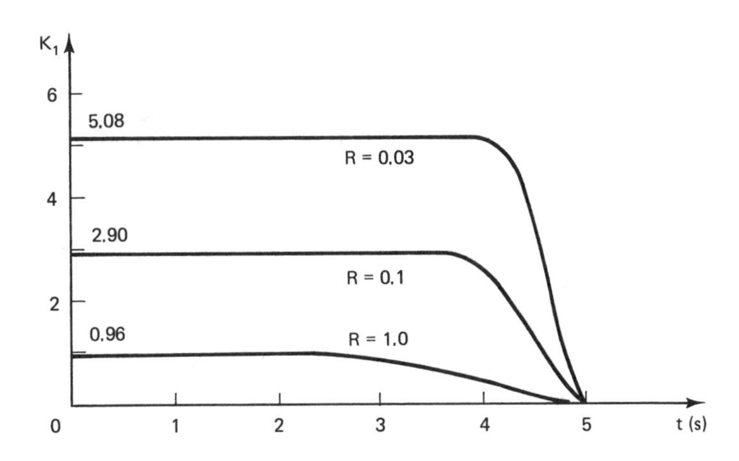

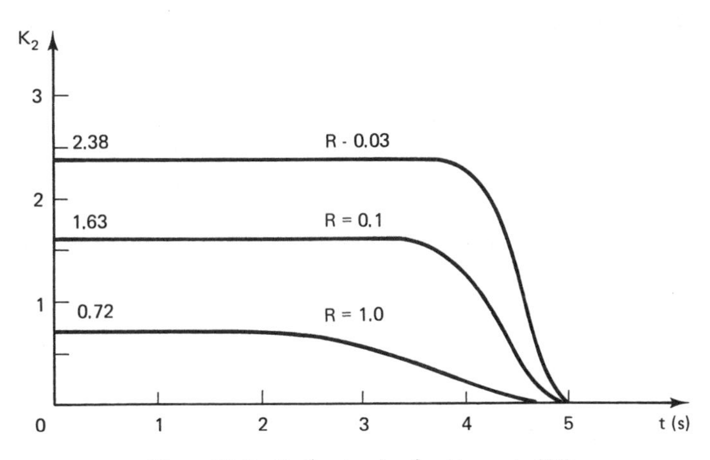

Figure 10-2 Optimal gains for Example 10.3.

Shown in Figure 10-3 is the initial-condition response, for $x_1(t)$, of this system for $R = 1.0$ and $R = 0.03$, with

$$\mathbf{x}(0) = \begin{bmatrix} 1 \\ 0 \end{bmatrix}$$

The effects of the choice of R on the transient response of the closed-loop system can be seen from this figure. Also, by comparing Figures 10-2 and 10-3, we see that the response has reached zero before the gains begin decreasing for the case that $R = 0.03$. This is not true for the case that $R = 1.0$.

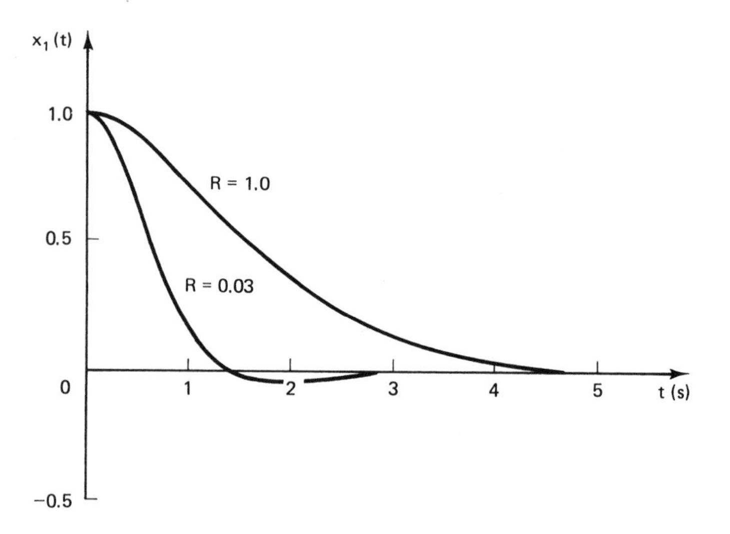

Figure 10-3 Time response for Example 10.3.

The initial-condition response for Example 9.1 is given in Figure 9-2. It is seen that the optimal system response settles faster for $R = 0.03$, but slower for $R = 1.0$.

10.5 THE MINIMUM PRINCIPLE

In the developments above, Bellman's principle was used in deriving the optimal control design equations (10-36) and (10-38). The equations may also be derived utilizing the minimum principle [4]. The minimum principle is presented in this section. This principle will prove to be useful when steady-state optimal control is considered in the following section.

The optimal control problem posed in Section 10.2 may be stated as follows:

For a linear discrete system described by (10-4),

$$\begin{aligned} \mathbf{x}(k+1) &= \mathbf{A}(k)\mathbf{x}(k) + \mathbf{B}(k)\mathbf{u}(k) \\ \mathbf{y}(k) &= \mathbf{C}(k)\mathbf{x}(k) \end{aligned} \tag{10-4}$$

determine the control in (10-7),

$$\mathbf{u}^o(k) = \mathbf{f}[\mathbf{x}(k)] \tag{10-7}$$

that minimizes the quadratic cost function of (10-3),

$$J_N = \frac{1}{2} \sum_{k=0}^{N} \mathbf{x}^T(k)\mathbf{Q}(k)\mathbf{x}(k) + \mathbf{u}^T(k)\mathbf{R}(k)\mathbf{u}(k) \tag{10-3}$$

where N is finite, $\mathbf{Q}(k)$ is positive semidefinite, and $\mathbf{R}(k)$ is positive definite. In (10-7), the superscript o denotes that the control is optimal.

Note that the factor $\frac{1}{2}$ added to (10-3) will not change the results, but will simplify the following derivations somewhat.

The minimum principle may be utilized to solve the optimal control problem.

Minimum Principle [4]. If the input $\mathbf{u}^o(k)$ and the corresponding trajectory $\mathbf{x}^o(k)$ are optimal, then there exists a nontrival vector sequence $\{\mathbf{p}^o(k)\}$ such that $\mathbf{u}^o(k)$ is the value of $\mathbf{u}(k)$ which minimizes the Hamiltonian

$$H = \frac{1}{2}[\mathbf{x}^{oT}(k)\mathbf{Q}(k)\mathbf{x}^o(k) + \mathbf{u}^T(k)\mathbf{R}(k)\mathbf{u}(k)] + [\mathbf{p}^o(k+1)]^T[\mathbf{A}(k)\mathbf{x}^o(k) + \mathbf{B}(k)\mathbf{u}(k)] \tag{10-42}$$

and the "costate vector" $\mathbf{p}^o(k)$ satisfies

$$\mathbf{p}^o(k) = \frac{\partial H}{\partial \mathbf{x}^o(k)}, \qquad \mathbf{p}^o(N) = \mathbf{Q}(N)\mathbf{x}^o(N) \tag{10-43}$$

for $k \leq N$.

From (10-43) we obtain

$$\mathbf{p}^o(k) = \mathbf{Q}(k)\mathbf{x}^o(k) + \mathbf{A}^T(k)\mathbf{p}^o(k+1) \tag{10-44}$$

and $\partial H/\partial \mathbf{u}(k) = 0$ yields

$$\mathbf{u}^o(k) = -\mathbf{R}^{-1}(k)\mathbf{B}^T(k)\mathbf{p}^o(k+1) \tag{10-45}$$

Hence this approach leads to the state equations (10-4),

$$\mathbf{x}(k+1) = \mathbf{A}(k)\mathbf{x}(k) + \mathbf{B}(k)\mathbf{u}(k) \tag{10-4}$$

and the costate equations (10-44), with the optimal control given by (10-45). The state equations will have an initial condition $\mathbf{x}(k_o)$, and are solved forward in time. The costate equations have the final state (10-43), and are solved backward in time. However, since the state and costate equations are coupled, these equations constitute a linear two-point boundary value problem. It is beyond the scope of this text to discuss the solution of two-point boundary value problems. However, it can be shown that the solution of (10-4), (10-44), and (10-45) yields the same optimal gain matrix as does the solution of (10-36) and (10-38).

10.6 STEADY-STATE OPTIMAL CONTROL

If we consider the design results of Example 10.3, given in Figure 10-2, we see that the optimal gains approach constant values with decreasing time. In general, this is true for time-invariant systems and cost functions.

Note also that if we increase N in Example 10.3, the optimal gains for the final 51 sample instants are the same as those calculated in this example. The values of the gains at earlier sampling instants will simply be those constant values indicated in Figure 10-2. Hence if we allow N to approach infinity, or if we allow the initial time to approach $-\infty$, we obtain a steady-state solution in which the optimal gains are constant values. We consider this steady-state solution in this section.

First we consider the difference equations, (10-36) and (10-38), employed in the design of optimal control systems.

$$\mathbf{K}(N-m) = [\mathbf{B}^T\mathbf{P}(N-m+1)\mathbf{B} + \mathbf{R}]^{-1}\mathbf{B}^T\mathbf{P}(N-m+1)\mathbf{A} \quad (10\text{-}36)$$

$$\mathbf{P}(N-m) = \mathbf{A}^T\mathbf{P}(N-m+1)[\mathbf{A} - \mathbf{B}\mathbf{K}(N-m)] + \mathbf{Q} \quad (10\text{-}38)$$

In this section **A, B, Q,** and **R** are assumed to be constant. If (10-36) is substituted into (10-38), the result is a difference equation for the matrix **P**.

$$\begin{aligned}\mathbf{P}(N-m) = \mathbf{A}^T\mathbf{P}(N-m+1)\{&\mathbf{A} - \mathbf{B}[\mathbf{B}^T\mathbf{P}(N-m+1)\mathbf{B} + \mathbf{R}]^{-1} \\ &\times \mathbf{B}^T\mathbf{P}(N-m+1)\mathbf{A}\} + \mathbf{Q}\end{aligned} \quad (10\text{-}46)$$

This equation is often written in a slightly different form,

$$\begin{aligned}\mathbf{P}(N-m) = \mathbf{A}^T\mathbf{P}(N-m+1)\mathbf{A} + \mathbf{Q} - \mathbf{A}^T\mathbf{P}(N-m+1)& \\ \times \mathbf{B}[\mathbf{B}^T\mathbf{P}(N-m+1)\mathbf{B} + \mathbf{R}]^{-1}\mathbf{B}^T\mathbf{P}(N-m+1)\mathbf{A}&\end{aligned} \quad (10\text{-}47)$$

and is referred to as the *discrete Riccati equation.* The inverse in (10-47) always exists since **R** is positive definite and $\mathbf{P}(N-m+1)$ is at least positive semidefinite [the cost function in (10-3) and (10-26) cannot be negative].

Before deriving the steady-state solution to (10-47), we will derive a nonrecursive solution for $\mathbf{P}(N-m)$ in the discrete Riccati equation for time-invariant systems. This solution will then yield the steady-state solution. We can express the state and costate equations of (10-4) and (10-44) as

$$\begin{bmatrix}\mathbf{x}^o(k)\\ \mathbf{p}^o(k)\end{bmatrix} = \mathcal{H}\begin{bmatrix}\mathbf{x}^o(k+1)\\ \mathbf{p}^o(k+1)\end{bmatrix} \quad (10\text{-}48)$$

where

$$\mathcal{H} = \begin{bmatrix}\mathbf{A}^{-1} & \mathbf{A}^{-1}\mathbf{R}_C \\ \mathbf{Q}\mathbf{A}^{-1} & \mathbf{A}^T + \mathbf{Q}\mathbf{A}^{-1}\mathbf{R}_C\end{bmatrix} \quad (10\text{-}49)$$

and

$$\mathbf{R}_C = \mathbf{B}\mathbf{R}^{-1}\mathbf{B}^T \quad (10\text{-}50)$$

It is shown in Ref. 5 that for a matrix

$$\mathcal{H} = \begin{bmatrix}\mathbf{D} & \mathbf{E}\\ \mathbf{F} & \mathbf{G}\end{bmatrix} \quad (10\text{-}51)$$

where the partitions **D, E, F,** and **G** are $n \times n$, then

$$\mathcal{H}^{-1} = \begin{bmatrix}\mathbf{D}^{-1} + \mathbf{D}^{-1}\mathbf{E}[\mathbf{G} - \mathbf{F}\mathbf{D}^{-1}\mathbf{E}]^{-1}\mathbf{F}\mathbf{D}^{-1} & -\mathbf{D}^{-1}\mathbf{E}[\mathbf{G} - \mathbf{F}\mathbf{D}^{-1}\mathbf{E}]^{-1} \\ -[\mathbf{G} - \mathbf{F}\mathbf{D}^{-1}\mathbf{E}]^{-1}\mathbf{F}\mathbf{D}^{-1} & [\mathbf{G} - \mathbf{F}\mathbf{D}^{-1}\mathbf{E}]^{-1}\end{bmatrix} \quad (10\text{-}52)$$

Hence, from (10-49), (10-51), and (10-52),

$$\mathcal{H}^{-1} = \begin{bmatrix}\mathbf{A} + \mathbf{R}_C\mathbf{A}^{-T}\mathbf{Q} & -\mathbf{R}_C\mathbf{A}^{-T} \\ -\mathbf{A}^{-T}\mathbf{Q} & \mathbf{A}^{-T}\end{bmatrix} \quad (10\text{-}53)$$

where

$$\mathbf{A}^{-T} = (\mathbf{A}^{-1})^T = (\mathbf{A}^T)^{-1} \quad (10\text{-}54)$$

If we denote $\mathbf{h}$ as an eigenvector of $\mathcal{H}$, and λ as an eigenvalue, then

$$\mathcal{H}\mathbf{h} = \lambda \mathbf{h} \tag{10-55}$$

Let

$$\mathbf{h} = \begin{bmatrix} \mathbf{f} \\ \mathbf{g} \end{bmatrix} \tag{10-56}$$

where $\mathbf{f}$ and $\mathbf{g}$ are n-vectors. Then, from (10-53) and (10-55),

$$\mathcal{H}^{-T}\begin{bmatrix} \mathbf{g} \\ -\mathbf{f} \end{bmatrix} = \lambda \begin{bmatrix} \mathbf{g} \\ -\mathbf{f} \end{bmatrix} \tag{10-57}$$

Thus λ is an eigenvalue of $\mathcal{H}^{-T}$ and hence $\mathcal{H}^{-1}$, and therefore $1/\lambda$ is an eigenvalue of $\mathcal{H}$. Thus the eigenvalues of $\mathcal{H}$ are such that the reciprocal of every eigenvalue is also an eigenvalue.

Next we define the similarity transformation $\mathbf{W}$ (see Section 2.9) such that

$$\begin{bmatrix} \mathbf{x}^o(k) \\ \mathbf{p}^o(k) \end{bmatrix} = \mathbf{W}\begin{bmatrix} \mathbf{v}(k) \\ \mathbf{s}(k) \end{bmatrix} \tag{10-58}$$

Then, for the case of distinct eigenvalues of $\mathcal{H}$, from (10-48) we write

$$\begin{bmatrix} \mathbf{v}(k) \\ \mathbf{s}(k) \end{bmatrix} = \begin{bmatrix} \mathbf{\Lambda} & 0 \\ 0 & \mathbf{\Lambda}^{-1} \end{bmatrix}\begin{bmatrix} \mathbf{v}(k+1) \\ \mathbf{s}(k+1) \end{bmatrix} = \mathbf{W}^{-1}\mathcal{H}\mathbf{W}\begin{bmatrix} \mathbf{v}(k+1) \\ \mathbf{s}(k+1) \end{bmatrix} \tag{10-59}$$

where $\mathbf{\Lambda}$ is a diagonal matrix of the eigenvalues of $\mathcal{H}$ that occur outside the unit circle. Then, in (10-58), the columns of $\mathbf{W}$ are the eigenvectors of $\mathcal{H}$, and $\mathbf{W}$ is the modal matrix.

Let the matrix $\mathbf{W}$ be partitioned into four $n \times n$ matrices.

$$\mathbf{W} = \begin{bmatrix} \mathbf{W}_{11} & \mathbf{W}_{12} \\ \mathbf{W}_{21} & \mathbf{W}_{22} \end{bmatrix} \tag{10-60}$$

Since, from (10-43),

$$\mathbf{p}^o(N) = \mathbf{Q}\mathbf{x}^o(N)$$

from (10-58) and (10-60),

$$[\mathbf{W}_{21}\mathbf{v}(N) + \mathbf{W}_{22}\mathbf{s}(N)] = \mathbf{Q}[\mathbf{W}_{11}\mathbf{v}(N) + \mathbf{W}_{12}\mathbf{s}(N)] \tag{10-61}$$

Thus

$$\mathbf{s}(N) = -[\mathbf{W}_{22} - \mathbf{Q}\mathbf{W}_{12}]^{-1}[\mathbf{W}_{21} - \mathbf{Q}\mathbf{W}_{11}]\mathbf{v}(N) = \mathbf{T}\mathbf{v}(N) \tag{10-62}$$

Also, for (10-59),

$$\mathbf{v}(N) = \mathbf{\Lambda}^{-k}\mathbf{v}(N-k) \tag{10-63}$$

$$\mathbf{s}(N) = \mathbf{\Lambda}^{k}\mathbf{s}(N-k) \tag{10-64}$$

Thus, from (10-62),

$$\mathbf{s}(N-k) = \mathbf{\Lambda}^{-k}\mathbf{T}\mathbf{\Lambda}^{-k}\mathbf{v}(N-k) = \mathbf{G}(k)\mathbf{v}(N-k) \tag{10-65}$$

From (10-58),

$$\mathbf{x}^o(N-k) = \mathbf{W}_{11}\mathbf{v}(N-k) + \mathbf{W}_{12}\mathbf{s}(N-k) \tag{10-66}$$

$$\mathbf{p}^o(N-k) = \mathbf{W}_{21}\mathbf{v}(N-k) + \mathbf{W}_{22}\mathbf{s}(N-k) \tag{10-67}$$

Eliminating $\mathbf{v}(N-k)$ and $\mathbf{s}(N-k)$ for (10-65), (10-66), and (10-67) yields

$$\begin{aligned}\mathbf{p}^o(N-k) &= [\mathbf{W}_{21} + \mathbf{W}_{22}\mathbf{G}(k)][\mathbf{W}_{11} + \mathbf{W}_{12}\mathbf{G}(k)]^{-1}\mathbf{x}^o(N-k) \\ &= \mathbf{M}(k)\mathbf{x}^o(N-k)\end{aligned} \tag{10-68}$$

This equation is a nonrecursive solution to the state and costate equations (10-4) and (10-44).

To find the optimal gain, $\mathbf{K}(N-m)$, of (10-36), we let $N-k = l+1$ in (10-68).

$$\mathbf{p}^o(l+1) = \mathbf{M}(N-l-1)\mathbf{x}^o(l+1) = \mathbf{M}[N-l-1][\mathbf{A}\mathbf{x}^o(l) + \mathbf{B}\mathbf{u}^o(l)] \tag{10-69}$$

Since the optimal control from (10-45) is

$$\mathbf{u}^o(l) = -\mathbf{R}^{-1}\mathbf{B}^T\mathbf{p}^o(l+1)$$

solving these two equations for $\mathbf{u}^o(l)$ yields

$$\mathbf{u}^o(l) = -[\mathbf{R} + \mathbf{B}^T\mathbf{M}(N-l-1)\mathbf{B}]^{-1}\mathbf{B}^T\mathbf{M}(N-l-1)\mathbf{A}\mathbf{x}^o(l)$$

Now, letting $l = N-m$,

$$\mathbf{u}^o(N-m) = -[\mathbf{R} + \mathbf{B}^T\mathbf{M}(m-1)\mathbf{B}]^{-1}\mathbf{B}^T\mathbf{M}(m-1)\mathbf{A}\mathbf{x}^o(N-m) \tag{10-70}$$

Comparing this equation with (10-31) and (10-36), we see that

$$\mathbf{M}(m-1) = \mathbf{P}(N-m+1) \tag{10-71}$$

Then, from (10-65) and (10-68),

$$\mathbf{P}(N-m+1) = [\mathbf{W}_{21} + \mathbf{W}_{22}\mathbf{G}(m-1)][\mathbf{W}_{11} + \mathbf{W}_{12}\mathbf{G}(m-1)]^{-1} \tag{10-72}$$

where

$$\mathbf{G}(m-1) = -\mathbf{\Lambda}^{-(m-1)}[\mathbf{W}_{22} - \mathbf{Q}\mathbf{W}_{12}]^{-1}[\mathbf{W}_{21} - \mathbf{Q}\mathbf{W}_{11}]\mathbf{\Lambda}^{-(m-1)} \tag{10-73}$$

The optimal gain is given by, from (10-70),

$$\mathbf{K}(N-m) = [\mathbf{R} + \mathbf{B}^T\mathbf{M}(m-1)\mathbf{B}]^{-1}\mathbf{B}^T\mathbf{M}(m-1)\mathbf{A} \tag{10-74}$$

For the steady-state solution such that the gains in (10-36) have become constant values, it must then be true in (10-47) that

$$\mathbf{P}(N-m) = \mathbf{P}(N-m+1) = \text{constant matrix} \tag{10-75}$$

We will denote this contant matrix as $\hat{\mathbf{P}}$. Then (10-47) becomes

$$\hat{\mathbf{P}} = \mathbf{A}^T\hat{\mathbf{P}}\mathbf{A} + \mathbf{Q} - \mathbf{A}^T\hat{\mathbf{P}}\mathbf{B}[\mathbf{B}^T\hat{\mathbf{P}}\mathbf{B} + \mathbf{R}]^{-1}\mathbf{B}^T\hat{\mathbf{P}}\mathbf{A} \tag{10-76}$$

This equation is referred to as the *algebraic Riccati equation.* Perhaps the simplest approach to finding the solution of this equation is that indicated in Example 10.3—set N to a large value and calculate the values of the $\mathbf{P}$ matrix (by computer) until the matrix elements become constant values. Then we have the solution to (10-76).

We may also find the solution to the algebraic Riccati equation from the foregoing nonrecursive solution to the discrete Riccati equation. In (10-68),

$$\mathbf{p}^o(N-k) = \mathbf{M}(k)\mathbf{x}^o(N-k) \tag{10-77}$$

The steady-state solution is obtained by allowing k to approach $-\infty$. From (10-65) and (10-68)

$$\lim_{k\to\infty} \mathbf{M}(k) = \mathbf{W}_{21}\mathbf{W}_{11}^{-1} \tag{10-78}$$

since $\mathbf{\Lambda}$ contains the eigenvalues outside the unit circle, and thus

$$\lim_{k\to\infty} \mathbf{\Lambda}^{-k} = \mathbf{0}$$

Hence, from (10-71), the solution to the algebraic Riccati equation, (10-76), is

$$\hat{\mathbf{P}} = \mathbf{W}_{21}\mathbf{W}_{11}^{-1} \tag{10-79}$$

Example 10.4

The nonrecursive method of solution of the optimal control problem will be illustrated using the first-order system of Example 10.1. We wish to design an optimal control law for the system

$$x(k+1) = 2x(k) + u(k)$$

with the cost function

$$J_N = \sum_{k=0}^{N} x^2(k) + u^2(k)$$

Thus the required parameters are

$$A = 2 \qquad Q = 1$$

$$B = 1 \qquad R = 1$$

and

$$R_C = BR^{-1}B^T = 1$$

Then in (10-49),

$$\mathcal{H} = \begin{bmatrix} A^{-1} & A^{-1}R_C \\ QA^{-1} & A^T + QA^{-1}R_C \end{bmatrix} = \begin{bmatrix} 0.5 & 0.5 \\ 0.5 & 2.5 \end{bmatrix}$$

The eigenvalues of $\mathcal{H}$ satisfy the equation

$$|\lambda\mathbf{I} - \mathcal{H}| = (\lambda - 0.5)(\lambda - 2.5) - 0.25 = \lambda^2 - 3\lambda + 1 = 0$$

Thus the eigenvalues are 2.618 and 0.382. The eigenvectors satisfy the equation

$$\mathcal{H}\mathbf{h} = \lambda\mathbf{h}$$

Thus the eigenvectors are

$$\mathbf{h}_1 = \begin{bmatrix} 1 \\ 4.237 \end{bmatrix}, \qquad \mathbf{h}_2 = \begin{bmatrix} 1 \\ -0.236 \end{bmatrix}$$

and the similarity transformation $\mathbf{W}$ is

$$\mathbf{W} = [\mathbf{h}_1 \quad \mathbf{h}_2] = \begin{bmatrix} 1 & 1 \\ 4.237 & -0.236 \end{bmatrix} = \begin{bmatrix} W_{11} & W_{12} \\ W_{21} & W_{22} \end{bmatrix}$$

Then, from (10-59),

$$\begin{bmatrix} \Lambda & 0 \\ 0 & \Lambda^{-1} \end{bmatrix} = \mathbf{W}^{-1}\mathcal{H}\mathbf{W} = \begin{bmatrix} 2.618 & 0 \\ 0 & 0.382 \end{bmatrix}$$

From (10-62),

$$T = -[W_{22} - QW_{12}]^{-1}[W_{21} - QW_{11}] = 2.619$$

and from (10-65),

$$G(k) = \Lambda^{-k}T\Lambda^{-k} = 2.619(0.382)^{2k}$$

From (10-71) and (10-72),

$$M(m-1) = \frac{4.237 - 0.618(0.382)^{2(m-1)}}{1 + 2.619(0.382)^{2(m-1)}}$$

and the optimal gain, from (10-74), is

$$K(N-m) = \frac{2M(m-1)}{1+M(m-1)}$$

For $N = 2$ and $m = 1$,

$$K(2-1) = \frac{2(1)}{1+1} = 1$$

and for $m = 2$,

$$K(0) = \frac{2(3)}{1+3} = 1.5$$

These values check those of Example 10.1. The steady-state value, $\hat{P}$, is obtained from (10-79).

$$\hat{P} = W_{21}W_{11}^{-1} = 4.237$$

and the steady-state gain is

$$\hat{K} = \frac{2(4.237)}{1+4.237} = 1.618$$

Some general results regarding the steady-state solution of the discrete Riccati equation, (10-47), will now be given. Note that the steady-solution of this equation is the solution with N finite and $m \longrightarrow \infty$.

First we define the term *stabilizable*.

Definition. The discrete-time system (10-4) is said to be stabilizable if there exists a matrix $\mathbf{K}$ such that the eigenvalues of $\mathbf{A} - \mathbf{BK}$ are all inside the unit circle.

Then we may state the following theorem.

Theorem 1. If the system (10-4) is either controllable or stabilizable, the discrete Riccati equation (10-47) has a limiting solution as $m \longrightarrow \infty$; that is,

$$\lim_{m\to\infty} \mathbf{P}(N-m) = \hat{\mathbf{P}}$$

with N finite.

We may prove this theorem in the following manner. First we must show that $\mathbf{P}(N-m)$ is bounded above. The optimal cost from the $(N-m)$ sample instant to termination is given by

$$S_m^o = \mathbf{x}^T(N-m+1)\mathbf{P}(N-m+1)\mathbf{x}(N-m+1) \qquad (10\text{-}26)$$

Since, if the system is either controllable or stabilizable, a control law can be chosen to make the system asymptotically stable (initial condition response goes to zero with increasing time), the optimal cost must be finite for the infinite-time case. Hence $\mathbf{P}(N-m)$ is bounded above.

Next we must show that

$$\mathbf{P}(N-m) \geq \mathbf{P}(N-m+1) \qquad (10\text{-}80)$$

Then the proof is complete. If the optimal control $\mathbf{u}^o(k)$ over the interval $N-m < k < N$ is used in evaluating the performance index, we may write

$$S_{m-1}^o \geq S_m^o \qquad (10\text{-}81)$$

Or, from (10-26) and (10-1),

$$\mathbf{x}^T(N-m)\mathbf{P}(N-m)\mathbf{x}(N-m) \geq \sum_{k=N-m}^{N} L[\mathbf{x}^o(k), \mathbf{u}^o(k)] \tag{10-82}$$

Let $\mathbf{u}^*(k)$ be optimal over the interval $N - m < k < N - 1$. Then

$$\sum_{k=N-m}^{N-1} L[\mathbf{x}^o(k), \mathbf{u}^o(k)] \geq \sum_{k=N-m}^{N-1} L[\mathbf{x}^*(k), \mathbf{u}^*(k)] \tag{10-83}$$

where $\mathbf{x}^o(k)$ denotes the trajectory for $\mathbf{u}^o(k)$, starting at $\mathbf{x}^o(n-m) = \mathbf{x}(n-m)$. The equivalent definition applies for $\mathbf{x}^*(k)$. But the right-hand side of (10-82) may be written as

$$\sum_{k=N-m}^{N-1} L[\mathbf{x}^*(k), \mathbf{u}^*(k)] = \mathbf{x}^T(N-m+1)\mathbf{P}(N-m+1)\mathbf{x}(N-m+1) \tag{10-84}$$

since the time invariance of the system and cost function allows a shift in time from the interval $[N-m, N-1]$ to $[N-m+1, N]$. Hence (10-80) is satisfied. This completes the proof.

Finally we present the following theorem [4].

Theorem 2. If the system (10-4) is observable, then $\hat{\mathbf{P}}$ is positive definite and the optimal closed-loop system is asymptotically stable.

The proof of this theorem [4] is beyond the scope of this text.

10.7 SUMMARY

Presented in this chapter were some basic results in linear quadratic optimal control design. Once the cost function has been chosen, the design involves the straightforward solution of a difference equation. Even though the basic formulation is for a finite-time problem, the design procedure is easily extended to the infinite-time problem. The design implementation requires full-state feedback, with the feedback gains time-varying for the finite-time problem and constant for the infinite-time problem.

REFERENCES AND FURTHER READING

1. P. M. DeRusso, R. J. Roy, and C. M. Close, *State Variables for Engineers*. New York: John Wiley & Sons, Inc., 1965.
2. R. Bellman, *Adaptive Control Process: A Guided Tour*. Princeton, N.J.: Princeton University Press, 1961.
3. *IEEE Trans. Autom. Control*, Bellman Special Issue, Vol. AC-26, Oct. 1981.
4. P. Dorato and A. H. Levis, "Optimal Linear Regulators: The Discrete Time Case," *IEEE Trans. Autom. Control*, Vol. AC-16, pp. 613–620, Dec. 1971.
5. T. E. Fortman, "A Matrix Inversion Identity," *IEEE Trans. Autom. Control*, Vol. AC-15, p. 599, Oct. 1970.

6. R. Gran and F. Kozin, *Applied Digital Control Systems*. George Washington University Short Course Notes, Washington, D.C., Aug. 1979.

7. D. R. Vaughan, "A Nonrecursive Algebraic Solution for the Discrete Riccati Equation," *IEEE Trans. Autom. Control*, Vol. AC-15, pp. 597–599, Oct. 1970.

PROBLEMS

10-1. Given that equations (10-21) and (10-22) are valid, derive equation (10-23).

10-2. Show that equation (10-37) can also be expressed as

$$\mathbf{P}(N-m) = \mathbf{A}^T[\mathbf{P}(N-m+1) - \mathbf{P}(N-m+1)\mathbf{B}\mathbf{D}\mathbf{B}^T\mathbf{P}(N-m+1)]\mathbf{A} + \mathbf{Q}$$

where $\mathbf{D} = [\mathbf{B}^T\mathbf{P}(N-m+1)\mathbf{B} + \mathbf{R}]^{-1}$.

10-3. Show that equation (10-37) can also be expressed as

$$\mathbf{P}(N-m) = \mathbf{A}^T\mathbf{P}(N-m+1)[\mathbf{A} - \mathbf{B}\mathbf{K}(N-m)] + \mathbf{Q}$$

10-4. Given a discrete system described by

$$x(k+1) = 0.9x(k) + u(k)$$

with the cost function

$$J_5 = \sum_{k=0}^{5} x^2(k) + 10u^2(k)$$

Calculate the feedback gains required to minimize the cost function.

10-5. For the system of Problem 10-4, find the feedback gain required to minimize the given J for the infinite-time problem. Find the closed-loop system characteristic equation.

10-6. Given a discrete system described by

$$x(k+1) = 0.9x(k) + u(k)$$

with the cost function

$$J_5 = \sum_{k=0}^{5} x^2(k)$$

Calculate the feedback gains required to minimize the cost function, and comment on the differences in the gains calculated for this design and for that of Problem 10-4.

10-7. Given a first-order time-invariant discrete system with a cost function

$$J_N = \sum_{k=0}^{N} Qx^2(k) + Ru^2(k)$$

Show that the optimal gains are a function of only the ratio

$$\alpha = \frac{Q}{R}$$

and not of Q and R singly.

10-8. Given the discrete system

$$\mathbf{x}(k+1) = \mathbf{A}\mathbf{x}(k) + \mathbf{B}\mathbf{u}(k)$$

with the cost function

$$J_N = \sum_{k=0}^{N} \mathbf{x}^T(k)\mathbf{Q}\mathbf{x}(k) + \mathbf{u}^T(k)\mathbf{R}\mathbf{u}(k)$$

show that the optimal gains which minimize J_N are unchanged if each element both in $\mathbf{Q}$ and in $\mathbf{R}$ are multiplied by the positive scalar β.

10-9. A satellite may be modeled as shown in Figure P10-9 (see Chapter 1).

(a) Develop a discrete model for this system.

(b) Determine the gains required to minimize the cost function

$$J_{10} = \sum_{k=0}^{10} \mathbf{x}^T(k)\begin{bmatrix} 1 & 0 \\ 0 & 0.1 \end{bmatrix}\mathbf{x}(k) + 10u^2(k)$$

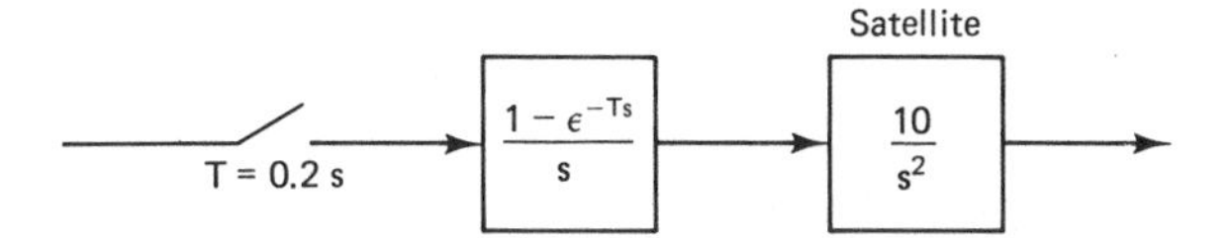

Figure P10-9 System for Problem 10-9.

10-10. For the satellite of Problem 10-9, the cost function to be minimized is given by

$$J_{10} = \sum_{k=0}^{10} \mathbf{x}^T(k)\begin{bmatrix} 1 & 0 \\ 0 & 0 \end{bmatrix}\mathbf{x}(k) + 10u^2(k)$$

Determine the optimal gains required.

10-11. Determine the gains required if Problem 10-9 is extended to an infinite-time problem. Find the closed-loop system characteristic equation.

10-12. Solve Problem 10-4 using the nonrecursive procedure of Section 10.6.

10-13. Solve Problem 10-5 using the infinite-time solution of Section 10.6. The results of Problem 10-12 are useful.

10-14. It is shown in Ref. 6 that, given the partitioned matrix

$$\mathcal{H} = \begin{bmatrix} \mathbf{D} & \mathbf{E} \\ \mathbf{F} & \mathbf{G} \end{bmatrix}$$

where each partition is $n \times n$, the determinant of $\mathcal{H}$ is given by

$$|\mathcal{H}| = |\mathbf{G}||\mathbf{D} - \mathbf{E}\mathbf{G}^{-1}\mathbf{F}| = |\mathbf{D}||\mathbf{G} - \mathbf{F}\mathbf{D}^{-1}\mathbf{E}|$$

Show that the determinant of $\mathcal{H}$ in (10-49) is equal to unity. Considering the development in Section 10.6, is this result expected?

11

Sampled-Data Transformation of Analog Filters

11.1 INTRODUCTION

In Chapters 8, 9, and 10 we presented methods for designing filters, for control systems, in the digital domain. Many times an application may require the transformation of an existing analog design to the digital domain. This requirement may result when an existing continuous control system is being replaced or updated with a digital version. Digital circuits eliminate many reliability problems and are less susceptible to electronic noise and electromagnetic radiation. In other cases the system designers may have more experience in designing continuous controllers, and wish to design their filters in the s-domain and then transform them into the z-domain.

Consequently, in this chapter we first present some basic ideas about sampled-data transformations. Next we review the fundamentals of designing Butterworth, Bessel, transitional, Chebyshev, and elliptic analog filters. These filter design methods are useful for implementing low-pass, high-pass, band-pass, and band-stop filters to be employed in control systems which have special requirements for processing sensor signals, eliminating noise frequency bands and the like. Finally, we apply the sampled-data transforms to a typical analog filter.

11.2 SAMPLED-DATA TRANSFORMATIONS

Sampled-data transformations are the techniques one uses to obtain numerical solutions to integral and differential equations. Any linear system's transfer function may be written as

$$G(s) = \frac{Y(s)}{X(s)}$$

$$Y(s) = \text{Laplace transform of the output}$$

$$X(s) = \text{Laplace transform of the input}$$

Alternatively, the relationship between input and output may be described as a differential or integral equation. Numerical methods may be employed to solve these equations; these methods approximate the integral and differential equations by difference equations. As we have seen previously, the difference equations may be represented by a discrete transfer function. The complete process is illustrated in Figure 11-1.

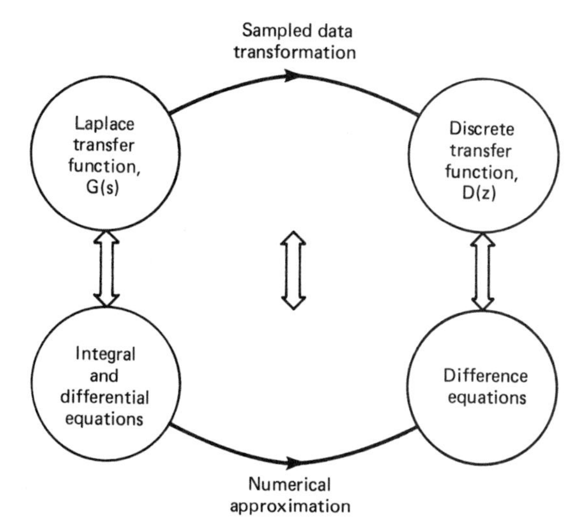

Figure 11-1 Relation between numerical approximations and sampled data transformations.

Numerical Approximations

Several numerical approximation techniques will now be presented, with some for differentiation and some for integration.

Backward difference. The backward difference is a simple technique that replaces the derivative of a function by

$$\frac{d}{dt}y(t) \doteq \frac{y(t) - y(t-T)}{T}$$

See Figure 11-2.

In the Laplace domain

$$sY(s) \doteq \frac{Y(s) - \epsilon^{-sT}Y(s)}{T} + y(0+)$$

If $y(0+)$ is small, then

$$s \doteq \frac{1 - \epsilon^{-sT}}{T}$$

$$s \doteq \frac{1 - z^{-1}}{T}$$

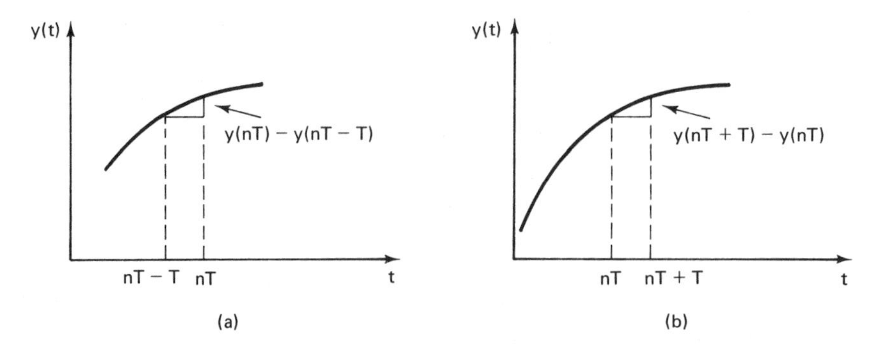

Figure 11-2 Difference approximations: (a) backward difference; (b) forward difference.

Hence,

$$D(z) = G(s)\big|_{s=(1-z^{-1})/T} \tag{11-1}$$

Example 11.1

Find a discrete approximation for

$$G(s) = \frac{s}{s+a}$$

$$Y(s) = G(s)X(s)$$

$$sY(s) + aY(s) = sX(s)$$

or

$$\frac{d}{dt}y(t) + ay(t) = \frac{d}{dt}x(t)$$

Now let

$$\frac{d}{dt}y(t) = \frac{y(t) - y(t-T)}{T}$$

$$\frac{d}{dt}x(t) = \frac{x(t) - x(t-T)}{T}$$

Therefore,

$$\frac{y(t) - y(t-T)}{T} + ay(t) = \frac{x(t) - x(t-T)}{T}$$

Evaluating at $t = nT$ yields

$$y(nT) = \frac{1}{1+Ta}(x(nT) - x(nT-T) + y(nT-T))$$

Employing the techniques of Chapter 2, we have

$$D(z) = \frac{1}{1+Ta}\,\frac{1 - z^{-1}}{1 - \dfrac{1}{1+Ta}z^{-1}}$$

An alternative solution employs equation (11-1) as follows:

$$D(z) = \left.\frac{s}{s+a}\right|_{s=(1-z^{-1})/T}$$

$$= \frac{\dfrac{1-z^{-1}}{T}}{a + \dfrac{1-z^{-1}}{T}}$$

$$= \frac{1-z^{-1}}{aT + 1 - z^{-1}}$$

$$= \frac{1}{1+aT} \frac{1-z^{-1}}{1 - \dfrac{1}{1+aT} z^{-1}}$$

Forward difference. A similar numerical technique approximates

$$\frac{d}{dt} y(t) \doteq \frac{y(t+T) - y(t)}{T}$$

See Figure 11-2.

This represents the equivalent Laplace domain approximation

$$sY(s) \doteq \frac{\epsilon^{sT} Y(s) - Y(s)}{T} + y(0+)$$

and, if $y(0+)$ is neglected,

$$s \doteq \frac{\epsilon^{sT} - 1}{T}$$

$$\doteq \frac{z-1}{T}$$

Hence,

$$D(z) = G(s)|_{s=(z-1)/T} \tag{11-2}$$

Example 11.2

Find a discrete version of $G(s)$ using the forward difference.

$$G(s) = \frac{s}{s+a}$$

$$D(z) = \left.\frac{s}{s+a}\right|_{s=(z-1)/T}$$

$$= \frac{(z-1)/T}{[(z-1)/T] + a}$$

$$= \frac{1 - z^{-1}}{1 + (aT - 1)z^{-1}}$$

Rectangular rule. Suppose now that we try some numerical approximations to integrals and compare results. The idea here is to represent $G(s)$ as

$$G(s) = \frac{\alpha_0 + \alpha_1 s^{-1} + \alpha_2 s^{-2} + \cdots + \alpha_n s^{-n}}{1 + \beta_1 s^{-1} + \beta_2 s^{-2} + \cdots + \beta_n s^{-n}} \tag{11-3}$$

Each s^{-1} represents an integrator in the s-domain. Hence, if we can replace each integrator by its digital equivalent

$$s^{-1} = f(z)$$

or

$$s = \frac{1}{f(z)} = g(z)$$

a digital equivalent of $G(s)$ will be produced.

Left-side rule. Let us determine the numerical approximation for

$$y(t) = \int_0^t x(t)\, dt$$

Assume that the upper limit of the integral is $t = nT$. Hence

$$y(nT) = \int_0^{nT} x(t)\, dt \tag{11-4}$$

Figure 11-3a illustrates the rectangular rule using the left side of the rectangles. Hence

$$y(nT) = T \sum_{i=0}^{n-1} x(iT)$$

$$y(nT + T) = T \sum_{i=0}^{n} x(iT) = T \sum_{i=0}^{n-1} x(iT) + Tx(nT)$$

$$= y(nT) + Tx(nT)$$

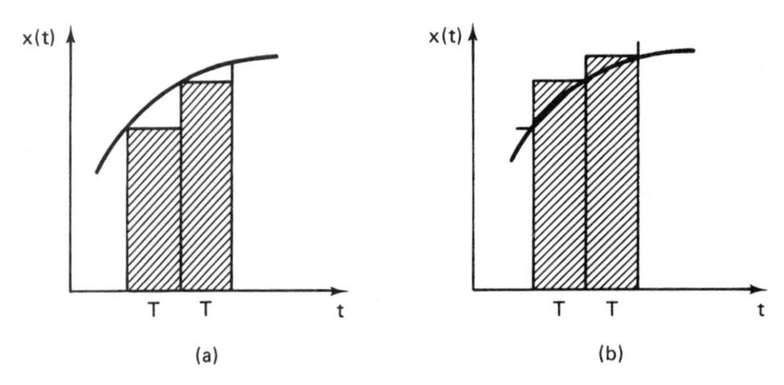

Figure 11-3 Rectangular rule: (a) left-side rule; (b) right-side rule.

Therefore, using the results of Chapter 2, the transfer function $Y(z)/X(z)$ is

$$D(z) = \frac{Tz^{-1}}{1 - z^{-1}}$$

$$= \frac{T}{z - 1}$$

Hence we have approximated the integration transfer function

$$\frac{1}{s} \doteq \frac{T}{z - 1} = f(z)$$

which gives the same results as equation (11-2) for the forward difference.

Right-side rule. Figure 11-3b illustrates the use of the right side of the rectangle in approximating equation (11-4). Therefore,

$$y(nT) = T \sum_{i=1}^{n} x(iT)$$

$$y(nT + T) = T \sum_{i=1}^{n+1} x(iT) = T \sum_{i=1}^{n} x(iT) + Tx(nT + T)$$

$$= y(nT) + Tx(nT + T)$$

Letting $n = n - 1$,

$$y(nT) = y(nT - T) + Tx(nT)$$

Hence, the transfer function is

$$D(z) = \frac{T}{1 - z^{-1}}$$

Consequently, we have approximated the integrator

$$\frac{1}{s} \doteq \frac{T}{1 - z^{-1}} = f(z)$$

which yields the identical result of equation (11-1) for the backward difference.

Trapezoidal rule. The trapezoidal rule takes the average of the left and right sides of the rectangles in Figure 11-3. Hence

$$y(nT) = \frac{T}{2} \sum_{i=0}^{n-1} [x(iT) + x(iT + T)]$$

$$= \frac{1}{2}\left[T \sum_{i=0}^{n-1} x(iT) + T \sum_{i=1}^{n} x(iT)\right]$$

Using the results of the rectangular rule, we see that the transfer function $Y(x)/X(z)$ is

$$D(z) = \frac{1}{2}\left[\frac{Tz^{-1}}{1 - z^{-1}} + \frac{T}{1 - z^{-1}}\right]$$

$$= \frac{T}{2} \frac{1 + z^{-1}}{1 - z^{-1}}$$

Thus we have approximated

$$\frac{1}{s} \doteq \frac{T}{2} \frac{1 + z^{-1}}{1 - z^{-1}} = f(z)$$

This approximation is the familiar bilinear z-transform.

Simpson's rule. Simpson's rule evaluates equation (11-4) by the formula

$$y(nT) = \frac{T}{3}[x(0) + 4x(T) + 2x(2T) + \cdots + 4x(nT - T) + x(nT)]$$

But

$$y(nT + 2T) = y(nT) + \frac{T}{3}[x(nT) + 4x(nT + T) + x(nT + 2T)]$$

Letting $n = n - 2$, the transfer function is

$$D(z) = \frac{T}{3} \frac{1 + 4z^{-1} + z^{-2}}{1 - z^{-2}}$$

Hence we have approximated

$$\frac{1}{s} \doteq \frac{T}{3} \frac{1 + 4z^{-1} + z^{-2}}{1 - z^{-2}} = f(z)$$

Impulse invariance [1]. Suppose that we want to find a discrete equivalent filter for the Laplace transfer function $G(s)$. Further suppose that we desire the impulse response of the discrete equivalent to match that of the analog filter as shown in Figure 11-4; that is, we desire impulse invariance:

$$g(nT) = d(nT)$$

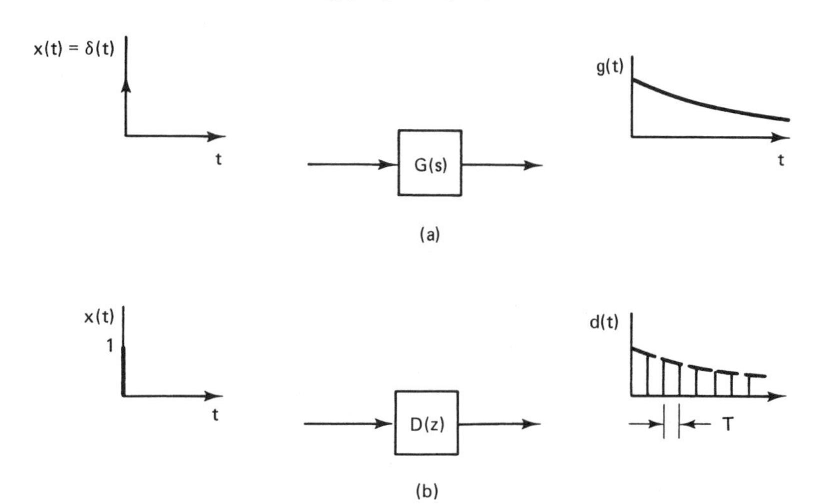

Figure 11-4 Impulse invariance: (a) analog filter; (b) digital filter.

Then

$$D(z) = \sum_{i=0}^{\infty} d(nT)z^{-n}$$

$$= \sum_{i=0}^{\infty} g(nT)z^{-n}$$

$$= G(z)$$

which is the standard z-transform. Hence, for impulse invariance

$$D(z) = \mathfrak{z}[G(s)] = G(z)$$

the digital approximation is just the standard z-transform of $G(s)$.

Impulse invariant integrator. Let us find the digital equivalent of an analog integrator using impulse invariance and the models of Figure 11-5. We know that

$$G(z) = \mathfrak{z}\left[\frac{1}{s}\right] = \frac{1}{1 - z^{-1}}$$

and that

$$G_{h0}(s) = \frac{1 - \epsilon^{-Ts}}{s} \doteq T$$

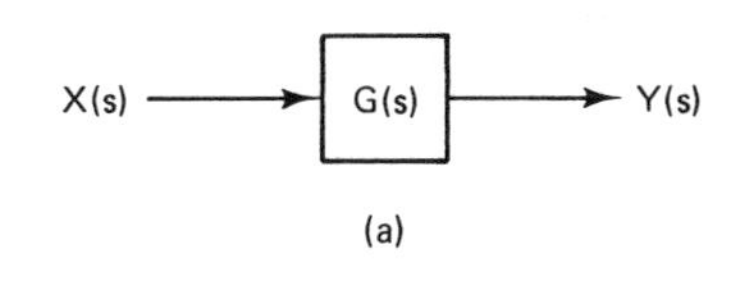

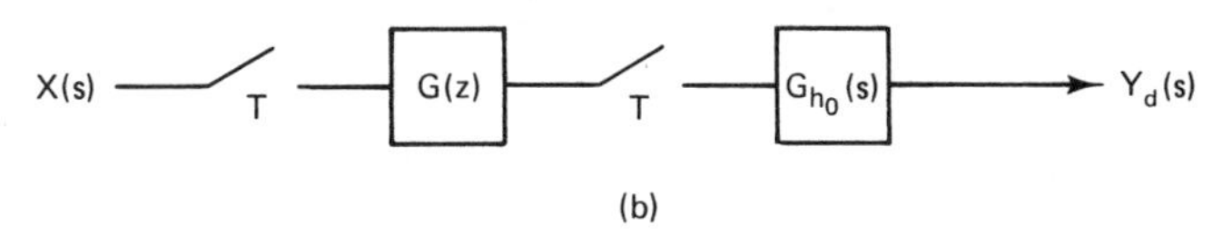

Figure 11-5 Impulse invariant integrator: (a) analog integrator; (b) digital integrator.

for small values of T. Hence

$$\frac{Y_d(z)}{X(z)} = \frac{T}{1 - z^{-1}}$$

and we have again approximated

$$\frac{1}{s} \doteq \frac{T}{1 - z^{-1}} = f(z)$$

Therefore, the backward difference, the right-side rectangular rule, and the impulse invariant integrator all indicate equation (11-1) as their equivalent sampled-data transformation.

Step invariance [2]. In Figure 11-4 a digital filter $D(z)$ which preserves the impulse response of an analog filter $G(s)$ was derived, and the result was the standard z-transform of $G(s)$. Suppose that instead of preserving the impulse response we preserve the step response. That is, the step response of $G(s)$ is set equal to the step response of $D(z)$ on a sample-by-sample basis. Then

$$\left(\frac{1}{1 - z^{-1}}\right) D(z) = \mathfrak{z}\left[\left(\frac{1}{s}\right) G(s)\right]$$

or

$$D(z) = (1 - z^{-1})\mathfrak{z}\left[\frac{G(s)}{s}\right]$$

Tables for $D(z)$ may be found in Ref. 2.

Example 11.3

Consider the step invariant integrator

$$G(s) = \frac{1}{s}$$

and

$$\begin{aligned} D(z) &= (1 - z^{-1})\mathfrak{z}\left[\frac{1}{s^2}\right] \\ &= (1 - z^{-1})\left[\frac{Tz^{-1}}{(1 - z^{-1})^2}\right] \\ &= \frac{Tz^{-1}}{1 - z^{-1}} \end{aligned}$$

which yields the same approximation derived earlier for the forward difference.

Mapping Functions Summary

As a result of our analysis of some elementary numerical approximation techniques we have identified several sampled-data mapping functions.

Standard *z*-transform. The standard z-transform yields an impulse-invariant filter. The mapping function for this transformation is

$$s = \frac{1}{T} \ln z \tag{11-5}$$

This mapping was defined in Chapter 4.

Backward difference [1]. The backward-difference approximation for the solution of differential equations provides the following mapping:

$$s = \frac{1 - z^{-1}}{T} \tag{11-6}$$

Solving for z yields

$$z = \frac{1}{1 - Ts}$$

Now substituting the frequency contour $s = j\omega$ produces

$$\begin{aligned} z &= \frac{1}{1 - j\omega T} \\ &= \frac{\frac{1}{2}(1 - j\omega T) + \frac{1}{2}(1 + j\omega T)}{1 - j\omega T} \\ &= \frac{1}{2}\left[1 + \frac{1 + j\omega T}{1 - j\omega T}\right] \\ &= \frac{1}{2}[1 + e^{j2\tan^{-1}(\omega T)}] \\ &= \frac{1}{2} + \frac{1}{2}e^{j\theta} \end{aligned}$$

Consequently, the left half of the s-plane maps inside the unit circle of the z-plane as shown in Figure 11-6. Hence stable analog filters will always result in stable digital equivalents. In fact, some unstable analog filters give stable digital ones. A major disadvantage of this mapping is seen in the frequency-response contour. The $j\omega$-axis in the s-plane does not map to the unit circle in the z-plane. Hence, as we get farther from $s = 0$ (or $z = 1$), the more degraded will be our desired frequency response. Thus we must decrease T (increase f_s) to improve the approximation.

Forward difference. The forward-difference approximation suggested the mapping

$$s = \frac{z - 1}{T}$$

Solving for z yields

$$z = 1 + Ts \tag{11-7}$$

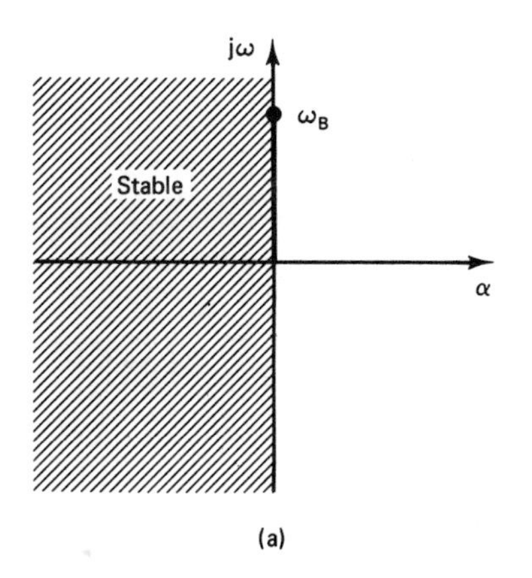

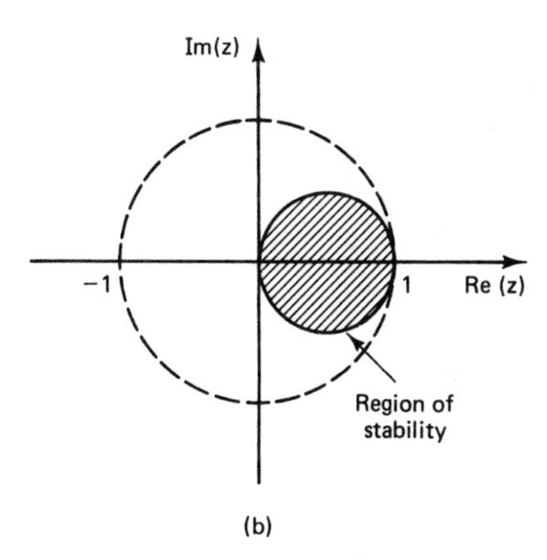

Figure 11-6 Mapping $z = 1/(1 - sT)$: (a) s-plane; (b) z-plane.

This mapping function is shown in Figure 11-7. Note that the left half-plane in the s-domain maps to the region to the left of $z = 1$ in the z-plane. But the interior of the unit circle represents the stability region in the z-plane. Consequently, some stable analog filters will give *unstable* digital ones. Unstable analog filters will also be unstable digital ones under this mapping. Yet a further disadvantage is that the frequency contour in the z-plane does not follow the unit circle. Hence, this is an undesirable mapping.

Bilinear *z*-transformation. The trapezoidal integration approximation led to the sampled-data mapping

$$s = \frac{2}{T}\frac{1 - z^{-1}}{1 + z^{-1}}$$

Solving for z yields

$$z = \frac{(2/T) + s}{(2/T) - s} \tag{11-8}$$

This mapping is illustrated in Figure 11-8. In Chapter 2 we employed this transform for a different purpose. Note here that the entire left half s-plane maps to the interior of the unit circle in the z-plane. Hence all stable analog filters will result in stable digital ones. Also, the $j\omega$-axis in the s-plane maps to the unit circle in the z-plane.

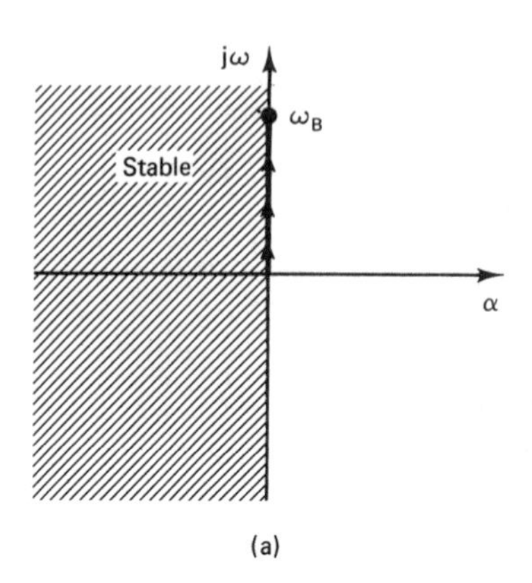

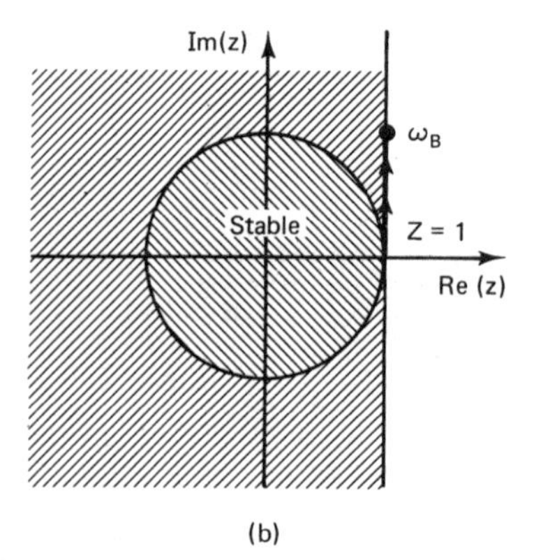

Figure 11-7 Mapping $z = 1 + Ts$: (a) s-plane; (b) z-plane.

However, the entire $j\omega$-axis maps *onto* the unit circle, which causes a mismatching of frequencies. This is a direct result of the characteristic that for a digital filter

$$z = 1 \longrightarrow \omega = 0$$

$$z = j1 \longrightarrow \omega = \frac{\omega_s}{4}$$

$$z = -1 \longrightarrow \omega = \frac{\omega_s}{2}$$

as required by equation (11-5). For the bilinear z-transform the frequencies in the z-plane (ω_D) are related to frequencies in the s-plane (ω_A) by

$$\frac{j\omega_A T}{2} = \frac{e^{j\omega_D T} - 1}{e^{j\omega_D T} + 1} = \frac{2j \sin(\omega_D T/2)}{2 \cos(\omega_D T/2)}$$

or

$$\omega_D = \frac{2}{T} \tan^{-1} \frac{\omega_A T}{2} \tag{11-9}$$

See Figure 11-9. Correction for this frequency-scale warping may be accomplished by redesigning (prewarping) the critical frequencies of the desired transfer function $G(s)$ before applying the bilinear z-transform.

This transformation maps circles and straight lines in the s-plane to circles in

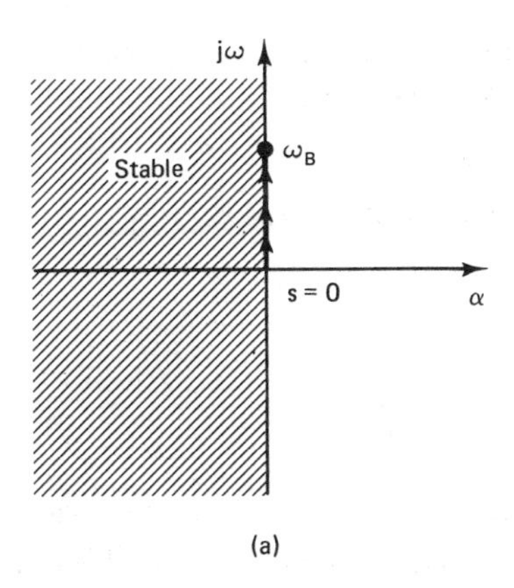

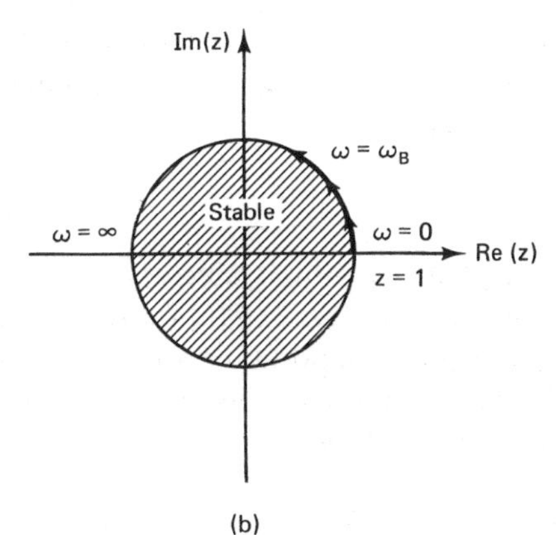

Figure 11-8 Bilinear z-transform: (a) s-plane; (b) z-plane.

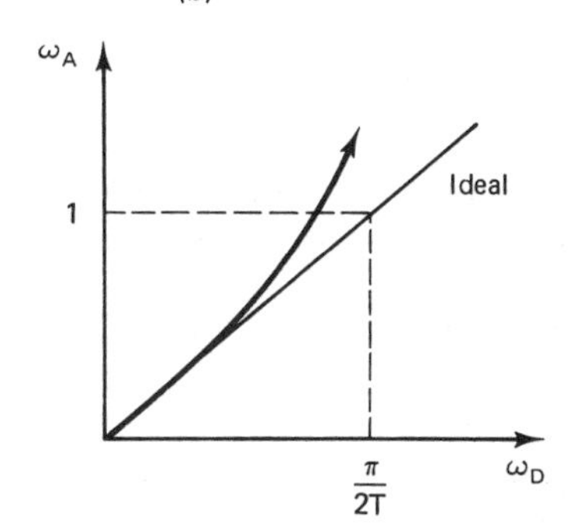

Figure 11-9 Change in frequency scale for the bilinear z-transform.

the z-plane. It works well for frequency characteristics that are piecewise linear. It also ensures that *no* frequency aliasing can occur in the transfer-function frequency characteristic because the $+j\omega$-axis does map onto the upper half of the unit circle. Hence the bilinear z-transform is quite popular.

Matched z-transforms [3]. The standard z-transform of $G(s)$ of equation (11-3) requires a partial-fraction expansion of $G(s)$ in order to complete the mapping

$$\frac{1}{s+u} \Longrightarrow \frac{1}{1-\epsilon^{-uT}z^{-1}}$$

For the purpose of simplifying the calculations, the matched z-transform maps the poles and zeros of $G(s)$ to the z-plane as follows:

$$s + \alpha \longrightarrow 1 - z^{-1}\epsilon^{-\alpha T} \tag{11-10}$$

Hence the matched z-transform of equation (11-3) is

$$G(z) = G(s)\Big|_{\substack{s+a_i = 1 - z^{-1}\epsilon^{-a_i T} \\ s+b_j = 1 - z^{-1}\epsilon^{-b_j T}}}$$

$$= K \frac{\prod_{i=1}^{m} (1 - z^{-1}\epsilon^{-a_i T})}{\prod_{j=1}^{n} (1 - z^{-1}\epsilon^{-b_j T})} \tag{11-11}$$

where K is adjusted to give the desired gain at dc ($z = 1$). This transform *matches* the poles and zeros in the s- and z-planes. Note that the poles of this transform are identical with those of the standard z-transform but that the zeros are different. Because of this difference, the matched z-transform may be used on nonbandlimited inputs. If $G(s)$ has no zeros, it is sometimes necessary to multiply $(1 + z^{-1})^N$, N an integer, times the expression (11-11).

Other Transforms

In general, any transformation that maps the stable region of the s-plane into the stable region of the z-plane may be used. It is helpful for the $j\omega$-axis in the s-plane to map to the z-plane's unit circle. Another important property is that rational functions $G(s)$ should be transformed into rational functions $D(z)$ so that the proper difference equations may be determined for realization.

Simpson's rule. The Simpson's rule approximation suggested that the mapping

$$s = \frac{3}{T}\frac{1 - z^{-2}}{1 + 4z^{-1} + z^{-2}} \tag{11-12}$$

be used as a transformation. Please note that a second-order function $G(s)$ will transform to a fourth-order $D(z)$. This is undesirable from an implementation viewpoint.

(w, v)-transform [4]. In some applications, the system transfer function $G(s, z, z^{\alpha})$ may be a function of s, $z = \epsilon^{Ts}$, and z^{α}, where $0 < \alpha < 1$. If all initial conditions are zero and

$$w = \frac{2}{T}\frac{1 - z^{-1}}{1 + z^{-1}}$$

$$v(\alpha) = 1 - \alpha(1 - z^{-1}) + \frac{\alpha(\alpha - 1)}{2}(1 - z^{-1})^2$$

then for a system described by

$$Y(s) = G(s, z, z^{-\alpha})X(s)$$

its z-transform will be

$$Y(z) = G(w, z, v(\alpha))\left[X(z) - \frac{x(0)}{1 + z^{-1}}\right]$$

If $x(0) = 0$, then

$$D(z) = G(s, z, z^{-\alpha})\Big|_{\substack{s=w \\ z^{-\alpha}=v(\alpha)}}$$

Example 11.4

Scott [5] has shown that a desirable phase-locked loop has the transfer function

$$G(s) = \frac{10}{s + 10z^{-0.5}}$$

Using the (w, v) transform to find a digital equivalent, if $x(0) = 0$,

$$D(z) = \frac{10}{s + 10z^{-0.5}}\Bigg|_{\substack{s=w=(2/T)[(1-z^{-1})/(1+z^{-1})] \\ z^{-0.5}=v(0.5)}}$$

$$v(0.5) = 1 - 0.5(1 - z^{-1}) + \frac{0.5(-0.5)}{2}(1 - z^{-1})^2$$

$$= 0.375 + 0.75z^{-1} - 0.125z^{-2}$$

$$D(z) = \frac{5T(1 + z^{-1})}{(1 + 1.875T) - (1 - 5.625T)z^{-1} + (3.125T)z^{-2} - (0.625T)z^{-3}}$$

11.3 REVIEW OF CONTINUOUS FILTER DESIGN [6]

The design of continuous filters can be accomplished by first designing several low-pass filter transfer functions $G(s)$, called prototype or normalized designs; the prototypes have a critical or break frequency of 1 rad/s. The prototype is used to realize a filter for a given specification by using the frequency transformations listed below:

$$\begin{aligned} &\text{Low pass:} \quad s \longrightarrow \frac{s}{\omega_u} \\ &\text{Band pass:} \quad s \longrightarrow \frac{s^2 + \omega_u\omega_l}{s(\omega_u - \omega_l)} \\ &\text{Band stop:} \quad s \longrightarrow \frac{s(\omega_u - \omega_l)}{s^2 + \omega_u\omega_l} \\ &\text{High pass:} \quad s \longrightarrow \frac{\omega_u}{s} \end{aligned} \tag{11-13}$$

where ω_u is the upper cutoff and ω_l is the low cutoff.

Five prototype filters are discussed in this section: Butterworth, Bessel, transitional, Chebyshev, and elliptical designs.

Butterworth

The Butterworth approximation to the ideal low-pass filter is defined by the squared frequency magnitude function

$$|G(\omega)|^2 = \frac{1}{1 + (\omega^2)^n} \tag{11-14}$$

where n is the order of the filter. The Laplace transfer function is given by

$$G(s)G(-s) = \frac{1}{1 + (-1)^n s^{2n}}$$

or

$$G(s) = \prod_{j=1}^{n} \frac{1}{s + b_j}$$

where

$$b_j = \epsilon^{-i\pi\left(\frac{1}{2}+\frac{2j-1}{2n}\right)}, \qquad i = \sqrt{-1}$$

Bessel

The Bessel filter approximation for the linear delay function $\epsilon^{-\tau s}$ may be written

$$G(s) = \frac{K_0}{B_n(s)} \tag{11-15}$$

where K_0 is a constant term and $B_n(s)$ are Bessel polynomials.

$$B_0 = 1$$
$$B_1 = s + 1$$
$$\vdots$$
$$B_n = (2n - 1)B_{n-1} + s^2 B_{n-2}$$

The roots of $B_n(s)$ are normalized using the factor $(K_0)^{1/n}$.

Transitional

The transitional filter combines roots of the nth-order Butterworth and normalized Bessel filters according to a transitional factor TF. Let

r_j = magnitude of jth transitional pole

r_{1j} = magnitude of jth Bessel pole

θ_j = angle of jth transitional pole

θ_{1j} = angle of jth Bessel pole

θ_{2j} = angle of jth Butterworth pole

the poles of the transitional filter are then described by

$$\begin{aligned} r_j &= r_{1j}\mathrm{TF} \\ \theta_j &= \theta_{2j} + \mathrm{TF}(\theta_{1j} - \theta_{2j}) \end{aligned} \tag{11-16}$$

Chebyshev

Chebyshev filters exhibit better cutoff characteristics for lower-order filters than do the designs above. Chebyshev type I and type II filters are defined by

$$|G_1(\omega)|^2 = \frac{1}{1 + \varepsilon^2 T_n^2(\omega)} \tag{11-17}$$

and

$$|G_2(\omega)|^2 = \frac{1}{1 + \varepsilon^2\left[\dfrac{T_n(\omega_r)}{T_n(\omega_r/\omega)}\right]^2} \tag{11-18}$$

where $T_n(\omega) = \cos(n \cos^{-1} \omega),\ 0 \leq \omega \leq 1$

$$= \cosh(n \cosh^{-1} \omega),\ \omega > 1$$

$$T_0(\omega) = 1$$
$$T_1(\omega) = \omega$$
$$T_2(\omega) = 2\omega^2 - 1$$
$$T_3(\omega) = 4\omega^3 - 3\omega$$

The order of the filter n is determined by specifying inband ripple E and the lowest frequency at which a loss of a decibels is achieved. Hence

$$\varepsilon = (10^{E/10} - 1)^{1/2} \tag{11-19}$$
$$A^2 = 10^{a/10}$$

and

$$n = \frac{\cosh^{-1}\sqrt{A^2 - 1/\epsilon}}{\cosh^{-1}(\omega_r)}$$

In equation (11-19), the variables E, a, or ω_r must be adjusted so that the n will be an integer. The type I filter differs from the type II in that the type I exhibits equiripple in the pass band while type II has equiripple in the stop band.

Elliptic

The elliptic filter has equiripple in both the pass and stop bands. Hence this type of design usually achieves the desired frequency response with a lower-order n than any of the types described above. The elliptic filter is determined by

$$|G(\omega)|^2 = \frac{1}{1 + \varepsilon^2 \psi_n^2(\omega)} \tag{11-20}$$

where

$$\psi_n = \begin{cases} \operatorname{sn}\left[n \dfrac{K(k_1)}{K(k)} \operatorname{sn}^{-1}(\omega; k); k_1\right], & n \text{ odd} \\ \operatorname{sn}\left[K(k_1) + N \dfrac{K(k_1)}{K(k)} \operatorname{sn}^{-1}(\omega; k); k_1\right], & n \text{ even} \end{cases}$$

with $\chi = \displaystyle\int_0^\omega \frac{d\omega}{[(1 - \omega^2)(1 - k^2\omega^2)]^{1/2}}$ = elliptic integral of the first kind

$\operatorname{sn}[\chi; k] = \omega$ = Jacobian elliptic function

$K(k)$ = complete elliptic integral of the first kind

$$= \int_0^{\pi/2} \frac{d\phi}{(1 - k^2 \sin^2 \phi)^{1/2}}$$

$$k = \frac{1}{\omega_r}$$
$$k_1 = \varepsilon(A^2 - 1)^{-1/2}$$
$$\varepsilon = (10^{E/10} - 1)^{1/2}$$
$$A^2 = 10^{a/10}$$

where e, a, and ω_r were defined for the Chebyshev filter; the order n is found by

$$n = \frac{K(k_1')K(k)}{K(k_1)K(k')}$$

with $k' = (1 - k^2)^{1/2}$

$k_1' = (1 - k_1^2)^{1/2}$

The result of any of the five design methods results in a Laplace transfer function $G(s)$ for the desired frequency response.

11.4 TRANSFORMING ANALOG FILTERS

Earlier in this chapter we described numerous sampled-data transformations that may be employed to achieve the goal of producing a digital approximation for an analog filter. However, we recommend that the standard z-transform, the billinear z-transform, or the matched z-transform be used in most practical applications.

Standard z-Transform

The problem of converting a continuous filter to a discrete one is presented in Figure 11-10. This figure is an expanded version of the ideas presented in Figure 11-5. Figure 11-10a illustrates the basic analog filter. Figure 11-10b shows the impulse invariance implications of the standard z-transform. If

$$Y_a^*(s) = Y_b^*(s)$$

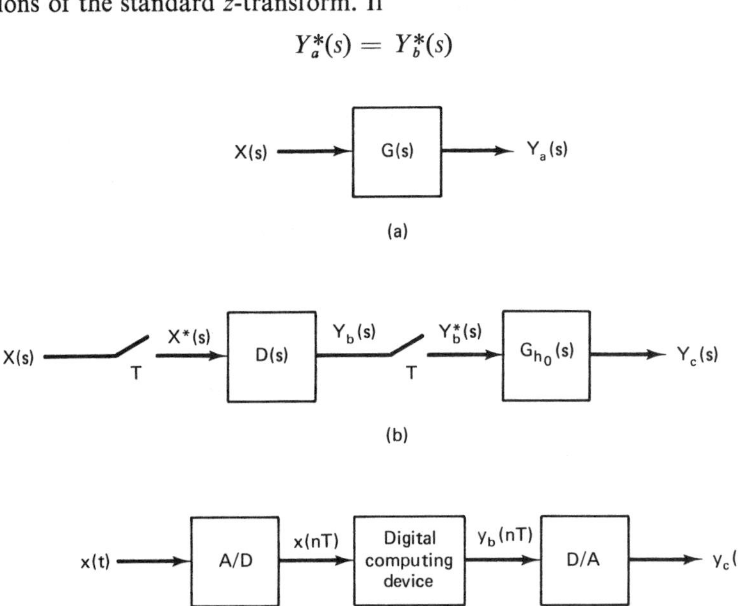

Figure 11-10 Transforming a continuous filter by the standard z-transform: (a) analog design; (b) impulse invariant model of an analog filter; (c) computational model.

then

$$Y_a^*(s) = D^*(s)X^*(s)$$

But

$$Y_a^*(s) = (G(s)X(s))^*$$

Impulse invariance implies that

$$D(z) \doteq G(z)$$

or

$$D^*(s) \doteq G^*(s)$$

Consequently,

$$(G(s)X(s))^* \doteq G^*(s)X^*(s)$$

This is the first assumption.

Next, consider Figure 11-10c regarding the digital implementation of the analog filter. First, it is desirable that

$$y_c(nT) = y_a(nT)$$

But

$$Y_c(s) = G_{h0}(s)\, Y_b^*(s)$$

where

$$G_{h0}(s) = \frac{1 - \epsilon^{-Ts}}{s}$$

$$Y_c(s) = D^*(s)X^*(s)G_{h0}(s)$$

Hence, since $G_{h0}^*(s) = 1$,

$$Y_c^*(s) = [D^*(s)X^*(s)G_{h0}(s)]^*$$
$$= D^*(s)X^*(s)G_{h0}^*(s) = D^*(s)X^*(s)$$

Consequently,

$$D(z) = G(z)$$

is the transfer function of the digital computer and we may write

$$D(z) = \sum_{k=1}^{n} \frac{R_k}{1 - z^{-1}e^{-Tb_k}} \tag{11-21}$$

An important point to remember is that if a zero-order-hold device is not being used, we adjust the gain of our filter by a factor T. Remember that:

1. Sampling increases the gain by a factor $1/T$ [see (3-33)].
2. The zero-order hold restores the gain by a factor T since

$$G_{h0}(s) = \frac{1 - e^{-Ts}}{s} \simeq \frac{1 - (1 - Ts)}{s} = T$$

for small T.

Consequently, for comparing the frequency responses, one should compute $G(j\omega)$ and $TD(e^{j\omega T})$. Note that the standard z-transform can be used only on bandlimited signals ($f < f_s/2$).

Bilinear z-Transform

The bilinear z-transform may be used to obtain a discrete equivalent of $G(s)$ as follows:

$$D(z) = G'(s)\big|_{s=(2/T)(1-z^{-1})/(1+z^{-1})} \tag{11-22}$$

where $G'(s)$ is a continuous filter whose critical frequencies differ from $G(s)$ by

$$f'_s = \frac{1}{\pi T \tan(\pi f_c T)} \tag{11-23}$$

Relation (11-23) is used *before* the continuous filter $G(s)$ is designed. The new filter $G'(s)$ is designed instead and then transformed to the z-plane by (11-22). The bilinear z-transform is a bandlimiting transformation with relatively flat magnitude characteristics in the pass and stop bands. However, the time response will be considerably different.

Matched z-Transform

The matched z-transform matches the poles and zeros of the discrete function to those of the continuous one. The digital equivalent of the $G(s)$ function is calculated as follows:

$$D(z) = G(s)\Big|_{\substack{s+a_i=1-z^{-1}\epsilon^{-a_iT} \\ s+b_j=1-z^{-1}\epsilon^{-b_jT}}} \tag{11-24}$$

If $G(s)$ has no zeros, it is sometimes necessary to multiply (11-24) by $(1 + z^{-1})^N$, N is an integer.

Summary

The standard z-transform is suitable only for bandlimited functions, while the bilinear and matched z-transforms are suitable for all filter types. The matched z-transform requires $G(s)$ in factored form; standard, in partial-fraction form; and bilinear, in prewarped frequency form. The standard z-transform preserves the shape of the impulse-time response; the matched, the shape of the frequency response; and the bilinear, the flat magnitude gain–frequency response characteristics. An example filter is designed and discretized in the following example.

Design example. In this section a digital filter will be designed using the techniques summarized above.

$$\omega_u = 200 = 2\pi(31.831)$$

$$\omega_l = 170 = 2\pi(27.056)$$

Cascaded with this filter will be a low-pass filter $G_2(s)$ with $\omega_\eta = 600 = 2\pi(95.493)$, with a dc gain of 1.356. The band-stop filter will be designed from Butterworth, Bessel, and Chebyshev I prototypes with $n = 2$. The low-pass filter will be designed with $n = 1$. The prototype of $G_2(s) = 1/(s + 1)$.

The prototype filters for $G_1(s)$ are given below.

Butterworth. The Butterworth filter is defined by

$$|G_1(\omega)|^2 = \frac{1}{1 + \omega^4}$$

$$G_1(s)G_1(-s) = \frac{1}{1 + s^4}$$

$$G_1(s) = \frac{1}{(s - \epsilon^{i3\pi/4})(s - \epsilon^{i5\pi/4})}$$

$$G_1(s) = \frac{1}{s^2 + \sqrt{2}\,s + 1}$$

Bessel. The Bessel prototype is defined by

$$G_1(s) = \frac{K_0}{B_2(s)} = \frac{3}{s^2 + 3s + 3}$$

Chebyshev I. The Chebyshev I filter is defined by

$$|G_1(\omega)|^2 = \frac{1}{1 + \varepsilon^2 T_2^2(\omega)}$$

$$T_2(\omega) = 2\omega^2 - 1$$

$$\varepsilon = (10^{E/10} - 1)^{1/2}$$

$$A^2 = 10^{a/10}$$

$$n = \frac{\cosh^{-1}(\sqrt{A^2 - 1}/\epsilon)}{\cosh^{-1}(\omega_r)} = 2$$

Let $E = 1.0$ dB. Then

$$\varepsilon = (10^{0.1} - 1)^{1/2} = 0.5$$

Let the filter gain be down 6 dB at ω_r.

$$a = 6$$

$$A^2 = 10^{0.6} = 4$$

$$n = \frac{\cosh^{-1}(\sqrt{A^2 - 1}/\varepsilon)}{\cosh^{-1}(\omega_r)} = 2$$

$$\cosh^{-1}(\omega_r) = \frac{1}{2}\cosh^{-1}(\sqrt{2}) = 0.44$$

$$\omega_r = 1.098$$

Hence

$$|G_1(\omega)|^2 = \frac{1}{\omega^4 - \omega^2 + 1.25}$$

$$G_1(s) = \frac{1}{(s + 1.057\underline{/58.28^\circ})(s + 1.057\underline{/-58.28^\circ})}$$

$$G_1(s) = \frac{1}{s^2 + 1.112s + 1.118}$$

The analog filters are designed from the prototypes by setting

$$G(s) = G_1(s)\,|_{s=[s(\omega_u-\omega_l)]/(s^2+\omega_u\omega_l)} \times G_2(s)\,|_{s=s/\omega_\eta}$$

and adjusting the dc gain to be 1.356. The resulting filter equations are given next.

Butterworth:

$$G(s) = 813.6 \frac{s^4 + 68{,}000s^2 + 1.156 \times 10^9}{(s^4 + 1272.8s^3 + 68{,}900s^2 + 4.3275 \times 10^7 s + 1.156 \times 10^9)(s + 600)}$$

Bessel:

$$G(s) = 2440.8$$

$$\times \frac{s^4 + 68{,}000s^2 + 1.156 \times 10^9}{(3s^4 + 27{,}000s^3 + 2.049 \times 10^5 s^2 + 9.18 \times 10^7 s + 3.465 \times 10^9)(s + 600)}$$

Chebyshev I:

$$G(s) = 909.6$$

$$\times \frac{s^4 + 68{,}000s^2 + 1.56 \times 10^9}{(1.118s^4 + 33.36s^3 + 76{,}924s^2 + 1.134 \times 10^6 s + 1.744 \times 10^9)(s + 600)}$$

The filter equations above were plotted for decibels and phase, ϕ, versus frequency, as shown in Figure 11-11 to 11-13. Since the plots are nearly identical, the Butterworth $G(s)$ was chosen to be discretized by the standard, bilinear, and matched z-transforms, with $T = 0.001$.

The Butterworth design for $G(s)$ may be written in partial-fraction expansion form as

$$G(s) = \frac{40.567}{s + 27.314} + \frac{2.4402 \times 10^{-4} + j7.5033 \times 10^{-5}}{s + 0.35375 + j185.39}$$

$$+ \frac{2.4402 \times 10^{-4} - j7.5033 \times 10^{-5}}{s + 0.35375 - j185.39}$$

$$+ \frac{1642.1}{s + 1244.8} + \frac{-869.03}{s + 600}$$

The standard z-transform is taken.

$$\frac{a}{s + u} \longrightarrow \frac{aT}{1 - \epsilon^{-uT} z^{-1}}$$

and

$$\frac{a + ib}{s + u + iv} + \frac{a - ib}{s + u - iv} \longrightarrow T \frac{[2a] + [2\epsilon^{-uT}(b \sin vT - a \cos vT)]z^{-1}}{1 + [-2\epsilon^{-uT} \cos vT]z^{-1} + [\epsilon^{-2uT}]z^{-2}}$$

Hence

$$D(z) = \frac{4.0567 \times 10^{-2}}{1 - 0.97306z^{-1}} + \frac{4.8803 \times 10^{-7} - 4.420 \times 10^{-7} z^{-1}}{1 - 1.9654z^{-1} + 0.99929z^{-2}}$$

$$+ \frac{1.6421}{1 - 0.28800z^{-1}} + \frac{-0.86903}{1 - 0.54881z^{-1}}$$

The frequency response of this function is found by letting $z = \epsilon^{j\omega T}$. The plot is shown in Figure 11-14. Note that this response is entirely inadequate. The standard z-transform is accurate only when $G(s)$ is limited to frequencies less than $1/2T$, or in this

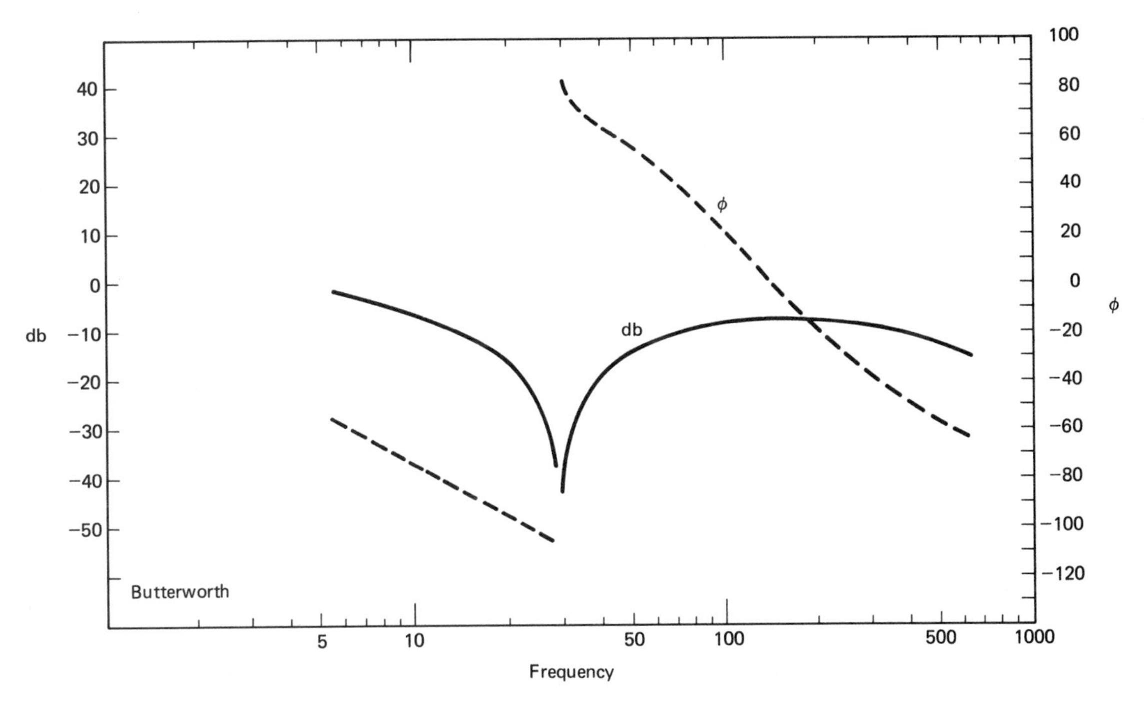

Figure 11-11 Butterworth prototype filter frequency response.

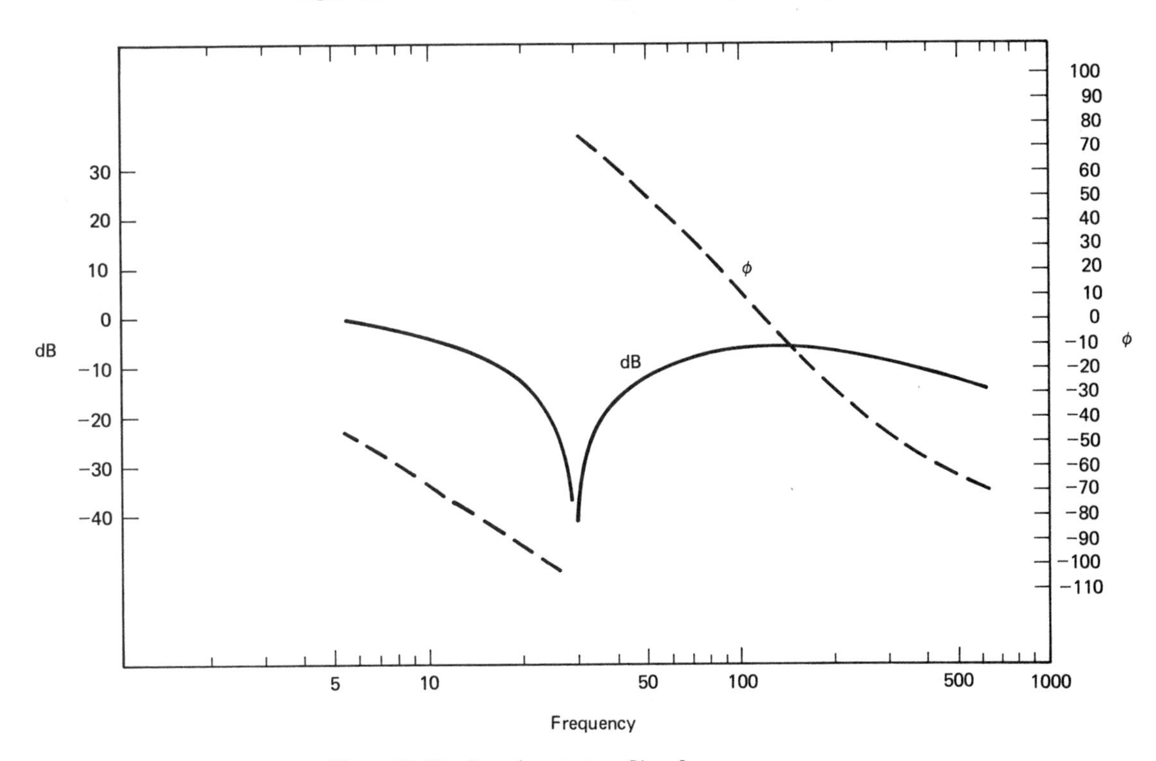

Figure 11-12 Bessel prototype filter frequency response.

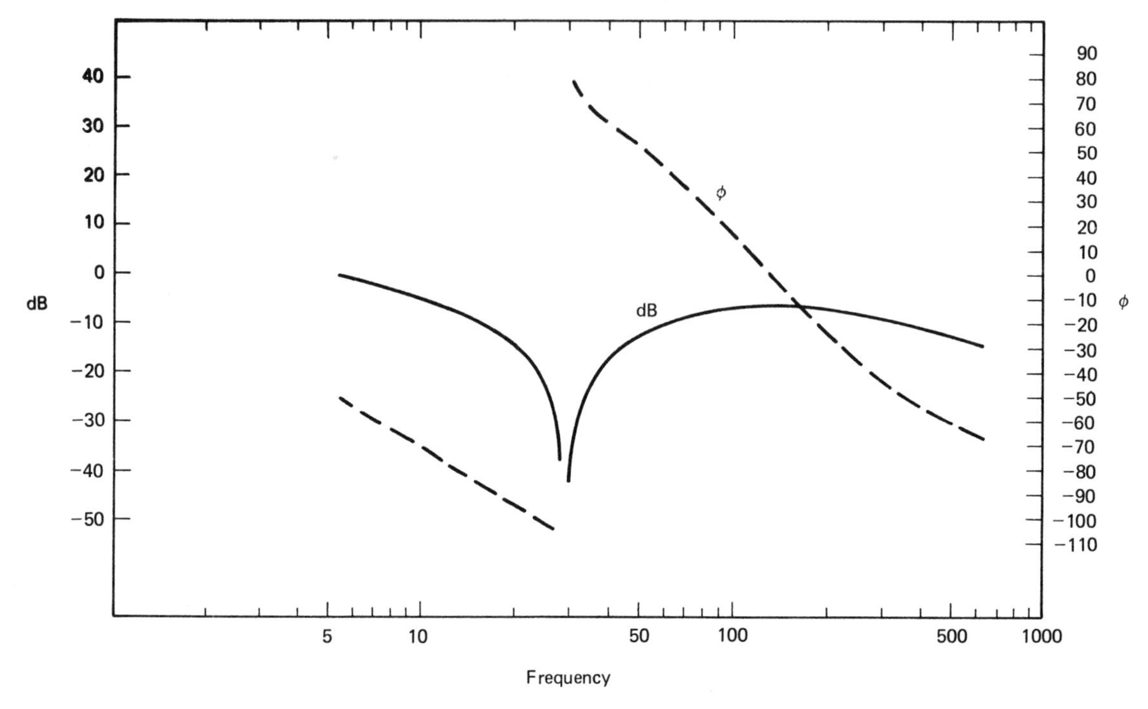

Figure 11-13 Chebyshev I prototype filter frequency response.

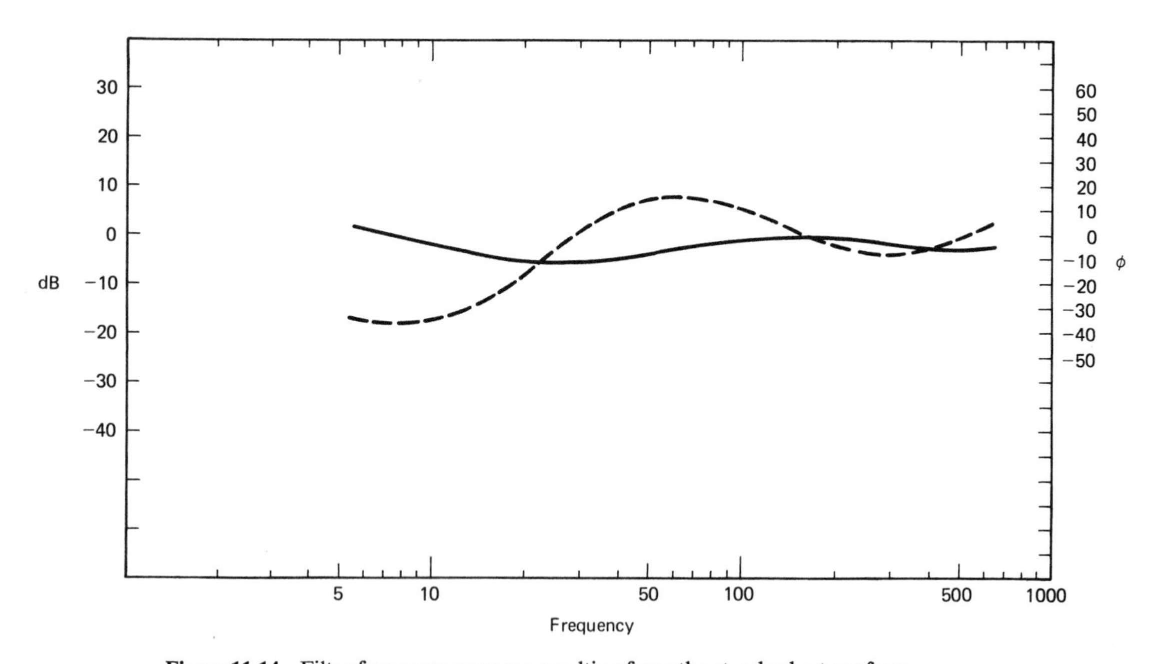

Figure 11-14 Filter frequency response resulting from the standard z-transform.

case, 500 Hz. This condition is violated, as is seen in the plot of the continuous Butterworth design $G(s)$.

The bilinear z-transform requires a prewarped frequency scale for the Butterworth $G(s)$ design, so

$$\omega_s = \frac{2}{T} \tan\left(\frac{\omega_z}{2} T\right)$$

	Unwarped	Warped
ω_u	200	200.67
ω_l	170	170.41
ω_n	600	618.67

The Butterworth design to be used in this case is

$$G(s) = \frac{1}{s^2 + \sqrt{2}\, s + 1}\bigg|_{s=[s(200.67-170.41)]/[s^2+(200.67)(170.41)]} \times \frac{1}{s+1}\bigg|_{s=s/618.67}$$

$$G(s) = 838.92 \frac{s^4 + 68{,}392s^2 + 1.1694 \times 10^9}{s^4 + 1294.8s^3 + 69{,}308s^2 + 4.479 \times 10^7 s + 1.169 \times 10^9)(s + 618.67)}$$

$$G(s) = 838.92 \frac{(s^2 + 3.4199 \times 10^4)^2}{(s + 26.987)(s^2 + 0.70708s + 34{,}199)(s + 1267.1)(s + 618.67)}$$

The bilinear z-transform is found by letting

$$D(z) = G(s)\,|_{s=(2/T)[(1-z^{-1})/(1+z^{-1})]}$$

$$D(z) = 0.19509$$

$$\times \frac{(1 - 1.9661z^{-1} + z^{-2})^2(1 + z^{-1})}{(1 - 0.97337z^{-1})(1 - 1.9654z^{-1} + 0.99930z^{-2})(1 - 0.22433z^{-1})(1 - 0.52749z^{-1})}$$

The frequency response for this function is found with $z = \epsilon^{j\omega T}$ and is plotted in Figure 11-15. Note that this plot closely matches Figure 11-11.

The Butterworth $G(s)$ may be factored as follows:

$$G(s) = 813.6$$

$$\times \frac{(s - j184.39)^2(s + j184.39)^2}{(s + 27.314)(s + 0.35375 + j184.39)(s + 0.35375 - j184.39)(s + 1244.8)(s + 600)}$$

The matched z-transform is given by

$$D(z) = G(s)\bigg|_{\substack{s+a=1-\epsilon^{-aT}z^{-1} \\ (s+u+iv)(s+u-iv)=1-2\epsilon^{-uT}\cos vT z^{-1}+\epsilon^{-2uT}z^{-2}}}$$

and $D(1)$ is set equal to 1.356, the dc gain. Hence

$$D(z) = 0.34607$$

$$\times \frac{(1 - 1.9661z^{-1} + z^{-2})^2}{(1 - 0.97035z^{-1})(1 - 1.9654z^{-1} + 0.99929z^{-2})(1 - 0.28800z^{-1})(1 - 0.54881z^{-1})}$$

The frequency response of this function is plotted in Figure 11-16. Note that the matched z-transform (like the bilinear) gives a good approximation to the response of Figure 11-11.

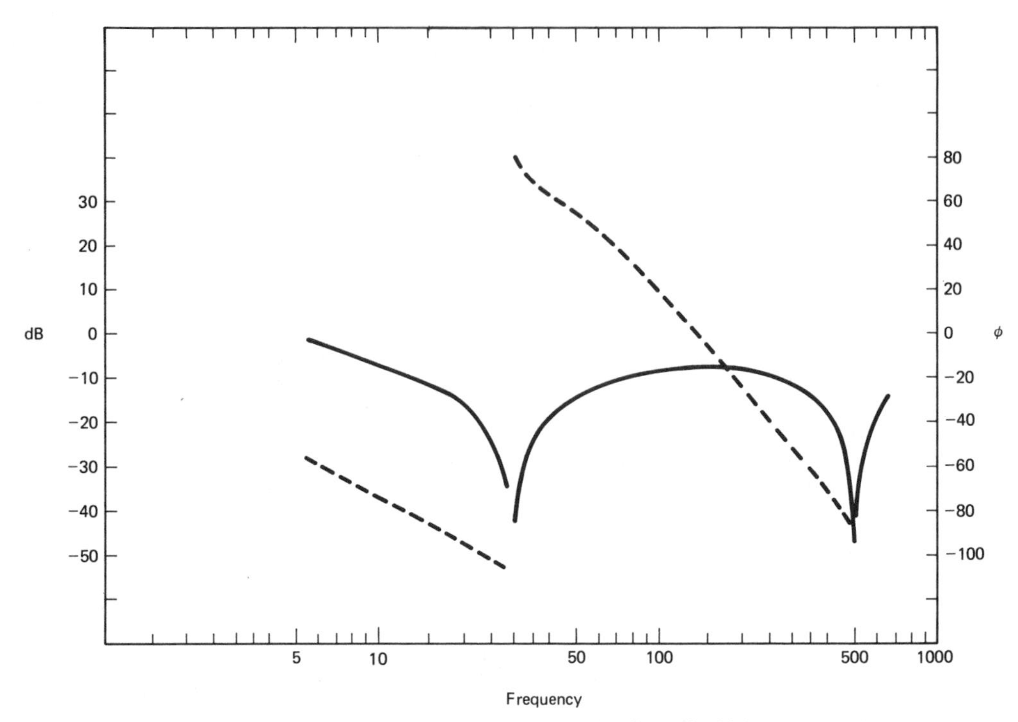

Figure 11-15 Filter frequency response resulting from the bilinear z-transform.

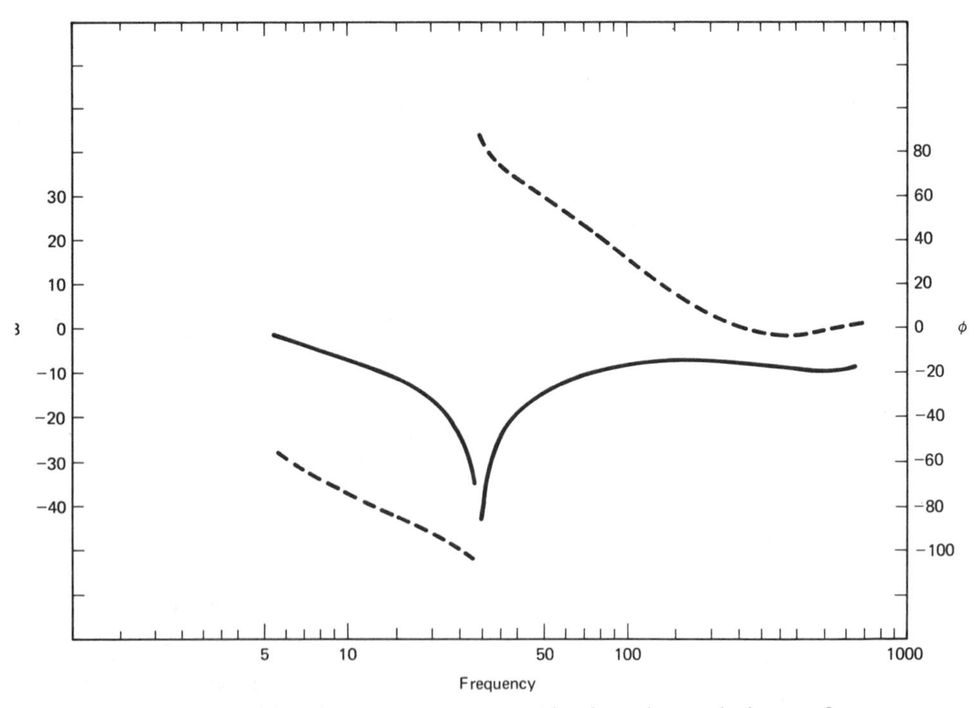

Figure 11-16 Filter frequency response resulting from the matched z-transform.

11.5 SUMMARY

In this chapter we have summarized most of the sampled-data transforms that are commonly used for transforming analog filters to digital filters. We have presented derivations of some of the transforms in order to give the reader a perspective on their relative accuracy and importance. A complete design example was presented to illustrate the most important principles.

REFERENCES

1. A. V. Oppenheim and R. W. Schafer, *Digital Signal Processing*. Englewood Cliffs, N.J.: Prentice-Hall, Inc., 1975.
2. C. P. Newman and C. S. Baradello, "Digital Transfer Functions for Microcomputer Control," *IEEE Trans. Syst. Man Cybern.*, Vol. SMC-9, No. 12, pp. 856–860, Dec. 1979.
3. R. M. Golden, "Designing Digital Filters, z-Transforms, and Fourier Analysis," *Proc. Natl. Electron. Conf.*, St. Charles, Ill., June 1969.
4. C. A. Halijak, "The (w, v)—Transform," *Proc. IEEE Region 3 Conv.*, Knoxville, Tenn., Apr. 10–12, 1972, pp. C4.1–3.
5. R. E. Scott, "An Improved Phase Lock Loop Derived from Ideal Single Sideband Modulation," Ph.D. dissertation, University of Denver, Denver, Colo., June, 1966, pp. 20–27.
6. A. S. Sedra and P. O. Brackett, *Filter Theory and Design: Active and Passive*. Beaverton, Oreg.: Matrix Publishers, Inc., 1978.

PROBLEMS

11-1. Given $G(s) = (s + 2)/(s^2 + 4s + 3)$ and $T = 1$ s, find the **(a)** standard z-transform; **(b)** bilinear z-transform; **(c)** matched z-transform.

11-2. Given $G(s) = [(s + 1)(s + 20)]/[(s + 2)(s + 10)]$, find an equivalent $D(z)$ using:

(a) $s = \dfrac{1 - z^{-1}}{T}$ **(b)** $s = \dfrac{z - 1}{T}$

(c) $s = \dfrac{2}{T}\left(\dfrac{1 - z^{-1}}{1 + z^{-1}}\right)$ **(d)** $s = e^{Ts}$

(e) $s + \alpha \Rightarrow 1 - z^{-1}e^{-\alpha T}$

11-3. Design a Chebyshev I prototype filter for $n = 3$, $a = 6$ dB, and $E = 1$ dB.

11-4. Using the prototype of Problem 11-3, design a low-pass analog filter with

$$\omega_u = 2\pi(100)$$

Use the bilinear z-transform to find a digital equivalent ($f_s = 1000$ Hz). Set the dc gain to 1.

11-5. Given $G(s) = G_1(s)G_2(s)$, where

$$G_1(s) = \frac{98{,}596s^2}{s^4 + 154.186s^3 + 491{,}994s^2 + 30{,}348{,}629s + 3.8884 \times 10^{10}}$$

$$G_2(s) = \frac{314s}{s^2 + 155.15s + 197{,}192}$$

(a) Plot the frequency response for $G(s)$.

(b) Find an equivalent $D(z)$ using the bilinear z-transform. Plot its frequency response.

(c) Repeat part (b) using the matched z-transform.

11-6. Repeat Problem 11-5 using **(a)** the backward difference; **(b)** the forward difference.

11-7. Compare the results of Problems 11-5 and 11-6.

11-8. Given a Chebyshev I prototype frequency function

$$|G_1(\omega)|^2 = \frac{1}{\omega^4 - \omega^2 + 1.25}$$

show that

$$G_1(s) = \frac{1}{s^2 + 1.112s + 1.118}$$

11-9. An analog PID controller may be written as

$$G(s) = K_P + K_I \frac{1}{s} + K_D s$$

Use the backward-difference mapping and find a digital equivalent.

11-10. Repeat Problem 11-9 using the trapezoidal mapping function.

11-11. Compare the answers to Problems 11-9 and 11-10 to the PID controller of (8-52). Which mapping functions are used in (8-52)?

12

Digital Filter Structures

12.1 INTRODUCTION

Up to this point we have been concerned with finding a transfer function $D(z)$ in the z-domain which is to perform digital filtering and control operations. The transfer function may be represented in general by

$$D(z) = \frac{a_0 + a_1 z^{-1} + \cdots + a_n z^{-n}}{1 + b_1 z^{-1} + \cdots + b_n z^{-n}} \tag{12-1}$$

where a_i and b_i are real coefficients, and n is the maximum of the orders of the denominator and numerator polynomials. Either polynomial may have zero coefficients in the higher-order terms so that (12-1) may describe the general case.

The purpose of this chapter is to describe block-diagram realizations of (12-1) using time-delay elements (represented by z^{-1}), adders, and multipliers. Each different block diagram is called a *filter structure*. Needless to say, there exist countless structures for (12-1) and we will describe only a few of the more important ones. In particular, we describe direct-form structures, second-order modules, cascaded modules, paralleled modules, and ladder structures, as well as suggest others.

12.2 DIRECT STRUCTURES

Direct structures for digital filters are those in which the real coefficients, a_i and b_i of (12-1), appear as multipliers in the block-diagram implementation.

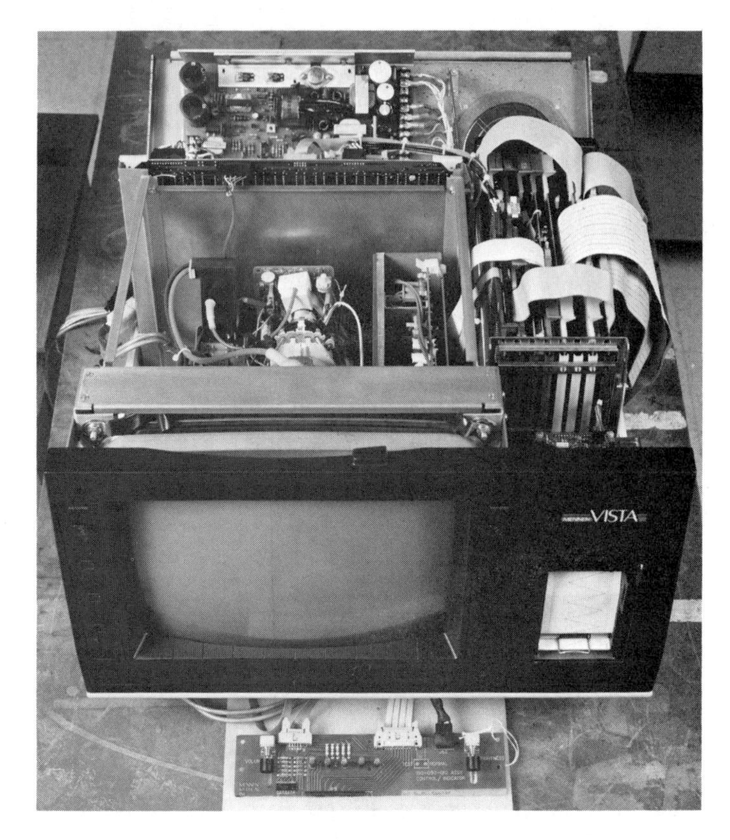

Nurse station for monitoring multipatient ECG, arrhythmia, and hemodynamic data, based on a single-board computer. (Courtesy of Digital Equipment Corporation.)

First Direct Structure

Suppose that we represent

$$D(z) = \frac{\sum_{i=0}^{n} a_i z^{-i}}{\sum_{i=0}^{n} b_i z^{-i}} \tag{12-2}$$

where $b_0 = 1$. If $X(z)$ is the filter input, and $Y(z)$, the output, then

$$\frac{Y(z)}{X(z)} = \frac{\sum_{i=0}^{n} a_i z^{-i}}{\sum_{i=0}^{n} b_i z^{-i}}$$

If an intermediate variable, say $M(z)$, is introduced,

$$\frac{Y(z)}{M(z)} \frac{M(z)}{X(z)} = \frac{\sum_{i=0}^{n} a_i z^{-i}}{\sum_{i=0}^{n} b_i z^{-i}}$$

such that

$$\frac{Y(z)}{M(z)} = \sum_{i=0}^{n} a_i z^{-i}$$

$$\frac{X(z)}{M(z)} = \sum_{i=0}^{n} b_i z^{-i}$$

Hence

$$X(z) = \sum_{i=0}^{n} b_i z^{-i} M(z)$$

or

$$M(z) = X(z) - \sum_{i=1}^{n} b_i z^{-i} M(z)$$

and

$$Y(z) = \sum_{i=0}^{n} a_i z^{-i} M(z)$$

In the time domain

$$\begin{aligned} m(k) &= x(k) - \sum_{i=1}^{n} b_i m(k-i) \\ y(k) &= \sum_{i=0}^{n} a_i m(k-i) \end{aligned} \tag{12-3}$$

Equations (12-3) define the first direct (1D) structure as shown in Figure 12-1a. In Figure 12-1 time delay (z^{-1}) is represented by rectangular boxes; multipliers, by labeled arrows; adders, by circles and elipses containing a plus (+); and signal distribution points, by dark dots at joining lines and dark bars. The 1D structure is called canonical because it possesses only n time-delay elements, the minimum number for an nth-order transfer function of (12-1).

Transpose Networks

The *transpose* structure of a digital filter structure is formed by reversing the signal flow in all branches of the block diagram [1]. Consequently, the summing junctions become signal distribution points, and vice versa. The input becomes the output, and vice versa. The transpose of a filter structure has the same transfer function as the original structure. Hence structures for digital filters exist as transpose pairs. Consequently, we may use these properties of structures to derive a second direct (2D) structure for (12-1).

Second Direct Structure

If we take the transpose of the 1D structure, we obtain the 2D structure shown in Figure 12-1b. It also implements (12-1), but it requires $n + 1$ difference equations (summing junctions), whereas the 1D structure required only 2. The 2D difference equations have the form

$$\begin{aligned} p_i(k) &= p_{i+1}(k-1) + a_i x(k) - b_i y(k), \qquad i = 1, n-1 \\ p_n(k) &= a_n x(k) - b_n y(k) \\ y(k) &= a_0 x(k) + p_1(k-1) \end{aligned} \tag{12-4}$$

This structure is also canonical.

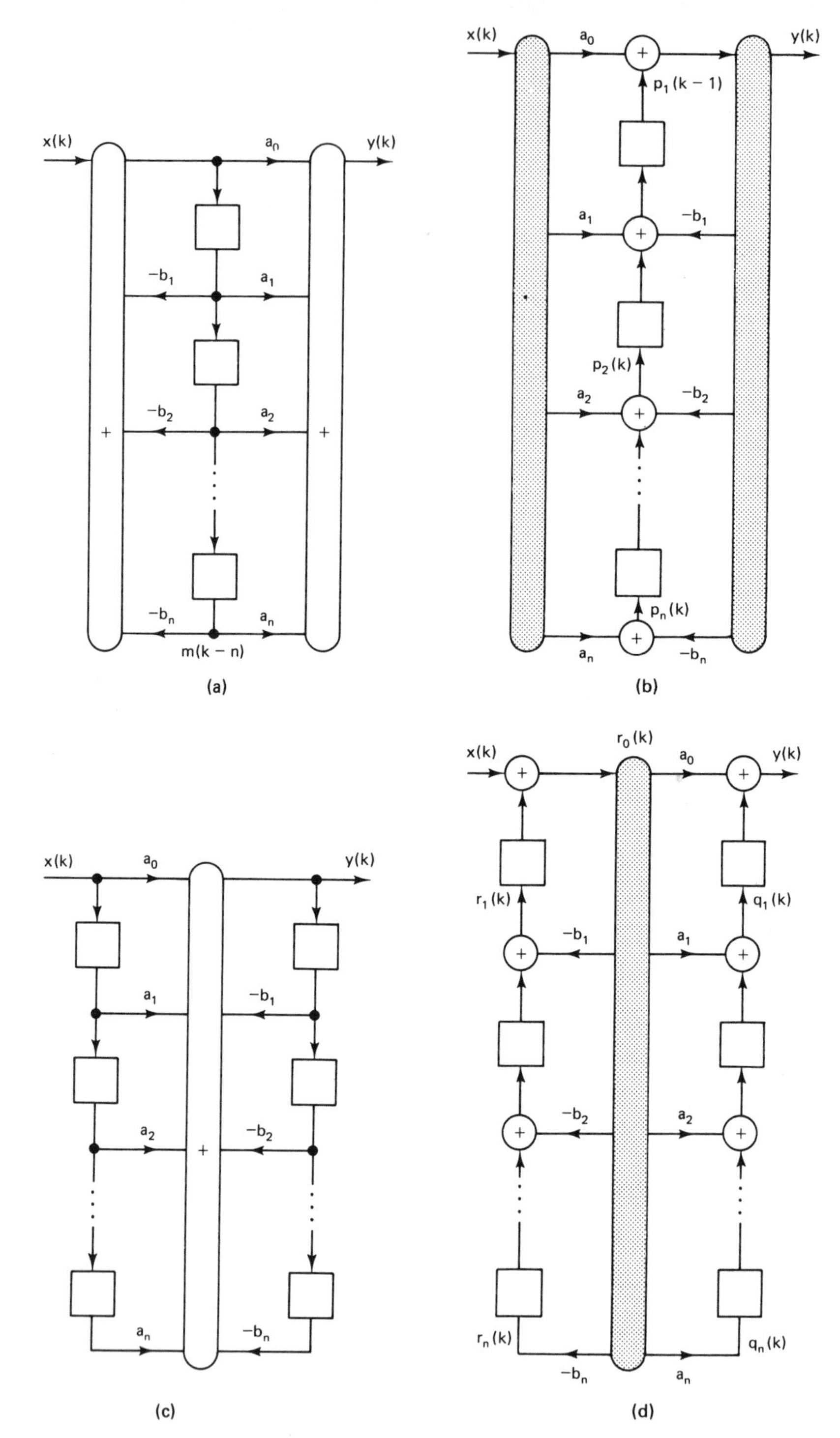

Figure 12-1 Direct structures: (a) 1D; (b) 2D; (c) 3D; (d) 4D. (From H. T. Nagle, Jr., and V. P. Nelson, "Digital Filter Implementation on 16-bit Microcomputers," *IEEE MICRO*, Vol. 1, No. 1, Fig. 1, p. 24, Feb. 1981, ©1981 IEEE.)

Third Direct Structure

Returning to (12-1) we may write

$$D(z) = \frac{Y(z)}{X(z)} = \frac{\sum_{i=0}^{n} a_i z^{-i}}{\sum_{i=0}^{n} b_i z^{-i}}$$

and thus

$$Y(z) \sum_{i=0}^{n} b_i z^{-i} = X(z) \sum_{i=0}^{n} a_i z^{-i}$$

Consequently,

$$Y(z) = \sum_{i=0}^{n} a_i z^{-i} X(z) - \sum_{i=1}^{n} b_i z^{-i} Y(z)$$

In the time domain

$$y(k) = \sum_{i=0}^{n} ax(k-i) - \sum_{i=1}^{n} b_i y(k-i) \tag{12-5}$$

This is the difference equation for the third direct (3D) structure, which is block diagrammed in Figure 12-1c. Notice that the structure has only one summing junction, but has $2n$ time-delay elements.

Fourth Direct Structure

The fourth direct (4D) structure is the transpose of the 3D structure, as shows in Figure 12-1d. This structure has only one signal distribution point, but has $2n$ difference equations, as expressed below:

$$\begin{aligned}
r_0(k) &= x(k) + r_1(k-1) \\
q_n(k) &= a_n r_0(k) \\
r_n(k) &= -b_n r_0(k) \\
q_i(k) &= a_i r_0(k) + q_{i+1}(k-1), \qquad i = 1, n-1 \\
r_i(k) &= -b_i r_0(k) + r_{i+1}(k-1) \\
y(k) &= a_0 r_0(k) + q_1(k-1)
\end{aligned} \tag{12-6}$$

Summary

Four direct structures for an nth-order digital filter have been derived. Table 12-1 summarizes their characteristics. Note that 1D and 2D conserve time-delay elements, while 1D and 3D conserve summing junctions. Conserving delay elements saves memory space in computer implementations, but memory is relatively inexpensive. Conserving summing junctions makes the control unit of a digital filter easier to design. Signal distribution usually has little impact, except in LSI applications, where routing is very important. All the direct structures suffer extreme coefficient sensitivity as n grows large. That is, a *small* change in a coefficient a_i or b_i, for n large, causes *large* changes in the zeros or poles of $D(z)$ of (12-1).

TABLE 12-1 PROPERTIES OF THE DIRECT STRUCTURES

	Structure			
	1D	2D	3D	4D
Time-delay elements	n	n	$2n$	$2n$
Multipliers	$2n + 1$	$2n + 1$	$2n + 1$	$2n + 1$
Summing junctions	2	$n + 1$	1	$2n$
Signal distribution points	$n + 1$	2	$2n$	1

Source: H. T. Nagle, Jr., and V. P. Nelson, "Digital Filter Implementation on 16-bit Microcomputers", *IEEE MICRO*, Vol. 1, No. 1, Table 1, p. 25, Feb. 1981, ©1981 IEEE.

12.3 SECOND-ORDER MODULES

To avoid the coefficient sensitivity problems, the transfer function $D(z)$ of (12-1) is usually implemented as a cascade or parallel of second-order modules of the form

$$D(z) = \frac{a_0 + a_1 z^{-1} + a_2 z^{-2}}{1 + b_1 z^{-1} + b_2 z^{-2}}$$

The structure of these second-order modules can themselves be of the direct format of Figure 12-1. Figure 12-2 illustrates the 1D, 2D, 3D, and 4D structures for second-order modules.

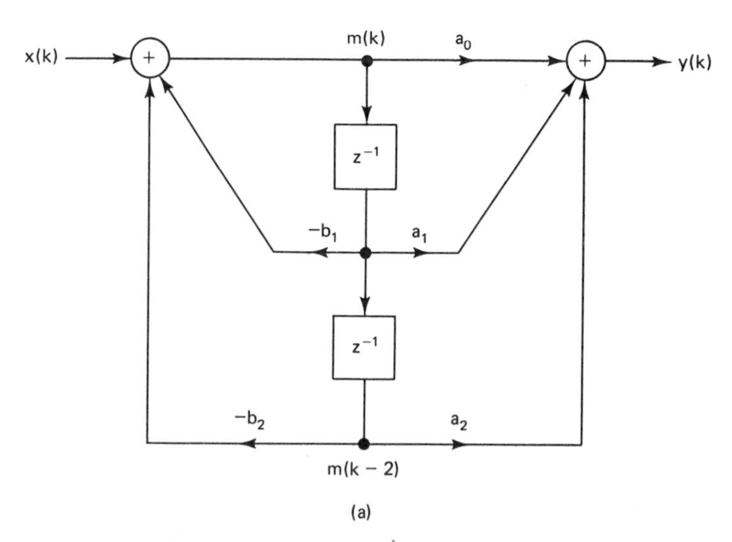

Figure 12-2 Direct second-order modules: (a) 1D; (b) 2D; (c) 3D; (d) 4D. (From H. T. Nagle, Jr., and V. P. Nelson, "Digital Filter Implementation on 16-bit Microcomputers," *IEEE MICRO*, Vol. 1, No. 1, Fig. 2, p. 25, Feb. 1981, ©1981 IEEE.)

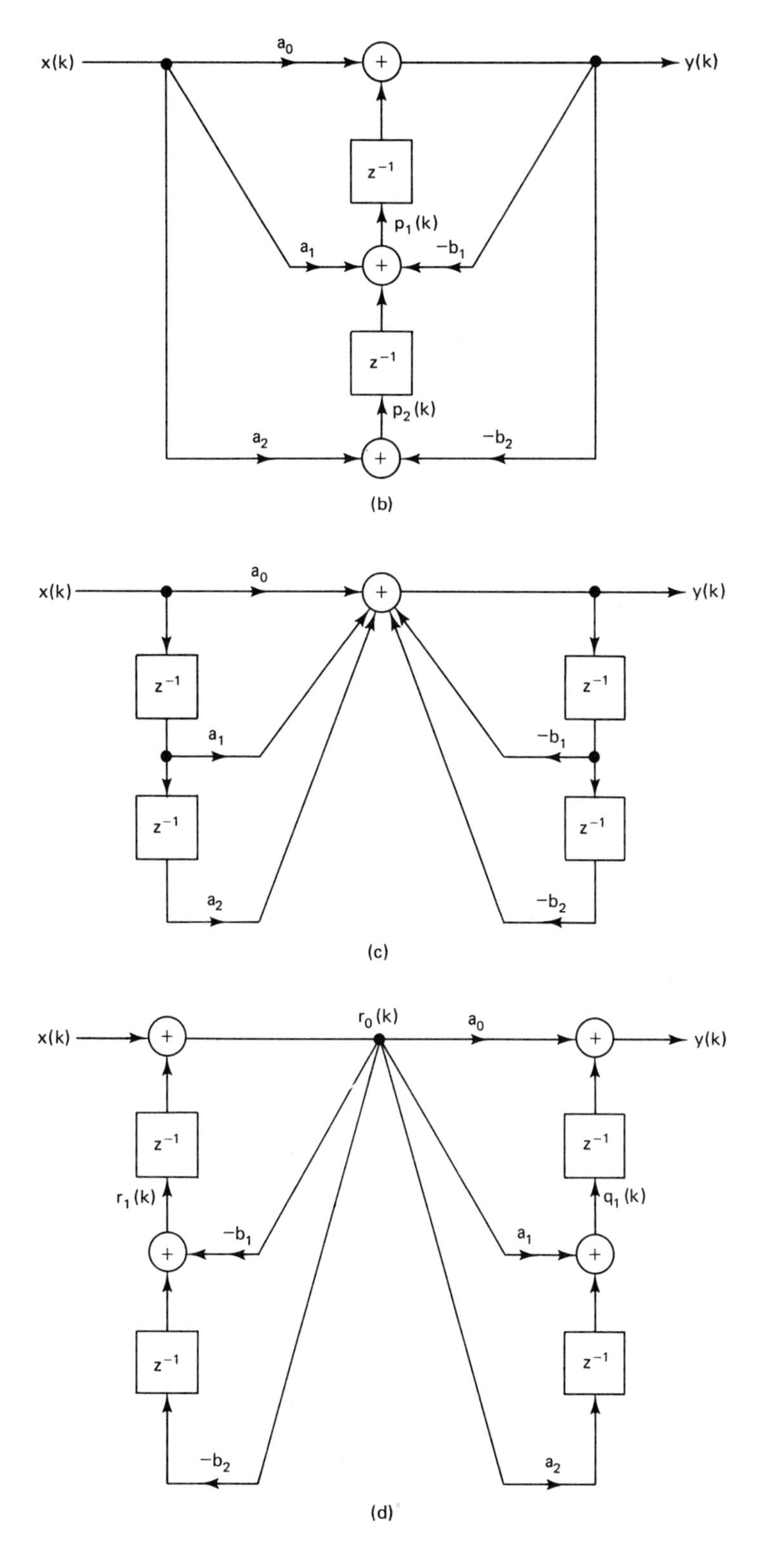

Figure 12-2 cont.

The difference equations describing each structure are:

$$\text{1D:}\quad \begin{aligned} m(k) &= x(k) - b_1 m(k-1) - b_2 m(k-2) \\ y(k) &= a_0 m(k) + a_1 m(k-1) + a_2 m(k-2) \end{aligned} \tag{12-7}$$

$$\text{2D:}\quad \begin{aligned} y(k) &= a_0 x(k) + p_1(k-1) \\ p_1(k) &= a_1 x(k) - b_1 y(k) + p_2(k-1) \\ p_2(k) &= a_2 x(k) - b_2 y(k) \end{aligned} \tag{12-8}$$

$$\text{3D:}\quad \begin{aligned} y(k) = a_0 x(k) + a_1 x(k-1) + a_2 x(k-2) \\ - b_1 y(k-1) - b_2 y(k-2) \end{aligned} \tag{12-9}$$

$$\text{4D:}\quad \begin{aligned} r_0(k) &= x(k) + r_1(k-1) \\ y(k) &= a_0 r_0(k) + q_1(k-1) \\ r_1(k) &= -b_1 r_0(k) - b_2 r_0(k-1) \\ q_1(k) &= a_1 r_0(k) + a_2 r_0(k-1) \end{aligned} \tag{12-10}$$

The equations should be calculated in the proper order. For example, in (12-7), $m(k)$ *must* first be obtained: in (12-10), $r_0(k)$. In (12-8) and (12-9) $y(k)$ *should* be calculated first to minimize the calculation time delay between the input sample $x(kT)$ and output generation $y(kT + \tau_c)$, where τ_c represents the filter calculation delay. Ideally, $\tau_c = 0$, but since this is unattainable we minimize τ_c by ordering our calculations. Practically, then, if $T \gg \tau_c$ we can neglect τ_c.

Other structures for second-order modules are possible. The cross-coupled structure of Figure 12-3 has often appeared in the literature [2–4] for the complex-pole-pair case. Here we call it the 1X structure. The difference equations are:

$$\begin{aligned} y(k) &= a_0 x(k) + s_2(k-1) \\ s_1(k) &= g_1 s_1(k-1) - g_2 s_2(k-1) + g_3 x(k) \\ s_2(k) &= g_1 s_2(k-1) + g_2 s_1(k-1) + g_4 x(k) \end{aligned} \tag{12-11}$$

where the g_i come from

$$D(z) = a_0 + \frac{A}{z+p} + \frac{A^*}{z+p^*}$$

and

$$\begin{aligned} g_1 &= -\text{Re}\,[p] \\ g_2 &= -\text{Im}\,[p] \\ g_3 &= 2\,\text{Im}\,[A] \\ g_4 &= 2\,\text{Re}\,[A] \end{aligned} \tag{12-12}$$

Note that the 1X structure is canonical.

The transpose of the 1X structure is shown in Figure 12-4 and is termed the 2X

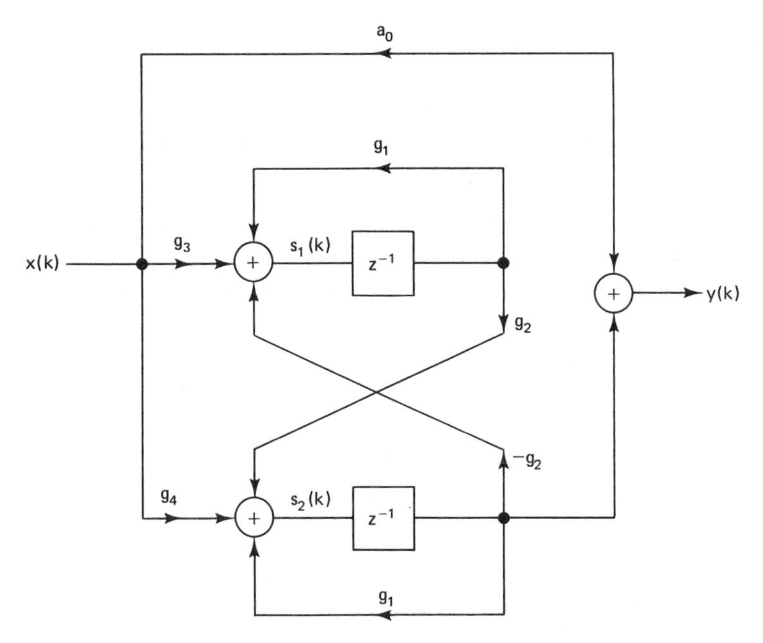

Figure 12-3 1X structure. (From H. T. Nagle, Jr., and V. P. Nelson, "Digital Filter Implementation on 16-bit Microcomputers," *IEEE MICRO*, Vol. 1, No. 1, Fig. 3, p. 26, Feb. 1981, ©1981 IEEE.)

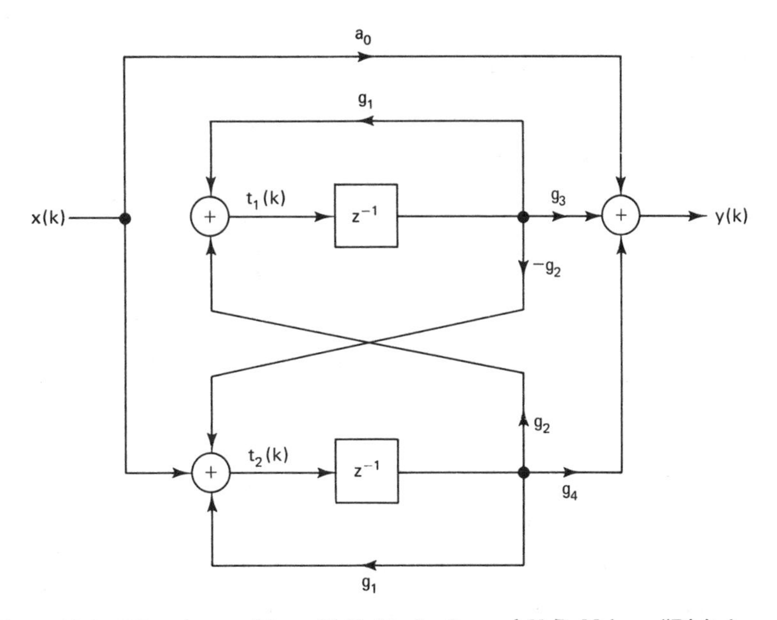

Figure 12-4 2X structure. (From H. T. Nagle, Jr., and V. P. Nelson, "Digital Filter Implementation on 16-bit Microcomputers," *IEEE MICRO*, Vol. 1, No. 1, Fig. 4, p. 26, Feb. 1981, ©1981 IEEE.)

structure. The difference equations are

$$
\begin{aligned}
y(k) &= a_0 x(k) + g_3 t_1(k-1) + g_4 t_2(k-1) \\
t_1(k) &= g_1 t_1(k-1) + g_2 t_2(k-1) \\
t_2(k) &= x(k) + g_1 t_2(k-1) - g_2 t_1(k-1)
\end{aligned}
\tag{12-13}
$$

where g_i are defined in (12-12).

These six structures will be used for second-order modules in the remainder of this text. Table 12-2 summarizes the structures. The 1X and 2X structures require two more multipliers than are required by the direct structures.

TABLE 12-2 SECOND-ORDER MODULES

	Structure					
	1D	2D	3D	4D	1X	2X
Time-delay elements	2	2	4	4	2	2
Multipliers	5	5	5	5	7	7
Summing junctions	2	3	1	4	3	3
Signal distribution points	3	2	4	1	3	3

Source: H. T. Nagle, Jr., and V. P. Nelson, "Digital Filter Implementation on 16-bit Microcomputers," *IEEE MICRO*, Vol. 1, No. 1, Table 2, p. 26, Feb. 1981, © 1981 IEEE.

12.4 CASCADE REALIZATION

To avoid coefficient-sensitivity problems, $D(z)$ of (12-1) may be implemented using a cascade of second-order modules. By factoring (12-1):

$$
D(z) = \frac{\prod_{i=1}^{m} (\alpha_{i0} + \alpha_{i1} z^{-1} + \alpha_{i2} z^{-2})}{\prod_{i-1}^{m} (1 + \alpha_{i3} z^{-1} + \alpha_{i4} z^{-2})} \tag{12-14}
$$

where m is the smallest integer greater than or equal to $n/2$. If the numerator and denominator factors are paired (the *pairing* problem) and the modules ordered in the cascade (the *ordering* problem), then

$$
D(z) = \prod_{i=1}^{m} A_i(z)
$$

where (12-15)

$$
A_i(z) = \frac{\alpha_{i0} + \alpha_{i1} z^{-1} + \alpha_{i2} z^{-2}}{1 + \alpha_{i3} z^{-1} + \alpha_{i4} z^{-2}}
$$

The pairing and ordering problems in cascaded second-order modules have been extensively studied in the literature [5–8]. In Chapter 14 we will examine these problems more closely. Figure 12-5 illustrates the cascade of (12-15); the second-order modules may be implemented in the direct or cross-coupled structures. If the direct structures are used, the cascade diagrams of Figure 12-6 result. These cascade structures are

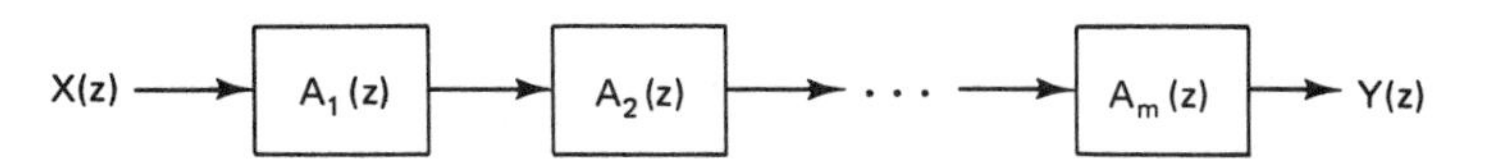

Figure 12-5 Cascaded second-order modules. (From H. T. Nagle, Jr., and V. P. Nelson, "Digital Filter Implementation on 16-bit Microcomputers," *IEEE MICRO,* Vol. 1, No. 1, Fig. 5, p. 27, Feb. 1981, ©1981 IEEE.)

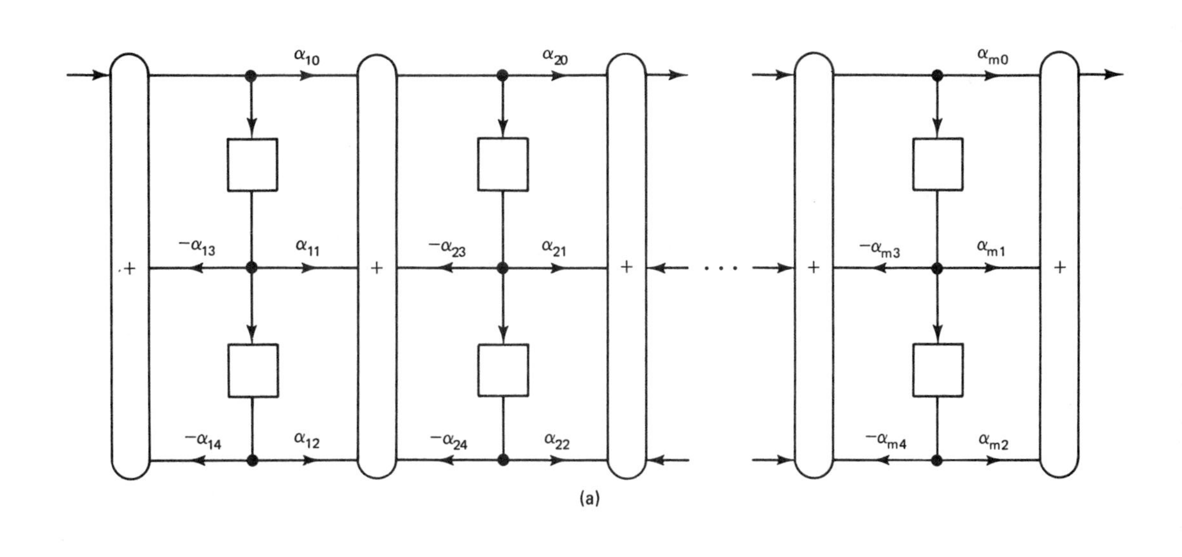

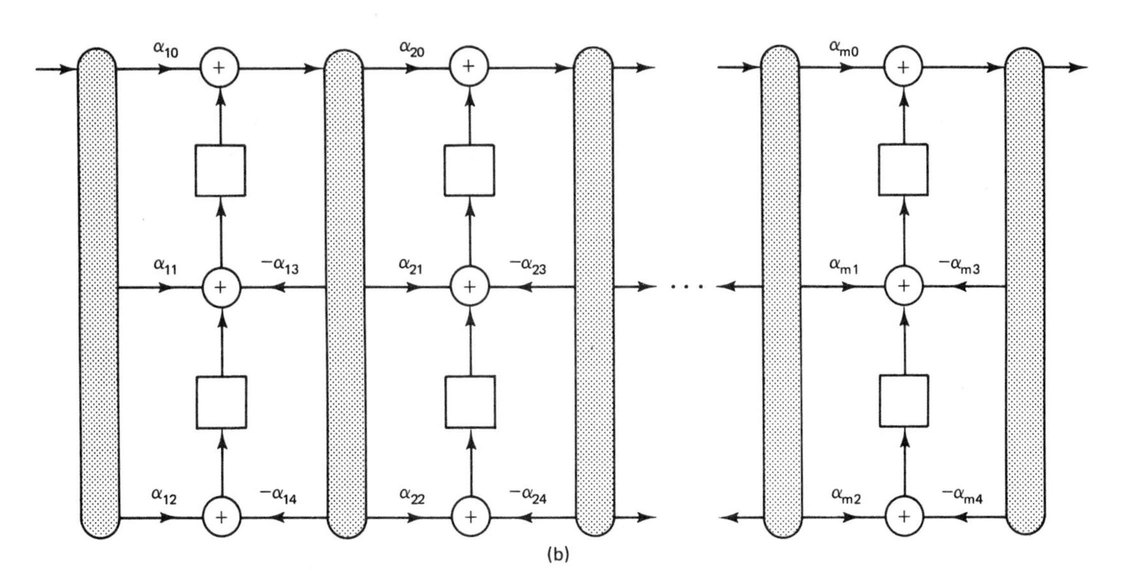

Figure 12-6 Cascade filter structures: (a) 1D; (b) 2D; (c) 3D; (d) 4D. (From H. T. Nagle, Jr., and V. P. Nelson, "Digital Filter Implementation on 16-bit Microcomputers," *IEEE MICRO*, Vol. 1, No. 1, Fig. 6, p. 28, Feb. 1981, ©1981 IEEE.)

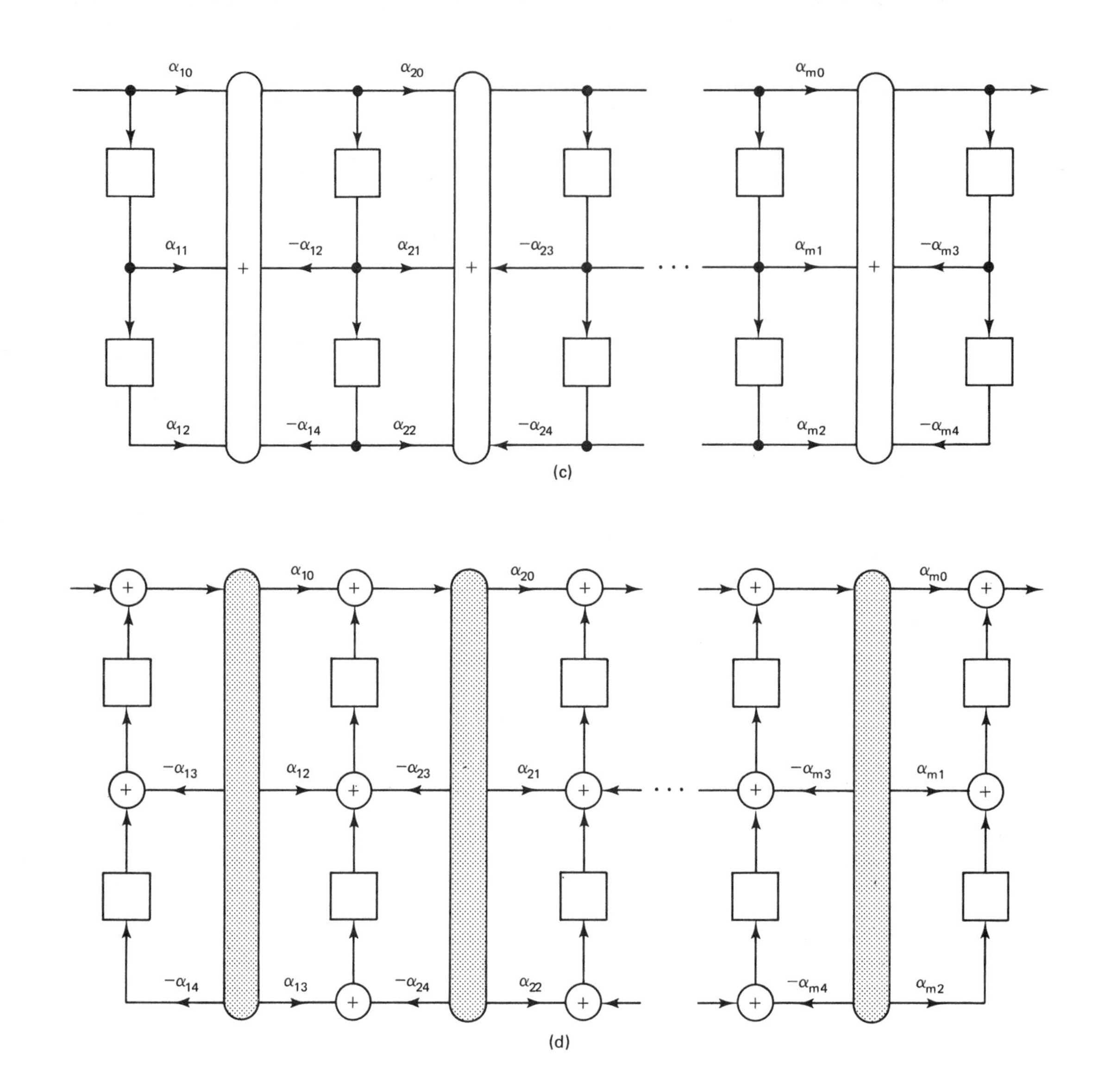

Figure 12-6 cont.

compared in Table 12-3. If one contrasts Tables 12-1 and 12-3, we see that cascading 3D and 4D modules saves $n - 2$ delay elements in each case. Cascading costs extra multipliers in every case. Cascaded direct modules require $m - 1$ extra summing junctions and signal distribution points over the direct structures.

TABLE 12-3 CASCADE STRUCTURES

	Structure[a]			
	1D	2D	3D	4D
Time-delay elements	$2m$	$2m$	$2m + 2$	$2m + 2$
	(n)	(n)	$(2n - (n - 2))$	$(2n - (n - 2))$
Multipliers	$5m$	$5m$	$5m$	$5m$
	$(n + n + m)$	$(n + n + m)$	$(2n + m)$	$(2n + m)$
Summing junctions	$m + 1$	$3m$	m	$3m + 1$
	$(2 + m - 1)$	$(n + m)$		$(2n - (m - 1))$
Signal distribution points	$3m$	$m + 1$	$3m + 1$	m
	$(n + m)$	$(2 + m - 1)$	$(2n - (m - 1))$	

[a] m is the smallest integer greater than or equal to $n/2$. The numbers in parentheses are for comparison with Table 12-1.

Source: H. T. Nagle, Jr., and V. P. Nelson, "Digital Filter Implementation on 16-bit Microcomputers," *IEEE MICRO*, Vol. 1, No. 1, Table 3, p. 27, Feb. 1981, © 1981 IEEE.

12.5 PARALLEL REALIZATION

A second method to avoid the coefficient-sensitivity problems of (12-1) is to factor the denominator of $D(z)$ and to perform a partial-fraction expansion to obtain (for distinct poles)

$$D(z) = \beta_0 + \sum_{i=1}^{m} B_i(z)$$

where (12-16)

$$B_i(z) = \frac{\beta_{i1} z^{-1} + \beta_{i2} z^{-2}}{1 + \beta_{i3} z^{-1} + \beta_{i4} z^{-2}}$$

Figure 12-7 depicts the parallel structure. Any of the six structures of Section 12.3

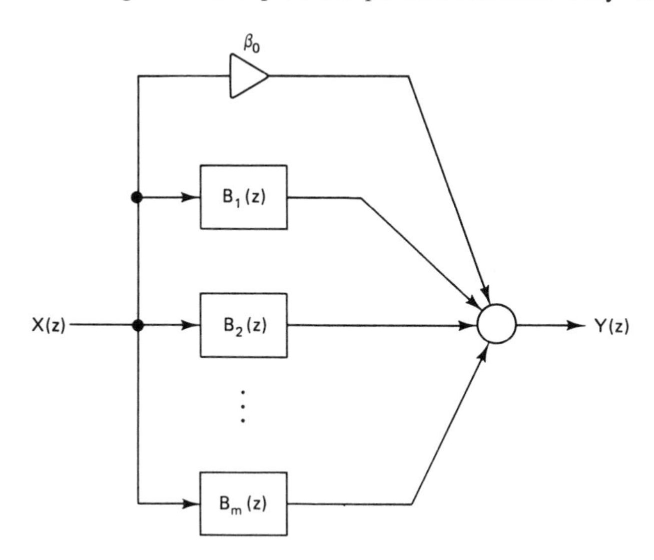

Figure 12-7 Parallel structure. (From H. T. Nagle, Jr., and V. P. Nelson, "Digital Filter Implementation on 16-bit Microcomputers," *IEEE MICRO*, Vol. 1, No. 1, Fig. 7, p. 28, Feb. 1981, ©1981 IEEE.)

may be used to implement the blocks of Figure 12-7. If the direct structures are used, some element sharing may be accomplished, as was the case in the cascade implementation. Figure 12-8 indicates the direct parallel structures. Table 12-4 compares the characteristics of the parallel structures with those of Table 12-1. Note that the 3D and 4D parallel structures save $n - 2$ time delays over the direct realizations. The number of multipliers is the same as in Table 12-1. The 1D parallel structure requires an additional $m - 1$ summing junctions; the 2D, $m - 1$ signal distribution points. The 3D requires m additional summing junctions and $n - 3$ fewer signal distribution points (and alternately for the 4D parallel case).

12.6 PID CONTROLLERS

In Chapter 8 we presented the concept of digital control using proportion (K_P), integration (K_I), and differentiation (K_D) in (8-52):

$$D(z) = K_P + \frac{K_I T}{2}\left(\frac{z+1}{z-1}\right) + \frac{K_D}{T}\left(\frac{z-1}{z}\right) \tag{12-17}$$

This transfer function may be implemented as shown in Figure 12-9. Here the proportional, integral, and differential terms are implemented separately and summed at the output. An alternative method would be to find a second-order transfer function for (12-17) and use the direct structures presented earlier. Consider

$$\begin{aligned} D(z) &= \frac{K_P(z-1)(z) + (K_I T/2)(z+1)(z) + (K_D/T)(z-1)(z-1)}{(z-1)(z)} \\ &= \frac{K_P(z^2 - z) + (K_I T/2)(z^2 + z) + (K_D/T)(z^2 - 2z + 1)}{z^2 - z} \\ &= \frac{(K_P + K_I T/2 + K_D/T)z^2 + (-K_P + K_I T/2 - 2K_D/T)z + K_D/T}{z^2 - z} \\ &= \frac{a_0 + a_1 z^{-1} + a_2 z^{-2}}{1 + b_1 z^{-1} + b_2 z^{-2}} \end{aligned}$$

where

$$\begin{aligned} a_0 &= K_P + \frac{K_I T}{2} + \frac{K_D}{T} \\ a_1 &= -K_P + \frac{K_I T}{2} - \frac{2K_D}{T} \\ a_2 &= \frac{K_D}{T} \\ b_1 &= -1 \\ b_2 &= 0 \end{aligned} \tag{12-18}$$

Consequently, a PID controller can be implemented by any of the second-order, direct, structures described earlier in the chapter.

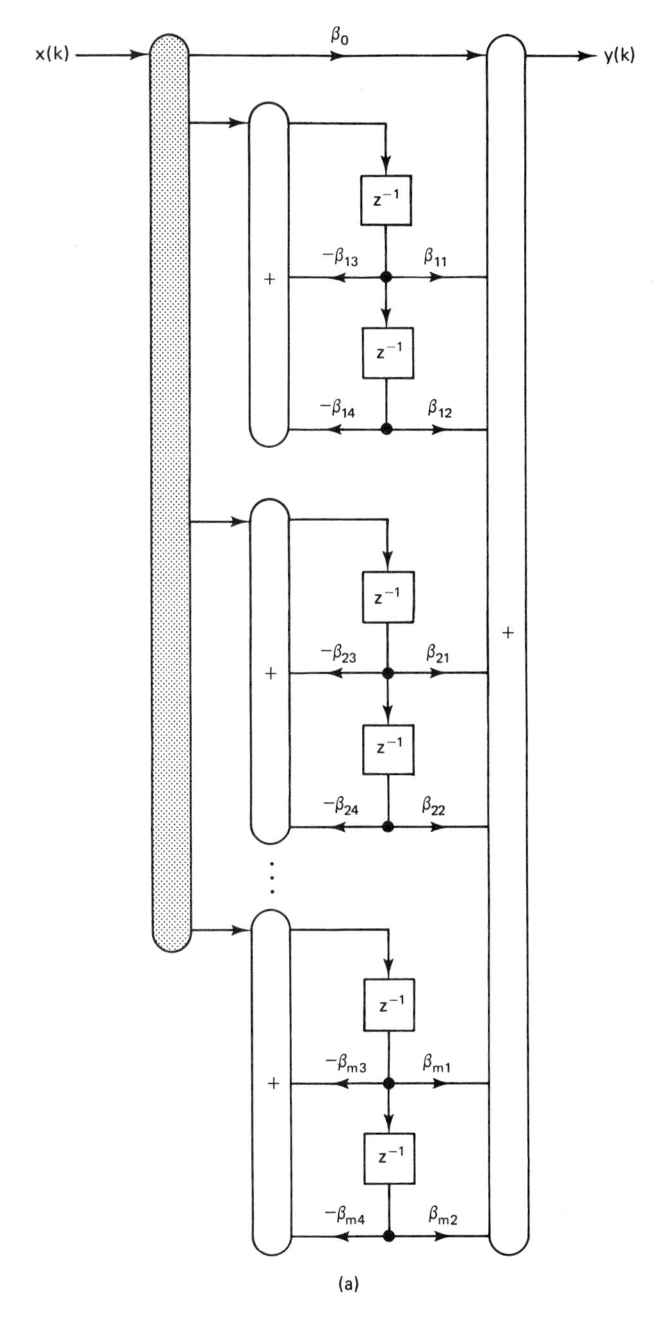

Figure 12-8 Paralleled second-order modules: (a) 1D; (b) 2D; (c) 3D; (d) 4D. (From H. T. Nagle, Jr., and V. P. Nelson, "Digital Filter Implementation on 16-bit Microcomputers," *IEEE MICRO*, Vol. 1, No. 1, Fig. 8, pp. 28–30, Feb. 1981, ©1981 IEEE.)

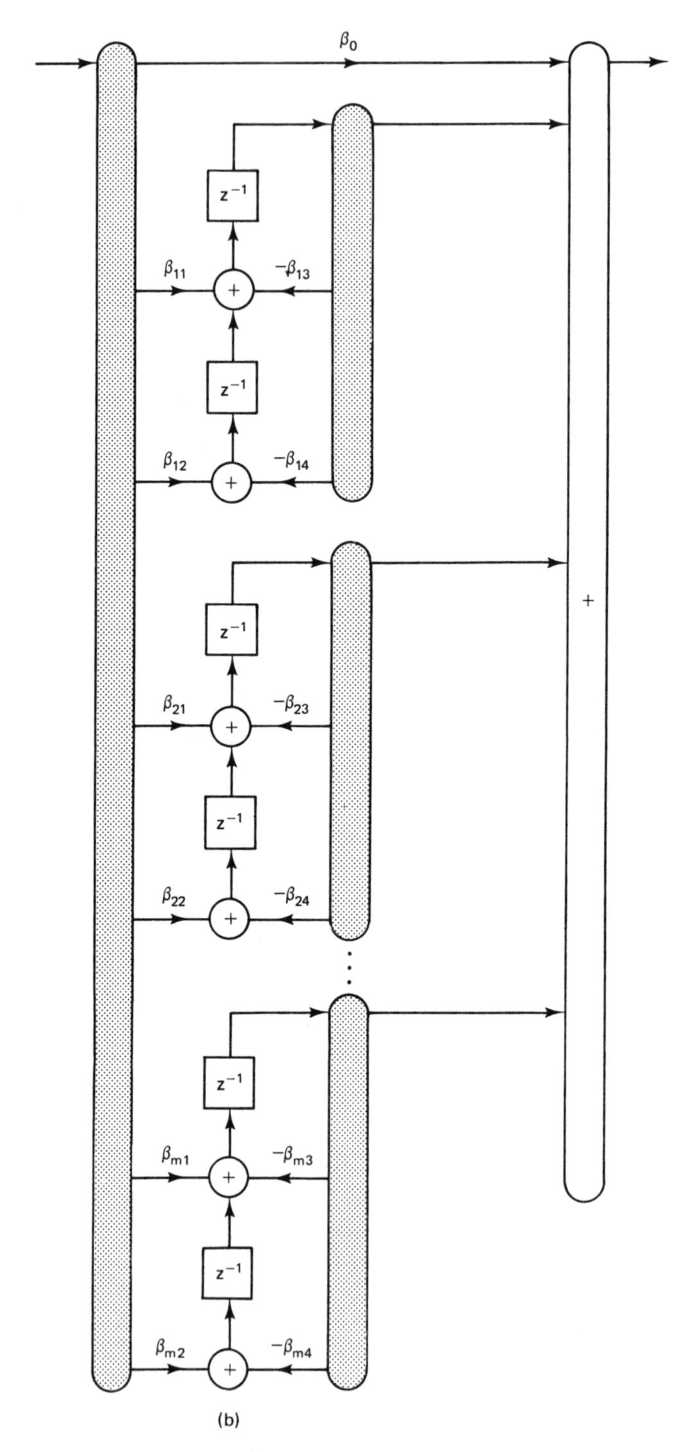

Figure 12-8 cont.

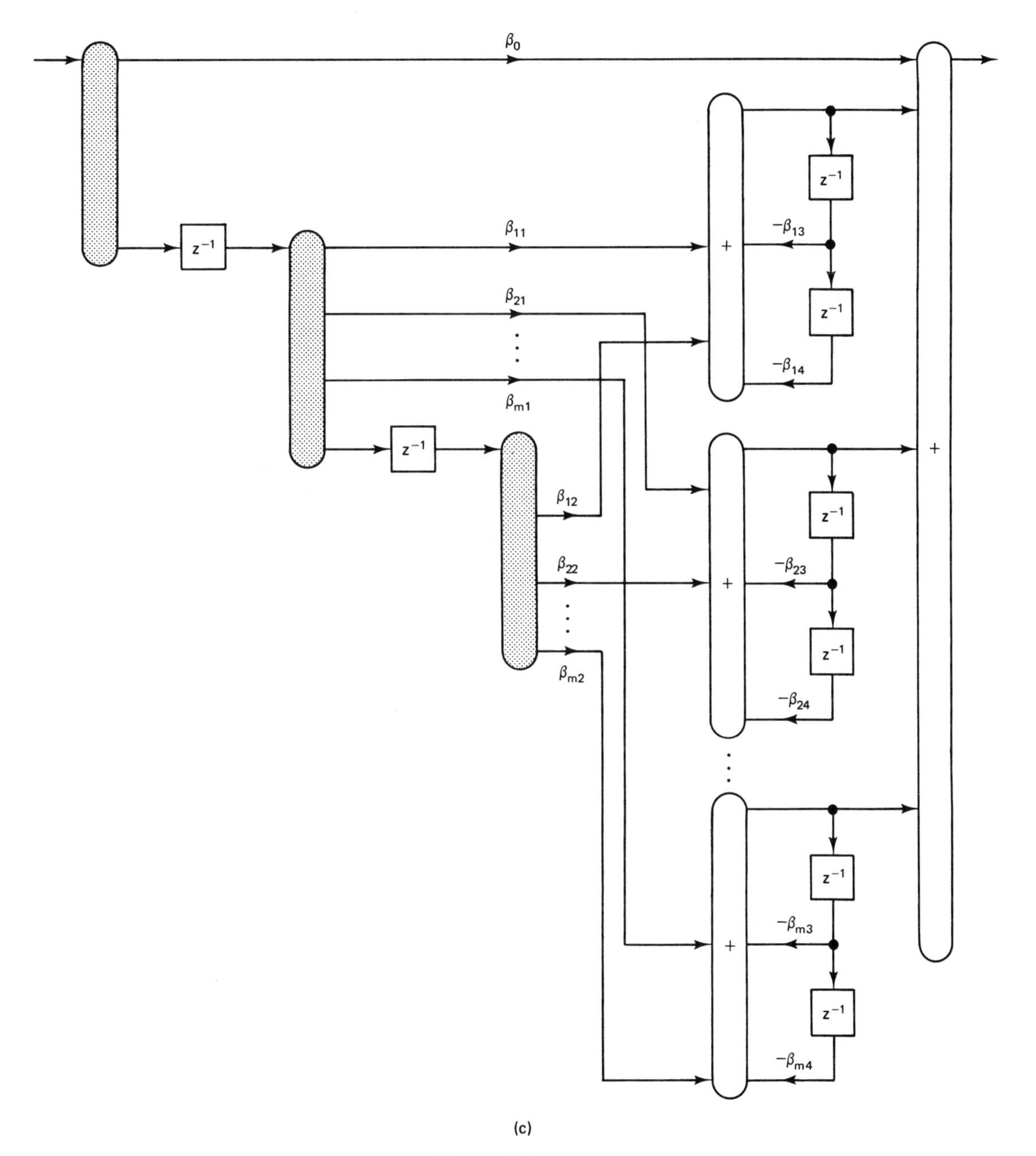

(c)

Figure 12-8 cont.

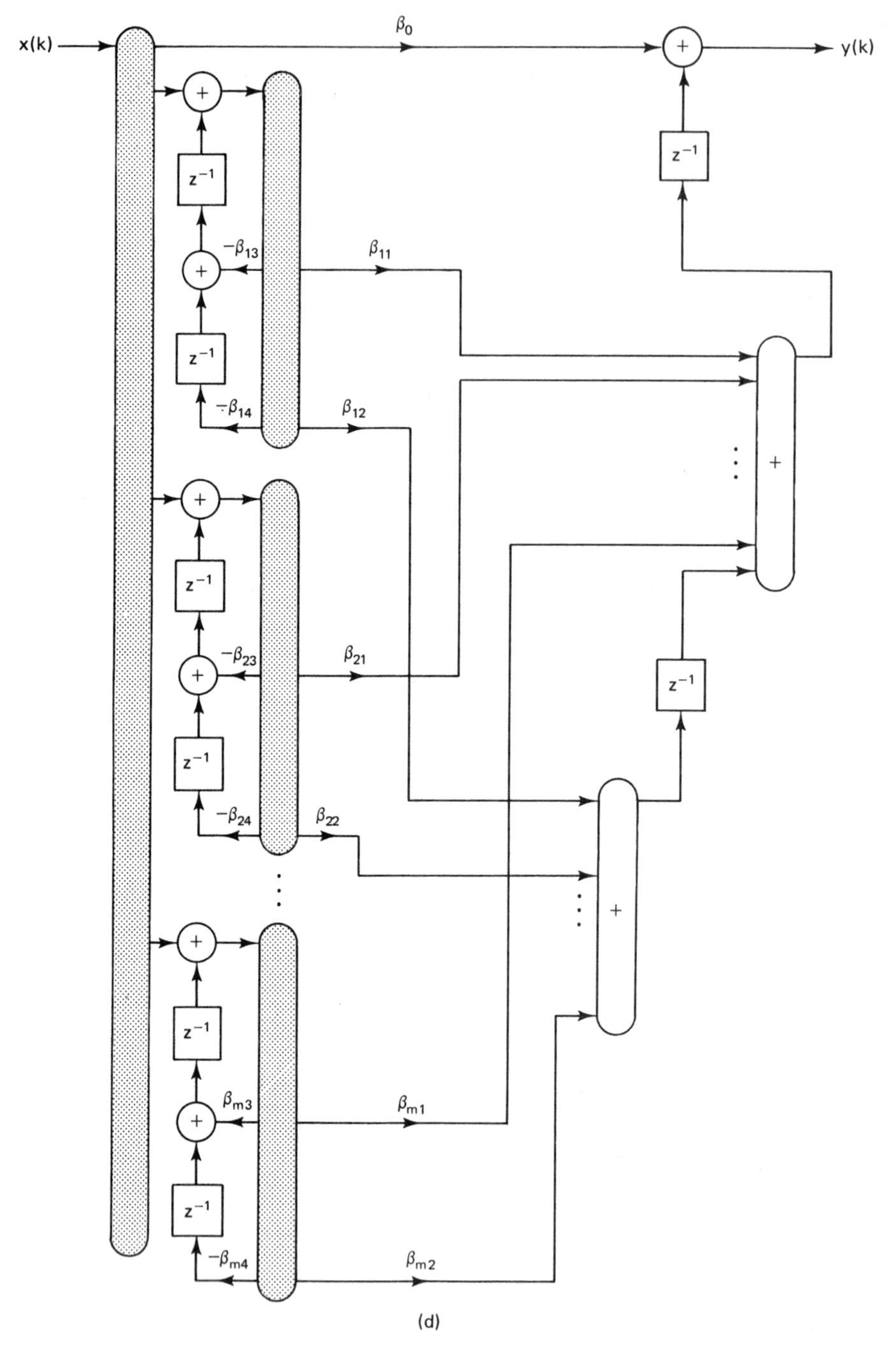

(d)

Figure 12-8 cont.

TABLE 12-4 PARALLEL STRUCTURES

	Structure[a]			
	1D	2D	3D	4D
Time-delay elements	$2m$ (n)	$2m$ (n)	$2m + 2$ $(2n - (n - 2))$	$2m + 2$ $(2n - (n - 2))$
Multipliers	$4m + 1$ $(2n + 1)$	$4m + 1$ $(2n + 1)$	$4m + 1$ $(2n + 1)$	$4m + 1$ $(2n + 1)$
Summing junctions	$m + 1$ $(2 + m - 1)$	$2m + 1$ $(n + 1)$	$m + 1$ $(1 + m)$	$2m + 3$ $(2n - (n - 3))$
Signal distribution points	$2m + 1$ $(n + 1)$	$m + 1$ $(2 + m - 1)$	$2m + 3$ $(2n - (n - 3))$	$m + 1$ $(1 + m)$

[a]m is the smallest integer greater than or equal $n/2$. Numbers in parentheses are for comparison with Table 12-1.

Source: H. T. Nagle, Jr., and V. P. Nelson, "Digital Filter Implementation on 16-bit Microcomputers," *IEEE MICRO*, Vol. 1, No. 1, Table 4, p. 31, Feb. 1981, © 1981 IEEE.

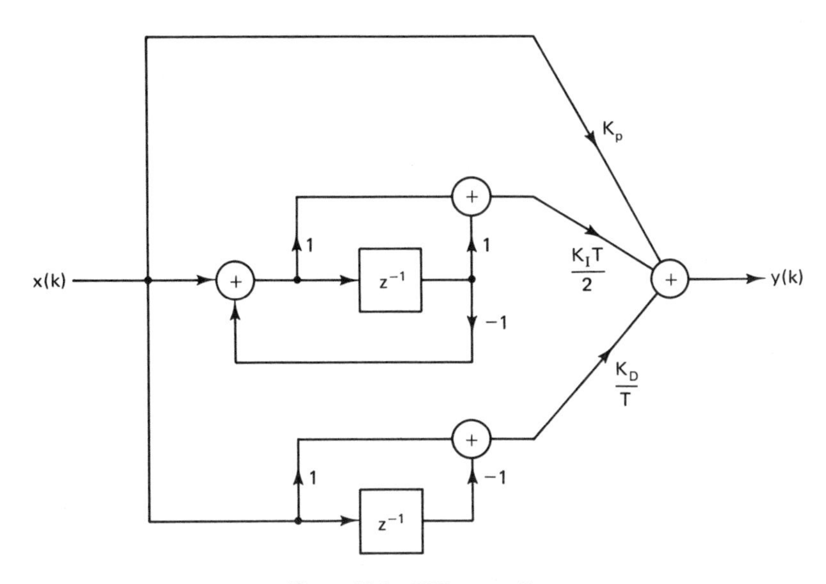

Figure 12-9 PID controller.

12.7 LADDER REALIZATION

A third method for improving the coefficient sensitivity of the direct structures for (12-1) is to implement a ladder network [9–11]. If $D(z)$ in (12-1) is expressed as

$$D(z) = \frac{a_0 + a_1 z^{-1} + a_2 z^{-2} + \cdots + a_n z^{-n}}{1 + b_1 z^{-1} + b_2 z^{-2} + \cdots + b_n z^{-n}}$$

where $a_n \neq 0$ and $b_n \neq 0$, then $D(z)$ can be expanded into continued-fraction form:

$$D(z) = A_0 + \cfrac{1}{B_1 z^{-1} + \cfrac{1}{A_1 + \cfrac{1}{B_2 z^{-1} + \cfrac{1}{\ddots + \cfrac{1}{B_n z^{-1} + \cfrac{1}{A_n}}}}}} \tag{12-19}$$

where A_i and B_i are real constants derived from a_i and b_i. If $a_n = 0$, then A_0 is zero. To implement (12-19) we must be able to implement

$$\begin{aligned} G_1(z) &= \frac{1}{Bz^{-1} + T(z)} \\ G_2(z) &= \frac{1}{A + T(z)} \end{aligned} \tag{12-20}$$

These functions may be implemented as shown in Figure 12-10.

The realization procedure is to first express (12-1) as

$$D(z) = A_0 + \frac{1}{B_1 z^{-1} + T_1(z)}$$

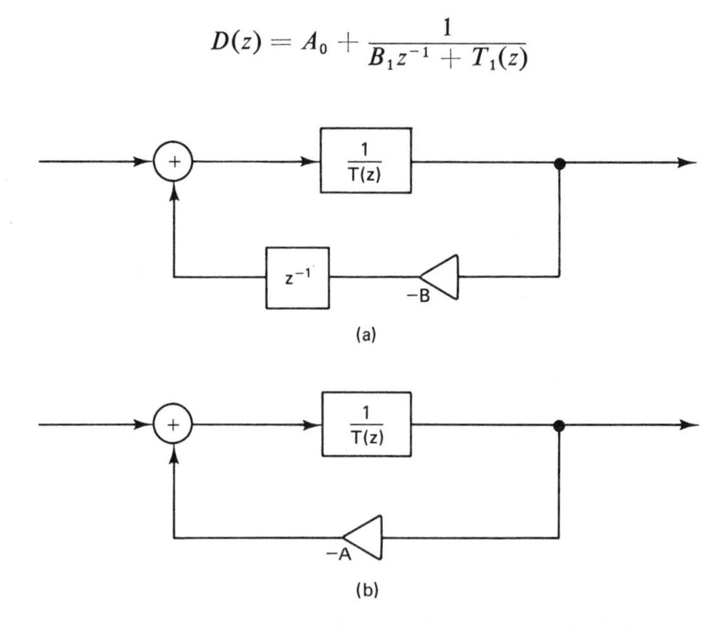

Figure 12-10 Continued Fraction Seconds: (a) $G_1(z)$; (b) $G_2(z)$. [From S. K. Mitra and R. J. Sherwood, "Canonic Realizations of Digital Filters Using the Continued Fraction Expansion," *IEEE Trans. Audio Electroacoust.*, Vol. AU-20, No. 3, Aug. 1972. Part (a) from Fig. 6, p. 188; part (b) from Fig. 2, p. 186, ©1972 IEEE.]

This part of the procedure is shown in Figure 12-11a. Next we express

$$T_1(z) = \frac{1}{A_1 + T_2(z)}$$

and

$$\frac{1}{T_1(z)} = A_1 + T_2(z)$$

This step is shown in Figure 12-11b. The third step is to express

$$T_2(z) = \frac{1}{B_2 z^{-1} + T_3(z)}$$

which is essentially repeating step 1, as shown in Figure 12-11c. The process is repeated until we arrive at the ladder structure of Figure 12-11d.

There are many different digital ladder networks. However, we only show one to illustrate the principles involved. If real-time response is required, the ladder of

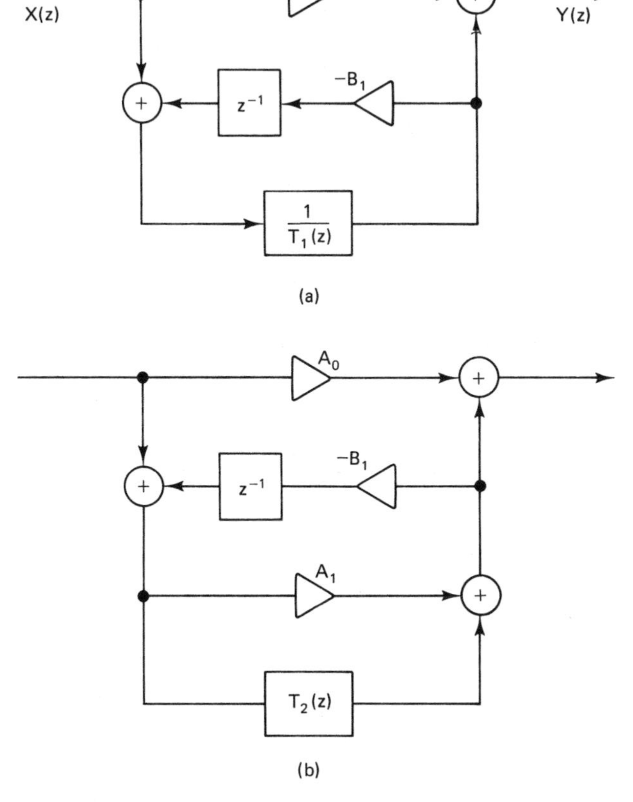

Figure 12-11 Ladder structure: (a) first step; (b) second step; (c) third step; (d) final step. [Part (d) from S. K. Mitra and R. J. Sherwood, "Canonic Realizations of Digital Filters Using the Continued Fraction Expansion," *IEEE Trans. Audio Electroacoust.*, Vol. AU-20, No. 3, Fig. 7, p. 188, Aug. 1972, ©1972 IEEE.]

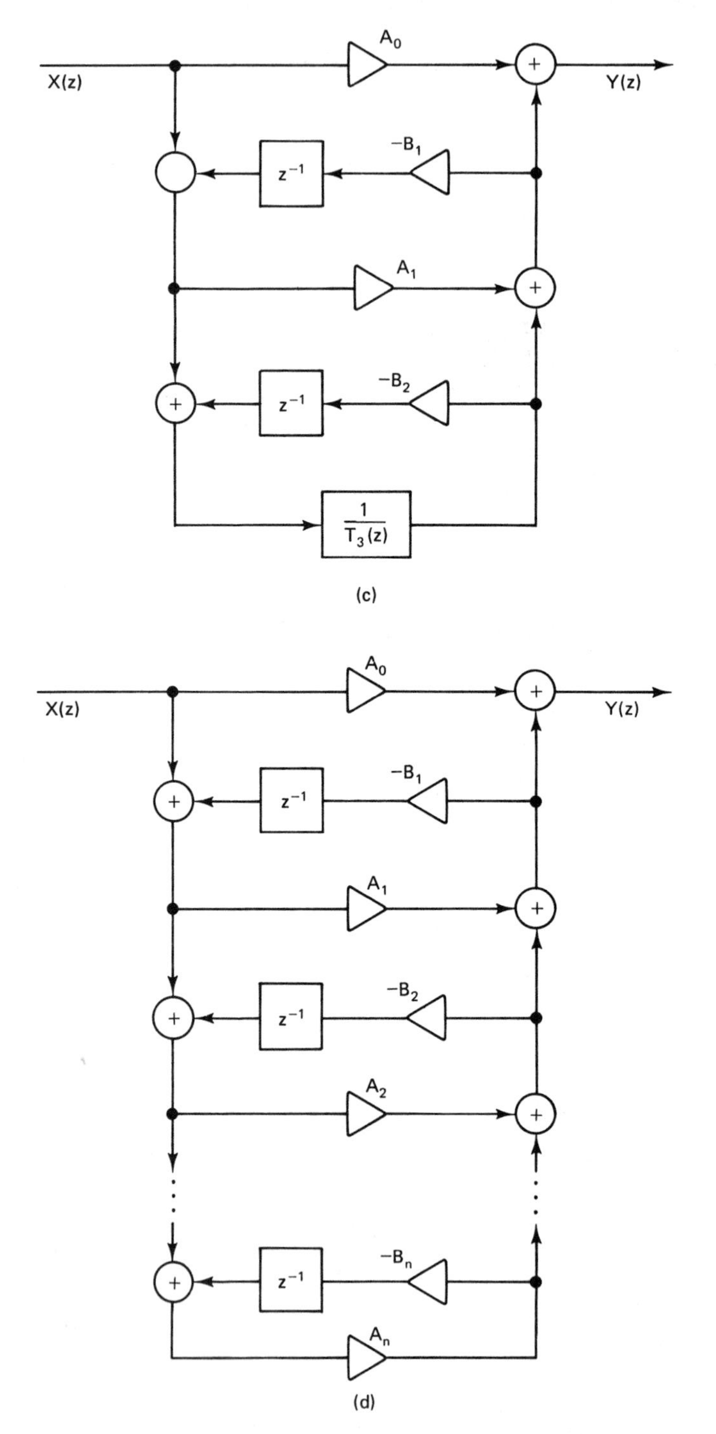

Figure 12-11 cont.

Figure 12-11d is not recommended because all of the $2n$ difference equations must be calculated before the output is available.

The ladder structure is canonical because it requires n time-delay elements. It also requires $2n + 1$ multipliers and has $2n$ signal distribution points.

Example 12.1

Find a ladder realization of the following:

$$D(z) = \frac{11/128 + 106/128z^{-1} + 224/128z^{-2} + z^{-3}}{1/128 + 34/128z^{-1} + 160/128z^{-2} + z^{-3}}$$

In another form:

$$D(z) = \frac{128z^{-3} + 224z^{-2} + 106z^{-1} + 11}{128z^{-3} + 160z^{-2} + 34z^{-1} + 1}$$

First, divide the denomination into the numerator:

$$
\begin{array}{rrrrr}
 & 1 & & & \\
128\ 160\ 34\ 1\,) & 128 & 224 & 106 & 11 \\
 & 128 & 160 & 34 & 1 \\ \hline
 & & 64 & 72 & 10
\end{array}
$$

Therefore,

$$D(z) = 1 + \cfrac{1}{\cfrac{128z^{-3} + 160z^{-2} + 34z^{-1} + 1}{64z^{-2} + 72z^{-1} + 10}}$$

Again we divide the denominator into the numerator:

$$
\begin{array}{rrrrr}
 & 2z^{-1} & & & \\
64\ 72\ 10\,) & 128 & 160 & 34 & 1 \\
 & 128 & 144 & 20 & \\ \hline
 & & 16 & 14 & 1
\end{array}
$$

Consequently,

$$D(z) = 1 + \cfrac{1}{2z^{-1} + \cfrac{1}{\cfrac{64z^{-2} + 72z^{-1} + 10}{16z^{-2} + 14z^{-1} + 1}}}$$

Repeating the procedure yields

$$
\begin{array}{rrrr}
 & 4 & & \\
16\ 14\ 1\,) & 64 & 72 & 10 \\
 & 64 & 56 & 4 \\ \hline
 & & 16 & 6
\end{array}
$$

The transfer function can then be written

$$D(z) = 1 + \cfrac{1}{2z^{-1} + \cfrac{1}{4 + \cfrac{1}{\cfrac{16z^{-2} + 14z^{-1} + 1}{16z^{-1} + 6}}}}$$

Again, repeating the division:

$$
\begin{array}{rrrr}
 & z^{-1} & & \\
16\ 6\,) & 16 & 14 & 1 \\
 & 16 & 6 & \\ \hline
 & & 8 & 1
\end{array}
$$

With each set, the order of the numerator or denominator is decreased by one.

$$D(z) = 1 + \cfrac{1}{2z^{-1} + \cfrac{1}{4 + \cfrac{1}{z^{-1} + \cfrac{16z^{-1} + 6}{8z^{-1} + 1}}}}$$

The final result is

$$D(z) = 1 + \cfrac{1}{2z^{-1} + \cfrac{1}{4 + \cfrac{1}{z^{-1} + \cfrac{1}{2 + \cfrac{1}{2z^{-1} + \cfrac{1}{4}}}}}}$$

The ladder structure is drawn in Figure 12-12.

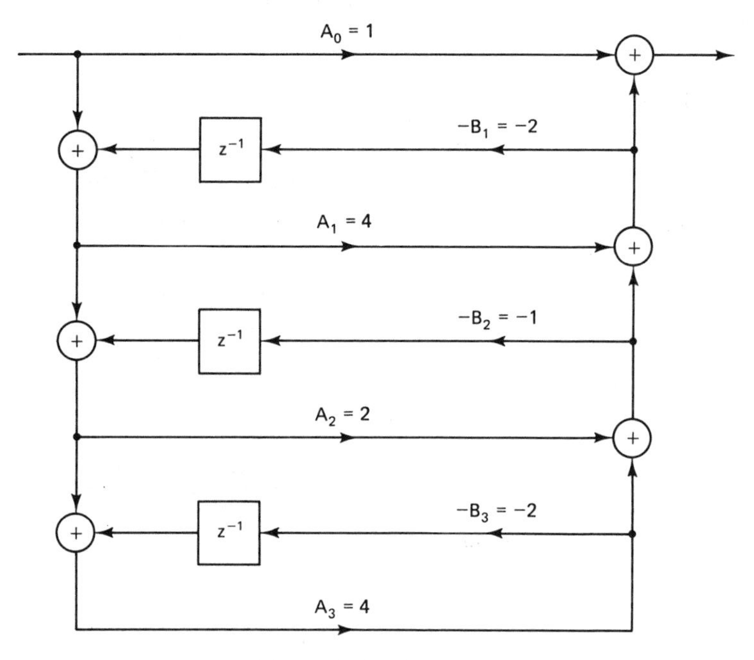

Figure 12-12 Example.

12.8 OTHER STRUCTURES

Researchers in the field of signal processing have explored many avenues in the realization of digital filter structures. Fettweis [12] designed wave digital filters which exhibited many desirable properties. However, the implementation of the wave digital filter is more complex than the examples of this chapter. Other researchers have produced special structure to accomplish specific goals. Ali and Constantinides [13] designed low-coefficient sensitive structures. Fam and Barnes [14] concentrated on structure to eliminate limit cycles. Chang [15] emphasized low round-off noise.

Abu-El-Haija et al. [16] have used a sampled-data transformation to represent an analog integrator, and has found filter structures for a digital incremental computer.

Nishimura and Hirano [17] have tackled the problem of multiple feedback (leapfrog) digital filters.

12.9 SUMMARY

In this chapter we have examined various structures for digital filters. The most commonly used in digital control applications are the direct, cascade, and parallel structures. If the direct structure can implement a specified $D(z)$ within its frequency tolerances, cascading or paralleling second-order modules is unnecessary. However, if coefficient round-off causes the poles and zeros of $D(z)$ to move beyond tolerable limits, cascading or paralleling second-order modules is recommended for solving the problem.

REFERENCES

1. R. E. Crochiere and A. V. Oppenheim, "Analysis of Linear Digital Networks," *Proc. IEEE*, Vol. 63, pp. 581–595, Apr. 1975.
2. A. E. Vereshkin et al., "Two New Structures for the Implementation of a Discrete Transfer Function with Complex Poles," *Autom. Remote Control*, pp. 1416–1422, Sept. 1968.
3. H. T. Nagle, Jr., "Survey of Digital Filtering," Final Technical Report, NASA-CR-124166, Contract NAS8-20163, Oct. 1972, Auburn University, Auburn, Ala., NASA Star CSCL09C, N73-20256.
4. L. B. Jackson, A. G. Lindgren, and Y. Kim, "Optimal Synthesis for Second-Order State-Space Structures for Digital Filters," *IEEE Trans. Circuits Syst.*, Vol. CAS-26, pp. 149–152, Mar. 1979.
5. L. B. Jackson, "Roundoff-Noise Analysis for Fixed-Point Digital Filters in Cascade or Parallel Form," *IEEE Trans. Audio Electroacoust.*, Vol. AU-18, pp. 107–122, June 1970.
6. S. Y. Hwang, "An Optimization of Cascade Fixed-Point Digital Filters," *IEEE Trans. Circuits Syst.* (Letters), Vol. CAS-21, pp. 163–166, Jan. 1974.
7. W. S. Lee, "Optimization of Digital Filters for Low Roundoff Noise," *IEEE Trans. Circuits Syst.*, Vol. CAS-21, pp. 424–431, May 1974.
8. B. Liu and A. Peled, "Heuristic Optimization of the Cascade Realization of Fixed-Point Digital Filters," *IEEE Trans. Acoust. Speech Signal Process.*, Vol. ASSP-23, pp. 464–473, Oct. 1975.
9. S. K. Mitra and R. J. Sherwood, "Canonic Realizations of Digital Filters Using the Continued Fraction Expansion," *IEEE Trans. Audio Electroacoust.*, Vol. AU-20, No. 3, pp. 185–194, Aug. 1972.
10. S. K. Mitra and R. J. Sherwood, "Digital Ladder Networks," *IEEE Trans. Audio Electroacoust.*, Vol. AU-21, pp. 30–36, Feb. 1973.
11. L. T. Bruton, "Low-Sensitivity Digital Ladder Filters," *IEEE Trans. Circuits Syst.*, Vol. CAS-22, pp. 168–176, Mar. 1975.

12. A. Fettweis, "Some Principles of Designing Digital Filters Imitating Classical Filter Structure," *IEEE Trans. Circuits Theory*, pp. 314–316, Mar. 1971.

13. A. M. Ali and A. G. Constantinides, "Design of Low Sensitivity and Complexity Digital Filter Structures," *1978 Eur. Conf. Circuit Theory Des.*, Lausanne, Switzerland, Sept. 1978, pp. 335–339.

14. A. T. Fam and C. W. Barnes, "Nonminimal Realizations of Fixed-Point Digital Filters That Are Free of All Finite Work-Length Limit Cycles," *IEEE Trans. Acoust. Speech Signal Process.*, Vol. ASSP-27, pp. 149–153, Apr. 1979.

15. T. L. Chang, "A Low Roundoff Noise Digital Filter Structure," *ISCAS. 78*, pp 1004–1008.

16. A. I. Abu-El-Haija, K. Shenoi, and A. M. Peterson, "Digital Filter Structures Having Low Errors and Simple Hardware Implementation," *IEEE Trans. Circuits Syst.*, Vol. CAS-25, pp. 593–599, Aug. 1978.

17. S. Nishimura and K. Hirano, "Realizations of Digital Filters Using GeneralizedMultiple-Feedback Structure," *ISCAS 78*, pp. 284–288.

PROBLEMS

12-1. Examine the structure of Figure P12-1. If $\alpha_1 = 0$, find α_0, α_2, and α_3 such that

$$D(z) = \frac{Y(z)}{X(z)} = \frac{a_0 + a_1 z^{-1} + a_2 z^{-2}}{1 + b_1 z^{-1} + b_2 z^{-2}}$$

Note: $\alpha_i = f_i(a_0, a_1, a_2, b_1, b_2)$.

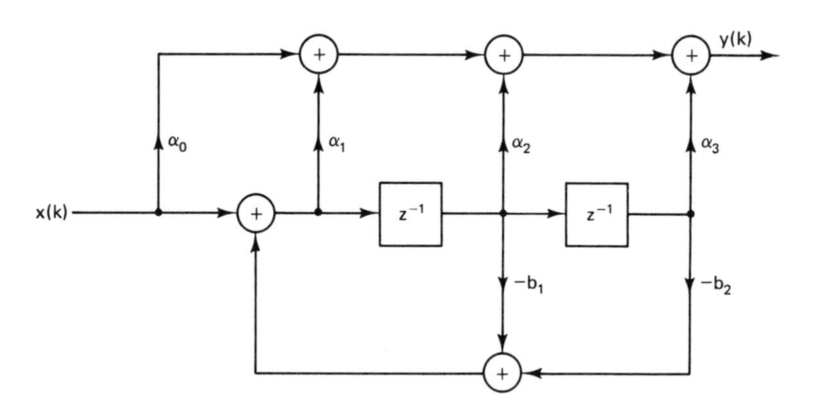

Figure P12-1

12-2. Repeat Problem 12-1 if $\alpha_2 = 0$.

12-3. Repeat Problem 12-1 if $\alpha_3 = 0$.

12-4. Examine Figure P12-4. Find α_0, α_1, α_2, such that

$$D(z) = \frac{a_0 + a_1 z^{-1} + a_2 z^{-2}}{1 + b_1 z^{-1} + b_2 z^{-2}}$$

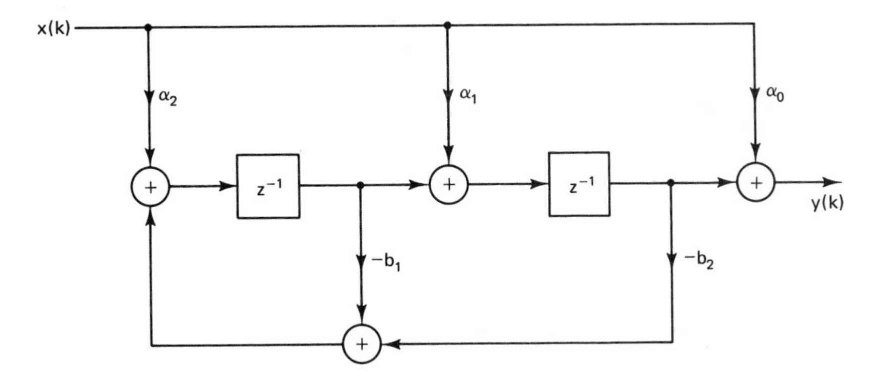

Figure P12-4

12-5. Consider the PID controller of Problem 11-9. Find a direct structure realization.

12-6. Consider the PID controller of Problem 11-10. Find a direct structure realization.

12-7. Can you find a 1X realization for the PID controllers of Problems 12-5 and 12-6?

13

Microcomputer Implementation of Digital Filters

13.1 INTRODUCTION

In Chapter 12 we examined various structures for realizing digital filters. Each structure may be described by a unique set of difference equations. The equations require the operations multiplication, addition, and time delay. In this chapter we explore the implementation of these difference equations by microcomputer. The microcomputers will be programmed in assembly language to calculate the difference equations for the parallel and cascade structures of Chapter 12. First we examine 16-bit microcomputers with on-chip multiplication. Next we explore microcomputers with special characteristics for digital filtering. The purpose of this chapter is to illustrate to the reader the many different microcomputers available.

13.2 THE INTEL 8086

The INTEL 8086, introduced in 1979, is a typical 16-bit microprocessor. We will describe it in some detail, and compare it to some other commercially available machines with similar features. The material in this section comes from Refs. 1 and 2. Now we describe the programming model of the INTEL 8086.

Register Architecture

The Intel 8086 [2] has been designed to provide high performance in applications requiring high-level language usage, extended memory addressing capabilities, 8- or

Manufacture of microcomputer components. (Courtesy of Intel Corporation.)

16-bit accuracy, and high throughput. These features are made possible by the register architecture of the 8086 as shown in Figure 13-1. These registers are divided into four groups, including (1) general register file, (2) pointer and index register file, (3) segment register file, and (4) instruction pointer (IP) and flag register. As shown in Figure 13-1,

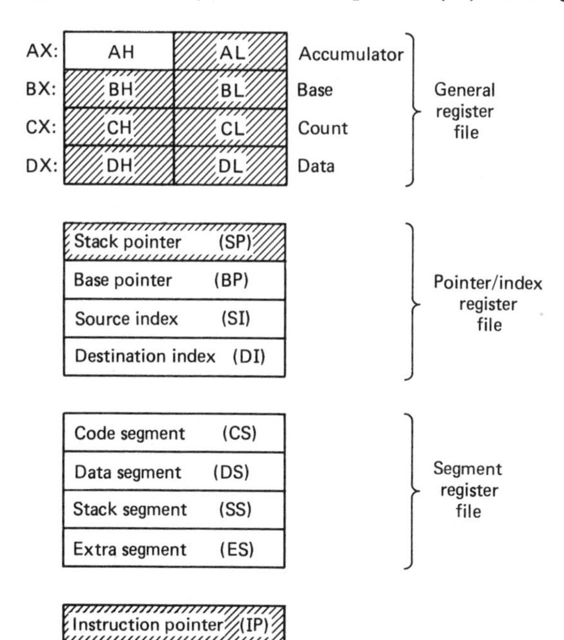

Figure 13-1 8086 register architecture. Shaded area indicates ⟹ 8080 register equivalent.

the basic registers of the Intel 8080 have been retained as a subset of the 8086 registers to facilitate upgrading of present 8080-based systems.

Unlike the 8080, arithmetic and logical operations are supported by all of the general, pointer, and index registers, although typically only the general registers (AX, BX, CX, DX) are used. The accumulator (AX) is the most efficient in most operations and is explicitly implied in a number of operations, such as multiply and divide. Any of the general registers may be addressed as words or bytes for source and destination operands. In addition to arithmetic and logical functions, a number of string operations have been provided which use the general registers as indicated by their mnemonics in Figure 13-1. Thus, for string operations, CX is assumed to be a counter, BX a base address register, and AX and DX are assumed to contain data.

The capabilities of the 8086 are further enhanced by its many modes of memory access. The CPU can directly address 1 megabytes of memory. Each 20-bit address is computed from two components as shown in Figure 13-2. The contents of a segment register is multiplied by 16 and then added to a 16-bit offset to determine the physical address in memory. Thus each of the four segment registers effectively points to one 64 kilobyte block of memory, resulting in separate 64K blocks for code (via CS), data (DS), and stack (SS), with the extra segment (ES) typically used as a second data-segment register. The offset of each address is computed using various combinations of constants and base or index register contents.

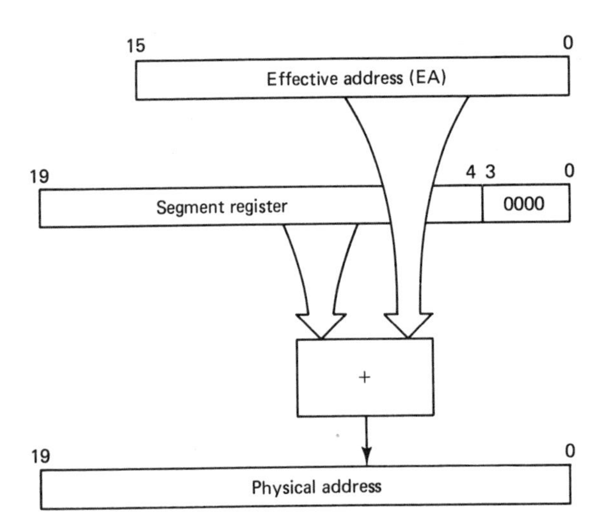

Figure 13-2 Physical address computation in the 8086.

In the case of instruction and stack operations, the segment and offset registers are explicitly implied; that is,

$$\text{Instruction PA} = (\text{CS}) + (\text{IP})$$
$$\text{Stack PA} = (\text{SS}) + (\text{SP})$$

where PA is the 20-bit physical address in memory. Operand addresses can be computed via a number of registers. The various operand addressing modes are summarized below:

1. Register: operand in {AX, BX, CX, DX, BP, SP, SI, DI}
2. Immediate: operand in instruction
3. Direct: PA = (DS) + DISP
4. Indirect:

$$\text{PA} = (\text{DS}) + (\text{BX}) + \text{DISP}$$

$$\text{PA} = (\text{SS}) + (\text{BP}) + \text{DISP}$$

$$\text{PA} = (\text{DS}) + \{(\text{SI}) \text{ or } (\text{DI})\} + \text{DISP}$$

$$\text{PA} = (\text{SS}) + (\text{BX}) + \{(\text{SI}) \text{ or } (\text{DI})\} + \text{DISP}$$

The segment register may be changed via a segment-override prefix. The displacement (DISP) may be 0, 8, or 16 bits.

8086 Instruction Set

The instruction set of the 8086 contains all of the 8080 primitive operations as a subset, with a number of extra instructions and features to improve the capabilities and throughput of the 8086 CPU. Those instructions common to the 8080 will not be discussed here; the reader is referred to Ref. 3. It should be noted, however, that a number of these 8080 operations can be used with the addressing modes described above.

1. IMUL/MUL: Signed and unsigned multiply (byte or word). Operand 1 is in AL or AX, and operand 2 is accessed via the addressing modes. The result is left in AX (byte op's) or DX, AX (word op's).
2. Loop control
   a. LOOP ADDR: The contents of CX are decremented by 1 and control is transferred to ADDR until (CX) = 0.
   b. LOOPZ (LOOPNZ): Similar to loop, but control is only transferred if ZF (zero flag) is set (not set).
3. String operations
   a. MOVS: Block move (byte or work operands) from $\langle(\text{SI})\rangle$ to $\langle(\text{DI})\rangle$, followed by an increment/decrement of both SI and DI.
   b. REP MOVS: Block move with CX decremented after each move and the move repeated until (CX) = 0.
   c. LODS/STOS: String load/store of one byte or work to/from AL to AX with (SI) as the operand address. SI is incremented/decremented after the transfer.
   d. CLD/STD: Set direction flag to zero (auto decrement mode) or one (auto increment mode) for string operations.
4. CBW/CWD: Convert byte (word) to word (double-word) by extending the sign of AL (AX) through AX (DX).
5. XCHG: Exchange contents of registers or register and memory.

The use of these 8086 instructions is expected to simplify greatly the implementation of the iterative calculations required in digital filtering as well as providing a minimum phase lag from filter input to the filter output.

13.3 IMPLEMENTING SECOND-ORDER MODULES

In Chapter 12 we examined six structures for second-order modules. Here we examine their implementation on the INTEL 8086.

Consider Figure 13-3. All programs for second-order modules may be modeled by this flowchart. This flowchart represents the processing required during one time interval ($kT < t < kT + T$). For example, consider the 1D structure of equation (12-7). If we precalculate (during the time interval $kT - T \leq t < kT$)

$$\begin{aligned} T_1 &= -b_1 m(k-1) - b_2 m(k-2) \\ T_2 &= a_1 m(k-1) + a_2 m(k-2) \end{aligned} \tag{13-1}$$

Then (during the time interval $kT \leq t < kT + T$) we may rapidly calculate the output $y(k)$ upon receipt of the input $x(k)$ as follows:

$$\begin{aligned} m(k) &= x(k) + T_1 \\ y(k) &= a_0 m(k) + T_2 \end{aligned} \tag{13-2}$$

This completes the processing of a 1D module so that no postprocessing is required.

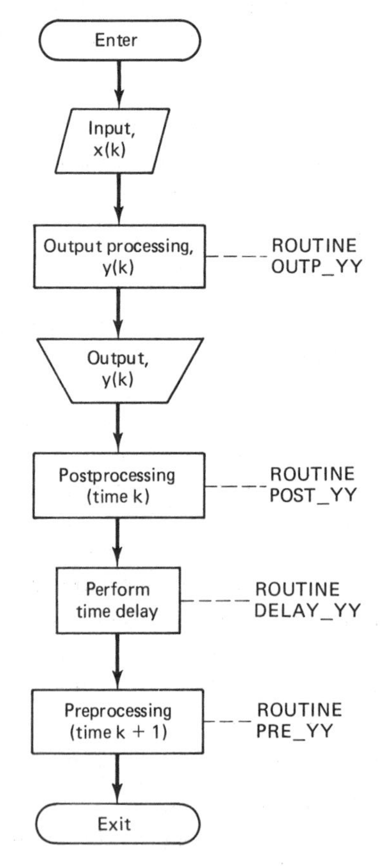

Figure 13-3 General second-order module. YY = 1D, 2D, 3D, 4D, 1X, 2X. (From H. T. Nagle, Jr., and V. P. Nelson, "Digital Filter Implementation on 16-bit Microcomputers," *IEEE MICRO*, Vol. 1, No. 1, Fig. 9, p. 31, Feb 1981, © 1981 IEEE.)

Equations (12-7) through (12-12) are reorganized below into the routines of Figure 13-3:

1D Structure:

$$\begin{aligned} &\text{OUTP} - 1\text{D:} && m(k) = x(k) + T_1 \\ & && y(k) = a_0 m(k) + T_2 \\ &\text{POST} - 1\text{D:} && \text{none} \\ &\text{PRE} - 1\text{D:} && T_1 = -b_1 m(k-1) - b_2 m(k-2) \\ & && T_2 = a_1 m(k-1) + a_2 m(k-2) \end{aligned} \tag{13-3}$$

2D Structure:

$$\begin{aligned} &\text{OUTP} - 2\text{D:} && y(k) = a_0 x(k) + p_1(k-1) \\ &\text{POST} - 2\text{D:} && p_1(k) = a_1 x(k) - b_1 y(k) + p_2(k-1) \\ & && p_2(k) = a_2 x(k) - b_2 y(k) \\ &\text{PRE} - 2\text{D:} && \text{none} \end{aligned} \tag{13-4}$$

3D Structure:

$$\begin{aligned} &\text{OUTP} - 3\text{D:} && y(k) = a_0 x(k) + T_3 \\ &\text{POST} - 3\text{D:} && \text{none} \\ &\text{PRE} - 3\text{D:} && T_3 = a_1 x(k-1) + a_2 x(k-2) \\ & && \qquad - b_1 y(k-1) - b_2 y(k-2) \end{aligned} \tag{13-5}$$

4D Structure:

$$\begin{aligned} &\text{OUTP} - 4\text{D:} && r_0(k) = x(k) + r_1(k-1) \\ & && y(k) = a_0 r_0(k) + q_1(k-1) \\ &\text{POST} - 4\text{D:} && r_1(k) = -b_1 r_0(k) - b_2 r_0(k-1) \\ & && q_1(k) = a_1 r_0(k) + a_2 r_0(k-1) \\ &\text{PRE} - 4\text{D:} && \text{none} \end{aligned} \tag{13-6}$$

1X Structure:

$$\begin{aligned} &\text{OUTP} - 1\text{X:} && y(k) = a_0 x(k) + s_2(k-1) \\ &\text{POST} - 1\text{X:} && s_1(k) = g_1 s_1(k-1) - g_2 s_2(k-1) + g_3 x(k) \\ & && s_2(k) = g_1 s_2(k-1) + g_2 s_1(k-1) + g_4 x(k) \\ &\text{PRE} - 1\text{X:} && \text{none} \end{aligned} \tag{13-7}$$

2X Structure:

$$\begin{aligned} &\text{OUTP} - 2\text{X:} && y(k) = a_0 x(k) + T_4 \\ &\text{POST} - 2\text{X:} && t_1(k) = g_1 t_1(k-1) + g_2 t_2(k-1) \\ & && t_2(k) = x(k) + g_1 t_2(k-1) - g_2 t_1(k-1) \\ &\text{PRE} - 2\text{X:} && T_4 = g_3 t_1(k-1) + g_4 t_2(k-1) \end{aligned} \tag{13-8}$$

The 1X and 2X structures require a pair of complex poles so that

$$D(z) = a_0 + \frac{A}{z + p} + \frac{A^*}{z + p^*} \tag{13-9}$$

and

$$g_1 = -\mathrm{Re}\,[p] \qquad g_3 = 2\,\mathrm{Im}\,[A]$$
$$g_2 = -\mathrm{Im}\,[p] \qquad g_4 = 2\,\mathrm{Re}\,[A]$$

Figures 13-4 to 13-10 demonstrate the INTEL 8086 assembly language programs which implement the six structures outlined above.

```
INPUT:       No parameters passed.
             Returns sample x(k) in AX.

OUTP_YY:     Pass x(k) in AX.
             Put number of cascaded stages in CX.
             Returns output y(k) in AX.

DELAY_YY:    Pass number of modules in CX
             to perform time delay.

PRE_YY:      Pass number of modules in CX
             for preprocessing.

POST_YY:     Pass number of modules in CX
             for postprocessing.

YY = 1D, 2D, 3D, 4D, 1X, or 2X.
```

Figure 13-4 Calling sequences for filter subroutine modules.

```
;
;OUTP_1D:   MO = X + T1  :  Y = AO*MO + T2
;X PASSED IN AX, Y RETURNED IN AX
;LOOP COUNT IN CX
OUTP_1D:   MOV   SI,#0          ;LOOP INDEX
           LEA   DI,MO          ;MEMORY POINTER
OLP_1D:    ADD   AX,T1[SI]      ;MO
           STOW                 ;STORE
           IMUL  AO[SI]         ;AO*MO/4 IN DX
           SAL   DX,2           ;AO*MO
           ADD   DX,T2[SI]      ;Y
           MOV   AX,DX          ;Y IN AX
           ADD   SI,#2          ;MOVE INDEX
           LOOP  OLP_1D         ;LOOP BACK
           RET
;
;
;
;DELAY_1D:   M2 = M1,  M1 = MO FOR TIME-DELAY
;LOOP COUNT IN CX
DELAY_1D:   LEA   DI,T1-2       ;POINT TO M2
            LEA   SI,M2-2       ;POINT TO M1
            STD                 ;SET AUTODECREMENT
            SAL   CX,1          ;DOUBLE LOOP COUNT
            REP
            MOVW                ;BLOCK MOVE
            CLD
            RET
;
;
```

Figure 13-5 1D module subroutines.

```
;
;PRE_1D:     T1 = -B1*M1 - B2*M2
             T2 = A1*M1 + A2*M2
;LOOP COUNT IN CX
PRE_1D:      LEA  SI,A1          ;POINT TO COEFS
             MOV  DI,#0          ;INDEX
PLP_1D:      LODW                ;A1/2
             IMUL M1[DI]         ;M1*A1/4 IN DX
             MOV  BX,DX          ;SAVE
             LODW                ;A2/2
             IMUL M2[DI]         ;M2*A2/4 IN DX
             ADD  BX,DX          ;T2/4
             SAL  BX,2           ;T2
             MOV  T2[DI],BX      ;STORE T2
             LODW                ;B1/2
             IMUL M1[DI]         ;M1*B1/4 IN DX
             MOV  BX,DX          ;SAVE
             LODW                ;B2/2
             IMUL M2[DI]         ;M2*B2/4
             ADD  BX,DX          ;-T1/4
             SAL  BX,2           ;-T1
             NEG  BX             ;T1
             MOV  T1[DI],BX      ;STORE T1
             ADD  DI,#2          ;MOVE INDEX
             LOOP PLP_1D         ;LOOP BACK
             RET
;
;
;
;POST_1D:    NOT USED IN A 1D MODULE
POST_1D:     RET
;
;
;
;1D DATA STORAGE FOR N STAGES
A0:          DW  A10, A20, ..., AN0        ;A0 COEFS
A1:          DW  A11, A12, A13, A14        ;STAGE 1 A1, A2, B1, B2
             DW  A21, A22, A23, A24        ;STAGE 2
             .
             .
             DW  AN1, AN2, AN3, AN4        ;STAGE N
M0:          DW  NDUP(0)                   ;M(k)
M1:          DW  NDUP(0)                   ;M(k-1)
M2:          DW  NDUP(0)                   ;M(k-2)
T1:          DW  NDUP(0)                   ;TEMP STORAGE
T2:          DW  NDUP(0)                   ;TEMP STORAGE
```

Figure 13-5 cont.

```
;
;
;OUTP_2D:   Y = A0*X + P1
;X PASSED IN AX, Y RETURNED IN AX
;LOOP COUNT IN CX
OUTP_2D:    MOV   SI,#0        ;INDEX
            LEA   DI,X         ;POINT TO X
OLP_2D:     STOW               ;SAVE X
            IMUL  A0[SI]       ;X*A0/4 IN DX
            SAL   DX,2         ;X*A0
            ADD   DX,P1[SI]    ;Y
            MOV   AX,DX        ;RETURN IN AX
            ADD   SI,#2        ;INDEX
            LOOP  OLP_2D       ;USE COUNT IN CX
            RET
;
;
```

Figure 13-6 2D module subroutines.

```
;
;PRE_2D AND DELAY_2D ARE NOT NEEDED
PRE_2D:    RET
DELAY_2D:  RET
;
;
;
;POST_2D:      P1 = A1*X - B1*Y + P2
               P2 = A2*X - B2*Y
;LOOP COUNT IN CX
POST_2D:       LEA    SI, A1        ;COEF POINTER
               LEA    BX, X         ;POINT TO INPUTS
               MOV    DI, #0        ; INDEX
POLP_2D:       LODW                 ;A1/2
               IMUL   [BX][DI]      ;X*A1/4 IN DX
               PUSH   DX            ;SAVE
               LODW                 ;B1/2
               IMUL   2[BX][DI]     ;B1*Y/4
               POP    AX
               SUB    AX, DX        ;(A1*X-B1*Y)/4
               SAL    AX, 2         ;A1*X-B1*Y
               ADD    AX, P2[DI]    ;COMPUTE P1
               MOV    P1[DI], AX    ;STORE
               LODW                 ;A2/2
               IMUL   [DX][DI]      ;X*A2/4 IN DX
               PUSH   DX
               LODW                 ;B2/2
               IMUL   2[DX][DI]     ;Y*B2/4
               POP    AX
               SUB    AX, DX        ;P2/2=(X*A2-Y*B2)/4
               SAL    AX, 2         ;P2
               MOV    P2[DI]        ;STORE P2
               ADD    DI, #2
               LOOP   POLP_2D
               RET
;
;
;


;2D CONSTANT STORAGE FOR N STAGES
A0:            DW     A10, . . . , AN0          ;A0 FOR N STAGES
A1:            DW     A11, B11, A12, B12        ;STAGE 1 COEFS
               DW     A21, B21, A22, B22        ;STAGE 2
               .
               .
               DW     AN1, BN1, AN2, BN2        ;STAGE N
;2D TEMPORARY STORAGE FOR N STAGES
X:             DW     (N+1)DUP(0)               ;INPUTS/OUTPUTS
P1:            DW     NDUP(0)
P2:            DW     NDUP(0)
```

Figure 13-6 cont.

```
;
;OUTP_3D:      Y = A0*X + T3
;PASS X IN AX, RETURN Y IN AX
;LOOP COUNT IN CX
OUTP_3D:       LEA    DI, X1        ;POINT TO X(k)
               MOV    SI, #0        ; INDEX
OLP_3D:        STOW                 ;SAVE X, Y
               IMUL   A0[SI]        ;A0*X/4 IN DX
               SAL    DX, 2         ;A0*X
               ADD    DX, T3[SI]    ;COMPUTE Y
```

Figure 13-7 3D module subroutines.

```
            MOV     AX,DX           ;RETURN Y IN AX
            ADD     SI,#2           ;INDEX
            LOOP    OLP_3D
            STOW                    ;SAVE LAST Y
            RET
;
;
;
;DELAY_3D:   X(k) TO X(k-1) OR X1 TO X2, Y1 TO Y2
;LOOP COUNT IN CX
DELAY_3D:   LEA     SI,X1           ;X(k)
            LEA     DI,X2           ;X(k-1)
            INC     CX              ;MOVE X AND Y VALUES
            REP
            MOVW                    ;BLOCK MOVE
            RET
;
;
;
;POST_3D:    NOT NEEDED
POST_3D:    RET
;
;
;
;PRE_3D:   T3 = A1*X1 + A2*X2 - B1*Y1 - B2*Y2
;LOOP COUNT IN CX
PRE_3D:     LEA     SI,A1           ;COEF POINTER
            MOV     DI,#0           ;INDEX
PRLP_3D     LODW                    ;A1/2
            IMUL    X1[DI]          ;X1*A1/4 IN DX
            MOV     BX,DX           ;PARTIAL SUM IN BX
            LODW                    ;A2/2
            IMUL    X2[DI]          ;X2*A2/4 IN DX
            ADD     BX,DX           ;PARTIAL SUM
            LODW                    ;B1/2
            IMUL    X1+2[DI]        ;Y1*B1/4 IN DX
            SUB     BX,DX           ;TOTAL
            LODW                    ;B2/2
            IMUL    X2+2[DI]        ;Y2*B2/4
            SUB     BX,DX           ;T3/4
            SAL     BX,2            ;T3
            MOV     T3[DI],BX       ;STORE
            ADD     DI,#2           ;INDEX
            LOOP    PRLP_3D
            RET
;
;
;
;3D CONSTANT STORAGE FOR N STAGES
A0:         DW      A10, ..., AN0           ;A0 COEFS FOR N STAGES
A1:         DW      A11, A12, B11, B12      ;COEFS FOR STAGE 1
            DW      A21, A22, B21, B22      ;STAGE 2
            DW      :               :         :
            DW      AN1, AN2, BN1, BN2      ;STAGE N
;
;3D STORAGE FOR N STAGES
X1:         DW      (N+1)DUP(0)             ;x(k), y(k)
X2:         DW      (N+1)DUP(0)             ;x(k-1), y(k-1)
T3:         DW      NDUP(0)
```

Figure 13-7 cont.

```
;
;OUTP_4D:   RO = X + R1; Y = AO*RO + Q1
;PASS X IN AX, RETURN Y IN AX
;LOOP COUNT IN CX
OUTP_4D:    LEA     DI,RO         ;POINT TO RO
            MOV     SI,#0         ;INDEX
OLP_4D:     ADD     AX,R1[SI]     ;RO
            STOW                  ;STORE
            IMUL    AO[SI]        ;RO*AO/4 IN DX
            SAL     DX,2          ;RO*AO
            ADD     DX,Q1[SI]     ;Y
            MOV     AX,DX         ;RETURN IN AX
            ADD     SI,#2         ;INDEX
            LOOP    OLP_4D
;           RET
;
;
PRE_4D:   RET ; NOT NECESSARY
;
;
;
;DELAY_4D:   DELAY RO(k) TO RO(k-1) OR RO TO R1
;LOOP COUNT IN CX
DELAY_4D:   LEA     SI,RO         ;R(k)
            LEA     DI,RO1        ;R(k-1)
            REP
            MOVW                  ;BLOCK MOVE
            RET
;
;POST_4D:   R1 = -B1*RO - B2*RO1
;           Q1 = A1*RO + A2*RO1
;LOOP COUNT IN CX
POST_4D:    LEA     SI,B1         ;COEF POINTER
            MOV     DI,#0         ;INDEX
POLP_4D:    LODW                  ;B1/2
            IMUL    RO[DI]        ;RO*B1/4 IN DX
            MOV     BX,DX
            LODW                  ;B2/2
            IMUL    RO1[DI]       ;RO1*B2/4
            ADD     BX,DX         ;-R1/4
            SAL     BX,2          ;-R1
            NEG     BX            ;R1
            MOV     R1[DI],BX     ;STORE R1
            LODW                  ;A1/2
            IMUL    RO[DI]        ;RO*A1/4
            MOV     BX,DX
            LODW                  ;A2/2
            IMUL    RO1[DI]       ;RO1*A2/4
            ADD     BX,DX         ;Q1/4
            SAL     BX,2          ;Q1
            MOV     Q1[DI],BX     ;STORE Q1
            ADD     DI,#2         ;INDEX
            LOOP    POLP_4D
            RET
;
;4D CONSTANT STORAGE FOR N STAGES
AO:         DW     A10, . . . , ANO             ;
B1:         DW     B11, B12, A11, A12           ;STAGE 1 COEFS
            DW     B21, B22, A21, A22           ;STAGE 2
            DW      :                           :
            DW     BN1, BN2, AN1, AN2           ;STAGE N
;4D TEMPORARY STORAGE FOR N STAGES
RO:         DW    NDUP(0)                       ;RO(k)
RO1:        DW    NDUP(0)                       ;RO(k-1)
R1:         DW    NDUP(0)                       ;R1(k)
Q1:         DW    NDUP(0)                       ;Q1(k)
```

Figure 13-8 4D module subroutines.

```
;
;OUTP_1X:   Y = A0*X + S2
;PASS X IN AX, RETURN Y IN AX
;LOOP COUNI IN CX
OUTP_1X:    LEA     DI,X         ;POINT TO X
            MOV     SI,#0        ;INDEX
OLP_1X:     STOW                 ;SAVE X
            IMUL    A0[SI]       ;X*A0/4
            SAL     DX,2         ;X*A0
            ADD     DX,S2[SI]    ;Y
            MOV     AX,DX        ;RET IN AX
            ADD     SI,#2        ;INDEX
            LOOP    OLP_1X
            RET
;
;
;
PRE_1X:   REI  : NOT NEEDED FOR 1X
;
;
;
;DELAY_1X:  DELAY S1(k) TO S1(k-1), S2(k) TO S2(k-1)
;LOOP COUNT IN CX
DELAY_1X:   LEA     SI,S1        ;SOURCE
            LEA     DI,S11       ;DESTINATION
            ADD     CX,CX        ;DOUBLE COUNT FOR
                                 ;S1 AND S2
            REP
            MOVW                 ;BLOCK MOVE
            RET
;
;
;
;POST_1X:   S1 = G1*S11 - G2*S21 + G3*X
;           S2 = G1*S21 + G2*S11 + G4*X
POST_1X:    LEA     SI,G1        ;COEF POINTER
            MOV     DI,#0        ;INDEX
POLP_1X:    LODW                 ;G1/2
            IMUL    S11[DI]      ;S11*G1/4
            MOV     BX,DX
            LODW                 ;G2/2
            IMUL    S21[DI]      ;S21*G2/4
            SUB     BX,DX
            LODW                 ;G3/2
            IMUL    X[DI]        ;X*G3/4
            ADD     BX,DX        ;S1/4
            SAL     BX,2         ;S1
            MOV     S1[DI],BX    ;STORE S1
            LODW                 ;G1/2
            IMUL    S21[DI]      ;S21*G1/4
            MOV     BX,DX
            LODW                 ;G2/2
            IMUL    S11[DI]      ;S11*G2/4
            ADD     BX,DX


            LODW                 ;G4/2
            IMUL    X[D1]        ;X*G4/4
            ADD     BX,DX        ;S2/4
            SAL     BX,2         ;S2
            MOV     S2[DI],BX    ;STORE S2
            ADD     DI,#2        ;INDEX
            LOOP    POLP_1X
            RET
```

Figure 13-9 1X module subroutines.

```
;
;
;
;1X CONSTANT STORAGE FOR N STAGES
A0:         DW    A10, . . . , AN0
G1:         DW    G11, G12, G13, G11, G12, G14    ;STAGE 1 COEFS
            DW    G21, G22, G23, G21, G22, G24    ;STAGE 2
            DW     .                                  .
            DW     .                                  .
            DW     .                                  .
            DW    GN1, GN2, GN3, GN1, GN2, GN4    ;STAGE N
;
;1X DATA STORAGE FOR N STAGES
X:          DW    NDUP(0)            ;INPUTS
S1:         DW    NDUP(0)            ;S1(k)
S2:         DW    NDUP(0)            ;S2(k)
S11:        DW    NDUP(0)            ;S1(k-1)
S21:        DW    NDUP(0)            ;S2(k-1)
```

Figure 13-9 cont.

```
;
;OUTP_2X:  Y = A0*X + T4
;PASS X IN AX, RETURN Y IN AX
;LOOP COUNT IN CX
OUTP_2X:  LEA     DI,X          ;POINT TO X
          MOV     SI,#0         ;INDEX
OLP_2X:   STOW                  ;SAVE INPUTS TO STAGES
          IMUL    A0[SI]        ;X*A0/4
          SAL     DX,2          ;X*A0
          ADD     DX,T4[SI]     ;COMPUTE Y
          MOV     AX,DX         ;RETURN IN AX
          ADD     SI,#2         ;INDEX
          LOOP    OLP_2X
          RET
;
;
;
;DELAY_2X:  T1(k) TO T1(k-1), T2(k) TO T2(k-1)
;LOOP COUNT IN     CX
DELAY_2X: LEA     SI, T1
          LEA     DI, T11
          ADD     CX, CX        ;DOULBE COUNT
          REP
          MOVW                  ;BLOCK MOVE
          RET
;
;
;
;PRE_2X:   T4 = G3*T1(k-1) + G4*T2(k-1)
;LOOP COUNT IN CX
PRE_2X:   LEA      SI,G3         ;COEF POINTER
          MOV      DI,#0         ;INDEX
PRLP_2X:  LODW                   ;G3/2
          IMUL     T11[DI]       ;G3*T11/4
          MOV      BX,DX
          LODW                   ;G4/2
          IMUL     T21[DI]       ;G4*T21/4
          ADD      BX,DX         ;T4/4
          SAL      BX,2          ;T4
          MOV      T4[DI],BX     ;STORE T3(k)
          ADD      DI,#2         ;INDEX
          LOOP     PRLP_2X
          RET
```

Figure 13-10 2X module subroutines.

```
;
;
;
;POST_2X:   T1  = G1*T1(k-1) + G2*T2(k-1)
;           T2  = X + G1*T2(k-1) - G2*T1(k-1)
;LOOP COUNT IN CX
POST_2X:   LEA      SI,G1
           MOV      DI,#0        ;INDEX
POLP_2X:   LODW                  ;G1/2
           IMUL     T11[DI]      ;T11*G1/4
           MOV      BX,DX

           LODW                  ;G2/2
           IMUL     T21[DI]      ;T21*G2/4
           ADD      BX,DX        ;T1/4
           SAL      BX,2         ;T1
           MOV      T1[DI],BX   ;STORE T1(k)
           SUB      SI,#4        ;BACK POINTER UP TO G1/2
           LODW                  ;G1/2
           IMUL     T21[DI]      ;G1*T21/4
           MOV      BX,DX
           LODW                  ;G2/2
           IMUL     T11[DI]      ;G2*T11/4
           SUB      BX,DX        ;PARTIAL SUM
           SAL      BX,2         ;T2-X
           ADD      BX,X[DI]     ;T2
           MOV      T2[DI],BX    ;STORE T2(k)
           ADD      DI,#2        ;INDEX
           LOOP     OLP_2X
           RET
;
;
;
;2X CONSTANT STORAGE FOR N STAGES
G1:        DW    G11, G12       ;STAGE 1
           DW    G21, G22       ;STAGE 2
           .
           .
           DW    GN1, GN2       ;STAGE N
G3:        DW    G13, G14       ;STAGE 1 COEFS
           DW    G23, G24       ;STAGE 2
           .
           .
           DW    GN3, GN4       ;STAGE N
A0:        DW    A01, ...., AON
;
;2X TEMPORARY STORAGE FOR N STAGES
X:         DW    NDUP(0)        ;INPUTS TO STAGES
T1:        DW    NDUP(0)        ;T1(k)
T2:        DW    NDUP(0)        ;T2(k)
T11:       DW    NDUP(0)        ;T1(k-1)
T21:       DW    NDUP(0)        ;T2(k-1)
T4:        DW    NDUP(0)        ;T4(k)
```

Figure 13-10 cont.

Note that the coefficient values are bounded by

$$0 \leq |a_0, a_2, b_2, g_1, g_2| \leq 1$$
$$0 \leq |a_1, b_1, g_3, g_4| \quad \leq 2 \qquad (13\text{-}10)$$

Since the two's-complement number system is used in the INTEL 8086, all numbers in the machine must be represented by

$$N = (SM_{14} \quad M_{13} \quad \cdots \quad M_1 \quad M_0 \cdot)_{2cns} \qquad (13\text{-}11)$$

Hence

$$-2^{15} \leq N \leq 2^{15} - 1 \tag{13-12}$$

If we consider all numbers to be scaled such that

$$N = (S \cdot M_{14} \quad M_{13} \quad \cdots \quad M_1 \quad M_0)_{2cns} \tag{13-13}$$

Thus

$$-1 \leq N \leq 1 - 2^{-15} \tag{13-14}$$

Consequently, coefficients in the range

$$1 \leq |N| < 2 \tag{13-15}$$

cannot be represented. Therefore, all coefficients will be stored as *half* their actual value, VALUE_STORED $= \lfloor \text{Value} * 2^{14} + .5 \rfloor$ and a left shift (multiply by 2) operation will be performed in each routine to compensate for this change. The symbol $\lfloor x \rfloor$ means the largest integer less than x.

Example 13.1

Suppose that coefficient SØ = 0.4383164 is to be stored in the INTEL 8086.

$$\begin{aligned} \text{VALUE\_STORED} &= \lfloor S\emptyset * 2^{14} + .5 \rfloor \\ &= \lfloor .4383164 * 16384 + .5 \rfloor \\ &= \lfloor 7181.8759 \rfloor = 7181 \end{aligned} \tag{13-16}$$

Next let us consider multiplication in the two's-complement number system. Figure 13-11 a illustrates the problem. An n-bit multiplicand (perhaps a signal variable)

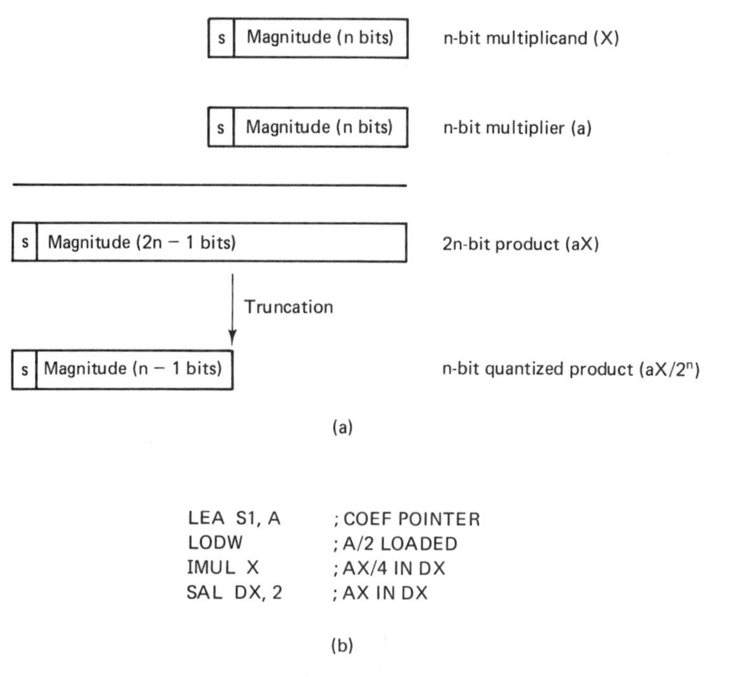

Figure 13-11 Two's-complement multiplication: (a) wordlength variation; (b) code sequence.

is multiplied by an n-bit multiplier (perhaps a filter coefficient) and the product has $2n$ bits. This product may be used as as another multiplicand in a later multiplication, so it is quantized (trucated here) back to n bits. The effect of this quantization is examined in great detail in Chapter 14.

Suppose that the multiplicand is X and the coefficient is a; then

$$\begin{array}{r} \mathrm{X} \\ \times\ \mathrm{a} \\ \hline \mathrm{aX} \end{array}$$

and the product is quantized Q[·]:

$$Q[aX] = \lfloor aX/2^n \rfloor \tag{13-17}$$

But from (13-11) and (13-13) we may see that the computer hardware actually handles the integers of (13-11) so that in hardware

$$\begin{array}{c} X * 2^{n-1} \\ a * 2^{n-1} \\ \hline aX * 2^{2n-2} \end{array}$$

is quantized

$$Q[aX] = \lfloor aX * 2^{2n-2}/2^n \rfloor = \lfloor aX * 2^{n-2} \rfloor \tag{13-18}$$

Consequently, the product must be multiplied by 2 (shifted one place left) so that the final truncated term is

$$Q[aX] = \lfloor aX * 2^{n-1} \rfloor \tag{13-19}$$

Earlier we explained that coefficients are actually stored as half-values; hence the computer actually performs the following operations:

1. Load the coefficient a into AX:

$$AX = a * 2^{n-2}$$

2. Multiply by the variable X:

$$\begin{aligned} DX, AX &= (a * 2^{n-2}) * (X * 2^{n-1}) \\ &= aX * 2^{2n-3} \\ &= (aX/4) * 2^{2n-1} \end{aligned}$$

   The product is now in the DX, AX register.

3. Shift DX left two places (quantize to 16 bits and multiply by 4):

$$\begin{aligned} DX &= \lfloor aX/4 * 2^{2n-1}/2^n \rfloor * 4 \\ &= \lfloor (aX/4) * 2^{n-1} \rfloor * 4 \\ &\doteq aX * 2^{n-1} \end{aligned}$$

   This operation left justifies the register DX and fills in two zeros in the least significant bits. The DX register now contains the truncated, properly scaled result. The instruction sequence is listed in Figure 13-11b. This sequence appears frequently in the computer programs of this chapter.

4. On computers with double register shifting, one would perform the double left shift first:

$$\text{double register} = (aX/4 * 2^{2n-1}) * 4 = aX * 2^{2n-1}$$

   and then truncate to the n most significant bits.

$$\text{Single register} = \lfloor (aX * 2^{2n-1})/2^n \rfloor$$
$$= \lfloor aX * 2^{n-1} \rfloor$$

which is more accurate than above.

Example 13.2

Consider multiplying

$$X = .7562867$$

and

$$a = .4383164$$

From (13-16),

$$\lfloor a * 2^{14} \rfloor = 7181$$

and from (13-13),

$$\lfloor X * 2^{15} \rfloor = 24782$$

After multiplication

$$aX * 2^{29} = 177959542$$

On the INTEL 8086, truncation must be accomplished before shifting left; hence

$$(aX * 2^{29})/2^{16} = 177959542/2^{16}$$
$$\lfloor aX * 2^{13} \rfloor = \lfloor 2715.4471 \rfloor = 2715$$

Finally, the shift left operation multiplies by 4:

$$\lfloor aX * 2^{13} \rfloor * 2^2 = 2715 * 4 \qquad (13\text{-}20)$$
$$= 10860$$

Hence

$$aX \doteq 10860/2^{15}$$
$$\doteq .3314209$$

The actual answer is .3314929.

On a computer with double register shifting, first we multiply by 4:

$$(aX * 2^{29}) * 4 = 177959542 * 4$$
$$aX * 2^{31} = 711838168$$

and then divide by 2^{16} and truncate:

$$(aX * 2^{31})/2^{16} = 711838168/2^{16}$$
$$\lfloor aX * 2^{15} \rfloor = \lfloor 10861.788 \rfloor \qquad (13\text{-}21)$$
$$= 10861$$

and

$$aX = 10861/2^{15}$$
$$= .3314514$$

which is more accurate. The INTEL 8086 can perform this more accurate computation by replacing the SAL DX, 2 instruction with the instruction sequence:

```
RCL   AX, 1
RCL   DX, 1
RCL   AX, 1
RCL   DX, 1
```

This replacement will slow down the filter's operating speed.

13.4 PARALLEL IMPLEMENTATION OF HIGHER-ORDER FILTERS

For implementing higher-order filters, we express $D(z)$ in the form of (12-16):

$$D(z) = \beta_0 + \sum_{i=1}^{m} B_i(z) \tag{13-22}$$

where

$$B_i(z) = \frac{\beta_{i1}z^{-1} + \beta_{i2}z^{-2}}{1 + \beta_{i3}z^{-1} + \beta_{i4}z^{-2}}$$

Figure 13-12 depicts the storage allocation for all modules. Coefficients and variable storage are grouped for more efficient addressing.

```
COEF:    DW                ;FOR ALL MODULES, HALF VALUES
                           ;ROM OR RAM
;
VAR:     DW       0        ;FOR ALL MODULES
                           ;RAM
;
TEMP:    DW       0        ;FOR ALL MODULES
                           ;RAM
```

Figure 13-12 Higher-order storage allocation.

Example 13.3

Consider the implementation of a fourth-order digital filter. This is a rate filter used in the vehicle control portion of the space shuttle [4]:

$$D(z) = \frac{1 + 0.390244z^{-1} - 1.24247z^{-2} + 0.344333z^{-3} + 1,977,044z^{-4}}{1 - 3.02828z^{-1} + 3.53682z^{-2} - 1.88867z^{-3} + 0.397506z^{-4}} \tag{13-23}$$

In the parallel form,

$$D(z) = \beta_0 + \beta_1\frac{\beta_{11}z^{-1} + \beta_{12}z^{-2}}{1 + \beta_{13}z^{-1} + \beta_{14}z^{-1}} + \beta_2\frac{\beta_{21}z^{-1} + \beta_{22}z^{-2}}{1 + \beta_{23}z^{-2} + \beta_{24}z^{-2}} \tag{13-24}$$

where

$$\begin{aligned} \beta_0 &= 1.0, & \beta_1 &= -8.0, & \beta_2 &= -4.0 \\ \beta_{11} &= 1.263998 & & & \beta_{21} &= 1.673365 \\ \beta_{12} &= -1.747506 & & & \beta_{22} &= -1.569220 \\ \beta_{13} &= -1.823002 & & & \beta_{23} &= -1.205277 \\ \beta_{14} &= 0.895895 & & & \beta_{24} &= 0.443701 \end{aligned}$$

Using the 1D modules, the structure of Figure 13-13 is obtained. Note that β_1 and β_2 represent shifting operations in the computer. Using the 1D routines of Figure 13-5, the parallel implementation of Figure 13-14 is obtained.

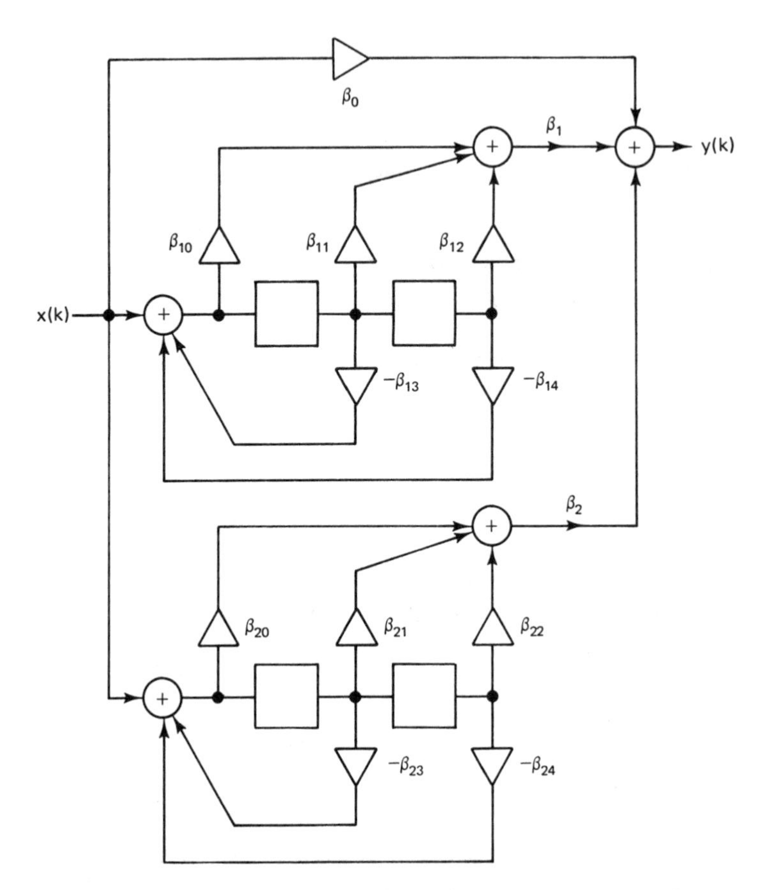

Figure 13-13 Fourth-order example. $\beta_{10} = \beta_{20} = 0$. (From H. T. Nagle, Jr., and V. P. Nelson, "Digital Filter Implementation on 16-bit Microcomputers," *IEEE MICRO*, Vol. 1, No. 1, Fig. 10, p. 32, Feb. 1981, ©1981 IEEE.)

```
;
;PARALLEL IMPLEMENTATION OF A FOURTH-ORDER FILTER
;
;
;
;MAIN PROGRAM-CALLS INTEL 8086 SUBROUTINES
FILTER 1:       CALL    INIT            ;INITIALIZE A/D, D/A
                                        ;AND VARIABLES
F1_LOOP:        CALL    INPUT           ;GET SAMPLE FROM A/D
                MOV     BX,AX           ;SAVE X
                IMUL    BO              ;BETAO*X/4
                SAL     DX,2            ;BETAO*X
                MOV     TEMP,DX         ;SAVE
                MOV     AX,BX           ;RESTORE X(k)
                MOV     CX,#1           ;CALCULATE 1 STAGE
                CALL    OUTP_1D         ;
                SAL     AX,3            ;BETA1 = -8
                NEG     AX
                ADD     AX,TEMP         ;INCLUDE IN OUTPUT
                MOV     TEMP,AX         ;SAVE
                MOV     AX,BX           ;RESTORE X(K)
```

Figure 13-14 Parallel implementation of a fourth-order filter.

```
              MOV       CX,#1
              CALL      OLP_1D          ;CALCULATE STAGE 2 WITH
                                        ;SI = 2 FROM 1st STAGE
              SAL       AX,2            ;BETA2 = -4
              NEG       AX
              ADD       AX,TEMP         ;CALCULATE Y
              CALL      OUTPUT          ;SEND Y TO D/A
              MOV       CX,#2
              CALL      DELAY_1D        ;TIME-DELAY
              MOV       CX,#2
              CALL      PRE_1D          ;PRE-PROCESS
              JMP       F1_LOOP         ;CONTINUE
;
;
;
;INIT:        CLEAR M1, M2, T1, T2
;             SET UP A/D, D/A DEVICES
INIT:         MOV       AX,#0
              MOV       CX,#11
              LEA       DI,M0
              REP
              STOW                      ;PUT 0 INTO ALL RAM
              RET
;
;
;
;INPUT-ASSUME MEMORY-MAPPED I/O
;A/D AT 60H, STATUS AT E0H
INPUT:        MOV       AX, #0000H      ;START-CONVERSION
              OUT       60H,AL
IN_LP:        IN        AL,0E0H         ;CHECK END-CONVERSION
              AND       AL,#04H
              JZ        IN_LP           ;WAIT UNTIL READY
              IN        AL,60H          ;GET SAMPLE FROM A/D
              RET
;
;
;
;OUTPUT TO D/A AT 80H
OUTPUT:       OUT       80H,AL          ;y(k) TO D/A
              RET
;
;
;
;COEFFICIENT STORAGE
B0:           DW        1                                    ;BETA_0
A0:           DW        0,0                                  ;BETA 10 = BETA 20 = 0
                                                             ;USED BY OUTP_1D
;HALF VALUES OF A1, A2, B1, B2 USED IN PRE_1D
A1:           DW        20709, -28631, -29868, 14678         ;A1, A2, B1, B2 FOR
                                                             ;PRE_1D-STAGE 1
              DW        27416, -25710, -19747, 7270          ;A1, A2, B1, B2 FOR
                                                             ;PRE_1D-STAGE 2
```

Figure 13-14 cont.

```
; VARIABLES
MO:             DW      2DUP(0)                 ;M(K)-BOTH STAGES
M1:             DW      2DUP(0)                 ;M(K-1)-BOTH STAGES
M2:             DW      2DUP(0)                 ;M(K-2)-BOTH STAGES
;
; TEMPORARY STORAGE
T1:             DW      2DUP(0)                 ;T1-BOTH STAGES
T2:             DW      2DUP(0)                 ;T2-BOTH STAGES
TEMP:           DW      0                       ;OUTPUT TEMP STORAGE
```

Figure 13-14 cont.

13.5 CASCADE IMPLEMENTATION OF HIGHER-ORDER FILTERS

For implementing filters as a cascade of second-order modules, we may write $D(z)$ in the form of (12-15):

$$D(z) = \prod_{i=1}^{m} A_i(z) \tag{13-25}$$

where

$$A_i(z) = \frac{\alpha_{i0} + \alpha_{i1}z^{-1} + \alpha_{i2}z^{-2}}{1 + \alpha_{i3}z^{-1} + {}_{i4}z^{-2}}$$

The storage allocation scheme of Figure 13-12 may also be used in this case.

Example 13.4

Consider the eighth-order digital filter designed in Ref. 5 and examined in Ref. 6:

$$D(z) = S_0 \prod_{i=1}^{4} A_i(z) \tag{13-26}$$

where

$$S_0 = 0.4383164$$

$$\alpha_{10} = \alpha_{12} = 0.7619852 \qquad \alpha_{13} = -0.90191$$

$$\alpha_{11} = -0.2727907 \qquad \alpha_{14} = 0.66204$$

$$\alpha_{20} = \alpha_{22} = 0.1963670 \qquad \alpha_{23} = -0.59971$$

$$\alpha_{21} = -0.0372331 \qquad \alpha_{24} = 0.96329$$

$$\alpha_{30} = \alpha_{32} = 0.2791792 \qquad \alpha_{33} = -1.17844$$

$$\alpha_{31} = 0.4518543 \qquad \alpha_{34} = 0.42357$$

$$\alpha_{40} = \alpha_{42} = 0.4660647 \qquad \alpha_{43} = -0.68404$$

$$\alpha_{41} = 0.1668512 \qquad \alpha_{44} = 0.85862$$

These coefficients were first designed in Ref. 5. Then Ref. 6 optimized the ordering and pairing to minimize the output round-off noise. Here we have changed the scaling constants (reflected in the numerator terms) to impose a signal limit of 90% of full scale at every internally constrained point in the filter for a full-scale step input at $x(k)$. The block diagram of the filter using 1D modules is shown in Figure 13-15.

The INTEL 8086 routines to implement Figure 13-15 are shown in Figure 13-16.

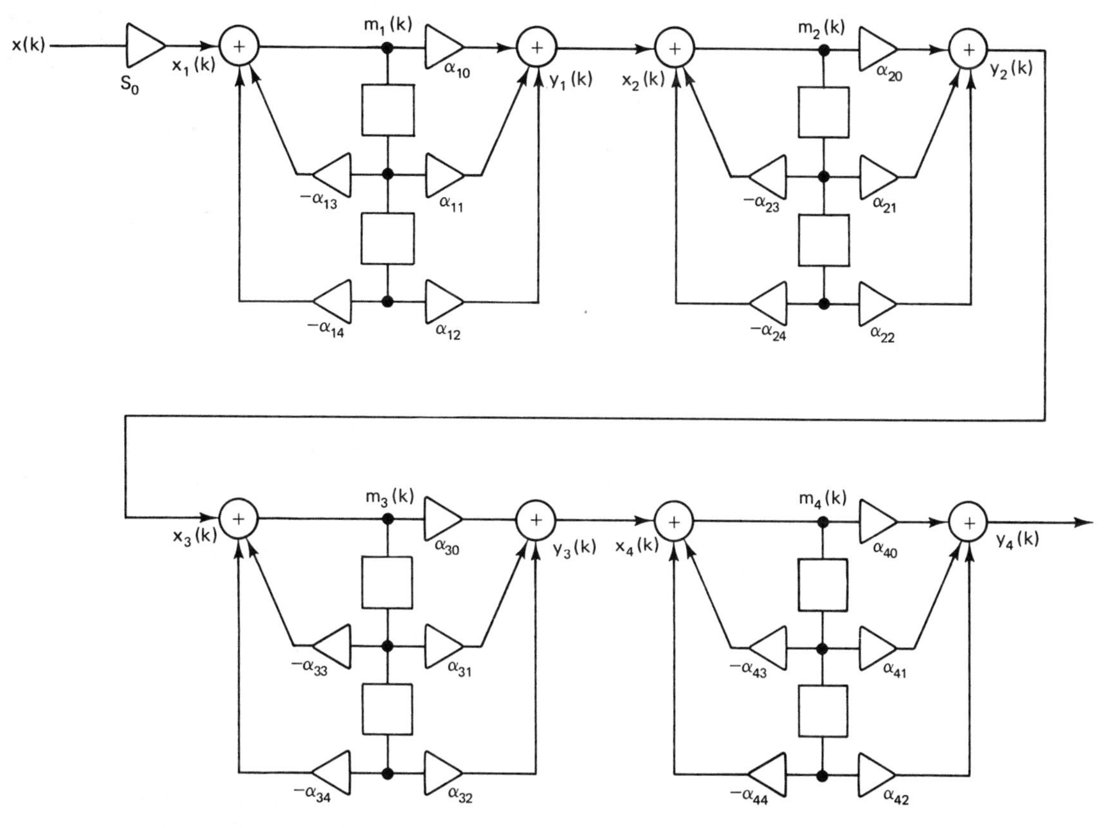

Figure 13-15 Eighth-order example. (From H. T. Nagle, Jr., and V. P. Nelson, "Digital Filter Implementation on 16-bit Microcomputers," *IEEE MICRO*, Vol. 1, No. 1, Fig. 1, p. 33, Feb. 1981, ©1981 IEEE.)

```
;
; INTEL 8086 MAIN
FILTER:         CALL    INIT            ; SEE PARALLEL EXAMPLE
FLOOP:          CALL    INPUT           ; GET X(K):  SEE PARALLEL EXAMPLE
                IMUL    SO              ; SO*X(K)/4
                SAL     DX, 2           ; SO*X(K)
                MOV     AX, DX
                MOV     CX, #4          ; DO 4 STAGES
                CALL    OUTP_1D         ; COMPUTE Y(K)
                CALL    OUTPUT          ; OUTPUT Y(K):  SEE PARALLEL EXAMPLE
                MOV     CX, #4
                CALL    DELAY_1D
                MOV     CX, #4          ; DO 4 STAGES
                CALL    PRE_1D          ; PREPROCESS FOR NEXT SAMPLE
                JMP     FLOOP           ; NEXT SAMPLE
;
;
;
; 1D FILTER COEFFICIENT STORAGE FOR 4 STAGES
SO:             DW      7181                            ; SO/2
AO:             DW      12484, 3217, 4574, 7636         ; ALPHA 10, 20, 30, 40
A1:             DW      -4469, 12484, -14777, 10847     ; ALPHA 11, 12, 13, 14
```

Figure 13-16 Cascade implementation of an eighth-order filter.

```
                DW      -610,3217,-9826,15783           ;ALPHA 21,22,23,24
                DW      7403,4574,-19308,6940           ;ALPHA 31,32,33,34
                DW      2734,7636,-11207,14068          ;ALPHA 41,42,43,44
;VARIABLE STORAGE
M0:             DW      4DUP(0)                         ;M(k)
M1:             DW      4DUP(0)                         ;M(k-1)
M2:             DW      4DUP(0)                         ;M(k-2)
;TEMPORARY STORAGE
T1:             DW      4DUP(0)                         ;TEMP STORAGE
T2:             DW      4DUP(0)                         ;TEMP STORAGE
```

Figure 13-16 cont.

13.6 COMPARISON OF STRUCTURES

Here we use the example of Section 13.5 to compare structures. Suppose that the cascade of four second-order modules were composed of 2D, 3D, 4D, 1X, or 2X structures. How does their sampling rates compare? Table 13-1 shows that for an

TABLE 13-1 COMPARISON OF EIGHTH-ORDER CASCADED FILTERS USING 1D, 2D, 3D, 4D, 1X, AND 2X STRUCTURES (FOR INTEL 8086 ROUTINES) (μs)

	Structure YY					
Routine	1D	2D	3D	4D	1X	2X
OUTP_YY	176.0	161.6	163.6	176.0	161.6	161.6
POST_YY	0	623.2	0	582.4	848.0	595.2
DELAY_YY	32.8	0	27.4	18.4	32.4	0
PRE_YY	582.4	0	565.6	0	0	296.8
TOTAL	791.2	784.8	756.6	776.8	1042.0	1053.6

Source: H. T. Nagle, Jr., and V. P. Nelson, "Digital Filter Implementation on 16-bit Microcomputers," *IEEE MICRO*, Vol. 1, No. 1, Table 5, p. 31, Feb. 1981, ©1981 IEEE.

eight-order filter, the 3D structure is fastest, while the 2X is slowest. The four direct structures are all approximately equal in sampling rate, while the cross-coupled structures are both considerably slower, requiring more multiply operations.

13.7 COMPARISON OF MACHINES

Other 16-bit microcomputers, with features similar to those of the 8086, could have been chosen to implement the examples of this chapter. For example, the Zilog Z8000, the Texas Instruments TMS 9995, the Fairchild-9445, and the Motorola MC 68000 subroutines are presented in Figures 13-17 to 13-20 to illustrate alternative realizations for the eighth-order filter example described earlier. Table 13-2 compares the results. Here we see that the 9445 outperforms the other machines, followed by

```
;
FILTER:     JSR      INIT              ;INIT A/D, D/A
FLOOP:      JSR      INPUT             ;GET SAMPLE INTO DO
            MULS     SO,DO             ;SO*X/4
            ASL.L    #2,DO             ;SO*X
            SWAP     DO                ;TRUNCATE
            MOVEQ    #4,D7             ;4 LOOPS
            JSR      OUTP_1D           ;OUTPUT PROCESS
            MOVE.B   DO,PORT2          ;OUTPUT TO D/A
            MOVEQ    #4,D7             ;4 LOOPS
            JSR      DELAY_1D          ;TIME-DELAY
            MOVEQ    #4,D7             ;4 LOOPS
            JSR      PRE_1D            ;PRE-PROCESS
            BRA      FLOOP             ;CONTINUE
;
;
;
;MC 68000 INPUT FROM A/D
INPUT       BSET     1,PORT4           ;RAISE START-CONVERT
            BCLR     1,PORT4           ;LOWER START-CONVERT
IN_LP:      BTST     1,PORT3           ;WAIT FOR END-CONVERT
            BNE      IN_LP
            MOVE     PORT1,DO          ;GET SAMPLE
            RTS
;
;
;
;MC 68000 1D OUTPUT PROCESSOR
;OUTP_1D:   MO = X + T1                ;Y = MO * AO + T2
;PASS X IN DO, RETURN Y IN DO
;LOOP COUNT IN D7
OUTP_1D:    LEA      T1,AO             ;POINT TO CONSTANTS
            LEA      MO,A1
            LEA      AO,A2
            LEA      T2,A3
OLP_1D:     ADD      @AO+,DO           ;COMPUTE MO
            MOVE.W   DO,@A1+           ;STORE
            MULS     @A2+,DO           ;MO*AO/4
            ASL.L    #2,DO             ;MO*AO
            SWAP     DO
            ADD.W    @A3+,DO           ;Y
            SUBQ.B   #1,D7             ;DECREMENT COUNTER
            BNE      OLP_1D            ;CONTINUE
            RTS
;
;
;
;MC 68000 1D TIME-DELAY
;DELAY_1D:   MOVE MO,M1, TO M1,M2, LOOP COUNT IN D7
DELAY_1D:   ADD.W    D7,D7             ;MOVE 2n WORDS
            LEA      T1,AO             ;POINT TO M(K-2)
            LEA      M2,A1             ;POINT TO M(K-1)
DEL_1D:     MOVE.W   @A1-,@AO-         ;M(K-2) = M(K-1), M(K-1) = M(K)
            SUBQ.B   #1,D7             ;DECREMENT COUNTER
            BNE      DEL_1D            ;CONTINUE
            RTS
;
;
;
;MC 68000 1D PREPROCESSOR
;PRE_1D:    T1 = -B1*M1 - B2*M2
;
;MC 68000 1D PREPROCESSOR
;PRE_1D:    T1 = -B1*M1 - B2*M2
;           T2 = A1*M1 + A2*M2
;LOOP COUNT IN D7
PRE_1D:     LEA      A1,AO             ;POINT TO CONSTANTS
            LEA      M1,A1
```

Figure 13-17 Motorola MC 68000 routines for the eighth-order cascaded filter.

```
                LEA      M2, A2
                LEA      T1, A3
                LEA      T2, A4
    PRLP_1D:    MOVE.W   @A1+, D0         ;M1
                MOVE.W   D0, D1           ;SAVE A COPY
                MOVE.W   @A2+, D2         ;M2
                MOVE.W   D2, D3           ;SAVE A COPY
                MULS     @A0+, D0         ;A1*M1/4
                MULS     @A0+, D2         ;A2*M2/4
                MULS     @A0+, D1         ;B1*M1/4
                MULS     @A0+, D3         ;B2*M2/4
                ADD.L    D2, D0           ;T2/4
                ADD.L    D3, D1           ;-T1/4
                ASL.L    #2, D1           ;-T1
                ASL.L    #2, D0           ;T2
                NEG.L    D1               ;T1
                SWAP     D0
                SWAP     D1
                MOVE.W   D1, @A3+         ;STORE T1
                MOVE.W   D0, @A4+         ;STORE T2
                SUBQ.B   #1, D7           ;DECREMENT COUNTER
                BNE      PRLP_1D          ;CONTINUE
                RTS
;
;
;1D DATA STORAGE
```

Figure 13-17 cont.

```
;
;Z8000 MAIN
FILTER:         CALL     INIT             ;INIT A/D, D/A
FLOOP:          CALL     INPUT            ;GET SAMPLE
                MULT     RR0, S0          ;X*S0/4
                SLAL     RR0, #2          ;X*S0
                LD       R7, #4           ;4 LOOPS
                CALL     OUTP_1D          ;OUTPUT PROCESS
                OUT      PORT2, R0        ;OUTPUT TO D/A
                LD       R7, #4           ;4 LOOPS
                CALL     DELAY_1D         ;TIME-DELAY
                LD       R10, #4          ;4 LOOPS
                CALL     PRE_1D           ;PREPROCESS
                JP       FLOOP            ;CONTINUE
;
;
;
;Z8000 INPUT FROM A/D
;INPUT:   PULSE START LINE, WAIT FOR EOC, THEN SAMPLE
;MEMORY-MAPPED IO
INPUT:          SETB     PORT4, 7         ;SET START-CONVERT
                RESB     PORT4, 7         ;CLEAR START-CONVERT
IN_LP:          BITB     PORT3, 1         ;WAIT FOR END-CONVERT
                JZ       IN_LP
                LD       R1, PORT1        ;GET SAMPLE
                RET
;
;
;
;Z8000 1D OUTPUT PROCESSOR
;OUTP_1D:   FOR Z8000, M0 = X + T1:   Y = A0*M0 + T2
;PASS X IN R0:   RETURN Y IN R0, LOOP COUNT IN R7
OUTP_1D:        LD       R8, #0           ;INDEX
OLP_1D:         ADD      R0, T1[R8]       ;COMPUTE M0
                LD       M0, [R8], R0     ;STORE
                LD       R1, R0           ;SET UP MULT
                MULT     RR0, A0[R8]      ;A0*M0/4
```

Figure 13-18 Zilog Z8000 routines for the eighth-order cascaded filter.

```
            SLAL    RR0,#2                ;A0*M0
            ADD     R0,T2[R8]             ;COMPUTE Y
            INC     R8,2                  ;MOVE INDEX
            DJNZ    R7,OLP_1D             ;CONTINUE
            RET
;
;
;
;Z8000 1D TIME-DELAY
;DELAY_1D - MOVE M0,M1, TO M1,M2, COUNT IN R7
DELAY_1D:   LDAR    R0,T1-2               ;POINT TO M2
            LDAR    R2,M2-2               ;POINT TO M1
            ADD     R7,R7                 ;DO 2N WORDS
            LDDR    @R0,@R2,R7            ;BLOCK MOVE
            RET
;
;
;Z8000 1D PREPROCESSOR
;PRE_1D:    T1 = -B1 * M1 - B2 * M2
;           T2 = A1 * M1 + A2 * M2        ;STAGES IN R10
PRE_1D:     LD      R8,#0                 ;INDEX
            LDA     R9,A1                 ;POINT TO COEFS
PLP_1D:     LD      R1,0H[R9]             ;A1
            LD      R3,4H[R9]             ;B1
            LD      R5,2H[R9]             ;A2
            LD      R7,6H[R9]             ;B2
            MULT    RR0,M1[R8]            ;A1*M1/4
            MULT    RR2,M1[R8]            ;B1*M1/4
            MULT    RR4,M2[R8]            ;A2*M2/4
            MULT    RR6,M2[R8]            ;B2*M2/4
            ADDL    RR0,RR4               ;T2/4
            SLAL    RR0,#2                ;T2
            ADDL    RR2,RR6               ;-T1/4
            SLAL    RR2,#2                ;-T1
            NEG     R2                    ;T1
            LD      T1[R8],R2             ;STORE T1
            LD      T2[R8],R0             ;STORE T2
            INC     R8,#2                 ;INDEX
            INC     R9,#8                 ;NEXT SET OF CONSTANTS
            DJNZ    R10,PLP_1D            ;CONTINUE
            RET
;
;
;
;1D DATA STORAGE
```

Figure 13-18 cont.

```
;
FILTER:     BL      @INIT                 ;INIT A/D, D/A
F_LOOP:     BL      @INIT                 ;SAMPLE X IN WR0
            MPY     R0,@S0                ;X*S0/4
            SLA     R0,2                  ;S*S0
            LI      R3,4                  ;4 STAGES
            BL      @OUTP_1D              ;OUTPUT PROCESS
            MOV     @PORT2,R0             ;OUTPUT TO D/A
            LI      R1,4                  ;4 STAGES
            BL      @DELAY_1D             ;TIME-DELAY
            LI      R3,4                  ;4 STAGES
```

Figure 13-19 Texas Instruments TMS 9995 routines for the eighth-order cascaded filter.

```
              BL      @PRE_1D              ;PREPROCESS
              B       @FLOOP               ;CONTINUE
;
;
;
;TMS 9995 INPUT FROM A/D
;RETURN SAMPLE IN WRO
INPUT:        SBO     SOC                  ;START CONVERT
              SBZ     SOC                  ;PULSE BIT
IN_LP:        TB      EOC                  ;WAIT FOR END-CONVERTER
              JNE     IN_LP
              MOV     RO,@A_TO_D           ;GET SAMPLE
              B       *R11                 ;RETURN
;
;
;
;TMS 9995 1D OUTPUT PROCESSOR
;OUTP_1D:  MO = X + T1;  Y = MO*AO + T2
;PASS X IN WRO, RETURN Y IN WRO
;LOOP COUNT IN WR14
OUTP_1D:      LI      R7,T1                ;POINT TO STORAGE
              LI      R8,MO
              LI      R9,AO
              LI      R10,T2
OLP_1D:       A       RO,*R7+              ;MO
              MOV     *R8+.RO              ;STORE MO
              MPY     RO.*R9+              ;MO*AO/4 IN WR1
              SLA     RO,2                 ;MO*AO
              A       RO,*R10+             ;Y
              DEC     R14                  ;CONTINUE
              JNE     OLP_1D
              B       *R11
;
;
;
;TMS 9995 1D TIME-DELAY
;DELAY_1D:  MOVE MO TO M1, M1 TO M2
;LOOP COUNTER IN R14, STORE M2, M1, MO IN ORDER
DELAY_1D:    LI      R1,M2                ;POINT TO M2
             LI      R2,M1                ;POINT TO M1
             A       R14,R14              ;DO 2N BYTES
OLP_1D:      MOV     *R2+,*R1+            ;MOVE 1 WORD
             DEC     R14                  ;STAGE COUNT 1
             JNE     DLP_1D               ;CONTINUE
             B       *R11                 ;RETURN
;
;
;
;TMS 9995 1D PREPROCESSOR
;PRE_1D:   T1 = -B1*M1 - B2*M2
;          T2 = A1*M1 + A2*M2
;LOOP COUNT IN R14
;
PRE_1D:      LI      R8,A1                ;POINT TO CONSTANTS
             LI      R9,M1                ;M1 POINTER
             LI      R10,M2               ;M2 POINTER
             LI      R12,T1               ;T1 POINTER
             LI      R13,T2               ;T2 POINTER
PRLP_1D:     MOV     RO,*R9+              ;M1
             MOV     R2,RO
             MOV     R4,*R10+             ;M2
             MOV     R6,R4
             MPY     RO,*R8+              ;M1*A1/4
```

Figure 13-19 cont.

```
                MPY         R4, *R8+            ; M2*A2/4
                MPY         R2, *R8+            ; M1*B1/4
                MPY         R6, *R8+            ; M2*B2/4
                A           R0, R4              ; T2/4
                SLA         R0, 2               ; T2
                MOV         *R13+, R0           ; STORE T2
                A           R2, R6              ; -T1/4
                SLA         R2, 2               ; -T1
                NEG         R2                  ; T1
                MOV         *R12+, R4           ; STORE T1
                DEC         R14                 ; STAGE COUNT-1
                JNE         PRLP_1D             ; CONTINUE
                B           *R11                ; RETURN
;
;
;
; 1D DATA STORAGE
```

Figure 13-19 cont.

```
;
; 9445 MAIN
FILTER:         JSR         INIT                ; INITIALIZE A/D, D/A
F_LOOP:         JSR         INPUT               ; GET SAMPLE TO AC1
                LDA         0, CONS_4           ; SET 4 LOOPS
                STA         0, COUNT
                LDA         2, S0               ; GET S0/2
                SUB         0, 0                ; IN AC0
                MULS                            ; X*S0/4
                LDA         2, CONS_2
                SALD                            ; X*S0
                JSR         OUTP_1D             ; PROCESS OUTPUT
                STA         0, PORT2            ; OUTPUT
                LDA         1, CONS_4
                STA         1, COUNT
                JSR         DELAY_1D            ; TIME-DELAY
                LDA         1, CONS_4
                STA         1, COUNT
                JSR         PRE_1D              ; PREPROCESS
                JMP         F_LOOP
;
;
;
; 9445 INPUT FROM A/D
; RETURNS SAMPLE IN AC0
INPUT:          NIOP        PORT0               ; PULSE TO START A/D
ILOOP:          SKPBZ       PORT0               ; WAIT FOR EOC
                JMP         ILOOP
                DIA         1, PORT0
                JMP         0, 3                ; RETURN
;
;
;
; 9445 1D OUIPUT PROCESSOR
; OUTP_1D:    M0 = X + T1;  Y = M0 * A0 + T2
; PASS SAMPLE X IN AC0, RETURN IN AC0
; LOOP COUNT STORED AT COUNT
OUTP_1D:        PSHA        3                   ; SAVE RETURN ADD
                SUB         3, 3                ; INDEX = 0
LOOP:           LDA         1, T1, 3            ; GET T1
                ADD         0, 1                ; COMPUTE M0
                STA         1, M0, 3            ; STORE
```

Figure 13-20 Fairchild 9445 (Microflame II) routines for the eighth-order cascaded filter.

```
          LDA     2,A0,3           ;A0/2
          SUB     0,0              ;CLEAR AC0
          MULS                     ;M0*A0/4
          LDA     2,CONS_2
          SALD                     ;M0*A0
          LDA     1,T2,3           ;GET T2
          ADD     1,0              ;Y
          INC     3                ;INDEX
          DSZ     COUNT            ;DO 4 LOOPS
          JMP     LOOP
          POPA    3                ;GET RETURN ADDRESS
          JMP     0,3              ;RETURN

;
;
;
;9445 1D TIME DELAY
;LOOP COUNT AT COUNT, M2=M1, M1=M0
DELAY_1D: LDA     0,COUNT
          ADD     0,0              ;DOUBLE COUNT
          STA     0,COUNT
          LDA     1,ADDR_T1        ;ONE ABOVE M2
          LDA     2,ADDR_M2        ;ONE ABOVE M1
          STA     1,48             ;AUTO DEC ADDRESSES
          STA     2,49
D_LOOP:   LDA     0,@49            ;GET NEW
          STA     0,@48            ;REPLACE OLD
          DSZ     COUNT            ;CONTINUE
          JMP     D_LOOP
          JMP     0,3
;
;
;
;9445 1D PREPROCESSOR
;PRE_1D:  T1 = -B1*M1 - B1*M2
;         T2 = A1*M1 + A2*M2
;LOOP COUNT AT COUNT
PRE_1D:   LDA     0,ADDR_A1-1,     ;FIRST CONSTANT
          STA     0,32             ;AUTO-INC ADDRESS
          PSHA    3                ;SAVE RETURN ADDR
          SUB     3,3              ;INDEX = 0
PRLP_1D:  SUB     0,0              ;SUM = 0
          LDA     1,@32            ;A1/2
          LDA     2,M1,3           ;M1
          MULS                     ;M1*A1/4 IN AC0
          LDA     1,@32            ;A2/2
          LDA     2,M2,3           ;M2
          MULS                     ;T2/4 IN AC0
          LDA     2,CONS_2
          SALD                     ;T2 IN AC0
          STA     0,T2,3           ;STORE
          SUB     0,0              ;CLEAR SUM
          LDA     1,@32            ;B1/2
          LDA     2,M1,3           ;M1
          MULS                     ;M1*B1/4
          LDA     1,@32            ;B2/2
          LDA     2,M2,3           ;M2
          MULS                     ;-T1/4 IN AC0
          LDA     2,CONS_2
          SALD                     ;-T1
          NEG     0,0              ;+T1
          STA     0,T1,3           ;STORE
          INC     3,3              ;INDEX
          DSZ     COUNT            ;4 LOOPS
```

Figure 13-20 cont.

```
            JMP     PRLP_1D
            POPA    3                   ;GET RETURN ADDR
            JMP     0,3                 ;RETURN
;
;
;
;1D DATA STORAGE:
```

Figure 13-20 cont.

the MC 68000, Z8000, 8086, and TMS 9900. It should be noted that these subroutines may not be optimum and thus give only approximate execution times. The largest factor contributing to these different execution times is the multiply time of the processor. These times are given in Table 13-3 for each of the five processors, together with the total time between samples spent doing multiply operations.

TABLE 13-2 PERFORMANCE TIMES (μs) OF 16-BIT MICROCOMPUTERS FOR AN EIGHTH-ORDER CASCADED FILTER

Program	MC68000	Z8000	8086	TMS 9995	9445
Filter (1 loop)	24.75	30.75	53.2	69.1	10.1
Input (no delay)	10.25	15.5	14.8	24.2	2.8
OUTP_1D	65.5	127.25	174.4	211.5	40.8
DELAY_1D	32.0	30.25	32.8	135.8	33.2
PRE_1D	194.5	380.25	582.4	559.4	108.0
Total sample time	327.0	594.0	855.6	1000.0	193.9
Time lag from input to output	82.25	156.25	212.8	253.9	47.8

Source: H. T. Nagle, Jr., and V. P. Nelson, "Digital Filter Implementation on 16-bit Microcomputers," *IEEE MICRO*, Vol. 1, No. 1, Table 1, p. 34, Feb. 1981, ©1981 IEEE.

TABLE 13-3 MULTIPLY TIMES (μs) OF 16-BIT MICROCOMPUTERS USING ADDRESSING MODES SELECTED FOR 1D MODULE SUBROUTINES

Processor	Clock rate (MHz)	16 × 16-bit multiply time (μs)	Total multiply time in four 1D modules (μs)
MC68000	8.0[a]	9.75	205.0
Z8000	4.0	18.0	378.0
8086	5.0	30.6	643.0
TMS 9995	3.3	18.2	382.0
9445	15.0	3.5	73.5

[a]One internal state = two clock cycles, giving an effective clock rate of 4 MHz.

Source: H. T. Nagle, Jr., and V. P. Nelson, "Digital Filter Implementation on 16-bit Microcomputers," *IEEE MICRO*, Vol. 1, No. 1, Table 2, p. 34, Feb. 1981, ©1981 IEEE.

13.8 THE INTEL 2920

The INTEL 2920 is a single-chip NMOS microcomputer designed especially for real-time digital signal processing applications. The 2920 performs all the functions shown in Figure 13-21. The processor employs a specialized canonical signed-digit code to implement multiplication by a minimum number of "shifts" and "adds." In presenting the 2920, we will first describe the canonical signed-digit code, then describe the processor itself, and finally examine an example filter program.

Computer-aided design of microcomputer components. (Courtesy of Intel Corporation.)

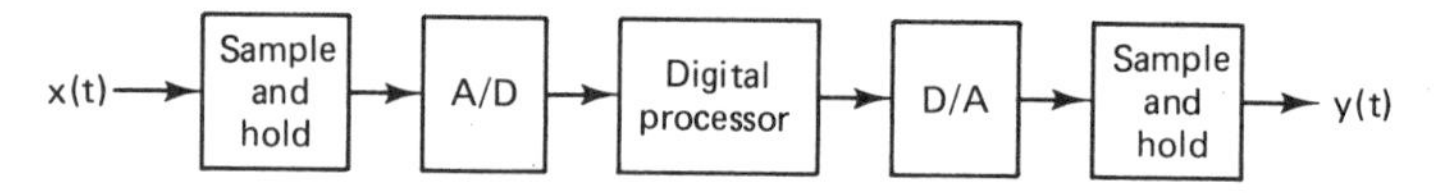

Figure 13-21 Functions of the 2920.

Canonical Signed-Digit Code (CSDC) [7]

The CSDC is a technique for representing a coefficient as

$$c_i = (c_{0i} \quad c_{1i} \quad \cdots \quad c_{B-1,i})_{\text{csdc}} \tag{13-27}$$

where the kth symbol c_{ki} takes on the values 1, 0, -1, and B is the coefficient word-length. One of the most important characteristics of this number representation is its property that approximately one-third of the digits of c_i are nonzero, and that no two adjacent digits c_{ki}, $c_{k+1,i}$ are both nonzero for all $k = 0, B-1$.

The algorithm for converting binary numbers from two's-complement notation to canonical signed-digit code is illustrated below.

In two's complement code

$$
\begin{aligned}
c_i &= (c_{0i} \quad c_{1i} \quad \cdots \quad c_{B-1,i})_{2\text{cns}} \\
&= (-c_{0i} + \sum_{k=1}^{B-1} c_{ki} \cdot 2^{-k}) 2^{B-1}
\end{aligned} \tag{13-28}
$$

where 2^{B-1} is a scaling constant added so that the coefficients c_i may be represented as integers for convenience. The sign bit is c_{0i} and $(c_{1i} \cdots c_{B-1,i})_2$ is the two's complement of the magnitude $|c_i|$. See Ref. 8 for details.

The canonical signed-digit code is based on the following property of binary numbers: A string of consecutive one-bits in a binary number from bit position i to j may be represented as $2^{i+1} - 2^j$.

Example 13.5

Assume that

$$
\begin{aligned}
& \quad\;\; 7 \quad 6 \quad 5 \quad 4 \quad 3 \quad 2 \quad 1 \quad 0 \\
c_i &= (0 \quad 0 \quad 1 \quad 1 \quad 1 \quad 1 \quad 0 \quad 0)_{2\text{cns}} \\
&= (0 \quad 1 \quad 0 \quad 0 \quad 0 \quad 0 \quad 0 \quad 0)_{2\text{cns}} \\
&\; - (0 \quad 0 \quad 0 \quad 0 \quad 1 \quad 0 \quad 0 \quad 0)_{2\text{cns}} \\
&= (2^6 - 2^2)_{10} \\
&= (64 - 4)_{10} \\
&= (60)_{10}
\end{aligned} \tag{13-29}
$$

A string of 1's in positions 5 to 2 is represented by a $(+1) \times 2^6$ and $(-1) \times 2^2$. In canonical signed-digit code

$$c_i = (0 \quad 1 \quad 0 \quad 0 \quad 0 \quad -1 \quad 0 \quad 0)_{\text{csdc}}$$

Example 13.6

Consider a second example.

$$
\begin{aligned}
c_i &= -(8)_{10} \\
&= -(0 \quad 0 \quad 0 \quad 0 \quad 1 \quad 0 \quad 0 \quad 0)_2 \\
&= (1 \quad 1 \quad 1 \quad 1 \quad 1 \quad 0 \quad 0 \quad 0)_{2\text{cns}}
\end{aligned} \tag{13-30}
$$

We may write

$$c_i = ((2^8 - 2^3) - 2^8)_{10}$$

Hence

$$c_i = (0 \quad 0 \quad 0 \quad 0 \quad -1 \quad 0 \quad 0 \quad 0)_{\text{csdc}}$$

Example 13.7

Consider a third example.

$$
\begin{aligned}
c_i &= -(34)_{10} \\
&= -(0 \quad 0 \quad 1 \quad 0 \quad 0 \quad 0 \quad 1 \quad 0)_2 \\
&= (1 \quad 1 \quad 0 \quad 1 \quad 1 \quad 1 \quad 1 \quad 0)_{2\text{cns}}
\end{aligned} \tag{13-31}
$$

This number may be decomposed to

$$
\begin{aligned}
c_i &= (1 \quad 1 \quad 0 \quad 0 \quad 0 \quad 0 \quad 0 \quad 0)_{2\text{cns}} \\
&\quad + (0 \quad 0 \quad 0 \quad 1 \quad 1 \quad 1 \quad 1 \quad 0)_{2\text{cns}}
\end{aligned}
$$

Then

$$
\begin{aligned}
c_i &= (1 \quad 1 \quad 0 \quad 0 \quad 0 \quad 0 \quad 0 \quad 0)_{2\text{cns}} \\
&\quad + (0 \quad 0 \quad 1 \quad 0 \quad 0 \quad 0 \quad 0 \quad 0)_{2\text{cns}} \\
&\quad + (0 \quad 0 \quad 0 \quad 0 \quad 0 \quad 0 \quad -1 \quad 0)_{\text{csdc}} \\
c_i &= (1 \quad 1 \quad 1 \quad 0 \quad 0 \quad 0 \quad 0 \quad 0)_{2\text{cns}} \\
&\quad + (0 \quad 0 \quad 0 \quad 0 \quad 0 \quad 0 \quad -1 \quad 0)_{\text{csdc}} \\
&= (0 \quad 0 \quad -1 \quad 0 \quad 0 \quad 0 \quad 0 \quad 0)_{\text{csdc}} \\
&\quad + (0 \quad 0 \quad 0 \quad 0 \quad 0 \quad 0 \quad -1 \quad 0)_{\text{csdc}}
\end{aligned}
$$

Finally,

$$c_i = (0 \quad 0 \quad -1 \quad 0 \quad 0 \quad 0 \quad -1 \quad 0)_{\text{csdc}}$$

These examples illustrate the following algorithm:

1. Scan the two's-complement representation of the coefficient from right to left.
2. Copy the bits until a string of one-bits is encountered.
3. Write the lowest-order bit of the string as -1.
4. Copy zeros in the positions of the rest of the string.
5. Insert a one-bit into the bit position that is in the next-most-significant position above the string. This position will always be a zero to define the upper limit of the string.
6. Go back to step 2 and repeat until the end of the string.

Example 13.8

$$
\begin{aligned}
c_i &= -(34)_{10} \\
&= (1 \quad 1 \quad \underset{1}{0} \quad 1 \quad 1 \quad 1 \quad 1 \quad 0)_{2\text{cns}} \qquad (13\text{-}32) \\
&= (0 \quad 0 \quad -1 \quad 0 \quad 0 \quad 0 \quad -1 \quad 0)_{\text{csdc}}
\end{aligned}
$$

Now that the canonical signed digit number system has been described, let us examine its use in calculating a sum of product terms. All digital filter structures of Chapter 12 may be implemented by computing devices which can calculate a sum of product terms:

$$y = \sum_{i=1}^{L} c_i X_i \tag{13-33}$$

If the coefficients c_i are implemented using the CSDC, the computing device can perform the multiplication operation by first shifting right by a power of 2 (for each nonzero digit of c_i), and then adding or subtracting the variable X_i. Consequently, each c_i determines the sequence of operations for each of the L terms of the sum.

The number N_{MUL} or microprocessor cycles required to carry out the sum of

L multiplications can be statistically estimated. It depends on the type of microprocessor (i.e., if it can perform add/subtract and shift in the same cycle). If so,

$$N_{MUL} = B + (L - 1)\left[\frac{1 - (-2^{-B})}{9} + \frac{B}{3}\right]$$
$$N_{MUL} \approx B + (L - 1)\frac{B}{3} \qquad (\text{for } B \geq 8) \tag{13-34}$$

where B = wordlength of the coefficients
($B \leq n$, the filter variable wordlength)
L = number of multipliers to the same adder

If one uses the CSDC for coefficients and implements them directly in microcode, increased sampling rates are achieved. This increase is gained at the expense of two important factors:

1. The first is increased microprogram storage. The cost of the increased storage may be diminished by using a relatively simple signal processor architecture with short microprogram wordlength, as in the INTEL 2920.
2. This method is suitable only for filters with fixed, time-invariant coefficients. The altering of a single coefficient would lead to regeneration of the whole microprogram.

Processor Description [9]

A functional block diagram of the INTEL 2920 is shown in Figure 13-22. Note that the four input lines and eight output lines may be multiplexed to allow the 2920 to implement several filters in parallel. The functions of the 2920 are under program control. Program instructions are stored in the 192-word (24-bits per word) erasable, programmable, read-only memory (EPROM). Each 24-bit instruction has the same execution time, and *no* conditional branching is allowed. Hence a program always executes in the same number of machine cycles, producing a constant sampling rate. The sampling rate is determined by the program length and the instruction cycle time, which is four clock cycles (400 ns at the maximum 10-MHz clock rate). Consequently, a full 192-instruction program (at 10 MHz) produces a 13-kHz sampling rate. A shorter program will generate a higher sampling rate.

Analog Operations

The 2920 input and output operations occur in parallel with the digital operations. To sample an input signal, one of the four input lines is selected and the sample and hold is activated. The resulting sample is digitized by the 9-bit successive approximation ADC and stored in register DAR. This register is the interface between the analog and digital sections of the 2920. Input signals are transferred from DAR to scratchpad memory for processing. Output signals (the 9 most significant bits) are transferred to DAR, which drives the DAC. The DAC is connected through the output demultiplexer to one of the eight output sample-and-hold circuits.

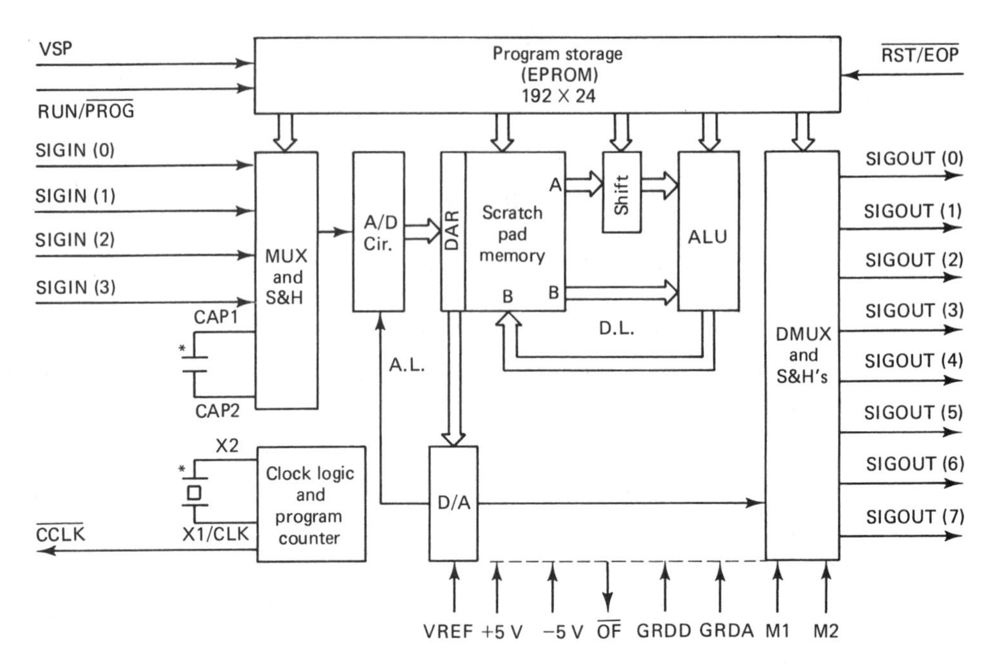

Figure 13-22 INTEL 2920 block diagram. (Courtesy of Intel Corporation.)

Digital Operations

The digital loop (D.L. in Figure 13-22) includes the two-port, 40-word scratchpad random access memory (RAM), a binary shifter, and the arithmetic unit (ALU). Two 25-bit words are fetched simultaneously from the two scratchpad output ports. The data from port A pass through the binary shifter for scaling from 2^2 (a 2-bit left shift) to 2^{-13} (a 13-bit right shift). The scaled A-value and unscaled B-value are furnished to the ALU, which produces a 25-bit result to be stored in the scratchpad location specified by the port B address.

Instruction Set

The 2920 instructions are 24 bits long and have five fields:

1. ALU instruction (opcode, 3 bits)
2. Destination (B address, 6 bits)
3. Source (A address, 6 bits)
4. Shift code (+2 to −13, 4 bits)
5. Analog instruction (5 bits)

Programs are implemented using a combination of analog and digital instructions from the instruction set of Table 13-4. Note that the arithmetic instructions are ADD

TABLE 13-4 INSTRUCTION SET AND OPERATIONS

Mnemonics		Operations			
Code	Condition	Digital instructions			
ADD		$(A \times 2^N)+B$	B[a]		
SUB		$B-(A \times 2^N)$	B		
LDA[b]		$(A \times 2^N)+0$	B		
XOR[b]		$(A \times 2^N) \oplus B$	B		
AND		$(A \times 2^N) \cdot B$	B		
ABS[b]		$[(A \times 2^N)]$	B		
ABA		$[(A \times 2^N)]+B$	B		
LIM		$\text{Sign}(A) \rightarrow \pm\text{F.S.}$	B[c]		
ADD	CND( )[d]	$(A \times 2)^N+B$	B		IFF DAR(K)=1
		B	B		IFF DAR(K)=0
SUB	CND( )[d,e]	$B-(A \times 2^N)$	B	& CY → DAR(K)	IFF $CY_P=1$
		$B+(A \times 2^N)$	B	& CY → DAR(K)	IFF $CY_P=0$[f]
LDA	CND( )[d]	$(A \times 2^N)$	B		IFF DAR(K)=1
		B	B		IFF DAR(K)=0
ABA	CND( )[g]	$(A \times 2^N)+B$	B		
XOR	CND( )[g]	$(A \times 2^N) \oplus B$	B		
		Analog Instructions			
IN(K)		Signal sample from input channel K			
OUT(K)		D/A to output channel K			
CVTS		Determine sign bit			
CVT(K)		Perform A/D on bit K			
EOP		Program counter to zero[h]			
NOP		No operation			
CND(K)		Select bit K for conditional instructions			
CNDS		Select sign bit for conditional instructions			

[a]Note that scaling of A always occurs before executing the digital operation.

[b]Clarification of CY_{OUT} sense for certain operations. For LDA, XOR, AND, ABS: $CY_{OUT} \rightarrow 0$.

[c]B is set to full scale (F.S.) amplitude with the same sign as the "A" port operand.

[d]CND( ) can be either CND(K) or CNDS testing amplitude bits or the sign bit of the DAR, respectively.

[e]For SUB CNDS operation $\overline{CY} \rightarrow$ DAR(S).

[f]The previous carry bit (CY_P) is tested to determine the operation. The present carry bit (CY) is loaded into the Kth bit location of the DAR. "Present carry (CY) is generated independent of overflow. It will represent the carry (CY) of a calculated 28-bit result."

[g]Does not affect DAR. In this case, CND is used with XOR/ABA to enable/disable the ALU overflow saturation algorithm. Use of either instruction causes the ALU output to roll over rather than go to full scale with sign bit preserved. An EOP instruction will also enable the ALU overflow saturation algorithm.

[h]EOP will also enable overflow correction if it was disabled during a program pass. The EOP must occur in ROM location 188.

Source: Intel Corporation.

and SUB with scaling of one operand, which is designed to implement the canonical signed-digit code described earlier.

Programming Example [10]

Suppose that we want to implement a second-order digital filter:

$$D(z) = \frac{0.00293}{1 - 1.7656z^{-1} + 0.99414z^{-2}} = \frac{a_0}{1 + b_1 z^{-1} + b_2 z^{-2}} \tag{13-35}$$

Hence

$$y(k) = a_0 x(k) - b_1 y(k-1) - b_2 y(k-2)$$

Since the INTEL 2920 executes instructions of the form

$$\begin{aligned} \text{B} &= \text{A} * 2^{\text{N}} \\ \text{B} &= \text{B} + \text{A} * 2^{\text{N}} \\ \text{B} &= \text{B} - \text{A} * 2^{\text{N}} \end{aligned} \tag{13-36}$$

where 2^{N} is the scaling (shifting) factor, the coefficients may be converted to CSDC as follows:

$$\begin{aligned} a_0 &= +(0.00293)_{10} \\ &= (0.000000001100000000000101\cdots)_2 \end{aligned}$$

Since the 2920 can shift from $+2$ to -13, the CSDC will have 16 digits, three to the left of the binary point. So we round a_0 to 16 bits:

$$\begin{aligned} a_0 &= +(000.0000000011000)_2 \\ &= (000.0000000011000)_{2\text{cns}} \\ &= (000.000000010\ -1000)_{\text{csdc}} \\ &= 2^{-8} - 2^{-10} \end{aligned} \tag{13-37}$$

In a like manner,

$$\begin{aligned} -b_1 &= +(1.7656)_{10} \\ &= +(1.11000011111111001011\cdots)_2 \\ &= +(001.1100010000000)_2 \\ &= (001.1100010000000)_{2\text{cns}} \\ &= (010.0\ -100010000000)_{\text{csdc}} \\ &= 2^1 - 2^{-2} + 2^{-6} \end{aligned} \tag{13-38}$$

and

$$\begin{aligned} -b_2 &= -(0.99414)_{10} \\ &= -(0.1111111001111111111110101\cdots)_2 \\ &= -(000.1111111010000)_2 \\ &= (111.0000000110000)_{2\text{cns}} \\ &= (00\ -1.00000010\ -10000)_{\text{csdc}} \\ &= -2^0 + 2^{-7} - 2^{-9} \end{aligned} \tag{13-39}$$

The INTEL 2920 program to this filter is given in Table 13-5. In the program

$$\begin{aligned} Y0 &= y(k - 0) \\ Y1 &= y(k - 1) \\ Y2 &= y(k - 2) \end{aligned} \tag{13-40}$$

TABLE 13-5 INTEL 2920 PROGRAM

Opcode (3 bits)	Destination (6 bits)	Source (6 bits)	Shift (4 bits)	Comments
;TIME DELAY				
LDA	Y2	Y1	0	;Y2 = Y1
LDA	Y1	Y0	0	;Y1 = Y0
;CALCULATE				
LDA	Y0	Y1	1	;$-b_1$
SUB	Y0	Y1	−2	
ADD	Y0	Y1	−6	;Y0 = $-b_1$Y1
SUB	Y0	Y2	0	;$-b_2$
ADD	Y0	Y2	−7	
SUB	Y0	Y2	−9	;Y0 = $-b_1$Y1$-b_2$Y2
ADD	Y0	X	−8	;a_0
SUB	Y0	X	−10	;Y0 COMPLETE

13.9 OTHER PROCESSORS

Many other microcomputers may be used to implement digital filters. Here we will examine briefly the AMD2901 bipolar bit-slice microprocessor, the Bell Laboratories Digital Signal Processor (DSP), the AMI S2811 signal processor, the Texas Instruments TMS 320, the NEC μPD7720, and three 8-bit microcomputers with A/D conversion (the AMI S2200, INTEL 8022, and the Motorola 6805).

Advanced Micro Devices 2901

The AMD2901 is a high-speed, bipolar, 4-bit slice microcomputer which can be used as a building block in high-speed signal processing applications. One example is the SDA processor [11], which uses four AMD2901s to implement a 16-bit signal processing machine. The SDA processor block diagram is shown in Figure 13-23. This processor has been constructed with 167 dual-in-line packages and consumes 48 W of power. The SDA processor can implement an eighth-order digital filter at a sampling rate of 35 kHz. Details of the implementation are given in Ref. 11.

Bell Laboratories Digital Signal Processor [12, 13]

The DSP is an NMOS integrated circuit signal processor designed for the Bell System. It is packaged in a 40-pin dual-in-line package, requires a single 5-V power supply, and runs at a 5-MHz rate. The machine cycle time is 800 ns. The circuit consists of about 45,000 transistors.

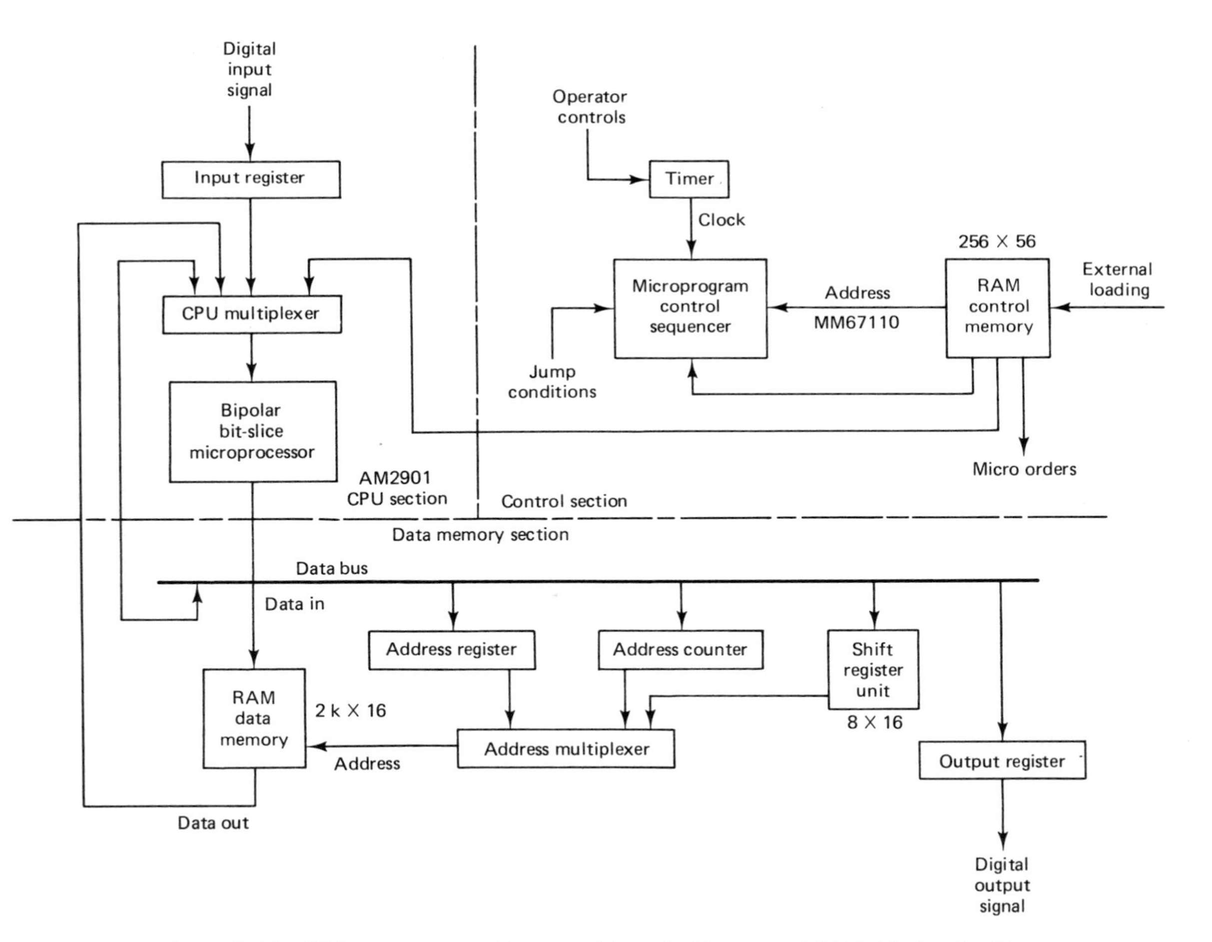

Figure 13-23 SDA processor architecture. (From J. Zeman and H. T. Nagle, Jr., "A High-Speed, Microprogrammable Digital Signal Processor Employing Distributed Arithmetic," *IEEE Trans. Comput.*, Vol. C-29, No. 2, Fig. 3, p. 137, Feb. 1980, © 1980 IEEE.)

The DSP architecture is shown in Figure 13-24. If features a 1024×16 ROM for instructions and fixed data, a 128×20 RAM for variable data, an address arithmetic unit (AAU), and an arithmetic unit (AU) which accepts 16-bit and 20-bit operands to form a 36-bit product and accumulates the product in a 40-bit register. The AU rounds the accumulator to a 20-bit result with overflow protection. The AU evaluates the expression

$$(x \cdot y) + a \longrightarrow a \longrightarrow w$$

where x is a 16-bit coefficient, y is a 20-bit variable, a is the 40-bit accumulator, and w is the 20-bit rounded result.

The DSP can implement up to 30 second-order filter modules at a sampling rate of 8 kHz.

American Microsystems S2811 [14]

The S2811 is a 16-bit microcomputer called the signal processing peripheral (SPP). The SSP architecture is shown in Figure 13-25. The instruction memory is a 256×17 ROM and data memory (ROM and RAM) is 256×16 bits. The SPP also implements the equation

$$(x \cdot y) + a \longrightarrow a$$

The multiplier is a parallel unit which multiplies two 12-bit inputs and rounds the result to 16 bits. The instruction cycle time is 300 ns. The SSP is a powerful signal processing microcomputer.

Texas Instruments TMS 320 [15]

The architecture of the Texas Instruments TMS 320 is demonstrated in Figure 13-26. The device is made from NMOS, 3-μm technology, has a power dissipation of 1 W, and is supplied in a 40-pin package. The process functions around two buses: a 16-bit program bus and a 16-bit data bus. A four-level stack is used to save subroutine call return links. Memory is organized as a ROM for program memory with 1.5K capacity, expandible to 4K externally. Data memory is a RAM of 144 words of 16 bits each. The ALU has a 16×16 bit parallel two's-complement multiplier. The accumulator is 32 bits and is used as the primary operand. No serial input/output (I/O) is provided. Most instructions are executed in a signal instruction cycle of 200 ns.

NEC μPD7720 [16]

The structure of the NEC μPD7720 is illustrated in Figure 13-27. It is constructed using NMOS, 2.5-μm technology and is supplied in a 28-pin package. The data RAM is organized as two 16-bit wide blocks, each consisting of four rows of 16 words, yielding a capacity of 128 words of 16 bits each. The data ROM (for coefficients) is 512 words of 13 bits each. The multiplier is 16×16 bits and used two's-complement code. The ALU uses 16 bit arithmetic and has two 16-bit accumulators. The unit has one serial I/O port as well as parallel I/O. Direct memory access (DMA) is provided A four-level stack for subroutine return linkage is also available. Full 32-bit precision from the ALU to memory is allowed for intermediate storage of double-precision

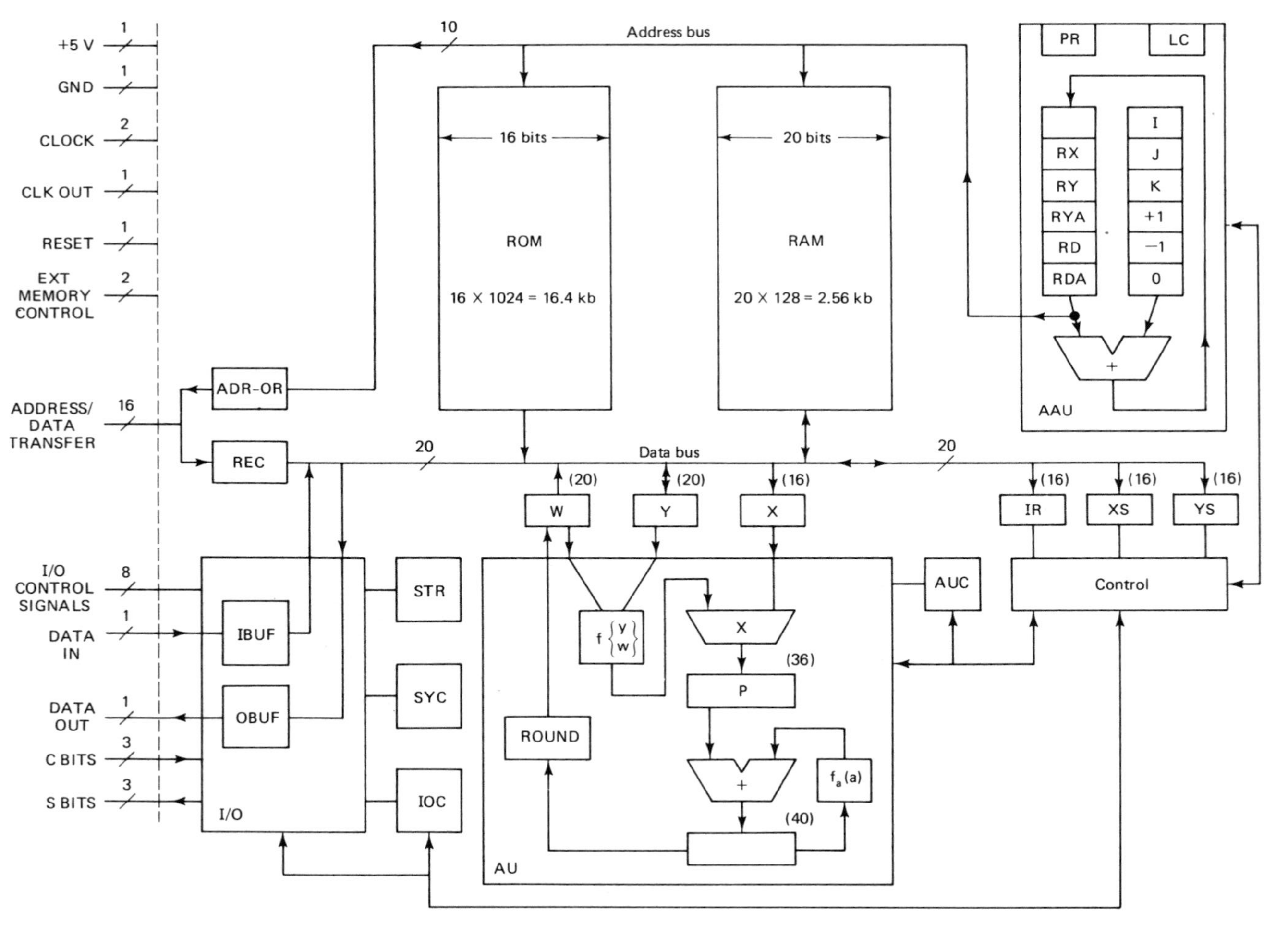

Figure 13-24 DSP architecture. (Reprinted with permission from J. Boddie, "Overview: The Device, Support Facilities, and Applications," *Bell Syst. Tech. J.*, Vol. 60, No. 7, Fig. 2, pp. 1431–1439, Sept. 1981. Copyright 1981, AT&T.)

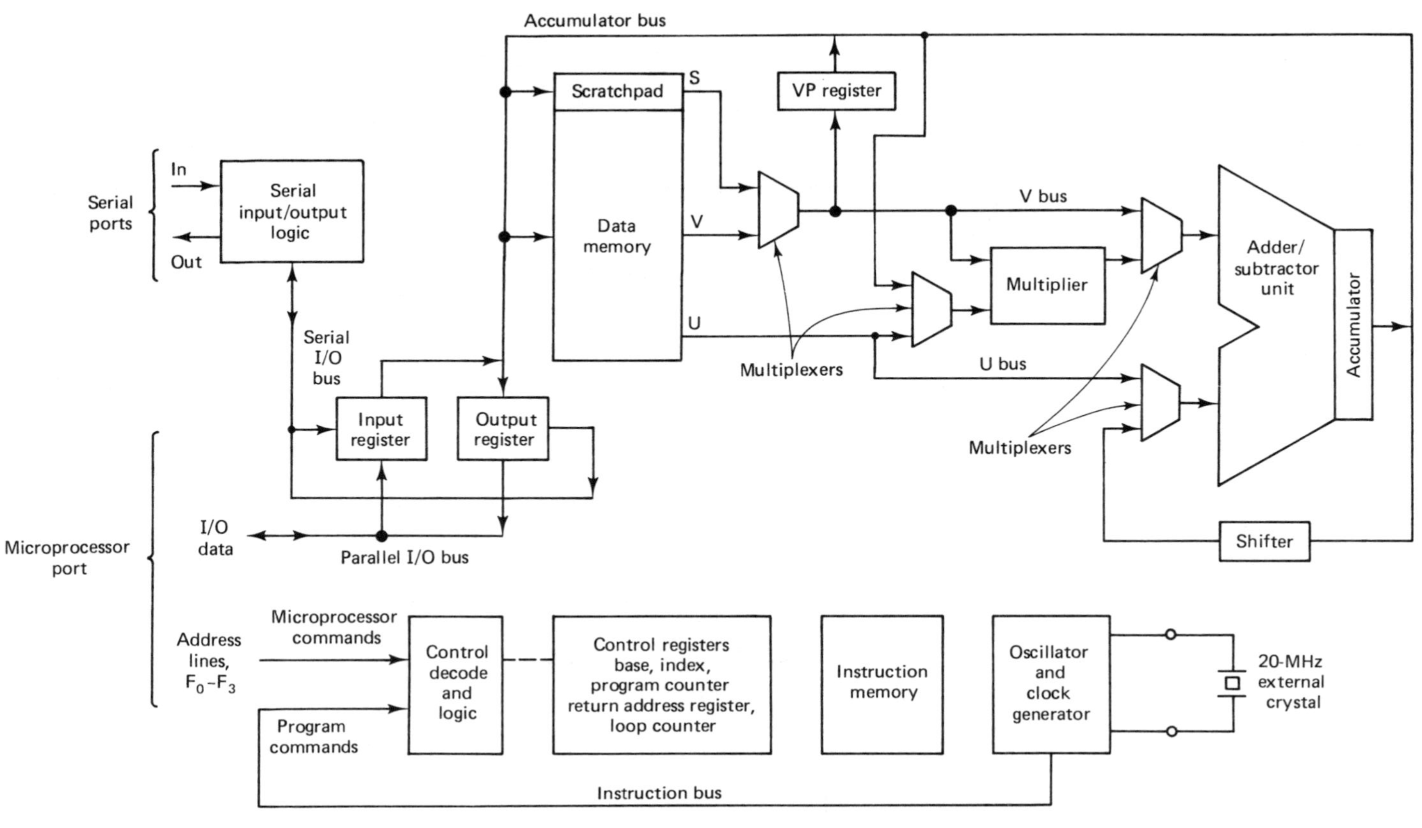

Figure 13-25 SSP architecture. (Reprinted from R. W. Blasco, "V-MOS Chip Joins Microprocessor to Handle Signals in Real Time," *Electronics*, pp. 131–138, Aug. 30, 1979.)

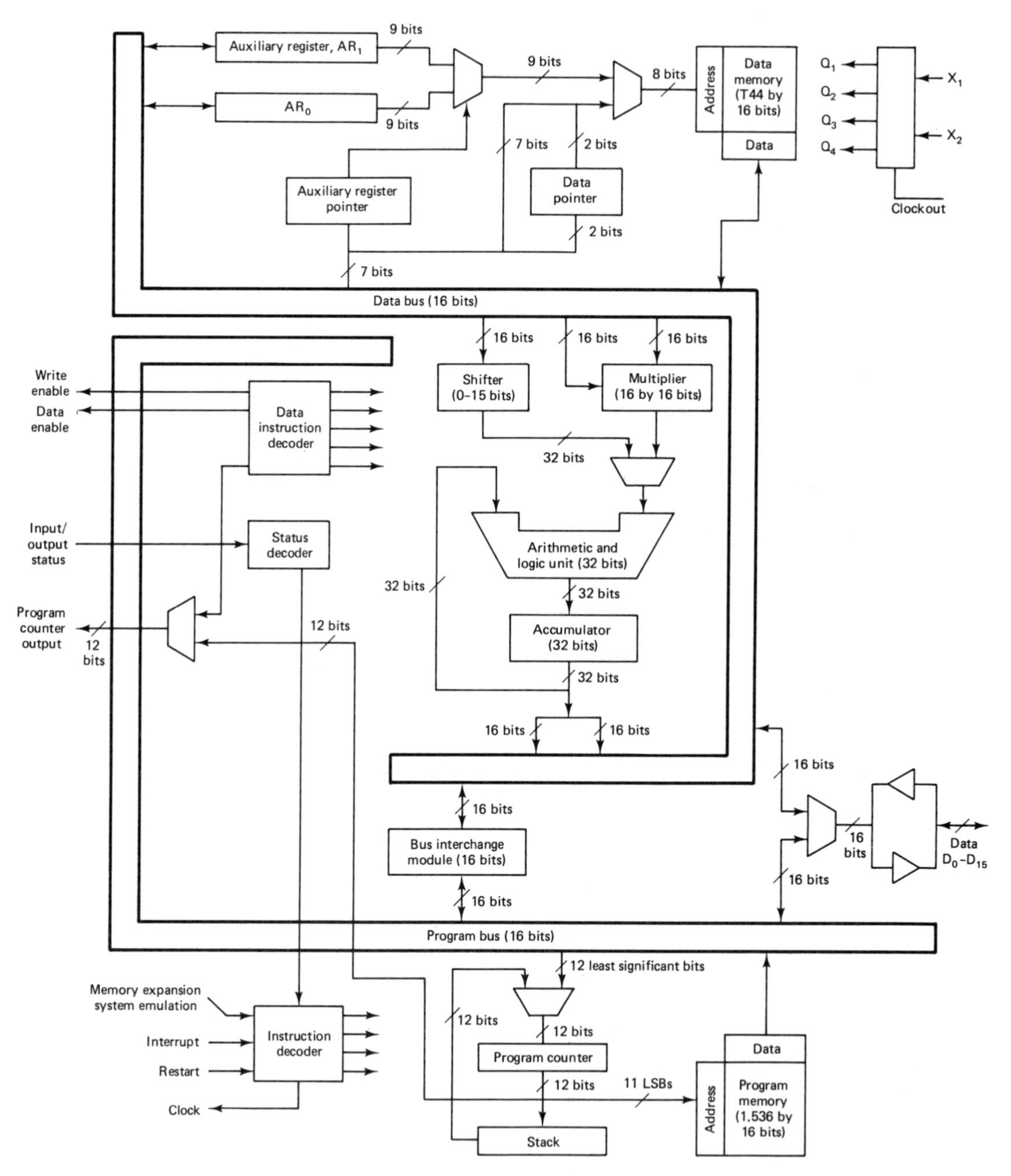

Figure 13-26 TI TMS 320. (Reprinted from K. McDonough et al., "Microcomputer with 32-Bit Arithmetic Does High Precision Number Crunching," *Electronics*, Vol. 55, No. 4, pp. 105–110, Feb. 24, 1982. Copyright © McGraw-Hill Inc. 1982. All rights reserved.)

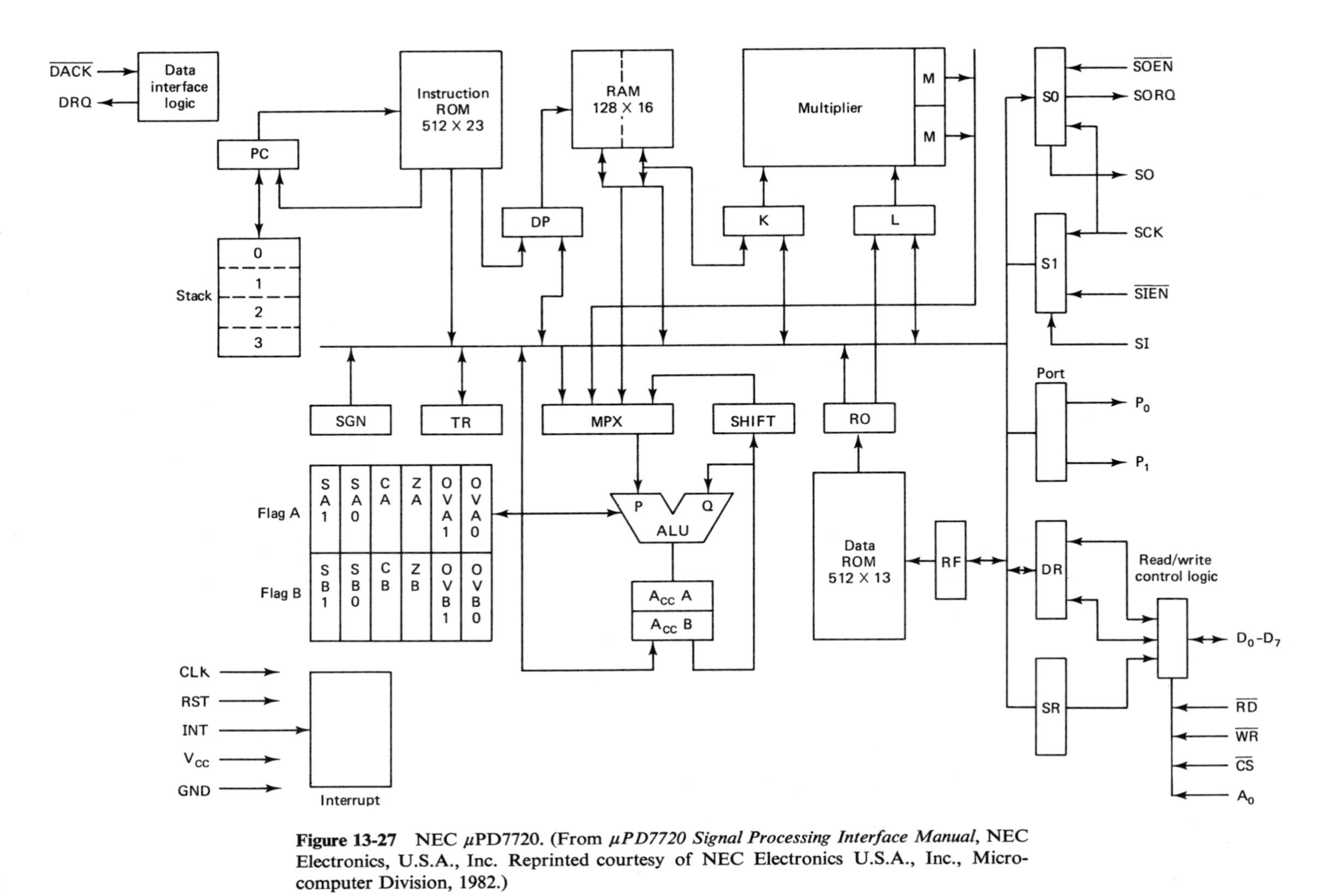

Figure 13-27 NEC μPD7720. (From *μPD7720 Signal Processing Interface Manual*, NEC Electronics, U.S.A., Inc. Reprinted courtesy of NEC Electronics U.S.A., Inc., Microcomputer Division, 1982.)

results. The main disadvantage of this device is that external expansion of the instruction memory is not allowed, and coefficients storage is limited to 13 bits. The device can implement a second-order digital filter in 2.25 μs, and calculate a 64-point complex FFT in 1.6 ms.

Microcomputers with A/D Conversion

In digital filtering applications in which low sampling rates (a few Hertz) and/or precision (8 bits) is adequate, the system designer may elect to employ an 8-bit microcomputer the built-in A/D conversion. Example machines are the AMI S2200 with a cycle of 4.5 μs, the INTEL 8022 with a cycle of 10 μs, and the Motorola 6805 with a cycle of 1 μs. These machines do not have multiply instructions, so that software multiplication routines must be used to bit test, shift, and add.

13.10 SUMMARY

In this chapter we have investigated microcomputer implementations of digital filters. Emphasis was placed on commerically available 16-bit machines with built-in hardware multiply operations. These machines offer the widest application since their wordlength is adequate for most digital filter applications. Special-purpose signal processing chips were also examined for high-speed applications. Eight-bit microcomputers with built-in ADC were suggested for low-speed applications.

REFERENCES

1. H. T. Nagle, Jr., and V. P. Nelson, "Digital Filter Implementation on 16-bit Microcomputers," *IEEE MICRO* Vol. 1, No. 1, pp. 23–41, Feb. 1981.
2. 8086 Reference Manual, Intel Corp., Santa Clara, Calif.
3. 8080 Reference Manual, Intel Corp., Santa Clara, Calif.
4. H. T. Nagle, Jr., and C. C. Carroll, "Realization of Digital Controllers," *Proc. IFAC Symp. Autom. Control Space*, Armenia, USSR, Aug. 1974.
5. B. Gold and C. M. Radar, *Digital Processing of Signals*. New York: McGraw-Hill Book Company, 1969.
6. S. Y. Huang, "On Optimization of Cascade Fixed-Point Digital Filters," *IEEE Trans. Circuits Syst.* (Letters), Vol. CAS-21, pp. 163–166, Jan. 1974.
7. A. Peled and B. Lia *Digital Signal Processing*. New York: John Wiley & Sons, Inc., 1976, pp. 177–178.
8. H. T. Nagle, Jr., B. D. Carroll, and J. D. Irwin, *An Introduction to Computer Logic*. Englewood Cliffs, N.J.: Prentice-Hall, Inc., 1975.
9. R. E. Holm and J. S. Rittenhouse, "Implementation of a Scanning Spectrum Analyzer Using the 2920 Signal Processor," Application Note, Intel Corp., Feb. 1980.
10. M. E. Hoff and M. Townsend, "Single-Chip n-MOS Microcomputer Processes Signals in Real Time," *Electronics*, pp. 105–110, Mar. 1, 1979.

11. J. Zeman and H. T. Nagle, Jr., "A High-Speed, Microprogrammable Digital Signal Processor Employing Distributed Arithmetic," *IEEE Trans. Comput.*, Vol. C-29, No. 2, pp. 134–144, Feb. 1980.
12. J. R. Boddie, "Digital Signal Processor Overview," *Bell Syst. Tech. J.*, Vol. 60, No. 7, Pt. 2, pp. 1431–1439, Sept. 1981.
13. J. R. Boddie et al., "Digital Signal Processor: Architecture and Performance," *Bell Syst. Tech. J.*, Vol. 60, No. 7, Pt. 2, pp. 1449–1462, Sept. 1981.
14. R. W. Blasco, "V-MOS Chip Joins Microprocessor to Handle Signals in Real Time," *Electronics*, pp. 131–138, Aug. 30, 1979.
15. K. McDonough et al., "Microcomputer with 32-Bit Arithmetic Does High Precision Number Crunching," *Electronics*, Vol. 55, No. 4, pp. 105–110, Feb. 24, 1982.
16. *μPD7720 Signal Processing Interface Manual*, NEC Electronics USA, Inc.

PROBLEMS

13-1. Given

$$D_1(z) = \frac{1 + z^{-1}}{1 - 0.97337z^{-1}}$$

Find the stored value of these coefficients for the INTEL 8086 1D Structure implementation.

13-2. Repeat Problem 13-1 for

$$D_2(z) = \frac{1 - 1.9661z^{-1} + z^{-2}}{1 - 1.9654z^{-1} + 0.99930z^{-2}}$$

13-3. Repeat Problem 13-1 for

$$D_3(z) = \frac{1 - 1.9661z^{-1} + z^{2}}{(1 - 0.22433z^{-1})(1 - 0.52749z^{-1})}$$

13-4. Compute the frequency response of

$$D(z) = D_1(z)D_2(z)D_3(z)$$

Compare the result with Figure 11-15.

13-5. Illustrate how the direct structure routines for the INTEL 8086 calculate

$$aX = (0.435102)(0.713001)$$

where a is a filter coefficient and X is a signal variable.

13-6. Repeat Problem 13-5 for the Z8000.

13-7. Repeat Problem 13-1 for the INTEL 2920.

13-8. Repeat Problem 13-2 for the INTEL 2920.

13-9. Repeat Problem 13-3 for the INTEL 2920.

13-10. Repeat Problem 13-4 using the results of Problems 13-7, 13-8, and 13-9.

14

Finite-Wordlength Effects

14.1 INTRODUCTION

In Chapters 4 through 12 we have presented the analysis and design of discrete-time linear systems. Signal variables and system coefficients were *real* numbers; that is, they were continuous (analog) variables and fixed constants without restriction on their specific values. However, in Chapter 13 we implemented in digital hardware the digital filters designed earlier. In these practical implementations, the values of signal variables and filter coefficients are restricted to a finite set of discrete magnitude values. In Chapter 13 we used a fixed-point number system (two's complement). Other researchers have described distributed arithmetic, signed-logarithm, canonical signed-digit code, input-scaled floating point, residue number systems, Fibonacci numbers, and Fermat transforms to implement digital filters. Floating point [1] has also been used in implementing digital systems. Fixed-point number systems are the most economical and generally applicable approach toward implementing digital filters. In this chapter we examine fixed-point number systems and analyze their effectiveness in implementing digital filters. Specifically, we will analyze coefficient quantization, filter input quantization, product quantization, round-off noise, limit cycles due to product quantization, overflow properties, signal dynamic range, and signal-to-noise ratios.

14.2 FIXED-POINT NUMBER SYSTEMS

The choice of a number system to implement a digital filter greatly affects the filter's performance. The accuracy with which coefficients and signal variables may be represented is directly related to the quantization properties, overflow characteristics, and dynamic range of the number system. In this section we describe two different number systems that have been used in implementing digital filters.

In what follows, we assume that a real number x is to be represented as a finite number of bits in a quantized version of x, say $Q(x)$. The accuracy of the representation may be measured by the error

$$e \triangleq Q(x) - x \tag{14-1}$$

To simplify the notation we will assume that each number x [and hence $Q(x)$] lie in the range

$$0 \leq |x| \leq 1 \tag{14-2}$$

In practice, if numbers are larger than 1, we may normalize them to this range by simply shifting the binary point by some L bits. So, if

$$0 \leq |x'| \leq C \tag{14-3}$$

then

$$0 \leq |Q(x')| \leq C$$

where $C > 1$, there is an L such that

$$Q(x) = 2^L Q(x') \tag{14-4}$$

and

$$0 \leq |Q(x)| \leq 1$$

Unless otherwise stated, we will always assume that numbers in each number system lie in this range.

Signed-Magnitude Number System

The signed-magnitude number system may be used to represent digital filter coefficients and signal variables. In general, any number in the signed-magnitude system may be expressed

$$Q^b(x) = (s.m_1 \quad m_2 \quad m_3 \quad \cdots \quad m_b)_{2\text{smns}}$$

where $Q^b(x)$ is a quantized version of a number x
- s is the sign bit
- $s = 0$ for x positive
- $s = 1$ for x negative
- m_i are the magnitude bits
  $(.m_1 \quad m_2 \quad \cdots \quad m_b)_2 = |Q(x)|$

That is, the magnitude of x is approximated by a binary fraction. Here all numbers are normalized such that

$$0 \leq |Q^b(x)| < 1$$

We may also use a series notation for $Q^b(x)$ as follows:

$$Q^b(x) = (1 - 2s) \sum_{i=1}^{b} m_i \cdot 2^{-i} \tag{14-5}$$

Now what remains is the matter of determining m_i, $i = 1, \ldots, b$ given x.

In this section we examine three quantizing methods which have been proposed for use in digital filters: truncation, round-off, and least significant bit 1 (LSB-1).

First let us consider the case of the *truncation* quantizer. In this case the $|x|$ is a positive real number and is converted to a binary fraction as an infinite series, and then the series is truncated at b bits, as shown below:

$$|x| = (\underbrace{.m_1 \quad m_2 \quad \cdots \quad m_b}_{|Q_t^b(x)|} \quad m_{b+1} \quad \cdots)_2 \tag{14-6}$$

and

$$|Q_t^b(x)| = (.m_1 \quad m_2 \quad \cdots \quad m_b)_2$$

The subscript t indicates trunaction and the superscript b indicates the number of bits. Now let us examine the error e_t introduced by the truncation. For $x \geq 0$, $|x| \geq |Q_t^b(x)|$, and

$$\begin{aligned} e_t &= Q_t^b(x) - x \\ &= |Q_t^b(x)| - |x| \\ &= -(.000 \quad \cdots \quad 0 \quad m_{b+1} \quad m_{b+2} \quad \cdots)_2 \\ &= -2^{-b}(.m_{b+1} \quad m_{b+2} \quad \cdots)_2 \end{aligned}$$

But $(.m_{b+1} \quad m_{b+2} \quad \cdots)_2$ is a real number bounded by

$$0 \leq (.m_{b+1} \quad m_{b+2} \quad \cdots)_2 < 1 \tag{14-7}$$

Therefore,

$$0 \geq e_t > -2^{-b}, \qquad x \geq 0 \tag{14-8}$$

For the case in which $x < 0$,

$$\begin{aligned} e_t &= Q_t^b(x) - x \\ &= -(||Q_t^b(x)| - |x||) \end{aligned}$$

so that

$$0 \leq e_t < 2^{-b}, \qquad x < 0 \tag{14-9}$$

The quantization characteristic for $Q_t^b(x)$ is illustrated in Figure 14-1a. This nonlinear transfer characteristic introduces many problems in digital filters.

Note that the sign of x determines the value of the truncation error. If we assume that x is a real number and is equally likely to have positive or negative values, the probability density function for e_t is continuous and may be depicted as shown in Figure 14-2a and the noise variance may be approximated by

$$\sigma_{e_t}^2 = \{E[e_t^2] - E^2[e_t]\}_{x\geq 0} + \{E[e_t^2] - E^2[e_t]\}_{x<0} \tag{14-10}$$

But

$$\begin{aligned} E[e_t^2]_{x\geq 0} &= \int_{-2^{-b}}^{0} e_t^2 \left(\frac{2^b}{2}\right) de_t \\ &= \frac{2^b}{2} \frac{e_t^3}{3}\Big|_{-2^{-b}}^{0} = \frac{2^b}{2} \frac{2^{-3b}}{3} = \frac{2^{-2b}}{6} \\ E[e_t]_{x\geq 0} &= \int_{-2^{-b}}^{0} e_t \left(\frac{2^b}{2}\right) de_t \\ &= \frac{2^b}{2} \frac{e_t^2}{2}\Big|_{-2^{-b}}^{0} = -\frac{2^b}{2} \frac{2^{-2b}}{2} = -\frac{2^{-b}}{4} \end{aligned}$$

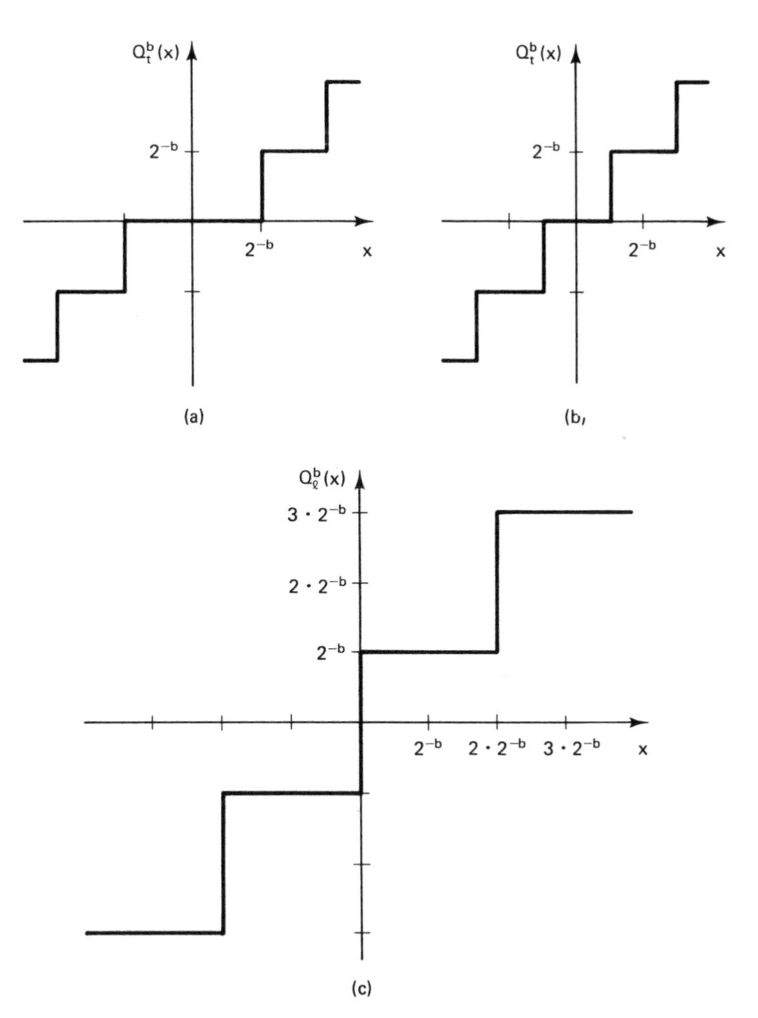

Figure 14-1 Quantizer characteristics for the signed-magnitude number system: (a) truncation; (b) round-off; (c) LSB-1.

$$E[e_t^2]_{x<0} = \int_0^{2^{-b}} e_t^2 \left(\frac{2^b}{2}\right) de_t$$

$$= \frac{2^b}{2} \frac{e_t^3}{3}\bigg|_0^{2^{-b}} = \frac{2^{-2b}}{6}$$

$$E[e_t]_{x<0} = \int_0^{2^{-b}} e_t \left(\frac{2^b}{2}\right) de_t$$

$$= \frac{2^b}{2} \frac{e_t^2}{2}\bigg|_0^{2^{-b}} = \frac{2^{-b}}{4}$$

Therefore,

$$\sigma_{e_t}^2 = \left\{\frac{2^{-2b}}{6} - \left(\frac{2^{-b}}{4}\right)^2\right\}_{x\geq 0} + \left\{\frac{2^{-2b}}{6} - \left(\frac{2^{-b}}{4}\right)^2\right\}_{x<0} = \frac{5 \cdot 2^{-2b}}{24} \tag{14-11}$$

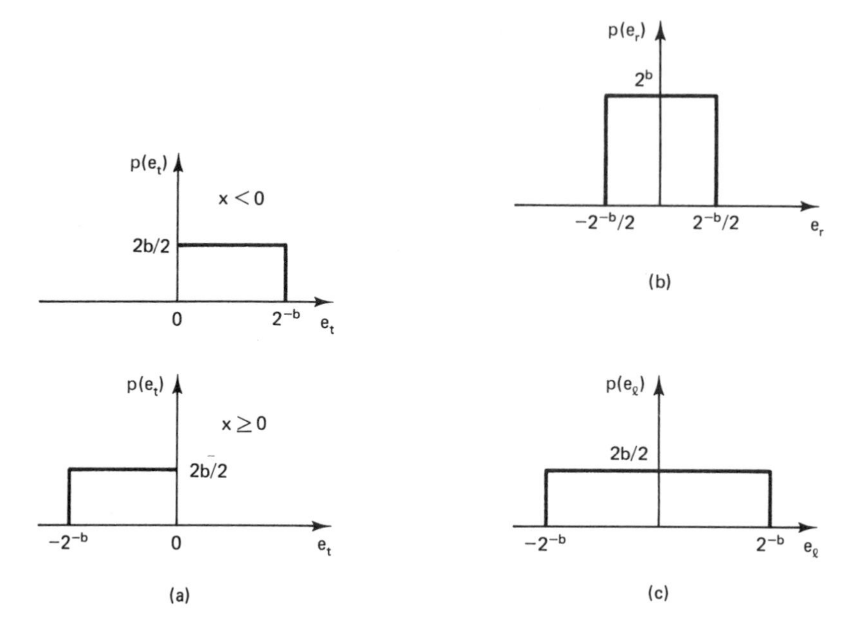

Figure 14-2 Quantification error probability density functions for the signed-magnitude number system: (a) truncation; (b) round-off; (c) LSB-1.

Next, let us consider the *round-off* quantizer. This quantizer takes a value $|x|$, which may be written as an infinite series in powers of 2, and rounds its value to the nearest term 2^{-b} as follows:

$$|x| = (.\,n_1 \quad n_2 \quad \cdots \quad n_b \quad n_{b+1} \quad \cdots)_2$$

Next we round the magnitude to b bits, which may be accomplished by adding 2^{-b-1} and then truncating the result to b bits:

$$\begin{array}{rl} x = & (s\,.\,n_1 \quad n_2 \quad \cdots \quad n_b \quad n_{b+1} \quad n_{b+2} \quad \cdots)_{2\text{smns}} \\ +\,2^{-b-1} = & (0.0 \quad 0 \quad \cdots \quad 0 \quad 1 \quad 0 \quad \cdots)_{2\text{smns}} \\ \hline x + 2^{-b-1} = & (\underbrace{s\,.\,m_1 \quad m_2 \quad \cdots \quad m_b}_{|Q_r^b(x)|} \quad n_{b+1} \oplus 1 \quad n_{b+2} \quad \cdots)_{2\text{smns}} \end{array} \tag{14-12}$$

Hence, if no overflow occurs,

$$Q_r^b(x) = (s\,.\,m_1 \quad m_2 \quad \cdots \quad m_b)_{2\text{smns}} \tag{14-13}$$

Now let us examine the error e_r introduced by the rounding process.

$$e_r = Q_r^b(x) - x$$

First, we may state that

$$x + 2^{-b-1} = (s\,.\,m_1 \quad m_2 \quad \cdots \quad m_b \quad n_{b+1} \oplus 1 \quad n_{b+2} \quad n_{b+3} \quad \cdots)_2$$

or

$$x + 2^{-b-1} = Q_r^b(x) + 2^{-b}(.\,n_{b+1} \oplus 1 \quad n_{b+2} \quad n_{b+3} \quad \cdots)_2$$

Thus

$$\underbrace{Q_r^b(x) - x}_{e_r} - 2^{-b-1} = -2^{-b}(.\,n_{b+1} \oplus 1 \quad n_{b+2} \quad n_{b+3} \quad \cdots)_2 \tag{14-14}$$

But the binary part of the right-hand side is bounded by

$$0 \leq (.\, n_{b+1} \oplus 1 \quad n_{b+2} \quad n_{b+3} \quad \cdots)_2 < 1 \tag{14-15}$$

and may assume any real number in the interval, so that

$$0 \geq e_r - 2^{-b-1} > -2^{-b} \tag{14-16}$$

Consequently,

$$\frac{2^{-b}}{2} \geq e_r > \frac{-2^{-b}}{2} \tag{14-17}$$

That is, the rounding error e_r is bounded by one-half the least significant bit. The rounding quantizer transfer characteristic is depicted in Figure 14-1b. If the number x may have any number in the range $0 \leq |x| \leq 1$, then e_r may be assumed to have a uniform probability density function as shown in Figure 14-2b. Hence e_r is considered to be a random variable. We may calculate the noise variance as follows:

$$\sigma_{e_r}^2 = E[e_r^2] - E^2[e_r] \tag{14-18}$$

where $E[e_r]$ is the expected value of e_r and

$$E[e_r^k] = \int_{-\infty}^{\infty} e_r^k p(e_r) de_r \tag{14-19}$$

Therefore,

$$\begin{aligned} E[e_r] &= \int_{-2^{-b}/2}^{2^{-b}/2} e_r \cdot 2^b \, de_r \\ &= \frac{e_r^2}{2} \cdot 2^{-b} \Bigg|_{-2^{-b}/2}^{2^{-b}/2} = 0 \\ E[e_r^2] &= \int_{-2^{-b}/2}^{2^{-b}/2} e_r^2 \cdot 2^b \, de_r \\ &= \frac{e_r^3}{3} \cdot 2^b \Bigg|_{-2^{-b}/2}^{2^{-b}/2} = \frac{2 \cdot 2^b}{3} \left(\frac{2^{-b}}{2}\right)^3 \\ &= \frac{2^{-2b}}{12} \end{aligned} \tag{14-20}$$

Hence

$$\sigma_{e_r}^2 = \frac{2^{-2b}}{12} \tag{14-21}$$

We note that

$$\sigma_{e_t}^2 = \tfrac{5}{2}\, \sigma_{e_r}^2 \tag{14-22}$$

Consequently, round-off gives about 1.6 times as much accuracy as truncation using the same number of bits.

Finally, let us examine the *LSB-1* quantization case. This quantizer simply forces the *l*east *s*ignificant *b*it to a value *1* in all numbers $Q_l^b(x)$. So if

$$x = (s.m_1 \quad m_2 \quad \cdots \quad m_{b-1} \quad m_b \quad m_{b+1} \quad \cdots)_{2\text{smns}} \tag{14-23}$$

then

$$Q_l^b(x) = (s.m_1 \quad m_2 \quad \cdots \quad m_{b-1} \quad 1 \quad 0 \quad \cdots)_{2\text{smns}} \tag{14-24}$$

Thus, the quantizing error is

$$\begin{aligned} e_l &= Q_l^b(x) - x \\ &= 2^{-b+1}(\tfrac{1}{2} - (.\, m_b \quad m_{b+1} \quad \cdots)_2) \end{aligned} \tag{14-25}$$

Since

$$0 \leq (.\, m_b \quad m_{b+1} \quad \cdots)_2 < 1$$

Then

$$-\tfrac{1}{2} < \tfrac{1}{2} - (.\, m_b \quad m_{b+1} \quad \cdots)_2 \leq \tfrac{1}{2} \tag{14-26}$$

Consequently, the error is bounded by

$$-2^{-b} < e_l \leq 2^{-b} \tag{14-27}$$

The quantizer characteristics and probability density function are illustrated in Figures 14-1c and 14-2c, respectively. From Figure 14-2c we may find that the noise variance in this case is

$$\sigma_{e_l}^2 = E[e_l^2] - E^2[e_l] \tag{14-28}$$

But

$$E[e_l] = \int_{-2^{-b}}^{2} e_l \left(\frac{2^b}{2}\right) de_l = \frac{2^b}{2} \frac{e_l^2}{2} \bigg|_{-2^{-b}}^{2^{-b}} = 0$$

$$E[e_l^2] = \int_{-2^{-b}}^{2^b} e_l^2 \left(\frac{2^b}{2}\right) de_l = \frac{2^b}{2} \frac{e_l^3}{3} \bigg|_{-2^{-b}}^{2^{-b}} \tag{14-29}$$

$$= \frac{2^b}{2}\left[\frac{2^{-3b}}{3} + \frac{2^{-3b}}{3}\right] = \frac{2^{-2b}}{3}$$

Hence

$$\sigma_{e_l}^2 = \frac{2^{-2b}}{3} \tag{14-30}$$

and

$$\sigma_{e_l}^2 = 4\sigma_{e_r}^2 \tag{14-31}$$

So, rounding yields twice as much accuracy as LSB-1 quantizing for the same number of bits.

As an interesting exercise, let us examine the relationship between the three quantizers Q_r^b, Q_t^b, and Q_l^b more closely.

In examining the round-off quantizer we actually calculated its error by using the truncation quantizer after adding a term $2^{-b}/2$. Or

$$Q_r^b(x) = Q_t^b\left(x + (1 - 2s)\frac{2^{-b}}{2}\right) \tag{14-32}$$

where the superscript has been added to indicate b bits. We may also express Q_l^b as

$$Q_l^b(x) = Q_t^{b-1}(x) + (1 - 2s)2^{-b} \tag{14-33}$$

A graphical interpretation of these expressions appears in Figure 14-3. In the figure the hexagons indicate a nonlinear transfer characteristic, as shown in Figure 14-1.

Next let us consider the dynamic range of the signed-magnitude number system. If we define *dynamic range* as

$$\text{D.R.} = \frac{\text{largest magnitude } (|\, Q^b(x)\,|_{\max})}{\text{smallest nonzero magnitude } (|\, Q^b(x)\,|_{\min} \neq 0)} \tag{14-34}$$

then

$$(\text{D.R.})_{2\text{smns}} = \frac{1 - 2^{-b}}{2^{-b}}$$

$$= -2^b - 1 \tag{14-35}$$

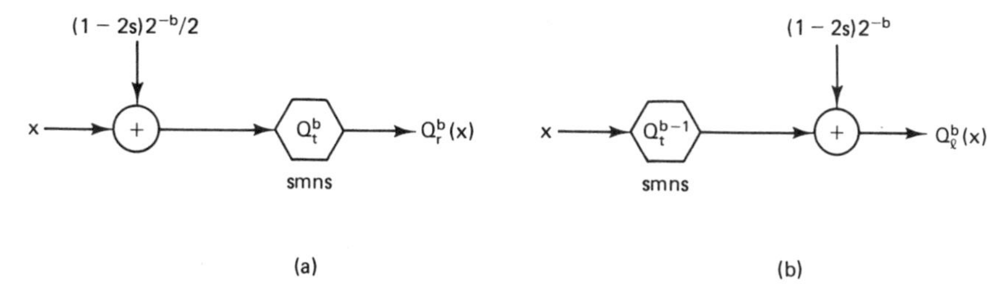

Figure 14-3 Equivalent quantizers: (a) round-off quantizer; (b) LSB-1 quantizer.

and numbers must fall in the range

$$-(2^b - 1) \leq 2^b \cdot Q^b(x) \leq (2^b - 1) \tag{14-36}$$

Suppose that we add two numbers in the binary signed-magnitude number system

$$\begin{array}{r} Q^b(x_1) = (s_1 . m_{11} \quad m_{12} \quad \cdots \quad m_{1b})_{2\text{smns}} \\ \underline{+Q^b(x_2) = (s_2 . m_{21} \quad m_{22} \quad \cdots \quad m_{2b})_{2\text{smns}}} \\ Q^b(x_3) = (s_3 . m_{31} \quad m_{32} \quad \cdots \quad m_{3b})_{2\text{smns}} \end{array} \tag{14-37}$$

If s_1 and s_2 are different, then

$$\begin{aligned} s_3 &= s_1 \qquad \text{if } |Q^b(x_1)| \geq |Q^b(x_2)| \\ s_3 &= s_2 \qquad \text{if } |Q^b(x_2)| > |Q^b(x_1)| \end{aligned} \tag{14-38}$$

and

$$|Q^b(x_3)| = \big||Q^b(x_1)| - |Q^b(x_2)|\big| \tag{14-39}$$

But if s_1 and s_2 are the same, then

$$s_3 = s_1 = s_2$$

and

$$|Q^b(x_3)| = |Q^b(x_1)| + |Q^b(x_2)| \tag{14-40}$$

and $|Q^b(x_3)|$ may exceed the range of the number system. If the addition of the number magnitudes is accomplished using a binary adder and the overflow bit is ignored, the resulting *overflow* characteristics for the signed-magnitude number system is given in Figure 14-4. This nonlinear overflow characteristic is important in analyzing the closed-loop effects of a digital filter's large-signal behavior in a discrete control system.

Two's-Complement Number System

The two's-complement number system is the number system used in most digital computers, and hence is commonly used to implement digital filters. Numbers are represented as

$$\begin{aligned} Q^b(x) &= (0 . m_1 \quad m_2 \quad \cdots \quad m_b)_{2\text{cns}}, \qquad 0 \leq x < 1 \\ &= (1 . n_1 \quad n_2 \quad \cdots \quad n_b)_{2\text{cns}}, \qquad -1 \leq x < 0 \end{aligned} \tag{14-41}$$

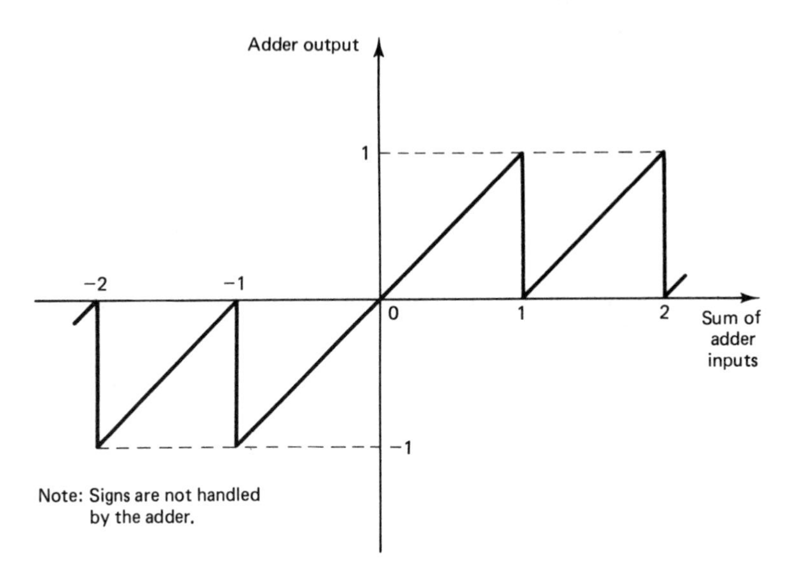

Figure 14-4 Signed-magnitude number system overflow characteristic.

where, for $x \geq 0$,

$$(.m_1 \quad m_2 \quad \cdots \quad m_b)_2 = |Q^b(x)| = \sum_{i=1}^{b} m_i \cdot 2^{-i} \tag{14-42}$$

as in the signed-magnitude number system, but for $x < 0$,

$$\begin{aligned}(.n_1 \quad n_2 \quad \cdots \quad n_b)_2 &= \text{two's complement of } |Q(x)| \\ &= 1.0 - |Q^b(x)| \\ &= 1.0 - \sum_{i=1}^{b} m_i \cdot 2^{-i}\end{aligned} \tag{14-43}$$

In series form

$$\begin{aligned}Q^b(x) &= 2.0 - \sum_{i=0}^{b} m_i \cdot 2^{-i}, \qquad x < 0 \\ Q^b(x) &= \sum_{i=0}^{b} m_i \cdot 2^{-i}, \qquad x \geq 0\end{aligned} \tag{14-44}$$

where m_0 is the sign bit. Here we should emphasize that positive numbers are identical to the signed-magnitude results presented earlier.

First let us consider the action of a truncation quantizer in generating $Q_t^b(x)$. For positive numbers we may use the results of the signed-magnitude analysis.

$$0 \geq e_t > -2^{-b}, \qquad x \geq 0 \tag{14-45}$$

For the negative case, however, the result is *not* identical with the signed-magnitude case.

Since

$$\begin{aligned}x &= (1.n_1 \quad n_2 \quad \cdots \quad n_b \quad n_{b+1} \quad \cdots)_{2cns} \\ Q_t^b(x) &= (1.n_1 \quad n_2 \quad \cdots \quad n_b \quad 0 \quad \cdots)_{2cns}\end{aligned} \tag{14-46}$$

Then

$$Q_t^b(x) - x = -2^{-b}(0.n_{b+1} \quad n_{b+2} \quad \cdots)_{2\text{cns}}$$

or

$$e_t = -2^{-b} \sum_{i=1}^{\infty} n_{b+i} \cdot 2^{-i} \tag{14-47}$$

Because

$$0 \leq \sum_{i=1}^{\infty} n_{b+i} \cdot 2^{-i} < 1 \tag{14-48}$$

Then

$$0 \geq e_t > -2^{-b}, \qquad x < 0 \tag{14-49}$$

which was the result for positive x. Hence for all x,

$$0 \geq e_t > -2^{-b} \tag{14-50}$$

The quantization characteristic and probability density function for the truncation error are shown in Figures 14-5a and 14-6a, respectively. Notice that the mean value will be nonzero:

$$\begin{aligned} E[e_t] &= \int_{-\infty}^{\infty} e_t p(e_t)\, de_t \\ &= \int_{-2^{-b}}^{0} e_t \cdot 2^b\, de_t = 2^b \frac{e_t^2}{2}\bigg|_{-2^{-b}}^{0} \\ &= -2^b \frac{2^{-2b}}{2} = \frac{-2^{-b}}{2} \end{aligned} \tag{14-51}$$

The expected value of e_t^2 is

$$\begin{aligned} E[e_t^2] &= \int_{-\infty}^{\infty} e_t^2 p(e_t)\, de_t \\ &= \int_{-2^{-b}}^{0} e_t^2 \cdot 2^b\, de_t = 2^b \frac{e_t^3}{3}\bigg|_{-2^{-b}}^{0} \\ &= 2^b \frac{2^{-3b}}{3} = \frac{2^{-2b}}{3} \end{aligned} \tag{14-52}$$

And we may calculate the variance

$$\begin{aligned} \sigma_{e_t}^2 &= E[e_t^2] - E^2[e_t] \\ &= \frac{2^{-2b}}{3} - \left(-\frac{2^{-b}}{2}\right)^2 = 2^{-2b}\left(\frac{1}{3} - \frac{1}{4}\right) \\ &= \frac{2^{-2b}}{12} \end{aligned} \tag{14-53}$$

Since two's-complement truncation has a nonzero mean, it can introduce dc biasing errors into the digital filter output. The use of this quantizer in a closed-loop control system requires that its effect be analyzed carefully in the closed-loop case.

Now suppose that we consider the case of the round-off quantizer in determining $Q_r^b(x)$. For positive numbers, round-off in the two's-complement number system duplicates the signed-magnitude case because the numbers have the same representation. Hence, from the signed-magnitude case, we have

$$\frac{2^{-b}}{2} \geq e_r > -\frac{2^{-b}}{2} \tag{14-54}$$

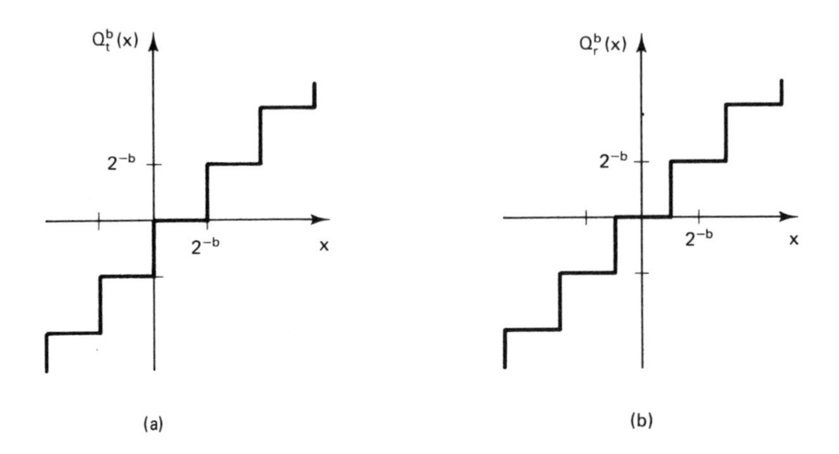

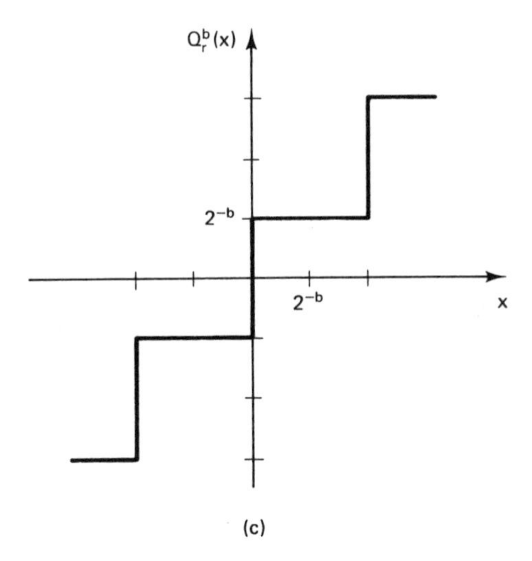

Figure 14-5 Quantizer characteristics for the two's-complement number system: (a) truncation; (b) round-off; (c) LSB-1.

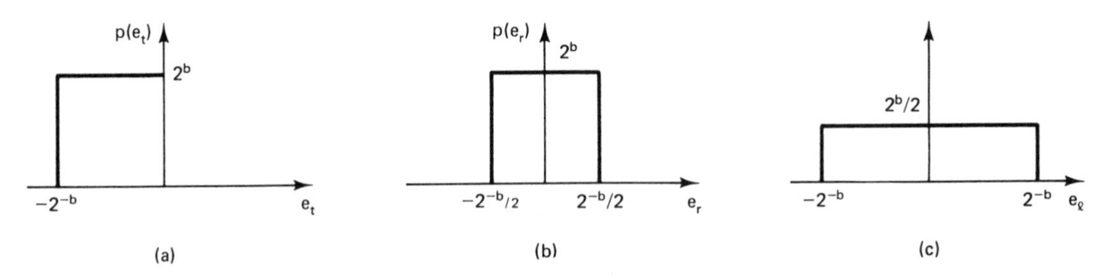

Figure 14-6 Quantizer error probability density functions: (a) truncation; (b) round-off; (c) LSB-1.

For the negative case, let

$$x = 2.0 - \sum_{i=0}^{\infty} n_i \cdot 2^{-i} \tag{14-55}$$

Recalling our development for the signed-magnitude case,

$$\begin{array}{rl} x = & (1.n_1 \quad n_2 \quad \cdots \quad n_b \quad n_{b+1} \qquad n_{b+2} \quad \cdots)_{\text{2cns}} \\ +\,2^{-b-1} = & (0.0 \quad 0 \quad \cdots \quad 0 \quad 1 \qquad 0 \quad \cdots)_{\text{2cns}} \\ \hline x + 2^{-b-1} = & (\underbrace{1.k_1 \quad k_2 \quad \cdots \quad k_b}_{Q_r^b(x)} \quad n_{b+1} \oplus 1 \quad n_{b+2} \quad \cdots)_{\text{2cns}} \end{array} \tag{14-56}$$

and

$$x + 2^{-b-1} = Q_r^b(x) + 2^{-b}(0.n_{b+1} \oplus 1 \quad n_{b+2} \quad \cdots)_{\text{2cns}} \tag{14-57}$$

But

$$0 \leq (0.n_{b+1} \oplus 1 \quad n_{b+2} \quad \cdots)_{\text{2cns}} < 1 \tag{14-58}$$

then

$$\frac{2^{-b}}{2} \geq e_r > -\frac{2^{-b}}{2}, \qquad x < 0 \tag{14-59}$$

Consequently, round-off for the two's-complement number system is the same as for signed-magnitude numbers, and the quantizer characteristic of Figure 14-5b holds. Therefore, the error distribution may be assumed to have the form of Figure 14-6b and

$$\sigma_{e_r}^2 = \frac{2^{-2b}}{12} \tag{14-60}$$

Finally, let us examine the LSB-1 quantizer for the two's-complement number system. Positive results are the same as the signed-magnitude case. Since negative numbers are expressed

$$x = (1.n_1 n_2 \quad \cdots \quad n_{b-1} \quad n_b \quad n_{b+1} \quad \cdots)_{\text{2cns}}, \qquad x < 0 \tag{14-61}$$

Then

$$x = -\left(2.0 - \sum_{i=0}^{\infty} n_i \cdot 2^{-i}\right)$$

$$Q_l^b(x) = -\left(2.0 - 2^{-b} - \sum_{i=0}^{b-1} n_i \cdot 2^{-i}\right) \tag{14-62}$$

Hence

$$\begin{aligned} e_l &= -2.0 + 2^{-b} + \sum_{i=0}^{b-1} n_i \cdot 2^{-i} + 2.0 - \sum_{i=0}^{\infty} n_i \cdot 2^{-i} \\ &= 2^{-b} - \sum_{i=b}^{\infty} n_i \cdot 2^{-i} \end{aligned} \tag{14-63}$$

But if we let $k = i - b$,

$$\begin{aligned} e_l &= 2^{-b} - \sum_{k=0}^{\infty} n_{b+k} \cdot 2^{-b-k} \\ &= 2^{-b}\left(1 - \sum_{k=0}^{\infty} n_{b+k} \cdot 2^{-k}\right) \end{aligned} \tag{14-64}$$

Since

$$0 \leq \sum_{k=0}^{\infty} n_{b+k} \cdot 2^{-k} < 2.0 \tag{14-65}$$

then

$$-1 < 1 - \sum_{k=0}^{\infty} n_{b+k} \cdot 2^{-k} \leq 1 \tag{14-66}$$

so that

$$-2^{-b} < e_l \leq 2^{-b}, \qquad x < 0 \tag{14-67}$$

which was our result for positive x and the relation thus holds for all x.

The quantizer characteristic is demonstrated in Figure 14-5c and the probability density function for e_l is plotted in Figure 14-6c. These curves were the same ones for the signed-magnitude case, so

$$\sigma_{e_l}^2 = \frac{2^{-2b}}{3} \tag{14-68}$$

We may compare the three quantizers by examining Figure 14-7. Here we have used the truncation quantizer to implement round-off and LSB-1 quantizers, because

$$Q_r^b(x) = Q_t^b\left(x + \frac{2^{-b}}{2}\right) \tag{14-69}$$

and

$$Q_l^b(x) = Q_t^{b-1}(x) + 2^{-b} \tag{14-70}$$

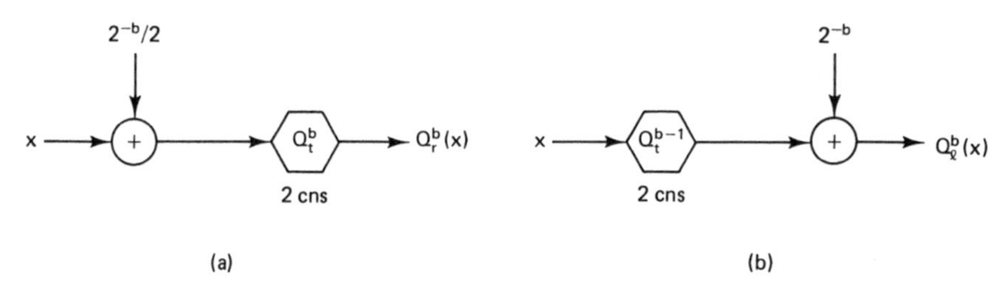

Figure 14-7 Equivalent quantizers for the two's-complement number system: (a) round-off quantizer; (b) LSB-1 quantizer.

Consider next the dynamic range of the two's-complement number system. Since

$$-1 \leq Q^b(x) \leq 1 - 2^{-b} \tag{14-71}$$

the dynamic range is given by

$$\begin{aligned}(\text{D.R.})_{2\text{cns}} &= \frac{|Q^b(x)|_{\max}}{|Q^b(x)|_{\min} \neq 0} \\ &= \frac{|-1|}{|2^{-b}|} \\ &= 2^b\end{aligned} \tag{14-72}$$

Finally, let us examine the overflow properties of the two's-complement number system. If we add two numbers to generate a third:

$$\begin{array}{rl} Q^b(x_1) = (m_{10} \,.\, m_{11} & m_{12} \quad \cdots \quad m_{1b})_{2\text{cns}} \\ + \; Q^b(x_2) = (m_{20} \,.\, m_{21} & m_{22} \quad \cdots \quad m_{2b})_{2\text{cns}} \\ \hline Q^b(x_3) = (m_{30} \,.\, m_{31} & m_{32} \quad \cdots \quad m_{3b})_{2\text{cns}} \end{array} \tag{14-73}$$

↑ sign position

the binary sum is calculated and the signs are automatically handled by the adder circuits. Any carry bit into the 2^1 position is ignored; this happens when two negative numbers are added. Overflow occurs when two positive numbers are added and a carry bit enters the sign position (2^0 position), or when two negative numbers are added and the carry bit into the sign position is absent. Hence the overflow characteristic for the two's-complement number system is shown in Figure 14-8.

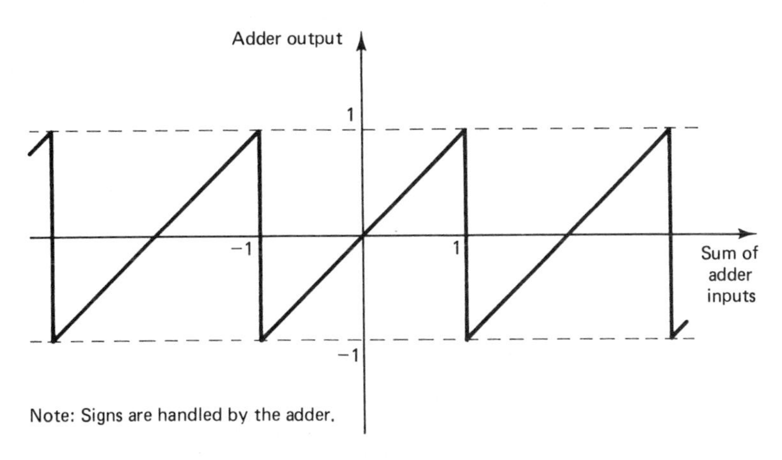

Figure 14-8 Two's-complement number system overflow characteristic.

14.3 COEFFICIENT QUANTIZATION

One effect of finite wordlengths in digital computers is that the filter's parameters, or coefficients, must be chosen from a finite set of allowable values. Classical design procedures yield filter transfer functions with coefficients of arbitrary precision which must be altered for implementation using digital computing devices. One approach to this problem is to select a filter structure that is not sensitive to coefficient inaccuracies. For example, realizing a filter directly allows a greater chance for instability than cascading or paralleling second-order modules because it is well known that the roots of polynomials become more sensitive to parameter changes as the order of the polynomial increases.

Pole/Zero Locations

Quantizing the coefficients of a digital filter effectively restricts the poles and zeros of the filter to lie on a finite number of discrete points in the z-plane. Consider the second-order filter

$$D(z) = \frac{a_0 + a_1 z^{-1} + a_2 z^{-2}}{1 + b_1 z^{-1} + b_2 z^{-2}} \tag{14-74}$$

implemented in the 1D structure (see Figure 14-9a). Since zeros cannot cause instability, let us examine only the poles. The poles are given

$$\begin{aligned}(z - re^{j\theta})(z - re^{-j\theta}) &= z^2 - (2r\cos\theta)z + r^2 \\ &= z^2 + b_1 z + b_2\end{aligned} \tag{14-75}$$

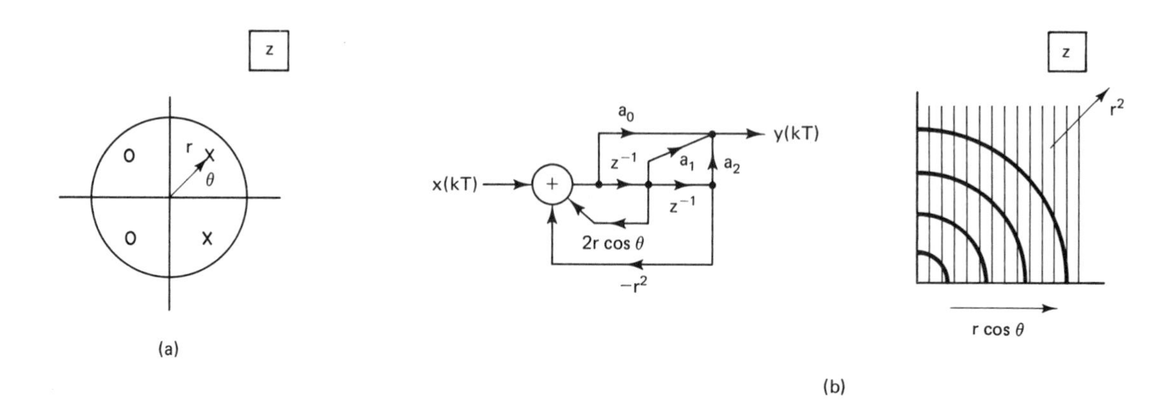

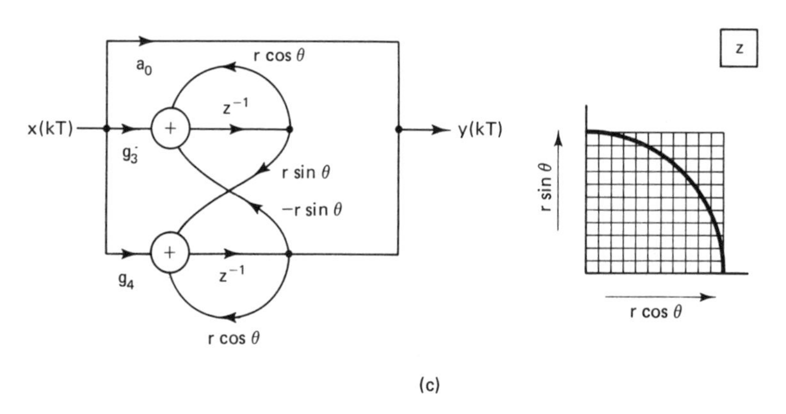

Figure 14-9 Coefficient quantization: (a) z-plane; (b) 1D structure; (c) 1X structure.

Hence

$$b_1 = -2r\cos\theta, \qquad b_2 = r^2$$

Suppose that we quantize b_1 to b_1^q and b_2 to b_2^q. Figure 14-9b illustrates that quantizing b_1 (i.e., $2r\cos\theta$) restricts the poles to lie on a finite number of vertical lines, $r\cos\theta = b_1^q/2$, while quantizing b_2 (i.e., r^2) further restricts the poles to lie on circles of radius $r = \sqrt{b_2^q}$.

Compare the 1D structure with the 1X structure of Figure 14-9c. The coefficients g_i are defined in Chapter 12. The poles are defined by

$$g_1 = r\cos\theta$$
$$g_2 = r\sin\theta$$

If we quantize g_1 to g_1^q and g_2 to g_2^q, then

$$Q(r\cos\theta) = g_1^q$$

and

$$Q(r\sin\theta) = g_2^q$$

restricts the poles to lie on a rectangular grid.

Consequently, one can see that the 1D structure is better suited to implement digital filters with poles near the unit circle, whereas the 1X structure gives a more uniform pattern of realizable locations throughout the unit circle.

Error Analysis

Sensitivity analysis may be employed to determine the effect of coefficient quantization on either the pole migration or change in transfer characteristic. Let us examine pole migration in the 1D filter structure. From Figure 14-9,

$$\begin{aligned} b_1 &= -2r\cos\theta \\ b_2 &= r^2 \end{aligned} \tag{14-76}$$

The pole migration is given by

$$\begin{aligned} \Delta r &= \frac{\partial r}{\partial b_1}\Delta b_1 + \frac{\partial r}{\partial b_2}\Delta b_2 \\ \Delta\theta &= \frac{\partial\theta}{\partial b_1}\Delta b_1 + \frac{\partial\theta}{\partial b_2}\Delta b_2 \end{aligned} \tag{14-77}$$

Using the relationship above,

$$\begin{aligned} \frac{\partial r}{\partial b_1} &= -\frac{1}{2\cos\theta} \\ \frac{\partial r}{\partial b_2} &= \frac{1}{2r} \\ \frac{\partial\theta}{\partial b_1} &= \frac{1}{2r\sin\theta} \\ \frac{\partial\theta}{\partial b_2} &= \frac{1}{2r^2\tan\theta} \end{aligned} \tag{14-78}$$

Consequently,

$$\begin{aligned} \Delta r &= \frac{-\Delta b_1}{2\cos\theta} + \frac{\Delta b_2}{2r} \\ \Delta\theta &= \frac{\Delta b_1}{2r\sin\theta} + \frac{\Delta b_2}{2r^2\tan\theta} \end{aligned} \tag{14-79}$$

These relations show us that, for a given r, as θ goes to zero, $\Delta\theta$ approaches infinity. Similarly for a constant θ, as r goes to zero, $\Delta\theta$ and Δr both go to infinity. These results agree with the grid pattern depicted in Figure 14-9b.

Consider now the effect of coefficient quantization on the transfer function of the digital filter [1] (consider Figure 14-10). The point a is the input mode for the filter; point b, the output mode. Points n and m are two arbitrary internal nodes and

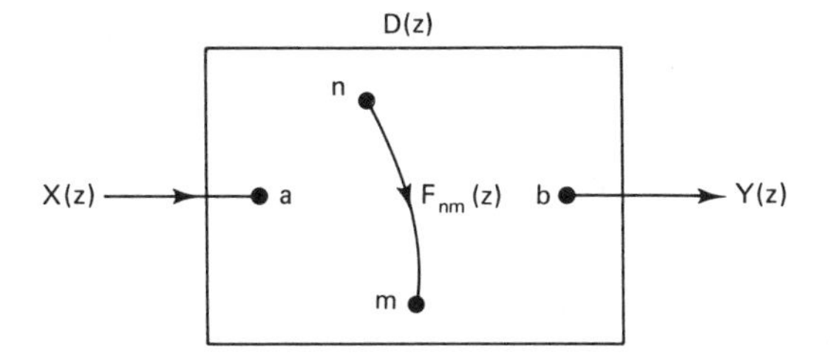

Figure 14-10 Generalized filter.

$F_{nm}(z)$ is the transfer function of a branch of the network from node n to node m. If we define $T_{ij}(z)$ to be the transfer function between any two network nodes i and j, then the sensitivity of $D(z)$ with respect to $F_{nm}(z)$ is given by

$$\frac{\partial D(z)}{\partial F_{nm}(z)} = T_{an}(z)T_{mb}(z) \tag{14-80}$$

the product of two transfer functions within the network [2]. If the network branch is a multiplicative coefficient

$$F_{nm}(z) = f_{cnm}$$

then

$$\frac{\partial |D(z)|}{\partial f_{cnm}} = \mathrm{Re}\left[\frac{|D(z)|}{D(z)} T_{an}(z)T_{mb}(z)\right] \tag{14-81}$$

If however, the branch includes a time-delay element

$$F_{nm}(z) = f_{dnm}z^{-1}$$

then

$$\frac{\partial |D(z)|}{\partial f_{dnm}} = \mathrm{Re}\left[\frac{|D(z)|}{D(z)} T_{an}(z)T_{mb}(z)z^{-1}\right] \tag{14-82}$$

Both f_{cnm} and f_{dnm} are real coefficients.

The sensitivities above may be used to calculate the change in the filter transfer function $D(z)$ due to a change in coefficient value. Consider the 1X structure of Figure 14-11. We may calculate the sensitivities mathematically:

$$\begin{aligned} D(z) &= \frac{a_0 + g_4z^{-1} - g_2g_3z^{-2}}{1 - 2g_1z^{-1} + (g_1 + g_2)z^{-2}} \\ &= a_0 + \frac{(g_4 - jg_3)/2}{z - g_1 + jg_2} + \frac{(g_4 + jg_3)/2}{z - g_1 - jg_2} \end{aligned} \tag{14-83}$$

and

$$\begin{aligned} \frac{\partial D(z)}{\partial a_0} &= 1 \\ \frac{\partial D(z)}{\partial g_1} &= \frac{-(g_4 - jg_3)/2}{(z - g_1 + jg_2)^2} + \frac{-(g_4 + jg_3)/2}{(z - g_1 - jg_2)^2} \\ \frac{\partial D(z)}{\partial g_2} &= \frac{-j(g_4 - jg_3)/2}{(z - g_1 + jg_2)^2} + \frac{-j(g_4 + jg_3)/2}{(z - g_1 - jg_2)^2} \\ \frac{\partial D(z)}{\partial g_3} &= \frac{-j/2}{z - g_1 + jg_2} + \frac{j/2}{z - g_1 - jg_2} = \frac{g_2}{(z - g_1)^2 + g_2^2} \\ \frac{\partial D(z)}{\partial g_4} &= \frac{1/2}{z - g_1 + jg_2} + \frac{1/2}{z - g_1 - jg_2} = \frac{z - g_1}{(z - g_1)^2 + g_2^2} \end{aligned} \tag{14-84}$$

Finally,

$$\Delta D(z) = \frac{\partial D(z)}{\partial a_0}\Delta a_0 + \sum_{i=1}^{4} \frac{\partial D(z)}{\partial g_i}\Delta g_i \tag{14-85}$$

Recall that we may calculate the sensitivities using transfer functions and (14-80). For example, from Figure 14-11,

$$\frac{\partial D(z)}{\partial g_3} = T_{aa}(z)T_{cb}(z) \tag{14-86}$$

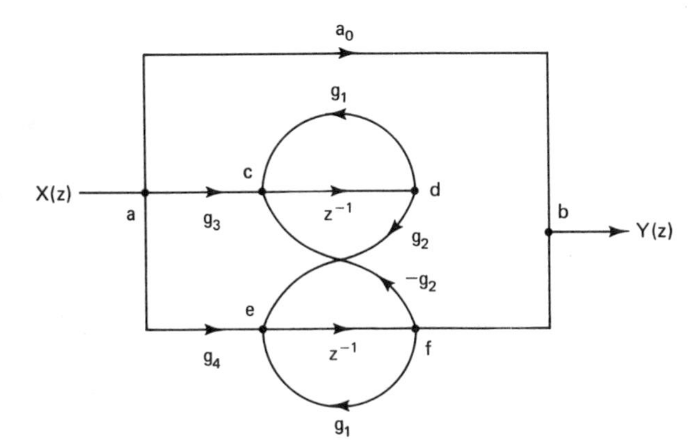

Figure 14-11 1X structure.

But $T_{aa}(z) = 1$. However, using Mason's gain formula,

$$
\begin{aligned}
T_{cb}(z) &= \frac{g_2 z^{-2}}{1 - 2g_1 z^{-1} + g_2^2 z^{-2} + g_1^2 z^{-2}} \\
&= \frac{g_2 z^{-2}}{1 - 2g_1 z^{-1} + g_1^2 z^{-2} + g_2^2 z^{-2}} \\
&= \frac{g_2}{(z - g_1)^2 + g_2^2}
\end{aligned}
\tag{14-87}
$$

Hence

$$\frac{\partial D(z)}{\partial g_3} = \frac{g_2}{(z - g_1)^2 + g_2^2}$$

which confirms our previous result in (14-84).

Observations

When digital filters are used in feedback control systems, the quantization of the filter coefficients can dramatically affect the system closed-loop performance in applications where pole and/or zero placement is critical. If the controller designer employs quantized coefficient values as he originally develops the digital filter transfer function, coefficient sensitivity problems may be avoided from the beginning of the design process.

When a digital filter must be quantized for some reason (say that a change in filter structure is desirable), the quantized filter should be returned to the controller designers to confirm that the resulting closed-loop characteristics (gain margin, phase margin, time response, etc.) are still within system specifications. The authors have found that, in most applications, coefficient quantization is rarely a problem.

14.4 SIGNAL QUANTIZATION ANALYSIS

In the preceding section we examined the effect of quantizing the coefficients of the filter transfer function. In this section we examine the quantization of the digital filter's signal variables, both at the input and internal nodes.

Filter Input Quantization

The input signal to a digital filter may come from an analog-to-digital (A/D) converter or from the output node of some other digital filter module. Consider first the case of the A/D as illustrated in Figure 14-12. The signal $x(t)$ is sampled and quantized into a sequence of time samples $\{Q(x(n))\}$ which are processed by the digital filter. Perhaps the most common A/D type is the bipolar (positive and negative values) successive-approximation converter, whose conversion time is proportional to the number of bits, $(b + 1)$. This converter is a truncation-type quantizer (see Figure 14-5a) whose output is in the two's-complement number system, so we may represent the A/D as shown in Figure 14-12b. If we bias the input by a small signal $2^{-b}/2$, we form the configuration of Figure 14-12c which, by Figure 14-7a, is equivalent to rounding the input values (Figure 14-12d). This is the A/D model which we will use in our results throughout this chapter. The A/D conversion introduces round-off noise into the digital filter as modeled in Figure 14-12e, where

$$e_r(n) = Q_r^b(x(n)) - x(n) \tag{14-88}$$

is the value of the round-off noise at time nT. The round-off noise is assumed to be uniformly distributed as shown in Figure 14-6b, with a variance of

$$\sigma_{e_r}^2 = \frac{2^{-2b}}{12} \tag{14-89}$$

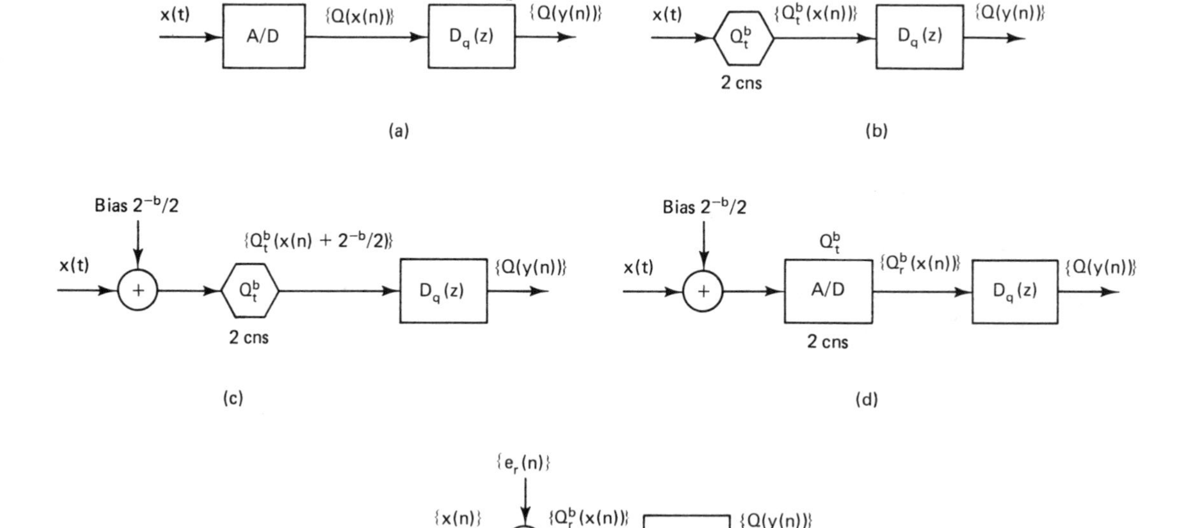

Figure 14-12 (a) General A/D; (b) successive approximation A/D; (c) biased successive approximation A/D; (d) A/D model; (e) noise model.

We also see that the noise is bounded by

$$\frac{2^{-b}}{2} \geq e_r > -\frac{2^{-b}}{2} \tag{14-90}$$

Another important point must be made about the successive approximation A/D of Figure 14-12d. This A/D usually exhibits saturation when the input signal $x(t)$ exceeds its dynamic range. Its overflow characteristic differs from the two's-complement addition characteristic of Figure 14-8. The A/D overflow characteristic is modeled in Figure 14-13.

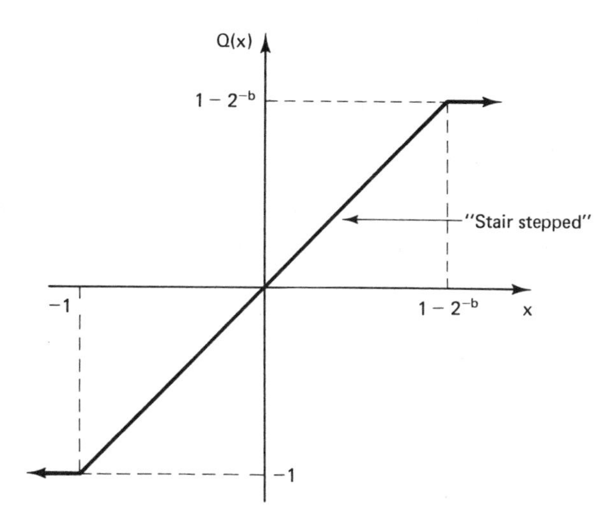

Figure 14-13 A/D overflow.

Internal Variable Quantization

In Chapter 12 we examined many different structures for digital filters. The internal nodes were always formed by summing product terms, with each product being generated by a filter coefficient and a signal variable. If $v_i(n)$ represents the internal variables and c_i the filter coefficients, an internal node $v_k(n)$ is generated by

$$v_k(n) = \sum_{i=1}^{L} c_i v_i(n) \tag{14-91}$$

the sum of several, say L, product terms. The process is illustrated in Figure 14-14a. This represents the ideal case with no quantization. If, however, the coefficients (c_i) are quantized (c_i^q) as discussed in the preceding section, and if the variables are represented in a finite-wordlength number system, say two's complement, the physically realizable case of Figure 14-14b results. Note that if the coefficients are quantized to a bits and the variables are quantized to b bits, the product terms $c_i^q Q^b(v_i(n))$ will have $a + b$ bits. The product terms may then be quantized to b bits by the quantizers labeled Q_1, or they may be summed in their entirety ($a + b$ bits) and the resulting sum quantized to b bits by quantizer Q_2. The choice of quantizing at location Q_1 versus location Q_2 must be made by evaluating the hardware required to compute $a + b$ bits versus the improved quantization noise performance. The noise model is depicted in Figure 14-14c. Note we have simplified the notation setting $v_i^q(n) = Q^b(v_i(n))$.

Assuming that

$$\sigma_{e_1}^2 = \sigma_{e_2}^2 \tag{14-92}$$

quantization at point Q_1 generates a noise variance, σ_ϵ^2, at $v_k^q(n)$ of

$$\sigma_\epsilon^2 = \sum_{i=1}^{L} \sigma_{e_1}^2 = L\sigma_{e_1}^2 = L\sigma_{e_2}^2 \tag{14-93}$$

whereas if quantization is delayed until point Q_2,

$$\sigma_\epsilon^2 = \sigma_{e_2}^2 \tag{14-94}$$

In other words, quantization at point Q_1 is L times as noisy as point Q_2. The final noise model is shown in Figure 14-14d.

Now let us examine the nature of the quantization error distributions for the error sources e_1 and e_2. Examine the product

$$\begin{array}{rl} v_i^q = & (m_{10}.m_{11} \quad m_{12} \quad \cdots \quad m_{1b})_{2cns} \\ \times\ c_i^q = & (m_{20}.m_{21} \quad m_{22} \quad \cdots \quad m_{2a})_{2cns} \\ \hline c_i^q v_i^q = & (m_{30}.m_{31} \quad m_{32} \quad \cdots \quad m_{3b} \quad m_{3,b+1} \quad \cdots \quad m_{3,b+a})_{2cns} \end{array} \tag{14-95}$$

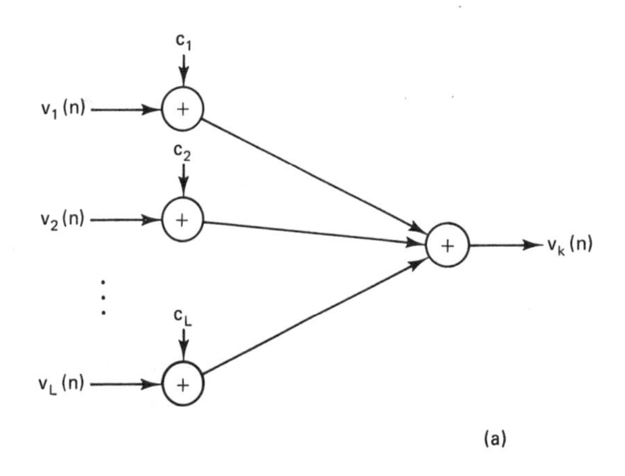

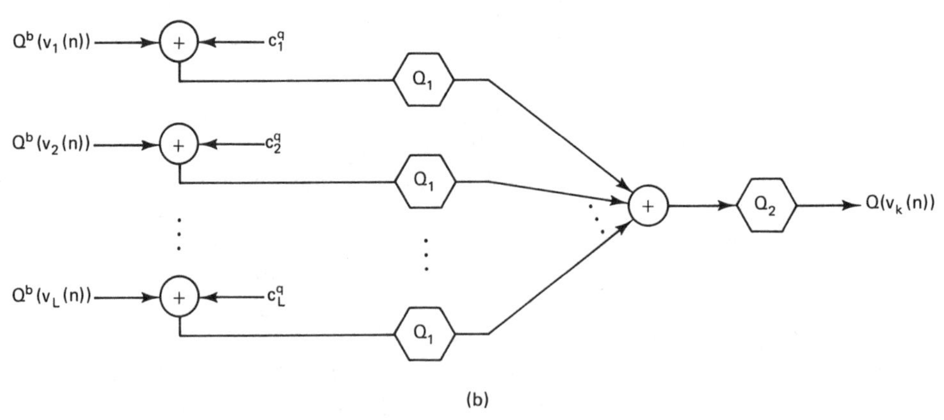

Figure 14-14 (a) Ideal case; (b) physically realizable case; (c) quantization noise model; (d) equivalent noise model.

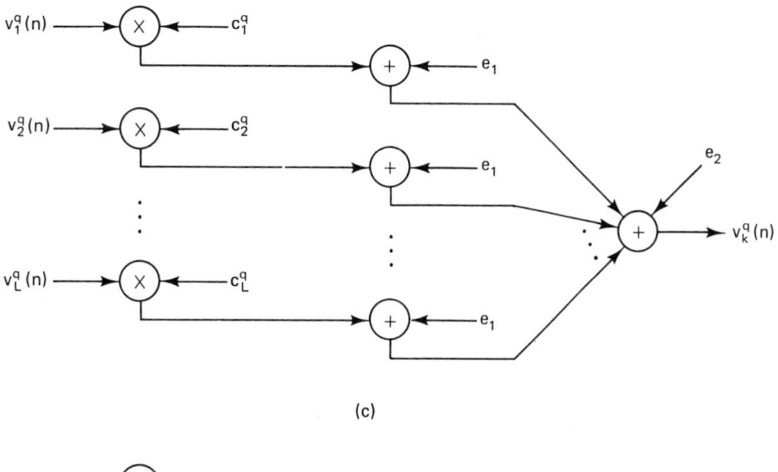

(c)

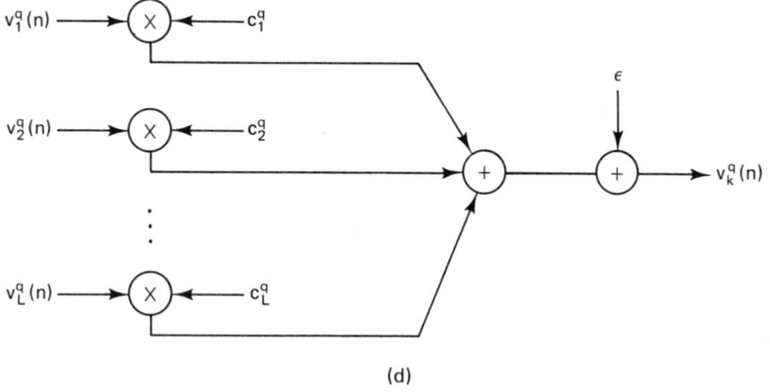

(d)

Figure 14-14 cont.

When this product is quantized (say round-off is used) to b bits, then

$$Q_r^b(c_i^q v_i^q) = (m_{40}.m_{41} \quad m_{42} \quad \cdots \quad m_{4b})_{2\text{cns}} \tag{14-96}$$

the resulting round-off error may be expressed as

$$e_r = Q_r^b(c_i^q v_i^q) - c_i^q v_i^q \tag{14-97}$$

But

$$\begin{array}{rl} c_i^q v_i^q = & (m_{30} \,.\, m_{31} \quad m_{32} \quad \cdots \quad m_{3b} \quad m_{3,b+1} \quad\quad m_{3,b+2} \quad \cdots \quad m_{3,b+a})_{2\text{cns}} \\ + \, 2^{-b-1} = & (0 \quad .\, 0 \quad 0 \quad \cdots \quad 0 \quad 1 \quad\quad 0 \quad \cdots \quad 0 \quad)_{2\text{cns}} \\ \hline 2^{-b-1} + c_i^q v_i^q = & (\underbrace{m_{40} \,.\, m_{41} \quad m_{42} \quad \cdots \quad m_{4b}}_{Q_r^b(c_i^q v_i^q)} \quad m_{3,b+1} \oplus 1 \quad m_{3,b+2} \quad \cdots \quad m_{3,b+a})_{2\text{cns}} \end{array} \tag{14-98}$$

Hence

$$\begin{aligned} e_r &= 2^{-b-1} - (0.00 \quad \cdots \quad 0 \quad m_{3,b+1} \oplus 1 \quad m_{3,b+2} \quad \cdots \quad m_{3,b+a})_{2\text{cns}} \\ &= 2^{-b}(2^{-1} - (.\, m_{3,b+1} \oplus 1 \quad m_{3,b+2} \quad \cdots \quad m_{3,b+a})_2) \end{aligned} \tag{14-99}$$

The binary number may be simplified by letting $m_1 = m_{3,b+1} \oplus 1$ and $m_i = m_{3,b+i}$ for $i = 2, a$, yielding

$$e_r = 2^{-b}(\tfrac{1}{2} - (.m_1 \quad m_2 \quad \cdots \quad m_a)_2) \tag{14-100}$$

Let us examine the binary number further. It is represented by a bits and may take on discrete value in the range

$$0 \leq (.m_1 \quad m_2 \quad \cdots \quad m_a)_2 \leq 1 - 2^{-a} \tag{14-101}$$

Then

$$0 \geq -(.m_1 \quad m_2 \quad \cdots \quad m_a)_2 \geq 1 - 2^{-a}$$

$$\frac{2^{-b}}{2} \geq \tfrac{1}{2} - (.m_1 \quad m_2 \quad \cdots \quad m_a)_2 \geq -(\tfrac{1}{2} - 2^{-a})2^{-b}$$

and hence the bounds on the error are

$$\frac{2^{-b}}{2} \geq e_r \geq -\frac{2^{-b}}{2} + 2^{-b-a} \tag{14-102}$$

But e_r may be expressed

$$\begin{aligned} 2^b e_r &= (.10 \quad \cdots \quad 0)_2 - (.m_1 \quad m_2 \quad \cdots \quad m_a)_2 \\ &= (n_0 . n_1 \quad n_2 \quad \cdots \quad n_a)_{2\text{cns}} \end{aligned} \tag{14-103}$$

and hence its smallest nonzero magnitude is

$$|e_r|_{\min \neq 0} = 2^{-b-a} \tag{14-104}$$

and its probability density function will be discrete and appear as illustrated in Figure 14-15. The density function consists of a series of a different impulse functions, each of weight 2^{-a}.

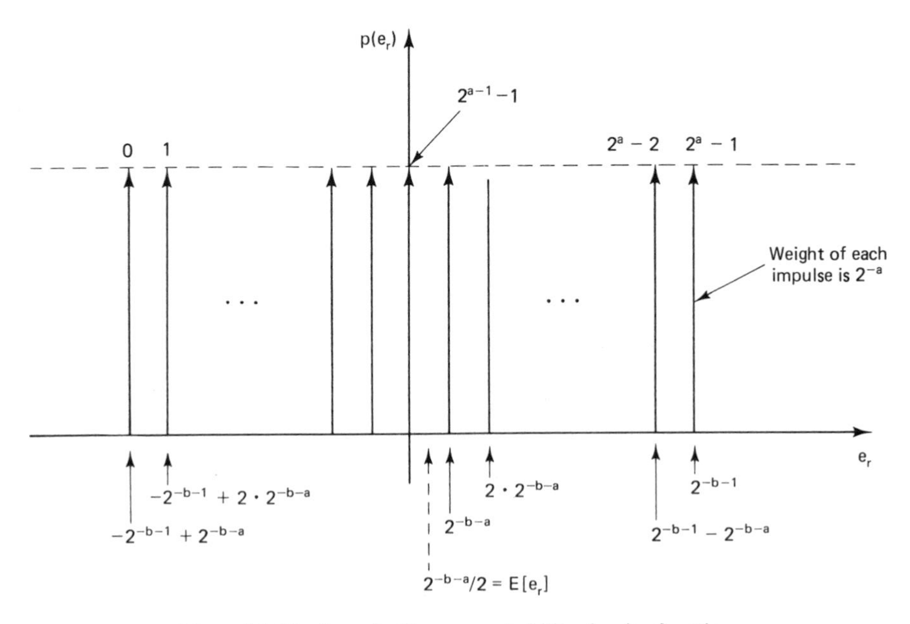

Figure 14-15 Round-off error probability density function.

The probability density function may be expressed as

$$p(e_r) = 2^{-a} \sum_{i=0}^{2^a-1} \delta[e_r - 2^{-b-1}(1 - i2^{-a+1})] \tag{14-105}$$

where $\delta(x)$ is the Dirac delta function. The expected value of e_r is thus

$$\begin{aligned} E[e_r] &= \int_{-\infty}^{\infty} e_r p(e_r)\, de_r \\ &= \int_{-\infty}^{\infty} e_r \left\{2^{-a} \sum_{i=0}^{2^a-1} \delta[e_r - 2^{-b-1}(1 - i2^{-a+1})]\right\} de_r \\ &= 2^{-a} \sum_{i=0}^{2^a-1} \int_{-\infty}^{\infty} e_r \delta[e_r - 2^{-b-1}(i - i2^{-a+1})]\, de_r \end{aligned} \tag{14-106}$$

But we know that

$$\int_{-\infty}^{\infty} x^r \delta(x - y)\, dx = y^r \tag{14-107}$$

so

$$\begin{aligned} E[e_r] &= 2^{-a} \sum_{i=0}^{2^a-1} [2^{-b-1}(1 - i2^{-a+1})] \\ &= 2^{-a-b-1}\left[\sum_{i=0}^{2^a-1} 1 - \sum_{i=0}^{2^a-1} i2^{-a+1}\right] \end{aligned} \tag{14-108}$$

But the sum of integers is

$$\begin{aligned} \sum_{i=0}^{n} 1 &= n + 1 \\ \sum_{i=0}^{n} i &= \frac{n(n+1)}{2} \end{aligned} \tag{14-109}$$

Hence

$$\begin{aligned} E[e_r] &= 2^{-a-b-1}\frac{2^a - 2^{-a+1}(2^a - 1)(2^a)}{2} \\ &= 2^{-a-b-1}[2^a - 2^a + 1] \\ &= \frac{2^{-b-a}}{2} \end{aligned} \tag{14-110}$$

Note this value is indicated on Figure 14-15.

Next let us calculate the expected value of e_r^2.

$$\begin{aligned} E[e_r^2] &= \int_{-\infty}^{\infty} e_r^2 p(e_r)\, de_r \\ &= \int_{-\infty}^{\infty} e_r^2 \left\{2^{-a} \sum_{i=0}^{2^a-1} \delta[e_r - 2^{-b-1}(1 - i2^{-a+1})]\right\} de_r \\ &= 2^{-a} \sum_{i=0}^{2^a-1} \int_{-\infty}^{\infty} e_r^2\, \delta[e_r - 2^{-b-1}(1 - i2^{-a+1})]\, de_r \\ &= 2^{-a} \sum_{i=0}^{2^a-1} [2^{-b-1}(1 - i2^{-a+1})]^2 \\ &= 2^{-a}2^{-2b-2} \sum_{i=0}^{2^a-1} [1 - 2i\, 2^{-a+1} + i^2 2^{-2a+2}] \\ &= 2^{-a-2b-2}\left[\left(\sum_{i=0}^{2^a-1} 1\right) - 2^{-a+2}\left(\sum_{i=0}^{2^a-1} i\right) + 2^{-2a+2}\left(\sum_{i=0}^{2^a-1} i^2\right)\right] \end{aligned} \tag{14-111}$$

But the sum

$$\sum_{i=0}^{n} i^2 = \frac{n(n+1)(2n+1)}{6} \tag{14-112}$$

Consequently,

$$\begin{aligned}
E[e_r^2] &= 2^{-a-2b-2}\left[2^a - \frac{2^{-a+2}(2^a-1)(2^a)}{2} + \frac{2^{-2a+2}(2^a-1)(2^a)(2^{a+1}-2+1)}{6}\right] \\
&= 2^{-a-2b-2}\left[2^a - 2(2^a-1) + \frac{2^{-a+1}(2^a-1)(2^{a+1}-1)}{3}\right] \\
&= 2^{-a-2b-2}\left[2^a - 2\cdot 2^a + 2 + 4\cdot\frac{2^a}{3} - 2 + 2\cdot\frac{2^{-a}}{3}\right] \\
&= 2^{-a}2^{-2b}2^{-2}\left[2^a\left(1-2+\frac{4}{3}\right) + 2^{-a}\left(\frac{2}{3}\right)\right] \\
&= 2^{-2b}2^{-2}\left[\left(\frac{1}{3}\right) + 2^{-2a}\left(\frac{2}{3}\right)\right] \\
&= \left(\frac{2^{-2b}}{12}\right)(1+2^{-2a+1})
\end{aligned} \tag{14-113}$$

The variance of e_r is thus

$$\begin{aligned}
\sigma_{e_r}^2 &= E[e_r^2] - E^2[e_r] \\
&= \left(\frac{2^{-2b}}{12}\right)(1+2^{-2a+1}) - \left(\frac{2^{-b-a}}{2}\right)^2 \\
&= \left(\frac{2^{-2b}}{12}\right)(1+2\cdot 2^{-2a} - 3\cdot 2^{-2a}) \\
&= \left(\frac{2^{-2b}}{12}\right)(1-2^{-2a})
\end{aligned} \tag{14-114}$$

Since in all practical cases, $a > 4$, then

$$2^{-2a} \ll 1 \tag{14-115}$$

and hence

$$\sigma_{e_r}^2 \doteq \frac{2^{-2b}}{12} \tag{14-116}$$

which is the same result for rounding of continuous signals. In the continuous case the expected value of the round-off error is zero. In a typical case, $b = 8$ and $a = 8$, then

$$\begin{aligned}
E[e_r] &= \frac{2^{-b-a}}{2} = 2^{-17} \\
&\doteq 7.62939 \times 10^{-6}
\end{aligned} \tag{14-117}$$

and

$$\begin{aligned}
\sigma_{e_r}^2 &= \frac{2^{-2b}}{12} = \frac{2^{-16}}{12} \\
&= 1.27157 \times 10^{-6} \\
\sigma_{e_r} &= 1.12764 \times 10^{-3} \gg E[e_r]
\end{aligned} \tag{14-118}$$

Consequently, we approximate the round-off of products by using the continuous results derived earlier.

Output Quantization Noise

In the preceding two sections we examined the digital filter input and product-term quantization effects. These quantization errors were modeled as additive input signals in Figures 14-12e and 14-14c. In this section we analyze the effect of these error sources on the output of the digital filter.

Consider the digital filter of Figure 14-16a. This model for a digital filter assumes Q summing junctions, each modeled after Figure 14-14d. In the model $e_0(n)$ represents the filter input quantizer. If we represent the transfer function from the filter input to the output of the ith summing junction by $F_i(z)$, and the transfer function from the output of the ith summing junction to the filter output by $G_i(z)$, we may determine the effect of each individual product term error as shown in Figure 14-16b. Using standard z-transform notation, we have

$$Y^q(z) = X(z)D(z) + E_o(z)D(z) + \sum_{i=1}^{Q} E_i(z)G_i(z) \tag{14-119}$$

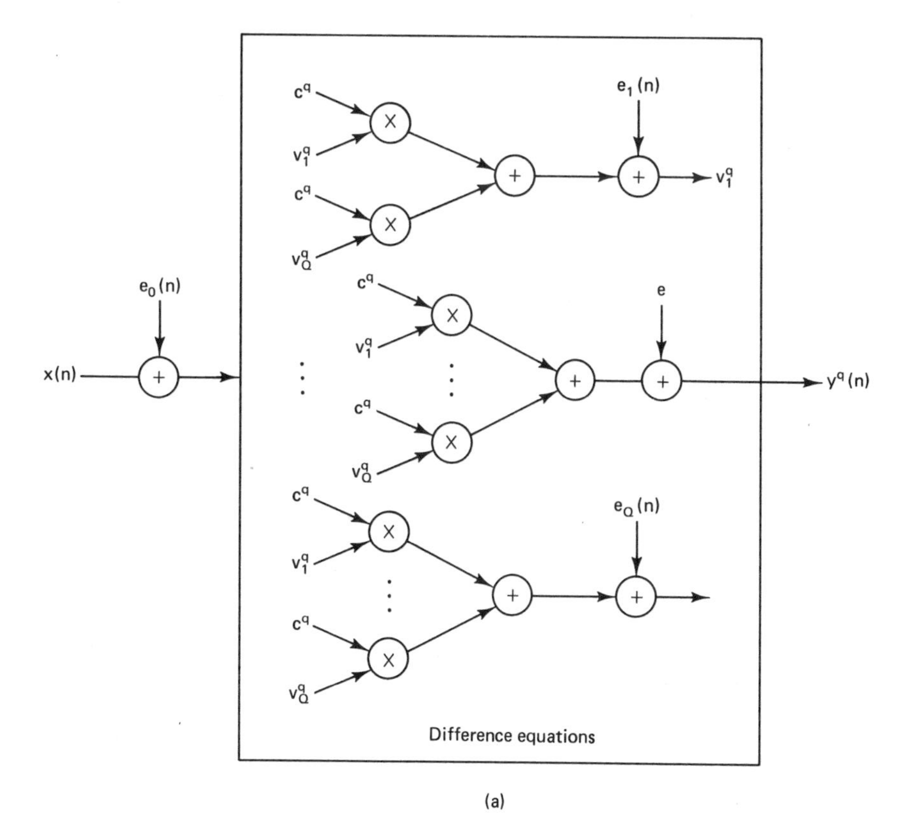

(a)

Figure 14-16 Filter noise models.

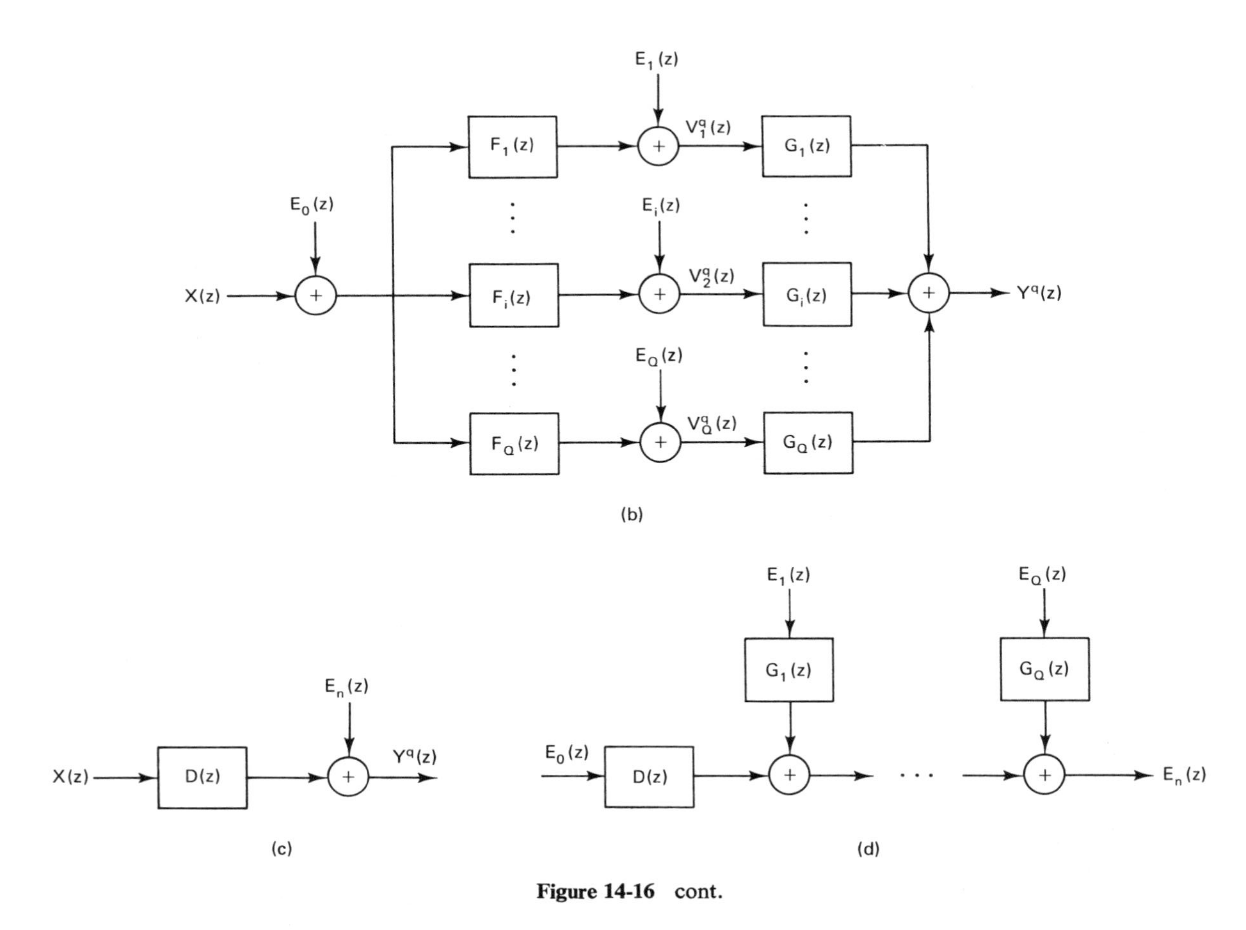

Figure 14-16 cont.

where $E_i(z) = z[e_i(n)]$. However, if we use the equivalent noise model of Figure 14-16c,

$$Y^q(z) = X(z)D(z) + E_n(z) \tag{14-120}$$

Hence the output noise is

$$E_n(z) = E_o(z)D(z) + \sum_{i=1}^{Q} E_i(z)G_i(z) \tag{14-121}$$

and is depicted in the quantization noise model of Figure 14-16d. Hence we may write

$$E_n(z) = \sum_{i=0}^{Q} E_{ni}(z) \tag{14-122}$$

where

$$E_{ni}(z) = E_i(z)G_i(z), \qquad i = 0, Q$$
$$G_0(z) = D(z)$$

We may examine $E_{ni}(z)$ in several ways. First let us write in the time domain

$$e_{ni}(n) = \sum_{j=0}^{\infty} g_i(j)e_i(n-j) \tag{14-123}$$

Hence

$$\begin{aligned} |e_{ni}(n)| &\leq \sum_{j=0}^{\infty} |g_i(j)||e_i(n-j)| \\ &\leq |e_i(n)|_{\max} \sum_{j=0}^{\infty} |g_i(j)| \end{aligned} \tag{14-124}$$

But we saw in an earlier section that if each product is rounded separately (see Figure 14-14c), then

$$\begin{aligned} |e_i(n)|_{\max} &= L_i |e_r|_{\max} \\ &= L_i \cdot \frac{2^{-b}}{2} \end{aligned} \tag{14-125}$$

Consequently,

$$|e_{ni}(n)| \leq \left(2^{-b} \cdot \frac{L_i}{2}\right) \sum_{j=0}^{\infty} |g_i(j)| \tag{14-126}$$

and

$$e_n(n) = \sum_{i=0}^{Q} e_{ni}(n)$$

$$\begin{aligned} |e_n(n)| &\leq \sum_{i=0}^{Q} |e_{ni}(n)| \\ &\leq \sum_{i=0}^{Q} \left(2^{-b} \cdot \frac{L_i}{2}\right) \sum_{j=0}^{\infty} |g_i(j)| \end{aligned}$$

So that

$$|e_n(n)| \leq \frac{2^{-b}}{2} \sum_{i=0}^{Q} \left(L_i \sum_{j=0}^{\infty} |q_i(j)|\right) \tag{14-127}$$

is an *absolute upper bound* on the magnitude of the round-off noise at the output of the digital filter. This is a very pessimistic bound because it assumes, at each sampling instance that the round-off noise at every quantizer is its maximum value and its worst-case sign, so that it drives the filter output to its *worst*-case value.

We may relax the condition of the worst-case *sign*, leaving the worst-case magnitude at each round-off quantizer, and we essentially apply a step input of magnitude $2^{-b}/2$ at each quantizer. Since

$$e_{ni}(n) = \sum_{j=0}^{\infty} g_i(j)e_i(n-j) \tag{14-128}$$

and

$$e_i(n-j) = L_i \cdot \frac{2^{-b}}{2}, \qquad n-j \geq 0$$
$$e_i(n-j) = 0, \qquad n-j < 0$$

or

$$e_{ni}(n) = \left(2^{-b} \cdot \frac{L_i}{2}\right) \sum_{j=0}^{n} g_i(j)$$

and, for large n,

$$|e_{ni}(\infty)| = \left(2^{-b} \cdot \frac{L_i}{2}\right) \left| \sum_{j=0}^{\infty} g_i(j) \right| \tag{14-129}$$

But

$$\sum_{j=0}^{\infty} g_i(j) = G(z)|_{z=1} = G(1)$$

Consequently,

$$|e_n(n)| \leq \frac{2^{-b}}{2} \sum_{i=0}^{Q} L_i |G_i(1)| \tag{14-130}$$

which is a bit more realistic than our absolute upper bound. We call this the *steady-state* bound.

An alternative derivation of this equation results after applying a step input error of magnitude $2^{-b}/2$ and then determining the steady-state error equation:

$$E_i(z) = \frac{L_i \cdot 2^{-b}/2}{1 - z^{-1}} \tag{14-131}$$

then

$$E_{ni}(z) = \frac{L_i \cdot 2^{-b}/2}{1 - z^{-1}} G_i(z) \tag{14-132}$$

But by the final-value theorem,

$$\begin{aligned} e_{ni}(\infty) &= \lim_{z \to 1} (1 - z^{-1}) E_{ni}(z) \\ &= L_i \cdot \frac{2^{-b}}{2} G_i(z)|_{z=1} \\ &= \left(L_i \cdot \frac{2^{-b}}{2}\right) G_i(1) \end{aligned} \tag{14-133}$$

So

$$\epsilon_n(\infty) = \frac{2^{-b}}{2} \sum_{i=0}^{Q} L_i G_i(1)$$

Hence

$$|\epsilon_n(\infty)| \leq \frac{2^{-b}}{2} \sum_{i=0}^{Q} L_i |G_i(1)| \tag{14-134}$$

Earlier in this chapter we examined the quantization characteristics of several number systems. There we noted that the quantization error for the round-off case could be modeled as a uniformly distributed random noise with zero mean value and variance

$$\sigma_{e_r}^2 = \frac{2^{-2b}}{12} \tag{14-135}$$

Consequently, here we abandon our deterministic approach used in our derivation of absolute upper and steady-state bounds and use statistical methods to analyze the output round-off error.

First, let us define some terminology. The autocovariance of a number sequence $\{x(n)\}$ is

$$Q_x(k - l) = E[x(k)x(l)] \tag{14-136}$$

Note that $Q_x(0) = E[x^2(n)]$. The spectral density can thus be defined as the z-transform of $\{Q(n)\}$

$$S_x(z) = \sum_{n=-\infty}^{\infty} Q_x(n)z^{-n} \tag{14-137}$$

so that

$$Q_x(n) = \frac{1}{2\pi j} \oint S_x(z)z^{n-1}\, dz \tag{14-138}$$

If we substitute $z = e^{j\omega T}$, then $dz = jTe^{j\omega T}\, d\omega$ and

$$Q_x(0) = \frac{1}{\omega_s} \int_0^{2\pi} S_x(e^{j\omega T})\, d\omega \tag{14-139}$$

Hence

$$E[x^2(n)] = \frac{1}{\omega_s} \int_0^{2\pi} S_x(e^{j\omega T})\, d\omega \tag{14-140}$$

It is well known [3] that a filter with transfer function $D(z)$ will have an output spectral density

$$S_y(z) = S_x(z)D(z)D\left(\frac{1}{z}\right) \tag{14-141}$$

Consequently,

$$\begin{aligned} E[y^2(n)] &= \frac{1}{\omega_s} \int_0^{2\pi} S_y(e^{j\omega T})\, d\omega \\ &= \frac{1}{\omega_s} \int_0^{2\pi} S_x(e^{j\omega T})D(e^{j\omega T})D(e^{-j\omega T})\, d\omega \\ &= \frac{1}{\omega_s} \int_0^{2\pi} S_x(e^{j\omega T})\,|\, D(e^{j\omega T})\,|^2\, d\omega \\ &= \sigma_y^2 \qquad \text{if } E[y(n)] = 0 \end{aligned} \tag{14-142}$$

For round-off noise analysis

$$Q_{e_r}(0) = E[e_r^2] = \frac{2^{-2b}}{12} \tag{14-143}$$

$$Q_{e_r}(n) = 0, \qquad n \neq 0$$

then

$$S_{e_r}(z) = \sum_{n=-\infty}^{\infty} Q_x(n)z^{-n} = \frac{2^{-2b}}{12} \tag{14-144}$$

If e_r is the input round-off error, then at the output of the filter

$$\begin{aligned} E[e_n^2(n)] &= \frac{1}{\omega_s}\int_0^{2\pi} S_{e_r}(e^{j\omega T})\,|\,D(e^{j\omega T})\,|^2\,d\omega \\ &= \frac{2^{-2b}}{12\omega_s}\int_0^{2\pi} |\,D(e^{j\omega T})\,|^2\,d\omega \end{aligned} \tag{14-145}$$

But

$$\begin{aligned} \frac{1}{\omega_s}\int_0^{2\pi} |\,D(e^{j\omega T})\,|^2\,d\omega &= \frac{1}{2\pi j}\oint D(z)D\left(\frac{1}{z}\right)\frac{dz}{z} \\ &= \sum_{m=0}^{\infty} d^2(m) \end{aligned}$$

Consequently,

$$\sigma_{\epsilon_n}^2 = E[e_n^2(n)] = \frac{2^{-2b}}{12}\sum_{m=0}^{\infty} d^2(m) \tag{14-146}$$

If we apply these results to the model of Figure 14-16,

$$\sigma_{e_{ni}}^2 = \sigma_{e_i}^2 \sum_{m=0}^{\infty} g_i^2(m) \tag{14-147}$$

But since the input random noise sources e_i have zero mean, then

$$\begin{aligned} \sigma_{e_n}^2 &= \sum_{i=0}^{Q}\sigma_{e_{ni}}^2 = \sum_{i=0}^{Q}\sigma_{e_i}^2\sum_{m=0}^{\infty} g_i^2(m) \\ &= \frac{2^{-2b}}{12}\sum_{i=0}^{Q}\sum_{m=0}^{\infty} g_i^2(m) \\ &= \frac{2^{-2b}}{12}\sum_{i=0}^{Q}\frac{1}{\omega_s}\int_0^{2\pi} |\,G_i(e^{j\omega T})\,|^2\,d\omega \\ &= \frac{2^{-2b}}{12}\sum_{i=0}^{Q}\frac{1}{2\pi j}\oint G_i(z)G_i\left(\frac{1}{z}\right)\frac{dz}{z} \end{aligned} \tag{14-148}$$

These relationships may be evaluated by a computing algorithm described in Ref. 4.

An important measure of a digital filter's performance is its signal-to-noise ratio. If we compute the ratio of the variance of the output signal to the output noise,

$$\begin{aligned} \frac{\sigma_y^2}{\sigma_{e_n}^2} &= \frac{(1/\omega_s)\int_0^{2\pi} S_x(e^{j\omega T})\,|\,G_0(e^{j\omega T})\,|^2\,d\omega}{(2^{-2b}/12)\sum_{i=0}^{Q}(1/\omega_s)\int_0^{2\pi} |\,G_i(e^{j\omega T})\,|^2\,d\omega} \\ &= 12\cdot 2^{2b}\,\frac{\int_0^{2\pi} S_x(e^{j\omega T})\,|\,G_0(e^{j\omega T})\,|^2\,d\omega}{\sum_{i=0}^{Q}\int_0^{2\pi} |\,G_i(e^{j\omega T})\,|^2\,d\omega} \end{aligned} \tag{14-149}$$

Note that the ratio is dependent upon the specific input signal. If we choose a random white noise input bounded by ± 1.0, then $S_x(e^{j\omega T}) = 1$ and

$$\frac{\sigma_y^2}{\sigma_{e_n}^2} = \frac{12\cdot 2^{2b}}{1 + \sum_{i=1}^{Q}\dfrac{\int_0^{2\pi} |\,G_i(e^{j\omega T})\,|^2\,d\omega}{\int_0^{2\pi} |\,G_0(e^{j\omega T})\,|^2\,d\omega}} \tag{14-150}$$

14.5 LIMIT CYCLES

Definitions

A *limit cycle* is a condition of sustained oscillation in a closed-loop system caused by nonlinearities within the loop. Consider the first-order digital filter of Figure 14-17.

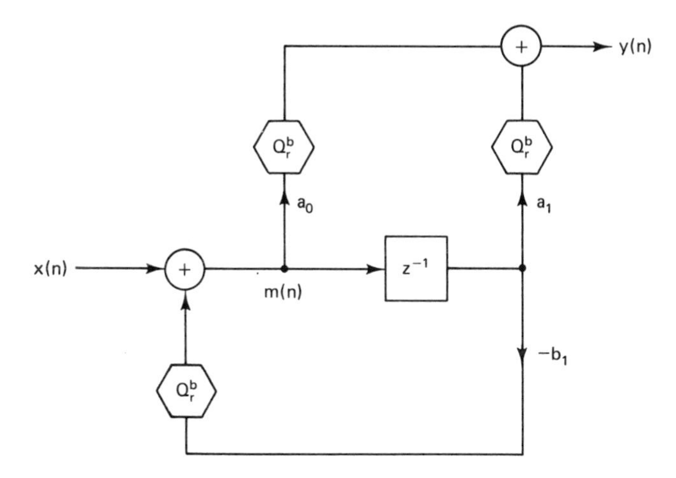

Figure 14-17 First-order 1D filter.

Let us examine the behavior of the filter if $b = 3$:

$$m(n) = (s.m_1 \quad m_2 \quad m_3)_{2cns} \tag{14-151}$$

and

$$b_1 = -0.6$$

Assume that at time zero ($n = 0$) an input pulse of value $m(n) = (0.100)_{2cns} = (0.5)_{10}$ is applied, and subsequent input values are zero. The filter signal $m(n)$ ideally would decay to zero and force the output $y(n)$ to zero as well. However, with the round-off quantizer in the loop:

n	$m(n-1)$	$-b_1m(n-1)$	$Q_r^3(b_1m(n-1))$	$m(n)$
0	0	0	0	4/8
1	4/8	2.4/8	2/8	2/8
2	2/8	1.2/8	1/8	1/8
3	1/8	0.6/8	1/8	1/8
4	1/8	0.6/8	1/8	1/8
5	etc.	etc.	etc.	etc.

Consequently, the signal $m(n)$ never reaches zero as in the ideal case. A truncation quantizer *would* have produced a zero value for $m(3)$. A digital filter which performs as shown above is said to possess a *deadband*. The deadband effectively *changes* the value of the feedback coefficient; in this case to $b_1 = -1.0$. We know that

$$\frac{M(z)}{X(z)} = \frac{1}{1 + b_1 z^{-1}} \tag{14-152}$$

and

$$m(n) = x(n) - b_1 m(n-1) \tag{14-153}$$

The impulse response is thus

$$\begin{aligned} m(n) &= \overset{0}{x(n)} - b_1(\overset{0}{x(n-1)} - b_1 m(n-2)) \\ &= (b_1)^n \end{aligned} \tag{14-154}$$

in the ideal case. In the case of the round-off quantizer,

$$m(n) = Q_r^b(-b_1 m(n-1)) \tag{14-155}$$

But we know that, for round-off,

$$|-b_1 m(n-1)| - |Q_r^b(-b_1 m(n-1))| \leq 2^{-b-1}$$

Consequently,

$$|-b_1 m(n-1)| - |m(n)| \leq 2^{-b-1}$$

If $b_1 > 0$, then its effective value in the deadband will be 1.0 and $m(n) = -m(n-1)$; if $b_1 < 0$, -1.0 and $m(n) = m(n-1)$. Hence

$$|m(n)| \leq \frac{2^{-b-1}}{1 - |b_1|} \tag{14-156}$$

Remember that $m(n)$ is an integer ($\times 2^{-b}$) and hence a deadband does not exist if $|b_1| < 0.5$.

Now let us consider the filter structure of Figure 14-18 under large-signal conditions. Suppose that we consider a 3-bit two's-complement number system with

$$b_1 = +1.5$$
$$b_2 = +0.5$$

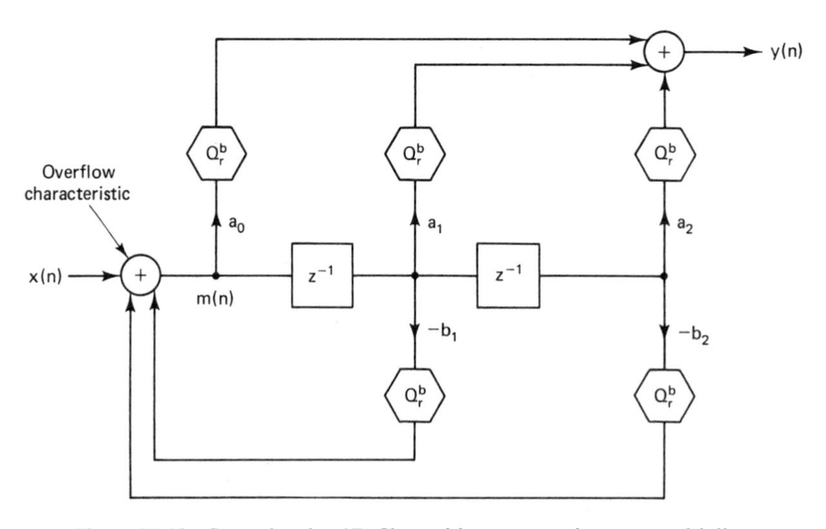

Figure 14-18 Second-order 1D filter with one quantizer per multiplier.

and let an input sequence of +5/8, 0, 0, 0, . . . be applied to the network. The poles of the filter are

$$(z^2 + 1.5z + 0.5) = (z + 1.0)(z + 0.5) \tag{14-157}$$

so that the ideal filter impulse response is stable:

n	$m(n-1)$	$Q_r^3(-b_1m(n-1))$	$m(n-2)$	$Q_r^3(-b_2m(n-2))$	$m(n)$
0	0	0	0	0	5/8
1	5/8	−7.5/8	0	0	−7.5/8
2	−7.5/8	11.25/8	5/8	−2.5/8	8.75/8
3	8.75/8	−13.125/8	−7/5.8	3.75/8	−9.375/8
4	−9.375/8	14.0625/8	8.75/8	−4.375/8	9.6875/8
5	9.6875/8	−14.53125/8	−9.375/8	4.6875/8	−9.84375/8
6	−9.84375/8	14.765625/8	9.6875/8	−4.84375/8	9.921765/8
7	9.921875/8	−14.8828125/8	−9.84375/8	4.921875/8	−9.9609375/8
8	−9.9609375/8	14.941406/8	9.921875/8	−4.9608375/8	9.9805688/8

Now suppose that we apply the round-off quantizers to the least significant bits (see Figure 14-5b) and the overflow characteristic to the most significant bits of the adder output (see Figure 14-8). The following sequence is produced:

n	$m(n-1)$	$Q_r^3(-b_1m(n-1))$	$m(n-2)$	$Q_r^3(-b_2m(n-2))$	$m(n)$
0	0	0	0	0	5/8
1	5/8	−8/8	0	0	−8/8
2	−8/8	12/8	5/8	−3/8	9/8 ⟶ −7/8
3	−7/8	11/8	−8/8	4/8	15/8 ⟶ −1/8
4	−1/8	2/8	−7/8	4/8	6/8
5	6/8	−9/8	−1/8	1/8	−8/8
6	−8/8	12/8	6/8	−3/8	9/8 ⟶ −7/8
7	−7/8	11/8	−8/8	4/8	15/8 ⟶ −1/8

Note that line $n = 7$ is identical to line $n = 3$. Hence a large limit cycle in the range

$$-1 \leq m(n) \leq +0.75 \tag{14-158}$$

has been produced, essentially by the overflow characteristic of the two's-complement number system adder as shown in the last column. For example, $n = 2$:

$$\begin{array}{rcrl} Q_r^3(-b_1m(n-1)) &=& 12/8 = & 1.100 \\ +\ \underline{Q_r^3(-b_2m(n-2))} &=& \underline{-3/8} = & \underline{1.101} \\ m(n) & & 9/8 & \underbrace{11.001}_{m(n)} \end{array} \tag{14-159}$$

Hence

$$\begin{aligned} m(n) &= (1.001)_{2\text{cns}} = -(0.111)_2 \\ &= -(7/8)_{10} \end{aligned} \tag{14-160}$$

Note that the quantizers Q_r^b in Figure 14-18 affect only the least significant bits, while the adder performs the overflow characteristic. The limit cycles produced by the overflow characteristic are called *overflow oscillations*.

Classification of Quantization Errors

The two cases shown above were examples of zero-input limit cycles. Consider the classification scheme of Table 14-1. The table illustrates the effect of signal amplitude

TABLE 14-1 CLASSIFICATION OF QUANTIZATION ERRORS

	Nonlinearity type	
Input Condition	Quantizer	Overflow
Zero input	Limit cycles	Overflow oscillations
Deterministic input		Overflow noise
Periodic	Limit cycles	
Nonperiodic	Quantization noise	
Stochastic input	Quantization noise	Overflow noise

quantizers and overflow characteristics on the output of a digital filter. The term *limit cycle* is commonly used to mean the small-signal limit cycles seen around the 2^{-b} signal level as described in the first-order example above. The term *quantization noise* refers to the error type described in Section 14.4. The term *overflow noise* has been included to mean cases in which an occasional overflow will added large noise "spike" into the filter output signal. In the remainder of this section we concentrate on analyzing the limit cycle and overflow oscillation phenomena.

Limit Cycles

Let us now examine a 3D filter of second order as shown in Figure 14-19. Note that double-precision product terms are added together and then quantized to form the output variable $y^q(k)$:

$$\begin{aligned} y(k) &= a_0 Q_r^b(x(k)) + a_1 Q_r^b(x(k-1)) \\ &\quad + a_2 Q_r^b(x(k-2)) - b_1 Q_r^b(y(k-1)) - b_2 Q_r^b(y(k-2)) \end{aligned} \tag{14-161}$$

For the zero-input limit cycle case

$$y(k) = -b_1 Q_r^b(y(k-1)) - b_2 Q_r^b(y(k-2))$$

And since $y^q(k) = Q_r^b(y(k))$,

$$|Q_r^b(y(k)) - y(k)| \leq 2^{-b-1} \tag{14-162}$$

Then

$$|Q_r^b(y(k)) + b_1 Q_r^b(y(k-1)) + b_2 Q_r^b(y(k-2))| \leq 2^{-b-1} \tag{14-163}$$

Case 1. Suppose that a constant nonzero output level is attained. Then

$$Q_r^b(y(k)) = Q_r^b(y(k-1)) = Q_r^b(y(k-2)) \tag{14-164}$$

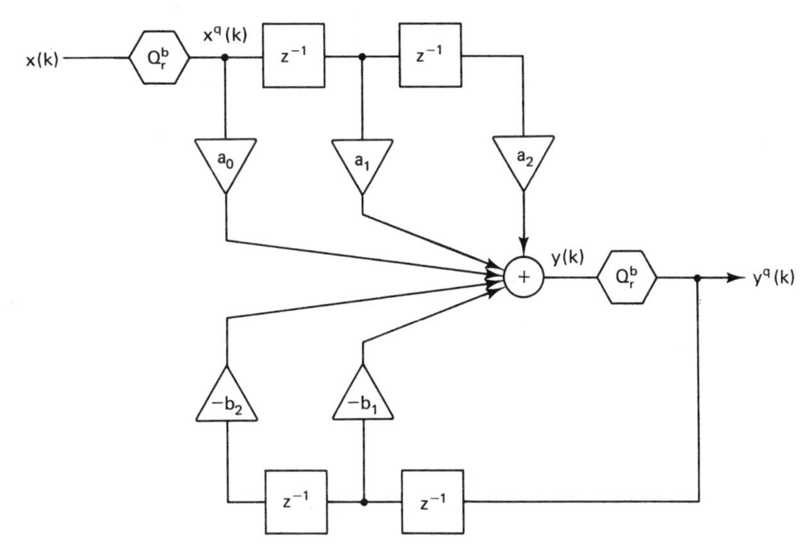

Figure 14-19 3D second-order filter.

and the deadband is

$$|Q_r^b(y(k))| \leq \frac{2^{-b-1}}{|1 + b_1 + b_2|} \tag{14-165}$$

Case 2. Suppose that a square-wave limit cycle is attained. Then

$$Q_r^b(y(k)) = -Q_r^b(y(k-1)) = Q_r^b(y(k-2)) \tag{14-166}$$

Consequently,

$$|Q_r^b(y(k))| \times |1 - b_1 + b_2| \leq 2^{-b-1}$$

and

$$|Q_r^b(y(k))| \leq \frac{2^{-b-1}}{|1 - b_1 + b_2|} \tag{14-167}$$

Case 3. Suppose that a sinusoidal limit cycle output is attained. Then the effective value of b_2 will be 1. That is,

$$Q_r^b(b_2 \times Q_r^b(y(k-2)) = Q_r^b(1 \times Q_r^b(y(k-2)) \tag{14-168}$$

Hence

$$|Q_r^b(y(k-2))| - |b_2 Q_r^b(y(k-2))| \leq 2^{-b-1}$$

and

$$|Q_r^b(y(k-2))| \leq \frac{2^{-b-1}}{1 - |b_2|}$$

Consequently,

$$|Q_r^b(y(k))| \leq \frac{2^{-b-1}}{1 - |b_2|} \tag{14-169}$$

In the cases presented above, we assumed a limit cycle waveform and solved for the magnitude bound of the filter output. Another way to view limit cycles is to consider the digital filter to be a finite-state, synchronous sequential circuit. The state of the circuit S_i is determined by the value of the variables in the delay elements (implemented by flip-flops or RAM cells). In Figure 14-19 there are four delay elements of $b + 1$ bits each so that the number of distinct states for the sequential circuit is

$$N_s = 2^{4(b+1)} \tag{14-170}$$

If $b = 15$ as in Chapter 11, then

$$N_s = 2^{64}$$

a very large number. For the zero-input case the subset of finite states is

$$N_s = 2^{32}$$

which is still over 1 billion.

In Table 14-1 we noted that limit cycles were always small-signal variations in the output signal. Practically speaking, only 3 or 4 bits are usually involved, so that

$$N_s \doteq 2^8 \tag{14-171}$$

which is a manageable number for analysis purposes.

Now examine Figure 14-20. The limit cycle $S_i \longrightarrow S_j \longrightarrow S_k \longrightarrow \cdots \longrightarrow S_i$ can have an even or odd number of states, M. In case 1 above we assumed $M = 1$; case 2,

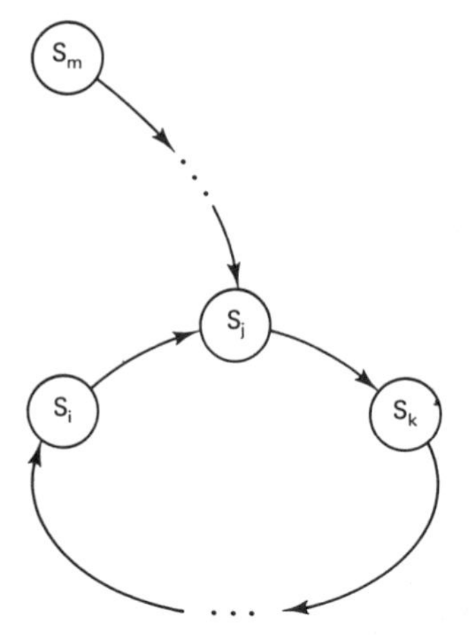

Figure 14-20 Limit cycle model.

$M = 2$. The limit cycle can be initiated by a starting state within the cycle (any of the limit cycle states) or from without the cycle as shown in state S_m in the model. Many researchers have sought to describe the behavior of the limit cycles modeled above [5–12]. In what follows we describe some of their results. The reader is referred to the literature for more advanced treatment of the limit cycle problem.

The limit cycle problem is essentially a feedback problem as modeled in Figure 14-21. The direct-form structures (Figure 14-21a–d) may all be described in the zero-input limit cycle case by Figure 14-21e.

$$m^q(k) = Q_2(Q_1(-b_1 m^q(k-1)) + Q_1(-b_2 m^q(k-2)) \tag{14-172}$$

The quantizers Q_1 or Q_2 are chosen for implementation. That is, if product terms are rounded after multiplication and before terms are added, Q_1 is active and Q_2 is absent. If, however, double-precision products are first added and the final result rounded, Q_2 is active and Q_1 is absent.

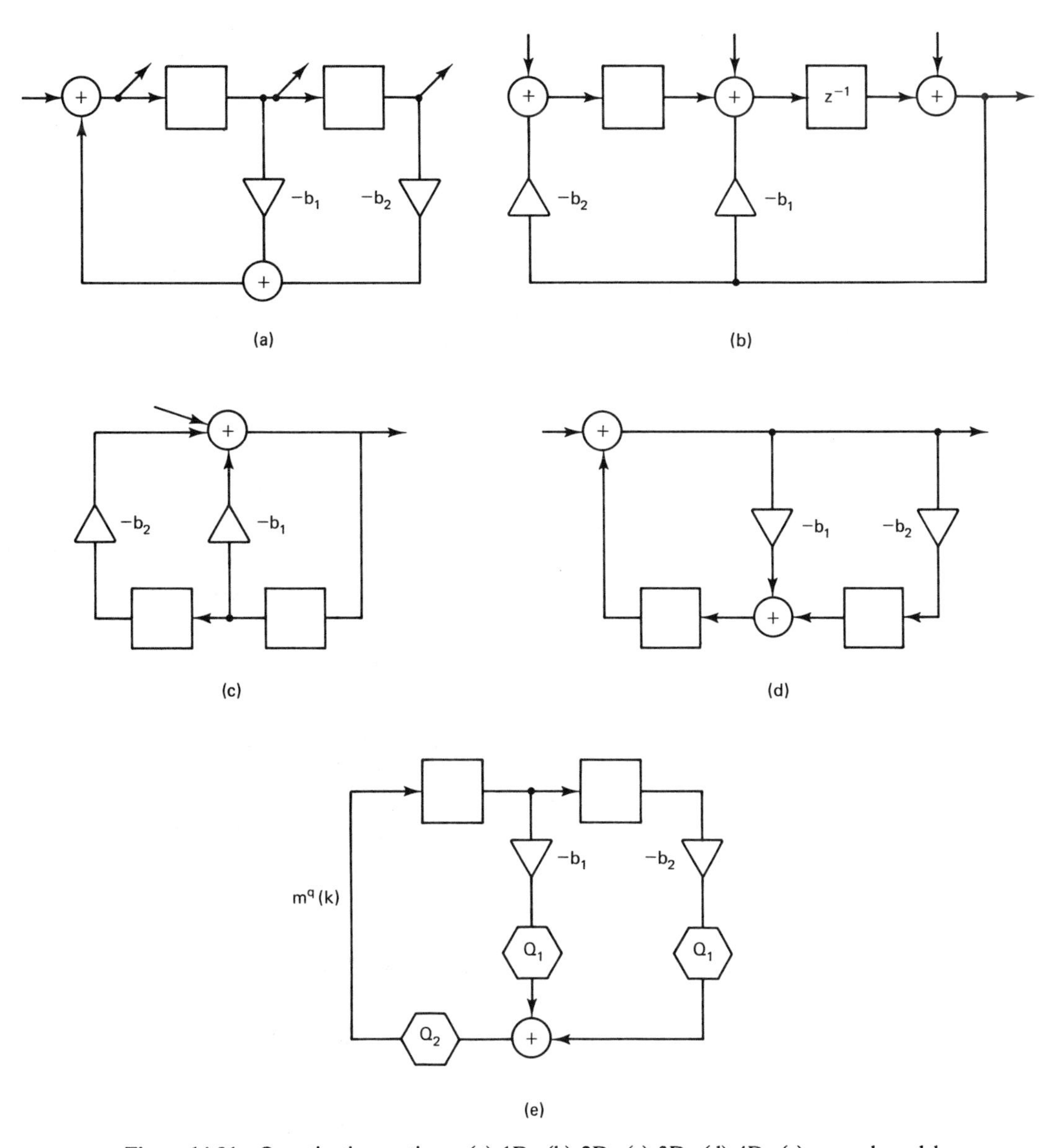

Figure 14-21 Quantization options: (a) 1D; (b) 2D; (c) 3D; (d) 4D; (e) general model for direct forms.

Limit cycle bounds. Long and Trick [9] have derived several bounds for limit cycle errors. Consider the 3D direct structure

$$y(k) = \sum_{i=0}^{n} a_i x(k-i) - \sum_{i=1}^{n} b_i y(k-i) \tag{14-173}$$

For the zero-input case

$$y(k) = -\sum_{i=1}^{n} b_i y(k-i) \tag{14-174}$$

Now adding quantizers at position Q_1 (see Figure 14-21)

$$y^q(k) = -\sum_{i=1}^{n} Q_r^b(b_i y^q(k-i)) \tag{14-175}$$

But if we define [see (14-1)]

$$Q_r^b(b_i y^q(k-i)) = b_i y^q(k-i) + e_i(k) \tag{14-176}$$

then

$$y^q(k) = -\sum_{i=1}^{n} (b_i y^q(k-i) + e_i(k)) \tag{14-177}$$

Hence

$$y^q(k) = -\sum_{i=1}^{n} b_i y^q(k-i) - \sum_{i=1}^{n} e_i(k) \tag{14-178}$$

So if we consider

$$e(k) = -\sum_{i=1}^{n} e_i(k) \tag{14-179}$$

to be an input signal into the digital filter, the limit cycle response will be

$$y^q(k) = \sum_{l=-\infty}^{k} h(k-l)e(l) \tag{14-180}$$

where $h(k)$ is the filter impulse response. We know that the limit cycle is periodic, say M cycles:

$$e(l) = e(l+M) \tag{14-181}$$

Hence

$$y^q(k) = \sum_{j=0}^{\infty}\left[\sum_{l=k-(j+1)M+1}^{k-jM} h(k-l)e(l)\right] \tag{14-182}$$

If $p = k - jM - l$,

$$\begin{aligned} y^q(k) &= \sum_{j=0}^{\infty}\left[\sum_{p=0}^{M-1} h(p+jM)e(k-p)\right. \\ &= \sum_{p=0}^{M-1} e(k-p)\left[\sum_{j=0}^{\infty} h(p+jM)\right] \end{aligned} \tag{14-183}$$

But

$$|e(k-p)| \leq n(2^{-b-1}) \tag{14-184}$$

so

$$y^q(k) = n \cdot 2^{-b-1} \sum_{p=0}^{M-1}\left[\sum_{j=0}^{\infty} h(p-jM)\right] \tag{14-185}$$

An absolute upper bound

$$|y^q(k)| \leq n \cdot 2^{-b-1} \sum_{p=0}^{\infty} |h(p)| \tag{14-186}$$

can be found which is the same bound found earlier for quantization noise in (14-127).

For second-order filters as modeled in Figure 14-21,

$$H(z) = \frac{1}{1 + b_1 z^{-1} + b_2 z^{-2}} \tag{14-187}$$

Hence

$$h(p) = \oint \frac{z^2}{z^2 + b_1 z + b_2} z^{p-1}\, dz \tag{14-188}$$

For distinct poles of the filter

$$|y^q(k)| \leq n \cdot 2^{-b-1} \sum_{p=0}^{M-1} \left| \frac{1}{2\left(\frac{b_1^2}{4} - b_2\right)} \right.$$

$$\times \left. \left[\frac{\left(\frac{-b_1}{2} + \sqrt{\frac{b_1^2}{4} - b_2}\right)^{p+1}}{1 - \left(-\frac{b_1}{2} + \sqrt{\frac{b_1^2}{4} - b_2}\right)^M} - \frac{\left(-\frac{b_1}{2} - \sqrt{\frac{b_1^2}{4} - b_1}\right)^{p+1}}{1 - \left(-\frac{b_1}{2} - \sqrt{\frac{b_1^2}{4} - b_2}\right)^M} \right] \right| \tag{14-189}$$

and for repeated poles

$$|y^q(k)| \leq n \cdot 2^{-b-1} \sum_{p=0}^{M-1} \left| \left(-\frac{b_1}{2}\right)^p \right.$$

$$\times \left. \left[\frac{k}{1 - \left(-\frac{b_1}{2}\right)^M} + \frac{1 + \left(-\frac{b_1}{2}\right)^M (M-1)}{\left[1 - \left(-\frac{b_1}{2}\right)^M\right]^2} \right] \right| \tag{14-190}$$

Hence the absolute bounds for limit cycles for second-order filters become:

Case 1. If $b_2 \leq 0$, or if $b_2 > 0$ and $2\sqrt{b_2} \leq |b_1|$,

$$|y^q(k)| \leq \frac{2^{-b}}{1 - |b_1| + b_2} \tag{14-191}$$

Case 2. If $b_2 > 0$ and $2b_2\sqrt{(2/\sqrt{b_2}) - 1} \leq |b_1| \leq 2\sqrt{b_2}$,

$$|y^q(k)| \leq \frac{2^{-b}}{(1 - \sqrt{b_2})^2} \tag{14-192}$$

Case 3. If $b_2 > 0$ and $|b_1| \leq 2b_2\sqrt{(2/\sqrt{b_2}) - 1}$,

$$|y^q(k)| \leq \frac{(1 + \sqrt{b_2})2^{-b}}{(1 - b_2)\sqrt{1 - b_1^2/4b_2}} \tag{14-193}$$

Absence of limit cycles [10]. Suppose that we model a digital filter with one quantizer as shown in Figure 14-22. Now

$$X(z) = W(z)\,Y(z) \tag{14-194}$$

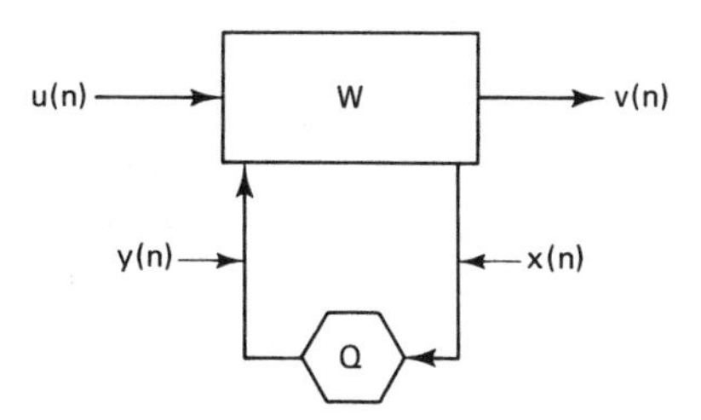

Figure 14-22 Digital filter with one quantizer.

If

$$Q(0) = 0$$

$$0 \le \frac{Q(x)}{x} \le k, \qquad x \ne 0 \tag{14-195}$$

$$\text{Re}\, W(z_l) - \frac{1}{k} < 0, \qquad l = 0, \quad \lfloor M/2 \rfloor$$

where $\lfloor x \rfloor$ indicates the integer part of x and

$$z_l = e^{j(2\pi/M)l}$$

then limit cycles of length M are *absent* from the digital filter [10].

If we apply these properties to the second-order digital filter of Figure 14-21e with the Q_2 quantizer, then

$$W(z) = b_1 z^{-1} + b_2 z^{-2} \tag{14-196}$$

and

$$\text{Re}\, W(z) = b_1 \cos\left[\left(\frac{2\pi}{M}\right)l\right] + b_2 \cos\left[\left(\frac{4\pi}{M}\right)l\right] \tag{14-197}$$

If $M = 1$, stationary limit cycles cannot exist if

$$b_1 + b_2 - \frac{1}{k} < 0 \tag{14-198}$$

Limit cycles of length $M = 2$ are absent if

$$-b_1 + b_2 - \frac{1}{k} < 0 \tag{14-199}$$

is also valid. Continuing this process, all limit cycles will be absent if

$$b_1 \cos \phi + b_2 \cos 2\phi - \frac{1}{k} < 0 \tag{14-200}$$

for $0 \le \phi \le \pi$.

These results are summarized in Figure 14-23. The quantizer characteristic must fall in the shaded area. If so, the foregoing conditions are shown in the coefficient space. If the coefficients fall within the shaded area, no limit cycles will exist. Note that for the round-off quantizer, $k = 2$; the truncation quantizer, $k = 1$.

Overflow Oscillations

Overflow oscillations must not be allowed to occur in a digital filter. Their avoidance can take three approaches:

1. Scale the input to the filter so that only small signal levels exist in the filter and overflow never occurs. This procedure will be discussed in Section 14.7.
2. Design the adder unit so that its overflow characteristic will not produce oscillations [13]. The overflow characteristic must lie in the shaded area of Figure 14-24. Note that the two's-complement and signed-magnitude adders do *not* satisfy this condition.

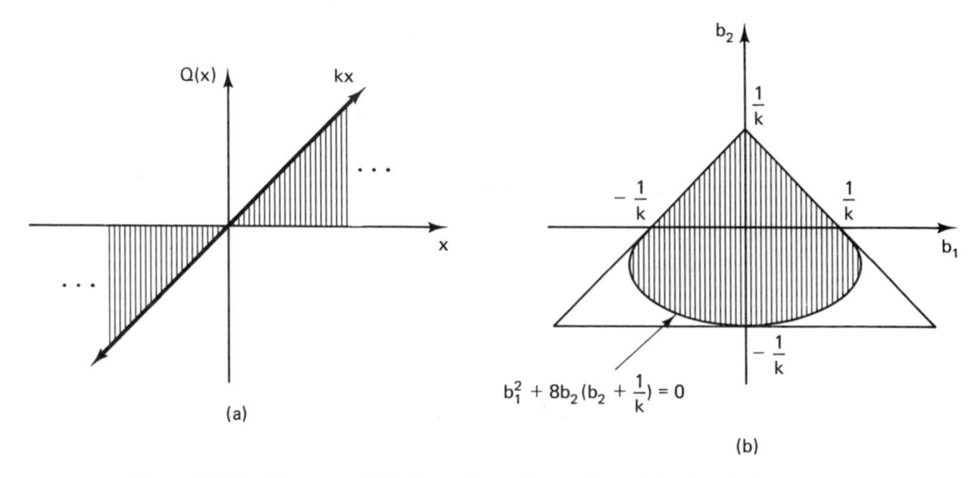

Figure 14-23 Absence of limit cycles: (a) nonlinear characteristic; (b) filter coefficient space.

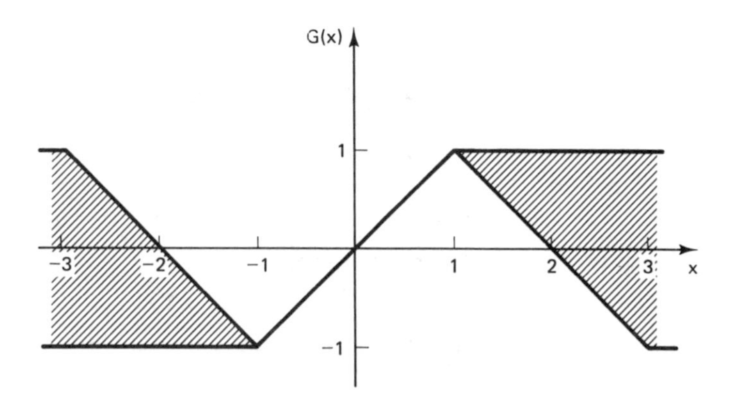

Figure 14-24 Absence of overflow oscillations.

3. Find a digital filter structure which is free of overflow oscillations, even when implemented with two's-complement arithmetic. This is the technique we will investigate here.

Conditions for absence of overflow oscillations [14]. Let us represent the digital filter in state-variable notation

$$\mathbf{x}(n+1) = A\mathbf{x}(n) + \mathbf{b}u(n) \tag{14-201}$$

The overflow characteristic may be imposed as follows:

$$\mathbf{x}(n+1) = G(A\mathbf{x}(n) + \mathbf{b}u(n)) \tag{14-202}$$

Theorem 1. If there is a diagonal matrix D with positive diagonal elements and $D - A^T DA$ is positive definite, overflow oscillations are impossible.

Theorem 2. Let A be a 2×2 matrix with eigenvalues $|\lambda| < 1$. There exists a positive definite diagonal matrix D for which $D - A^T DA$ is positive definite if and only if

$$a_{12}a_{21} \geq 0 \tag{14-203}$$

or

$$a_{12}a_{21} < 0 \quad \text{and} \quad |a_{11} - a_{22}| + \det(A) < 1 \tag{14-204}$$

Example 14.1

Consider the second-order filter of Figure 14-18.

$$A = \begin{bmatrix} 0 & 1 \\ -b_2 & -b_1 \end{bmatrix}, \qquad \mathbf{b} = \begin{bmatrix} 0 \\ 1 \end{bmatrix}$$
$$\mathbf{x}(n) = \begin{bmatrix} m(n-2) \\ m(n-1) \end{bmatrix} \tag{14-205}$$

For stability it is known that

$$\begin{aligned} 1 - b_2 &> 0 \\ 1 + b_1 + b_2 &> 0 \\ 1 - b_1 + b_2 &> 0 \end{aligned} \tag{14-206}$$

Using Theorem 2, overflow oscillations are absent if

$$a_{12}a_{21} = -b_2 \geq 0 \tag{14-207}$$

or if

$$-b_2 < 0 \quad \text{and} \quad |a_{11} - a_{22}| + \det(A) = b_1 + b_2 < 1 \tag{14-208}$$

Hence overflow oscillations are absent if

$$|b_1| + |b_2| < 1 \tag{14-209}$$

Unfortunately, this condition cannot always be met when designing second-order modules.

Structure with absent overflow oscillations. For any stable filter there is always a two's-complement implementation in which overflow oscillations are absent. The Liapunov stability theory states that the unique solution P to

$$P = A^T PA + I \tag{14-210}$$

is positive definite. If T is a symmetric square root of P^{-1}, then

$$P^{-1} = I - (T^{-1}AT)^T I(T^{-1}AT) \tag{14-211}$$

This T is a coordinate transform that will produce a new A which meets the conditions of Theorem 1. However, this solution will require a large number of multiplications and may not be practical in some applications.

The treatment of the overflow oscillation problem is an important step in the design of a digital filter. The three approaches above may not offer an optimal solution, but their application to a specific design problem can indeed find a practical solution.

14.6 IMPACT OF FINITE WORDLENGTH ON FILTER IMPLEMENTATION

In Chapters 8 through 11, we have discussed the design of digital filter transfer functions:

$$D(z) = \frac{\sum_{i=0}^{n} a_i z^{-i}}{1 + \sum_{i=1}^{n} b_i z^{-i}} \tag{14-212}$$

where a_i and b_i are constant, real numbers. In Chapter 12 we displayed several direct and cross-coupled digital filter structures suitable for implementing (14-212). In Chapter 13, we presented techniques for realizing the structures in Chapter 12. These realization methods impose finite-wordlength constraints on (14-212). The nature of these finite-wordlength constraints was examined earlier in this chapter. Here we explore the impact of these finite-wordlength constraints on the filter design and implementation process.

In Chapter 12 it was noted that higher-order filters ($n \geq 4$) are usually implemented as cascaded or paralleled second-order modules in order to avoid the pole-sensitivity problem described in Section 14.3. In avoiding the coefficient-sensitivity problem, we introduce other problems; specifically, pole–zero pairing, module scaling, and module ordering. In what follows we examine each of these new problems and give practical design guidelines for handling them.

14.7 CASCADED SECOND-ORDER MODULES

In implementing (14-212) as cascaded second-order modules, $D(z)$ may be factored into second-order numerator and denominator terms:

$$D(z) = \frac{\prod_{i=1}^{m} \alpha_i(z)}{\prod_{i=1}^{m} \beta_i(z)} = \prod_{i=1}^{m} A_i(z) \tag{14-213}$$

where, from (12-14),

$$\begin{aligned} \alpha_i(z) &= \alpha_{i0} + \alpha_{i1} z^{-1} + a_{i2} z^{-2} \\ \beta_i(z) &= 1 + \alpha_{i3} z^{-1} + \alpha_{i4} z^{-2} \end{aligned} \tag{14-214}$$

and m is the smallest integer greater than $n/2$. Figure 14-25 displays the cascaded realization. In order to form (14-213), second-order numerator and denominator terms must first be paired; then the pairs must be ordered in cascade. Each second-order module may then be implemented by one of the structures of Chapter 12. Consider the four direct structures of Figure 12-2. These structures have been redrawn in Figure 14-26a–d showing the signal amplitude error sources as modeled in Figure

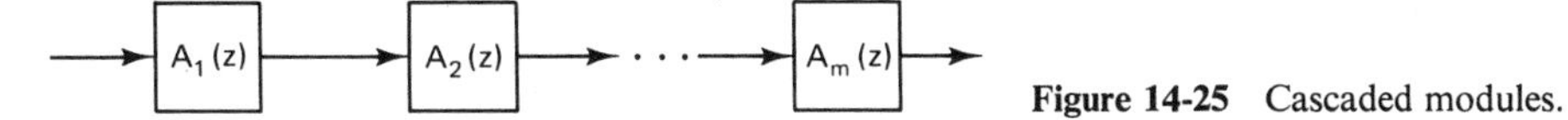

Figure 14-25 Cascaded modules.

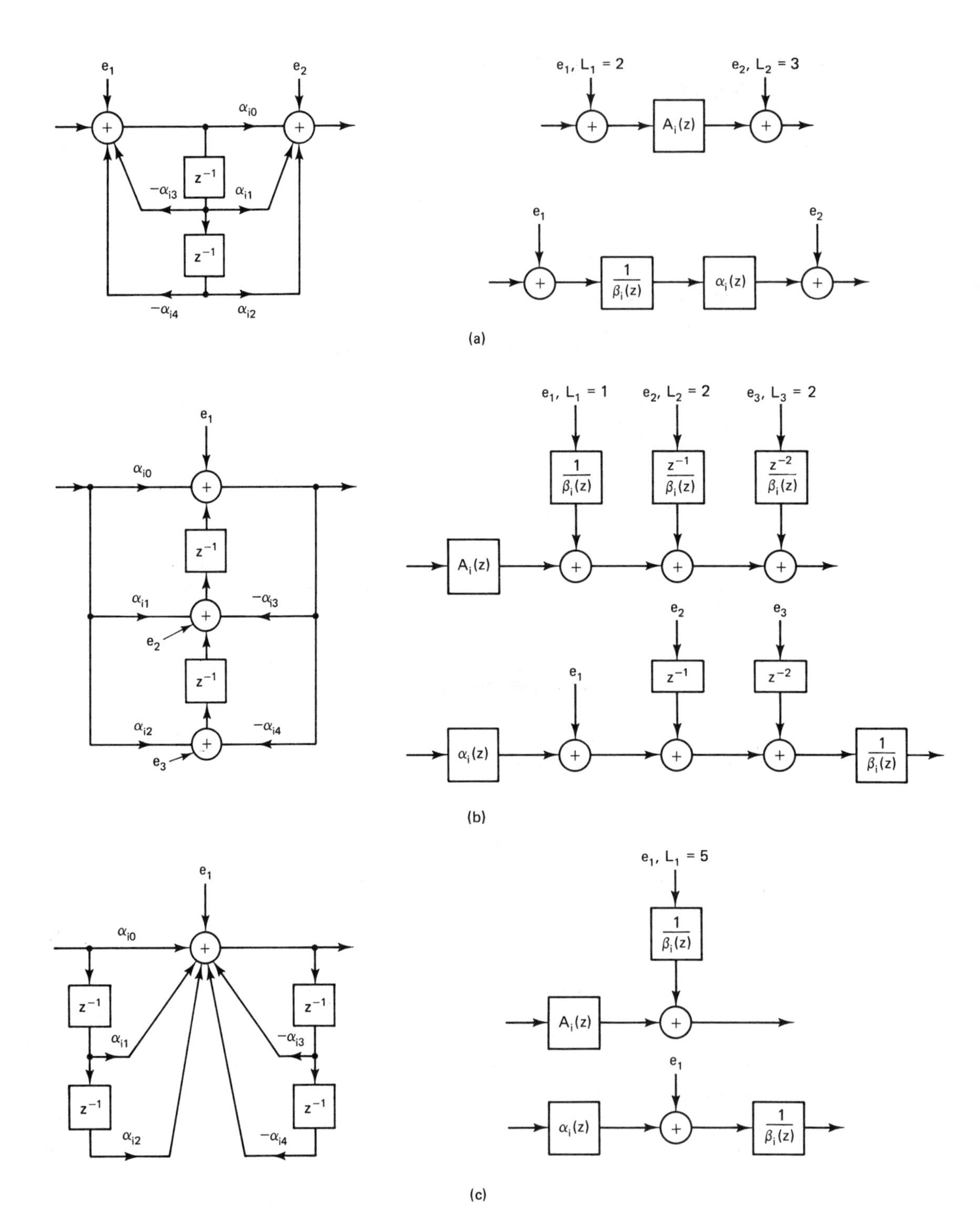

Figure 14-26 Direct structure error models: (a) 1D; (b) 2D; (c) 3D; (d) 4D; (e) 1D, 4D composite model; (f) 2D, 3D composite model.

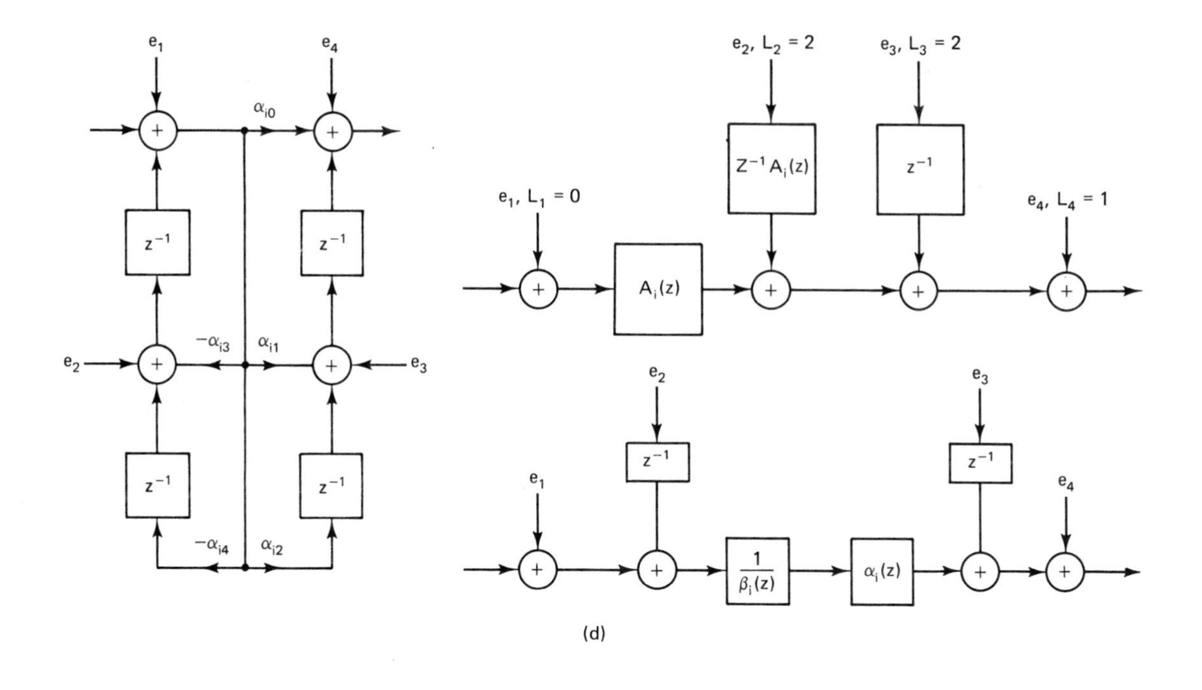

(d)

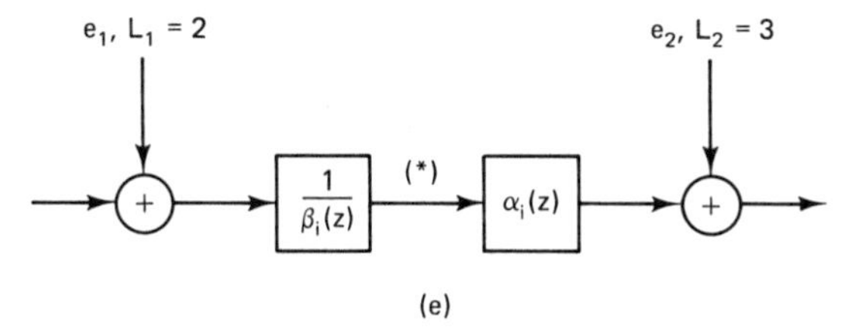

(e)

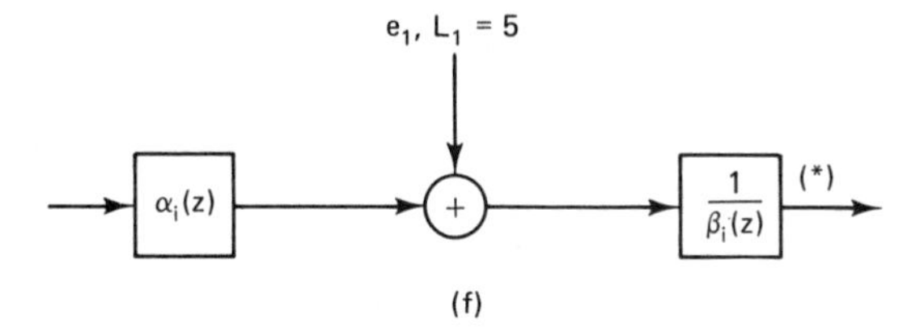

(f)

Figure 14-26 cont.

14-16. Please note that the 1D structure has two error sources per module; and the 3D, one. However, the 1D and 4D structures have similar error behavior, as do the 2D and 3D. Composite error models for these structures are shown in Figure 14-26e and f.

Earlier we emphasized that overflow oscillations can produce disastrous results. In practice, *scaling factors* are introduced into the cascade in order to limit signal amplitudes so that overflow does not occur. Let us examine a cascade of 1D (or 4D) second-order modules with scaling coefficients included in the cascade (see Figure

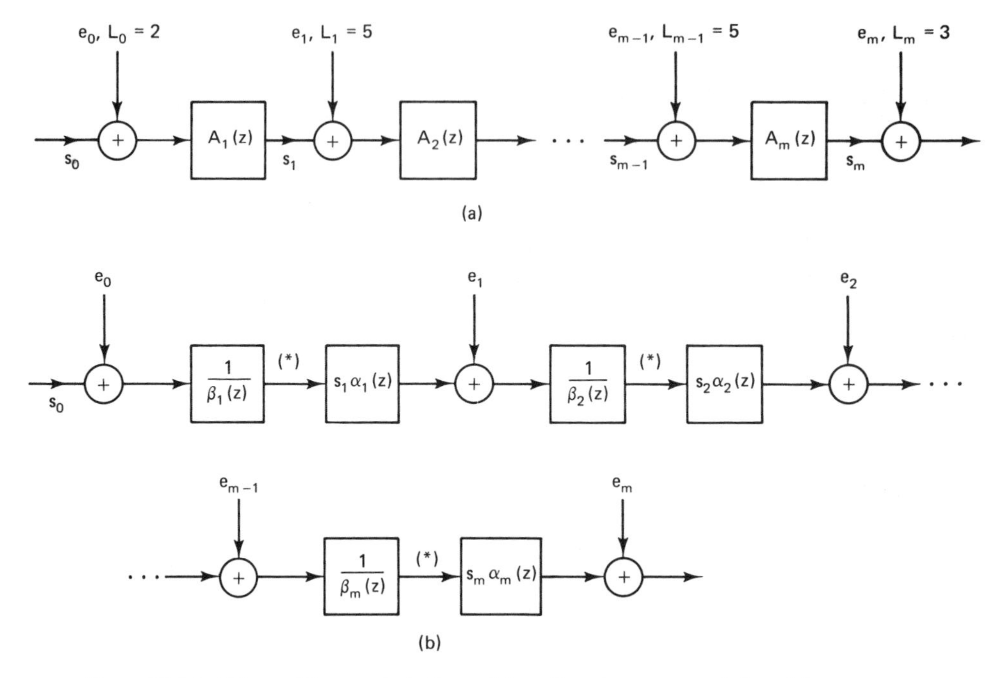

Figure 14-27 1D or 4D cascaded filter: (a) scaling between modules; (b) numerator scaling.

14-27a). We may simplify the notation by factoring a_0 from (14-212):

$$D(z) = a_0 \prod_{i=1}^{m} A_i(z) \tag{14-215}$$

where

$$A_i(z) = \frac{1 + \alpha_{i1} z^{-1} + \alpha_{i2} z^{-2}}{1 + \alpha_{i3} z^{-1} + \alpha_{i4} z^{-2}}$$

Consequently,

$$\prod_{i=0}^{m} s_i = a_0 \tag{14-216}$$

Previously, we derived several estimates of the output noise of a digital filter as a function of its amplitude error sources. For example, (14-148) states that the variance of the output noise may be computed by

$$\sigma_{e_n}^2 = \frac{2^{-2b}}{12} \sum_{i=0}^{Q} \frac{1}{2\pi j} \oint G_i(z) G_i\left(\frac{1}{z}\right) \frac{dz}{z} \tag{14-217}$$

where Q is the number of error sources (round-off quantizers) and $G_i(z)$ is the transfer function from the error source e_i to the filter output. Hence in Figure 14-27b

$$\begin{aligned}\sigma_{e_n}^2 = \frac{2^{-2b}}{12}\Bigg[L_m &+ \frac{L_{m-1}}{2\pi j} \oint \frac{s_m\alpha_m(z) s_m\alpha_m(1/z)}{\beta_m(z)\beta_m(1/z)} \frac{dz}{z} \\ &+ \frac{L_{m-2}}{2\pi j} \oint \frac{s_m\alpha_m(z) s_{m-1}\alpha_{m-1}(z) s_m\alpha_m(1/z) s_{m-1}\alpha_{m-1}(1/z)}{\beta_m(z)\beta_{m-1}(z)\beta_m(1/z)\beta_{m-1}(1/z)} \frac{dz}{z} \\ &+ \cdots + \frac{L_0}{2\pi j} \oint \prod_{i=1}^{m} \frac{s_i\alpha_i(z) s_i\alpha_i(1/z)}{\beta_i(z)\beta_i(1/z)} \frac{dz}{z}\Bigg]\end{aligned} \tag{14-218}$$

or by rearranging terms

$$\sigma_{e_n}^2 = \frac{2^{-2b}}{12}\left[L_m + \sum_{l=1}^{m} \frac{L_{l-1}}{2\pi j} \oint \prod_{i=l}^{m} s_i^2 A_i(z) A_i\left(\frac{1}{z}\right) \frac{dz}{z}\right] \tag{14-219}$$

where $L_0 = 2$, $L_m = 3$; otherwise, $L_i = 5$.

The goal of design in a cascade of second-order modules is to minimize (14-219). The parameters may be varied by pairing, scaling, and ordering.

Next let us examine the behavior of a cascade of 2D (or 3D) modules (see Figure 14-28). Using the model in Figure 14-26 and (14-217), we have

$$\begin{aligned}\sigma_{e_n}^2 = \frac{2^{-2b}}{12}\Bigg[&\frac{L_m}{2\pi j} \oint \frac{s_m^2(dz/z)}{\beta_m(z)\beta_m(1/z)} \\ &+ \frac{L'_{m-1}}{2\pi j} \oint \frac{s_m^2 s_{m-1}^2 A_m(z) A_m(1/z)}{\beta_{m-1}(z)\beta_{m-1}(1/z)} \frac{dz}{z} \\ &+ \frac{L_{m-2}}{2\pi j} \oint \frac{s_m^2 s_{m-1}^2 s_{m-2}^2 A_m(z) A_{m-1}(z) A_m(1/z) A_{m-1}(1/z)}{\beta_{m-2}(z)\beta_{m-2}(1/z)} \frac{dz}{z} \\ &+ \cdots + \frac{L_1}{2\pi j} \oint \frac{s_m^2 s_{m-1}^2 \cdots s_1^2 A_m(z) \cdots A_2(z) A_m(1/z) \cdots A_2(1/z)}{\beta_1(z)\beta_1(1/z)} \frac{dz}{z}\Bigg]\end{aligned} \tag{14-220}$$

Hence

$$\sigma_{e_n}^2 = \frac{2^{-2b}}{12} \sum_{l=1}^{m} \frac{L_l}{2\pi j} \oint \frac{\prod_{i=l}^{m} s_i^2 A_{i+1}(z) A_{i+1}(1/z)}{\beta_l(z)\beta_l(1/z)} \frac{dz}{z} \tag{14-221}$$

where $A_{m+1}(z) = 1$. Again the design goal is to minimize (14-221) by properly pairing, scaling, and ordering the terms of (14-213).

Signal Scaling

In Section 14.5 an example of a digital filter with overflow oscillations was presented. The filter was a 1D structure of second-order employing two's-complement arithmetic. In this example the internal signal $m(n)$ in Figure 14-18 exceeded the dynamic range $(1 > m(n) \geq -1)$ of the fixed-point number system and hence overflow occurred and introduced large-scale oscillations in the structure. The overflow oscillations, in general, may be eliminated by reducing (*scaling*) the input signal to a digital filter. However, if one scales the signals down to very small values, quantization errors become a significant part of the internal signals, and the signal-to-noise ratio is degraded. Consequently, an important design problem is to choose scaling factors that reduced the probability of overflow while maintaining signals at significant levels of the dynamic range.

The insertion of scaling coefficients between cascaded modules has been illustrated in Figures 14-27 and 14-28. These scaling coefficients are chosen such that the magnitude of the internal signal values [$V_i^q(z)$ of Figure 14-16] does not exceed 1, the overflow limit. Several methods of scaling have been presented in the literature [15–21]. Here we discuss a few of them.

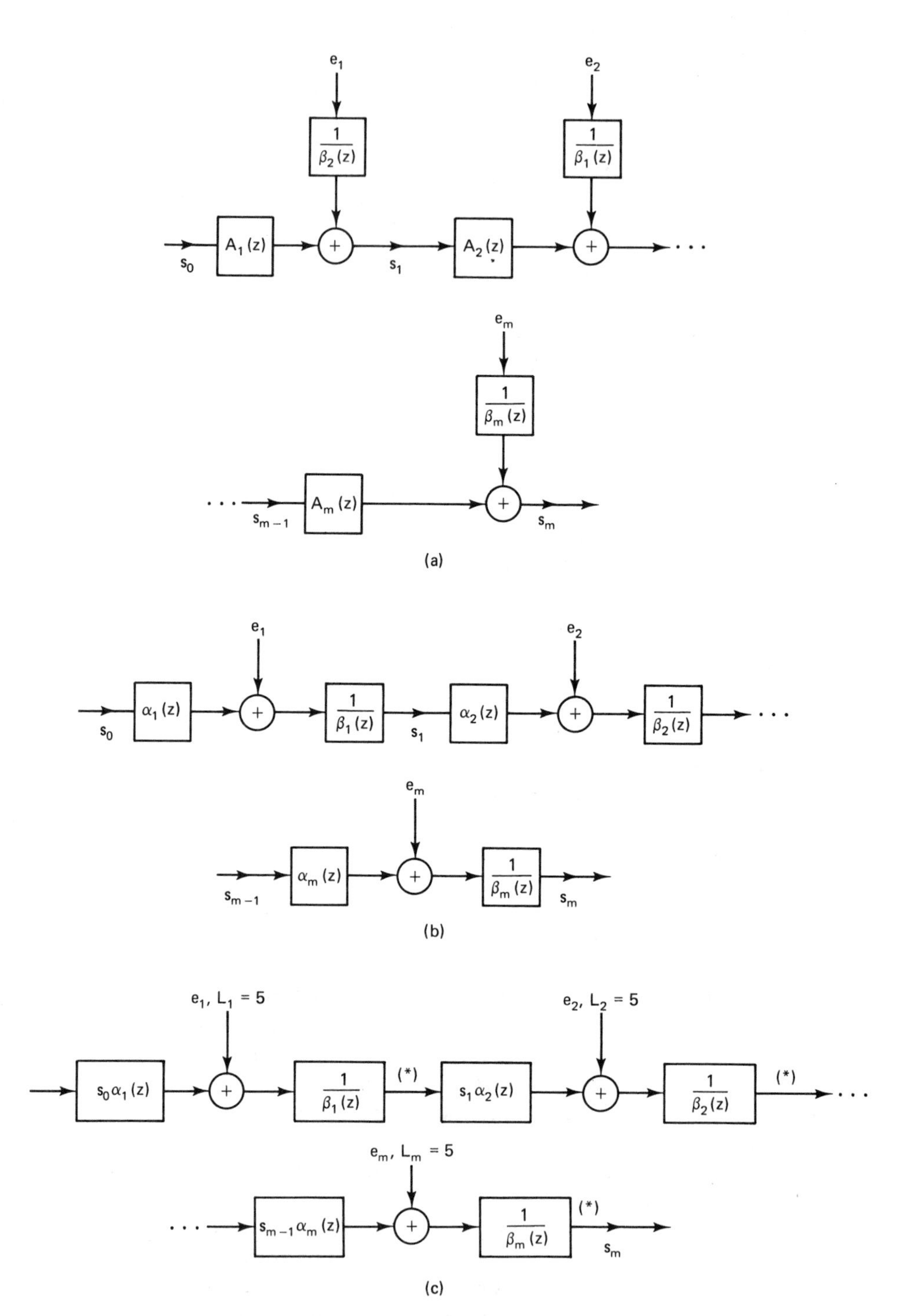

Figure 14-28 2D or 3D cascaded filter: (a) scaling between modules; (b) rearranging terms; (c) numerator scaling.

Upper-bound scaling. Examine Figure 14-16. In order to limit the signal at any point V_i^q to 1, we may select an input scaling factor λ_i such that

$$V_i^q(z) = \lambda_i X(z) F_i(z) \tag{14-222}$$

Hence

$$v_i^q(n) = \sum_{k=0}^{\infty} \lambda_i f_i(k) x(n-k) \tag{14-223}$$

But

$$|v_i^q(n)| \leq \sum_{k=0}^{\infty} |\lambda_i f_i(k)| |x(n-k)| \tag{14-224}$$

and since $|x(n-k)| \leq 1$ by definition,

$$|v_i^q(n)| < \sum_{k=0}^{\infty} |\lambda_i f_i(k)| \tag{14-225}$$

The object of scaling is to ensure that

$$|v_i^q(n)| \leq 1 \tag{14-226}$$

Consequently, we may ensure (14-226) by forcing

$$|v_i^q(n)| \leq \sum_{k=0}^{\infty} |\lambda_i f_i(k)| = 1 \tag{14-227}$$

or

$$\lambda_i = \frac{1}{\sum_{k=0}^{\infty} |f_i(k)|} \tag{14-228}$$

This scaling procedure produces a set of λ_i, $i = 1, Q$, where Q is the number of internal quantizers in the structure. If one chooses the scaling constant s:

$$s = \min(\lambda_i), \qquad i = 1, Q \tag{14-229}$$

overflow can *never* occur in the filter structure. Equation (14-228) is an absolute upper bound case, and is too conservative in most applications. The signal levels are quite restricted, leaving much of the filter dynamic range unused.

L_p-norm scaling. The inverse z-transform of (14-222) gives the expression

$$v_i^q(n) = \frac{1}{2\pi j} \oint \lambda_i X(z) F_i(z) z^{n-1} \, dz \tag{14-230}$$

If the contour of integration is the unit circle, $z = e^{j\omega T}$, and

$$v_i^q(n) = \frac{1}{\omega_s} \int_0^{\omega_s} \lambda_i X(e^{j\omega T}) F_i(e^{j\omega T}) e^{jn\omega T} \, d\omega \tag{14-231}$$

Using the Schwarz inequality [15]

$$\begin{aligned} |v_i^q(n)| \leq & \left[\frac{1}{\omega_s} \int_0^{\omega_s} |\lambda_i F_i(e^{j\omega T})|^2 \, d\omega \right]^{1/2} \\ & \times \left[\frac{1}{\omega_s} \int_0^{\omega_s} |X(e^{j\omega T})|^2 \, d\omega \right]^{1/2} \end{aligned} \tag{14-232}$$

Another way to write (14-230) is

$$|v_i^q(n)| \leq \frac{1}{\omega_s}\int_0^{\omega_s} |X(e^{j\omega T})||\lambda_i F_i(e^{j\omega T})|\,d\omega$$

or

$$|v_i^q(n)| \leq [\max_{\omega=[0,\omega_s]} |X(e^{j\omega T})|]\frac{1}{\omega_s}\int_0^{\omega_s} |\lambda_i F_i(e^{j\omega T})|\,d\omega \tag{14-233}$$

Similarly,

$$|v_i^q(n)| \leq [\max_{\omega=[0,\omega_s]} |\lambda_i F(e^{j\omega T})|]\frac{1}{\omega_s}\int_0^{\omega_s} |X(e^{j\omega T})|\,d\omega \tag{14-234}$$

Equations (14-231)–(14-234) may be conveniently expressed in terms of L_p norms. The L_p norm of a periodic function $G(\omega)$ with period ω_s is

$$\|G\|_p = \left[\frac{1}{\omega_s}\int_0^{\omega_s} |G(\omega)|^p\,d\omega\right]^{1/P}, \qquad p \geq 1 \tag{14-235}$$

If $G(\omega)$ is continuous, then

$$\begin{aligned}\lim_{p\to\infty} \|G\|_p = \|G\|_\infty &= \max_{\omega=[0,\omega_s]} |G(\omega)| \\ \|G\|_\infty &\geq \|G\|_p\end{aligned} \tag{14-236}$$

Substituting (14-235) and (14-236) into (14-231)–(14-234) produces

$$|v_i^q(n)| \leq \|\lambda_i F_i\|_2 \|X\|_2 \tag{14-237}$$

$$|v_i^q(n)| \leq \|X\|_\infty \|\lambda_i F_i\|_1 \tag{14-238}$$

$$|v_i^q(n)| \leq \|\lambda_i F_i\|_\infty \|X\|_1 \tag{14-239}$$

In general [21],

$$|v_i^q(n) \leq \|X\|_p \|\lambda_i F_i\|_q \tag{14-240}$$

where $1/p + 1/q = 1$.

These relations may be used in scaling since (14-240) holds for all X and F_i; let $\lambda_i F_i = 1$. Then

$$|v_i^q(n)| = |x(n)| \leq \|x\|_p \|1\|_q$$

or

$$|x(n)| = \|X\|_p \tag{14-241}$$

We may add the absolute upper bound

$$|x(n)| \leq \|x\|_p \leq 1 \tag{14-242}$$

But

$$|v_i^q(n)| \leq \|X\|_p \|\lambda_i F_i\|_q \leq 1 \|\lambda_i F_i\|_q \tag{14-243}$$

If we choose λ_i such that

$$|v_i^q(n)| \leq 1 \tag{14-244}$$

then

$$\|\lambda_i F_i\|_q = 1$$

or

$$\lambda_i = \frac{1}{\|F_i\|_q}, \qquad q \geq 1 \tag{14-245}$$

These values of λ_i may be used in (14-229). When $q = \infty$, then

$$\lambda_i = \frac{1}{\max\limits_{\omega=[0,\omega_s]} F_i(e^{j\omega T})} \tag{14-246}$$

which represents sinusoidal scaling. That is, a unit sine-wave input at a frequency that produces max $F_i(e^{j\omega T})$ will give a unit sine wave out of the filter.

When $q = 2$, then

$$\lambda_i = \frac{1}{\|F_i\|_2} = \left[\frac{1}{\omega_s}\int_0^{\omega_s} |F_i(e^{j\omega T})|^2 \, d\omega\right]^{-1/2}$$

and

$$\lambda_i^2 = \left[\frac{1}{\omega_s}\int_0^{\omega_s} F(e^{j\omega T})F(e^{-j\omega T}) \, d\omega\right]^{-1}$$

but

$$\sum_{k=0}^{n} (v_i^q(k))^2 = \left(\sum_{k=0}^{n} x^2(k)\right)\left(\frac{1}{\omega_s}\int_0^{\omega_s} F(e^{j\omega T})F(e^{-j\omega T}) \, d\omega\right)$$

or

$$\lambda_i^2 = \frac{\sum\limits_{k=0}^{n} x^2(k)}{\sum\limits_{k=0}^{n} (v_i^q(k))^2} \tag{14-247}$$

and the scaling constant relates the mean-squared values of the input and internal variable (energy scaling).

Unit step scaling. If the input $x(n)$ to the filter is a unit step (step of maximum amplitude), then (14-223) becomes

$$V_i^q(n) = \sum_{k=0}^{n} \lambda_i f_i(k) \tag{14-248}$$

and

$$|v_i^q(n)| \leq \max_{n=[0,\infty)} \left|\sum_{k=0}^{n} \lambda_i f_i(k)\right| \leq 1 \tag{14-249}$$

Hence we may choose

$$\lambda_i = \left[\max_{n=[0,\infty)} \left|\sum_{k=0}^{n} f_i(k)\right|\right]^{-1} \tag{14-250}$$

Averaging method [22]. This scaling method attempts to avoid overflow and maximize the signal-to-noise ratio. Let

$$F_{k\max} = \max_{\omega=[0,\omega_s]} |F_{ki}(e^{j\omega T})| \tag{14-251}$$

for the ith signal variable (constraint point) of the kth module in a cascade. Also, let

$$F_{k\min} = \min_{\omega=[0,\omega_s]} |F_{ki}(e^{j\omega T})| \tag{14-252}$$

To avoid overflow one would like to have

$$s_0 F_{1\max} = s_1 F_{2\max} = s_2 F_{3\max} = \cdots = s_{m-1} F_{m\max} \tag{14-253}$$

To maximize the signal-to-noise ratio, however,

$$s_0 F_{1\min} = s_1 F_{2\min} = \cdots = s_{m-1} F_{m\min} \tag{14-254}$$

would be more appropriate. A compromise solution is to average the two:

$$a = \frac{F_{k\max} + F_{k\min}}{2} s_{k-1} \tag{14-255}$$

where a is average for all k. Consequently,

$$\prod_{k=1}^{m} \left(\frac{F_{k\max} + F_{k\min}}{2}\right) s_{k-1} = a^m \tag{14-256}$$

But by (14-216)

$$\prod_{k=0}^{m} s_k = a_0$$

and

$$\prod_{k=1}^{m} s_{k-1} = \frac{a_0}{s_m} \tag{14-257}$$

Therefore, and scale factor s_{i-1} may be calculated by

$$a = \frac{F_{i\max} + F_{i\min}}{2} s_{i-1}$$

$$\left(\frac{F_{i\max} + F_{i\min}}{2} s_{i-1}\right)^m = \prod_{k=1}^{m} \left(\frac{F_{k\max} + F_{k\min}}{2}\right) s_{k-1}$$

$$= \frac{\prod_{k=1}^{m} s_{k-1} \prod_{k=1}^{m} (F_{k\max} + F_{kmin})}{2^m}$$

or

$$S_{i-1} = \frac{\left[(a_0/s_m) \prod_{k=1}^{m} (F_{k\max} + F_{k\min})\right]^{1/m}}{F_{i\max} + F_{i\min}} \tag{14-258}$$

for $i = 1, m$.

Optimization

Equation (14-219) expresses the output noise of a cascade of 1D second-order modules. In order to minimize the output noise, several researchers have used optimization techniques [16–20]. The scaling method is first selected. For example, if L_2-norm scaling is used in $q = 2$. In Figure 14-27, (14-245) is applicable with $q = 2$. In Figure 14-27, points at which overflow must be contrained are labeled (*). So, for the lth module,

$$s_0 s_1 s_2 \cdots s_{l-1} = \frac{1}{\| A_1 A_2 \cdots A_{l-1}/\beta_l \|_2} \tag{14-259}$$

or

$$(s_0 \cdots s_{l-1}) \left\| \frac{A_1 \cdots A_{l-1}}{\beta_l} \right\|_2 = 1 \tag{14-260}$$

$$(s_0 \cdots s_{l-1})^2 \left[\frac{1}{\omega_s} \int_0^{\omega_s} \left| \frac{A_1 \cdots A_{l-1}}{\beta_l} \right|^2 d\omega \right] = 1$$

Consequently,

$$\frac{1}{2\pi j} \oint \prod_{l=1}^{l} s_{i-1}^2 \frac{\alpha_{i-1}(z)\alpha_{i-1}(1/z)}{\beta_i(z)\beta_i(1/z)} \frac{dz}{z} = 1 \tag{14-261}$$

where $\alpha_0(z) = 1$. Finally, we may incorporate (14-261) into (14-219):

$$\sigma_{e_n}^2 = \frac{2^{-2b}}{12}\left[L_m + \sum_{l=1}^{m} \frac{L_{l-1}}{2\pi j} \oint \prod_{i=1}^{l} s_{i-1}^2 \frac{\alpha_{i-1}(z)\alpha_{i-1}(1/z)}{\beta_i(z)\beta_i(i/z)} \frac{dz}{z} \right. \\ \left. \times \frac{1}{2\pi j} \oint \prod_{i=l}^{m} s_i^2 \frac{\alpha_i(z)\alpha_i(1/z)}{\beta_i(z)\beta_i(1/z)} \frac{dz}{z}\right] \tag{14-262}$$

This relation is the one used in optimization.

Most designers of real-time digital filters do not have optimization programs [23] available. Consequently, in what follows we present design quidelines that usually give "good" results, but not optimal. One should always remember that an optimal solution to (14-219) involves selecting a particular scaling method as we did in (14-262). The selection of an "optimal" scaling method will depend on the application. For example, in closed-loop control systems, signal levels into a digital controller can be very low. In these cases L_∞ or absolute upper bound scaling would not be wise. The L_2, L_1, unit step, or averaging methods would be better.

Pole–Zero Pairing

In optimizing (14-219), a search of all possible pairing of numerator and denominator terms requires $(m!)^2$ evaluations of (14-219). Jackson [24] has suggested that good results are obtained if

$$\rho_1 < \rho_2 < \cdots < \rho_m \tag{14-263}$$

where

$$\rho_i = \frac{\|A_i\|_\infty}{\|A_i\|_2}$$

The authors have found that the following algorithm provides similar results, and may be calculated by graphical methods.

1. Plot poles and zeros graphically in the z-plane.
2. Pair real poles with each other. Find the real pole nearest the $z = +1$ point. Pair it with the real pole farthest from the $z = +1$ point. Continue until all real poles have been paired.
3. Pair poles and zeros. Begin by finding the pole nearest the $z = +1$ point. Match it with the zero nearest its location. If the zero is real, match the other pole of the pole pair in the same manner. Repeat step 3 until all poles and zeros have been matched.

The algorithm above tends to balance the dc gain of each module:

$$A_1(1) \doteq A_2(1) \doteq \cdots \doteq A_m(1) \tag{14-264}$$

Balancing the dc gain gives results similar to (14-263).

Example 14.2

Consider the fourth-order digital filter of Figure 14-29. Here the real poles are first paired. Then the real pair of poles is matched with the two real zeros. Finally, the complex poles are matched with the complex zeros.

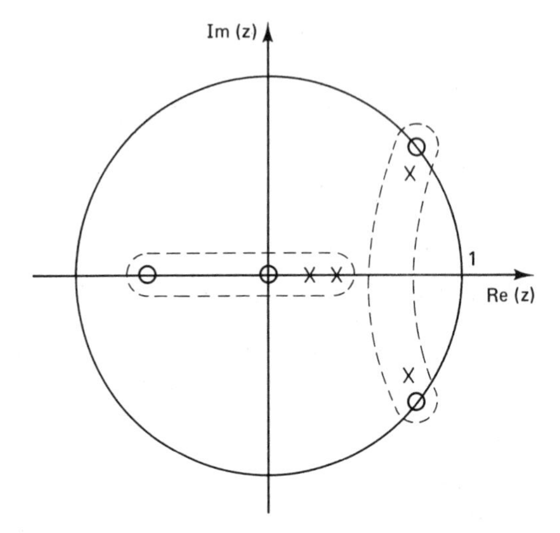

Figure 14-29 Pole–zero pairing.

Example 14.3

Consider the sixth-order filter [24] in cascaded 2D form:

$$\begin{aligned}
\alpha_1 &: \; 1 - 1.8118373z^{-1} + z^{-2} \\
\beta_1 &: \; 1 - 1.7636952z^{-1} + 0.90352914z^{-2} \\
\alpha_2 &: \; 1 - 1.6545862z^{-1} + z^{-2} \\
\beta_2 &: \; 1 - 1.4427789z^{-1} + 0.84506679z^{-2} \\
\alpha_3 &: \; 1 - 1.7442502z^{-1} + z^{-2} \\
\beta_3 &: \; 1 - 1.5334490z^{-1} + 0.75829007z^{-2}
\end{aligned} \tag{14-265}$$

	$\|\alpha_i/\beta_i\|_\infty$	$\|\alpha_i/\beta_i\|_2$	ρ_i	s_i
$i = 1$	1.65	1.056	1.56	1/1.65
$i = 2$	1.37	1.076	1.27	1/1.37
$i = 3$	1.137	1.066	1.07	1/1.137

Hence by (14-263),

$$\rho_1 < \rho_2 < \rho_3$$

The algorithm above achieves the same pairing as Ref. 24.

Example 14.4

Lee [17] has optimized the cascade realization of a cascaded 1D and 3D seventh-order filter:

$$
\begin{aligned}
\alpha_1&: \ 1 + 1.12368z^{-1} + z^{-2}\\
\beta_1&: \ 1 + 0.05358156z^{-1} + 0.9403549z^{-2}\\
\alpha_2&: \ 1 + 0.216853z^{-1} + z^{-2}\\
\beta_2&: \ 1 + 0.019248z^{-1} + 0.74403z^{-2}\\
\alpha_3&: \ 1 + 0.42371z^{-1} + z^{-2}\\
\beta_3&: \ 1 + 0.19405z^{-1} + 0.33908z^{-2}\\
\alpha_4&: \ 1 + z^{-1}\\
\beta_4&: \ 1 + 0.16535z^{-1}
\end{aligned}
\tag{14-266}
$$

Consider the pole–zero plot of Figure 14-30 and the following algorithm.

1. Real pole β_4 is matched with α_4.
2. Complex poles β_3 are matched with complex zeros α_3.
3. Complex poles β_2 are matched with complex zeros α_2.
4. Complex poles β_1 are matched with complex zeros α_1.

This pairing is the same as obtained by optimization in Ref. 17.

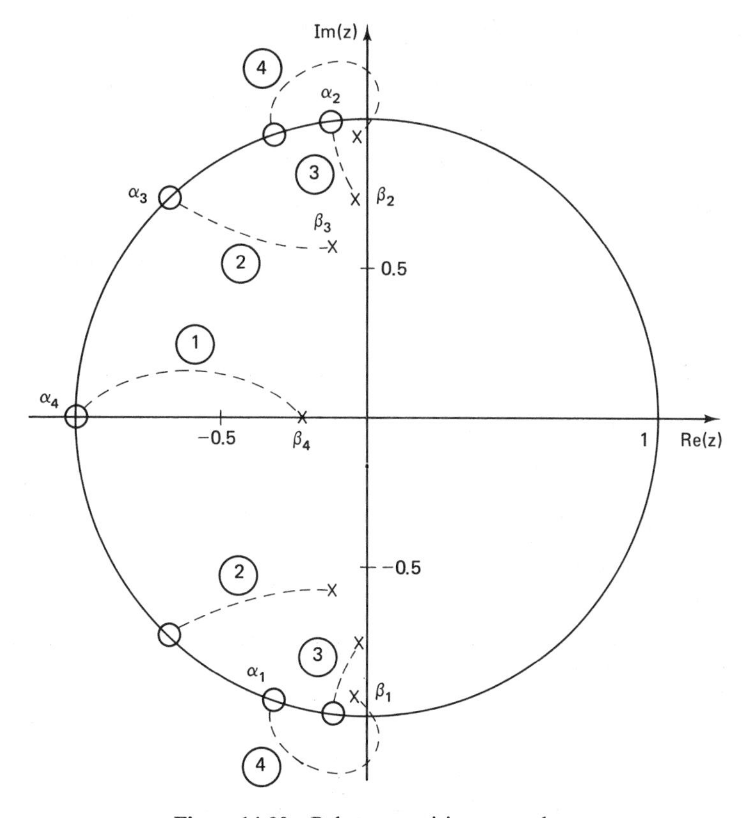

Figure 14-30 Pole–zero pairing example.

Ordering

Once the second-order modules have been paired, they must be ordered to minimize the output noise and limit cycle response. Ordering algorithms have been proposed in the literature [24–27]. Here we will derive a method based on (14-219) for cascaded 1D modules. The technique may be extended to other direct structures.

Since

$$\frac{1}{2\pi j}\oint A_i(z)A_i(1/z)\,dz/z = \frac{1}{\omega_s}\int_0^{\omega_s} |A_i(e^{j\omega T})|^2\,d\omega = \sum_{m=0}^{\infty} a_i^2(m) \tag{14-267}$$

we may write (14-219) in the form

$$\sigma_{e_n}^2 = \frac{2^{-2b}}{12}\left[L_m + \sum_{l=1}^{m} L_{l-1}\frac{1}{\omega_s}\int_0^{\omega_s}\left|\prod_{i=l}^{m} s_i A_i(e^{j\omega T})\right|^2 d\omega\right] \tag{14-268}$$

Consequently,

$$\sigma_{e_n}^2 = \frac{2^{-2b}}{12}\left[L_m + \sum_{l=1}^{m} L_{l-1}\frac{1}{\omega_s}\int_0^{\omega_s}\prod_{i=l}^{m} s_i^2 |A_i(e^{j\omega T})|^2\,d\omega\right] \tag{14-269}$$

But we may interchange the integral and product and create an inequality:

$$\begin{aligned}\sigma_{e_n}^2 &\le \frac{2^{-2b}}{12}\left[L_m + \sum_{l=1}^{m} L_{l-1}\prod_{i=l}^{m}\frac{1}{\omega_s}\int_0^{\omega_s} s_i^2 |A_i(e^{j\omega T})|^2\,d\omega\right]\\ &\le \frac{2^{-2b}}{12}\left[L_m + \sum_{l=1}^{m} L_{l-1}\prod_{i=l}^{m} C_i\right]\end{aligned} \tag{14-270}$$

where

$$C_i = \frac{1}{\omega_s}\int_0^{\omega_s} s_i^2 |A_i(e^{j\omega T})|^2\,d\omega = \sum_{m=0}^{\infty} s_i^2 a_i^2(m)$$

is a constant for each module. Equation (14-270) may be expanded:

$$\sigma_{e_n}^2 \le \frac{2^{-2b}}{12}[L_m + L_{m-1}C_m + L_{m-2}C_{m-1}C_m + \cdots + L_0C_1C_2\cdots C_m] \tag{14-271}$$

and may also be written as

$$\sigma_{e_n}^2 \le (C_m \cdots (C_2(C_1L_0 + L_1) + L_2)\cdots + L_m)\frac{2^{-2b}}{12} \tag{14-272}$$

The result in (14-272) may be drawn graphically as shown in Figure 14-31. Since $L_0 = 2$, $L_m = 3$ and $L_i = 5$, $\sigma_{e_n}^2$ may be reduced in value by requiring that

$$C_1 \ge C_2 \ge C_3 \cdots \ge C_m \tag{14-273}$$

These constants may be calculated for each module once the pairing and scaling have been accomplished.

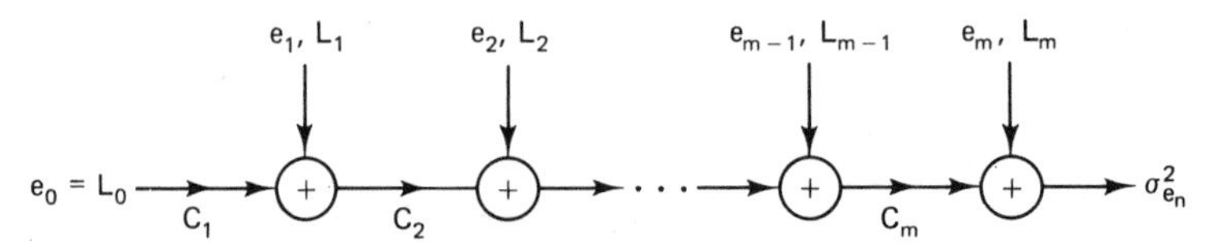

Figure 14-31 Ordering model for 1D modules.

Design Guidelines

When a designer is faced with implementing a cascade of second-order modules, the following guidelines for the cascade procedure may be used to achieve a "good" suboptimal implementation.

1. Factor $D(z)$ of (14-212) to obtain (14-213).
2. Quantize the coefficients for the desired implementation wordlength (e.g., 16 bits for the INTEL 8086).
3. Verify that (14-213) with quantized coefficients meets the system specifications.
4. Employ the pole–zero pairing algorithm.
5. Choose a structure for the modules (e.g., 1D).
6. Scaling may now be applied to each module *independently*. We suggest the unit step method (except for integrators). However, any of the other methods may be used.
7. Apply the ordering algorithm. This algorithm uses constants which may be calculated from each independent module. The implementation is now complete.
8. Simulate the implementation in the open-loop case and test its step response to assure that the dynamic range of the internal variables is mid range.
9. Simulate the implementation in the closed-loop case to ensure that system specifications are met.

The key to the foregoing procedure is that calculations are done on independent modules to determine scaling and ordering. This drastically reduces the computations needed to implement a filter, and can be done by hand without computer-aided design programs.

14.8 PARALLEL SECOND-ORDER MODULES

In implementing (14-212) as a parallel connection of second-order modules, $D(z)$ may be expressed, using a partial-fraction expansion, as

$$D(z) = \beta_0 + \sum_{i=1}^{m} B_i(z) \tag{14-274}$$

where

$$B_i(z) = \frac{\beta_{i1} z^{-1} + \beta_{i2} z^{-2}}{1 + \beta_{i3} z^{-1} + \beta_{i4} z^{-2}} = \frac{\alpha_i(z)}{\beta_i(z)}$$

Figure 14-32a depicts the parallel realization. The modules B_i may be implemented using the direct structures of Figure 14-26. The 1D (or 4D) direct structure has been used in the implementation of Figure 14-32b. The scaling constants have been added to avoid overflow in the second-order modules. The inverse of the scaling constant is used to restore the signal level before the output adder is reached, as is required in

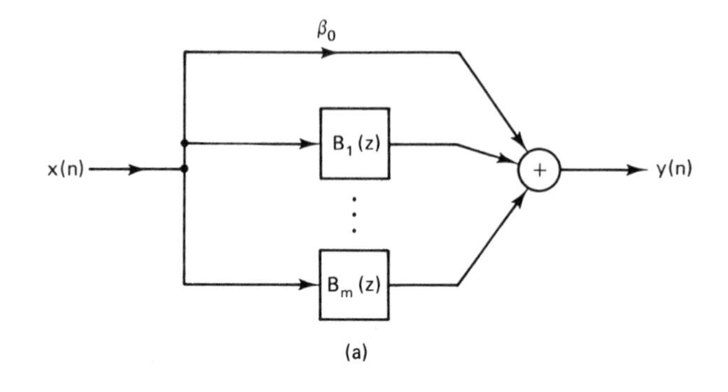

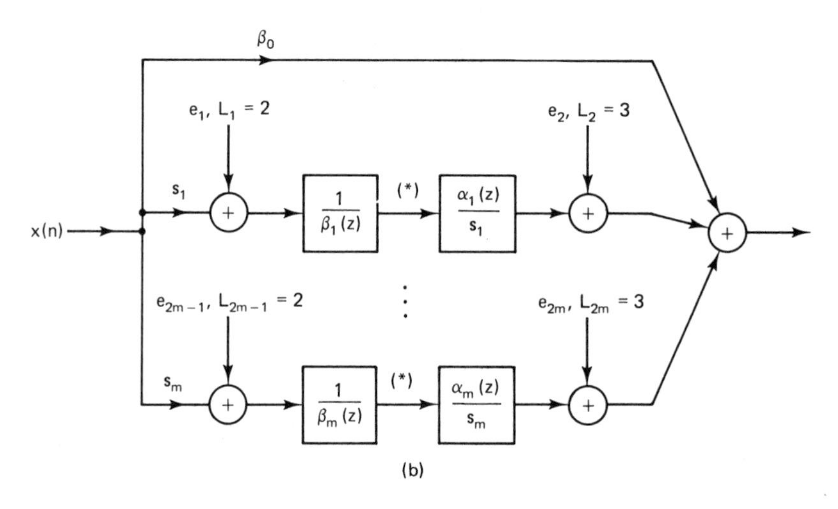

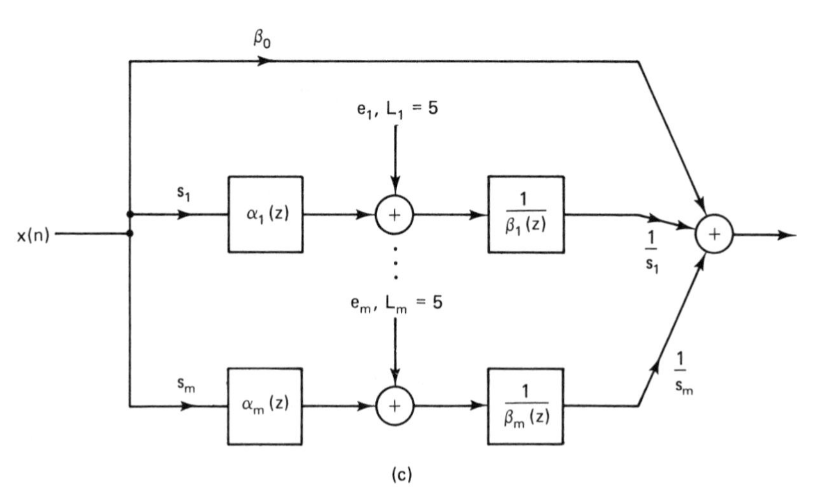

Figure 14-32 Parallel implementation models: (a) general model; (b) 1D, 4D implementation; (c) 2D, 3D implementations.

(14-274). Here the error variance may be expressed, by using (14-217),

$$\sigma_{e_n}^2 = \frac{2^{-2b}}{12}\left[3m + \sum_{i=1}^{m} \frac{2m}{2\pi j} \oint \frac{B_i(z)B_i(1/z)}{s_i^2} \frac{dz}{z}\right] \tag{14-275}$$

and by (14-267),

$$\sigma_{e_n}^2 = \frac{2^{-2b}}{12}\left[3m + \sum_{i=1}^{m} \frac{2m}{s_i^2} \sum_{l=0}^{\infty} b_i^2(l)\right] \tag{14-276}$$

Implementing (14-274) in the 2D or 3D structures produces the error configuration shown in Figure 14-32c. The output noise variance in this case is computed by

$$\sigma_{e_n}^2 = \frac{2^{-2b}}{12}\left[\sum_{i=1}^{m} \frac{L_m}{2\pi j} \oint \frac{1}{s_i^2 \beta_i(z)\beta_i(1/z)} \frac{dz}{z}\right] \tag{14-277}$$

where $L_m = 5$.

The parallel implementation of a digital filter is a rather straightforward process.

1. Perform a partial-fraction expansion on (14-212) to obtain (14-274).
2. Quantize the coefficients to the realizable wordlength.
3. Verify that (14-274) with quantized coefficients meets the system specifications.
4. Choose a structure for the second-order modules.
5. Scaling may be applied to each module (e.g., the unit step method). The implementation is now complete.
6. Simulate the open-loop digital filter and test its step, impulse, and sinusoidal responses to assure that the dynamic range of all variables is appropriate.
7. Insert the digital filter simulation into the total system simulation to ensure that system specifications have been satisfied.

Note that the design procedure for the parallel case is similar to the cascade case, with the exception that pairing and ordering are not necessary.

14.9 SUMMARY

In this chapter we have discussed the finite-wordlength complications of implementing digital filters. Signal amplitude quantization was shown to add low-level noise to signal variables. The resulting output effect is described by random noise or limit cycles. Coefficient quantization is shown to change the transfer characteristics of the digital filter. Overflow has been shown to impose disastrous consequences on the digital filter; hence overflow oscillations must be avoided.

We have also discussed the cascade and parallel implementations of digital filters. In the cascade case we factored $D(z)$ and used coefficient quantization, pole–zero pairing, direct structures for second-order modules, module scaling, module ordering, and simulation to achieve and verify our implementation. In the parallel case we used a partial-fraction expansion of $D(z)$ and omitted the pole–zero pairing and ordering steps. The overall design process is summarized in Figure 14-33.

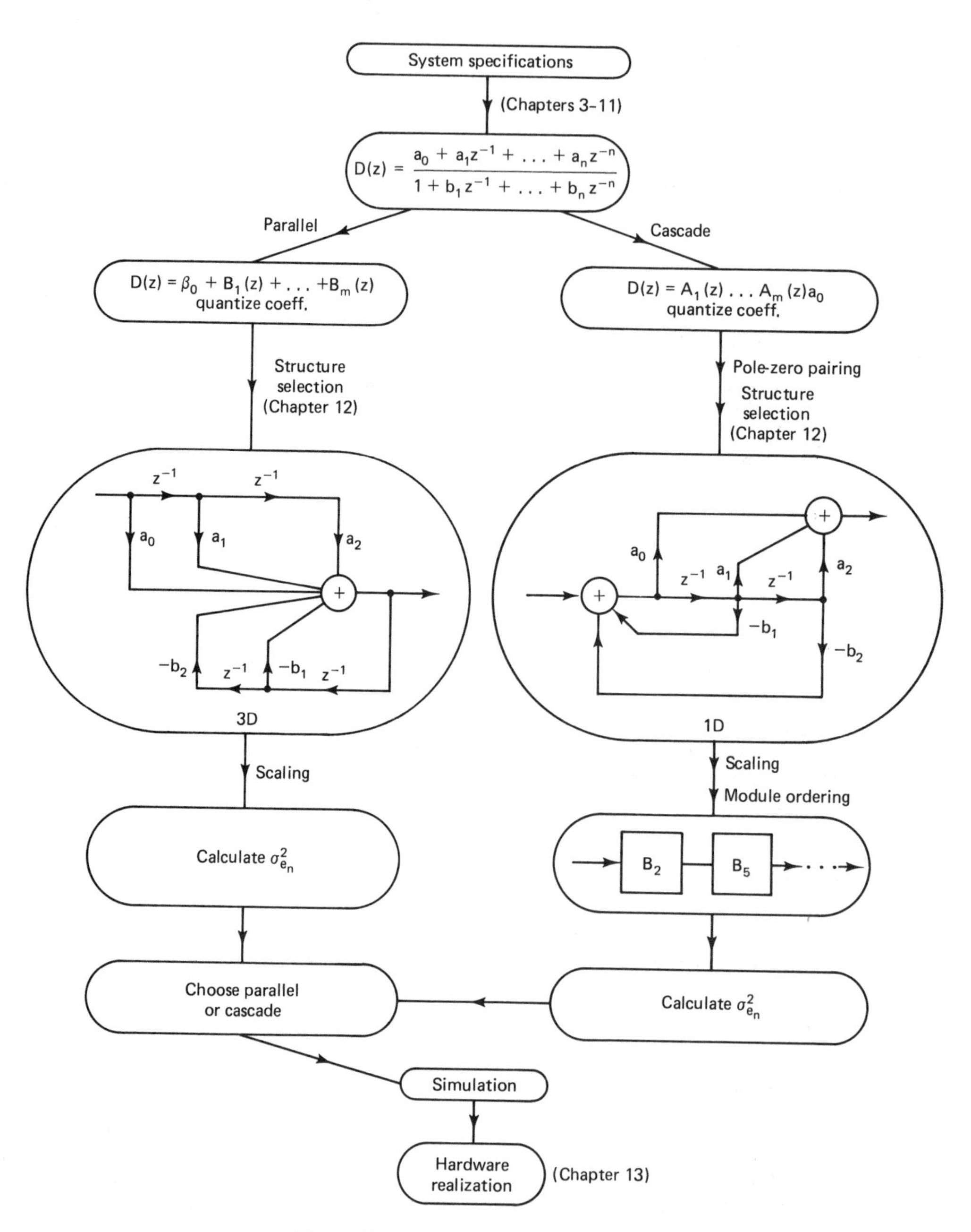

Figure 14-33 General design procedure.

REFERENCES

1. A. B. Sripad and D. L. Snyder, "Quantization Errors in Floating-Point Arithmetic," *IEEE Trans. Acoust., Speech Signal Process.*, Vol. ASSP-26, pp. 456–463, Oct. 1978.
2. R. E. Crochiere and A. V. Oppenheim, "Analysis of Linear Digital Networks," *Proc. IEEE*, Vol. 63, pp. 581–595, Apr. 1975.
3. B. Gold and C. M. Rudar, *Digital Processing of Signals.* New York: McGraw-Hill Book Company, 1969.
4. S. K. Mitra, K. Hirano, and H. Sakaguchi, "A Simple Method of Computing the Input Quantization and Multiplication Roundoff Errors in a Digital Filter," *IEEE Trans. Acoust. Speech Signal Process.*, Vol. ASSP-22, pp. 326–329, Oct. 1974.
5. L. B. Jackson, "An Analysis of Limit Cycles Due to Multiplication Rounding in Recursive (Sub) Filters," *Proc. 7th Annu. Allerton Conf. Circuit Syst. Theory*, 1969, pp. 69–78.
6. S. R. Parker and S. F. Hess, "Limit-Cycle Oscillations in Digital Filters," *IEEE Trans. Circuit Theory*, Vol. CT-18, pp. 687–697, Nov. 1971.
7. I. W. Sandberg and J. F. Kaiser, "A Bound on Limit Cycles in Fixed-Point Implementations of Digital Filters," *IEEE Trans. Audio Electroacoust.*, Vol. AU-20, pp. 110–112, June 1972.
8. T. A. C. M. Claasen, W. F. G. Mecklenbraeuker, and J. B. H. Peek, "Some Remarks on the Classification of Limit Cycles in Digital Filters," *Phillips Res. Rep.*, Vol. 28, pp. 297–305, Aug. 1973.
9. J. L. Long and T. N. Trick, "An Absolute Bound on Limit Cycles Due to Roundoff Errors in Digital Filters," *IEEE Trans. Audio Electroacoust.*, Vol. AU-21, pp. 27–30, Feb. 1973.
10. T. Claasen, W. F. G. Meckenbraeuker, and J. B. H. Peek, "Frequency Domain Criteria for the Absence of Zero-Input Limit Cycles in Nonlinear Discrete-Time Systems, with Applications to Digital Filters," *IEEE Trans. Circuits Syst.*, Vol. CAS-22, 232–239, Mar. 1975.
11. V. B. Lawrence and K. V. Mina, "Control of Limit Cycle Oscillations in Second-Order Recursive Digital Filters Using Constrainted Random Quantization," *IEEE Trans. Acoust. Speech Signal Process.*, Vol. ASSP-26, pp. 127–134, Apr. 1978.
12. L. B. Jackson, "Limit Cycles in State-Space Structures for Digital Filters," *IEEE Trans. Circuits Syst.*, Vol. CAS-26, pp. 67–68, Jan. 1979.
13. P. M. Ebert, J. E. Mazo, and M. G. Taylor, "Overflow Oscillations in Digital Filters," *Bell Syst. Tech. J.*, Vol. 48, pp. 2999–3020, Nov. 1969.
14. W. L. Mills, C. T. Mullis, and R. A. Roberts, "Digital Filter Realizations without Overflow Oscillations," *IEEE Trans. Acoust. Speech Signal Process.*, Vol. ASSP-24, pp. 334–338, Aug. 1978.
15. L. B. Jackson, "On the Interaction of Roundoff Noise and Dynamic Range in Digital Filters," *Bell Syst. Tech. J.*, Vol. 49, pp. 159–184, Feb. 1970.
16. S. Y. Hwang, "On Optimization of Cascade Fixed-Point Digital Filters," *IEEE Trans. Circuits Syst.*, Vol. CAS-21, pp. 163–166, Jan. 1974.
17. W. S. Lee, "Optimization of Digital Filters for Low Roundoff Noise," *IEEE Trans. Circuits Syst.*, Vol. CAS-22, pp. 424–431, May 1974.

18. B. Liu and A. Peled, "Heuristic Optimization of the Cascade Realization of Fixed-Point Digital Filters," *IEEE Trans. Acoust. Speech Sig. Process.*, Vol. ASSP-23, pp. 464–473, Oct. 1975.

19. E. Lueder, H. Hug, and W. Wolf, "Minimizing the Roundoff Noise in Digital Filters by Dynamic Programming," *Frequenz*, Vol. 29, pp. 211–214, 1975.

20. C. T. Mullis and R. A. Roberts, "Synthesis of Minimum Roundoff Noise Fixed Point Digital Filters," *IEEE Trans. Circuits Syst.*, Vol. CAS-23, pp. 551–562, Sept. 1976.

21. B. P. Gaffney and J. N. Gowdy, "A Symmetry Relationship for "Between" Scaling in Cascade Digital Filters," *IEEE Trans. Acoust. Speech Signal Process.*, Vol. ASSP-25, pp. 350–351, Aug. 1977.

22. G. Moschytz, *Linear Integrated Networks Fundamentals.* New York: Van Nostrand Reinhold Company, 1974.

23. A. Peled and B. Liu, *Digital Signal Processing: Theory, Design, and Implementation.* New York: John Wiley & Sons, Inc., 1976.

24. L. B. Jackson, "Roundoff Noise Analysis for Fixed Point Digital Filters Realized in Cascade or Parallel Form," *IEEE Trans. Audio Electroacoust.*, Vol. AU-18, pp. 107–122, June 1970.

25. E. P. F. Kan and J. K. Aggarwal, "Minimum-Deadband Design of Digital Filters," *IEEE Trans. Audio Electroacoust.*, Vol. AU-19, pp. 292–296, Dec. 1971.

26. G. A. Maria and M. F. Fahmy, "Limit Cycle Oscillations in a Cascade of First- and Second-Order Digital Sections," *IEEE Trans. Circuits Syst.*, Vol. CAS-22, pp. 131–134, Feb. 1975.

27. K. Steiglitz and B. Liu, "An Improved Algorithm for Ordering Poles and Zeros of Fixed-Point Recursive Digital Filters," *IEEE Trans. Acoust. Speech Signal Process.*, Vol. ASSP-24, pp. 341–343, Aug. 1976.

PROBLEMS

14-1. For the 1D structure for second-order modules:

(a) Find $\partial D(z)/\partial a_i$, $\partial D(z)/\partial b_i$.

(b) Check the results in part (a) using (14-80).

(c) Find $\Delta D(z)$.

14-2. Assume that

$$D(z) = \frac{1 + 1.858741z^{-1} + z^{-2}}{1 - 0.748386z^{-1} + 0.213374z^{-2}}$$

is to be implemented using the 1D structure and an 8-bit two's-complement number system:

$$\text{coefficients:}\quad (s \quad m\,.\,m \quad m \quad m \quad m \quad m \quad m)_{2\text{cns}}$$

(a) Find Δa_i, Δb_i.

(b) Find $\Delta D(z)$. *Hint:* Use the results of Problem 14-1.

(c) Plot $\Delta D(e^{j\omega T})$.

14-3. Repeat Problem 14-2 for a 16-bit two's-complement number system.

14-4. For the direct structures for second-order modules, assume an 8-bit two's-complement number system

$$\text{coefficient:} \quad (s \;\; m \,.\, m \;\; m \;\; m \;\; m \;\; m \;\; m)_{2\text{cns}}$$

$$\text{variables:} \quad (s \,.\, m \;\; m \;\; m \;\; m \;\; m \;\; m \;\; m)_{2\text{cns}}$$

Calculate the output steady-state error bound if each summing node is implemented using double-precision accumulation before round-off (Q_2 quantizers in Figures 14-14b and 14-21e). Use the $D(z)$ of Problem 14-2. Which structure exhibits the best steady-state error performance?

14-5. For the digital filters below, assume that two's-complement round-off is used to implement each filter in a direct structure. Are overflow oscillations possible? Are limit cycles possible? If limit cycles are possible, find their bounds.

(a) $D_1(z) = 1/(1 + 0.98381z^{-1} + 0.09443z^{-2})$

(b) $D_2(z) = 1/(1 + 0.17819z^{-1} - 0.17187z^{-2})$

(c) $D_3(z) = 1/(1 - 1.49513z^{-1} + 0.56292z^{-2})$

15

Case Studies

15.1 INTRODUCTION

This chapter presents three case studies of digital control systems that have been implemented. Two of the control systems were designed using the frequency-response techniques of Chapter 8, and the design of the third, even though empirical, was based on the techniques of Chapter 8.

The first case study is a second-order position control system (servomotor) for which both a phase-lead and a phase-lag controller were designed. The responses are compared. Sampling rate selection is discussed, and certain significant effects from plant nonlinearities are noted. The digital controller is implemented using a Texas Instruments TI9900 microprocessor.

The second system studied is an environmental chamber control system, which is composed of a temperature control system, a carbon dioxide control system, a chamber water-loss monitor, an outside rainfall monitor, and a data acquisition system. The temperature system controller is a PID compensator; the carbon dioxide system controller is a quasi-proportional compensator. Both control systems were designed empirically, and all operations are implemented via a time-shared TI9900 microprocessor. The microcomputer software is discussed.

The third case study is the lateral control system of an automatic aircraft landing system for U. S. Marine fighter aircraft. The plant for this control system is the aircraft lateral dynamics, including the bank autopilot, and is ninth order. The digital controller generates bank commands into the autopilot, and the aircraft position is determined by a phased array radar. The system contains significant noise and disturbance inputs, which must be considered in the controller design. The digital controller is a PID compensator plus added filtering to reduce noise effects.

15.2 SERVOMOTOR SYSTEM

The design of a digitally controlled servomotor system [1] is presented in this section. This system is low order and presents no particular design nor implementation difficulties. However, the system does contain nonlinearities which have an observable influence on system response. These nonlinear effects will be discussed as they are encountered.

The control-system block diagram is shown in Figure 15-1 and the system

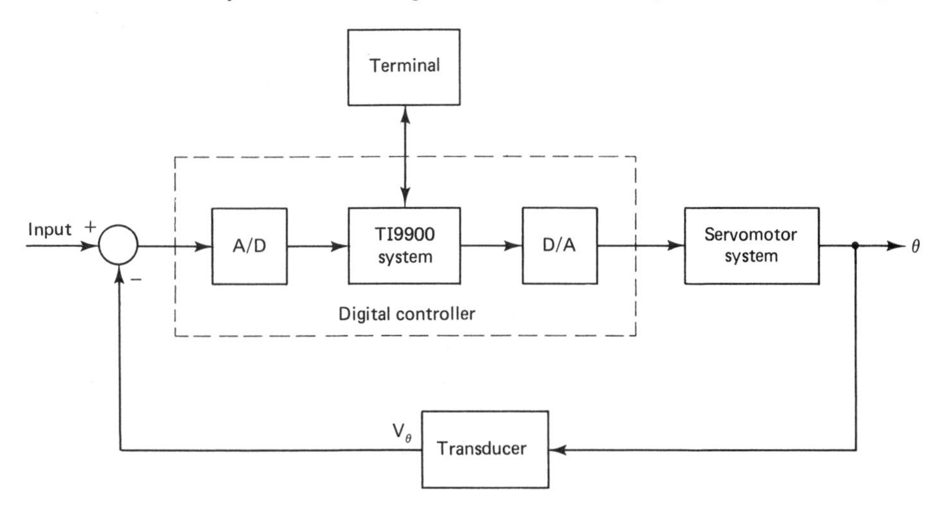

Figure 15-1 System block diagram.

hardware configuration is given in Figure 15-2. The Texas Instruments TI9900 microprocessor system [2] was chosen for the implementation of the digital controller. At the time of the system design, this microprocessor system was one of the few 16-bit processors available. In addition, the processor has hardware multiplication, and software support is available. The terminal indicated in Figure 15-1 was chosen to be a Texas Instruments Microterminal [2] and is used to initialize the system, change filter parameters if desired, test system operation, and so on.

The data (number) format for data internal to the computer is shown in Figure 15-3 and is fixed-point format. The magnitudes of both the fractional part and the integer part of numbers are represented by 16 bits. Thus the dynamic range is from a (nonzero) minimum of [3]

$$2^{-16} = 0.000015$$

to a maximum of

$$(2^{16} - 1)_{\text{integer}} + (1 - 2^{-16})_{\text{fraction}} = 65{,}535.999985$$

The 32-bit precision is not needed for this application. It was chosen to facilitate data manipulations, since processing time is abundant. In addition, the sign flag was placed

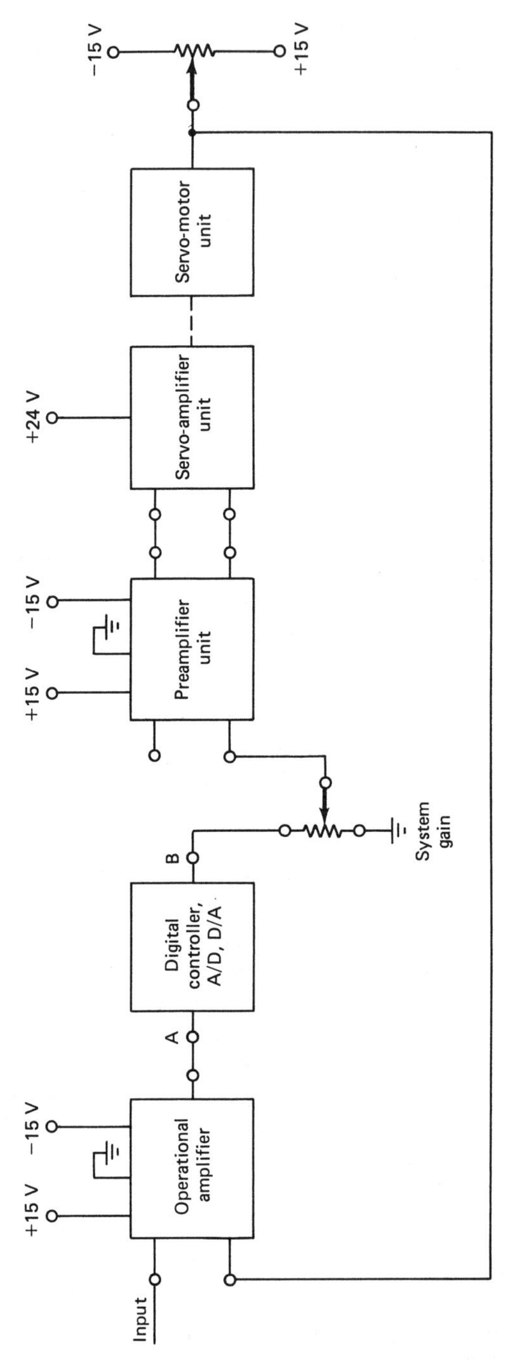

Figure 15-2 System hardware.

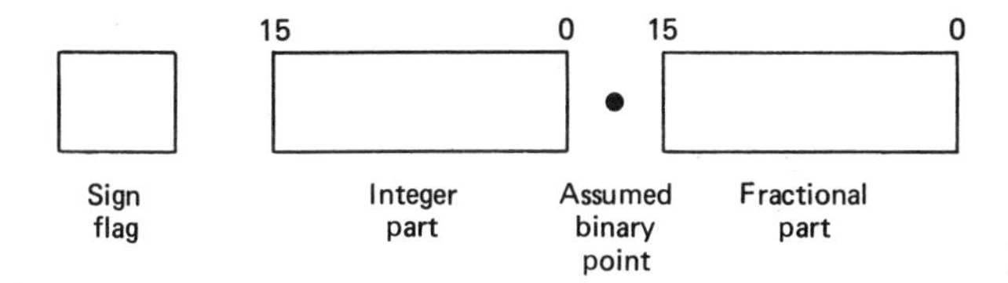

Figure 15-3 Computer data format.

external to the integer part of the number to facilitate multiplication. The filter coefficients are stored in sign-magnitude format, which is the form required for the multiplication hardware.

System Model

The control system can be mathematically modeled as shown in Figure 15-4. A first step in the design process was to determine the plant transfer function $G_P(s)$. This transfer function was obtained experimentally by removing the digital controller,

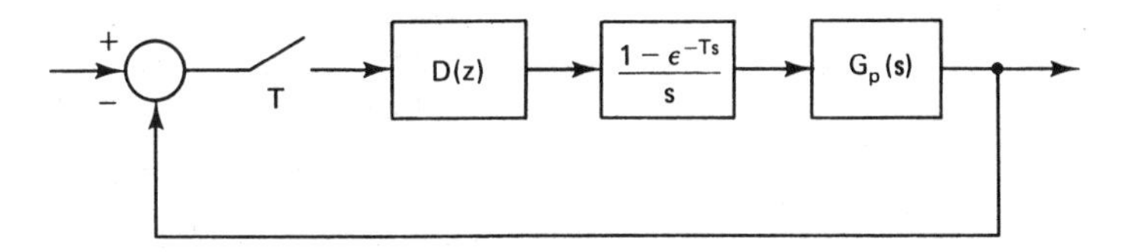

Figure 15-4 Control system model.

the A/D, and the D/A shown in Figure 15-2. Thus, in this figure, points A and B were connected. Since the servomotor is dc and armature-controlled, the plant transfer function was assumed to be [4]

$$G_P(s) = \frac{\omega_n^2}{s(s + 2\delta\omega_n)} \tag{15-1}$$

giving a closed-loop transfer function of

$$T(s) = \frac{G_P(s)}{1 + G_P(s)} = \frac{\omega_n^2}{s^2 + 2\delta\omega_n s + \omega_n^2} \tag{15-2}$$

Thus the system frequency response is given by

$$T(j\omega) = \frac{\omega_n^2}{(\omega_n^2 - \omega^2) + j2\delta\omega_n\omega} \tag{15-3}$$

To obtain the system model experimentally, a sinusoid was applied to the system and the frequency varied until the system output lagged the input by 90° in phase. As seen from (15-3), this frequency is ω_n. The amplitude of the response at this frequency was then used to calculate δ. The resulting transfer function was found to be

$$T(s) = \frac{36}{s^2 + 3.6s + 36} \tag{15-4}$$

Next a value for the sample period, T, was determined experimentally. One criterion that has been used successfully is to choose T as approximately one-tenth of the system rise time [5, 6]. The system step response is recorded in Figure 15-5,

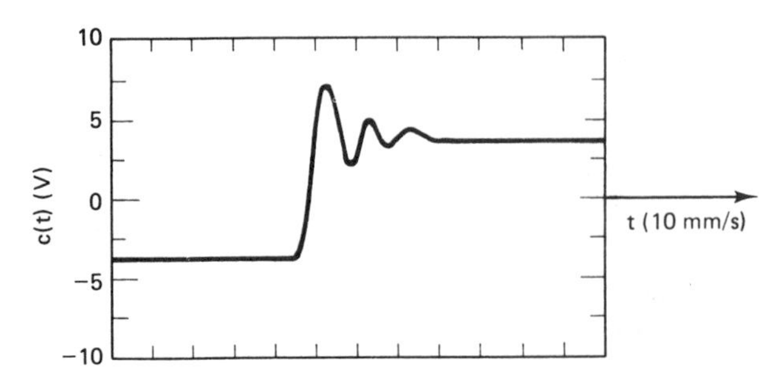

Figure 15-5 Step response of the servomotor system.

indicating a rise time of approximately 0.3 s. Thus a sample period of 30 ms is a value for consideration.

To test this sample period, the system was connected with the microcomputer in the loop, as shown in Figure 15-1. The computer was programmed to be a simple gain of unity (i.e., the A/D, computer, and D/A performed the function of a sampler and zero-order hold). The step response was then run for several values of T in the vicinity of T equal to 30 ms. The results are given in Figure 15-6.

First, note that the response for a sample period of 5 ms. is quite close to that of the analog system (see Figure 15-5). Next note that the system is approaching instability for $T = 40$ ms. However, the small amplitude of the oscillation indicates a nonlinear effect, since generally a linear system oscillation triggered by a step input will have approximately the amplitude of the step. This nonlinear oscillation is the limit cycle as discussed in Chapter 14. An investigation of this system determined that a jump resonance nonlinearity [7], usually caused by saturation, is present in the system. Jump resonance has been shown to produce limit cycles in sampled-data systems of this type [8]. Thus, for this system, the sampling rate is determined not by linear sampling theory, but by a system nonlinearity. It should be pointed out that a servo system is often designed such that the servoamplifier is saturated for a large percentage of the time. Thus the motor input voltage is maximum during this time, ensuring maximum speed of response.

Consideration of the step responses in Figure 15-6 resulted in a choice for T of 5 ms. With this value of T, we can calculate $G(z)$. From (15-2) and (15-4),

$$G_P(s) = \frac{36}{s(s + 3.6)} \tag{15-5}$$

Then

$$G(z) = \mathscr{z}\left[\frac{36(1 - \epsilon^{-Ts})}{s^2(s + 3.6)}\right] = \frac{0.00044731z + 0.00044739}{z^2 - 1.982161z + 0.982161} \tag{15-6}$$

The closed-loop transfer function is then

$$\frac{G(z)}{1 + G(z)} = \frac{0.00044731z + 0.00044739}{z^2 - 1.981714z + 0.982608} \tag{15-7}$$

Note the numerical problems. At low frequencies (in the vicinity of $z = 1$), the frequency response is determined principally by the last digit of the numerator coefficients, and thus numerical inaccuracies will be present. These numerical problems are caused by using a sampling frequency that is very large compared to the system natural frequencies. To calculate the frequency response from (15-6) and

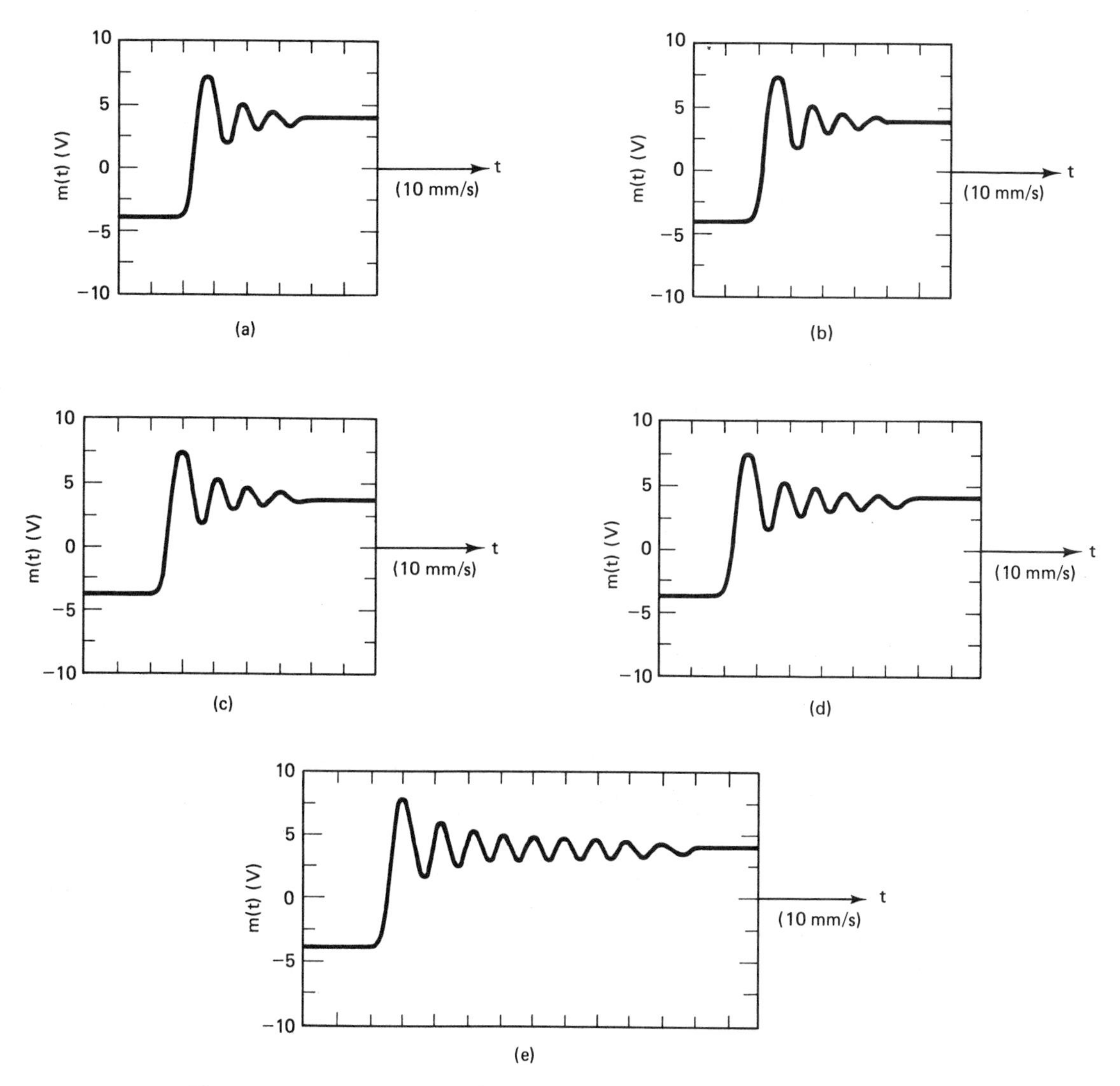

Figure 15-6 Step response of the servomotor for different sampling periods: (a) $T = 5$ ms; (b) $T = 10$ ms; (c) $T = 20$ ms; (d) $T = 30$ ms; (e) $T = 40$ ms.

(15-7), numerical accuracy greater than that shown must be employed. However, use of (7-23) to calculate the frequency response circumvents these problems.

Shown in Figure 15-7 is the measured system frequency response and also the frequency response of an analog (linear) computer simulation of the system. The system frequency response in the vicinity of the resonance peak is actually double valued, because of the jump resonance effects [7], but is shown as being single valued. Since both of these curves were obtained experimentally, they are plotted versus real, or s-plane, frequency. The w-plane frequency is obtained from (7-10):

$$\omega_w = \frac{2}{T} \tan \frac{\omega T}{2} \tag{15-8}$$

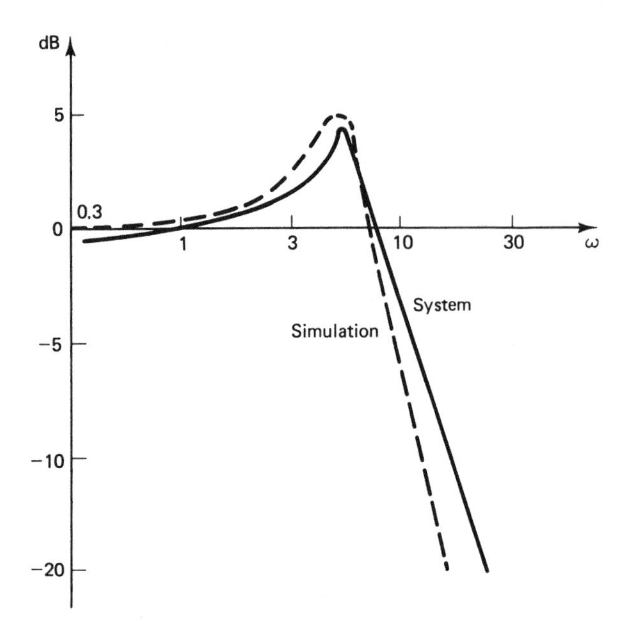

Figure 15-7 Closed-loop frequency response. $D(z) = 1$, $T = 0.005$ s.

Design

The frequency response for $G(z)$, obtained using (7-23) (program 3 in Appendix I), is plotted in Figure 15-8.

As described in Chapter 8, phase lead design is principally trial and error. Use of the procedure of Chapter 8 resulted in a phase-lead filter transfer function of

$$D(w) = \frac{0.693(1 + w/2.21)}{1 + w/44.8} \tag{15-9}$$

and thus, from (8-14),

$$D(z) = \frac{12.74z - 12.6}{z - 0.798} \tag{15-10}$$

The computer calculates

$$y(nT) = 12.74x(nT) - 12.6x(nT - T) + 0.798y(nT - T)$$

The resulting phase margin was 80°. The system closed-loop frequency response is given in Figure 15-9. Both responses were obtained experimentally. The analog computer simulation, with filter, was linear; thus the closeness of the two responses indicates little nonlinear effects in the system. The system step response is given in Figure 15-10.

A phase lag filter was also designed, using the procedure given in Chapter 8. The resultant filter transfer function is

$$D(w) = \frac{1.36(1 + w/0.20)}{1 + w/0.044} \tag{15-11}$$

and thus

$$D(z) = \frac{0.3z - 0.2997}{z - 0.99978} \tag{15-12}$$

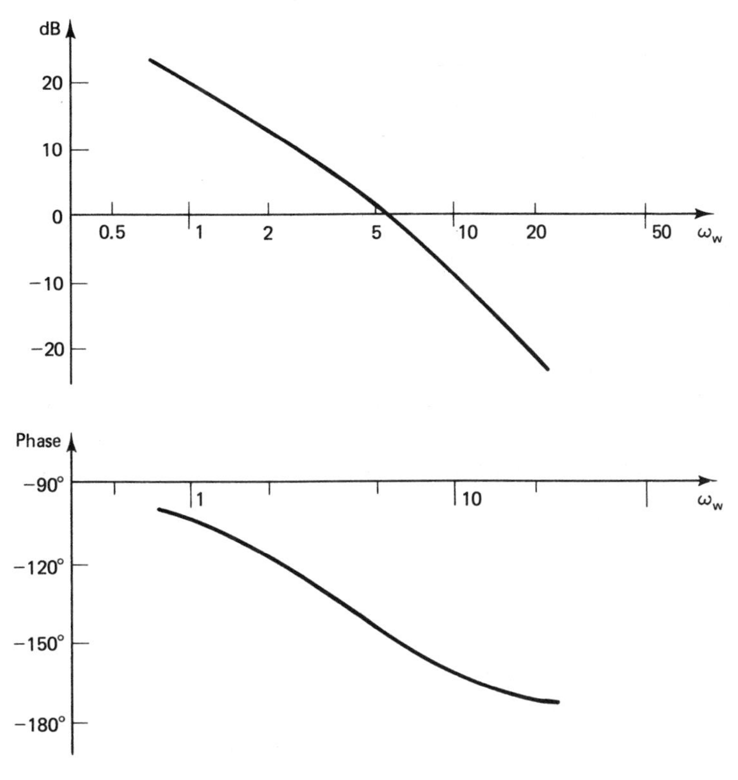

Figure 15-8 Open-loop frequency response. $T = 0.005$ s.

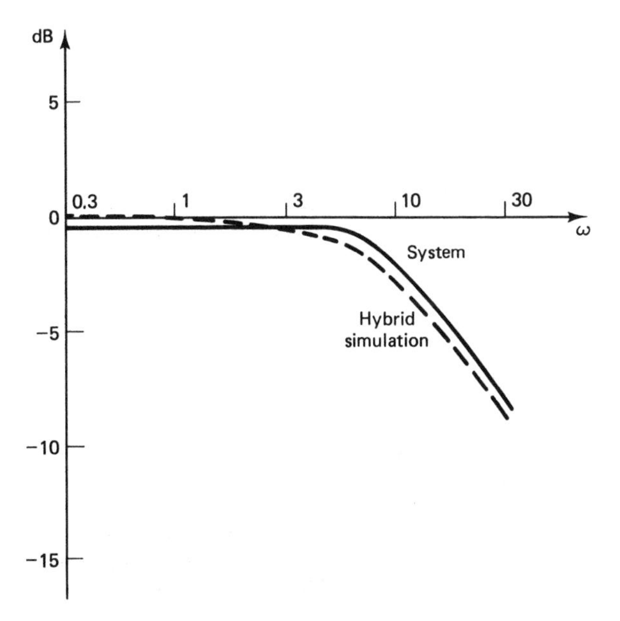

Figure 15-9 Closed-loop frequency response with phase-lead compensation ($T = 5$ ms).

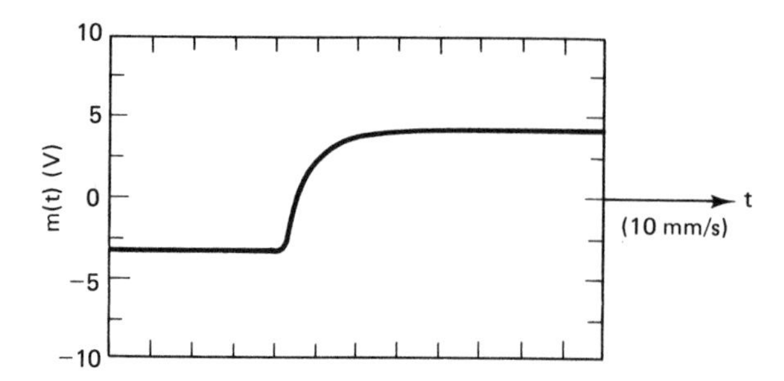

Figure 15-10 Step response of the servomotor system with a phase-lead compensator ($T = 5$ ms).

Here the computer calculates

$$y(nT) = 0.3x(nT) - 0.2997x(nT - T) + 0.99978y(nT - T)$$

The phase margin for this system is 60°. However, when the frequency responses were obtained experimentally, the curves in Figure 15-11 resulted. Here the differences in the frequency responses indicate gross nonlinear effects. The system step response is given in Figure 15-12.

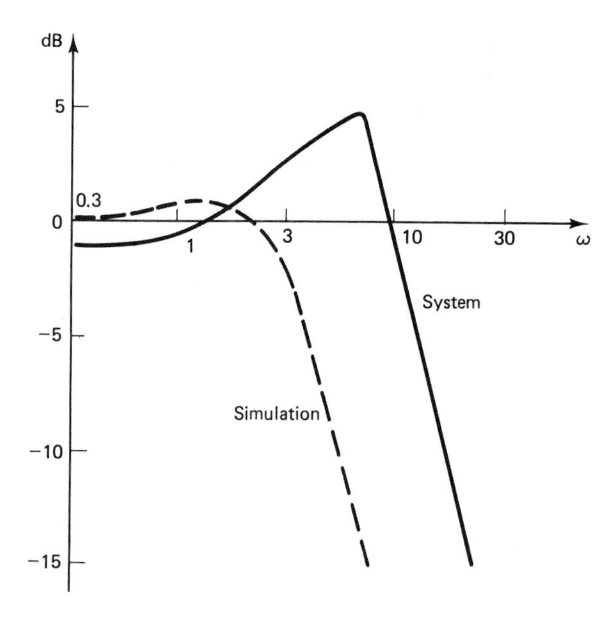

Figure 15-11 Closed-loop frequency response with phase-lag compensation ($T = 5$ ms).

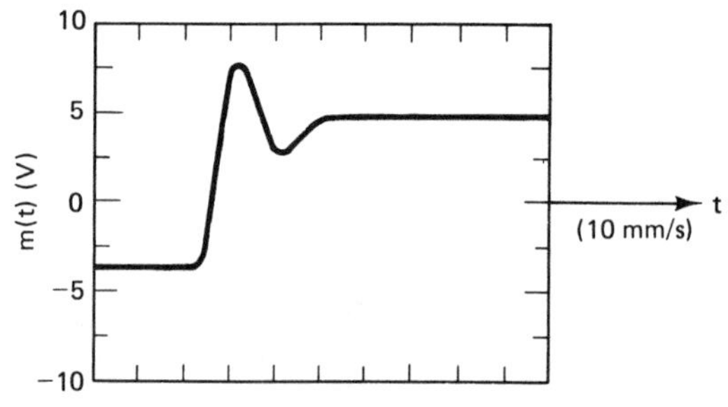

Figure 15-12 Step response of the servomotor system with a phase-lag compensator ($T = 5$ ms).

15.3 ENVIRONMENTAL CHAMBER CONTROL SYSTEM

This section presents the case study of a digital control system for an environmental chamber designed for the study of plant growth [9]. Two control systems were implemented: one to control dry-bulb temperature and the other to control the carbon dioxide (CO_2) content in the chamber atmosphere.

The chamber is constructed of plexiglass with measurements of approximately $0.7 \times 1.3 \times 2$ m. The chamber system is hermetically sealed such that both moisture content and carbon dioxide content can be accurately monitored.

Figure 15-13 gives a hardware description of the temperature control system. The chamber is cooled (air conditioned) to the extent that heaters are required to

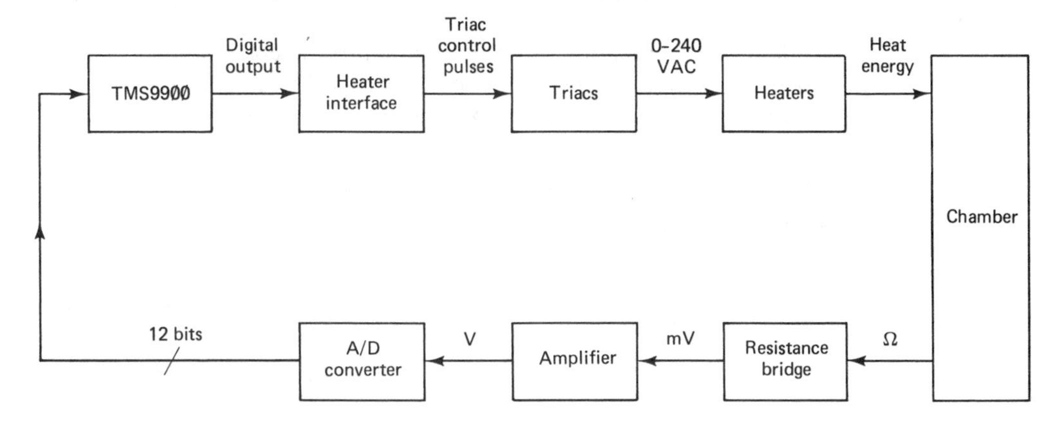

Figure 15-13 Chamber temperature control hardware diagram.

maintain the desired temperature. Thus the heaters are the controlling elements in the closed-loop system. The resistance bridge is dictated by the temperature sensor used, and has an output in the millivolt range. An operational amplifier increases the amplitude of this signal to the ± 5-V range required by the A/D converter. The measured temperature is then subtracted from the desired temperature, which is stored in the TMS 9900 microcomputer system. Next this error signal is processed by the TMS 9900 (the system compensation), resulting in an output signal to the heater interface. The heater interface is a complex logic circuit [9] which converts the computer output signal into the triacs' [10] control pulses. The triacs control the electrical energy into the heaters by in effect controlling the rms voltage applied to the heaters.

The hardware description of the CO_2 control system is given in Figure 15-14. The CO_2 content of the chamber atmosphere is measured by the gas analyzer in parts per million (ppm). This signal is compared to the desired set point, and if the error is negative (a CO_2 deficit), the computer opens a solenoid value from a CO_2 supply for a length of time dependent on the error magnitude. The control system has no capability to remove excess CO_2. Because of the time lag required for the gas analysis, the sampling rate for this control system cannot be greater than 0.0222 Hz ($T = 45$ s). In addition, no compensation is employed. However, as will be shown later, satisfactory control occurs.

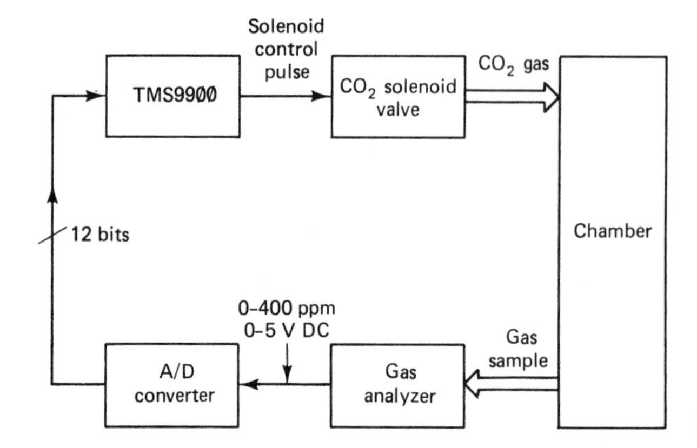

Figure 15-14 CO_2 control hardware.

The data (number) format for data internal to the computer is given in Figure 15-15 and is a two's-complement fixed-point format [3]. For both control systems, the 16-bit accuracy was found to be sufficient, and convenient.

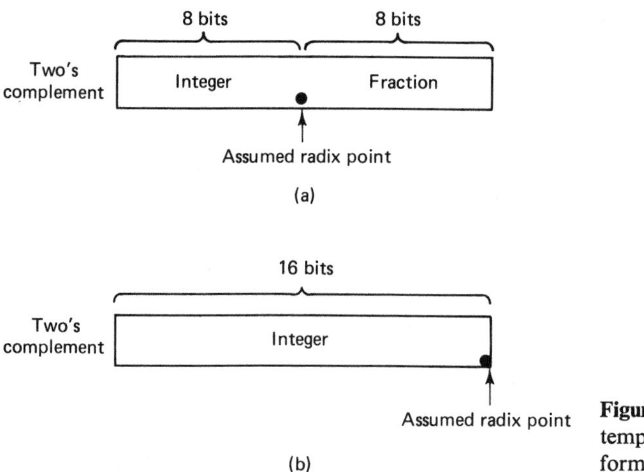

Figure 15-15 (a) Binary format for temperature software; (b) binary format for CO_2 software.

Temperature Control System

The temperature control system was designed empirically. Because of the complexity of the system, no attempt was made to derive a plant model. Instead, a step response was run between the heater input and the temperature sensor output. The temperature rise was found to be approximately exponential with a time constant of 60 s. Thus, as a first try, the sample period was set to 6 s.

To aid in the empirical design, a proportional-plus-integral-plus-derivative (PID) controller was chosen for implementation (see Chapter 8). The analog version of the PID filter is

$$m(t) = K_P e(t) + K_I \int e(t)\,dt + K_D \frac{de(t)}{dt} \tag{15-13}$$

where $e(t)$ is the controller input and $m(t)$ is the controller output. The discrete controller implementation of the integrator was chosen to be

$$m_i(k) = \frac{T}{2}[e(k) + e(k-1)] + m_i(k-1) \tag{15-14}$$

and for the differentiator,

$$m_d(k) = \frac{1}{T}[e(k) - e(k-1)] \tag{15-15}$$

[see (8-34) and (8-37)]. Hence the discrete controller implementation of (15-13) is

$$m(k) = K_P e(k) + K_I m_i(k) + K_D m_d(k) \tag{15-16}$$

It is evident from these equations that four parameters are to be determined: T, K_P, K_I, and K_D. However, if an appropriate choice is made for T such that accurate differentiation and integration occur, the parameters K_P, K_I, and K_D will be independent of T.

The hardware for the temperature control system was chosen prior to design of the controller. In many applications the control system designer will find predefined hardware constraints, and must learn to adapt his design techniques, from ideal to practical. Two heaters were specified: one with a slow response (large mass) and one with a faster response (smaller mass). Thus the slow-response heater adds phase lag to the system, which is undesirable. Subsequently, experimentation on the physical system resulted in a choice for the form of the controller as shown in Figure 15-16. Different PID gains (P1, I1, Di, equivalent to K_P, K_I, K_D, respectively) are implemented for the fast-response heater [controlled by the signal $c_1(k)$] from those employed with the slow-response heater, which is controlled by $c_2(k)$. This form for the con-

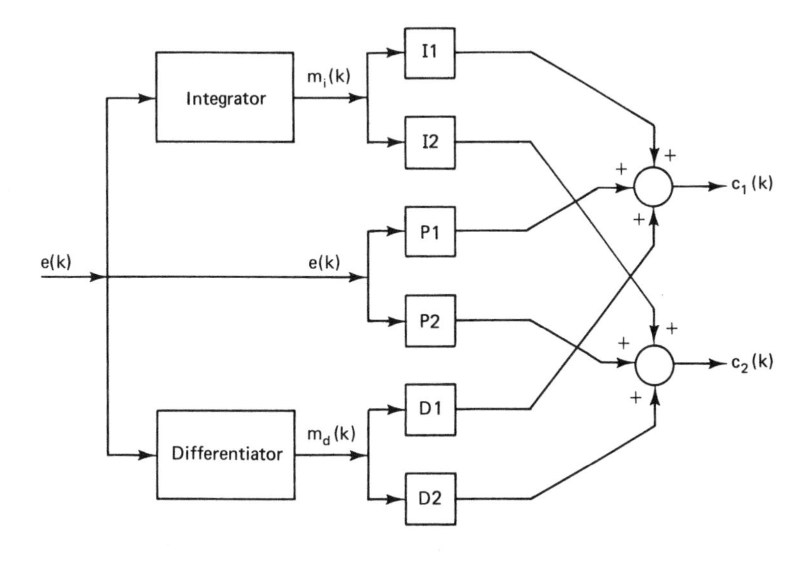

where e(k) = error @ t = kT
$m_i(k) = T/2*e(k) + T/2*e(k-1) + m_i(k-1)$
$m_d(k) = 1/T*e(k) - 1/T*e(k-1)$

Figure 15-16 Block diagram of PID filter implementation.

troller results in a better system response than that obtained by controlling both heaters with a single signal.

The controller gains were determined by first implementing a proportional-only controller, and varying P1 and P2 in Figure 15-16. A typical step response is given in Figure 15-17, with P1 and P2 both equal to 25. The small oscillation visible in this response is caused by the air conditioner cycling on and off. The control system was commanded to a 10° Celsius step input, from 25°C to 35°C. Note the steady-state error in Figure 15-17. During these tests, the sample period T (Ts in Figure 15-17) was also varied, and a value of $T = 1$ s was chosen; that is, choosing T less than 1 s caused no observable improvement in the system response.

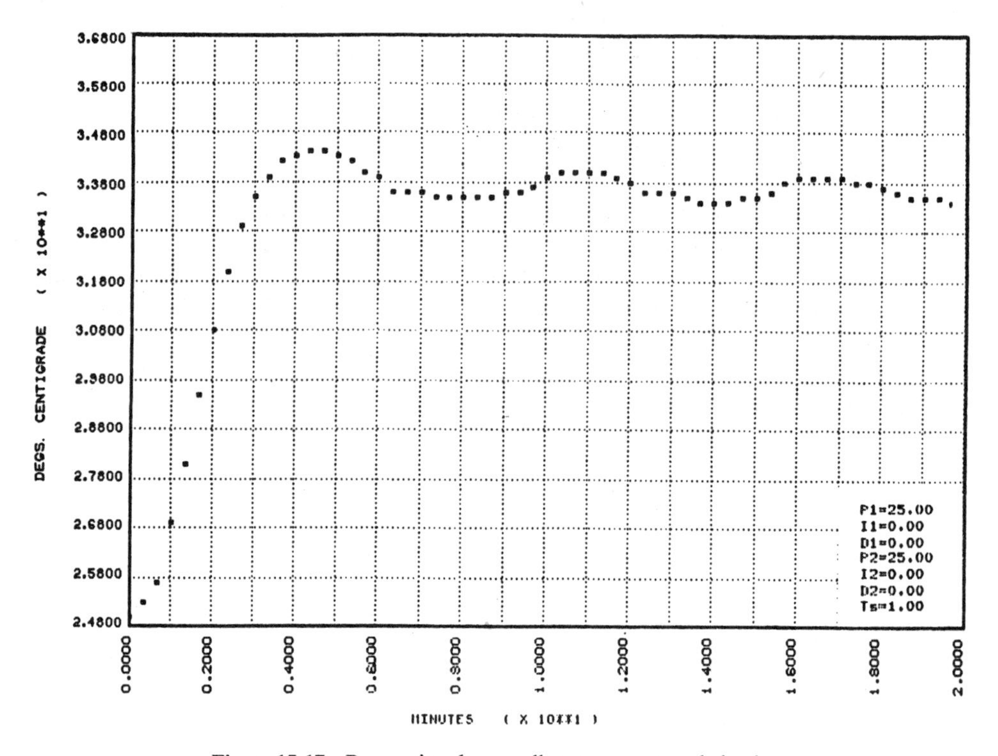

Figure 15-17 Proportional controller step response behavior.

Next the integral term was added to the controller, and a typical step response is given in Figure 15-18. Note the elimination of the steady-state error, as expected (see Section 8.9). Finally, the derivative term was added to the controller, resulting in a typical response as shown in Figure 15-19. Note that no improvement occurs in the rise time for the PID controller, because the heaters are full on for large error signals for both the P controller and the PI controller (a nonlinear effect). Thus the system cannot respond any faster for large errors and this effect is independent of the form of the controller. However, for small errors (e.g., small overshoot), the derivative term in the controller has a definite and predictable effect (normal linear operation).

In all the tests above, both heaters were controlled by the same signal. Attempts

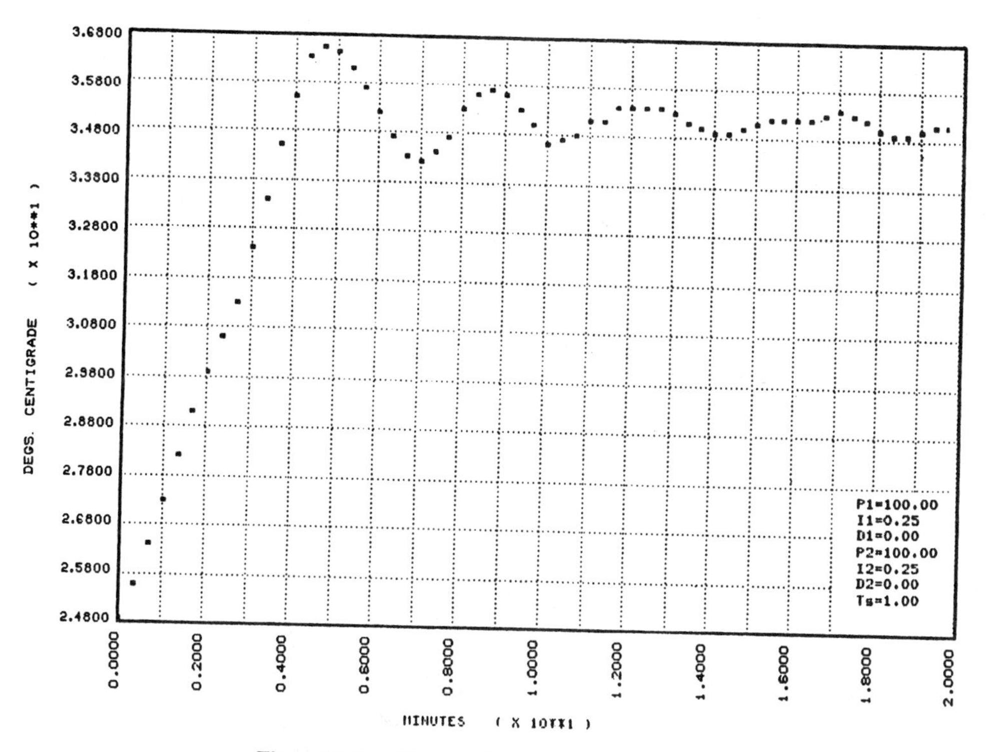

Figure 15-18 PI controller step response behavior.

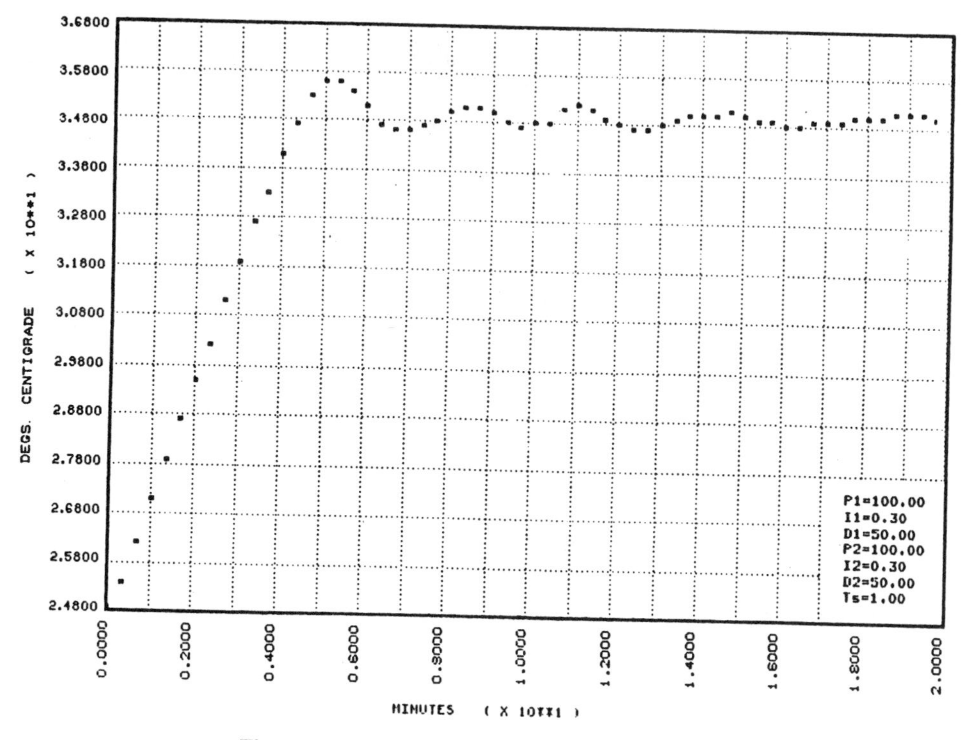

Figure 15-19 PID controller step response behavior.

to increase the derivative signal beyond certain values while using this strategy resulted in a tendency toward instability. This effect is predictable, since the maximum phase lead attainable from a PID controller is 90°, but the gain may be increased without limit (theoretically). At this point, the PID gains for the fast heater were varied from those for the slow heater in order to reduce overshoot and settling time. The final values for the filter gains were chosen to be: P1 = 100, I1 = 0.3, D1 = 100, P2 = 25, I2 = 0.25, D2 = 25, and T = 1 s.

A typical 24-h system response is given in Figure 15-20. Time equal to zero is midnight. The carbon dioxide can be controlled only during daylight hours, since plants emit CO_2 at night. For this figure, the temperature was commanded to follow a triangular wave which is an approximation to actual temperature variations during a typical 24-h period.

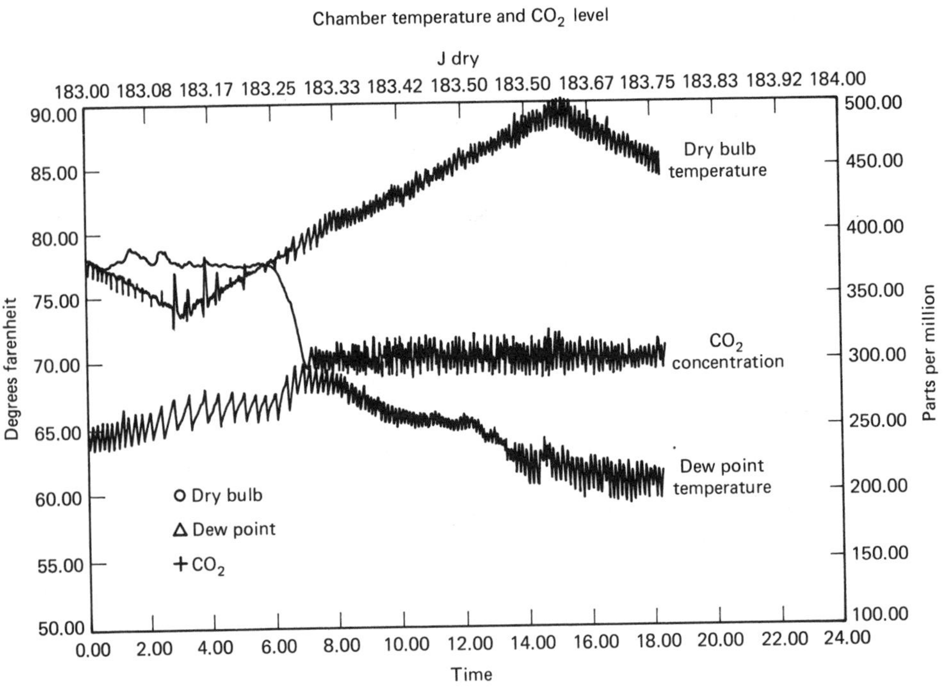

Figure 15-20 Chamber behavior for normal operating conditions.

Software [9]

Real-time operating software was written to control the complete operation of this system. The computer operation is based on interrupts [11], which are in effect subroutine calls whose execution is prompted by hardware signals. For example, the microcomputer system constructed for the control system contains an interval timer. Once each 250 ms this timer outputs a hardware signal to a specific terminal of the microprocessor, causing the microprocessor to interrupt normal processing and update a real-time clock. The real-time clock is composed of memory locations which contain a numerical description of the present second, minute, hour, and day of the year (Julian day).

A flowchart of the operating software is given in Figure 15-21. Note that the principal program operation is through the interrupts. After system initialization, the

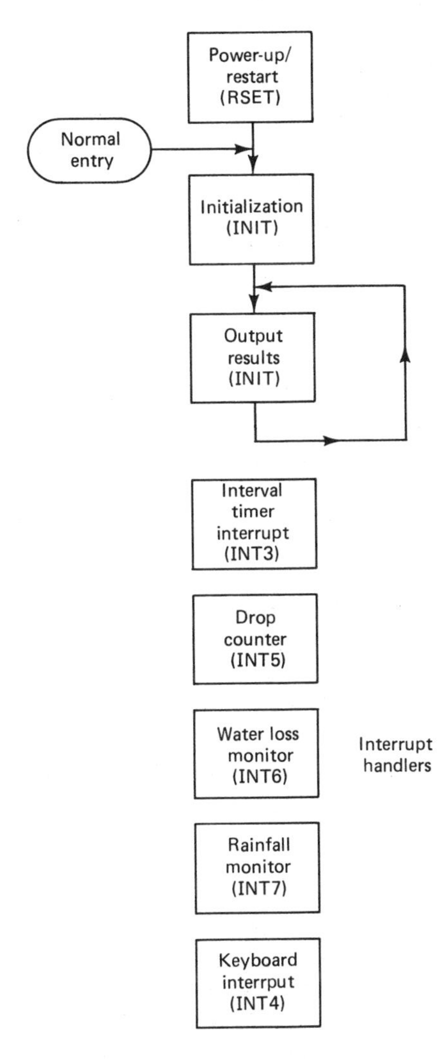

Figure 15-21 Program flowchart.

computer is in a wait loop except when printing is required. The interrupt system for the TI 9900 microprocessor is based on priorities (i.e., certain interrupts take precedence over others). The interrupt priorities for this system's operating software is given in Table 15-1. The various interrupt service routines will now be described.

INTO. This routine provides an automatic restart of the system following a power failure, and points a message stating that a power failure has occured.

TABLE 15-1 OPERATING SYSTEM PRIORITIES

Priority ranking	Task description	How implemented
1	Power-up/restart	$\overline{\text{INT0}}$
2	Program initialization	Main program (system not operating)
3	Real-time clock; solenoid valve operation; data output; temperature set point update; control functions	$\overline{\text{INT3}}$
4	Water drop counting	$\overline{\text{INT5}}$
5	Water volume measurement	$\overline{\text{INT6}}$
6	Rainfall measurement	$\overline{\text{INT7}}$
7	Parameter entry	$\overline{\text{INT4}}$
8	Printed results	Main program (system operating)

INT3. This interrupt is caused by the interval timer decrementing to zero (250 ms has passed), and the service routine is responsible for five tasks: CO_2 value operation, real-time clock service, data output into a data acquisition system, temperature set point update, and sample-interval timing. The sample-interval timing initiates the control processing for the CO_2 controller and the temperature controller at the appropriate sample instants.

INT4, INT5, INT6. These three interrupt service routines allow devices for water measurement to be monitored. Water drop counting and water volume measurement measure the water removed from the chamber atmosphere by the air conditioner. Rainfall measurement records rainfall outside the chamber. All three routines simply count the total number of interrupts occurring (each representing a known volume of water) and reset the devices. The totals are output for data acquisition purposes by the INT3 routine.

15.4 AIRCRAFT LANDING SYSTEM

This section presents the design of an automatic aircraft landing system. The particular control system described is that of the Marine Air Traffic Control and Landing System (MATCALS) [12, 13]. The system is illustrated in Figure 15-22. The radar (the control system sensor) employs a phased array antenna, which was chosen because the radar beam can be rapidly repositioned to control a number of planes at sampling rates of up to 40 Hz. However, the positional information provided by the radar system is corrupted with significant noise, which presents problems to the control system designer.

The control operation is composed of two independent (decoupled) control systems: the vertical control system, which keeps the plane on a 3.5° glide slope, and the lateral control system, which maintains the plane on the extended centerline of

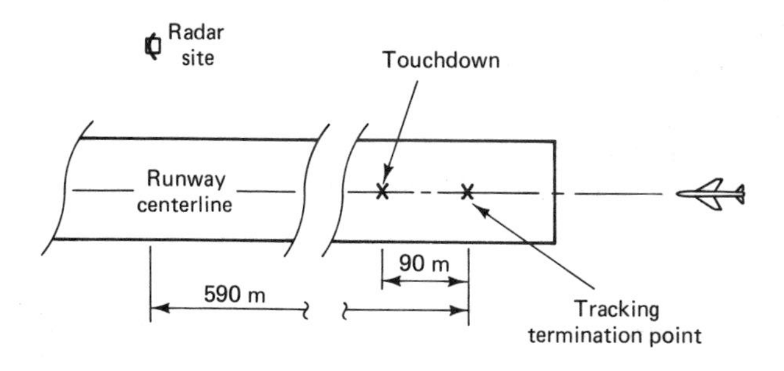

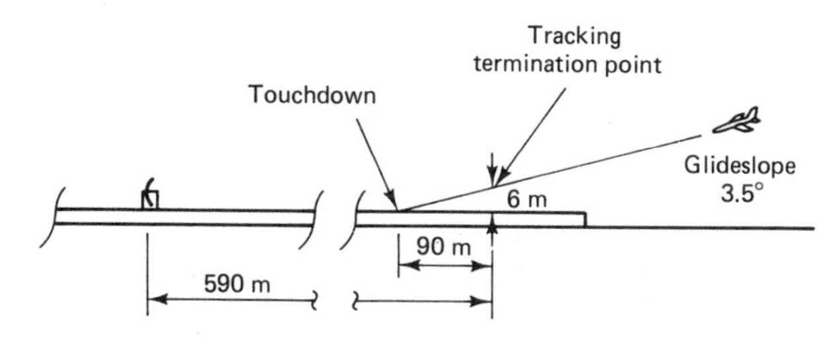

Figure 15-22 Aircraft landing system.

the runway. These two control systems are illustrated in Figure 15-23. Two separate computer-processing algorithms are utilized in each control system. The first algorithm processes the antenna return signals to determine the aircraft position (centroid position) with respect to the runway touchdown point. The second algorithm comprises the difference equations that describe the digital controllers. The outputs of the

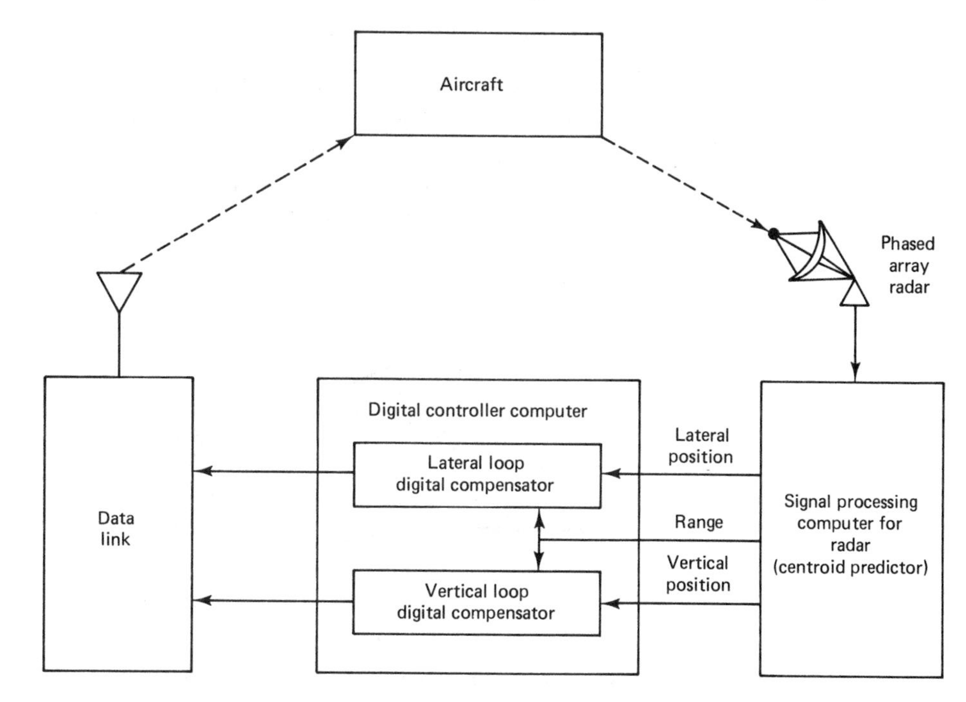

Figure 15-23 MATCALS automatic landing control loops.

controllers are transmitted via a data link to the aircraft and applied to the appropriate autopilot inputs: the pitch autopilot for the vertical control system and the bank autopilot for the lateral control system.

In the following development, only the lateral control system will be discussed. The design of the vertical control system follows similar paths of development.

Plant Model

A block diagram of the lateral control system is given in Figure 15-24. The lateral aircraft dynamics, described by the transfer function $G_L(s)$, include the bank autopilot

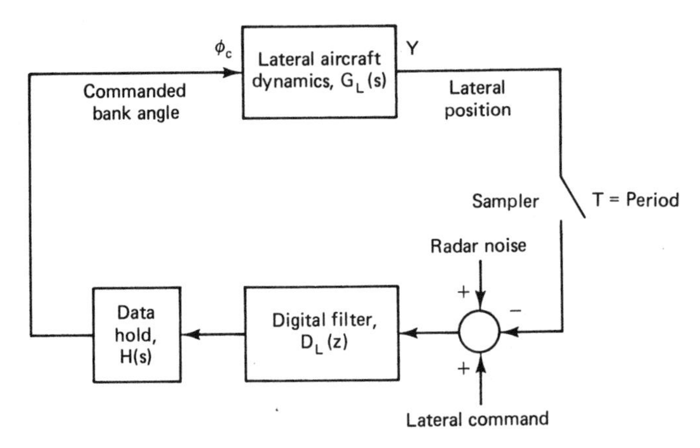

Figure 15-24 Lateral control loop.

and typically have a frequency response as shown in Figure 15-25. These dynamics are of course aircraft dependent; the frequency response of the McDonnell Aircraft Company F4J aircraft is given in Figure 15-26, and was obtained from measurements on the aircraft in flight [14]. The lateral aircraft equations of motion will not be developed here (see Ref. 15), but are ninth order for the F4J aircraft with the auto-

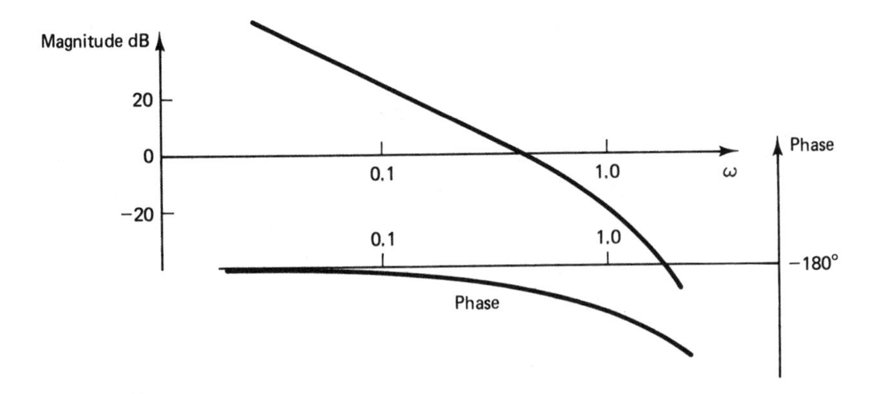

Figure 15-25 Typical frequency response of $G_L(s)$.

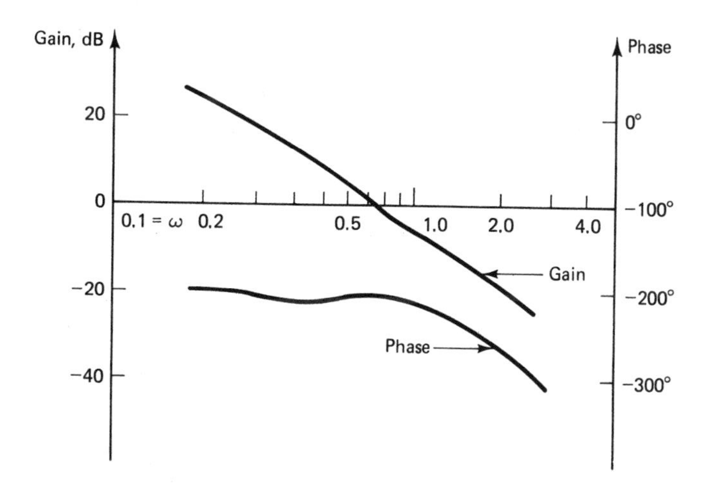

Figure 15-26 F4J lateral frequency response (Y/ϕ_c).

pilot included. Note from the −40-dB slope on the magnitude curve of $G_L(j\omega)$ in Figure 15-26 that $G_L(s)$ has a second-order pole at the origin.

A sketch of the Nyquist diagram for the F4J aircraft lateral control system is given in Figure 15-27, as can be seen from Figure 15-26. Since the aircraft with autopilot must be (open-loop) stable, it is evident that phase-lead compensation must be employed to eliminate the encirclement of the -1 point, in order to stabilize the system.

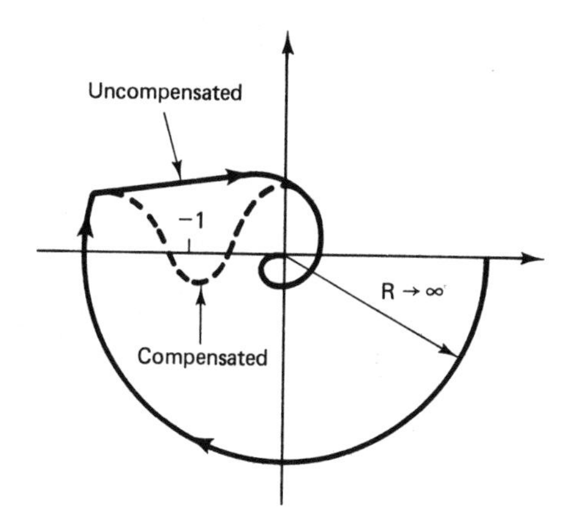

Figure 15-27 Nyquist diagram of $G_L(s)$, and $G_L(s)$ with possible compensation.

Design

There are three significant disturbance sources in the lateral control system which must be considered in the system design. The first of these is the radar noise, and is indicated in Figure 15-24. The other two disturbances, which will be considered first, are direct inputs to the aircraft lateral dynamics. Thus each of these disturbances may be modeled as shown in Figure 15-28. Hence the influence of the disturbance on the

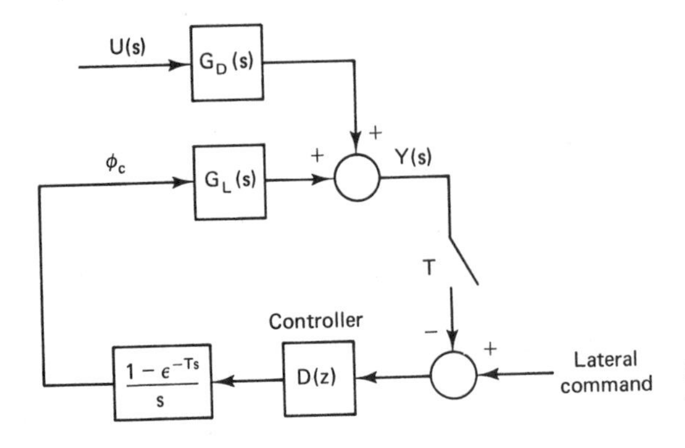

Figure 15-28 Lateral control system with a disturbance.

lateral position $y(k)$ is given by

$$Y(z) = \frac{\overline{G_D U}(z)}{1 + D(z)G(z)} \tag{15-17}$$

where

$$G(z) = \mathfrak{z}\left[\frac{1 - \epsilon^{-Ts}}{s} G_L(s)\right] \tag{15-18}$$

and, as stated previously, $G_L(s)$ has two poles at $s = 0$, resulting in two poles of $G(z)$ at $z = 1$.

One of the disturbances to be modeled as $U(s)$ in (15-17) is a relatively constant error (called a dc bias) in the output of a rate gyroscope in the autopilot. This error can be treated as a system input. Note that, in general, a constant input for the disturbance in Figure 15-28 will result in a constant output $y(k)$ in the steady state. The aircraft would then land to one side of the runway centerline, which is unacceptable. Thus a system requirement is that the final-value theorem when applied to (15-17) yield a value of zero; that is,

$$\lim_{k \to \infty} y(k) = \left. \frac{(z-1)\overline{G_D U}(z)}{1 + D(z)G(z)} \right|_{z=1} = 0 \tag{15-19}$$

with $U(s) = 1/s$. Now both $G_D(s)$ and $G_L(s)$ of Figure 15-28 have two poles at $s = 0$. Thus $\overline{G_D U}(z)$ has three poles at $z = 1$, and $G(z)$ has two poles at $z = 1$. If $D(z)$ is given a pole at $z = 1$, (15-19) is satisfied. Hence a proportional-plus-integral (PI) controller is a system requirement. As was seen from Figure 15-27, phase lead is required for stability. Since the derivative term in the PID controller contributes phase lead [see (8-47)], the PID controller was chosen to compensate this system. The compensated-system Nyquist diagram then appears as shown in Figure 15-29.

The second disturbance source of the type illustrated in Figure 15-28 is the wind. It is desirable that the aircraft not respond to wind inputs, but of course this is not possible. So the controller is designed to reduce the effects of wind. Consider again the effect of the wind disturbance as modeled in Figure 15-28 and equation (15-17). Of course, $G_D(s)$ is different for wind disturbances from that for the dc bias disturbance. In (15-17), the response is reduced if $D(z)G(z)$ is made large in the frequency range of the wind (turbulence) input signal. Since the frequency response of $G(z)$ is fixed, $D(z)$ should be designed to increase the system bandwidth. As we have seen in Chapter 8,

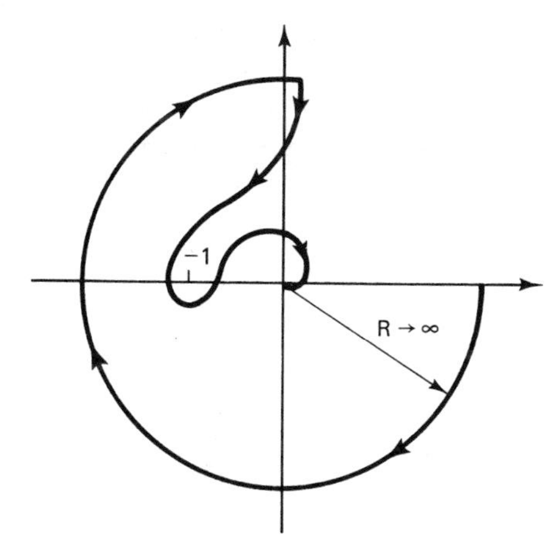

Figure 15-29 Lateral control system Nyquist diagram.

the derivative path of the PID controller not only introduces phase lead, but also increases system bandwidth. Hence we require a high gain in the derivative path.

The third disturbance source, which is the noise present in the radar system output signal, is modeled as shown in Figure 15-24. Note that the radar noise enters the system at the same point as does the lateral position command signal. Thus a system design that yields a good response to the command input also yields a good response to the radar noise. This noise response is a major problem in the design of this system. For aircraft carrier-based systems [16], a parabolic radar antenna system is used. The signal from this type of antenna system is relatively noise-free, and the noise problem is not as critical as in the land-based system.

α-β filter. To reduce the effects of the radar noise, an α-β tracking filter [17] was chosen to filter the radar signal at the controller input. (The α-β filter equations are given in Problem 2-22.) This filter is designed to estimate $y(k)$, the aircraft lateral position, and $\dot{y}(k)$, the aircraft lateral velocity, given a noisy radar system output signal. The estimate $y(k)$ is then transmitted to the position path and the integrator path, and the estimate $\dot{y}(k)$ to the derivative path, of the PID compensator. Thus the PID compensator has the basic configuration shown in Figure 15-30.

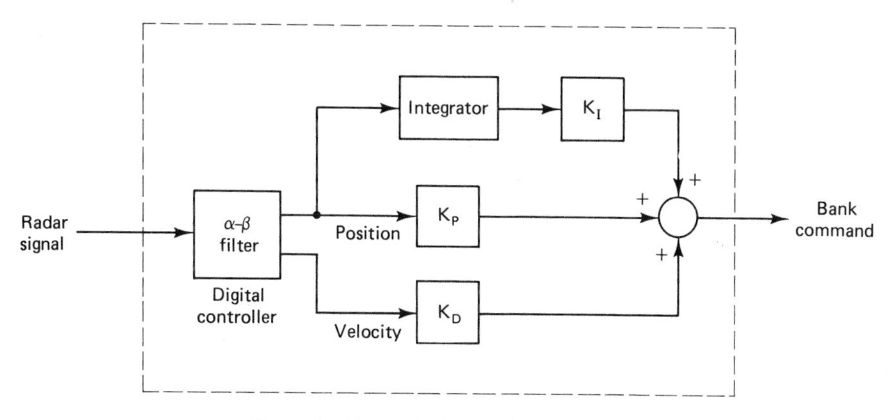

Figure 15-30 Basic form of the controller.

The frequency responses plotted versus real frequency ω of an α-β filter are given in Figure 15-31. The sample period is $T = 0.1$ s; thus $\omega_s/2$ is 31.4 rad/s. Note that at low frequencies the position filter (Figure 15-31a) has unity gain and no phase shift, but it attenuates high-frequency noise. At low frequencies the velocity filter response (Figure 15-31b) has a slope of -20 dB/decade, which is equal to that of an

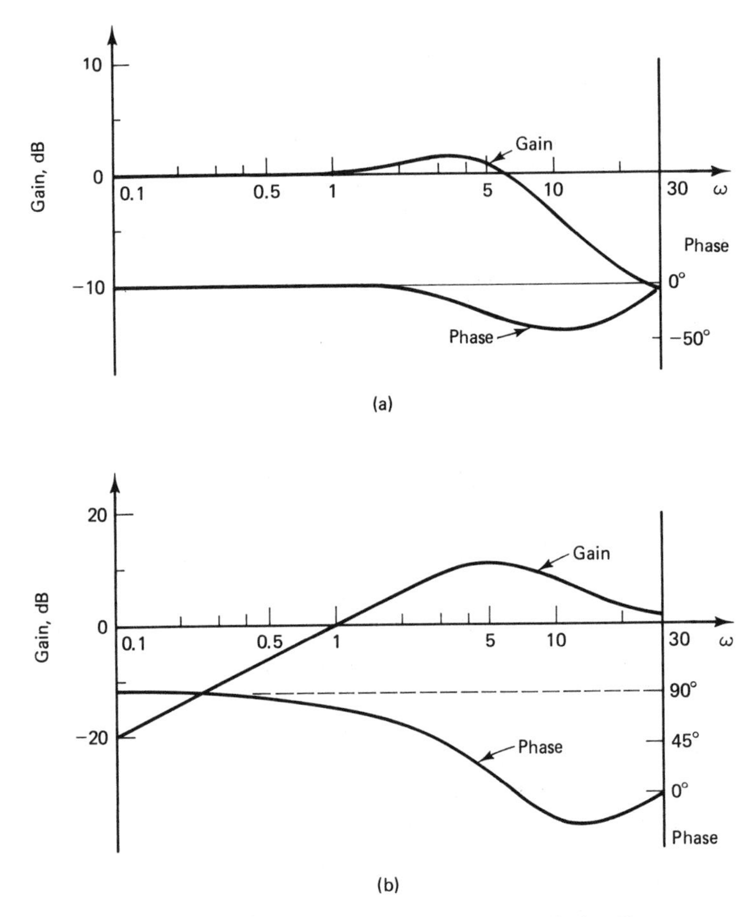

Figure 15-31 (a) α-β position filter response; (b) α-β velocity filter response.

exact differentiator. However, the phase characteristic varies from that of a differentiator, which has a constant phase of 90°. This filter also attenuates high-frequency noise. For the filter plotted, $\alpha = 0.51$ and $\beta = 0.1746$, which are the values used in the F4J controller.

α-filters. The noise-reduction characteristics of the α-β filter do not adequately attenuate the radar noise in this system. To reduce the high-frequency noise further, low-pass filters (at times called α-filters [16]) are added at various points in the PID controller. The α-filter has a transfer function given by

$$D_\alpha(z) = \frac{\alpha z}{z - (1 - \alpha)} \tag{15-20}$$

The frequency response of an α-filter for $\alpha = 0.234$ and $T = 0.1$ s is given in Figure 15-32. This filter is also employed in the F4J controller. As is normal in low-pass filtering, phase lag is added to the system. It is seen from the system Nyquist diagram in Figure 15-29 that phase lag is already a major design problem. Thus the final design must include a trade-off between desired stability margins and radar noise rejection.

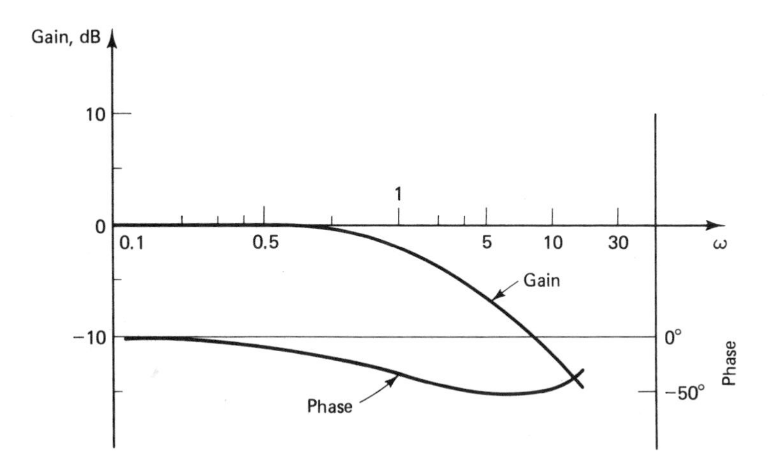

Figure 15-32 α-filter frequency response.

The foregoing system requirements resulted in a final filter design illustrated in Figure 15-33. An additional differentiator is added to the $\dot{y}(k)$ path to produce additional phase lead, since that produced by the derivative term was insufficient. The differentiator has the transfer function

$$D_a(z) = \frac{z - 1}{Tz} \tag{15-21}$$

[see (8-38)]. Because of the high-frequency noise amplification from this differentiation of a noisy signal, two sections of α-filters are required for this path. The nonlinearities shown in the filter are not exercised during normal system operation. The limits ϕ_{lim} and Y_{lim}, and the various gains shown, are all functions of range, with the gains increasing as the aircraft nears touchdown. The gains are constant for the final 1500 m of flight. The frequency response (versus real frequency ω) of the lateral controller for the F4J aircraft is given in Figure 15-34; the standard PID filter response is seen, except for the high-frequency attenuation added to reduce noise effects.

The PID path gains used, for range less than 1500 m, are $K_P = 0.1$, $K_I = 0.0033$, $K_D = 0.75$, and (acceleration path) $K_A = 0.75$. The resultant phase margin is 60°, and the gain margin is 8 dB. Two significant nonlinearities in the lateral control

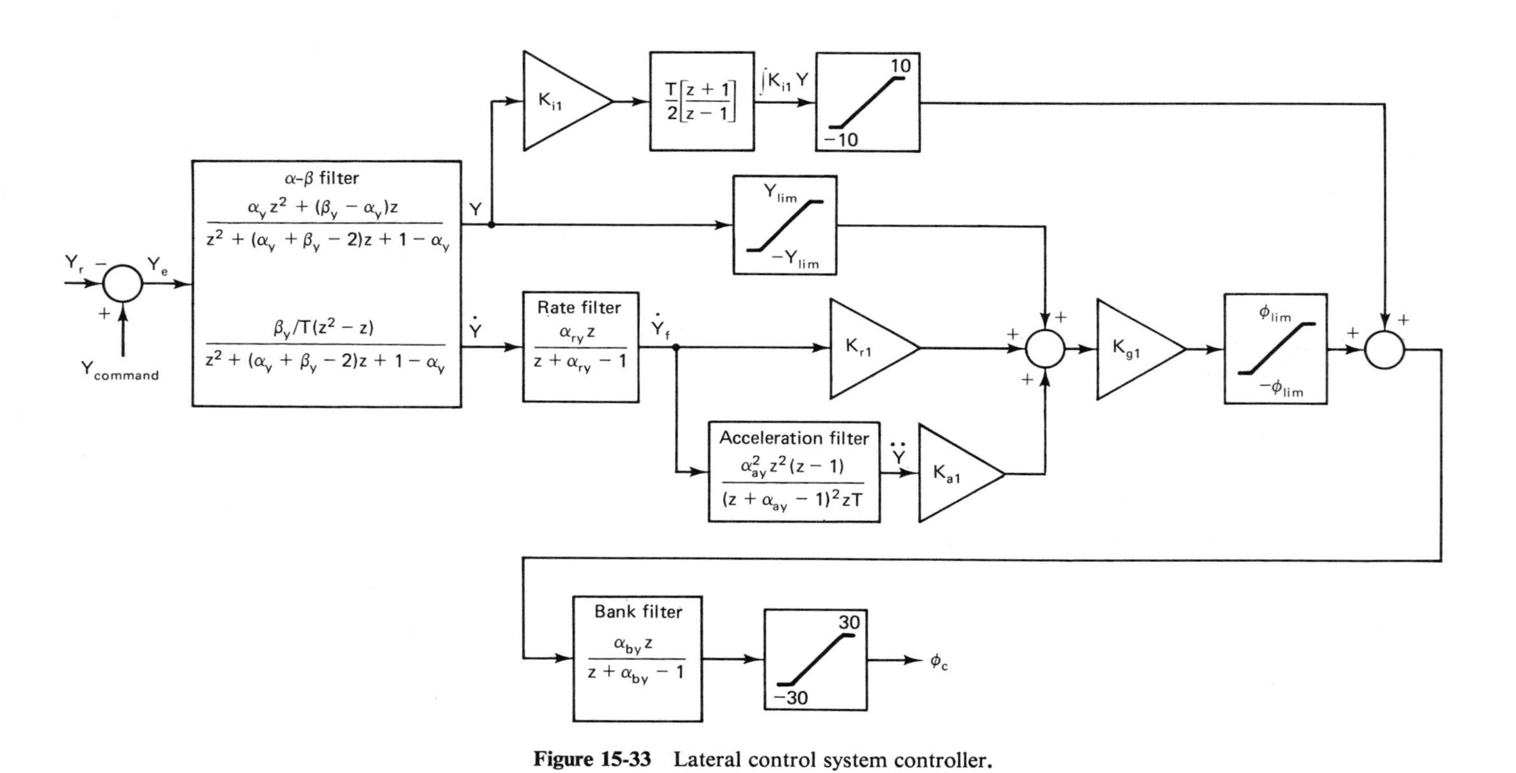

Figure 15-33 Lateral control system controller.

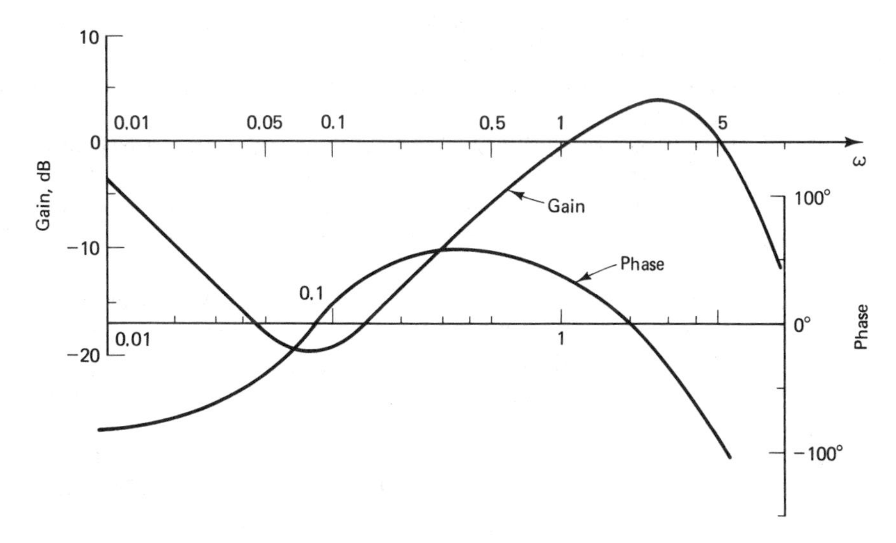

Figure 15-34 F4J lateral controller frequency response, $T = 0.1$ s.

system are the mechanical limits on the rotations of the ailerons and of the rudder. For a landing in high wind turbulence, these nonlinearities are exercised and the stability characteristics are degraded.

REFERENCES

1. M. S. Paranjape, "Microprocessor Controller for a Servomotor System," M.S. thesis, Auburn University, Auburn, Ala., 1980.
2. ———, *9900 Family Systems Design.* Dallas, Tex.: Texas Instruments Learning Center, 1978.
3. H. T. Nagle, Jr., B. D. Carroll, and J. D. Irwin, *An Introduction to Computer Logic.* Englewood Cliffs, N.J.: Prentice-Hall, Inc., 1975.
4. S. M. Shinners, *Modern Control System Theory and Application.* Reading, Mass.: Addison-Wesley Publishing Company, Inc., 1978.
5. J. D. Powell and P. Katz, "Sample Rate Selection for Aircraft Digital Control," *AIAAJ.*, Vol. 13, pp. 975–979, Aug. 1975.
6. G. F. Franklin and J. D. Powell, *Digital Control of Dynamic Systems.* Reading, Mass.: Addison-Wesley Publishing Company, Inc., 1980.
7. J. C. Hsu and A. U. Meyer, *Modern Control Principles and Applications.* New York: McGraw-Hill Book Company, 1968.
8. A. Gelb and W. E. Vander Velde, *Multiple-Input Describing Functions and Nonlinear Design.* New York: McGraw-Hill Book Company, 1968.
9. R. E. Wheeler, "A Digital Control System for Plant Growth Chambers," M.S. thesis, Auburn University, Auburn, Ala., 1980.
10. R. Boylestad and L. Nashelsby, *Electronic Devices and Circuit Theory.* Englewood Cliffs, N.J.: Prentice-Hall, Inc., 1978.

11. L. A. Leventhal, *Introduction to Microprocessors: Software, Hardware, Programming.* Englewood Cliffs, N.J.: Prentice-Hall, Inc., 1978.
12. "AN/TPN-22 Mode 1 Final Report," Contract N00039-75-C-0021, ITT Gilfillan, Van Nuys, Calif., 1979.
13. "Software Implementation ALS Computer Program," Contract N00421-75-C-0058, Bell Aerospace Corporation, Buffalo, N.Y., Mar. 1975.
14. A. P. Schust, Jr., "Determination of Aircraft Response Characteristics in Approach/Landing Configuration for Microwave Landing System Program," Report FT-61R-73, Naval Air Test Center, Patuxeut River, Md., 1973.
15. C. L. Phillips, E. R. Graf, and H. T. Nagle, Jr., "MATCALS Error and Stability Analysis," Report AU-EE-75-2080-1, Auburn University, Auburn, Ala., 1975.
16. R. F. Wigginton, "Evaluation of OPS-II Operational Program for the Automatic Carrier Landing System," Naval Electronic Systems Test and Evaluation Facility, Saint Inigoes, Md., 1971.
17. T. R. Benedict and G. W. Bordner, "Synthesis of an Optimal Set of Radar Track-While-Scan Smoothing Equations," *IRE Trans. Autom. Control*, pp. 27–31, July 1962.

APPENDIX I

Computer Programs

PROGRAM 1

This program, written in BASIC, solves the state equations

$$\mathbf{x}(k+1) = \mathbf{A}\mathbf{x}(k) + \mathbf{B}\mathbf{u}(k)$$
$$\mathbf{y}(k) = \mathbf{C}\mathbf{x}(k) + \mathbf{D}\mathbf{u}(k)$$

```
10   REM  THIS PROGRAM SOLVES DISCRETE STATE EQUATIONS
20   REM           X(K+1) = AX(K) + BU(K)
30   REM             Y(K) = CX(K) + DU(K)
40   DIM A[10,10],B[10,10],C[10,10],D[10,10],X[10],Y[10]
50   DIM U[10],V[10],W[10],Z[10]
60   REM  MATRICES ARE IDENTIFIED IN STATE EQUATIONS
70   REM  X=X(K+1) , V=X(K) , Y=Y(K) , U=U(K)
80   REM  W AND Z ARE WORKING MATRICES
90   PRINT "INPUT SYSTEM ORDER, # OF INPUTS, AND # OF OUTPUTS"
100   INPUT N,R,M
110   PRINT "INPUT FINAL VALUE OF K DESIRED"
120   INPUT S
130   PRINT "INPUT INITIAL VALUES OF STATES"
140   MAT  INPUT V[N]
150   PRINT "INPUT A MATRIX, BY ROWS"
160   MAT  INPUT A[N,N]
170   PRINT "INPUT B MATRIX, BY ROWS"
180   MAT  INPUT B[N,R]
190   PRINT "INPUT C MATRIX, BY ROWS"
200   MAT  INPUT C[M,N]
210   PRINT "INPUT D MATRIX, BY ROWS"
220   MAT  INPUT D[M,R]
230   PRINT  USING 240;N
240   IMAGE "SYSTEM ORDER =",3D
250   PRINT  USING 260;R,M
260   IMAGE "NUMBER OF INPUTS=",3D3X,"NUMBER OF OUTPUTS=",3D/
```

```
270  PRINT "INITIAL STATES ARE"
280  MAT  PRINT V
290  PRINT "A MATRIX IS"
300  MAT  PRINT A
310  PRINT "B MATRIX IS"
320  MAT  PRINT B
330  PRINT "C MATRIX IS"
340  REM  EACH INPUT MUST BE ENTERED, STARTING WITH STATEMENT #470
350  REM  EXAMPLE:     U(1) = (.5)**K
360  REM  EXAMPLE:     U(2) = COS(.3*K)
370  MAT  PRINT C
380  PRINT "D MATRIX IS"
390  MAT  PRINT D
400  REM  TO SET DIMENSIONS OF W,Y,AND Z
410  MAT W=ZER[N]
420  MAT Z=ZER[M]
430  MAT Y=ZER[M]
440  MAT X=ZER[N]
450  MAT U=ZER[R]
460  FOR K=1 TO S
470  U[1]=1
480  MAT X=A*V
490  MAT W=B*U
500  MAT X=X+W
510  MAT Y=C*V
520  MAT Z=D*U
530  MAT Y=Y+Z
540  PRINT  USING 550;K-1
550  IMAGE "K=",5D
560  PRINT "OUTPUT VARIABLES ARE"
570  MAT  PRINT Y
580  MAT V=X
590  NEXT K
600  END
```

PROGRAM 2

This program calculates the discrete system matrices from the continuous system matrices for a sampled-data system. The program is based on the development in Section 4.10. Let the continuous system be described by

$$\dot{\mathbf{x}}(t) = \mathbf{A}_c\mathbf{x}(t) + \mathbf{B}_c\mathbf{u}(t)$$
$$\mathbf{y}(t) = \mathbf{C}_c\mathbf{x}(t) + \mathbf{D}_c\mathbf{u}(t)$$

where $\mathbf{u}(t)$ are the outputs of zero-order holds. Also, let the discrete system be described by

$$\mathbf{x}(k+1) = \mathbf{A}\mathbf{x}(k) + \mathbf{B}\mathbf{m}(k)$$
$$\mathbf{y}(k) = \mathbf{C}\mathbf{x}(k) + \mathbf{D}\mathbf{m}(k)$$

where $\mathbf{m}(k) = \mathbf{u}(t)$ and $kT \leq t < (k+1)T$. Then

$$\mathbf{A} = \mathbf{I} + \mathbf{A}_c T + \frac{\mathbf{A}_c^2 T^2}{2!} + \frac{\mathbf{A}_c^3 T^3}{3!} + \cdots$$

$$\mathbf{B} = \left(\mathbf{I}T + \mathbf{A}_c\frac{T^2}{2!} + \mathbf{A}_c^2\frac{T^3}{3!} + \cdots\right)\mathbf{B}_c$$

$$\mathbf{C} = \mathbf{C}_c$$

$$\mathbf{D} = \mathbf{D}_c$$

```
10  REM  THIS PROGRAM CALCULATES DISCRETE SYSTEM MATRICES FROM
20  REM  CONTINUOUS SYSTEM MATRICES
30  DIM A[10,10],B[10,10],R[10,10],S[10,10]
40  DIM K[10,10],L[10,10],G[10,10],H[10,10]
50  REM  A AND B ARE CONTINUOUS SYSTEM MATRICES
60  REM  R IS DISCRETE SYSTEM A MATRIX
70  REM  S IS DISCRETE SYSTEM B MATRIX
80  REM  G,H,K,L ARE WORKING MATRICES
90  PRINT "INPUT T, SYSTEM ORDER, AND NUMBER OF INPUTS"
100  INPUT T,N,P
110  PRINT "INPUT NUMBER OF TERMS TO BE USED IN SERIES"
120  INPUT F
130  PRINT "INPUT CONTINUOUS A MATRIX, BY ROWS"
140  MAT  INPUT A[N,N]
150  PRINT "INPUT CONTINUOUS B MATRIX, BY ROWS"
160  MAT  INPUT B[N,P]
170  PRINT
180  PRINT  USING 190;T,N,P
190  IMAGE "T=",2D.3D3X,"SYSTEM ORDER=",3D3X,"# OF INPUTS=",3D/
200  PRINT "CONTINUOUS SYSTEM A MATRIX IS"
210  MAT  PRINT A
220  PRINT "CONTINUOUS B MATRIX IS"
230  MAT  PRINT B
240  REM TO INITIALIZE AND SET DIMENSIONS ON MATRICES
250  MAT R=IDN[N,N]
260  MAT S=IDN[N,N]
270  MAT L=IDN[N,N]
280  MAT G=ZER[N,N]
290  MAT H=ZER[N,N]
300  MAT K=ZER[N,N]
310  FOR J=1 TO F
320  MAT G=L
330  MAT L=(T/(J+1))*G
340  MAT G=L
350  MAT L=G*A
360  MAT S=S+L
370  MAT K=(J+1)*L
380  MAT R=R+K
390  NEXT J
400  MAT H=S
410  MAT S=(T)*H
420  MAT G=ZER[N,P]
430  MAT G=S*B
440  PRINT "DISCRETE A MATRIX IS"
450  MAT  PRINT R
460  PRINT "DISCRETE B MATRIX IS"
470  MAT  PRINT G
480  END
```

PROGRAM 3

This program calculates the frequency response of a sampled-data system containing a digital filter. Let $G(s)$ be the plant transfer function, which includes the data-hold transfer function, and let $D(z)$ be the transfer function of the digital filter. Then this program calculates

$$G^*(j\omega)D(\epsilon^{j\omega T}) = \frac{1}{T}[G(j\omega) + G(j\omega + j\omega_s) + G(j\omega - j\omega_s) + \cdots]D(\epsilon^{j\omega T})$$

```
C       THIS PROGRAM CALCULATES THE PLANT FREQUENCY RESPONSE BY SERIES
C       OMEGA = S-PLANE FREQUENCY
C       OMEGA1 = SHIFTED S-PLANE FREQUENCY
C       OMEGAS = SAMPLING FREQUENCY
C       S = S-PLANE VARIABLE
C       DELTA = S-PLANE FREQUENCY INCREMENT
C       OMEGAF = FINAL S-PLANE FREQUENCY
C       OMEGAW = W-PLANE FREQUENCY
C       T = SAMPLING PERIOD
C       NTERMS IS USED TO DETERMINE THE NUMBER OF TERMS IN G*(S)
C       NX = NUMBER OF TERMS IN G*(S)
C       GS = PLANT TRANSFER FUNCTION
C       HOLD = DATA HOLD TRANSFER FUNCTION
C       DZ = THE DIGITAL COMPENSATOR TRANSFER FUNCTION
C       SG = (HOLD*GS)*
C       TF = D(Z)*(HOLD*GS)*
C
        COMPLEX TF,HOLD,GS,TERM,CEXP,CMPLX,S,Z,DZ,SG
        DATA T/.1/,DELTA/.1/,NTERMS/5/,OMEGAF/3.14/,OMEGA/.005/
        NX=2*NTERMS+1
        OMEGAS=2.*3.14159/T
    4   SG = CMPLX(0.0, 0.0)
        DO 24 J=1,NX
        XJ=-NTERMS+J-1
        OMEGA1=OMEGA+XJ*OMEGAS
        S = CMPLX(0.,OMEGA1)
        GS=.8/(S+1.)/(S+.5)/(S+.2)
        HOLD=(1.-CEXP(-T*S))/S
        TERM=1./T*HOLD*GS
   24   SG=SG+TERM
        Z=CEXP(S*T)
        DZ=11.44*(Z-.9417)/(Z-.333)
        TF=SG*DZ
        ABSVAL = CABS(TF)
        DB = 20.0*ALOG10 (ABSVAL)
        PHASE = 57.29578*ATAN2(AIMAG(TF), REAL(TF))
        IF(PHASE) 30,30,31
   31   PHASE=PHASE-360.
   30   CONTINUE
        OMEGAW=TAN(OMEGA*T/2.)*2./T
        PRINT 25,OMEGA,DB,PHASE,OMEGAW
        OMEGA=OMEGA+DELTA
        IF(OMEGA-OMEGAF) 4,4,7
    7   CONTINUE
   25   FORMAT(5X,6HOMEGA=F9.5,5X,3HDB=F9.3,5X,6HPHASE=F9.3,5X,
       1 7HOMEGAW=F10.6)
        STOP
        END
```

PROGRAM 4

This program calculates the frequency response of a system whose transfer function is $D(z)G(z)$, with $z = \epsilon^{j\omega T}$.

```
C       THIS PROGRAM CALCULATES THE FREQUENCY RESPONSE OF G(Z)*D(Z)
C       OMEGA = S-PLANE FREQUENCY
C       DELTA = S-PLANE FREQUENCY INCREMENT
C       OMEGAF = FINAL S-PLANE FREQUENCY
C       OMEGAW = W-PLANE FREQUENCY
C       T = SAMPLING PERIOD
C       S = S-PLANE VARIABLE
C       Z = Z-PLANE VARIABLE
C       GZ = G(Z)
C       DZ = D(Z)
C       TF = D(Z)*G(Z)
C
```

```
      COMPLEX GZ,S,Z,CMPLX,CEXP,DZ,TF
      DATA T/.1/,DELTA/.1/,OMEGAF/3.14/,OMEGA/.005/
 4    CONTINUE
      OMEGAW=TAN(OMEGA*T/2.)*2./T
      S=CMPLX(0.,OMEGA)
      Z=CEXP(S*T)
      GZ=(Z-1.)/Z*(8.*Z/(Z-1.)-2.*Z/(Z-.9048)+10.664*Z/(Z-.9512)
     1  -16.664*Z/(Z-.9802))
      DZ=11.44*(Z-.9417)/(Z-.333)
      TF=DZ*GZ
      ABSVAL=CABS(TF)
      DB = 20.0*ALOG10 (ABSVAL)
      PHASE = 57.29578*ATAN2(AIMAG(TF), REAL(TF))
      IF (PHASE) 31,31,30
 30   PHASE=PHASE-360.
 31   CONTINUE
      PRINT 25,OMEGA,DB,PHASE,OMEGAW
      OMEGA=OMEGA+DELTA
      IF(OMEGA-OMEGAF) 4,4,7
 7    CONTINUE
 25   FORMAT(5X,6HOMEGA=F9.5,5X,3HDB=F9.3,5X,6HPHASE=F9.3,5X,
     1 7HOMEGAW=F10.6)
      STOP
      END
```

PROGRAM 5

```
C      THIS PROGRAM CALCULATES THE SCALAR TRANSFER FUNCTION TF, WHERE
C
C                                        -1
C                  TF = H (ZI-F)    G
C
C      WHERE THE (N ORDER) SYSTEM EQUATIONS ARE
C
C                  X(K+1) = F X(K) + G U(K)
C
C                  Y(K)= H X(K)
C
C      NUMERICAL ACCURACY IS INDICATED BY THE 'RE (N+1) MATRIX',
C      WHICH SHOULD BE THE NULL MATRIX.  THE POLES AND ZEROS
C      OF THE TRANSFER FUNCTION ARE ALSO CALCULATED.
C
C      THE LEVERRIER ALGORITHM IS USED TO CALCULATE THE MATRIX INVERSE.
C
       IMPLICIT REAL*8(A-H),REAL*8(O-Z)
       DIMENSION F(10,10),G(10,10),H(10,10),CO(11),RE(10,10,10),CA(11)
       DIMENSION POR(10),POI(10),F1(10,10),E1(10,10),R1(10,10)
       DIMENSION R2(10,10),R3(10,10),U(10),V(10),CAD(11,3,3)
       REAL NCO(11),NCA(11)
       READ(5,101)NP,NU,NY
 101   FORMAT(3I2)
       CALL MXIN(F,NP,NP)
       CALL MXIN(G,NP,NU)
       CALL MXIN(H,NY,NP)
       NP1=NP+1
 103   FORMAT(14X,16H NO. OF STATES =,I4/14X,16H NO. OF INPUTS =,
      *I4/14X,17H NO. OF OUTPUTS =,I4)
       WRITE(6,103)NP,NU,NY
 105    FORMAT(17H F MATRIX FOLLOWS)
       WRITE(6,105)
       CALL MXOUT(F,NP,NP)
 107   FORMAT(17H G MATRIX FOLLOWS)
       WRITE(6,107)
       CALL MXOUT(G,NP,NU)
```

```
 109  FORMAT(17H H MATRIX FOLLOWS)
      WRITE(6,109)
      CALL MXOUT(H,NY,NP)
      DO 1 I=1,NP
      DO 1 J=1,NP
 1    RE(1,I,J)=0.
      DO 3 I=1,NP
 3    RE(1,I,I)=1.
 114  FORMAT(1H0,'RE1 MATRIX FOLLOWS')
      WRITE(6,114)
 116  FORMAT((1H ,1P9D14.4))
      DO 8 I=1,NP
 8    WRITE(6,116)(RE(1,I,J),J=1,NP)
      CO(1)=1.
      DO 5 I=1,NP
      DO 5 J=1,NP
 5    F1(I,J)=F(I,J)
      DO 7 K=2,NP1
      CALL MTRACE(NP,F1,W)
      CO(K)=-W/(K-1)
      DO 9 I=1,NP
      DO 9 J=1,NP
 9    E1(I,J)=-RE(1,I,J)*W/(K-1)
      CALL MADD(NP,NP,F1,E1,R1)
 115  FORMAT(1H0, 'RE',I2,' MATRIX FOLLOWS')
      WRITE(6,115)K
      CALL MXOUT(R1,NP,NP)
      DO 11 I=1,NP
      DO 11 J=1,NP
 11   RE(K,I,J)=R1(I,J)
      CALL MMUL(NP,NP,NP,F,R1,F1)
 7    CONTINUE
 117  FORMAT(52H THE CHAR POLY COEF IN DESCENDING POWERS OF S FOLLOW//)
      WRITE(6,117)
 119  FORMAT(D16.7/)
      WRITE(6,119)(CO(I),I=1,NP1)
      IR=-1
      CALL PROOT(NP,CO,U,V,IR)
      DO 91 I=1,NP
 91    POR(I)=U(I)
      DO 92 I=1,NP
 92   POI(I)=-V(I)
 127  FORMAT(1H0, 23HTHE EQUATION ROOTS ARE ,13X, 9HREAL PART,
     *10X,    14HIMAGINARY PART//)
      WRITE(6,127)
 123  FORMAT(25X,1P2D20.7)
      WRITE(6,123)(POR(I),POI(I),I=1,NP)
      DO 10 K=1,NP
      DO 12 I=1,NP
      DO 12 J=1,NP
 12   R1(I,J)=RE(K,I,J)
      CALL MMUL(NP,NP,NU,R1,G,R2)
      CALL MMUL(NY,NP,NU,H,R2,R3)
      CA(K)=R3(1,1)
 10   CONTINUE
 167  FORMAT(52H THE NUMR POLY COEF IN DESCENDING POWERS OF S FOLLOW//)
      WRITE(6,167)
 169  FORMAT(D16.7/)
      WRITE(6,169)(CA(I),I=1,NP)
      IF(NP.EQ.2) GO TO 5034
      CA(NP1)=0.
      IR=-1
      N1=NP-1
      CALL PROOT(N1,CA,U,V,IR)
      DO 37 I=1,NP
 37    POR(I)=U(I)
      DO 38 I=1,NP
 38   POI(I)=-V(I)
      WRITE(6,127)
```

```
      WRITE(6,123)(POR(I),POI(I),I=1,N1)
 5034 CONTINUE
      DO 1011 I=1,NP1
 1011 NCO(I)=CO(I)
      DO 1012 I=1,NP
 1012 NCA(I)=CA(I)
      STOP
      END
      SUBROUTINE MATRIX
C     THIS ROUTINE IS USED FOR MATRIX OPERATIONS.
      IMPLICIT REAL*8(A-H),REAL*8(O-Z)
       DIMENSION A(10,10),B(10,10),C(10,10)
      ENTRY       MADD (M,N,A,B,C)
      DO 29 I=1,M
      DO 29 J=1,N
   29 C(I,J)=A(I,J) + B(I,J)
      RETURN
      ENTRY       MMUL (MP,NP,NU,A,B,C)
      DO 11 K=1,MP
      DO 11 I=1,NU
      SUM = 0.0
      DO 31 J=1,NP
   31 SUM = SUM + A(K,J)*B(J,I)
   11 C(K,I) = SUM
      RETURN
      ENTRY       MTRACE(N,A,W)
      W=0.0
      DO 7 I=1,N
    7 W=W+A(I,I)
      RETURN
      ENTRY       MXIN (A,M,N)
      DO 50 I=1,M
   50 READ(5,100)  (A(I,J), J=1,N)
  100 FORMAT (8D10.3)
      RETURN
      ENTRY       MXOUT (A,L,M)
      DO 103 I=1,L
  103 WRITE (6,109) (A(I,J),J=1,M)
  109 FORMAT ((1H ,1P9D14.4))
      RETURN
      END
      SUBROUTINE PROOT(NP,AR,U,V,IR)
C     THIS SUBROUTINE USES A MODIFIED BARSTOW METHOW TO FIND THEROOTS
C     OF A POLYNOMINAL SEE HAMMMINGNEKSA P111
      IMPLICIT REAL*8(A-H),REAL*8(O-Z)
      DIMENSION AR(10),U(10),V(10),H(11),BB(11),CC(11)
       IREV=IR
      NC=NP+1
      DO 1 I=1,NC
 1    H(I)=AR(I)
      P=0.
      Q=0.
      R=0.
 3    IF(H(1))4,2,4
 2    NC=NC-1
      V(NC)=0.
      U(NC)=0.
      DO 1002 I=1,NC
 1002   H(I)=H(I+1)
      GO TO 3
 4    CONTINUE
      IF(NC-1)5,100,5
 5    IF(NC-2)7,6,7
 6    R=-H(1)/H(2)
      GO TO 50
 7    IF(NC-3)9,8,9
 8    P=H(2)/H(3)
      Q=H(1)/H(3)
      GO TO 70
```

```
9      IF(DABS(H(NC-1)/H(NC))-DABS(H(2)/H(1)))10,19,19
10     IREV=-IREV
       M=NC/2
       DO 11 I=1,M
       NL=NC+1-I
       F=H(NL)
       H(NL)=H(I)
11     H(I)=F
       IF(Q)13,12,13
12     P=0.
       GO TO 15
13     P=P/Q
       Q=1./Q
15     IF(R)16,19,16
16     R=1./R
19     E=5.E-10
       BB(NC)=H(NC)
       CC(NC)=H(NC)
       BB(NC+1)=0.
       CC(NC+1)=0.
       NQ=NC-1
20     DO 49 J=1,1000
       DO 21 I1=1,NQ
       I=NC-I1
       BB(I)=H(I)+R*BB(I+1)
21     CC(I)=BB(I)+R*CC(I+1)
       IF(DABS(BB(1)/H(1))-E)50,50,24
24     IF(CC(2))23,22,23
22     R=R+1.
       GO TO 30
23     R=R-BB(1)/CC(2)
30     DO 37 I1=1,NQ
       I=NC-I1
       BB(I)=H(I)-P*BB(I+1)-Q*BB(I+2)
37     CC(I)=BB(I)-P*CC(I+1)-Q*CC(I+2)
       IF(H(2))32,31,32
31     IF(DABS(BB(2)/H(1))-E)33,33,34
32     IF(DABS(BB(2)/H(2))-E)33,33,34
33     IF(DABS(BB(1)/H(1))-E)70,70,34
34     CBAR=CC(2)-BB(2)
       D=CC(3)**2-CBAR*CC(4)
       IF(D)36,35,36
35     P=P-2.
       Q=Q*(Q+1.)
       GO TO 49
36     P=P+(BB(2)*CC(3)-BB(1)*CC(4))/D
49     CONTINUE
       Q=Q+(-BB(2)*CBAR+BB(1)*CC(3))/D
       E=E*10.
       GO TO 20
50     NC=NC-1
       V(NC)=0.
       IF(IREV)51,52,52
51     U(NC)=1./R
       GO TO 53
52     U(NC)=R
53     DO 54 I=1,NC
54     H(I)=BB(I+1)
       GO TO 4
70     NC=NC-2
       IF(IREV)71,72,72
71     QP=1./Q
       PP=P/(Q*2.)
       GO TO 73
72     QP=Q
       PP=P/2.
73     F=(PP)**2-QP
       IF(F)74,75,75
74     U(NC+1)=-PP
```

```
      U(NC)=-PP
      V(NC+1)=DSQRT(-F)
      V(NC)=-V(NC+1)
      GO TO 76
75    U(NC+1)=-(PP/DABS(PP))*(DABS(PP)+DSQRT(F))
      V(NC+1)=0.
      U(NC)=QP/U(NC+1)
      V(NC)=0.
      IF(NC.EQ.1) GO TO 100
 76   DO 77 I=1,NC
 77   H(I)=BB(I+2)
      GO TO 4
 100  RETURN
      END
```

PROGRAM 6

This program solves Ackermann's formula of the form of equation (9-25).

$$\mathbf{K} = [0 \quad 0 \quad \cdots \quad 0 \quad 1][\mathbf{B} \quad \mathbf{AB} \quad \cdots \quad \mathbf{A}^{n-2}\mathbf{B} \quad \mathbf{A}^{n-1}\mathbf{B}]^{-1}\alpha_c(\mathbf{A}) \qquad (9\text{-}25)$$

By replacing **A** with $\mathbf{A}^T$ and **B** with **C**, the program also solves equation (9-48).

```
POLEAS

10  REM THIS PROGRAM SOLVES ACKERMAN'S FORMULA,EQUATIONS (9-25) AND
20  REM (9-48). FOR (9-25),DATA STATEMENTS BEGIN IN LINE 730 WITH
30  REM             N=SYSTEM ORDER
40  REM             A MATRIX, BY ROWS
50  REM             B MATRIX
60  REM             CHARACTERISTIC POLYNOMIAL COEFFICIENTS,
70  REM                  IN DESCENDING ORDER
80  REM THE GAIN MATRIX K IS CALCULATED. FOR (9-48), DATA STATEMENTS
90  REM BEGIN IN LINE 730 WITH
100  REM             N=SYSTEM ORDER
110  REM             A MATRIX, BY COLUMNS
120  REM             C MATRIX
130  REM             CHARACTERISTIC POLYNOMIAL COEFFICIENTS,
140  REM                  IN DESCENDING ORDER
150  REM THE GAIN MATRIX G IS CALCULATED.
160  DIM A[10,10],B[10,4],C[1,10]
170  DIM F[10,10],L[10,1]
180  DIM W[10,10],R[10,10],T[10,10],Q[10,10]
190  DIM U[1,10],H[10,1],P[10],S[1,11]
200  READ N
210  N1=1
220  MAT A=ZER[N,N]
230  MAT B=ZER[N,N1]
240  MAT C=ZER[1,N]
250  MAT F=ZER[N,N]
260  MAT L=ZER[1,N]
270  MAT W=ZER[N,N]
280  MAT R=IDN[N,N]
290  MAT T=ZER[N,N]
300  MAT Q=ZER[N,N]
310  MAT U=ZER[N,1]
320  MAT H=ZER[1,N]
330  MAT P=ZER[N]
340  N2=N+1
350  MAT S=ZER[1,N2]
360  MAT  READ A,B,S
```

```
370  FOR I=1 TO N
380  J=I+1
390  P[I]=(-1)*S[1,J]
400  NEXT I
410  PRINT 'MATRICES OF THE SYSTEM'
420  PRINT 'MATRIX A'
430  MAT  PRINT A
440  PRINT 'MATRIX B'
450  MAT  PRINT B
460  PRINT 'COEFFICIANTS OF DESIRED CHARACTERISTIC POLYNOMIAL'
470  MAT  PRINT S;
480  FOR I=1 TO N
490  M=N-(I-1)
500  MAT U=R*B
510  X=P[M]
520  MAT W=(X)*R
530  MAT T=T+W
540  FOR K=1 TO N
550  Q[K,I]=U[K,1]
560  NEXT K
570  MAT W=A*R
580  MAT R=W
590  NEXT I
600  MAT W=R
610  MAT T=W-T
620  MAT W=INV(Q)
630  H[1,N]=1
640  MAT Q=W*T
650  MAT L=H*Q
660  MAT W=B*L
670  MAT F=A-W
680  PRINT 'MATRICES OF THE CLOSED-LOOP SYSTEM'
690  PRINT 'MATRIX A-BK'
700  MAT  PRINT F
710  PRINT 'MATRIX K'
720  MAT  PRINT L
730  DATA 2
740  DATA 1,.0952,0,.905
750  DATA .00484,.0952
760  DATA 1,-1.5167,.6703
770  END
```

PROGRAM 7

This program computes the gain matrix **K** which minimizes the quadratic cost function

$$J_N = \sum_{k=0}^{N} \mathbf{x}^T(k)\mathbf{Q}\mathbf{x}(k) + \mathbf{u}^T(k)\mathbf{R}\mathbf{u}(k)$$

for the system

$$\mathbf{x}(k+1) = \mathbf{A}\mathbf{x}(k) + \mathbf{B}\mathbf{u}(k)$$

with $\mathbf{u}(k) = -\mathbf{K}(k)\mathbf{x}(k)$.

```
10  REM THIS PROGRAM COMPUTES THE OPTIMUM GAIN MATRIX WHICH MINIMIZES
20  REM THE QUADRATIC COST FUNCTION GIVEN BY
30  REM                                   T             T
40  REM     COST=(FROM K=1 TO N) SUM OF X (K)QX(K) + U (K)RU(K)
50  REM
60  REM WHERE :
70  REM           X IS THE STATE OF THE SYSTEM
80  REM           U IS THE CONTROL SYSTEM
```

```
90  REM            Q AND R ARE POSITIVE DEFINITE WEIGHTING MATRICES
100  REM THE STATE VARIABLES ARE RELATED BY THE DISCRETE TIME DIFFERENCE
110  REM EQUATION
120  REM               X(K+1) = AX(K) + BU(K)
130  REM
140  REM THE PROGRAM SOLVES THE DISCRETE TIME MATRIX RICATTI EQUATION
150  REM IN REVERSE TIME. DATA STATEMENTS BEGIN IN LINE 640 WITH
160  REM            NUMBER OF INPUTS
170  REM            ORDER OF SYSTEM
180  REM            N FROM COST FUNCTION
190  REM            NUMBER OF ITERATIONS BETWEEN PRINTED OUTPUTS
200  REM            A MATRIX, BY ROWS
210  REM            B MATRIX, BY ROWS
220  REM            Q MATRIX, BY ROWS
230  REM            R MATRIX, BY ROWS
240  REM THE GAIN MATRIX IS CALLED K. THE P MATRIX IS ALSO PRINTED.
250  REM
260  DIM A[12,12],B[12,5],Q[12,12],R[5,5],P[12,12],K[5,12]
270  DIM T[12,12],U[5,12],V[12,5],W[5,5],X[5,5],Y[5,12],Z[12,12]
280  LET J=0
290  READ M,N,S1,P2
300  LET J2=0
310  MAT T=ZER[N,N]
320  MAT U=ZER[M,N]
330  MAT V=ZER[N,M]
340  MAT W=ZER[M,M]
350  MAT X=ZER[M,M]
360  MAT Y=ZER[M,N]
370  MAT Z=ZER[N,N]
380  MAT K=ZER[M,N]
390  MAT P=ZER[N,N]
400  MAT  READ A[N,N],B[N,M],Q[N,N],R[M,M]
410  MAT P=Q
420  MAT U=TRN(B)
430  MAT Y=U*P
440  MAT W=Y*B
450  MAT W=W+R
460  MAT X=INV(W)
470  MAT K=Y*A
480  MAT Y=X*K
490  MAT K=(-1)*Y
500  MAT T=B*K
510  MAT T=T+A
520  MAT Z=P*T
530  MAT T=TRN(A)
540  MAT P=T*Z
550  MAT P=P+Q
560  LET J=J+1
570  IF INT((J-J2)/P2)=0 THEN 610
580  LET J2=J
590  PRINT "P AND K SUB";J;"ARE : "
600  MAT  PRINT P,K
610  IF J<S1 THEN 430
620  GOTO 280
630  REM BEGIN DATA STATEMENTS HERE.........
640  DATA 1,2,50,10
650  DATA 1,.0952,0,.905
660  DATA .00484,.0952
670  DATA 1,0,0,0,1
999  END
```

APPENDIX II

The Signal Flow Graph Technique

In order to understand the terminology employed in the use of signal flow graphs, consider the diagram shown in Figure A2-1. The following terms are used with reference to the graph.

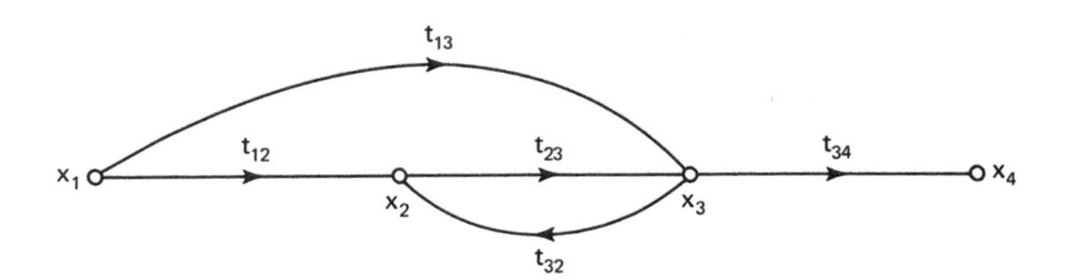

Figure A2-1 Signal flow graph illustration.

Nodes: represent variables $x_1, x_2, \ldots$.

Branches: represent unidirectional paths between variables.

Transmissions: functional relationships between pairs of variables, (e.g. t_{12}, $t_{23}, \ldots$).

Input node (*source*)*:* has outgoing branches only (e.g., x_1).

Output node (*sink*)*:* has incoming branches only (e.g., x_4).

Path: a continuous unidirectional succession of branches traversed in the arrowhead direction.

Forward path: a path from input to output nodes in which no node is encountered more than once.

Feedback loop: a closed path that returns to the starting node, in which no node is encountered more than once.

Path transmission: transmission along a continuous path.

Loop transmission: transmission around a closed feedback loop.

The algebra involved in signal flow graph reduction is illustrated below.

1. Addition:

$$x_k = \sum_{i=1}^{n} t_{ik} x_i, \qquad k \leq n$$

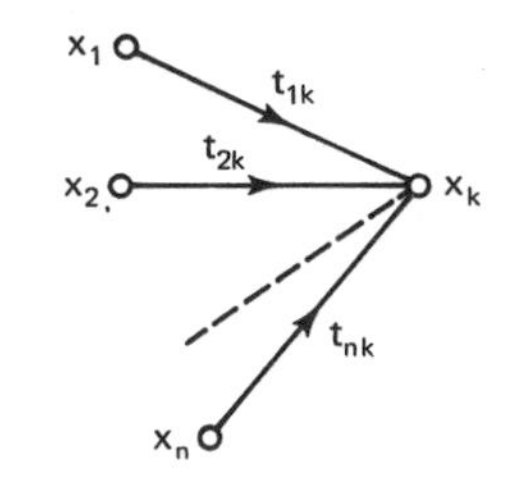

2. Multiplication:

$$x_3 = t_{12} t_{23} x_1$$

t12 t23
x1 x2 x3

3. Parallel branches—transmissions added:

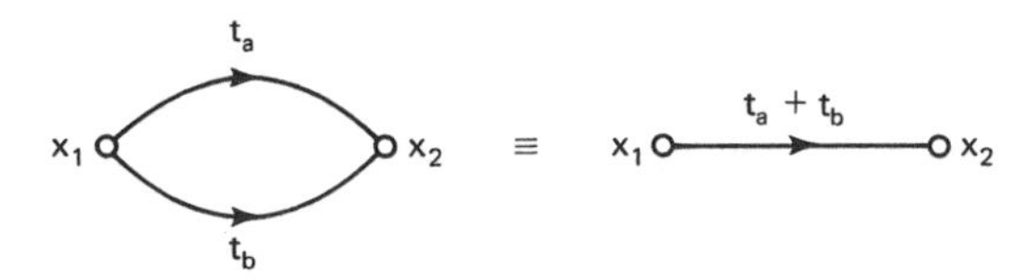

4. Series branches—transmissions multiplied:

t12 t23 t34 ≡ t12 t23 t34
x1 x2 x3 x4 x1 x4

5. Change of termination:

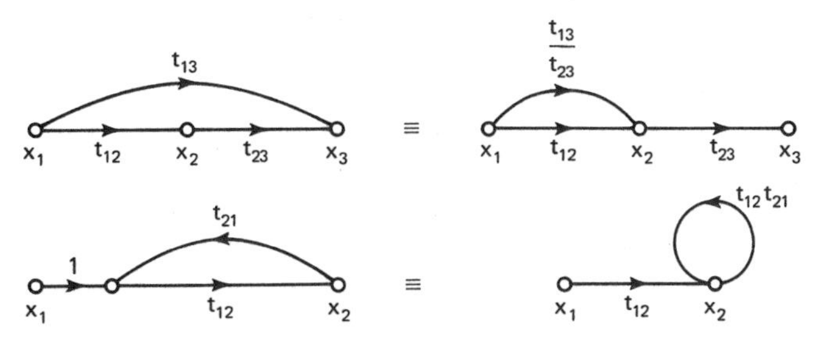

6. Removal of a self-loop:

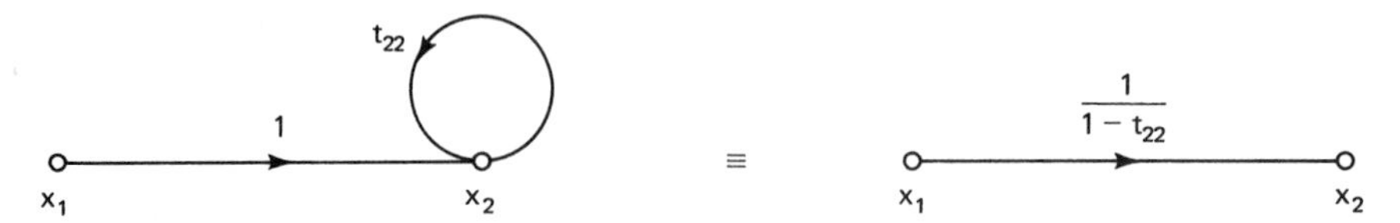

7. Removal of a node:

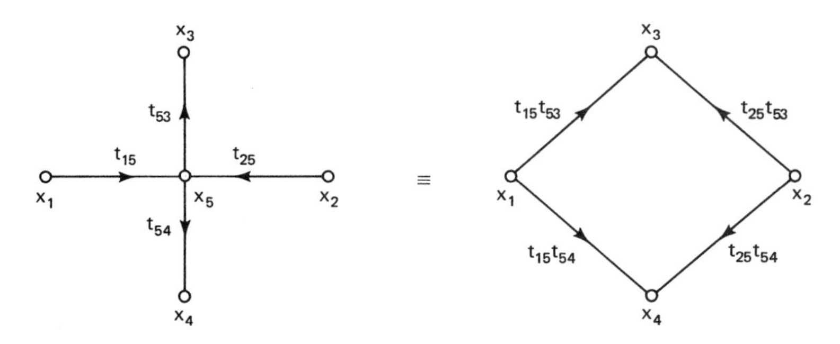

Although the techniques shown above for signal flow graph simplification are useful, in general it is the gain or transfer function of the flow graph which is of primary importance. The transfer function can usually be written by inspection by means of Mason's gain formula. This formula is given by

$$M = \frac{1}{\Delta} \sum_k M_k \Delta_k$$

where $\Delta = 1$ − (sum of all *individual* loop transmissions)
+ (sum of the product of the loop transmission of all nontouching loops taken *two* at a time)
− (sum of the products of the loop transmission of all nontouching loops taken *three* at a time)
+ $(\cdots)$

M_k = path transmission of kth forward path

Δ_k = value of Δ for that part of the graph not touching the kth forward path

Although signal flow graphs can be extremely useful in many areas, such as in active and passive circuit analysis, we concentrate here only on the determination of block diagram transfer functions. The following examples have been carefully chosen to illustrate the salient features of Mason's gain formula.

Example A2.1

Consider the block diagram shown in Figure A2-2a. The corresponding signal flow graph is shown in Figure A2-2b. First we note that there is one forward path and that there are two loops. These loops have the path G_2 in common and therefore there are no nontouching loops.

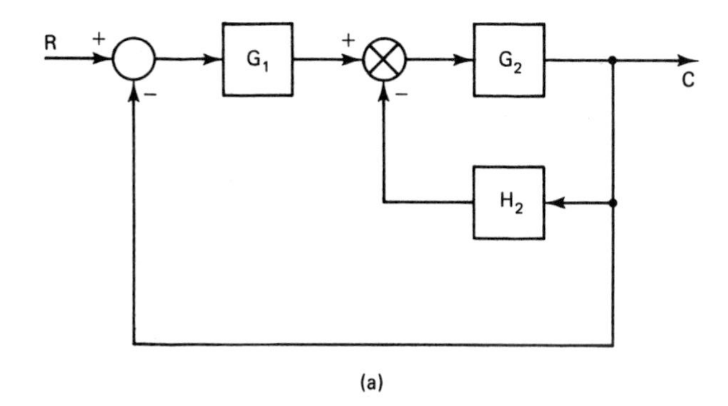

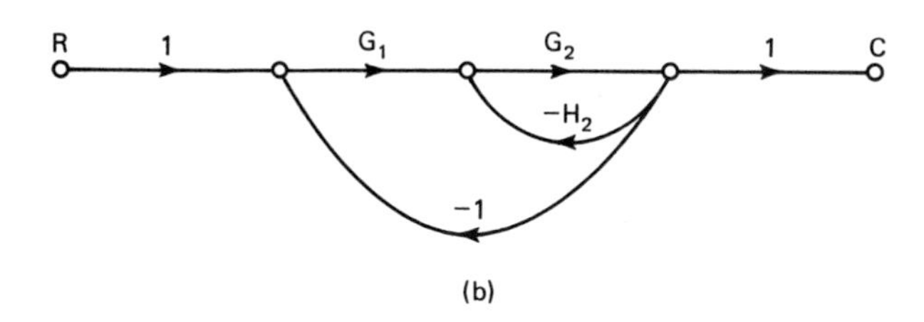

Figure A2-2 Diagrams for Example A2.1: (a) block diagram; (b) signal flow graph.

Thus from the definitions

$$\Delta = 1 - (-G_1G_2 - G_2H_2)$$
$$M_1 = G_1G_2$$
$$\Delta_1 = 1$$

Therefore,

$$\frac{C}{R} = \frac{G_1G_2}{1 + G_1G_2 + G_2H_2}$$

Example A2.2

The block diagram and corresponding signal flow graph for this example are shown in Figure A2-3. In determining the transfer function for this system we note that there is only one forward path and there are three loops. Note, however, that two of the loops are nontouching loops.

Therefore,

$$\Delta = 1 - (-G_1H_1 - G_2H_2 - G_1G_2H_3) + (-G_1H_1)(-G_2H_2)$$
$$M_1 = G_1G_2$$
$$\Delta_1 = 1$$

Hence

$$\frac{C}{R} = \frac{G_1G_2}{1 + G_1H_1 + G_2H_2 + G_1G_2H_3 + G_1G_2H_1H_2}$$

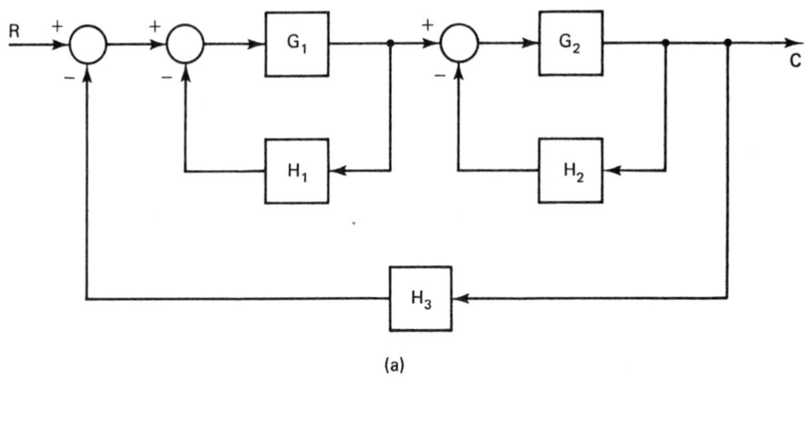

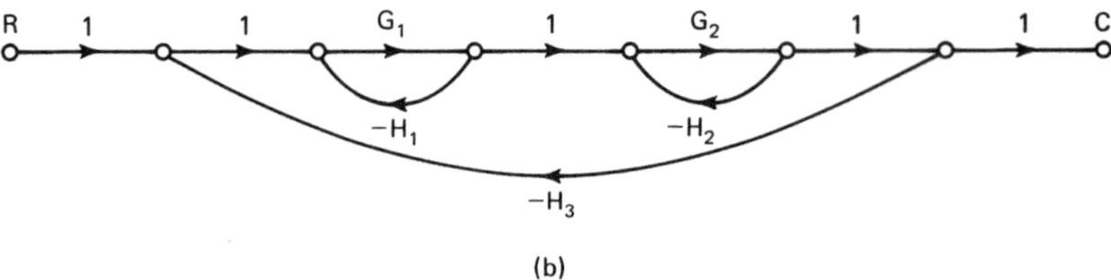

Figure A2-3 Diagrams for Example A2.2: (a) block diagram; (b) signal flow graph.

Example A2.3

We wish to determine the transfer function of the block diagram shown in Figure A2-4a. The signal flow graph for this diagram is shown in Figure A2-4b. Note carefully that there are two forward paths, one of which does not touch the loop $(-G_2H_2)$. There are four loops in the flow graph and the outside loop $(-G_1G_4G_3)$ does not touch the inner loop $(-G_2H_2)$. Therefore,

$$\Delta = 1 - (-G_1 - G_2H_2 - G_1G_2G_3 - G_1G_4G_3) + (-G_2H_2)(-G_1G_4G_3) + (-G_1)(-G_2H_2)$$

$$M_1 = G_1G_2G_3$$

$$\Delta_1 = 1$$

$$M_2 = G_1G_4G_3$$

$$\Delta_2 = 1 - (-G_2H_2)$$

Hence

$$\frac{C}{R} = \frac{G_1G_2G_3 + G_1G_4G_3(1 + G_2H_2)}{1 + G_1 + G_2H_2 + G_1G_2G_3 + G_1G_4G_3 + G_2H_2G_1G_4G_3 + G_1G_2H_2}$$

For simplicity, we have described the signal flow graph techniques using continuous systems. However, the methods are equally applicable to discrete systems, as shown in Chapter 5.

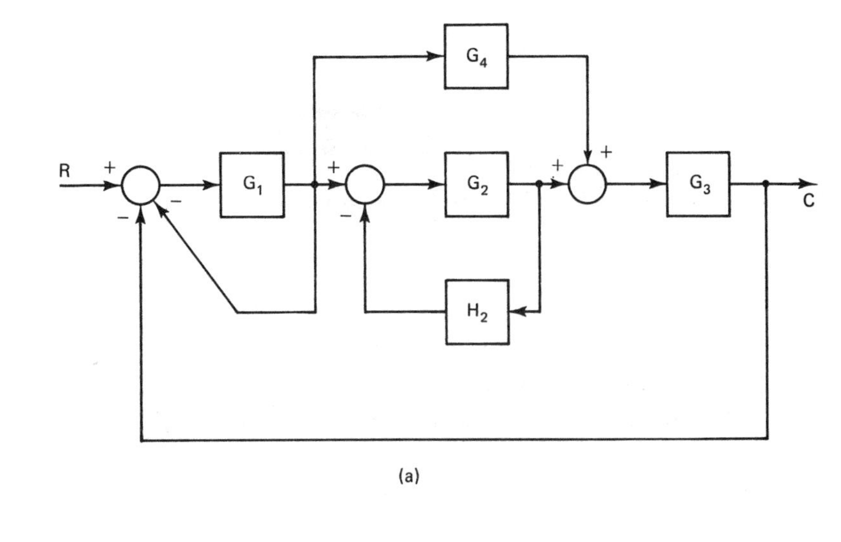

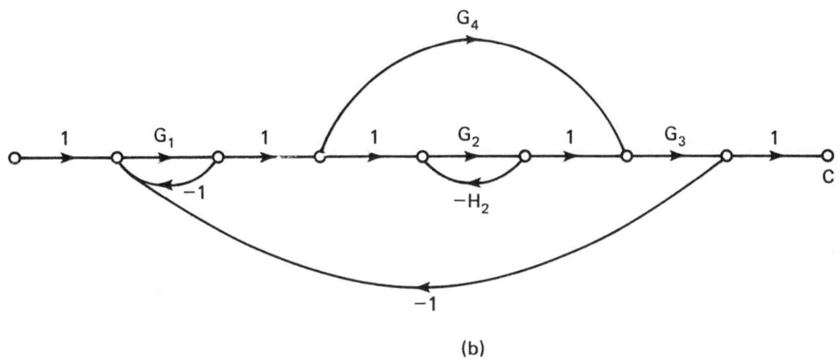

Figure A2-4 Diagrams for Example A2.3: (a) block diagram; (b) signal flow graph.

APPENDIX III

Evaluation of $E^*(s)$

$E^*(s)$, as defined by (3-7), has limited usefulness in an analysis because it is expressed as an infinite series. However, for many useful time functions $E^*(s)$ can be expressed in closed form. This closed form of $E^*(s)$ will now be derived.

From (3-6),

$$e^*(t) = e(t)\delta_T(t) \tag{A3-1}$$

However, this function is the inverse Laplace transform of $E^*(s)$ only for cases in which $e(t)$ is continuous at all sampling instants. Problems arise if $e(t)$ is discontinuous at any sampling instant, since the inverse Laplace transform evaluated at a discontinuity will give the average value of the discontinuity. As defined in Section 3.3, however, if $e(t)$ is discontinuous at a sampling instant, then at that instant $E^*(s)$ assumes the value of $e(t)$ from the right. For example, if $e(t)$ is discontinuous at the origin, the calculation of $E^*(s)$ from (A3-1) would yield a value for the function at $t = 0$ of $\frac{1}{2}e(0)$. Thus, if $e(0) \neq 0$, (A3-1) must be expressed as

$$e^*(t) = e(t)\delta_T(t) + e(0)\delta(t) \tag{A3-2}$$

If $e(t)$ is discontinuous at other sampling instants, then impulse functions with values $\Delta e(kT)\ \delta(t - kT)$ must be added to (A3-2), where $\Delta e(kT)$ is the amplitude of the discontinuity of $e(t)$ at $t = kT$; that is

$$\Delta e(kT) = e(kT^+) - e(kT^-)$$

where $e(kT^-) = e(t)$ evaluated at $t = kT - \epsilon$, and where ϵ is arbitrarily small.

The following derivation applies for the case in which $e(t)$ is continuous at all sampling instants. From (A3-1),

$$E^*(s) = E(s) * \Delta_T(s) \tag{A3-3}$$

where $*$ denotes complex convolution [1] and $\Delta_T(s)$ is the Laplace transform of $\delta_T(t)$. From (3-5),

$$\Delta_T(s) = 1 + \epsilon^{-Ts} + \epsilon^{-2Ts} + \cdots = \frac{1}{1 - \epsilon^{-Ts}} \tag{A3-4}$$

Therefore, the poles of $\Delta_T(s)$ occur at values of s for which

$$\epsilon^{-Ts} = 1 \tag{A3-5}$$

Equation (A3-5) is satisfied for $s = j(2\pi n/T) = jn\omega_s$, $n = 0, \pm 1, \pm 2, \ldots$, where ω_s is the sampling frequency expressed in radians per second. The poles of $\Delta_T(s)$ are shown on the s-plane plot of Figure A3-1.

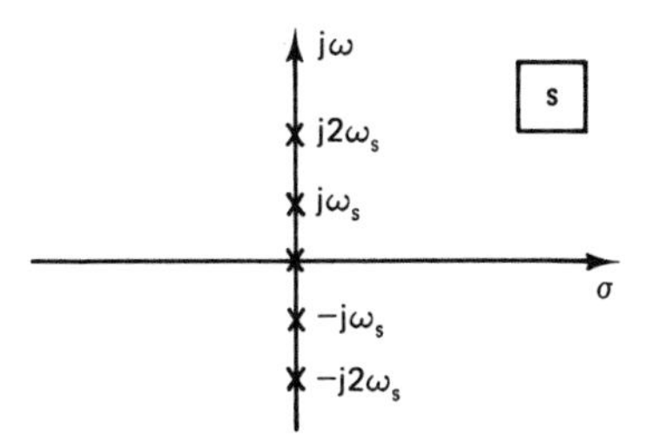

Figure A3-1 Poles of $\Delta_T(s)$.

By definition, equation (A3-3) can be expressed as

$$\begin{aligned} E^*(s) &= \frac{1}{2\pi j}\int_{c-j\infty}^{c+j\infty} E(\lambda)\,\Delta_T(s-\lambda)\,d\lambda \\ &= \frac{1}{2\pi j}\int_{c-j\infty}^{c+j\infty} E(\lambda)\frac{1}{1-\epsilon^{-T(s-\lambda)}}\,d\lambda \end{aligned} \tag{A3-6}$$

where the poles of the integrand of (A3-6) occur as shown in Figure A3-2. The value of c must be chosen such that the poles of $E(\lambda)$ are to the left of the path of integration, and the value of s must be selected so that the poles of $\Delta_T(s-\lambda)$ are to the right of the path of integration [1]. An examination of (A3-6) and Figure A3-2 indicates that $E^*(s)$ can be expressed as

$$E^*(s) = \frac{1}{2\pi j}\oint_\gamma E(\lambda)\,\Delta_T(s-\lambda)\,d\lambda - \frac{1}{2\pi j}\oint E(\lambda)\,\Delta_T(s-\lambda)\,d\lambda \tag{A3-7}$$

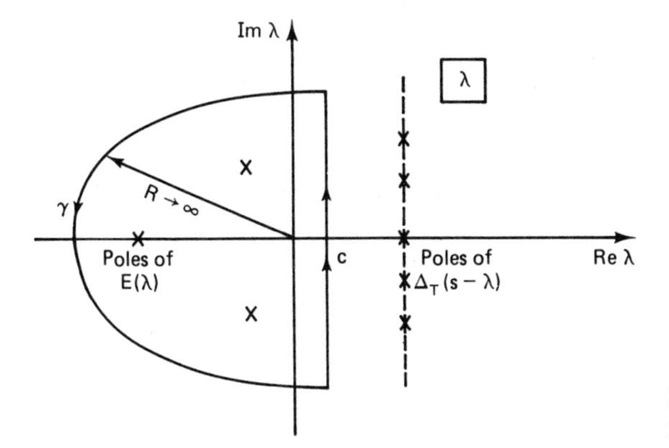

Figure A3-2 Pole locations for the integrand of (A3-6).

Consider the second integral of (A3-7). This term is zero if $\lim_{\lambda\to\infty} \lambda E(\lambda) = 0$ [i.e., if $e(0) = 0$]. If $e(0)$ is not zero, the second integral can be shown [2] to be equal to $e(0)/2$. However, from (A3-2), if $e(0)$ is not zero, an additional term equal to $e(0)/2$ must be added to (A3-7). Therefore, in either case,

$$E^*(s) = \frac{1}{2\pi j} \oint_{\gamma} E(\lambda)\Delta_T(s - \lambda)\, d\lambda \tag{A3-8}$$

The reader interested in a more detailed mathematical development of (A3-8) is referred to Refs. 1–3.

The theorem of residues [3] can be used to evaluate (A3-8).

Theorem of Residues. If C is a closed curve and if $f(z)$ is analytic within and on C except at finite number of singular points in the interior of C, then

$$\oint_C f(z)\, dz = 2\pi j[r_1 + r_2 + \cdots + r_n]$$

where $r_1, r_2, \ldots, r_n$ are the residues of $f(z)$ at the singular points within C.

Using this theorem, we can express (A3-8) as

$$E^*(s) = \sum_{\substack{\text{at poles}\\ \text{of } E(\lambda)}} \left[\text{residues of } E(\lambda)\frac{1}{1 - \epsilon^{-T(s-\lambda)}}\right] \tag{A3-9}$$

Recall that the residues are evaluated at the poles of the function $E(\lambda)$, since as shown in Figure A3-2, the singular points that lie within the closed contour are derived from this function. For the case in which $E(s)$ has only simple poles, we may use (2-28). Or, by letting

$$E(s) = \frac{N(s)}{D(s)} \tag{A3-10}$$

where $N(s)$ and $D(s)$ are polynomials in s, (A3-9) can be expressed as [4]

$$E^*(s) = \sum_n \frac{N(\lambda_n)}{D'(\lambda_n)} \frac{1}{1 - \epsilon^{-T(s-\lambda_n)}} \tag{A3-11}$$

where λ_n are the locations of simple poles of $E(\lambda)$ and

$$D'(\lambda) = \frac{dD(\lambda)}{d\lambda}$$

For multiple-ordered poles, the residue may be found using the expression illustrated in (2-29).

It is of interest to consider the case in which the function $e(t)$ contains a time delay. For example, consider a delayed signal of the type

$$e(t) = e_1(t - t_0)u(t - t_0) \tag{A3-12}$$

Then

$$E(s) = \epsilon^{-t_0 s}\mathfrak{L}[e_1(t)] = \epsilon^{-t_0 s}E_1(s) \tag{A3-13}$$

For this case, in general $\lim_{\lambda\to\infty} \lambda E(\lambda)$ is not finite in the second integral in (A3-7) (see Figure A3-2), and thus (A3-9) does not apply. Special techniques are required to find the starred transform of a delayed signal in closed form, and these techniques are presented in Chapter 4, where the modified z-transform is developed. However, for the special case in which the time signal is delayed a whole number of sampling periods, (A3-9) can be applied in a slightly different form,

$$[\epsilon^{-kTs}E_1(s)]^* = \epsilon^{-kTs} \sum_{\substack{\text{at poles} \\ \text{of } E_1(\lambda)}} \left[\text{residues of } E_1(\lambda)\frac{1}{1-\epsilon^{-T(s-\lambda)}}\right] \qquad \text{(A3-14)}$$

where k is a positive integer.

Equation (A3-6) can also be evaluated using the path α shown in Figure A3-3. In this case, the poles of $\Delta_T(s - \lambda)$ are enclosed by α. For the case in which

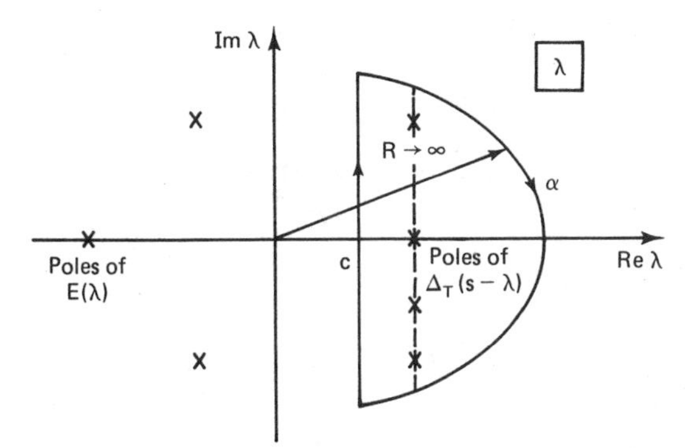

Figure A3-3 Integration path in the λ-plane.

$\lim_{\lambda\to\infty} \lambda E(\lambda)$ is zero, the integral around the infinite semicircular portion of α is zero and

$$E^*(s) = -\sum_{\substack{\text{poles of} \\ \Delta_T(s-\lambda)}} [\text{residues of } E(\lambda)\Delta_T(s-\lambda)] \qquad \text{(A3-15)}$$

This relationship is derived using the same steps that were used in the derivation of (A3-9). Recall that

$$\Delta_T(s-\lambda) = \frac{1}{1-\epsilon^{-T(s-\lambda)}} = \frac{N(\lambda)}{D(\lambda)} \qquad \text{(A3-16)}$$

has simple poles located at periodic intervals along a fixed line in the λ-plane as shown in Figure A3-3, and $E^*(s)$ can be expressed as

$$E^*(s) = -\sum_{n=-\infty}^{\infty} \frac{N(\lambda_n)}{D'(\lambda_n)} E(\lambda_n) \qquad \text{(A3-17)}$$

The poles of $\Delta_T(s - \lambda)$ occur at

$$s - \lambda_n = j\frac{2\pi n}{T} = jn\omega_s, \qquad n = 0, \pm 1, \pm 2, \ldots \qquad \text{(A3-18)}$$

or, solving for λ_n,

$$\lambda_n = s - jn\omega, \qquad n = 0, \pm 1, \pm 2, \ldots \qquad \text{(A3-19)}$$

Now in this case

$$D'(\lambda) = -T\epsilon^{-T(s-\lambda)} \qquad \text{(A3-20)}$$

Thus

$$D'(\lambda_n) = -T\epsilon^{-j2\pi n} = -T \qquad \text{(A3-21)}$$

Then using (A3-16) and (A3-21), (A3-17) can be expressed as

$$E^*(s) = \frac{1}{T} \sum_{n=-\infty}^{\infty} E(s + jn\omega_s) \tag{A3-22}$$

For the case in which $\lim_{\lambda\to\infty} E(\lambda) = 0$, but $e(0) \neq 0$, the integral around the infinite semicircular portion of α is also zero. However, because of the additional term in (A3-2), (A3-22) becomes

$$E^*(s) = \frac{1}{T} \sum_{n=-\infty}^{\infty} E(s + jn\omega_s) + \frac{e(0^+)}{2} \tag{A3-23}$$

Therefore, the general expression for (A3-23) is

$$E^*(s) = \frac{1}{T} \sum_{n=-\infty}^{\infty} E(s + jn\omega_s) + \frac{1}{2} \sum_{n=0}^{\infty} \Delta e(nT)\epsilon^{-nTs} \tag{A3-24}$$

where $\Delta e(nT)$ is the amplitude of the discontinuity of $e(t)$ at $t = nT$.

In summary, there are three expressions of $E^*(s)$. These are:

$$E^*(s) = \sum_{n=0}^{\infty} e(nT)\epsilon^{-nTs} \tag{A3-25}$$

$$E^*(s) = \sum_{\substack{\text{at poles} \\ \text{of } E(\lambda)}} \left[\text{residues of } E(\lambda) \frac{1}{1 - \epsilon^{-T(s-\lambda)}} \right] \tag{A3-26}$$

$$E^*(s) = \frac{1}{T} \sum_{n=-\infty}^{\infty} E(s + jn\omega_s) + \frac{1}{2} \sum_{n=0}^{\infty} \Delta e(nT)\epsilon^{-nTs} \tag{A3-27}$$

Equation (A3-25) is the defining equation for $E^*(s)$, and (A3-26) and (A3-27) are derived from (A3-25).

REFERENCES

1. M. F. Gardner and J. L. Barnes, *Transients in Linear Systems*, Vol. I. New York: John Wiley & Sons, Inc., 1942.
2. C. L. Phillips, D. L. Chenoweth, and R. K. Cavin III, "*z*-Transform Analysis of Sampled-Data Control Systems without Reference to Impulse Functions," *IEEE Trans. Educ.*, Vol. E-11, pp. 141–144, June 1968.
3. C. R. Wylie, Jr., *Advanced Engineering Mathematics*. New York: McGraw-Hill Book Company, 1951.
4. E. A. Guillemin, *The Mathematics of Circuit Analysis*. New York: John Wiley & Sons, Inc., 1949.

APPENDIX IV

Review of Matrices

Presented in this appendix is a brief review of the algebra of matrices. Those readers interested in more depth are referred to Refs. 1–4.

The study of matrices originated in linear equations. As an example, consider the equations

$$\begin{aligned} x_1 + x_2 + \ \ x_3 &= 3 \\ x_1 + x_2 - \ \ x_3 &= 1 \\ 2x_1 + x_2 + 3x_3 &= 6 \end{aligned} \tag{A4-1}$$

In a *vector-matrix* format we write these equations as

$$\begin{bmatrix} 1 & 1 & 1 \\ 1 & 1 & -1 \\ 2 & 1 & 3 \end{bmatrix} \begin{bmatrix} x_1 \\ x_2 \\ x_3 \end{bmatrix} = \begin{bmatrix} 3 \\ 1 \\ 6 \end{bmatrix} \tag{A4-2}$$

We define the following:

$$\mathbf{A} = \begin{bmatrix} 1 & 1 & 1 \\ 1 & 1 & -1 \\ 2 & 1 & 3 \end{bmatrix}, \qquad \mathbf{x} = \begin{bmatrix} x_1 \\ x_2 \\ x_3 \end{bmatrix}, \qquad \mathbf{u} = \begin{bmatrix} 3 \\ 1 \\ 6 \end{bmatrix} \tag{A4-3}$$

Then (A4-2) may be expressed as

$$\mathbf{Ax} = \mathbf{u} \tag{A4-4}$$

In this equation $\mathbf{A}$ is a 3×3 (3 rows, 3 columns) *matrix*, $\mathbf{x}$ is a 3×1 *matrix*, and $\mathbf{u}$ is a 3×1 *matrix*. Usually matrices that contain only one row or only one column are called *vectors*. A matrix of only one row and one column is a *scalar*. In (A4-1), x_1 is a scalar.

The general matrix $\mathbf{A}$ is written as

$$\mathbf{A} = \begin{bmatrix} a_{11} & a_{12} & \cdots & a_{1n} \\ a_{21} & a_{22} & \cdots & a_{2n} \\ \cdots & \cdots & \cdots & \cdots \\ a_{m1} & a_{m2} & \cdots & a_{mn} \end{bmatrix} \tag{A4-5}$$

This matrix has m rows and n columns, and thus is an $m \times n$ matrix. The element a_{ij} is the element common to the ith row and jth column.

Some special definitions will now be given.

Identity Matrix. The identity matrix is an $n \times n$ matrix with all main diagonal elements equal to 1 and all off-diagonal elements equal to zero. For example, the 3×3 identity matrix is

$$\mathbf{I} = \begin{bmatrix} 1 & 0 & 0 \\ 0 & 1 & 0 \\ 0 & 0 & 1 \end{bmatrix} \tag{A4-6}$$

If $\mathbf{A}$ is also $n \times n$,

$$\mathbf{AI} = \mathbf{IA} = \mathbf{A} \tag{A4-7}$$

Diagonal Matrix. A diagonal matrix is an $n \times n$ (square) matrix with all off-diagonal elements equal to zero.

$$\mathbf{D} = \begin{bmatrix} d_{11} & 0 & 0 \\ 0 & d_{22} & 0 \\ 0 & 0 & d_{33} \end{bmatrix} \tag{A4-8}$$

Transpose of a Matrix. To take the transpose of a matrix, interchange rows and columns. For example,

$$\mathbf{A} = \begin{bmatrix} 1 & 1 & 1 \\ 1 & 1 & -1 \\ 2 & 1 & 3 \end{bmatrix}, \qquad \mathbf{A}^T = \begin{bmatrix} 1 & 1 & 2 \\ 1 & 1 & 1 \\ 1 & -1 & 3 \end{bmatrix} \tag{A4-9}$$

where $\mathbf{A}^T$ denotes the transpose of $\mathbf{A}$. A property of the transpose is

$$(\mathbf{AB})^T = \mathbf{B}^T\mathbf{A}^T \tag{A4-10}$$

Minor. The minor m_{ij} of element a_{ij} of a square matrix $\mathbf{A}$ is the determinant of the array remaining when the ith row and jth column are deleted from $\mathbf{A}$. For example, m_{21} for $\mathbf{A}$ of (A4-9) is

$$m_{21} = \begin{vmatrix} 1 & 1 \\ 1 & 3 \end{vmatrix} = 3 - 1 = 2 \tag{A4-11}$$

Cofactor. The cofactor c_{ij} of element a_{ij} of the matrix $\mathbf{A}$ is given by

$$c_{ij} = (-1)^{i+j} m_{ij} \tag{A4-12}$$

For (A4-11),

$$c_{21} = (-1)^{2+1}(2) = -2 \tag{A4-13}$$

Adjoint. The matrix of cofactors, when transposed, is called the adjoint of **A** (adj **A**). For **A** of (A4-9),

$$\text{adj } \mathbf{A} = \begin{bmatrix} c_{11} & c_{12} & c_{13} \\ c_{21} & c_{22} & c_{23} \\ c_{31} & c_{32} & c_{33} \end{bmatrix}^T = \begin{bmatrix} 4 & -5 & -1 \\ -2 & 1 & 1 \\ -2 & 2 & 0 \end{bmatrix}^T \tag{A4-14}$$

Inverse. The inverse of **A** is given by

$$\mathbf{A}^{-1} = \frac{\text{adj } \mathbf{A}}{|\mathbf{A}|} \tag{A4-15}$$

where $\mathbf{A}^{-1}$ denotes the inverse of **A** and $|\mathbf{A}|$ denotes the determinant of **A**. For **A** of (A4-9) and (A4-14)

$$|\mathbf{A}| = -2$$

and

$$\mathbf{A}^{-1} = \begin{bmatrix} -2 & 1 & 1 \\ \frac{5}{2} & -\frac{1}{2} & -1 \\ \frac{1}{2} & -\frac{1}{2} & 0 \end{bmatrix} \tag{A4-16}$$

Two properties of matrix inverses are

$$\mathbf{A}\mathbf{A}^{-1} = \mathbf{A}^{-1}\mathbf{A} = \mathbf{I} \tag{A4-17}$$

$$(\mathbf{AB})^{-1} = \mathbf{B}^{-1}\mathbf{A}^{-1} \tag{A4-18}$$

ALGEBRA OF MATRICES

The algebra of matrices must be defined such that the operations indicated in (A4-2), and any additional operation we may wish to perform, lead us back to (A4-1).

Addition. To form the sum of matrices **A** and **B**, we add corresponding elements a_{ij} and b_{ij}, for each ij. For example,

$$\begin{bmatrix} 1 & 2 \\ 3 & 4 \end{bmatrix} + \begin{bmatrix} 5 & 6 \\ 7 & 8 \end{bmatrix} = \begin{bmatrix} 6 & 8 \\ 10 & 12 \end{bmatrix} \tag{A4-19}$$

Multiplication by a Scalar. To multiply a matrix **A** by a scalar k, multiply each element of **A** by k.

Multiplication of Vectors. The multiplication of the $1 \times n$ (row) vector with an $n \times 1$ (column) vector is defined as

$$[x_1 \; x_2 \; \cdots \; x_n] \begin{bmatrix} y_1 \\ y_2 \\ \cdots \\ y_n \end{bmatrix} = x_1 y_1 + x_2 y_2 + \cdots + x_n y_n \tag{A4-20}$$

Multiplication of Matrices. An $n \times p$ matrix $\mathbf{A}$ may be multiplied by only a $p \times m$ matrix $\mathbf{B}$; that is, the number of columns of $\mathbf{A}$ must equal the number of rows of $\mathbf{B}$. Let

$$\mathbf{AB} = \mathbf{C}$$

Then the ijth element of $\mathbf{C}$ is equal to the multiplication (as vectors) of the ith row of $\mathbf{A}$ with the jth column of $\mathbf{B}$. As an example, consider the produce $\mathbf{AA}^{-1}$ from (A4-9) and (A4-16).

$$\begin{aligned}\mathbf{AA}^{-1} &= \begin{bmatrix} 1 & 1 & 1 \\ 1 & 1 & -1 \\ 2 & 1 & 3 \end{bmatrix}\begin{bmatrix} -2 & 1 & 1 \\ \frac{5}{2} & -\frac{1}{2} & -1 \\ \frac{1}{2} & -\frac{1}{2} & 0 \end{bmatrix} \\ &= \begin{bmatrix} 1 & 0 & 0 \\ 0 & 1 & 0 \\ 0 & 0 & 1 \end{bmatrix} = \mathbf{I}\end{aligned} \tag{A4-21}$$

OTHER RELATIONSHIPS

Other important matrix relationships will now be given.

Differentiation. The derivative of a matrix is obtained by differentiating the matrix element by element. For example, let

$$\mathbf{x} = \begin{bmatrix} x_1 \\ x_2 \end{bmatrix} \tag{A4-22}$$

Then

$$\frac{d\mathbf{x}}{dt} = \begin{bmatrix} \dfrac{dx_1}{dt} \\ \dfrac{dx_2}{dt} \end{bmatrix} \tag{A4-23}$$

Integration. The integral of a matrix is obtained by integrating the matrix element by element. In (A4-22),

$$\int \mathbf{x}\, dt = \begin{bmatrix} \int x_1\, dt \\ \int x_2\, dt \end{bmatrix} \tag{A4-24}$$

Trace. The trace of a matrix is equal to the sum of its diagonal elements. Given an $n \times n$ matrix $\mathbf{A}$,

$$\text{trace of } \mathbf{A} = \text{tr } \mathbf{A} = a_{11} + a_{22} + \cdots + a_{nn} \tag{A4-25}$$

Quadratic Forms. The scalar

$$F = \mathbf{x}^T\mathbf{Q}\mathbf{x} \tag{A4-26}$$

is called a quadratic form. For example, if $\mathbf{x}$ is second order,

$$F = \mathbf{x}^T\mathbf{Q}\mathbf{x} = [x_1 \quad x_2]\begin{bmatrix} q_{11} & q_{12} \\ q_{21} & q_{22} \end{bmatrix}\begin{bmatrix} x_1 \\ x_2 \end{bmatrix} \tag{A4-27}$$

$$= q_{11}x_1^2 + (q_{12} + q_{21})x_1x_2 + q_{22}x_2^2$$

Note that $\mathbf{Q}$ can be assumed symmetric with no loss of generality. Now

$$\frac{\partial F}{\partial \mathbf{x}} = \begin{bmatrix} \dfrac{\partial F}{\partial x_1} \\ \dfrac{\partial F}{\partial x_2} \end{bmatrix} = \begin{bmatrix} 2q_{11}x_1 + (q_{12} + q_{21})x_2 \\ (q_{12} + q_{21})x_1 + 2q_{22}x_2 \end{bmatrix} = 2\mathbf{Q}\mathbf{x} \tag{A4-28}$$

Bilinear Forms. The scalar

$$G = \mathbf{x}^T\mathbf{Q}\mathbf{y} \tag{A4-29}$$

is called a bilinear form. For example, for $\mathbf{x}$ and $\mathbf{y}$ second order,

$$G = \mathbf{x}^T\mathbf{Q}\mathbf{y} = [x_1 \quad x_2]\begin{bmatrix} q_{11} & q_{12} \\ q_{21} & q_{22} \end{bmatrix}\begin{bmatrix} y_1 \\ y_2 \end{bmatrix} \tag{A4-30}$$

$$= q_{11}x_1y_1 + q_{12}x_1y_2 + q_{21}x_2y_1 + q_{22}x_2y_2$$

Note that $\mathbf{Q}$ cannot be assumed symmetric. Then

$$\frac{\partial G}{\partial \mathbf{x}} = \begin{bmatrix} \dfrac{\partial G}{\partial x_1} \\ \dfrac{\partial G}{\partial x_2} \end{bmatrix} = \begin{bmatrix} q_{11}y_1 + q_{12}y_2 \\ q_{21}y_1 + q_{22}y_2 \end{bmatrix} = \mathbf{Q}\mathbf{y} \tag{A4-31}$$

and

$$\frac{\partial G}{\partial \mathbf{y}} = \begin{bmatrix} \dfrac{\partial G}{\partial y_1} \\ \dfrac{\partial G}{\partial y_2} \end{bmatrix} = \begin{bmatrix} q_{11}x_1 + q_{21}x_2 \\ q_{12}x_1 + q_{22}x_2 \end{bmatrix} = \mathbf{Q}^T\mathbf{x} \tag{A4-32}$$

REFERENCES

1. F. R. Gantmacher, *Theory of Matrices*, Vols. I and II. New York: Chelsea Publishing Company, Inc., 1959.
2. P. M. DeRusso, R. J. Roy, and C. M. Close, *State Variables for Engineers*. New York: John Wiley & Sons, Inc., 1965.
3. K. Ogata, *State Space Analysis of Control Systems*. Englewood Cliffs, N.J.: Prentice-Hall, Inc., 1967.
4. G. Strang, *Linear Algebra and Its Applications*. New York: Academic Press, Inc., 1976.

APPENDIX V

Design Equations

In this appendix, (8-32) and (8-33) are shown to be solutions to (8-30) and (8-31), by direct substitution. For convenience, $D(j\omega_{w1})$ will be written as D, and similarly for G. Then, from (8-23), (8-32), and (8-33),

$$D = \frac{1 + j\dfrac{\omega_{w1}}{\omega_{w0}}}{1 + j\dfrac{\omega_{w1}}{\omega_{wp}}} = \frac{1 + j\dfrac{\omega_{w1}}{\omega_{w1}\sin\theta}\left(\dfrac{1}{|G|} - \cos\theta\right)}{1 + j\dfrac{\omega_{w1}}{\omega_{w1}\sin\theta}(\cos\theta - |G|)}$$

$$= \frac{\sin\theta + j\left(\dfrac{1}{|G|} - \cos\theta\right)}{\sin\theta + j(\cos\theta - |G|)} \tag{A5-1}$$

Then

$$|D|^2 = \frac{\sin^2\theta + \dfrac{1}{|G|^2} - \dfrac{2\cos\theta}{|G|} + \cos^2\theta}{\sin^2\theta + |G|^2 - 2|G|\cos\theta + \cos^2\theta}$$

$$= \frac{1 - \dfrac{2\cos\theta}{|G|} + \dfrac{1}{|G|^2}}{1 - 2|G|\cos\theta + |G|^2} = \frac{1}{|G|^2} \tag{A5-2}$$

Thus

$$|D||G| = 1 \tag{A5-3}$$

and (8-30) is satisfied.

To show that the angle of $D(\omega_{w1})$ is θ, let the numerator angle and the denominator angle in (A5-1) be denoted as θ_1 and θ_2, respectively. Then, from (A5-1),

$$\tan(\theta_1 - \theta_2) = \frac{\tan\theta_1 - \tan\theta_2}{1 + \tan\theta_1 \tan\theta_2}$$

$$= \frac{\dfrac{(1/|G|) - \cos\theta}{\sin\theta} - \dfrac{\cos\theta - |G|}{\sin\theta}}{1 + \dfrac{(1/|G|)\cos\theta - 1 - \cos^2\theta + |G|\cos\theta}{\sin^2\theta}} \tag{A5-4}$$

$$= \frac{\sin\theta\,(|G| + (1/|G|) - 2\cos\theta)}{|G|\cos\theta + (\cos\theta/|G|) - 2\cos^2\theta} = \tan\theta$$

Thus the angle of $D(j\omega_{w1})$ is θ, and (8-31) is satisfied.

APPENDIX VI

z-Transform Tables

Laplace transform $E(s)$	Time function $e(t)$	z-Transform $E(z)$	Modified z-transform $E(z, m)$
$\frac{1}{s}$	$u(t)$	$\frac{z}{z-1}$	$\frac{1}{z-1}$
$\frac{1}{s^2}$	t	$\frac{Tz}{(z-1)^2}$	$\frac{mT}{z-1}+\frac{T}{(z-1)^2}$
$\frac{1}{s^3}$	$\frac{t^2}{2}$	$\frac{T^2z(z+1)}{2(z-1)^3}$	$\frac{T^2}{2}\left[\frac{m^2}{z-1}+\frac{2m+1}{(z-1)^2}+\frac{2}{(z-1)^3}\right]$
$\frac{(k-1)!}{s^k}$	t^{k-1}	$\lim_{a\to 0}(-1)^{k-1}\frac{\partial^{k-1}}{\partial a^{k-1}}\left[\frac{z}{z-\epsilon^{-aT}}\right]$	$\lim_{a\to 0}(-1)^{k-1}\frac{\partial^{k-1}}{\partial a^{k-1}}\left[\frac{\epsilon^{-amT}}{z-\epsilon^{-aT}}\right]$
$\frac{1}{s+a}$	ϵ^{-at}	$\frac{z}{z-\epsilon^{-aT}}$	$\frac{\epsilon^{-amT}}{z-\epsilon^{-aT}}$
$\frac{1}{(s+a)^2}$	$t\epsilon^{-at}$	$\frac{Tz\epsilon^{-aT}}{(z-\epsilon^{-aT})^2}$	$\frac{T\epsilon^{-amT}[\epsilon^{-aT}+m(2-\epsilon^{-aT})]}{(z-\epsilon^{-aT})^2}$
$\frac{(k-1)!}{(s+a)^k}$	$t^k\epsilon^{-at}$	$(-1)^k\frac{\partial^k}{\partial a^k}\left[\frac{z}{z-\epsilon^{-aT}}\right]$	$(-1)^k\frac{\partial^k}{\partial a^k}\left[\frac{\epsilon^{-amT}}{z-\epsilon^{-aT}}\right]$
$\frac{a}{s(s+a)}$	$1-\epsilon^{-at}$	$\frac{z(1-\epsilon^{-aT})}{(z-1)(z-\epsilon^{-aT})}$	$\frac{1}{z-1}-\frac{\epsilon^{-amT}}{z-\epsilon^{-aT}}$
$\frac{a}{s^2(s+a)}$	$t-\frac{1-\epsilon^{-at}}{a}$	$\frac{z[(aT-1+\epsilon^{-aT})z+(1-\epsilon^{-aT}-aT\epsilon^{-aT})]}{a(z-1)^2(z-\epsilon^{-aT})}$	$\frac{T}{(z-1)^2}+\frac{amT-1}{a(z-1)}+\frac{\epsilon^{-amT}}{a(z-\epsilon^{-aT})}$
$\frac{a^2}{s(s+a)^2}$	$1-(1+at)\epsilon^{-at}$	$\frac{z}{z-1}-\frac{z}{z-\epsilon^{-aT}}-\frac{aT\epsilon^{-aT}z}{(z-\epsilon^{-aT})^2}$	$\frac{1}{z-1}-\left[\frac{1+amT}{z-\epsilon^{-aT}}+\frac{aT\epsilon^{-aT}}{(z-\epsilon^{-aT})^2}\right]\epsilon^{-amT}$

Laplace transform $E(s)$	Time function $e(t)$	z-Transform $E(z)$	Modified z-transform $E(z, m)$
$\frac{b-a}{(s+a)(s+b)}$	$\epsilon^{-at} - \epsilon^{-bt}$	$\frac{(\epsilon^{-aT} - \epsilon^{-bT})z}{(z-\epsilon^{-aT})(z-\epsilon^{-bT})}$	$\frac{\epsilon^{-amT}}{z-\epsilon^{-aT}} - \frac{\epsilon^{-bmT}}{z-\epsilon^{-bT}}$
$\frac{a}{s^2+a^2}$	$\sin(at)$	$\frac{z\sin(aT)}{z^2 - 2z\cos(aT) + 1}$	$\frac{z\sin(amT) + \sin(1-m)aT}{z^2 - 2z\cos(aT) + 1}$
$\frac{s}{s^2+a^2}$	$\cos(at)$	$\frac{z(z - \cos(aT))}{z^2 - 2z\cos aT + 1}$	$\frac{z\cos(amT) - \cos(1-m)aT}{z^2 - 2z\cos(aT) + 1}$
$\frac{1}{(s+a)^2+b^2}$	$\frac{1}{b}\epsilon^{-at}\sin bt$	$\frac{1}{b}\left[\frac{z\epsilon^{-aT}\sin bT}{z^2 - 2z\epsilon^{-aT}\cos(bT) + \epsilon^{-2aT}}\right]$	$\frac{1}{b}\left[\frac{\epsilon^{-amT}[z\sin bmT + \epsilon^{-aT}\sin(1-m)bT]}{z^2 - 2z\epsilon^{-aT}\cos bT + \epsilon^{-2aT}}\right]$
$\frac{s+a}{(s+a)^2+b^2}$	$\epsilon^{-at}\cos bt$	$\frac{z^2 - z\epsilon^{-aT}\cos bT}{z^2 - 2z\epsilon^{-aT}\cos bT + \epsilon^{-2aT}}$	$\frac{\epsilon^{-amT}[z\cos bmT + \epsilon^{-aT}\sin(1-m)bT]}{z^2 - 2z\epsilon^{-aT}\cos bT + \epsilon^{-2aT}}$
$\frac{a^2+b^2}{s[(s+a)^2+b^2]}$	$1 - \epsilon^{-at}\left(\cos bt + \frac{a}{b}\sin bt\right)$	$\frac{z(Az+B)}{(z-1)(z^2 - 2z\epsilon^{-aT}\cos bT + \epsilon^{-2aT})}$ $A = 1 - \epsilon^{-aT}\left(\cos bT + \frac{a}{b}\sin bT\right)$ $B = \epsilon^{-2aT} + \epsilon^{-aT}\left(\frac{a}{b}\sin bT - \cos bT\right)$	$\frac{1}{z-1}$ $- \frac{\epsilon^{-amT}[z\cos bmT + \epsilon^{-aT}\sin(1-m)bT]}{z^2 - 2z\epsilon^{-aT}\cos bT + \epsilon^{-2aT}}$ $\frac{+\frac{a}{b}\{\epsilon^{-amT}[z\sin bmT - \epsilon^{-aT}\sin(1-m)bT]\}}{z^2 - 2z\epsilon^{-aT}\cos bT + \epsilon^{-2aT}}$
$\frac{1}{s(s+a)(s+b)}$	$\frac{1}{ab} + \frac{\epsilon^{-at}}{a(a-b)}$ $+ \frac{\epsilon^{-bt}}{b(b-a)}$	$\frac{(Az+B)z}{(z-\epsilon^{-aT})(z-\epsilon^{-bT})(z-1)}$	$A = \frac{b(1-\epsilon^{-aT}) - a(1-\epsilon^{-bT})}{ab(b-a)}$ $B = \frac{a\epsilon^{-aT}(1-\epsilon^{-bT}) - b\epsilon^{-bT}(1-\epsilon^{-aT})}{ab(b-a)}$

Index

A

B

C

D

E

F

N

O

P

Q

R

S

T

U

W

Z